FOUNDATIONS OF EARTH SCIENCE

SEVENTH EDITION

FOUNDATIONS OF EARTH SCIENCE

SEVENTH EDITION

FREDERICK K. LUTGENS

EDWARD J. TARBUCK

ILLUSTRATED BY

DENNIS TASA

Boston   Columbus   Indianapolis   New York   San Francisco   Upper Saddle River
Amsterdam   Cape Town   Dubai   London   Madrid   Milan   Munich   Paris   Montréal   Toronto
Delhi   Mexico City   São Paulo   Sydney   Seoul   Singapore   Taipei   Tokyo

Geology Editor: Andrew Dunaway
Senior Marketing Manager: Maureen McLaughlin
Senior Project Editor: Crissy Dudonis
Assistant Editor: Michelle White
Senior Marketing Assistant: Nicola Houston
Director of Development: Jennifer Hart
Executive Development Editor: Jonathan Cheney
Media Producer: Tod Regan
Managing Editor, Geosciences and Chemistry:
   Gina M. Cheselka
Production Project Manager: Edward Thomas
Full Service/Composition: Element, LLC.

Full Service Project Manager: Heidi Allgair; Element, LLC.
Design Manager: Derek Bacchus
Cover and Interior Design: Gary Hespenheide
Illustrations: International Mapping
Photo Manager: Maya Melenchuk
Photo Researcher: Marta Johnson
Text Permissions Manager: Michael Farmer
Text Permissions Researcher: Katherine Benzer
Operations Specialist: Michael Penne
Cover Image Credit: Gordon Wiltsie/National
   Geographic Stock

Credits and acknowledgments for materials borrowed from other sources and reproduced, with permission, in this textbook appear on the appropriate page within the book.

Library of Congress Cataloging-in-Publication Data is available from the publisher upon request.

2 3 4 5 6 7 8 9 10—DOW—17 16 15 14 13

Student Edition
ISBN-10: 0-321-81179-8
ISBN-13: 978-0-321-81179-0

Instructor's Review Copy
ISBN-10: 0-321-82195-5
ISBN-13: 978-0-321-82195-9

www.pearsonhighered.com

To Our Grandchildren

Shannon, Amy, Andy, Ali, and Michael

Allison and Lauren

Each is a bright promise for the future.

Brief Table of Contents

Contents

UNIT I   EARTH MATERIALS   22

1

Matter and Minerals   22

UNIT II   SCULPTURING EARTH'S SURFACE 76

3

Landscapes Fashioned by Water 76

4

Glacial and Arid Landscapes   118

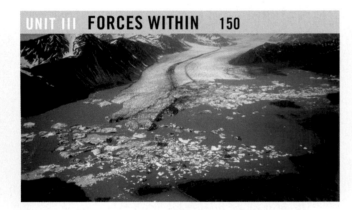

UNIT III **FORCES WITHIN** 150

5
Plate Tectonics: A Scientific Revolution Unfolds 150

6
Restless Earth: Earthquakes, Geologic Structures, and Mountain Building 188

7

Volcanoes and Other Igneous Activity   230

**UNIT IV   DECIPHERING EARTH'S
HISTORY   270**

8

Geologic Time   270

Contents

UNIT VI   **EARTH'S DYNAMIC ATMOSPHERE   352**

11

Heating the Atmosphere   352

12

Moisture, Clouds, and Precipitation   386

13
The Atmosphere in Motion 420

14
Weather Patterns and Severe Weather 442

UNIT VII **EARTH'S PLACE IN THE UNIVERSE**

15
The Nature of the Solar System 472

Preface

Foundations of Earth Science, seventh edition, is a college-level text designed for an introductory course in Earth science. It consists of seven units that emphasize broad and up-to-date coverage of basic topics and principles in geology, oceanography, meteorology, and astronomy. The book is intended to be a meaningful, nontechnical survey for undergraduate students who have little background in science. Usually these students are taking an Earth science class to meet a portion of their college's or university's general requirements.

In addition to being informative and up-to-date, *Foundations of Earth Science,* seventh edition, strives to meet the need of beginning students for a readable and user-friendly text and a highly usable "tool" for learning basic Earth science principles and concepts.

New to the Seventh Edition

- **SmartFigures—art that teaches.** Inside every chapter are several *SmartFigures. Foundations of Earth Science,* seventh edition, has more than 80 of these figures. Just use your mobile device to scan the Quick Response (QR) code next to SmartFigure and the art comes alive. Each 2- to 3-minute feature, prepared and narrated by Professor Callan Bentley, is a mini-lesson that examines and explains the concepts illustrated by the figure. It is truly *art that teaches*.
- **New active learning path.** *Foundations of Earth Science,* seventh edition, is designed for learning. Every chapter begins with *Focus on Concepts*. Each numbered learning objective corresponds to a major section in the chapter. The statements identify the knowledge and skills students should master by the end of the chapter, helping students prioritize key concepts. Within the chapter, each major section concludes with *Concept Checks* that allow students to check their understanding and comprehension of important ideas and terms before moving on to the next section. Chapters conclude with a new section called *Give It Some Thought*. These questions and problems challenge learners by involving them in activities that require higher-order thinking skills such as application, analysis, and synthesis of material in the chapter.
- **Concepts in Review.** This all new end-of-chapter feature is an important part of the book's active learning path. Each review is coordinated with the *Focus on Concepts* at the beginning of the chapter and with the numbered sections within the chapter. It is a readable and concise overview of key ideas, which makes it a valuable review tool for students.

- **An unparalleled visual program.** In addition to more than 200 new, high-quality photos and satellite images, dozens of figures are new or have been redrawn by renowned geoscience illustrator Dennis Tasa. Maps and diagrams are frequently paired with photographs for greater effectiveness. Further, many new and revised figures have additional labels that narrate the process being illustrated and guide students as they examine the figures. The result is a visual program that is clear and easy to understand.
- **MasteringGeology™.** MasteringGeology delivers engaging, dynamic learning opportunities—focused on course objectives and responsive to each student's progress—that are proven to help students absorb course material and understand difficult concepts. Assignable activities in MasteringGeology include Encounter Earth activities using Google Earth™, SmartFigure activities, GeoTutor activities, GigaPan® activities, Geoscience Animation activities, GEODe activities, and more. MasteringGeology also includes all instructor resources and a robust Study Area with resources for students.
- **Significant updating and revision of content.** A basic function of a college science textbook is to provide clear, understandable presentations that are accurate, engaging, and up-to-date. Our number-one goal is to keep *Foundations of Earth Science* current, relevant, and highly readable for the beginning students. Every part of this book has been examined carefully with this goal in mind. Many discussions, case studies, and examples have been revised. The seventh edition represents perhaps the *most extensive and thorough revision* in the long history of this textbook.

Distinguishing Features

Readability

The language of this book is straightforward and *written to be understood*. Clear, readable discussions with a minimum of technical language are the rule. The frequent headings and subheadings help students follow discussions and identify the important ideas presented in each chapter. In the seventh edition, we have continued to improve readability by examining chapter organization and flow and by writing in a more personal style. Significant portions of several chapters were substantially rewritten in an effort to make the material easier to understand.

Focus on Basic Principles and Instructor Flexibility

Although many topical issues are treated in the seventh edition of *Foundations of Earth Science*, it should be emphasized that the main focus of this new edition remains the same as the focus of each of its predecessors: to promote student understanding of basic Earth science principles. As much as possible, we have attempted to provide the reader with a sense of the observational techniques and reasoning processes that constitute the Earth sciences.

A Strong Visual Component

Earth science is highly visual, and art and photographs play a critical role in an introductory textbook. As in all previous editions, Dennis Tasa, a gifted artist and respected geoscience illustrator, has worked closely with the authors to plan and produce the diagrams, maps, graphs, and sketches that are so basic to student understanding. The result is art that is clearer and easier to understand than ever before.

Our aim is to get *maximum effectiveness* from the visual component of the book. Michael Collier, an award-winning geologist-photographer aided greatly in this quest. As you page through this book, you will see dozens of his extraordinary aerial photographs. His contribution truly helps bring geology alive for the reader.

The Teaching and Learning Package

MasteringGeology™ with Pearson eText

MasteringGeology delivers engaging, dynamic learning opportunities—focused on course objectives and responsive to each student's progress—that are proven to help students absorb course material and understand difficult concepts:

- MasteringGeology provides a rich and flexible set of course materials to get instructors started quickly, including prebuilt assignments that instructors can use as-is or customize to fit their needs.

- MasteringGeology provides quick and easy access to information on student performance against learning outcomes. Instructors can quickly add their own learning outcomes or use publisher-provided ones to track student performance.

- The MasteringGeology gradebook and diagnostic tools capture the step-by-step work of every student, providing unique insight into class performance.

- Assignable items in MasteringGeology include:
  - SmartFigures bring key chapter illustrations to life! Found throughout the book, SmartFigures are sophisticated, annotated illustrations that are also narrated videos. The SmartFigure videos are accessible on mobile devices via scannable Quick Response (QR) codes printed in the text and through the Study Area in MasteringGeology.
  - GigaPan® activities take advantage of the GigaPan high-resolution panoramic picture technology developed by Carnegie Mellon University in conjunction with NASA. Photos and accompanying questions correlate with concepts in the student book.
  - Encounter Earth activities provide interactive explorations of Earth science concepts using Google Earth™. Students work through the activities in Google Earth and then test their knowledge by answering the assessment questions, which include hints and specific wrong-answer feedback.
  - Animation activities illuminate the most difficult-to-understand topics in Earth science. Each Animation activity includes audio narration, a text transcript, and assignable multiple-choice questions with specific wrong-answer feedback.
  - GeoTutor activities address the most frequently cited thoughest topics in Earth science and consist of sophisticated, high impact visuals that ask students to demonstrate their knowledge by synthesizing and analyzing core concepts using higher-order thinking skills.
  - Concept Check questions, Give It Some Thought questions, Reading questions, and Test Bank questions, are all assignable in MasteringGeology.

- MasteringGeology includes a Study Area for students to access all their study resources. The Study Area includes geoscience animations, GEODe activities, *In the News* RSS feeds, Self-Study Quizzes, Web Links, Glossary, Flashcards, and an optional Pearson eText.

See www.masteringgeology.com.

Instructor's Resource Center

The authors and publisher were pleased to work with a number of talented people who produced an excellent supplements package.

Instructor Resource Center (IRC) on DVD The IRC puts all your lecture resources in one easy-to-reach place:
- *All* of the line art, tables, and photos from the text in .jpg files
- More than 100 animations of key geologic processes
- An *Images of Earth* photo gallery
- PowerPoint presentations
- An *Instructor Manual* in Microsoft Word
- A *Test Bank File* in Microsoft Word
- TestGen® test-generation and management software
- Electronic transparency acetates

Animations The Geoscience Animation Library, fifth edition, includes animations that illuminate many difficult-to-visualize topics of Earth science. Created through a unique collaboration among five of Pearson's leading geoscience authors, these animations represent a significant leap forward in lecture presentation aids. They are provided both as Flash files and, for your convenience, preloaded into PowerPoint slides.

PowerPoint Presentations The IRC provides three PowerPoint files for each chapter. They help you cut down on your preparation time, no matter what your lecture needs:

- *Exclusively art*—All the photos, art, and tables from the text, in order, loaded into PowerPoint slides.
- *Lecture outline*—This set averages 50 slides per chapter and includes customizable lecture outlines with supporting art.
- *Classroom Response System (CRS) questions*— Authored for use in conjunction with the CRS, these PowerPoints allow you to electronically poll your class for responses to questions, pop quizzes, attendance, and more.

Instructor Manual

The Instructor Manual is designed to help seasoned and new professors alike. It offers an introduction and outline, as well as learning objectives, teaching strategies, teacher resources, and answers to the *Concept Checks* and *Give It Some Thought* questions for each chapter.

TestGen® Computerized Test Bank (Download Only)

TestGen® is a computerized test generator that lets instructors view and edit *Test Bank* questions, transfer questions to tests, and print the test in a variety of customized formats. This *Test Bank* includes more than 800 multiple-choice, true/false, short-answer, matching, visualization, and essay questions. Questions are correlated to Bloom's Taxonomy, each chapter's learning objectives, the Earth Science Learning Objectives, and the Pearson Science Global Outcomes to help instructors better map the assessments against both broad and specific teaching and learning objectives. The *Test Bank* is also available in Microsoft Word® and is importable into Blackboard. www.pearsonhighered.com/irc.

Blackboard Already have your own website set up? The *Test Bank* is available in Blackboard for importation. Additional course resources are available on the IRC and are available for use with permission.

Electronic Acetates Provided electronically, every table and most of Dennis Tasa's illustrations in *Foundations of Earth Science*, seventh edition, are available to be printed out as full-color, projection-enhanced transparencies.

Acknowledgments

Writing a college textbook requires the talents and cooperation of many people. It is truly a team effort and the authors are fortunate to be part of an extraordinary team at Pearson. In addition to being great people to work with, all are committed to producing the best textbooks possible. Special thanks to our geology editor, Andy Dunaway, who invested a great deal of time, energy, and effort in this project. We appreciate his enthusiasm, hard work, and quest for excellence. We also appreciate our conscientious project manager, Crissy Dudonis, whose job it was to keep track of all that was going on—and a lot was going on. We spent many hours having conversations with Derek Bacchus and Gary Hespenheide, two talented men who are largely responsible for the new design of this seventh edition. We think it is a job well done. As always, our marketing manager, Maureen McLaughlin, provided helpful advice and many good ideas. *Foundations of Earth Science,* seventh edition, was truly improved with the help of our developmental editor, Jonathan Cheney. Many thanks. The production team was led by Ed Thomas at Pearson and by Heidi Allgair at Element LLC. It was their job to make this book into a finished product. We think they did a great job. All are true professionals with whom we are very fortunate to be associated.

The authors owe a special thanks to three people who are very important parts of this project.

- Working with Dennis Tasa, who is responsible for all of the text's outstanding illustrations, is always special for us. He has been a part of our team for more than 30 years. We not only value his artistic talents, hard work, patience, and imagination, but his friendship as well.

- As you turn the pages of this book, you will see many extraordinary photographs by Michael Collier. Most are aerial shots taken from his nearly 60-year-old Cessna 180. Michael is an award-winning geologist, author, and photographer. Among his many awards is the American Geological Institute Award for Outstanding Contribution to the Public Understanding of the Geosciences. John S. Shelton, a pioneering geologist-pilot-photographer, wrote of Michael, "Collier's combination of skill and talents—writer, photographer, geologist, and pilot . . . is truly impressive. He is indeed master of both the medium and the message." We agree. Michael's photographs are the next best thing to being there. We were fortunate to have had Michael's assistance on *Foundations of Earth Science,* seventh edition. Thanks, Michael.

- Callan Bentley has been an important addition to the *Foundations* team. Callan is an assistant professor of geology at Northern Virginia Community College in Annandale, where he has been honored many times as an outstanding teacher. He is a frequent contributor to *Earth* magazine and is author of the popular geology blog *Mountain Beltway*. Callan was responsible for preparing the dozens of SmartFigures that appear throughout this book. As you take advantage of these outstanding learning aids, you will hear his voice explaining the concepts and ideas related to each illustration.

Great thanks also go to those colleagues who prepared in-depth reviews. Their critical comments and thoughtful input helped guide our work and clearly strengthened the text. Special thanks to:

Winston Crausaz, Lorain County Community College
Chris Hansen, Southern Adventist University
Miriam Helen Hill, Jacksonville State University
Paul Horton, Indian River State College
Adam Kalkstein, United States Military Academy
Kody Kuehnl, Franklin University

Judith Lemons, Hannibal LaGrange University
Randal Mandock, Clark Atlanta University
Kevin Marty, Imperial Valley University
Gustavo Morales, Valencia College
Malcolm Skinner, Brigham Young University

Last, but certainly not least, we gratefully acknowledge the support and encouragement of our wives, Nancy Lutgens and Joanne Bannon. Preparation of *Foundations of Earth Science,* seventh edition, would have been far more difficult without their patience and understanding.

Fred Lutgens
Ed Tarbuck

About Our Sustainability Initiatives

Pearson recognizes the environmental challenges facing this planet, as well as acknowledges our responsibility in making a difference. This book is carefully crafted to minimize environmental impact. The binding, cover, and paper come from facilities that minimize waste, energy consumption, and the use of harmful chemicals. Pearson closes the loop by recycling every out-of-date text returned to our warehouse.

Along with developing and exploring digital solutions to our market's needs, Pearson has a strong commitment to achieving carbon-neutrality. As of 2009, Pearson became the first carbon- and climate-neutral publishing company. Since then, Pearson remains strongly committed to measuring, reducing, and offsetting our carbon footprint.

The future holds great promise for reducing our impact on Earth's environment, and Pearson is proud to be leading the way. We strive to publish the best books with the most up-to-date and accurate content, and to do so in ways that minimize our impact on Earth. To learn more about our initiatives, please visit www.pearson.com/responsibility.

MasteringGeology™   www.masteringgeology.com

MasteringGeology delivers engaging, dynamic learning opportunities—focused on course objectives and responsive to each student's progress—that are proven to help students absorb course material and understand difficult concepts.

Encounter Activities provide rich, interactive explorations of Earth science concepts using the dynamic features of Google Earth™ to visualize and explore Earth's physical landscape. Dynamic assessment includes questions related to core Earth science concepts. All explorations include corresponding Google Earth KMZ media files, and questions include hints and specific wrong-answer feedback to help coach students towards mastery of the concepts.

NEW! Inquiry-based interactive simulations, developed to allow students to manipulate Earth processes, assist students in mastering the most difficult Earth science processes as identified by today's instructors.

NEW! GigaPan® Activities take advantage of the GigaPan high-resolution panoramic picture technology developed by Carnegie Mellon University in conjunction with NASA. Photos and accompanying questions correlate with concepts in the student book.

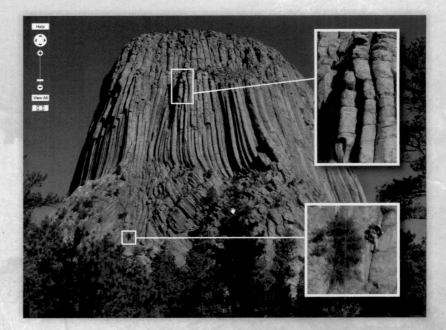

Geoscience Animations and Activities illuminate the most difficult-to-visualize topics from across the physical geosciences. MasteringGeology allows instructors to easily assign the animations and corresponding assessment questions, all of which include hints and specific wrong-answer feedback.

GEODe Activities provide an interactive visual walkthrough of each chapter's core content through animations, videos, illustrations, photographs, and narration. Activities include assessment questions to test those concepts with hints and specific wrong-answer feedback.

Give It Some Thought questions and problems relate back to each chapter's learning objectives, challenge learners by involving them in activities that require higher-order thinking skills such as synthesis, analysis, and application.

MasteringGeology™   www.masteringgeology.com

Quickly monitor and display student results

With the Mastering gradebook and diagnostics, instructors will be better informed about students' progress than ever before. Mastering captures the step-by-step work of every student—including wrong answers submitted, hints requested, and time taken at every step of every problem—all providing unique insight into the most common misconceptions of the class.

The **Gradebook** records all scores for automatically graded assignments. Shades of red highlight struggling students and challenging assignments.

Diagnostics provide unique insight into class and student performance. With a single click, charts summarize the most difficult items, vulnerable students, grade distribution, and score improvement over the duration of the course.

With a single click, **Individual Student Performance Data** provides **at-a-glance statistics** into each individual student's performance, including time spent on the item, number of hints opened, and number of wrong and correct answers submitted.

Easily measure student performance against learning outcomes

Learning Outcomes

MasteringGeology provides quick and easy access to information on student performance against learning outcomes and makes it easy for instructors to share those results.

Instructors can:

- Quickly add learning outcomes or use publisher-provided ones to track student performance and report it to administration.
- View class and individual student performance against specific learning outcomes.
- Effortlessly export results to a spreadsheet and further customize and/or share with chairs, deans, administrators, and/or accreditation boards.

Easy to Customize

Instructors can customize publisher-provided problems or quickly add their own. MasteringGeology makes it easy for instructors to edit any questions or answers, import their own questions, and quickly add images, links, and files to further enhance the student experience.

Instructors can upload their own video and audio files from their hard drives to share with students, as well as record video from their computer's webcam directly into MasteringGeology—no plug-ins required. Students can download video and audio files to their local computer or launch them in Mastering to view the content.

Pearson eText gives students access to *Foundations of Earth Science, Seventh Edition* whenever and wherever they can access the Internet. The eText pages look exactly like the printed text, and include powerful interactive and customization functions. Users can create notes, highlight text in different colors, create bookmarks, zoom, click hyperlinked words and phrases to view definitions, and view as a single page or as two pages. Pearson eText links students to associated media files, enabling them to view an animation as they read the text, and offers a full-text search and the ability to save and export notes. The Pearson eText also includes embedded URLs in the chapter text with active links to the Internet.

NEW! The Pearson eText app is a great companion to Pearson's eText browser-based book reader. It allows existing subscribers who view their Pearson eText titles on a Mac or PC to additionally access their titles in a bookshelf on the iPad™ and Android™ device either online or via download.

FOUNDATIONS OF EARTH SCIENCE

SEVENTH EDITION

Introduction to
Earth Science

Each statement represents the primary learning objective for the corresponding major heading within the Introduction. After you complete the Introduction, you should be able to:

I.1 List and describe the sciences that collectively make up Earth science.

I.2 Describe the four "spheres" that comprise Earth's natural environment.

I.3 Define *system* and explain why Earth is considered to be a system.

I.4 Discuss the scales of space and time in Earth science.

I.5 Summarize some important connections between people and the physical environment.

I.6 Discuss the nature of scientific inquiry and distinguish between a hypothesis and a theory.

An afternoon rain storm near Muddy Creek in southern Utah. (Photo by Michael Collier)

The spectacular eruption of a volcano, the magnificent scenery of a rocky coast, and the destruction created by a hurricane are all subjects for the Earth scientist. The study of Earth science deals with many fascinating and practical questions about our environment. What forces produce mountains? Why is our daily weather so variable? Is climate really changing? How old is Earth, and how is it related to the other planets in the solar system? What causes ocean tides? What was the Ice Age like? Will there be another? Can a successful well be located at this site?

The subject of this text is *Earth science.* To understand Earth is not an easy task because our planet is not a static and unchanging mass. Rather, it is a dynamic body with many interacting parts and a long and complex history.

I.1 What Is Earth Science?

List and describe the sciences that collectively make up Earth science.

Earth science is the name for all the sciences that collectively seek to understand Earth and its surrounding space. It includes geology, oceanography, meteorology, and astronomy.

In this text, Units 1 through 4 focus on the science of **geology**, a word that literally means "study of Earth." Geology is traditionally divided into two broad areas—physical and historical.

Physical geology examines the materials comprising Earth and seeks to understand the many processes that operate beneath and upon its surface. Earth is a dynamic, ever-changing planet. Internal forces create earthquakes, build mountains, and produce volcanic structures. At the surface, external processes break rock apart and sculpt a broad array of landforms (**Figure I.1**). The erosional effects of water, wind, and ice result in a great diversity of landscapes. Because rocks and minerals form in response to Earth's internal and external processes, the study of Earth materials is basic to an understanding of our planet.

In contrast to physical geology, the aim of *historical geology* is to understand the origin of Earth and the evolution of the planet through its 4.6-billion-year history. It strives to establish an orderly, chronological arrangement of the multitude of physical and biological changes that have occurred in the geologic past. Logically, the study of physical geology precedes the study of Earth's history because we must first understand how Earth works before we attempt to unravel its past.

Unit 5, *The Global Ocean*, is devoted to **oceanography**. Oceanography is actually not a separate and distinct science. Rather, it involves the application of all sciences in a comprehensive and interrelated study of the oceans in all their aspects and relationships. Oceanography integrates chemistry, physics, geology, and biology. It includes the study of the composition and movements of seawater, as well as coastal processes, seafloor topography, and marine life.

Unit 6, *Earth's Dynamic Atmosphere*, examines the mixture of gases that is held to the planet by gravity and thins rapidly with altitude. Acted on by the combined effects of Earth's motions and energy from the Sun, the formless and invisible atmosphere reacts by producing an infinite variety of weather, which in turn creates the basic pattern of global climates. **Meteorology** is the study of the atmosphere and the processes that produce weather and climate. Like oceanography, meteorology involves the application of other sciences in an integrated study of the thin layer of air that surrounds Earth.

Figure I.1 Alaska's Mount Foraker Internal forces created this mountain and external processes are sculpting it. This mountain in Denali National Park is the fourth highest peak in the United States. (Photo by Michael Collier)

Unit 7, *Earth's Place in the Universe,* demonstrates that an understanding of Earth requires that we study our planet in relation to the larger universe. Because Earth is related to all of the other objects in space, the science of **astronomy**—the study of the universe—is useful in determining the origins of our own environment. As we are so closely acquainted with the planet on which we live, it is easy to forget that Earth is just a tiny object in a vast universe. Indeed, Earth is subject to the same physical laws that govern the many other objects populating the great expanse of space. Thus, to understand explanations of our planet's origin, it is useful to learn something about the other members of our solar system. Moreover, it is helpful to view our solar system as part of the great assemblage of stars that comprise our galaxy, which is only one of many galaxies.

Understanding Earth science is challenging because our planet is a dynamic body with many interacting parts and a complex history. Throughout its long existence, Earth has been changing. In fact, it is changing as you read this page and will continue to do so into the foreseeable future. Sometimes the changes are rapid and violent, as when severe storms, landslides, or volcanic eruptions occur. Just as often, change takes place so gradually that

it goes unnoticed during a lifetime. Scales of size and space also vary greatly among the phenomena studied in Earth science.

Earth science is often perceived as science that is done outdoors, and rightly so. A great deal of what Earth scientists study is based on observations and experiments performed in the field. But Earth science is also conducted in the laboratory, where the study of various Earth materials provides insights into many basic processes. For example, the creation of complex computer models in the laboratory allows for the simulation of our planet's complicated climate system. Frequently, Earth scientists require an understanding and application of knowledge and principles from physics, chemistry, and biology. Geology, oceanography, meteorology, and astronomy are sciences that seek to expand our knowledge of the natural world and our place in it

---

I.1 CONCEPT CHECKS

1. List and briefly describe the sciences that collectively make up Earth science.
2. Name the two broad subdivisions of geology and distinguish between them.

I.2 Earth's Spheres

Describe the four "spheres" that comprise Earth's natural environment.

The images in **Figure I.2** are considered classics because they let humanity see Earth differently than ever before. These early views profoundly altered our conceptualizations of Earth and remain powerful images decades after they were first viewed. Such images remind us that our home is, after all, a planet—small, self-contained, and in some ways even fragile.

As we look closely at our planet from space, it becomes clear that Earth is much more than rock and soil. In fact, the most conspicuous features in

This image taken from *Apollo 17* **in December 1972 is perhaps the first to be called "The Blue Marble". The dark blue ocean and swirling cloud patterns remind us of the importance of the oceans and atmosphere.**

View called "Earthrise" that greeted *Apollo 8* **astronauts as their spacecraft emerged from behind the Moon in December 1968. This classic image let people see Earth differently than ever before.**

Figure I.2 Two classic views Earth from space. (NASA)

Figure I.3 Interactions among Earth's spheres The shoreline is considered an *interface*—a common boundary where different parts of a system interact. In this scene, ocean waves (hydrosphere) that were created by the force of moving air (atmosphere) break against a rocky shore (geosphere). The force of the water can be powerful, and the erosional work that is accomplished can be great. (Photo by Michael Collier)

Figure I.2 are not continents, but swirling clouds suspended above the surface and the vast global ocean. These features emphasize the importance of air and water to our planet.

The closer view of Earth from space shown in Figure I.2 helps us appreciate why traditionally the physical environment is divided into three major parts: the water portion of our planet, called the hydrosphere; Earth's gaseous envelope, called the atmosphere; and, of course, the solid Earth, or geosphere.

It should be emphasized that our environment is highly integrated and not dominated by water, air, or rock alone. Rather, it is characterized by continuous interactions as air comes in contact with rock, rock with water, and water with air. Moreover, the biosphere, the totality of life forms on our planet, extends into each of the three physical realms and is an equally integral part of the planet. Thus, Earth can be thought of as consisting of four major

spheres: the hydrosphere, atmosphere, geosphere, and biosphere.

The interactions among Earth's four spheres are incalculable. **Figure I.3** provides an easy-to-visualize example. The shoreline is an obvious meeting place for rock, water, and air. In this scene, ocean waves that were created by the drag of air moving across the water are breaking against the rocky shore. The force of the water can be powerful, and the erosional work that is accomplished can be great.

Hydrosphere

Earth is sometimes called the *blue planet*. Water, more than anything else, makes Earth unique. The **hydrosphere** is a dynamic mass of water that is continually moving, evaporating from the oceans to the atmosphere, precipitating to the land, and flowing back to the ocean. The global ocean is certainly the most prominent feature of the hydrosphere, blanketing nearly 71 percent of Earth's surface to an average depth of about 3800 meters (12,500 feet). It accounts for more than 96 percent of Earth's water (**Figure I.4**). The hydrosphere also includes the freshwater found underground and in streams, lakes, and glaciers. Moreover, water is an important component of all living things.

Although these latter sources constitute just a tiny fraction of the total, they are much more important than their meager percentage indicates. In addition to

Hydrosphere

Fresh water 2.56%

Saline water 97.44%

Oceans 96.5%

Saline groundwater and lakes 0.94%

Groundwater 0.77%

Glaciers 1.76%

Streams, lakes, soil moisture, atmosphere, etc. account for 0.03% (3/100 of 1%)

Michael Collier

Stream channel

Nearly 69% of Earth's fresh water is locked up in glaciers.

Bernhard Edmaier/ Photo Researchers, Inc.

Glaciers

Although fresh groundwater represents less than 1% of the hydrosphere, it accounts for 30% of all fresh water and about 96% of all liquid fresh water.

Michael Collier

Groundwater (spring)

Figure I.4 The water planet Distribution of water in the hydrosphere.

providing the freshwater that is so vital to life on the continents, streams, glaciers, and groundwater are responsible for sculpturing and creating many of our planet's varied landforms. Water in the atmosphere, in the form of clouds and water vapor, plays a critical role in weather and climate processes.

Atmosphere

Earth is surrounded by a life-giving gaseous envelope called the **atmosphere** (**Figure I.5**). When we watch a high-flying jet plane cross the sky, it seems that the atmosphere extends upward for a great distance. However, when compared to the thickness (radius) of the solid Earth (about 6400 kilometers or 4000 miles), the atmosphere is a very shallow layer. Despite its modest

dimensions, this thin blanket of air is nevertheless an integral part of the planet. It not only provides the air that we breathe, but it also acts to protect us from the dangerous radiation emitted by the Sun. The energy exchanges that continually occur between the atmosphere and Earth's surface, and between the atmosphere and space, produce the effects we call *weather* and *climate*. Climate has a strong influence on the nature and intensity of Earth's surface processes. When climate changes, these processes respond.

If, like the Moon, Earth had no atmosphere, our planet would be lifeless because many of the processes and interactions that make Earth's surface such a dynamic place could not operate. Without weathering and erosion, the face of our planet might more closely resemble the surface of the Moon, which has not changed appreciably in nearly 3 billion years.

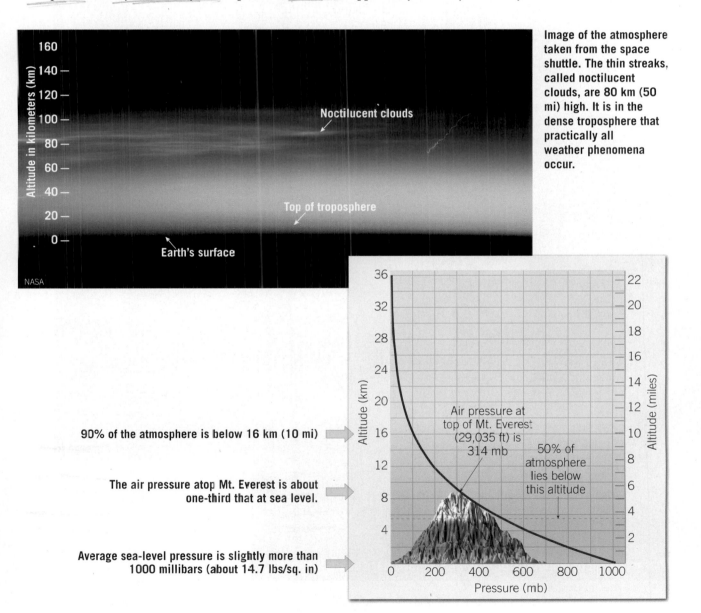

Image of the atmosphere taken from the space shuttle. The thin streaks, called noctilucent clouds, are 80 km (50 mi) high. It is in the dense troposphere that practically all weather phenomena occur.

90% of the atmosphere is below 16 km (10 mi)

The air pressure atop Mt. Everest is about one-third that at sea level.

Average sea-level pressure is slightly more than 1000 millibars (about 14.7 lbs/sq. in)

Air pressure at top of Mt. Everest (29,035 ft) is 314 mb

50% of atmosphere lies below this altitude

Figure I.5 A shallow layer The atmosphere is an integral part of the planet. (NASA)

Biosphere

The **biosphere**, includes all life on Earth (**Figure I.6**). Ocean life is concentrated in the sunlit waters of the sea. Most life on land is also concentrated near the surface, with tree roots and burrowing animals reaching a few meters underground and flying insects and birds reaching a kilometer or so above the surface. A surprising variety of life forms are also adapted to extreme environments. For example, on the deep ocean floor where pressures are extreme and no light penetrates, there are places where vents spew hot, mineral-rich fluids that support communities of exotic life forms. On land,

some bacteria thrive in rocks as deep as 4 kilometers (2.5 miles) and in boiling hot springs. Moreover, air currents can carry microorganisms many kilometers into the atmosphere. But even when we consider these extremes, life still must be thought of as being confined to a narrow band very near Earth's surface.

Plants and animals depend on the physical environment for the basics of life. However, organisms do more than just respond to this environment. Through countless interactions, life forms help maintain and alter their physical environment. Without life, the makeup and nature of the geosphere, hydrosphere, and atmosphere would be very different.

The ocean contains a significant portion of Earth's biosphere. Modern coral reefs are unique and complex examples and are home to about 25% of all marine species. Because of this diversity, they are sometimes referred to as the ocean equivalent of a rain forest.

A.

Tropical rain forests are characterized by hundreds of different species per square kilometer.

Figure I.6 The biosphere The biosphere, one of Earth's four spheres, includes all life. (Coral reef photo by Darryl Leniuk/AGE Fotostock; rain forest photo by AGE Fotostock/SuperStock)

Geosphere

Lying beneath the atmosphere and the ocean is the solid Earth or **geosphere**. The geosphere extends from the surface to the center of the planet, a depth of nearly 6400 kilometers (4000 miles), making it by far the largest of Earth's spheres. Much of our study of the solid Earth focuses on the more accessible surface and near-surface features, but it is worth noting that many of these features are linked to the dynamic behavior of Earth's interior. **Figure I.7** provides a diagrammatic view of Earth's interior. As you can see, it is not uniform but is divided into various layers. Based on compositional differences, three principal layers are identified: a dense inner sphere called the **core**; the less dense **mantle**; and the **crust**, which is the light and very thin outer skin of Earth. The crust is not a layer of uniform thickness. It is thinnest beneath the oceans and

thickest where continents exist. Although the crust may seem insignificant when compared with the other layers of the geosphere, which are much thicker, it was created by the same general processes that formed Earth's present structure. Thus, the crust is important in understanding the history and nature of our planet.

The solid Earth is also divided based on how materials behave when various forces and stresses are applied. The term **lithosphere** refers to the rigid outer layer that includes the crust and uppermost mantle. Beneath the rigid rocks that compose the lithosphere, the rocks of the **asthenosphere** are weak and are able to slowly flow in response to the uneven distribution of heat deep within Earth.

The two principal divisions of Earth's surface are the continents and the ocean basins. The most obvious difference between these two diverse provinces is their relative levels. The average elevation of the continents above sea

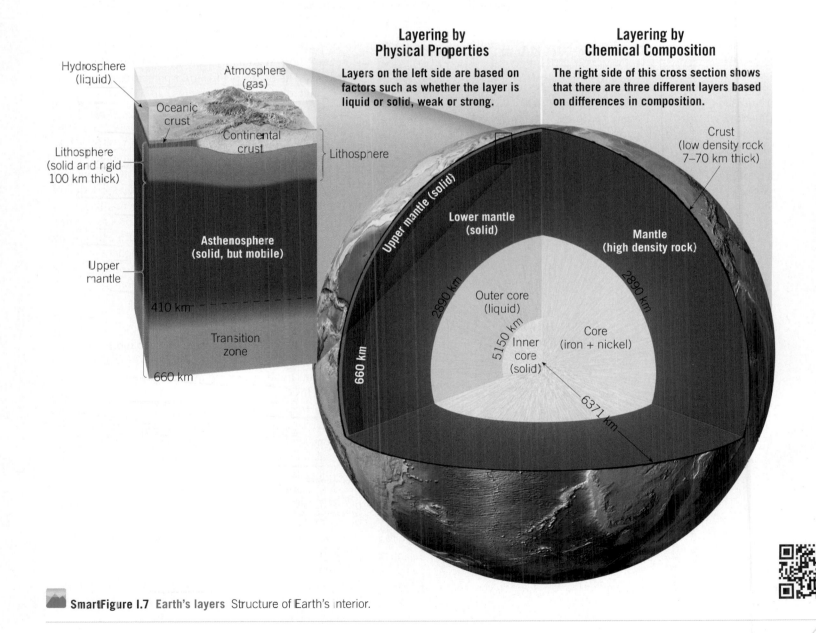

SmartFigure I.7 Earth's layers Structure of Earth's interior.

level is about 840 meters (2750 feet), whereas the average depth of the oceans is about 3800 meters (12,500 feet). Thus, the continents stand on the average 4640 meters (about 4.6 kilometers or nearly 3 miles) above the level of the ocean floor.

Soil, the thin veneer of material at Earth's surface that supports the growth of plants, may be thought of as part of all four spheres. The solid portion is a mixture of weathered rock debris (geosphere) and organic matter from decayed plant and animal life (biosphere). The decomposed and disintegrated rock debris is the product of weathering processes that require air (atmosphere) and water (hydrosphere).

Air and water also occupy the open spaces between solid particles.

I.2 CONCEPT CHECKS

1. List and briefly define the four "spheres" that constitute our environment.
2. Compare the height of the atmosphere to the thickness of the geosphere.
3. How much of Earth's surface do oceans cover? How much of the planet's total water supply do oceans represent?
4. Briefly summarize Earth's layered structure.
5. To which sphere does soil belong?

I.3 Earth as a System

Define *system* and explain why Earth is considered to be a system.

Figure I.8 Deadly debris flow This image provides an example of interactions among different parts of the Earth system. On January 10, 2005, extraordinary rains triggered this debris flow (popularly called a *mudslide*) in the coastal community of La Conchita, California. (AP Wideworld Photo)

Anyone who studies Earth soon learns that our planet is a dynamic body with many separate but interacting parts or spheres. The hydrosphere, atmosphere, biosphere, geosphere, and all of their components can be studied separately. However, the parts are not isolated. Each is related in many ways to the others to produce a complex and continuously interacting whole that we call the **Earth system**.

A simple example of the interactions among different parts of the Earth system occurs every winter as moisture evaporates from the Pacific Ocean and subsequently falls as rain in the hills of southern California, triggering destructive landslides (**Figure I.8**). The processes that move water from the hydrosphere to the atmosphere and then to the geosphere have a profound impact on the physical landscape and on the plants and animals (including humans) that inhabit the affected regions.

What Is a System?

Scientists have recognized that in order to more fully understand our planet, they must learn how its individual components (land, water, air, and life forms) are interconnected. This endeavor, called *Earth system science,* aims to study Earth as a *system* composed of numerous interacting parts, or *subsystems.* Using an interdisciplinary approach, those engaged in Earth system science attempt to achieve the level of understanding necessary to comprehend and solve many of our global environmental problems.

A **system** is a group of interacting, or interdependent, parts that form a complex whole. Most of us hear and use the term frequently. We may service our car's cooling *system,* make use of the city's transportation *system,* and participate in the political *system.* A news report might inform us of an approaching weather *system.* Further, we know that Earth is just a small part of a large system known as the *solar system,* which, in turn, is a subsystem of an even larger system called the Milky Way galaxy.

The Parts Are Linked

The parts of the Earth system are linked so that a change in one part can produce changes in any or all of the other parts. For example, when a volcano erupts, lava from Earth's interior may flow out at the surface and block a nearby valley. This new obstruction influences the region's drainage system by creating a lake or causing streams to change course. The large quantities of volcanic ash and gases that can be emitted during an eruption might be blown high into the atmosphere and influence the amount of solar energy that can reach Earth's surface. The result could be a drop in air temperatures over the entire hemisphere.

Where the surface is covered by lava flows or a thick layer of volcanic ash, existing soils are buried. This causes the soil-forming processes to begin anew to transform the new surface material into soil (**Figure I.9**). The soil that eventually forms will reflect the interactions among many parts of the Earth system—the volcanic parent material, the type and rate of weathering, and the impact of biological activity. Of course, there would also be significant changes in the biosphere. Some organisms and their habitats would be eliminated by the lava and ash, whereas new settings for life, such as the lake, would be created.

The potential climate change could also affect sensitive life forms.

Time and Space Scales The Earth system is characterized by processes that vary on spatial scales from fractions of millimeters to thousands of kilometers. Time scales for Earth's processes range from milliseconds to billions of years. As we learn about Earth, it becomes increasingly clear that despite significant separations in distance or time, many processes are connected and a change in one component can influence the entire system.

Energy for the Earth System The Earth system is powered by energy from two sources. The Sun drives external processes that occur in the atmosphere, hydrosphere, and at Earth's surface. Weather and climate, ocean circulation, and erosional processes are driven by energy from the Sun. Earth's interior is the second source of energy. Heat remaining from when our planet formed and heat that is continuously generated by radioactive decay power the internal processes that produce volcanoes, earthquakes, and mountains.

People and the Earth System Humans are *part of* the Earth system; a system in which the living and nonliving components are entwined and interconnected. Therefore,

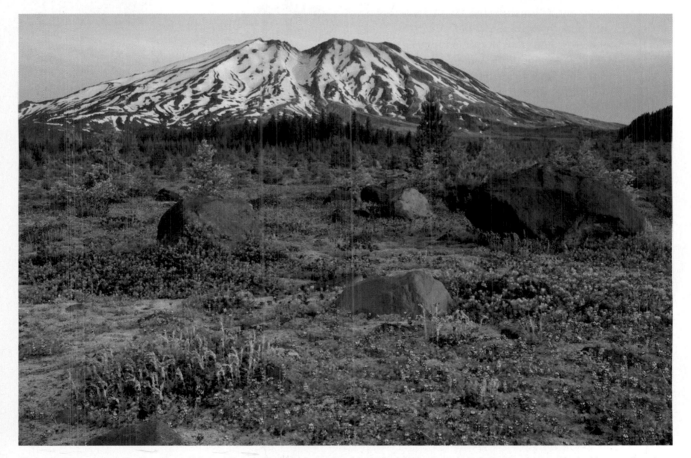

Figure I.9 Change is a constant. When Mount St. Helens erupted in May 1980, the area shown here was buried by a volcanic mudflow. Now plants are reestablished and new soil is forming. (Photo by Terry Donnelly/Alamy Images)

our actions produce changes in all of the other parts. When we burn gasoline and coal, build seawalls along the shoreline, dispose of our wastes, and clear the land, we cause other parts of the system to respond, often in unforeseen ways. Throughout this text, you will learn about many of Earth's subsystems: the hydrologic system, the tectonic (mountain-building) system, and the climate system, to name a few. Remember that these components *and we humans* are all part of the complex interacting whole that we call the Earth system.

I.4 Scales of Space and Time in Earth Science

Discuss the scales of space and time in Earth science.

When we study Earth, we must contend with a broad array of space and time scales (**Figure I.10**). Some phenomena are relatively easy for us to imagine, such as the size and duration of an afternoon thunderstorm or the dimensions of a sand dune. Other phenomena are so vast or so small that they are difficult to imagine. The number of stars and distances in our galaxy (and beyond!) or the internal arrangement of atoms in a mineral crystal are examples of such phenomena.

Some of the events we study occur in fractions of a second. Lightning is an example. Other processes extend over spans of tens or hundreds of millions of years. The lofty Himalaya Mountains began forming nearly 50 million years ago, and they continue to develop today.

The concept of **geologic time**, the span of time since the formation of Earth, is new to many nonscientists. People are accustomed to dealing with increments of time that are measured in hours, days, weeks, and years. Our history books often examine events over spans of centuries, but even a century is difficult to appreciate fully. For most of us, someone or something that is 90 years old is *very old,* and a 1000-year-old artifact is *ancient.*

By contrast, those who study Earth science must routinely deal with vast time periods—millions or billions (thousands of millions) of years. When viewed in the context of Earth's nearly 4.6-billion-year history, an event that occurred 100 million years ago may be characterized as "recent" by a geologist, and a rock sample that has been dated at 10 million years may be called "young."

An appreciation for the *magnitude of geologic time* is important in the study of our planet because many processes are so gradual that vast spans of time are needed before significant changes occur. How long is 4.6 billion years? If you were to begin counting at the rate of one number per second and continued 24 hours a day, seven days a week and never stopped, it would take about two lifetimes (150 years) to reach 4.6 billion!

The previous is just one of many analogies that have been conceived in an attempt to convey the magnitude of geologic time. Although helpful, all of them, no matter

Figure I.10   From atoms to galaxies Earth science involves investigations of phenomena that range in size from atoms to galaxies and beyond.

What if we compress the 4.6 billion years of Earth history into a single year?

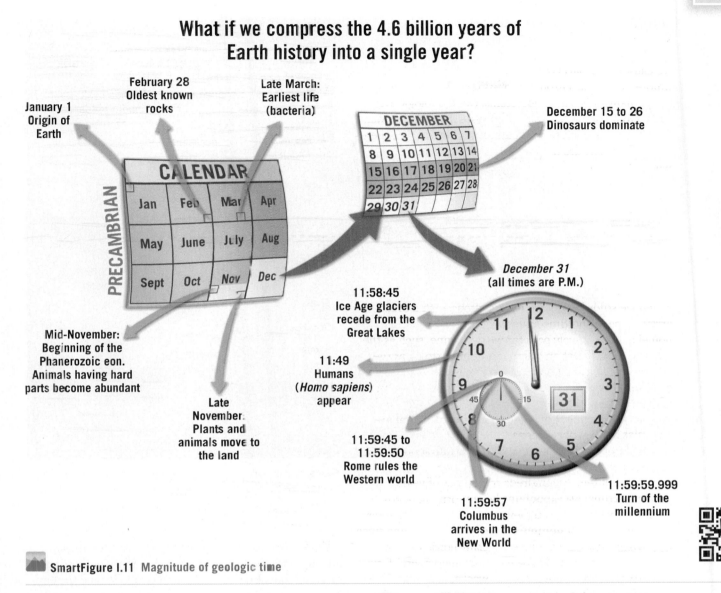

SmartFigure I.11 Magnitude of geologic time

how clever, only begin to help us comprehend the vast expanse of Earth history. **Figure I.11** provides another interesting way of viewing the age of Earth.

Over the past 200 years or so, Earth scientists have developed a *geologic time scale* of Earth history. It subdivides the 4.6-billion-year history of Earth into many different units and provides a meaningful time frame within which the events of the geologic past are arranged (see Figure 8.22, page 288). The geologic time scale and the principles used to develop it are examined at some length in Chapter 8.

I.4 CONCEPT CHECKS

1. List two examples of size/space scales in Earth science that are at opposite ends of the spectrum.
2. How old is Earth?
3. When all of geologic time is compressed into a single year, how much time has elapsed since Columbus arrived in the New World?

I.5 Resources and Environmental Issues

Summarize some important connections between people and the physical environment.

Environment refers to everything that surrounds and influences an organism. Some of these things are biological and social, but others are nonliving. The factors in this latter category are collectively called our physical environment. The **physical environment** encompasses water, air, soil, and rock, as well as conditions such as

Figure I.12 Surface coal mine Coal remains an important source of energy worldwide. In the United States it represents about 20 percent of our total energy consumption. Most coal is used to generate electricity. In the United States, coal was responsible for generating 46 percent of our electricity in 2011.

temperature, humidity, and sunlight. The phenomena and processes studied by the Earth sciences are basic to an understanding of the physical environment. In this sense, most of Earth science may be characterized as environmental science.

However, when the term *environmental* is applied to Earth science today, it usually means relationships between people and the physical environment. Application of the Earth sciences is necessary to understand and solve problems that arise from these interactions.

We can dramatically influence natural processes. For example, river flooding is natural, but the magnitude and frequency of flooding can be changed significantly by human activities such as clearing forests, building cities, and constructing dams. Unfortunately, natural systems do not always adjust to artificial changes in ways we can anticipate. Thus, an alteration to the environment that was intended to benefit society may have the opposite effect.

Resources

Resources are an important focus of the Earth sciences that is of great value to people. Resources include water and soil, a great variety of metallic and nonmetallic minerals, and energy. Together, they form the very foundation of modern civilization (**Figure I.12**). The Earth sciences deal not only with the formation and occurrence of these vital resources but also with maintaining supplies and the environmental impact of their extraction and use.

Few people who live in highly industrialized nations realize the quantity of resources needed to maintain their present standard of living. **Figure I.13** shows the annual per capita consumption of several important metallic and nonmetallic mineral resources for the United States. This is each person's prorated share of the materials required by industry to provide the vast array of products modern society demands. Figures for other highly industrialized countries are comparable.

Resources are commonly divided into two broad categories. Some are classified as **renewable**, which means that they can be replenished over relatively short time spans. Common examples are plants and animals for food, natural fibers for clothing, and forest products for lumber and paper. Energy from flowing water, wind, and the Sun is also considered renewable (**Figure I.14**).

By contrast, many other basic resources are classified as **nonrenewable**. Important metals such as iron, aluminum, and copper fall into this category, as do our most important fuels: oil, natural gas, and coal. Although these and other resources continue to form, the processes that

Nonmetallic Resources

5713 kg (12695 lbs)
Stone

4025 kg (8945 lbs)
Sand and gravel

360 kg (790 lbs)
Cement

137 kg (304 lbs)
Clays

178 kg (395 lbs)
Salt

162 kg (361 lbs)
Phosphate rock

302 kg (672 lbs)
Other nonmetals

Metallic Resources

35 kg (77 lbs)
Aluminum

6 kg (14 lbs)
Lead

6 kg (13 lbs)
Manganese

249 kg (553 lbs)
Iron

11 kg (25 lbs)
Copper

5 kg (11 lbs)
Zinc

9 kg (20 lbs)
Other metals

Energy Resources

3500 kg (7700 lbs)
Petroleum

3700 kg (8140 lbs)
Coal

3850 kg (8470 lbs)
Natural gas

Figure I.13 How much do each of us use? The annual per capita consumption of nonmetallic and metallic resources for the United States is about 11,000 kilograms (12 tons). About 97 percent of the materials used are nonmetallic. The per capita use of oil, coal, and natural gas exceeds 11,000 kilograms. (U.S. Geological Survey)

Figure I.14 Solar energy is renewable. These solar collectors focus sunlight onto collection pipes filled with a fluid. The heat is used to make steam that drives turbines used to generate electricity. (Photo by Jim West/Alamy Images)

Figure I.15 Urban air pollution A severe air pollution episode at Beijing, China, on March 18, 2008. Fuel combustion by factories, power plants, and motor vehicles provided a high proportion of the pollutants. Meteorological factors determine whether pollutants remain "trapped" in the city or are dispersed. (Photo by Ng Han Guar/Associated Press)

create them are so slow that significant deposits take millions of years to accumulate. In essence, Earth contains *fixed* quantities of these substances. When the present supplies are mined or pumped from the ground, there will be no more. Although some nonrenewable resources, such as aluminum, can be used over and over again, others, such as oil, cannot be recycled.

How long will the remaining supplies of basic resources last? How long can we sustain the rising standard of living in today's industrial countries and still provide for the growing needs of developing regions? How much environmental deterioration are we willing to accept in pursuit of basic resources? Can alternatives be found? If we are to cope with an increasing demand and a growing world population, it is important that we have some understanding of our present and potential resources.

Environmental Problems

In addition to the quest for adequate mineral and energy resources, the Earth sciences must also deal with a broad array of environmental problems. Some

are local, some are regional, and still others are global in extent. Serious difficulties face developed and developing nations alike. Urban air pollution, acid rain, ozone depletion, and global climate change are just a few that pose significant threats (**Figure I.15**). Other problems involve the loss of fertile soils to erosion, the disposal of toxic wastes, and the contamination

Figure I.16 Tornadoes are natural hazards. A devastating tornado struck Joplin, Missouri, on Sunday, May 22, 2011. There were more than 150 tornado deaths and nearly 1000 injured. (Photo by c51/ZUMA Press/Newscom)

Earthquakes, landslides, volcanic eruptions, floods, and hurricanes are just five of the many risks. Others, such as drought, although not as spectacular, are nevertheless equally important environmental concerns. It should be emphasized that *natural hazards* are *natural phenomena* that only become hazards when people try to live where these phenomena occur. In many cases, the threat of natural hazards is aggravated by increases in population as more people crowd into places where an impending danger exists.

Complicating all environmental issues is rapid world population growth and everyone's aspiration to a better standard of living. This means ballooning demand for resources *and* growing pressure for people to dwell in environments having significant natural hazards. Therefore, an understanding of Earth is essential not only for the location and recovery of basic resources, but also for dealing with the human impact on the environment and minimizing the effects of natural hazards. Knowledge about our planet and how it works is necessary to our survival and well-being. Earth is the only suitable habitat we have, and its resources are limited.

and depletion of water resources. The list continues to grow.

In addition to human-induced and human-accentuated problems, people must also cope with the many **natural hazards** posed by the physical environment (**Figure I.16**).

I.5 CONCEPT CHECKS

1. Define *physical environment*.
2. Distinguish between renewable and nonrenewable resources. What are some examples of each?
3. List at least four phenomena that can be regarded as natural hazards.

I.6 The Nature of Scientific Inquiry

Discuss the nature of scientific inquiry and distinguish between a hypothesis and a theory.

As members of a modern society, we are constantly reminded of the benefits derived from science. But what exactly is the nature of scientific inquiry? Science is a process of producing knowledge. The process depends both on making careful observations and on creating explanations that make sense of the observations. Developing an understanding of how science is done and how scientists work is an important theme that appears throughout this text. You will explore the difficulties in gathering data and some of the ingenious methods that have been developed to overcome these difficulties. You will also see examples of how hypotheses are formulated and tested, as well as learn about the evolution and development of some major scientific theories.

All science is based on the assumption that the natural world behaves in a consistent and predictable manner that is comprehensible through careful, systematic study. The overall goal of science is to discover the underlying

patterns in nature and then to use this knowledge to make predictions about what should or should not be expected, given certain facts or circumstances. For example, by understanding the processes and conditions that produce certain cloud types, meteorologists are often able to predict the approximate time and place of their formation.

The development of new scientific knowledge involves some basic logical processes that are universally accepted. To determine what is occurring in the natural world, scientists collect scientific *facts* through observation and measurement (**Figure I.17**). The types of facts that are collected often seek to answer a well-defined question about the natural world. Because some error is inevitable, the accuracy of a particular measurement or observation is always open to question. Nevertheless, these data are essential to science and serve as the springboard for the development of scientific theories.

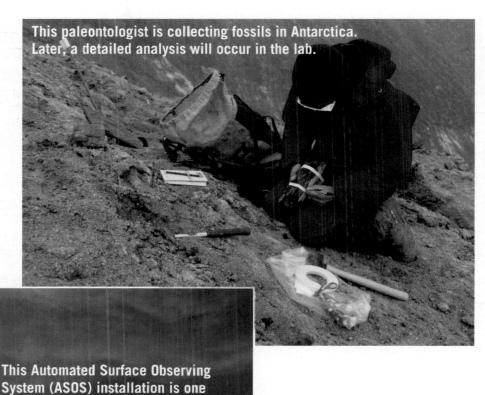

This paleontologist is collecting fossils in Antarctica. Later, a detailed analysis will occur in the lab.

This Automated Surface Observing System (ASOS) installation is one of nearly 900 in use for data gathering as part of the U.S. primary surface observing network.

Figure I.17 Observation and measurement Gathering data and making careful observations are a basic part of scientific inquiry. (Instrument photo by Bobbé Christopherson; paleontologist photo by British Antarctic Survey/Photo Researchers, Inc.)

Hypothesis

Once facts have been gathered and principles have been formulated to describe a natural phenomenon, investigators try to explain how or why things happen in the manner observed. They often do this by constructing a tentative (or untested) explanation, which is called a scientific **hypothesis**. It is best if an investigator can formulate more than one hypothesis to explain a given set of observations. If an individual scientist is unable to devise multiple hypotheses, others in the scientific community will almost always develop alternative explanations. A spirited debate frequently ensues. As a result, extensive research is conducted by proponents of opposing hypotheses, and the results are made available to the wider scientific community in scientific journals.

Before a hypothesis can become an accepted part of scientific knowledge, it must pass objective testing and analysis. If a hypothesis cannot be tested, it is not scientifically useful, no matter how interesting it might seem. The verification process requires that *predictions* be made based on the hypothesis being considered and that the predictions be tested by comparing them against objective observations of nature. Put another way, hypotheses must fit observations other than those used to formulate them in the first place. Those hypotheses that fail rigorous testing are ultimately discarded. The history of science is littered with discarded hypotheses. One of the best known is the Earth-centered model of the universe—a proposal that was supported by the apparent daily motion of the Sun, Moon, and stars around Earth. As the mathematician Jacob Bronowski so ably stated, "Science is a great many things, but in the end they all return to this: Science is the acceptance of what works and the rejection of what does not."

Theory

When a hypothesis has survived extensive scrutiny and when competing ones have been eliminated, a hypothesis may be elevated to the status of a scientific **theory**. In everyday language we may say, "That's only a theory." But a scientific theory is a well-tested and widely accepted view that the scientific community agrees best explains certain observable facts. Some theories that are extensively documented and extremely well supported are comprehensive in scope. One example, is the *nebular theory* discussed in Chapter 15, which explains the formation of our solar system. Another, called the *theory of plate tectonics* provides the framework for understanding the origin of mountains, earthquakes, and volcanic activity—ideas that are explored in some detail in Chapters 5, 6, and 7.

Scientific Methods

The process just described, in which researchers gather facts through observations and formulate scientific hypotheses and theories, is often referred to as the *scientific method*. Contrary to popular belief, the scientific method is not a standard recipe that scientists apply in a routine manner to unravel the secrets of our natural world. Rather, it is an endeavor that involves creativity and insight. Rutherford and Ahlgren put it this way: "Inventing hypotheses or theories to imagine how the world works and then figuring out how they can be put to the test of reality is as creative as writing poetry, composing music, or designing skyscrapers."[*]

[*]F. James Rutherford and Andrew Ahlgren, *Science for All Americans* (New York: Oxford University Press, 1990), p. 7.

There is not a fixed path that scientists always follow that leads unerringly to scientific knowledge. Nevertheless, many scientific investigations involve the following:

- a question is raised about the natural world.
- scientific data are collected that relate to the question.
- questions are posed that relate to the data and one or more working hypotheses are developed that may answer these questions.
- observations and experiments are developed to test the hypotheses.
- the hypotheses are accepted, modified, or rejected based on extensive testing.
- data and results are shared with the scientific community for critical examination and further testing.

Other scientific discoveries may result from purely theoretical ideas, which stand up to extensive examination. Some researchers use high-speed computers to create models that simulate what is happening in the "real" world. These models are useful when dealing with natural processes that occur on very long time scales or take place in extreme or inaccessible locations. Still other scientific advancements are made when a totally unexpected happening occurs during an experiment. These serendipitous discoveries are more than pure luck, for as Louis Pasteur said, "In the field of observation, chance favors only the prepared mind."

Scientific knowledge is acquired through several avenues, so it might be best to describe the nature of scientific inquiry as the methods of science rather than the scientific method. In addition, it should always be remembered that even the most compelling scientific theories are still simplified explanations of the natural world.

In this text, you will discover the results of centuries of scientific work. You will see the end product of millions of observations, thousands of hypotheses, and hundreds of theories. We have distilled all of this to give you a "briefing" in Earth science.

But realize that our knowledge of Earth is changing daily, as thousands of scientists worldwide make satellite observations, analyze drill cores from the seafloor, study earthquake waves, develop computer models to predict climate, examine the genetic codes of organisms, and discover new facts about our planet's long history. This new knowledge often updates hypotheses and theories. Expect to see many new discoveries and changes in scientific thinking in your lifetime.

1.6 CONCEPT CHECKS

1. How is a scientific hypothesis different from a scientific theory?
2. Summarize the basic steps followed in many scientific investigations.

CONCEPTS IN REVIEW: INTRODUCTION TO EARTH SCIENCE

1.1 What Is Earth Science?

List and describe the sciences that collectively make up Earth science.

KEY TERMS: Earth science, geology, oceanography, meteorology, astronomy

- Earth science includes geology, oceanography, meteorology, and astronomy.
- There are two broad subdivisions of geology. Physical geology studies Earth materials and the internal and external processes that create and shape Earth's landscape. Historical geology examines Earth history.
- The other Earth sciences seek to understand the oceans, the atmosphere's weather and climate, and Earth's place in the universe.

1.2 Earth's Spheres

Describe the four "spheres" that comprise Earth's natural environment.

KEY TERMS: hydrosphere, atmosphere, biosphere, geosphere, core, mantle, crust, lithosphere, asthenosphere

- Earth's physical environment is traditionally divided into three major parts: the solid Earth, called the geosphere; the water portion of our planet, called the hydrosphere; and Earth's gaseous envelope, called the atmosphere.
- A fourth Earth sphere is the biosphere, the totality of life on Earth. It is concentrated in a relatively thin zone that extends a few kilometers into the hydrosphere and geosphere, and a few kilometers up into the atmosphere.
- Of all the water on Earth, more than 96 percent is in the oceans, which cover nearly 71 percent of the planet's surface.

? The diagram represents Earth's layered structure. Is it showing layering based on physical properties or layering based on composition? Identify the lettered layers.

A.

B.

C.

2900 km

D.

5150 km

E.

6371 km

1.3 Earth as a System

Define *system* and explain why Earth is considered to be a system.

KEY TERMS: Earth system, system

- Although each of Earth's four spheres can be studied separately, they are all related in a complex and continuously interacting whole that is called the Earth system.
- Earth system science uses an interdisciplinary approach to integrate the knowledge of several academic fields in the study of our planet and its global environmental problems.
- The two sources of energy that power the Earth system are (1) the Sun, which drives the external processes that occur in the atmosphere, hydrosphere, and at Earth's surface, and (2) heat from Earth's interior that powers the internal processes that produce volcanoes, earthquakes, and mountains.

1.4 Scales of Space and Time in Earth Science

Discuss the scales of space and time in Earth science.

KEY TERMS: geologic time

- Earth science must deal with processes and phenomena that vary from the subatomic scale of matter to the nearly infinite scale of the universe.
- The time scales of phenomena studied in Earth science range from tiny fractions of a second to many billions of years.
- Geologic time, the span of time since the formation of Earth, is about 4.6 billion years, a number that is difficult to comprehend.

1.5 Resources and Environmental Issues

Summarize some important connections between people and the physical environment.

KEY TERMS: physical environment, renewable resource, nonrenewable resource, natural hazard

- Environment refers to everything that surrounds and influences an organism. Nonliving components, such as air, water, soil, and rock are collectively called the physical environment.
- Important relationships between people and the environment include the quest for resources, the impact of people on the natural environment, and the effects of natural hazards.
- Two broad categories of resources are (1) renewable, which means that they can be replenished over relatively short time spans, and (2) nonrenewable.

? As the photo reminds us, large quantities of aluminum are recycled. Does that mean that aluminum is a renewable resource? Explain.

I.6 The Nature of Scientific Inquiry

Discuss the nature of scientific inquiry and distinguish between a hypothesis and a theory.

KEY TERMS: hypothesis, theory

- Scientists make careful observations, construct tentative explanations for those observations (hypotheses), and then test those hypotheses with field investigations and laboratory work. In science, a theory is a well-tested and widely accepted explanation that the scientific community agrees best fits certain observable facts.

- As failed hypotheses are discarded, scientific knowledge moves closer to a correct understanding, but we can never be fully confident that we know all the answers. Scientists must always be open to new information that forces change in our model of the world.

GIVE IT SOME THOUGHT

1. As you enter a dark room, you turn on a wall switch, but the ceiling light does not come on. Formulate at least three hypotheses that might explain this observation. Once you have formulated your hypotheses, what should your next step be?

2. The circumference of Earth is slightly more than 40,000 kilometers (nearly 25,000 miles). How long would it take a jet plane traveling at 1000 kilometers (620 miles) per hour to circle our planet? The circumference of the Sun is 109 times that of Earth. How long would it take the jet mentioned above to circle the Sun?

3. The accompanying photo provides an example of interactions among different parts of the Earth system. It is a view of a mudflow that was triggered by extraordinary rains. Which of Earth's four "spheres" were involved in this natural disaster that buried a small town on the Philippine island of Leyte? Describe how each contributed to the mudflow.

AP Photo/Pat Roque

4. Examine **Figure I.4** to answer these questions.
   a. Where is most of Earth's fresh water stored?
   b. Where is most of Earth's liquid fresh water found?

5. Refer to the graph in **Figure I.5** to answer the following questions.
   a. If you were to climb to the top of Mount Everest, how many breaths of air would you have to take at that altitude to equal one breath at sea level?
   b. If you were flying in a commercial jet at an altitude of 12 kilometers (about 39,000 feet), about what percentage of the atmosphere would be below the plane?

6. Look at the concept map linking the four spheres of the Earth system. Between each sphere are arrows representing processes by which these spheres interact and influence each other. For each arrow, describe at least one interaction.

MasteringGeology™

1

Matter and Minerals

Each statement represents the primary learning objective for the corresponding major heading within the chapter. After you complete the chapter, you should be able to:

1.1 List the main characteristics that an Earth material must possess to be considered a mineral and describe each.

1.2 Compare and contrast the three primary particles contained in atoms.

1.3 Distinguish among ionic bonds, covalent bonds, and metallic bonds.

1.4 List and describe the properties that are used in mineral identification.

1.5 List the common silicate and nonsilicate minerals and describe what characterizes each group.

The Cave of Crystals, Chihuahua, Mexico, contains giant gypsum crystals, some of the largest natural crystals ever found. (Photo by Carsten Peter/ Speleoresearch & Films/National Geographic Stock)

Earth's crust and oceans are home to a wide variety of useful and essential minerals. Most people are familiar with the common uses of many basic metals, including aluminum in beverage cans, copper in electrical wiring, and gold and silver in jewelry. But some people are not aware that pencil "lead" contains the greasy-feeling mineral graphite and that bath powders and many cosmetics contain the mineral talc. Moreover, many do not know that dentists use drill bits impregnated with diamonds to drill through tooth enamel or that the common mineral quartz is the source of silicon for computer chips. In fact, practically every manufactured product contains materials obtained from minerals.

In addition to rocks and minerals having economic uses, all the processes that geologists study are in some way dependent on the properties of these basic Earth materials. Events such as volcanic eruptions, mountain building, weathering and erosion, and even earthquakes involve rocks and minerals. Consequently, a basic knowledge of Earth materials is essential to understanding all geologic phenomena.

1.1 Minerals: Building Blocks of Rocks

List the main characteristics that an Earth material must possess to be considered a mineral and describe each.

We begin our discussion of Earth materials with an overview of **mineralogy** (*mineral* = mineral, *ology* = study of) because minerals are the building blocks of rocks. In addition, humans have used minerals for both practical and decorative purposes for thousands of years (**Figure 1.1**). The first minerals mined were flint and chert, which were fashioned into weapons and cutting tools. As early as 3700 B.C., Egyptians began mining gold, silver, and copper; by 2200 B.C., humans had discovered how to combine copper with tin to make bronze, a strong, hard alloy. Later, a process was developed to extract iron from minerals such as hematite—a discovery that marked the decline of the Bronze Age. During the Middle Ages, mining of a variety of minerals became common, and the impetus for the formal study of minerals was in place.

The term *mineral* is used in several different ways. For example, those concerned with health and fitness extol the benefits of vitamins and minerals. The mining industry typically uses the word to refer to anything taken out of the ground, such as coal, iron ore, or sand and gravel. The guessing game known as *Twenty Questions* usually begins with the question "Is it animal, vegetable, or mineral?" What criteria do geologists use to determine whether something is a mineral?

Defining a Mineral

Geologists define **mineral** as *any naturally occurring inorganic solid that possesses an orderly crystalline structure and a definite chemical composition that allows for some variation.* Thus, Earth materials that are classified as minerals exhibit the following characteristics:

Figure 1.1 Quartz crystals Well-developed quartz crystals found near Hot Springs, Arkansas. (Photo by Jeff Scovil)

1. **Naturally occurring.** Minerals form by natural geologic processes. Synthetic materials, meaning those produced in a laboratory or by human intervention, are not considered minerals.

2. **Generally inorganic.** Inorganic crystalline solids, such as ordinary table salt (halite), that are found naturally in the ground are considered minerals. (Organic compounds, on the other hand, are generally not. Sugar, a crystalline solid like salt but that comes from sugarcane or sugar beets, is a common example of such an organic compound.) Many marine animals secrete inorganic compounds, such as calcium carbonate (calcite), in the form of shells and coral reefs. If these materials are buried and become part of the rock record, geologists consider them minerals.

3. **Solid substance.** Only solid crystalline substances are considered minerals. The exception is mercury, which is found in its liquid form in nature. In addition, ice (frozen water) fits this criterion and is considered a mineral, whereas liquid water and water vapor do not.

4. **Orderly crystalline structure.** Minerals are crystalline substances, which means their atoms (ions) are arranged in an orderly, repetitive manner (**Figure 1.2**). This orderly packing of atoms is reflected in regularly shaped objects called crystals. Some naturally occurring solids, such as volcanic glass (obsidian), lack a repetitive atomic structure and are not considered minerals.

5. **Definite chemical composition that allows for some variation.** Most minerals are chemical compounds having compositions that can be expressed by a chemical formula. For example, the common mineral quartz has the formula SiO_2, which indicates that quartz consists of silicon (Si) and oxygen (O) atoms, in a ratio of one-to-two. This proportion of silicon to oxygen is true for any sample of pure quartz, regardless of its origin. However, the compositions of some minerals vary *within specific, well-defined limits*. This occurs because certain elements can substitute for others of similar size without changing the mineral's internal structure.

A. Sodium and chlorine ions.

B. Basic building block of the mineral halite.

C. Collection of basic building blocks (crystal).

D. Crystals of the mineral halite.

Figure 1.2 Arrangement of sodium and chloride ions in the mineral halite The arrangement of atoms (ions) into basic building blocks that have a cubic shape results in regularly shaped cubic crystals. (Photo by Dennis Tasa)

What Is a Rock?

In contrast to minerals, rocks are more loosely defined. Simply, a **rock** is any solid mass of mineral, or mineral-like, matter that occurs naturally as part of our planet. Most rocks, like the sample of granite shown in **Figure 1.3**, occur as aggregates of several different minerals. The term *aggregate* implies that the minerals are joined in such a way that their individual properties are

Granite
(Rock)

Quartz
(Mineral)

Hornblende
(Mineral)

Feldspar
(Mineral)

SmartFigure 1.3 Most rocks are aggregates of minerals Shown here is a hand sample of the igneous rock granite and three of its major constituent minerals. (Photos by E. J. Tarbuck)

Did You Know?

Archaeological discoveries show that more than 2000 years ago, the Romans transported water in lead pipes. In fact, Roman smelting of lead and copper ores between 500 B.C. and A.D. 300 caused a small rise in atmospheric pollution that was recorded in Greenland ice cores.

retained. Note that the mineral constituents of granite can be easily identified. However, some rocks are composed almost entirely of one mineral. A common example is the sedimentary rock *limestone*, which is an impure mass of the mineral calcite.

Some rocks are composed of nonmineral matter. These include the volcanic rocks *obsidian* and *pumice*, which are noncrystalline glassy substances, and *coal*, which consists of solid organic debris.

Although this chapter deals primarily with the nature of minerals, keep in mind that most rocks are simply aggregates of minerals. Because the properties of rocks are determined largely by the chemical composition and crystalline structure of the minerals contained within them, we will first consider these Earth materials.

1.1 CONCEPT CHECKS

1. List five characteristics an Earth material must have in order to be considered a mineral.
2. Based on the definition of *mineral*, which of the following materials are not classified as minerals and why: gold, water, synthetic diamonds, ice, and wood.
3. Define the term *rock*. How do rocks differ from minerals?

1.2 Atoms: Building Blocks of Minerals

Compare and contrast the three primary particles contained in atoms.

When minerals are carefully examined, even under optical microscopes, the innumerable tiny particles of their internal structures are not discernible. Nevertheless, scientists have discovered that all matter, including minerals, is composed of minute building blocks called **atoms**—the smallest particles that cannot be chemically split. Atoms, in turn, contain even smaller particles—*protons* and *neutrons* located in a central **nucleus** that is surrounded by *electrons* (**Figure 1.4**).

Properties of Protons, Neutrons, and Electrons

Protons and **neutrons** are very dense particles with almost identical masses. By contrast, **electrons** have a negligible mass, about 1/2000 that of a proton. For comparison, if a proton or a neutron had the mass of a baseball, an electron would have the mass of a single grain of rice.

Both protons and electrons share a fundamental property called *electrical charge*. Protons have an electrical charge of +1, and electrons have a charge of −1. Neutrons, as the name suggests, have no charge. The charges of protons and electrons are equal in magnitude but opposite in polarity, so when these two particles are paired, the charges cancel each other out. Since matter typically contains equal numbers of positively charged protons and negatively charged electrons, most substances are electrically neutral.

In illustrations, electrons are sometimes shown orbiting the nucleus in a manner that resembles the planets of our solar system orbiting the Sun (see Figure 1.4A). However, electrons do not actually behave this way. A more realistic depiction shows electrons as a cloud of negative charges surrounding the nucleus (see Figure 1.4B). Studies of the arrangements of electrons show that they move about the nucleus in regions called *principal shells*, each with an associated energy level. In addition, each shell can hold a specific number of electrons, with the outermost shell generally containing **valence electrons** that interact with other atoms to form chemical bonds.

Figure 1.4 Two models of an atom A. A very simplified view of an atom. The central nucleus consists of protons and neutrons encircled by high-speed electrons. **B.** This model of an atom shows spherically shaped electron clouds (shells) surrounding a central nucleus. The nucleus contains virtually all of the mass of the atom. The remainder of the atom is the space occupied by negatively charged electrons. (The relative sizes of the nuclei shown are greatly exaggerated.)

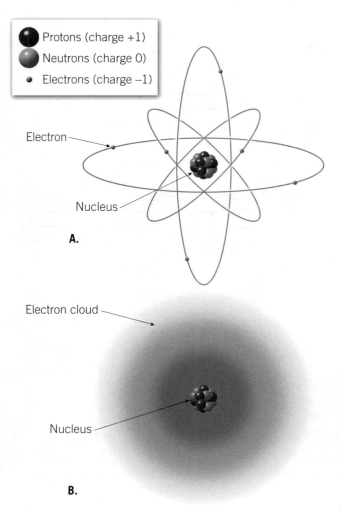

Protons (charge +1)
Neutrons (charge 0)
Electrons (charge −1)

Electron

Nucleus

A.

Electron cloud

Nucleus

B.

Most of the atoms in the universe (except hydrogen and helium) were created inside massive stars by nuclear fusion and released into interstellar space during hot, fiery supernova explosions. As this ejected material cooled, the newly formed nuclei attracted electrons to complete their atomic structure. At the temperatures found at Earth's surface, all free atoms (those not bonded to other atoms) have a full complement of electrons—one for each proton in the nucleus.

Elements: Defined by Their Number of Protons

The simplest atoms have only 1 proton in their nuclei, whereas others have more than 100. The number of protons in the nucleus of an atom, called the **atomic number**, determines its chemical nature. All atoms with the same number of protons have the same chemical and physical properties. Together, a group of the same kind of atoms is called an **element**. There are about 90 naturally occurring elements, and several more have been synthesized in the laboratory. You are probably familiar with the names of many elements, including carbon, nitrogen, and oxygen. All carbon atoms have six protons, whereas all nitrogen atoms have seven protons, and all oxygen atoms have eight.

Elements are organized so that those with similar properties line up in columns, referred to as groups. This arrangement, called the **periodic table**, is shown in **Figure 1.5**. Each element has been assigned a one- or two-letter symbol. The atomic numbers and masses are also included for each element.

Atoms of the naturally occurring elements are the basic building blocks of Earth's minerals. A few minerals,

Figure 1.5 Periodic table of the elements

A. Gold on quartz   **B. Sulfur**   **C. Copper**

Figure 1.6 Minerals composed of a single element (Photos by Dennis Tasa)

such as native copper, diamonds, sulfur, and gold, are made entirely of atoms of only one element (**Figure 1.6**). However, most elements tend to join with atoms of other elements to form **chemical compounds**. Most minerals are chemical compounds composed of atoms of two or more elements.

1.2 CONCEPT CHECKS

1. List the three main particles of an atom and explain how they differ from one another.
2. Make a simple sketch of an atom and label its three main particles.
3. What is the significance of valence electrons?

1.3 Why Atoms Bond

Distinguish among ionic bonds, covalent bonds, and metallic bonds.

Except for a group of elements known as the noble gases, atoms bond to one another under the conditions (temperatures and pressures) that occur on Earth. Some atoms bond to form *ionic compounds*, some form *molecules*, and still others form *metallic substances*. Why does this happen? Experiments show that electrical forces hold atoms together and bond them to each other. These electrical attractions lower the total energy of the bonded atoms, which, in turn, generally makes them more stable. Consequently, atoms that are bonded in compounds tend to be more stable than atoms that are free (not bonded).

The Octet Rule and Chemical Bonds

As was noted earlier, valence (outer shell) electrons are generally involved in chemical bonding. **Figure 1.7** shows a shorthand way of representing the number of valence electrons for selected elements in each group. Notice that the elements in Group I have one valence electron, those in Group II have two valence electrons, and so on, up to eight valence electrons in Group VIII.

The noble gases (except helium) have very stable electron arrangements with eight valence electrons and, therefore, tend to lack chemical reactivity. Many other atoms gain, lose, or share electrons during chemical reactions, ending up with electron arrangements of the noble gases. This observation led to a chemical guideline known as the **octet rule**: *Atoms tend to gain, lose,*

or share electrons until they are surrounded by eight valence electrons. Although there are exceptions to the octet rule, it is a useful *rule of thumb* for understanding chemical bonding.

When an atom's outer shell does not contain eight electrons, it is likely to chemically bond to other atoms to fill its shell. A **chemical bond** is a transfer or sharing of electrons that allows each atom to attain a full valence shell of electrons. Some atoms do this by transferring all their valence electrons to other atoms so that an inner shell becomes the full valence shell.

When the valence electrons are transferred between the elements to form ions, the bond is an *ionic bond*. When the electrons are shared between the atoms, the

Electron Dot Diagrams for Some Representative Elements							
I	II	III	IV	V	VI	VII	VIII
H •							He :
Li •	• Be •	• Ḃ •	• Ċ •	• N̈ •	:Ö •	:F̈ •	:N̈e:
Na •	• Mg •	• A̋l •	• S̈i •	• P̈ •	:S̈ •	:C̈l •	:Är:
K •	• Ca •	• Ga •	• Ge •	• Äs •	:S̈e •	:B̈r •	:K̈r:

Figure 1.7 Dot diagrams for certain elements Each dot represents a valence electron found in the outermost principal shell.

bond is a *covalent bond*. When the valence electrons are shared among all the atoms in a substance, the bonding is *metallic*.

Ionic Bonds: Electrons Transferred

Perhaps the easiest type of bond to visualize is the *ionic bond*, in which one atom gives up one or more of its valence electrons to another atom to form **ions**—*positively and negatively charged atoms*. The atom that loses electrons becomes a positive ion, and the atom that gains electrons becomes a negative ion. Oppositely charged ions are strongly attracted to one another and join to form ionic compounds.

Consider the ionic bonding that occurs between sodium (Na) and chlorine (Cl) to produce the solid ionic compound sodium chloride—the mineral halite (common table salt). Notice in **Figure 1.8A** that a sodium atom gives up its single valence electron to chlorine and, as a result, becomes a positively charged sodium ion. Chlorine, on the other hand, gains one electron and becomes a negatively charged chloride ion. We know that ions having unlike charges attract. Thus, an **ionic bond** is an attraction of oppositely charged ions to one another, producing an electrically neutral ionic compound.

Figure 1.8B illustrates the arrangement of sodium and chlorine ions in ordinary table salt. Notice that salt consists of alternating sodium and chlorine ions, positioned in such a manner that each positive ion is attracted to and surrounded on all sides by negative ions and vice versa. This arrangement maximizes the attraction between ions with opposite charges while minimizing the repulsion between ions with identical charges. Thus, ionic compounds consist of an orderly arrangement of oppositely charged ions assembled in a definite ratio that provides overall electrical neutrality.

The properties of a chemical compound are dramatically different from the properties of the various elements comprising it. For example, sodium is a soft silvery metal that is extremely reactive and poisonous. If you were to consume even a small amount of elemental sodium, you would need immediate medical attention. Chlorine, a green poisonous gas, is so toxic that it was used as a chemical weapon during World War I. Together, however, these elements produce sodium chloride, a harmless flavor enhancer that we call table salt. Thus, when elements combine to form compounds, their properties change significantly.

Covalent Bonds: Electron Sharing

Sometimes the forces that hold atoms together cannot be understood on the basis of the attraction of oppositely charged ions. One example is the hydrogen molecule (H_2), in which the two hydrogen atoms are held together tightly and no ions are present. The strong attractive force that holds two hydrogen atoms together results from a **covalent bond**, *a chemical bond formed by the sharing of a pair of electrons between atoms.*

A. The transfer of an electron from a sodium (Na) to a chlorine (Cl) atom leads to the formation of a Na⁺ ion and a Cl⁻ ion.

B. The arrangement of Na⁺ and Cl⁻ in the solid ionic compound sodium chloride (NaCl), table salt.

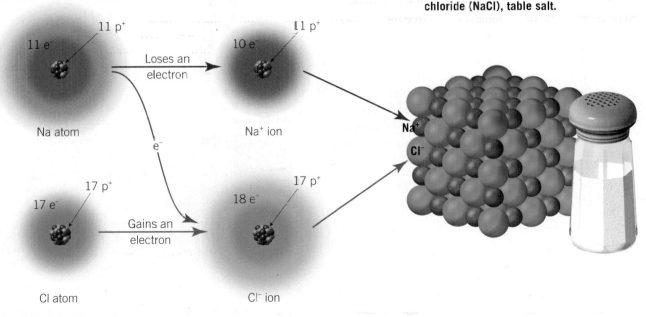

Figure 1.8 Formation of the ionic compound sodium chloride

Imagine two hydrogen atoms (each with one proton and one electron) approaching one another as shown in **Figure 1.9**. Once they meet, the electron configuration will change so that both electrons will primarily occupy the space between the atoms. In other words, the two electrons are shared by both hydrogen atoms and are attracted simultaneously by the positive charge of the proton in the nucleus of each atom. Although hydrogen atoms do not form ions, the force that holds these atoms together arises from the attraction of oppositely charged particles—positively charged protons in the nuclei and negatively charged electrons that surround these nuclei.

Figure 1.9 Formation of a covalent bond Covalent bonding of two hydrogen atoms (H) to form a hydrogen molecule (H_2). When hydrogen atoms bond, the negatively charged electrons are shared by both hydrogen atoms and attracted simultaneously by the positive charge of the proton in the nucleus of each atom.

Two hydrogen atoms combine to form a hydrogen molecule, held together by the attraction of oppositely charged particles—positively charged protons in each nuclei and negatively charged electrons that surround these nuclei.

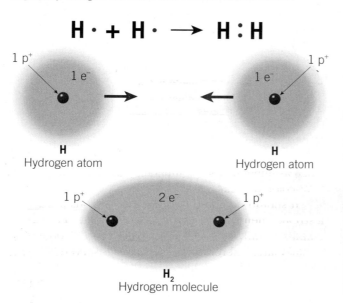

Metallic Bonds: Electrons Free to Move

In **metallic bonds**, the valence electrons are free to move from one atom to another so that all atoms share the available valence electrons. This type of bonding is found in metals such as copper, gold, aluminum, and silver and in alloys such as brass and bronze. Metallic bonding accounts for the high electrical conductivity of metals, the ease with which metals are shaped, and numerous other special properties.

1.3 CONCEPT CHECKS

1. What is the difference between an atom and an ion?
2. What occurs in an atom to produce a positive ion? A negative ion?
3. Briefly distinguish between ionic and covalent bonding and the role that electrons play in both.

1.4 Physical Properties of Minerals

List and describe the properties that are used in mineral identification.

Minerals have definite crystalline structures and chemical compositions that give them unique sets of physical and chemical properties shared by all samples of that mineral. For example, all specimens of halite have the same hardness, have the same density, and break in a similar manner. Because a mineral's internal structure and chemical composition are difficult to determine without the aid of sophisticated tests and equipment, the more easily recognized physical properties are frequently used in identification.

Optical Properties

Of the many optical properties of minerals, their luster, their ability to transmit light, their color, and their streak are most frequently used for mineral identification.

Luster The appearance or quality of light reflected from the surface of a mineral is known as **luster**. Minerals that have the appearance of metals, regardless of color, are said to have a *metallic luster* (**Figure 1.10**). Some metallic minerals, such as native copper and galena, develop a dull coating or tarnish when exposed to the atmosphere.

The freshly broken sample of galena (right) displays a metallic luster, while the sample on the left is tarnished and has a submetallic luster.

Figure 1.10 Metallic versus submetallic luster (Photo courtesy of E. J. Tarbuck)

Because they are not as shiny as samples with freshly broken surfaces, these samples are often said to exhibit a *submetallic luster*.

Most minerals have a *nonmetallic luster* and are described using adjectives such as *vitreous* or *glassy*. Other nonmetallic minerals are described as having a *dull*, or *earthy*, *luster* (a dull appearance like soil) or a *pearly luster* (such as a pearl or the inside of a clamshell). Still others exhibit a *silky luster* (like satin cloth) or a *greasy luster* (as though coated in oil).

Ability to Transmit Light Another optical property used in the identification of minerals is the ability to transmit light. When no light is transmitted, the mineral is described as *opaque*; when light, but not an image, is transmitted through a mineral sample, the mineral is said to be *translucent*. When both light and an image are visible through the sample, the mineral is described as *transparent*.

Color Although **color** is generally the most conspicuous characteristic of any mineral, it is considered a diagnostic property of only a few minerals. Slight impurities in the common mineral quartz, for example, give it a variety of tints, including pink, purple, yellow, white, gray, and even black (**Figure 1.11**). Other minerals, such as tourmaline, also exhibit a variety of hues, with multiple colors sometimes occurring in the same sample. Thus, the use of color as a means of identification is often ambiguous or even misleading.

Streak The color of a mineral in powdered form, called **streak**, is often useful in identification A mineral's streak is obtained by rubbing it across a *streak plate* (a piece of unglazed porcelain) and observing the color of the mark it leaves (**Figure 1.12**). Although the color of a mineral may vary from sample to sample, its streak is usually consistent in color.

Streak can also help distinguish between minerals with metallic luster and those with nonmetallic luster. Metallic minerals generally have a dense, dark streak,

Although the color of a mineral is not always helpful in identification, the streak, which is the color of the powdered mineral, can be very useful.

Figure 1.12 Streak

whereas minerals with nonmetallic luster typically have a light-colored streak.

It should be noted that not all minerals produce a streak when rubbed across a streak plate. For example, the mineral quartz is harder than a porcelain streak plate. Therefore, no streak is observed for quartz.

Crystal Shape or Habit

Mineralogists use the term **crystal shape**, or **habit**, to refer to the common or characteristic shape of individual crystals or aggregates of crystals. Some minerals tend to grow equally in all three dimensions, whereas others tend to be elongated in one direction or flattened if growth in one dimension is suppressed. A few minerals have crystals that exhibit regular polygons that are helpful in their identification. For example, magnetite crystals sometimes occur as octahedrons, garnets often form dodecahedrons, and halite and fluorite crystals tend to grow as cubes or near cubes. While minerals tend to have one common crystal shape, a few have two or more characteristic crystal shapes, such as the pyrite samples shown in **Figure 1.13**.

Figure 1.13 Common crystal shapes of pyrite

A. Fluorite   **B. Quartz**

SmartFigure 1.11 Color variations in minerals Some minerals, such as fluorite and quartz, occur in a variety of colors. (Photo A by Dennis Tasa; photo B by E. J. Tarbuck)

Although most minerals exhibit only one common crystal shape, some, such as pyrite, have two or more characteristic habits.

Dennis Tasa

A. Fibrous

B. Bladed

C. Banded

D. Cubic crystals

SmartFigure 1.14 Common crystal habits A. Thin, rounded crystals that break into fibers. **B.** Elongated crystals that are flattened in one direction. **C.** Minerals that have stripes or bands of different color or texture. **D.** Groups of crystals that are shaped like a cube.

In addition, some mineral samples consist of numerous intergrown crystals that exhibit characteristic shapes that are useful for identification. Terms commonly used to describe these and other crystal habits include *equant* (equidimensional), *bladed*, *fibrous*, *tabular*, *prismatic*, *platy*, *blocky*, *banded*, *granular*, and *botryoidal*. Some of these habits are pictured in **Figure 1.14**.

Mineral Strength

How easily minerals break or deform under stress is determined by the type and strength of the chemical bonds that hold the crystals together. Mineralogists use terms including *tenacity*, *hardness*, *cleavage*, and *fracture* to describe mineral strength and how minerals break when stress is applied.

Tenacity The term **tenacity** describes a mineral's toughness, or its resistance to breaking or deforming. Minerals that are ionically bonded, such as fluorite and halite, tend to be *brittle* and shatter into small pieces when struck. By contrast, minerals with metallic bonds, such as native copper, are *malleable*, or easily hammered into different shapes. Minerals, including gypsum and talc, that can be cut into thin shavings are described as *sectile*. Still others, notably the micas, are *elastic* and will bend and snap back to their original shape after stress is released.

Hardness One of the most useful diagnostic properties is **hardness**, a measure of the resistance of a mineral to abrasion or scratching. This property is determined by rubbing a mineral of unknown hardness against one of known hardness or vice versa. A numerical value of hardness can be obtained by using the **Mohs scale** of hardness, which consists of 10 minerals arranged in order from 1 (softest) to 10 (hardest), as shown in **Figure 1.15A**. It should be noted that the Mohs scale is a relative ranking, and it does not imply that mineral number 2, gypsum, is twice as hard as mineral 1, talc. In fact, gypsum is only slightly harder than talc, as **Figure 1.15B** indicates.

In the laboratory, other common objects can be used to determine the hardness of a mineral. These include a human fingernail, which has a hardness of about 2.5, a copper penny (3.5), and a piece of glass (5.5). The mineral gypsum, which has a hardness of 2, can be easily scratched with a fingernail. On the other hand, the

SmartFigure 1.15 Hardness scales A. The Mohs scale of hardness, with the hardness of some common objects. **B.** Relationship between the Mohs relative hardness scale and an absolute hardness scale.

A. Mohs scale (Relative hardness)

INDEX MINERALS		COMMON OBJECTS
Diamond	10	
Corundum	9	
Topaz	8	
Quartz	7	
Orthoclase	6	— Streak plate (6.5)
Apatite	5	— Glass & knife blade (5.5)
Fluorite	4	— Wire nail (4.5)
Calcite	3	— Copper penny (3.5)
Gypsum	2	— Fingernail (2.5)
Talc	1	

B. Comparison of Mohs scale and an absolute scale

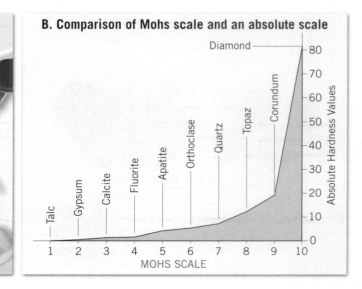

mineral calcite, which has a hardness of 3, will scratch a fingernail but will not scratch glass. Quartz, one of the hardest common minerals, will easily scratch glass. Diamonds, hardest of all, scratch anything, including other diamonds.

Cleavage In the crystal structure of many minerals, some atomic bonds are weaker than others. It is along these weak bonds that minerals tend to break when they are stressed. **Cleavage** (*kleiben* = carve) is the tendency of a mineral to break (cleave) along planes of weak bonding. Not all minerals have cleavage, but those that do can be identified by the relatively smooth, flat surfaces that are produced when the mineral is broken.

The simplest type of cleavage is exhibited by the micas (**Figure 1.16**). Because these minerals have very weak bonds in one direction, they cleave to form thin, flat sheets. Some minerals have excellent cleavage in one, two, three, or more directions, whereas others exhibit fair or poor cleavage, and still others have no cleavage at all. When minerals break evenly in more than one direction, cleavage is described by *the number of cleavage directions and the angle(s) at which they meet* (**Figure 1.17**).

Each cleavage surface that has a different orientation is counted as a different direction of cleavage. For example, some minerals cleave to form six-sided cubes. Because cubes are defined by three different sets of parallel planes that intersect at 90-degree angles,

Figure 1.16 Micas exhibit perfect cleavage. The thin sheets shown here exhibit one plane of cleavage. (Photo by Chip Clark)

Knife blade
Weak bonds
Strong bonds

cleavage is described as *three directions of cleavage that meet at 90 degrees.*

Do not confuse cleavage with crystal shape. When a mineral exhibits cleavage, it will break into pieces that all have the same geometry. By contrast, the smooth-sided quartz crystals shown in Figure 1.1 do not have cleavage. If broken, they fracture into shapes that do not resemble one another or the original crystals.

Did You Know?

Clay minerals are sometimes used as an additive to thicken milkshakes in fast-food restaurants.

A. Cleavage in one direction.
Example: Muscovite

B. Cleavage in two directions at 90° angles.
Example: Feldspar

Fracture not cleavage
90°

C. Cleavage in two directions not at 90° angles.
Example: Hornblende

Fracture not cleavage
120°

D. Cleavage in three directions at 90° angles.
Example: Halite
90°

E. Cleavage in three directions not at 90° angles.
Example: Calcite
75°

F. Cleavage in four directions.
Example: Fluorite

SmartFigure 1.17 Cleavage directions exhibited by minerals (Photos by E. J. Tarbuck and Dennis Tasa)

A. Irregular fracture   **B. Conchoidal fracture**

Figure 1.18 Irregular versus conchoidal fracture (Photo by E. J. Tarbuck)

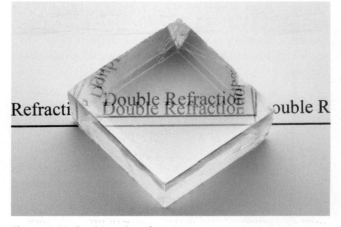

Figure 1.19 Double refraction This sample of calcite exhibits double refraction. (Photo by Chip Clark)

Fracture Minerals that have chemical bonds that are equally, or nearly equally, strong in all directions exhibit a property called **fracture**. When minerals fracture, most produce uneven surfaces and are described as exhibiting *irregular fracture*. However, some minerals, such as quartz, break into smooth, curved surfaces resembling broken glass. Such breaks are called *conchoidal fractures* (**Figure 1.18**). Still other minerals exhibit fractures that produce splinters or fibers that are referred to as *splintery fracture* and *fibrous fracture*, respectively.

Density and Specific Gravity

Density, an important property of matter, is defined as mass per unit volume. Mineralogists often use a related measure called **specific gravity** to describe the density of minerals. Specific gravity is a number representing the ratio of a mineral's weight to the weight of an equal volume of water.

Most common minerals have a specific gravity between 2 and 3. For example, quartz has a specific gravity of 2.65. By contrast, some metallic minerals, such as pyrite, native copper, and magnetite, are more than twice as dense and thus have more than twice the specific gravity of quartz. Galena, an ore of lead, has a specific gravity of roughly 7.5, whereas the specific gravity of 24-karat gold is approximately 20.

With a little practice, you can estimate the specific gravity of a mineral by hefting it in your hand. Does this mineral feel about as "heavy" as similar sized rocks you have handled? If the answer is "yes," the specific gravity of the sample will likely be between 2.5 and 3.

Other Properties of Minerals

In addition to the properties discussed thus far, some minerals can be recognized by other distinctive properties. For example, halite is ordinary salt, so it can be quickly identified through taste. Talc and graphite both have distinctive feels: Talc feels soapy, and graphite feels greasy. Further, the streaks of many sulfur-bearing minerals smell like rotten eggs. A few minerals, such as magnetite, have a high iron content and can be picked up with a magnet, while some varieties (such as lodestone) are natural magnets and will pick up small iron-based objects such as pins and paper clips (see Figure 1.30F).

Moreover, some minerals exhibit special optical properties. For example, when a transparent piece of calcite is placed over printed text, the letters appear twice. This optical property is known as *double refraction* (**Figure 1.19**).

One very simple chemical test involves placing a drop of dilute hydrochloric acid from a dropper bottle onto a freshly broken mineral surface. Using this technique, certain minerals, called carbonates, will effervesce (fizz) as carbon dioxide gas is released (**Figure 1.20**). This test is especially useful in identifying the common carbonate mineral calcite.

Figure 1.20 Calcite reacting with a weak acid (Photo by Chip Clark)

1.4 CONCEPT CHECKS

1. Define *luster*.
2. Why is color not always a useful property in mineral identification? Give an example of a mineral that supports your answer.
3. What do we mean when we refer to a mineral's *tenacity*? List three terms that describe tenacity.
4. What differentiates cleavage from fracture?
5. What simple chemical test is useful in the identification of the mineral calcite?

1.5 Mineral Groups

List the common silicate and nonsilicate minerals and describe what characterizes each group.

More than 4000 minerals have been named, and several new ones are identified each year. Fortunately for students who are beginning to study minerals, no more than a few dozen are abundant! Collectively, these few make up most of the rocks of Earth's crust and, as such, they are often referred to as the **rock-forming minerals**.

Although less abundant, many other minerals are used extensively in the manufacture of products and are called **economic minerals**. However, rock-forming minerals and economic minerals are not mutually exclusive groups. When found in large deposits, some rock-forming minerals are economically significant. One example is the mineral calcite, which is the primary component of the sedimentary rock limestone and has many uses, including being used in the production of cement.

It is worth noting that *only eight elements* make up the vast majority of the rock-forming minerals and represent more than 98 percent (by weight) of the continental crust (**Figure 1.21**). These elements, in order of abundance from most to least, are oxygen (O), silicon (Si), aluminum (Al), iron (Fe), calcium (Ca), sodium (Na), potassium (K), and magnesium (Mg). As shown in Figure 1.21, silicon and oxygen are by far the most common elements in Earth's crust. Furthermore, these two elements readily combine to form the basic "building block" for the most common mineral group, the **silicates**. More than 800 silicate minerals are known, and they account for more than 90 percent of Earth's crust.

Because other mineral groups are far less abundant in Earth's crust than the silicates, they are often grouped together under the heading **nonsilicates**. Although not as common as silicates, some nonsilicate minerals are very important economically. They provide us with iron and aluminum to build automobiles, gypsum for plaster and drywall for home construction, and copper wire that carries electricity and connects us to the Internet. Some common nonsilicate mineral groups include the carbonates, sulfates, and halides. In addition to their economic importance, these mineral groups include members that are major constituents in sediments and sedimentary rocks.

Silicate Minerals

Each of the silicate minerals contains oxygen and silicon atoms. Except for a few silicate minerals such as quartz, most silicate minerals also contain one or more additional elements in their crystalline structure. These elements give rise to the great variety of silicate minerals and their varied properties.

All silicates have the same fundamental building block, the **silicon–oxygen tetrahedron** (*tetra* = four, *hedra* = a base). This structure consists of four oxygen atoms surrounding a much smaller silicon atom, as shown in **Figure 1.22**. In some minerals, the tetrahedra are joined into chains, sheets, or three-dimensional networks by sharing oxygen atoms (**Figure 1.23**). These larger silicate structures are then connected to one another by other

Figure 1.21 The eight most abundant elements in the continental crust

A. Silicon–oxygen tetrahedron

B. Expanded view of silicon-oxygen tetrahedron

Figure 1.22 Two representations of the silicon–oxygen tetrahedron elements. The primary elements that join silicate structures are iron (Fe), magnesium (Mg), potassium (K), sodium (Na), and calcium (Ca).

Major groups of silicate minerals and common examples are given in Figure 1.23. The *feldspars* are by far the most plentiful group, comprising over 50 percent of Earth's crust. *Quartz*, the second-most-abundant mineral in the continental crust, is the only common mineral made completely of silicon and oxygen.

Figure 1.23 Common silicate minerals Note that the complexity of the silicate structure increases from top to bottom. (Photos by Dennis Tasa and E. J. Tarbuck)

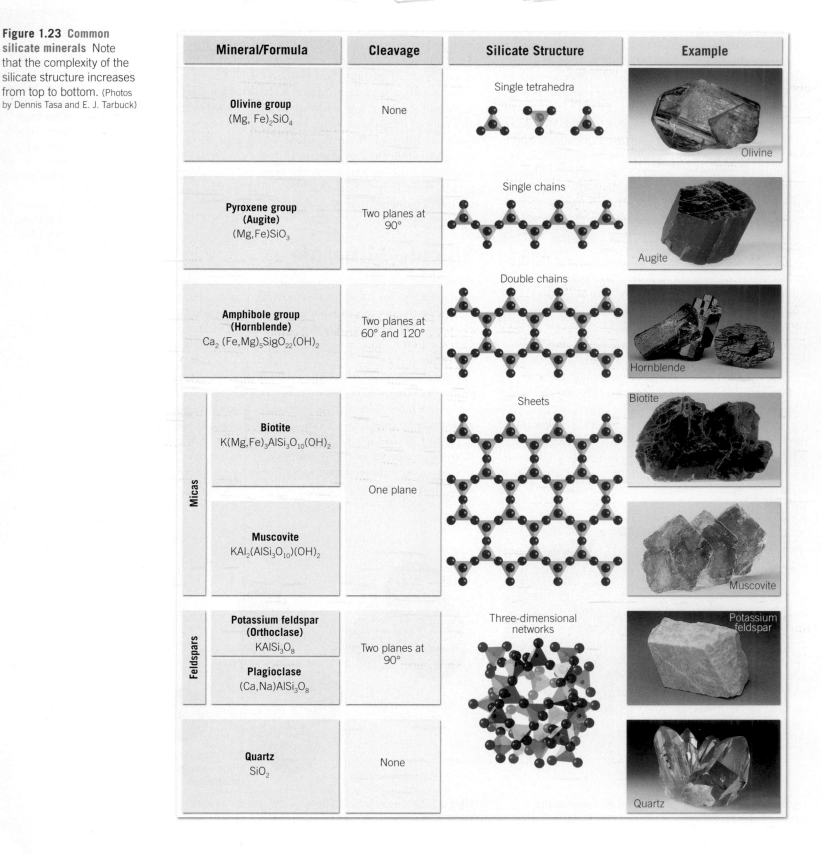

Notice in Figure 1.23 that each mineral *group* has a particular silicate *structure*. A relationship exists between this internal structure of a mineral and the *cleavage* it exhibits. Because the silicon–oxygen bonds are strong, silicate minerals tend to cleave between the silicon–oxygen structures rather than across them. For example, the micas have a sheet structure and thus tend to cleave into flat plates (see muscovite in Figure 1.17). Quartz, which has equally strong silicon–oxygen bonds in all directions, has no cleavage but fractures instead.

How do silicate minerals form? Most of them crystallize from molten rock as it cools. This cooling can occur at or near Earth's surface (low temperature and pressure) or at great depths (high temperature and pressure). The *environment* during crystallization and the *chemical composition of the molten rock* mainly determine which minerals are produced. For example, the silicate mineral olivine crystallizes at high temperatures (about 1200°C [2200°F]), whereas quartz crystallizes at much lower temperatures (about 700°C [1300°F]).

In addition, some silicate minerals form at Earth's surface from the weathered products of other silicate minerals. Clay minerals are an example. Still other silicate minerals are formed under the extreme pressures associated with mountain building. Each silicate mineral, therefore, has a structure and a chemical composition that *indicate the conditions under which it formed*. Thus, by carefully examining the mineral makeup of rocks, geologists can often determine the circumstances under which the rocks formed.

We will now examine some of the most common silicate minerals, which are divided into two major groups based on their chemical makeup.

Common Light Silicate Minerals

The most common light silicate minerals include the feldspars, quartz, muscovite, and the clay minerals. Generally light in color and having a specific gravity of about 2.7, **light silicate minerals** contain varying amounts of aluminum, potassium, calcium, and sodium.

The most abundant mineral group, the *feldspars*, are found in many igneous, sedimentary, and metamorphic rocks (**Figure 1.24**). One group of feldspar minerals contains potassium ions in its crystalline structure and is referred to as *potassium feldspar*. The other group, called *plagioclase feldspar*, contains calcium and/or sodium ions (see Figure 1.24). All feldspar minerals have two directions of cleavage that meet at 90-degree angles and are relatively hard (6 on the Mohs scale). The only reliable way to physically differentiate the feldspars is to look for striations that are present on some cleavage surfaces of plagioclase feldspar (see Figure 1.24) but do not appear in potassium feldspar.

Quartz is a major constituent of many igneous, sedimentary, and metamorphic rocks. Found in a wide

Potassium Feldspar

A. Potassium feldspar crystal (orthoclase)

B. Potassium feldspar showing cleavage (orthoclase)

Plagioclase Feldspar

C. Sodium-rich plagioclase feldspar (albite)

D. Plagioclase feldspar showing striations (labradorite)

Figure 1.24 Feldspar minerals A. Characteristic crystal form of potassium feldspar. **B.** Like this sample, most salmon-colored feldspar belongs to the potassium feldspar subgroup. **C.** Most sodium-rich plagioclase feldspar is light colored and has a porcelain luster. **D.** Calcium-rich plagioclase feldspar tends to be gray, blue-gray, or black in color. Labradorite, the variety shown here, exhibits striations on one of its crystal faces.

variety of colors (caused by impurities), quartz is quite hard (7 on the Mohs scale) and exhibits conchoidal fracture when broken (**Figure 1.25**). Pure quartz is clear, and if allowed to grow without interference, it will develop hexagonal crystals with pyramid-shaped ends (see Figure 1.1).

Another light silicate mineral, *muscovite*, is an abundant member of the mica family and has excellent cleavage in one direction. Muscovite is relatively soft (2.5 to 3 on the Mohs scale).

Clay minerals are light silicates that typically form as products of chemical weathering of igneous rocks. They make up much of the surface material we call

A. Smoky quartz

B. Rose quartz

C. Milky quartz

D. Jasper

Figure 1.25 Quartz is one of the most common minerals and has many varieties. A. Smoky quartz is commonly found in coarse-grained igneous rocks. **B.** Rose quartz owes its color to small amounts of titanium. **C.** Milky quartz often occurs in veins that occasionally contain gold. **D.** Jasper is a variety of quartz composed of minute crystals.

Figure 1.26 Kaolinite
Kaolinite is a common clay mineral formed by weathering of feldspar minerals.

Kaolinite

Figure 1.29 Well-formed garnet crystal Garnets come in a variety of colors and are commonly found in mica-rich metamorphic rocks. (Photo by E. J. Tarbuck)

Did You Know?

Most "scoopable" kitty litter sold today contains a naturally occurring material called *bentonite*. Bentonite, largely composed of highly absorbent clay minerals that swell and clump in the presence of moisture, allows kitty waste to be isolated and scooped out, leaving behind clean litter.

soil, and nearly half of the volume of sedimentary rocks is composed of clay minerals. Kaolinite is a common clay mineral formed from the weathering of feldspar (**Figure 1.26**).

Common Dark Silicate Minerals The **dark silicate minerals** contain iron and magnesium in their crystalline structures and include the pyroxenes, amphiboles, olivine, biotite, and garnet. Iron gives the dark silicates their color and contributes to their high specific gravity, which is between 3.2 and 3.6, significantly greater than the specific gravity of the light silicate minerals.

Olivine is an important group of dark silicate minerals that are major constituents of dark-colored igneous rocks. Abundant in Earth's upper mantle, olivine is black to olive green in color, has a glassy luster, and often forms small crystals which gives it a granular appearance (**FIGURE 1.27**).

The *pyroxenes* are a group of dark silicate minerals that are important components of dark-colored igneous rocks. The most common member, *augite*, is a black, opaque mineral with two directions of cleavage that meet at nearly 90-degree angles (**Figure 1.28A**).

The *amphibole group*, the most common of which is *hornblende*, is usually dark green to black in color (**Figure 1.28B**). Except for its cleavage angles, which are about 60 degrees and 120 degrees, hornblende is very similar in appearance to augite. Found in igneous rocks, hornblende makes up the dark portion of otherwise light-colored rocks.

Biotite is the dark, iron-rich member of the mica family. Like other micas, biotite possesses a sheet structure that gives it excellent cleavage in one direction. Biotite's shiny appearance helps distinguish it from other dark silicate minerals. Like hornblende, biotite is a common constituent of most light-colored igneous rocks, including granite.

Another dark silicate is *garnet* (**Figure 1.29**). Similar to olivine, garnet has a glassy luster, lacks cleavage, and exhibits conchoidal fracture. Although the colors of garnet are varied, the mineral is most often brown to deep red, and in transparent form, it is used as a gemstone.

Important Nonsilicate Minerals

Nonsilicate minerals are typically divided into groups, based on the negatively charged ion or complex ion that the members have in common. For example, the *oxides* contain the negative oxygen ions (O^{2-}), which are bonded to one or more kinds of positive ions. Thus, within each mineral group, the basic structure and type of bonding is similar. As a result, the minerals in each group have similar physical properties that are useful in mineral identification.

Although the nonsilicates make up only about 8 percent of Earth's crust, some minerals, such as gypsum, calcite, and halite, occur as constituents in sedimentary rocks in significant amounts. Furthermore, many others are important economically. **Table 1.1** lists some of the nonsilicate mineral groups and a few examples of each. A brief discussion of a few of the most common nonsilicate minerals follows.

Olivine-rich peridotite (variety dunite)

Figure 1.27 Olivine Commonly black to olive green in color, olivine has a glassy luster and a granular appearance. Olivine is commonly found in the igneous rock basalt.

A. Augite

B. Hornblende

Figure 1.28 Augite and hornblende These dark-colored silicate minerals are common constituents of a variety of igneous rocks. (Photos by E. J. Tarbuck)

TABLE 1.1   Common Nonsilicate Mineral Groups

Mineral Groups (key ion[s] or element[s])	Mineral Name	Chemical Formula	Economic Use
Carbonates (CO_3^{2-})	Calcite	$CaCO_3$	Portland cement, lime
	Dolomite	$CaMg(CO_3)_2$	Portland cement, lime
Halides (Cl^{1-}, F^{1-}, Br^{1-})	Halite	$NaCl$	Common salt
	Fluorite	CaF_2	Used in steelmaking
	Sylvite	KCl	Fertilizer
Oxides (O^{2-})	Hematite	Fe_2O_3	Ore of iron, pigment
	Magnetite	Fe_3O_4	Ore of iron
	Corundum	Al_2O_3	Gemstone, abrasive
	Ice	H_2O	Solid form of water
Sulfides (S^{2-})	Galena	PbS	Ore of lead
	Sphalerite	ZnS	Ore of zinc
	Pyrite	FeS_2	Sulfuric acid production
	Chalcopyrite	$CuFeS_2$	Ore of copper
	Cinnabar	HgS	Ore of mercury
Sulfates (SO_4^{2-})	Gypsum	$CaSO_4 \cdot 2H_2O$	Plaster
	Anhydrite	$CaSO_4$	Plaster
	Barite	$BaSO_4$	Drilling mud
Native elements (single elements)	Gold	Au	Trade, jewelry
	Copper	Cu	Electrical conductor
	Diamond	C	Gemstone, abrasive
	Sulfur	S	Sulfa drugs, chemicals
	Graphite	C	Pencil lead, dry lubricant
	Silver	Ag	Jewelry, photography
	Platinum	Pt	Catalyst

Some of the most common nonsilicate minerals belong to one of three classes of minerals: the carbonates (CO_3^{2-}), the sulfates (SO_4^{2-}), and the halides (Cl^{1-}, F^{1-}, Br^{1-}). The carbonate minerals are much simpler structurally than the silicates. This mineral group is composed of the carbonate ion (CO_3^{2-}) and one or more kinds of positive ions. The two most common carbonate minerals are *calcite*, $CaCO_3$ (calcium carbonate) and *dolomite*, $CaMg(CO_3)_2$ (calcium/magnesium carbonate) (**Figure 1.30A,B**). Calcite and dolomite are usually found together as the primary constituents in the sedimentary rocks limestone and dolostone. When calcite is the dominant mineral, the rock is called *limestone*, whereas *dolostone* results from a predominance of dolomite. Limestone has many uses, including as road aggregate, as building stone, and as the main ingredient in Portland cement.

A. Calcite B. Dolomite C. Halite
D. Gypsum E. Hematite F. Magnetite
G. Galena H. Chalcopyrite I. Fluorite

Figure 1.30 Important nonsilicate minerals (Photos by Dennis Tasa and E. J. Tarbuck)

Figure 1.31 Thick bed of halite exposed in an underground mine Halite (salt) mine in Grand Saline, Texas. Note the person for scale. (Photo by Tom Bochsler)

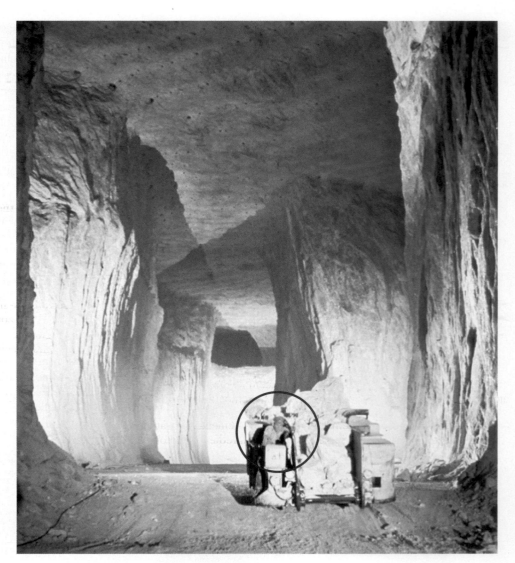

Two other nonsilicate minerals frequently found in sedimentary rocks are *halite* and *gypsum* (Figure 1.30C,D). Both minerals are commonly found in thick layers that are the last vestiges of ancient seas that have long since evaporated (**Figure 1.31**). Like limestone, both halite and gypsum are important nonmetallic resources. Halite is the mineral name for common table salt (NaCl). Gypsum ($CaSO_4 \cdot 2H_2O$), which is calcium sulfate with water bound into the structure, is the mineral of which plaster and other similar building materials are composed.

Most nonsilicate mineral classes contain members that are prized for their economic value. This includes the oxides, whose members *hematite* and *magnetite* are important ores of iron (Figure 1.30E,F). Also significant are the sulfides, which are basically compounds of sulfur (S) and one or more metals. Examples of important sulfide minerals include galena (lead), sphalerite (zinc), and chalcopyrite (copper). In addition,

native elements—including gold, silver, and carbon (diamonds)—plus a host of other nonsilicate minerals—fluorite (flux in making steel), corundum (gemstone, abrasive), and uraninite (a uranium source)—are important economically.

1.5 CONCEPT CHECKS

1. List the eight most common elements in Earth's crust, in order of abundance (most to least).
2. Explain the difference between the terms *silicon* and *silicate*.
3. Draw a sketch of the silicon–oxygen tetrahedron.
4. What is the most abundant mineral in Earth's crust?
5. List six common nonsilicate mineral groups. What key ion(s) or element(s) define each group?
6. What is the most common carbonate mineral?
7. List eight common nonsilicate minerals and their economic uses.

CONCEPTS IN REVIEW: MATTER AND MINERALS

1.1 Minerals: Building Blocks of Rocks

List the main characteristics that an Earth material must possess to be considered a mineral and describe each.

KEY TERMS: mineralogy, mineral, rock

- In Earth science, the word mineral refers to naturally occurring inorganic solids that possess an orderly crystalline structure and a characteristic chemical composition. The study of minerals is mineralogy.
- Minerals are the building blocks of rocks. Rocks are naturally occurring masses of minerals or mineral-like matter, such as natural glass or organic material.

1.2 Atoms: Building Blocks of Minerals

Compare and contrast the three primary particles contained in atoms.

KEY TERMS: atom, nucleus, proton, neutron, electron, valence electron, atomic number, element, periodic table, chemical compound

- Minerals are composed of atoms of one or more elements. The atoms of any element consist of the same three basic ingredients: protons, neutrons, and electrons.
- The number of protons in an atom is its atomic number. For example, an oxygen atom has eight protons, so its atomic number is eight. Protons and neutrons are approximately the same size and mass, but while protons are positively charged, neutrons have no charge.

- Electrons are much smaller than both protons and neutrons, and they weigh about 2000 times less. Each electron has a negative charge, equal in magnitude to the positive charge of a proton. Electrons swarm around an atom's nucleus at a distance, generally in several distinctive energy levels called principal shells. The electrons in the outermost principal shell, called valence electrons, are important when one atom bonds with other atoms to form chemical compounds.
- Elements with similar numbers of valence electrons tend to behave in similar ways. The periodic table displays these similarities in its graphical arrangement of the elements.

? Use the periodic table to identify these geologically important elements by their number of protons: (A) 14, (B) 6, (C) 13, (D) 17, and (E) 26.

1.3 Why Atoms Bond

Distinguish among ionic bonds, covalent bonds, and metallic bonds.

KEY TERMS: octet rule, chemical bond, ion, ionic bond, covalent bond, metallic bond

- When atoms attract to other atoms, they can form chemical bonds, which generally involve the transfer or sharing of valence electrons. For most atoms, the most stable arrangement is to have eight electrons in the outermost principal shell. This idea is called the *octet rule.*

- Ionic bonds involve atoms of one element giving up electrons to atoms of another element, forming positively and negatively charged atoms called *ions.* Positively charged ions bond with negative ions to form ionic bonds.
- Covalent bonds involve the sharing of electrons between two adjacent atoms. In metallic bonds, the sharing is more extensive: Electrons can freely move from one atom to the next throughout the entire mass.

? Which of the situations in the diagram shows ionic bonding? What are its distinguishing characteristics?

1.4 Physical Properties of Minerals

List and describe the properties that are used in mineral identification.

KEY TERMS: luster, color, streak, crystal shape (habit), tenacity, hardness, Mohs scale, cleavage, fracture, density, specific gravity

- The composition and internal crystalline structure of a mineral give it specific physical properties. These properties can be used to tell minerals apart from each another and make certain minerals useful for specific human tasks.
- Luster is a mineral's ability to reflect light. The degree to which a mineral can transmit light may be described as transparent, translucent, or opaque. Color is an unreliable characteristic for identification, as slight impurities can "stain" minerals into misleading colors. A trustworthy characteristic is streak, the color of the powder generated by scraping a mineral against a porcelain streak plate.
- The shape a crystal assumes as it grows is often useful for identifying it.
- Variations in the strength of chemical bonds give minerals properties such as tenacity (whether a mineral breaks in a brittle fashion or bends when stressed) and hardness (resistance to being scratched). Cleavage, the preferential breakage of a mineral along planes of weakly bonded atoms, is often useful in identifying minerals.
- The amount of matter packed into a given volume determines a mineral's density. Mineralogists find it simplest to measure a similar quantity, specific gravity, which is the ratio between a mineral's density and the density of water.
- Other properties are diagnostic for certain minerals but rare in most others. Smell, taste, feel, reaction to hydrochloric acid, magnetism, and double refraction are examples.

? **Research the minerals *quartz* and *calcite*. List three physical properties that may be used to distinguish one from the other.**

Dennis Tasa

Quartz

Dennis Tasa

Calcite

1.5 Mineral Groups

List the common silicate and nonsilicate minerals and describe what characterizes each group.

KEY TERMS: rock-forming mineral, economic mineral, silicate, nonsilicate, silicon–oxygen tetrahedron, light silicate mineral, dark silicate mineral

- Silicate minerals have a basic building block in common: a small pyramid-shaped structure that consists of one silicon atom surrounded by four oxygen atoms. Because this structure has four sides, it is called the silicon-oxygen tetrahedron. Neighboring tetrahedra can share some of their oxygen atoms, causing them to develop long chains or sheet structures.
- Silicate minerals make up the most common mineral class on Earth. They are subdivided into minerals that contain iron and/or magnesium (dark silicates) and those that do not (light silicates). The light silicate minerals are generally light in color and of relatively low specific gravity. Feldspar, quartz, muscovite, and clay minerals are examples. The dark silicate minerals are generally dark in color and relatively dense. Olivine, pyroxene, amphibole, biotite, and garnet are examples.
- Nonsilicate minerals include oxides, which contain oxygen ions that bond to other elements (usually metals); carbonates, which have CO_3 as a critical part of their crystal structure; sulfates, which have SO_4 as their basic building block; and halides, which contain a nonmetal ion such as chlorine, bromine, or fluorine that bonds to a metal ion such as sodium or calcium.

GIVE IT **SOME THOUGHT**

1. Using the geologic definition of *mineral* as your guide, determine which of the items in this list are minerals and which are not. If something in this list not a mineral, explain.
   a. Gold nugget
   b. Seawater
   c. Quartz
   d. Cubic zirconia
   e. Obsidian
   f. Ruby
   g. Glacial ice
   h. Amber

   Refer to the periodic table of the elements (Figure 1.5) to help you answer Questions 2 and 3.

2. Assume that the number of protons in a neutral atom is 92 and its mass number is 238.
   a. What is the name of the element?
   b. How many electrons does it have?
   c. How many neutrons does it have?

3. Which of the following elements is more likely to form chemical bonds: xenon (Xe) or sodium (Na)? Explain why.

4. Referring to the accompanying photos of five minerals, determine which of these specimens exhibit a metallic luster and which have a nonmetallic luster. (Photos by Dennis Tasa)

A.   B.   C.

D.   E.

5. Gold has a specific gravity of almost 20. A 5-gallon bucket of water weighs 40 pounds. How much would a 5-gallon bucket of gold weigh?

6. Examine the accompanying photo of a mineral that has several smooth, flat surfaces that resulted when the specimen was broken.

   Cleaved sample

   a. How many flat surfaces are present on this specimen?
   b. How many different directions of cleavage does this specimen have?
   c. Do the cleavage directions meet at 90-degree angles?

7. Each of the following statements describes a silicate mineral or mineral group. In each case, provide the appropriate name:
   a. The most common member of the amphibole group
   b. The most common light-colored member of the mica family
   c. The only common silicate mineral made entirely of silicon and oxygen
   d. A silicate mineral with a name that is based on its color
   e. A silicate mineral that is characterized by striations
   f. A silicate mineral that originates as a product of chemical weathering

8. What mineral property is illustrated in the accompanying photo?

9. Do an Internet search to determine what minerals are extracted from the ground during the manufacturing of the following products.
   a. Stainless steel utensils
   b. Cat litter
   c. Tums brand antacid tablets
   d. Lithium batteries
   e. Aluminum beverage cans

Dennis Tasa

10. Most states have designated a state mineral, rock, or gemstone to promote interest in the state's natural resources. Describe your state mineral, rock, or gemstone and explain why it was selected. If your state does not have a state mineral, rock, or gemstone, complete the exercise by selecting one from a state adjacent to yours.

2

Rocks: Materials of the Solid Earth

FOCUS ON CONCEPTS

Each statement represents the primary learning objective for the corresponding major heading within the chapter. After you complete the chapter, you should be able to:

2.1 Sketch, label, and explain the rock cycle.

2.2 Describe the two criteria used to classify igneous rocks and explain how the rate of cooling influences the crystal size of minerals.

2.3 Compare and contrast mechanical and chemical weathering and give examples of each.

2.4 List and describe the different categories of sedimentary rocks and discuss the processes that change sediment into sedimentary rock.

2.5 Define *metamorphism*, explain how metamorphic rocks form, and describe the agents of metamorphism.

These sedimentary rocks are located north of Capital Reef National Park, Utah. Dikes, a type of igneous intrusion, were feeders for the dark igneous rocks at the top. (Photo by Michael Collier)

Why study rocks? You have already learned that some rocks and minerals have great economic value. In addition, all Earth processes depend in some way on the properties of these basic Earth materials. Events such as volcanic eruptions, mountain building, weathering, erosion, and even earthquakes involve rocks and minerals. Consequently, a basic knowledge of Earth materials is essential to understanding most geologic phenomena.

Every rock contains clues about the environment in which it formed. For example, some rocks are composed entirely of small shell fragments. This tells Earth scientists that the rock likely originated in a shallow marine environment. Other rocks contain clues which indicate that they formed from a volcanic eruption or deep in the Earth during mountain building. Thus, rocks contain a wealth of information about events that have occurred over Earth's long history.

2.1 Earth as a System: The Rock Cycle

Sketch, label, and explain the rock cycle.

Earth as a system is illustrated most vividly when we examine the rock cycle. The **rock cycle** allows us to see many of the interactions among the many components and processes of the Earth system (**Figure 2.1**). It helps us understand the origin of igneous, sedimentary, and metamorphic rocks and how they are connected. In addition, the rock cycle demonstrates that any rock type, under the right circumstances, can be transformed into any other type.

The Basic Cycle

We begin our discussion of the rock cycle with molten rock, called *magma*, which forms by melting that occurs primarily within Earth's crust and upper mantle (Figure 2.1). Once formed, a magma body often rises toward the surface because it is less dense than the surrounding rock. Occasionally, magma reaches Earth's surface, where it erupts as *lava*. Eventually, molten rock cools and solidifies, a process called *crystallization*, or *solidification*. Molten rock may solidify either beneath the surface or, following a volcanic eruption, at the surface. In either situation, the resulting rocks are called *igneous rocks*.

If igneous rocks are exposed at the surface, they undergo *weathering*, in which the daily influences of the atmosphere slowly disintegrate and decompose rocks. The loose materials that result are often moved downslope by gravity and then picked up and transported by one or more erosional agents—running water, glaciers, wind, or waves. Eventually, these particles and dissolved substances, called *sediment*, are deposited. Although most sediment ultimately comes to rest in the ocean, other sites of deposition include river floodplains, desert basins, lakes, inland seas, and sand dunes.

Next, the sediments undergo *lithification*, a term meaning "conversion into rock." Sediment is usually lithified into *sedimentary rock* when compacted by the weight of overlying materials or when cemented as percolating groundwater fills the pores with mineral matter.

If the resulting sedimentary rock becomes deeply buried or is involved in the dynamics of mountain building, it will be subjected to great pressures and intense heat. The sedimentary rock may react to the changing environment by turning into the third rock type, *metamorphic rock*. If metamorphic rock is subjected to still higher temperatures, it may melt, creating magma, and the cycle begins again.

Although rocks may appear to be stable, unchanging masses, the rock cycle shows that they are not. The changes, however, take time—sometimes millions or even billions of years. In addition, the rock cycle operates continuously around the globe, but in different stages, depending on the location. Today, new magma is forming under the island of Hawaii, whereas the rocks that comprise the Colorado Rockies are slowly being worn down by weathering and erosion. Some of this weathered debris will eventually be carried to the Gulf of Mexico, where it will add to the already substantial mass of sediment that has accumulated there.

Alternative Paths

Rocks do not necessarily go through the cycle in the order just described. Other paths are also possible. For example, rather than being exposed to weathering and erosion at Earth's surface, igneous rocks may remain deeply buried (Figure 2.1). Eventually, these masses may

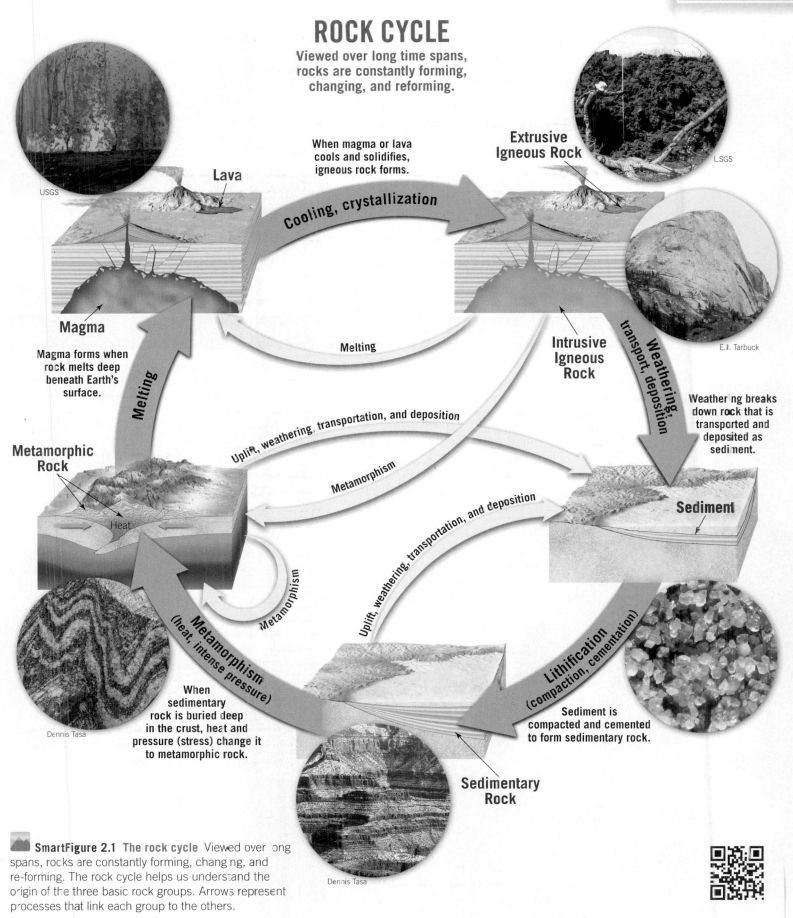

ROCK CYCLE

Viewed over long time spans, rocks are constantly forming, changing, and reforming.

USGS

When magma or lava cools and solidifies, igneous rock forms.

Lava

Extrusive Igneous Rock

LSGS

Cooling, crystallization

Magma

Magma forms when rock melts deep beneath Earth's surface.

Melting

Melting

Intrusive Igneous Rock

E.J. Tarbuck

Weathering, transport, deposition

Weathering breaks down rock that is transported and deposited as sediment.

Metamorphic Rock

Uplift, weathering, transportation, and deposition

Metamorphism

Heat

Sediment

Uplift, weathering, transportation, and deposition

Metamorphism

Dennis Tasa

Metamorphism (heat, intense pressure)

When sedimentary rock is buried deep in the crust, heat and pressure (stress) change it to metamorphic rock.

Lithification (compaction, cementation)

Sediment is compacted and cemented to form sedimentary rock.

Sedimentary Rock

Dennis Tasa

SmartFigure 2.1 The rock cycle Viewed over long spans, rocks are constantly forming, changing, and re-forming. The rock cycle helps us understand the origin of the three basic rock groups. Arrows represent processes that link each group to the others.

be subjected to the strong compressional forces and high temperatures associated with mountain building. When this occurs, they are transformed directly into metamorphic rocks.

Metamorphic and sedimentary rocks, as well as sediment, do not always remain buried. Rather, overlying layers may be eroded away, exposing the once-buried rock. When this happens, the material is attacked by weathering processes and turned into new raw materials for sedimentary rocks.

In a similar manner, igneous rocks that formed at depth can be uplifted, weathered, and turned into sedimentary rocks. Alternatively, igneous rocks may remain at depth, where the high temperatures and forces associated with mountain building may metamorphose or even melt them. Over time, rocks may be transformed into any other rock type, or even into a different form of the original type. Rocks may take many paths through the rock cycle.

What drives the rock cycle? Earth's internal heat is responsible for the processes that form igneous and metamorphic rocks. Weathering and the transport of weathered material are external processes, powered by energy from the Sun. External processes produce sedimentary rocks.

2.1 CONCEPT CHECKS

1. Sketch and label the rock cycle. Make sure your sketch includes alternative paths.
2. Use the rock cycle to explain the statement "One rock is the raw material for another."

2.2 Igneous Rocks: "Formed by Fire"

Describe the two criteria used to classify igneous rocks and explain how the rate of cooling influences the crystal size of minerals.

Figure 2.2 Fluid basaltic lava emitted from Hawaii's Kilauea Volcano Kilauea, on the Big Island, is one of the most active volcanoes on Earth. (Photo courtesy of USGS)

In the discussion of the rock cycle, we pointed out that **igneous rocks** form as *magma* or *lava* cools and crystallizes. **Magma** is molten rock that is most often generated by melting of rocks in Earth's mantle, although melting of crustal rock generates some magma. Once formed, a magma body buoyantly rises toward the surface because it is less dense than the surrounding rocks.

When magma reaches the surface, it is called **lava** (**Figure 2.2**). Sometimes, lava is emitted as fountains produced when escaping gases propel molten rock skyward. On other occasions, magma is explosively ejected from vents, producing a spectacular eruption such as the 1980 eruption of Mount St. Helens. However, most eruptions are not violent; rather, volcanoes most often emit quiet outpourings of lava.

When molten rock solidifies *at the surface*, the resulting igneous rocks are classified as **extrusive** or **volcanic** (after the Roman fire god, Vulcan). Extrusive igneous rocks are abundant in western portions of the Americas, including the volcanic cones of the Cascade Range and the extensive lava flows of the Columbia Plateau. In addition, many oceanic islands, including the Hawaiian Islands, are composed almost entirely of volcanic igneous rocks.

Most magma, however, loses its mobility before reaching the surface and eventually crystallizes deep below the surface. Igneous rocks that *form at depth* are termed **intrusive** or **plutonic** (after Pluto, the god of the underworld in classical mythology). Intrusive igneous rocks remain at depth unless portions of the crust are uplifted and the overlying rocks are stripped away by erosion. Exposures of intrusive igneous rocks occur in many places, including Mount Washington, New Hampshire; Stone Mountain, Georgia; Mount Rushmore in the Black Hills of South Dakota; and Yosemite National Park, California (**Figure 2.3**).

Figure 2.3 Mount Rushmore National Memorial, Black Hills of South Dakota This memorial is carved from the intrusive igneous rock granite. This massive igneous body cooled very slowly at depth and has since been uplifted, with the overlying rocks stripped away by erosion. (Photo by Barbara A. Harvey/Shutterstock)

From Magma to Crystalline Rock

Magma is molten rock (*melt*) composed of ions of the elements found in silicate minerals, mainly silicon and oxygen that move about freely. Magma also contains gases, particularly water vapor, that are confined within the magma body by the weight (pressure) of the overlying rocks, and it may contain some solids (mineral crystals). As magma cools, the once-mobile ions begin to arrange themselves into orderly patterns—a process called *crystallization*. As cooling continues, numerous small crystals develop, and ions are systematically added to these centers of crystal growth. When the crystals grow large enough for their edges to meet, their growth ceases for lack of space. Eventually, all the liquid is transformed into a solid mass of interlocking crystals.

The rate of cooling strongly influences crystal size. If magma cools very slowly, ions can migrate over great distances. Consequently, *slow cooling results in the formation of fewer, larger crystals*. On the other hand, if cooling occurs rapidly, the ions lose their motion and quickly combine. This results in a large number of tiny crystals that all compete for the available ions. Therefore, *rapid cooling results in the formation of a solid mass of small intergrown crystals*.

If the molten material is cooled almost instantly, there is insufficient time for the ions to arrange themselves into a crystalline network. Solids produced in this manner consist of randomly distributed atoms. Such rocks are called *glass* and are quite similar to ordinary manufactured glass. "Instant" quenching sometimes occurs during violent volcanic eruptions that produce tiny shards of glass called *volcanic ash*.

In addition to the rate of cooling, the composition of a magma and the amount of dissolved gases influence crystallization. Because magmas differ in each of these aspects, the physical appearance and mineral composition of igneous rocks vary widely.

What Can Igneous Textures Tell Us?

Geologists describe the overall appearance of a rock, based on the *size*, *shape*, and *arrangement* of its mineral grains as its **texture**. Texture is an important property because it allows geologists to make inferences about a rock's origin, based on careful observations of crystal size and other characteristics. You learned that rapid cooling produces small crystals, whereas very slow cooling produces much larger crystals. As you might expect, the rate of cooling is slow in magma chambers that lie deep within the crust, whereas a thin layer of lava extruded upon Earth's surface may chill to form solid rock in a matter of hours. Small molten blobs ejected from a volcano during a violent eruption can solidify in mid-air.

Igneous rocks that form rapidly at the surface or as small masses within the upper crust have a **fine-grained texture**, with individual crystals too small to be seen with the unaided eye (**Figure 2.4D**). Common in many fine-grained igneous rocks are voids, called *vesicles*, left by gas bubbles that formed as the lava solidified. Rocks containing these voids are said to display a **vesicular texture** (**Figure 2.5**).

When a large mass of magma solidify far below the surface, the surrounding rock insulates it, and heat is lost more slowly. This allows time for the magma's ions to organize into larger crystals, resulting in igneous rocks that exhibit a **coarse-grained texture**. Coarse-grained rocks have the appearance of a mass of intergrown crystals, roughly equal in size and large enough that individual

minerals can be identified with the unaided eye. Granite is a classic example (**Figure 2.4E**).

A large mass of magma located at depth may require tens of thousands, or even millions, of years to solidify. Because materials crystallize under different environmental conditions (such as temperature and pressure), it is possible for crystals of one mineral to become quite large before other minerals begin forming. If molten rock containing some large crystals moves to a different environment—for example, by erupting at the surface—the remaining molten portion of the lava will cool more quickly. The resulting rock, which has a two-stage cooling history, will be composed of large crystals embedded in a matrix of smaller crystals. Rocks of this type exhibit a **porphyritic texture** (**Figure 2.4C**).

During some volcanic eruptions, molten rock is ejected into the atmosphere, where it is quenched quickly. In this situation, rapid cooling may generate rock

A. Glassy texture
Composed of unordered atoms and resembles dark manufactured glass. (Obsidian is a natural glass that usually forms when highly silica-rich magmas solidify.)

B. Vesicular texture
Extrusive rock containing voids left by gas bubbles that escape as lava solidifies. (Pumice is a frothy volcanic glass that displays a vesicular texture.)

C. Porphyritic texture
Composed of two distinctly different crystal sizes.

D. Fine-grained texture
Composed of crystals that are too small for the individual minerals to be identified without a microscope.

E. Coarse-grained texture Composed of mineral grains that are large enough to be identified without a microscope.

SmartFigure 2.4 Igneous rock textures (Photos courtesy of E. J. Tarbuck)

Figure 2.5 Vesicular texture
Vesicles form as gas bubbles escape near the top of a lava flow.

Scoria, a volcanic rock with a vesicular texture.

having a **glassy texture** (**Figure 2.4A**). Glass results when unordered atoms are "frozen in place" before they are able to unite into an orderly crystalline structure. In addition, magmas containing large amounts of silica (SiO_2) are more likely to form rocks that exhibit a glassy texture than are those with a low silica content.

Igneous Compositions

Igneous rocks are composed mainly of silicate minerals. Chemical analysis shows that silicon and oxygen—usually expressed as the silica (SiO_2) content of a magma—are by far the most abundant constituents of igneous rocks. These two elements, plus ions of aluminum (Al), calcium (Ca), sodium (Na), potassium (K), magnesium (Mg), and iron (Fe), make up roughly 98 percent by weight of most magmas. In addition, magma contains small amounts of many other elements, including titanium and manganese, and trace amounts of much rarer elements, such as gold, silver, and uranium.

As magma cools and solidifies, these elements combine to form two major groups of silicate minerals. The *dark silicates* are rich in iron and/or magnesium and are comparatively low in silica (SiO_2). *Olivine*, *pyroxene*, *amphibole*, and *biotite mica* are the common dark silicate minerals of Earth's crust. By contrast, the *light silicates* contain greater amounts of potassium, sodium, and calcium and are richer in silica than dark silicates. Light silicates include *quartz, muscovite mica*, and the most abundant mineral group, the *feldspars*. Feldspars make up at least 40 percent of most igneous rocks. Thus, in addition to feldspar, igneous rocks contain some combination of the other light and/or dark silicates listed earlier.

Classifying Igneous Rocks

Igneous rocks are classified by their texture and mineral composition. The texture of an igneous rock is mainly a result of its cooling history, whereas its mineral composition is largely a result of the chemical makeup of the parent magma and the environment of crystallization.

Despite their great compositional diversity, igneous rocks can be divided into broad groups according to their proportions of light and dark minerals. A general classification scheme based on texture and mineral composition is provided in **Figure 2.6**.

Granitic (Felsic) Rocks
Near one end of the continuum are rocks composed almost entirely of light-colored silicates—quartz and potassium feldspar. Igneous rocks in which these are the dominant minerals have a **granitic composition**. Geologists also refer to granitic rocks as being **felsic**, a term derived from *feldspar* and *silica* (quartz). In addition to quartz and feldspar, most granitic rocks contain about 10 percent dark silicate minerals, usually biotite mica and amphibole. Granitic rocks are rich in silica (about 70 percent) and are major constituents of the continental crust.

Granite is a coarse-grained igneous rock that forms where large masses of magma slowly solidify at depth. During episodes of mountain building, granite and related crystalline rocks may be uplifted, with the processes of weathering and erosion stripping away the overlying crust. Areas where large quantities of granite are exposed at the surface include Pikes Peak in the Rockies, Mount Rushmore in the Black Hills, Stone Mountain in Georgia, and Yosemite National Park in the Sierra Nevada (**Figure 2.7**).

Granite is perhaps the best-known igneous rock, in part because of its natural beauty, which is enhanced

IGNEOUS ROCK CLASSIFICATION CHART

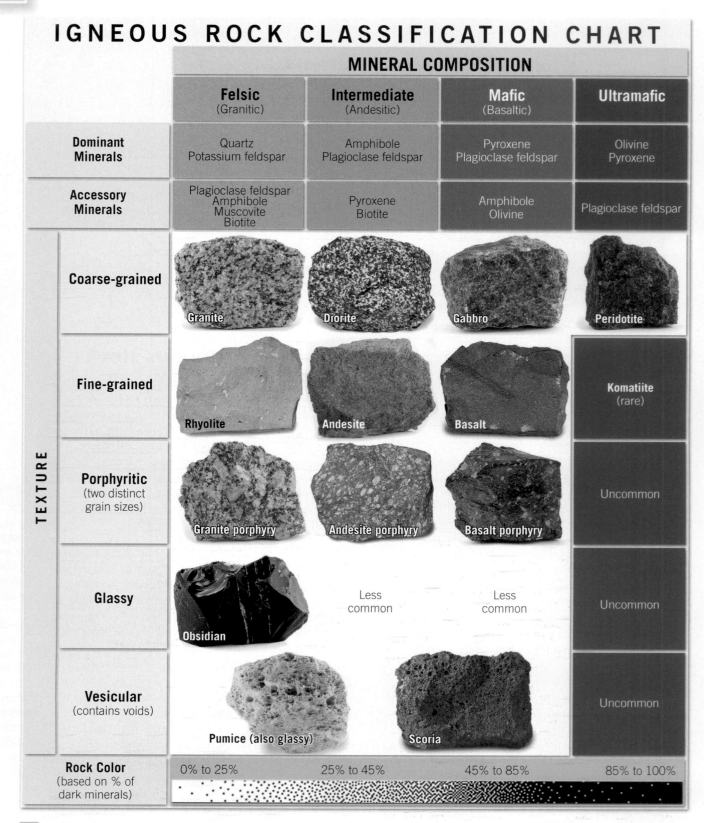

	MINERAL COMPOSITION			
	Felsic (Granitic)	**Intermediate** (Andesitic)	**Mafic** (Basaltic)	**Ultramafic**
Dominant Minerals	Quartz Potassium feldspar	Amphibole Plagioclase feldspar	Pyroxene Plagioclase feldspar	Olivine Pyroxene
Accessory Minerals	Plagioclase feldspar Amphibole Muscovite Biotite	Pyroxene Biotite	Amphibole Olivine	Plagioclase feldspar
TEXTURE Coarse-grained	Granite	Diorite	Gabbro	Peridotite
Fine-grained	Rhyolite	Andesite	Basalt	Komatiite (rare)
Porphyritic (two distinct grain sizes)	Granite porphyry	Andesite porphyry	Basalt porphyry	Uncommon
Glassy	Obsidian	Less common	Less common	Uncommon
Vesicular (contains voids)	Pumice (also glassy)		Scoria	Uncommon
Rock Color (based on % of dark minerals)	0% to 25%	25% to 45%	45% to 85%	85% to 100%

SmartFigure 2.6 Classification of igneous rocks, based on their mineral composition and texture Coarse-grained rocks are plutonic, solidifying deep underground. Fine-grained rocks are volcanic, or solidify as shallow, thin plutons. Ultramafic rocks are dark, dense rocks, composed almost entirely of minerals containing iron and magnesium. Although relatively rare on Earth's surface, these rocks are major constituents of the upper mantle. (Photos by E.J. Tarbuck and Dennis Tasa)

Corey Rich/
Getty Images

Henrik Lehneren/
Glow Images

Granite

Figure 2.7 Granitic rock exposed in California's Yosemite National Park This rock formed from magma that crystallized deep beneath the surface.

when polished, and partly because of its abundance. Slabs of polished granite are commonly used for tombstones, monuments, and countertops.

Rhyolite is the extrusive equivalent of granite and, likewise, is composed essentially of light-colored silicates (see Figure 2.6). This fact accounts for its color, which is usually buff to pink or light gray. Rhyolite is fine grained and frequently contains glass fragments and voids, indicating rapid cooling in a surface environment. In contrast to granite, which is widely distributed as large intrusive masses, rhyolite deposits are less common and generally less voluminous. Yellowstone Park is one well-known exception where extensive lava flows and thick ash deposits of rhyolitic composition are found.

Obsidian is a common type of natural glass that is similar in appearance to a dark chunk of manufactured glass (**Figure 2.8**). Although dark in color, obsidian usually has a higher silica content and a chemical composition similar to that of light-colored igneous rocks. Obsidian's dark color results from small amounts of metallic ions in an otherwise clear, glassy substance. Another silica-rich volcanic rock that exhibits a glassy and also vesicular texture is *pumice*. Often found with obsidian, pumice forms when large amounts of gas escape from molten rock to generate a gray, frothy mass (**Figure 2.9**). In some samples,

the vesicles are quite noticeable, whereas in others, the pumice resembles fine shards of intertwined glass. Because of the large volume of air-filled voids, many samples of pumice float in water (Figure 2.9).

Basaltic (Mafic) Rocks Rocks that contain substantial dark silicate minerals and calcium-rich plagioclase feldspar (but no quartz) are said to have a **basaltic composition**

← 2 cm →

Figure 2.9 Pumice, a vesicular glassy rock Pumice is very lightweight because it contains numerous vesicles. (Photo by E. J. Tarbuck; inset photo by Chip Clark)

Figure 2.8 Obsidian, a natural glass Native Americans used obsidian to make arrowheads and cutting tools. (Photo by Jeffrey Scovil)

(see Figure 2.6). Basaltic rocks contain a high percentage of dark silicate minerals, so geologists also refer to them as **mafic** (from *magnesium* and *ferrum*, the Latin name for iron). Because of their iron content, basaltic rocks are typically darker and denser than granitic rocks.

Basalt, the most common extrusive igneous rock, is a very dark green to black, fine-grained volcanic rock composed primarily of pyroxene, olivine, and plagioclase feldspar. Many volcanic islands, such as the Hawaiian Islands and Iceland, are composed mainly of basalt (**Figure 2.10**). Furthermore, the upper layers of the oceanic crust consist of basalt. In the United States, large portions of central Oregon and Washington were the sites of extensive basaltic outpourings.

The coarse-grained, intrusive equivalent of basalt is *gabbro* (see Figure 2.6). Although gabbro is not commonly exposed at the surface, it makes up a significant percentage of the oceanic crust.

Andesitic (Intermediate) Rocks   As you can see in Figure 2.6, rocks with a composition between granitic and basaltic rocks are said to have an **andesitic composition**, after the common volcanic rock *andesite*, or **intermediate composition**. Andesitic rocks contain a mixture of both light- and dark-colored minerals, mainly amphibole and plagioclase feldspar. This important category of igneous rocks is associated with volcanic activity typically confined to continental margins. When magma of intermediate composition crystallizes at depth, it forms the coarse-grained rock called *diorite* (Figure 2.6).

Ultramafic Rocks   Another important igneous rock, *peridotite*, contains mostly the dark-colored minerals olivine and pyroxene and, thus, falls on the opposite side of the compositional spectrum from granitic rocks (Figure 2.6). Because peridotite is composed almost entirely of dark silicate minerals, its chemical composition is referred to as **ultramafic**. Although ultramafic rocks are rare at Earth's surface, peridotite is believed to be the main constituent of the upper mantle.

How Different Igneous Rocks Form

Because a large variety of igneous rocks exist, it is logical to assume that an equally large variety of magmas also exist. However, geologists have observed that a single volcano may extrude lavas exhibiting quite different compositions. Data of this type led them to examine the possibility that magma might change (evolve) and thus become the parent to a variety of igneous rocks. To explore this idea, a pioneering investigation into the crystallization of magma was carried out by N. L. Bowen in the first quarter of the twentieth century.

Bowen's Reaction Series   In a laboratory setting, Bowen demonstrated that magma, with its diverse

Figure 2.10 Basaltic lava flowing from Kilauea Volcano, Hawaii (Photo by David Reggie/Getty Images)

chemistry, crystallizes over a temperature range of at least 200°C, unlike simple compounds (such as water), which solidify at specific temperatures. As magma cools, certain minerals crystallize first at relatively high temperatures. At successively lower temperatures, other minerals begin to crystallize. This arrangement of minerals, shown in **Figure 2.11**, became known as **Bowen's reaction series**.

Bowen discovered that the first mineral to crystallize from a body of magma is *olivine*. Further cooling results in the formation of *pyroxene*, as well as *plagioclase feldspar*. At intermediate temperatures, the minerals *amphibole* and *biotite* begin to crystallize.

During the last stage of crystallization, after most of the magma has solidified, the minerals *muscovite* and *potassium feldspar* may form (Figure 2.11). Finally, *quartz* crystallizes from any remaining liquid. Olivine and quartz are seldom found in the same igneous rock because quartz crystallizes at much lower temperatures than olivine.

Analysis of igneous rocks provides evidence that this crystallization model approximates what can happen in nature. In particular, we find that minerals that form in the same general range on Bowen's reaction series are

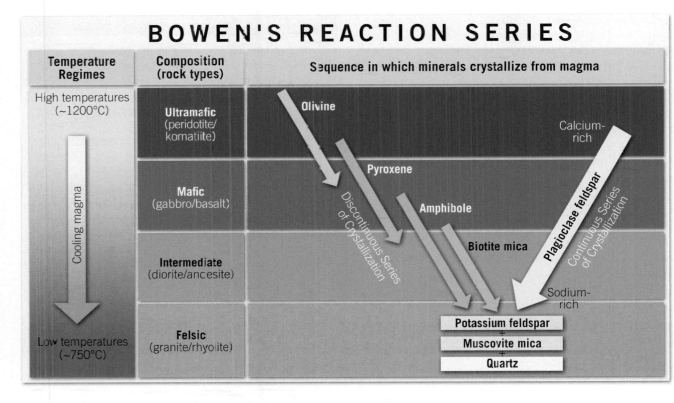

BOWEN'S REACTION SERIES

Temperature Regimes	Composition (rock types)	Sequence in which minerals crystallize from magma
High temperatures (~1200°C)	Ultramafic (peridotite/komatiite)	Olivine
Cooling magma	Mafic (gabbro/basalt)	Pyroxene Amphibole
	Intermediate (diorite/andesite)	Biotite mica
Low temperatures (~750°C)	Felsic (granite/rhyolite)	Potassium feldspar + Muscovite mica + Quartz

Discontinuous Series of Crystallization

Plagioclase feldspar — Continuous Series of Crystallization — Calcium-rich / Sodium-rich

Figure 2.11 Bowen's reaction series This diagram shows the sequence in which minerals crystallize from a magma. Compare this figure to the mineral composition of the rock groups in Figure 2.6. Note that each rock group consists of minerals that crystallize in the same temperature range.

found together in the same igneous rocks. For example, notice in Figure 2.11 that the minerals quartz, potassium feldspar, and muscovite, located in the same region of Bowen's diagram, are typically found together as major constituents of the igneous rock *granite*.

Magmatic Differentiation Bowen demonstrated that different minerals crystallize from magma according to a predictable pattern. But how do Bowen's findings account for the great diversity of igneous rocks? During the crystallization process, the composition of magma

continually changes. This occurs because as crystals form, they selectively remove certain elements from the magma, which leaves the remaining liquid portion (melt) depleted in these elements. Occasionally, separation of the solid and liquid components of magma occurs during crystallization, which creates different sets of minerals and thus, different types of igneous rocks. One such scenario, called **crystal settling**, occurs when the earlier formed minerals are more dense (heavier) than the liquid portion and sink toward the bottom of the magma chamber, as shown in **Figure 2.12**. When the

Figure 2.12 Crystal settling results in a change in the composition of the remaining melt.

A. A magma having a mafic (basaltic) composition erupts fluid basaltic lavas.

B. Cooling of the magma body causes crystals of olivine, pyroxene, and calcium-rich plagioclase to form and settle out, or crystallize along the magma body's cool margins.

C. The remaining melt will be enriched with silica, and should a subsequent eruption occur, the rocks generated will be more silica-rich and closer to the felsic (granitic) end of the compositional range than the initial magma.

remaining molten material solidifies—either in place or in another location, if it migrates into fractures in the surrounding rocks—it will form a rock with a chemical composition much different from the parent magma (Figure 2.12). The formation of one or more secondary magmas from a single parent magma is called **magmatic differentiation**.

At any stage in the evolution of a magma, the solid and liquid components can separate into two chemically distinct units. Furthermore, magmatic differentiation within the secondary magma can generate other chemically distinct masses of molten rock. Consequently, magmatic differentiation and separation of the solid and liquid components at various stages of crystallization can produce several chemically diverse magmas and, ultimately, a variety of igneous rocks.

2.3 Weathering of Rocks to Form Sediment

Compare and contrast mechanical and chemical weathering and give examples of each.

All materials are susceptible to weathering. Consider, for example, the synthetic rock we call concrete. A newly poured concrete sidewalk is smooth, but many years later, the same sidewalk will appear chipped, cracked, and rough, with pebbles exposed at the surface. If a tree is nearby, its roots may grow under the sidewalk, heaving and buckling the concrete. The same natural processes that eventually break apart a concrete sidewalk also act to disintegrate natural rocks, regardless of their type or strength.

Why does rock weather? Simply, weathering is the natural response of Earth materials to a new environment.

For instance, after millions of years of erosion, the rocks overlying a large body of intrusive igneous rock may be removed. This exposes the igneous rock to a whole new environment at the surface. This mass of crystalline rock, which formed deep below ground, where temperatures and pressures are high, is now subjected to very different and comparatively hostile surface conditions. In response, this rock mass will gradually change until it is once again in equilibrium, or balance, with its new environment. Such transformation of rock is what we call *weathering*.

In the following sections, we discuss the two basic categories of weathering—mechanical and chemical. Mechanical weathering (disintegration) is the physical breaking up of rocks. Chemical weathering (decomposition) actually alters what a rock is, changing it into different substances. Although we will consider these two processes separately, keep in mind that they usually work simultaneously in nature. Furthermore, the activities of erosional agents—wind, water, and glaciers—that transport weathered rock particles are important. As these mobile agents move rock debris, they relentlessly disintegrate it further.

Mechanical Weathering

When a rock undergoes **mechanical weathering**, it is broken into smaller and smaller pieces. Each piece retains the characteristics of the original material. The end result is many small pieces from a single large one. **Figure 2.13** shows that breaking a rock into smaller pieces increases the surface area available for chemical attack. An example is adding sugar to water. A chunk of rock candy will dissolve much more slowly than will an equal volume of sugar

As mechanical weathering breaks rock into smaller pieces, more surface area is exposed to chemical weathering.

2
2
4 square units

4 square units ×
6 sides ×
1 cube =
24 square units

1
1
1 square unit

1 square unit ×
6 sides ×
8 cubes =
48 square units

.5 .5

Increase in surface area

.25 square unit ×
6 sides ×
64 cubes =
96 square units

SmartFigure 2.13

Mechanical weathering increases surface area. Mechanical weathering adds to the effectiveness of chemical weathering because chemical weathering occurs mainly on exposed surfaces.

granules because of the vast difference in surface area. Hence, by breaking rocks into smaller pieces, mechanical weathering increases the amount of surface area available for chemical weathering.

In nature, three important physical processes break rocks into smaller fragments: frost wedging, sheeting, and biological activity.

Frost Wedging If you leave a glass bottle full of water in a freezer too long, you will find the bottle fractured. The bottle breaks because liquid water has the unique property of expanding about 9 percent upon freezing. This is also the reason that poorly insulated and exposed water pipes rupture during frigid weather. You might also expect this process to fracture rocks in nature. After water works its way into cracks in the rock, it freezes, expands, and enlarges the openings (**Figure 2.14**). After many freeze–thaw cycles, the rock is broken into pieces.

This process is appropriately called **frost wedging** (see Figure 2.14). Frost wedging is most pronounced in mountainous regions in the middle latitudes, where a daily freeze–thaw cycle often exists. Here, sections of

Frost wedging

Slightly tilted sedimentary beds

Falling rock debris

Falling rock debris

Patches of snow

Talus slope composed of angular rock fragments

SmartFigure 2.14 Frost wedging In mountainous areas frost wedging creates angular rock fragments that accumulate to form talus slopes. (Photo by Marli Miller)

rock are wedged loose and may tumble into large piles called *talus* or *talus slopes* that often form at the base of steep rock outcrops (see Figure 2.14).

Sheeting When large masses of intrusive igneous rock are exposed by erosion, entire slabs begin to break loose, like the layers of an onion. This process, called **sheeting**, is thought to occur because of the great reduction in pressure that accompanies the removal of the overlying rock (**Figure 2.15**). The outer layers expand more than the

Confining pressure

Deep pluton

This large igneous mass formed deep beneath the surface where confining pressure is great.

Joints

Expansion and sheeting

Uplift

As erosion removes the overlying bedrock (unloading), the outer parts of the igneous mass expand. Joints form parallel to the surface. Continued weathering causes thin slabs to separate and fall off.

SmartFigure 2.15 Unloading leads to sheeting Sheeting leads to the formation of an exfoliation dome. (Photo by Gary Moon/AGE Fotostock America, Inc.)

The summit of Half Dome in California's Yosemite National Park is an exfoliation dome and illustrates the onion-like layers created by sheeting.

rock below and thus separate from the rock body. Granite is particularly prone to sheeting.

Continued weathering eventually causes the slabs to separate and fall off, creating **exfoliation domes**. Excellent examples of exfoliation domes include Stone Mountain, Georgia, and Half Dome in Yosemite National Park (see Figure 2.15).

Biological Activity Weathering is also accomplished by the activities of organisms, including plants, burrowing animals, and humans. Plant roots in search of water grow into fractures, and as the roots grow, they wedge the rock apart (**Figure 2.16**). Burrowing animals further break down

the rock by moving fresh material to the surface, where physical and chemical processes can more effectively attack it.

Chemical Weathering

Chemical weathering alters the internal structure of minerals by removing and/or adding elements. During this transformation, the original rock is altered into substances that are stable in the surface environment.

Oxidation Water is the most important agent of chemical weathering. Oxygen dissolved in water will *oxidize* some materials in a process called *oxidation*. For example, when an iron nail is found in the soil, it will have a coating of rust (iron oxide), and if the time of exposure has been long, the nail will be so weak that it can be broken as easily as a toothpick. When rocks containing iron-rich minerals (such as hornblende) oxidize, a yellow to reddish-brown rust will appear on the surface.

Carbonic Acid Carbon dioxide (CO_2) dissolved in water (H_2O) forms *carbonic acid* (H_2CO_3). This is the same weak acid produced when soft drinks are carbonated. Rain dissolves some carbon dioxide as it falls through the atmosphere, so normal rainwater is mildly acidic. Water in the soil also dissolves carbon dioxide released by decaying organic matter. The result is that acidic water is everywhere on Earth's surface.

How does rock decompose when attacked by carbonic acid? Consider the weathering of the common igneous rock granite. Recall that granite is composed mainly of quartz and potassium feldspar. As the weak acid slowly reacts with crystals of potassium feldspar, potassium ions are displaced. *This changes the mineral's crystalline structure.*

The most abundant products of the chemical breakdown of feldspar are clay minerals. Because clay minerals are the end product of chemical weathering, they are very stable under surface conditions. Consequently, clay minerals make up a high percentage of the inorganic material in many soils.

In addition to the formation of clay minerals, some silica (SiO_2) is removed from the feldspar structure and carried

Figure 2.16 Plants can break rock. Root wedging near Boulder, Colorado. (Photo by Kristin Piljay)

Plant roots can extend into joints and grow in diameter and length. This process enlarges fractures and breaks rock.

away by groundwater. The dissolved silica will eventually precipitate to produce a hard, dense sedimentary rock (chert), fill pore spaces between mineral grains, or be carried to the ocean, where microscopic animals will build silica shells from it.

Quartz, the other main component of granite, is very resistant to chemical weathering. Because it is durable, quartz remains substantially unaltered when attacked by weak acid. As granite weathers, the feldspar crystals become dull and slowly turn to clay, releasing the once-interlocked quartz grains, which still retain their fresh, glassy appearance. Although some quartz remains in the soil, much is eventually transported to the ocean and other sites, where it becomes sandy beaches and sand dunes.

Products of Chemical Weathering **Table 2.1** lists the weathered products of some of the most common silicate minerals. Remember that silicate minerals make up most of Earth's crust and are composed primarily of just eight elements: oxygen, silicon, aluminum, iron, magnesium, potassium, sodium, and calcium. When chemically weathered, the silicate minerals yield sodium, calcium, potassium, and magnesium ions that may be used by plants or removed by groundwater. The element iron readily combines with oxygen to produce iron-oxide compounds that give soil a reddish-brown or yellowish color. The three remaining elements—aluminum, silicon, and oxygen—join with water and

TABLE 2.1 Products of Weathering

Mineral	Residual Products	Material in Solution
Quartz	Quartz grains	Silica
Feldspars	Clay minerals	Silica, K^+, Na^+, Ca^{2-}
Amphibole	Clay minerals, Iron oxides	Silica, Ca^{2+}, Mg^{2+}
Olivine	Iron oxides	Silica, Mg^{2+}

other elements to produce clay minerals that become an important part of the soil.

Ultimately, the products of physical and chemical weathering form the raw materials for building sedimentary rocks, which we consider next.

2.3 CONCEPT CHECKS

1. What are the two basic categories of weathering?
2. When a rock is mechanically weathered, how does its surface area change? How does this influence chemical weathering?
3. How can water cause mechanical weathering? Chemical weathering?
4. How does biological activity contribute to weathering?
5. How is carbonic acid formed in nature?
6. What products result when granite is chemically weathered by carbonic acid?

2.4 Sedimentary Rocks: Compacted and Cemented Sediment

List and describe the different categories of sedimentary rocks and discuss the processes that change sediment into sedimentary rock.

Recall the rock cycle, which shows the origin of **sedimentary rocks**. Weathering begins the process. Next, gravity and erosional agents (running water, wind, waves, and glacial ice) remove the products of weathering and carry them to a new location, where they are deposited. Usually, the particles are broken down further during this transport phase. Following deposition, this **sediment** may become lithified, or "turned to rock." Commonly, *compaction* and *cementation* transform the sediment into solid sedimentary rock.

The word *sedimentary* indicates the nature of these rocks, for it is derived from the Latin *sedimentum*, which means "settling," a reference to a solid material settling out of a fluid. Most sediment is deposited in this fashion. Weathered debris is constantly being swept from bedrock and carried away by water, ice, or wind. Eventually, the material is deposited in lakes, river

valleys, seas, and countless other places. The particles in a desert sand dune, the mud on the floor of a swamp, the gravels in a streambed, and even household dust are examples of sediment produced by this never-ending process.

The weathering of bedrock and the transport and deposition of the weathering products are continuous. Therefore, sediment is found almost everywhere. As piles of sediment accumulate, the materials near the bottom are compacted by the weight of the overlying layers. Over long periods, these sediments are cemented together by mineral matter deposited from water in the spaces between particles. This forms solid sedimentary rock.

Geologists estimate that sedimentary rocks account for only about 5 percent (by volume) of Earth's outer 16 kilometers (10 miles). However, the importance of this

Figure 2.17 Sedimentary rocks exposed along the Vermillion Cliffs, Arizona Sedimentary rocks occur in layers called *strata*. About 75 percent of all rock exposures on continents are sedimentary rocks. (Photo by Michael Collier)

group of rocks is far greater than this percentage implies. If you sampled the rocks exposed at Earth's surface, you would find that the great majority are sedimentary (**Figure 2.17**). Indeed, about 75 percent of all rock outcrops on the continents are sedimentary. Therefore, we can think of sedimentary rocks as comprising a relatively thin and somewhat discontinuous layer in the uppermost portion of the crust. This makes sense because sediment accumulates at the surface.

It is from sedimentary rocks that geologists reconstruct many details of Earth's history. Because sediments are deposited in a variety of different settings at the surface, the rock layers that they eventually form hold many clues to past surface environments. They may also exhibit characteristics that allow geologists to decipher information about the method and distance of sediment transport. Furthermore, sedimentary rocks contain fossils, which are vital evidence in the study of the geologic past.

Finally, many sedimentary rocks are important economically. Coal, which is burned to provide a significant portion of U.S. electrical energy, is classified as a sedimentary rock. Other major energy resources (such as petroleum and natural gas) occur in pores within

sedimentary rocks. Other sedimentary rocks are major sources of iron, aluminum, manganese, and fertilizer, plus numerous materials essential to the construction industry.

Classifying Sedimentary Rocks

Materials that accumulate as sediment have two principal sources. First, sediments may originate as solid particles from weathered rocks, such as the igneous rocks described earlier. These particles are called *detritus*, and the sedimentary rocks they form are called **detrital sedimentary rocks** (**Figure 2.18**).

Detrital Sedimentary Rocks Though a wide variety of minerals and rock fragments may be found in detrital rocks, clay minerals and quartz dominate. As you learned earlier, clay minerals are the most abundant product of the chemical weathering of silicate minerals, especially the feldspars. Quartz, on the other hand, is abundant because it is extremely durable and

Figure 2.18 Detrital sedimentary rocks

very resistant to chemical weathering. Thus, when igneous rocks such as granite are weathered, individual quartz grains are set free.

Geologists use particle size to distinguish among detrital sedimentary rocks. Figure 2.18 presents the four size categories for particles making up detrital rocks. When gravel-size particles predominate, the rock is called *conglomerate* if the sediment is rounded and *breccia* if the pieces are angular (Figure 2.18). Angular fragments indicate that the particles were not transported very far from their source prior to deposition and so have not had corners and rough edges abraded. *Sandstone* is the name given to rocks when sand-size grains prevail. *Shale*, the most common sedimentary rock, is made of very fine-grained sediment and composed mainly of clay minerals (Figure 2.18). *Siltstone*, another rather fine-grained rock, is composed of clay-sized sediment intermixed with slightly larger silt-sized grains.

Particle size is not only a convenient method of dividing detrital rocks; the sizes of the component grains also provide useful information about the environment in which the sediment was deposited. Currents of water or air sort the particles by size. The stronger the current, the larger the particle size carried. Gravels, for example, are moved by swiftly flowing rivers, rockslides, and glaciers. Less energy is required to transport sand; thus, it is common in windblown dunes, river deposits, and beaches. Because silts and clays settle very slowly, accumulations of these materials are generally associated with the quiet waters of a lake, lagoon, swamp, or marine environment.

Although detrital sedimentary rocks are classified by particle size, in certain cases, the mineral composition is also part of naming a rock. For example, most sandstones are predominantly quartz rich, and they are often referred to as quartz sandstone. In addition, rocks consisting of detrital sediments are rarely composed of grains of just one size. Consequently, a rock containing quantities of both sand and silt can be correctly classified as sandy siltstone or silty sandstone, depending on which particle size dominates.

Detrital Sedimentary Rocks

ClasticTexture (particle size)	Sediment Name	Rock Name
Coarse (over 2 mm)	Gravel (Rounded particles)	Conglomerate
Coarse (over 2 mm)	Gravel (Angular particles)	Breccia
Medium (1/16 to 2 mm)	Sand	Sandstone
Medium (1/16 to 2 mm)	Sand	Arkose*
Fine (1/16 to 1/256 mm)	Silt	Siltstone
Very fine (less than 1/256 mm)	Clay	Shale or Mudstone

*If abundant feldspar is present the rock is called Arkose.

Chemical and Biochemical Sedimentary Rocks

In contrast to detrital rocks, which form from the solid products of weathering, **chemical** and **biochemical sedimentary rocks** are derived from material (ions) that is carried in solution to lakes and seas (**Figure 2.19**). This material does not remain dissolved in water indefinitely. Under certain conditions, it precipitates (settles out) to form *chemical sediments* as a result of physical processes. An example of chemical sediments resulting from physical processes is the salt left behind as a body of saltwater evaporates.

Precipitation may also occur indirectly through life processes of water-dwelling organisms that form

Did You Know?

The most important and common material used for making glass is silica, which is usually obtained from the quartz in "clean," well-sorted sandstones.

Chemical and Organic Sedimentary Rocks

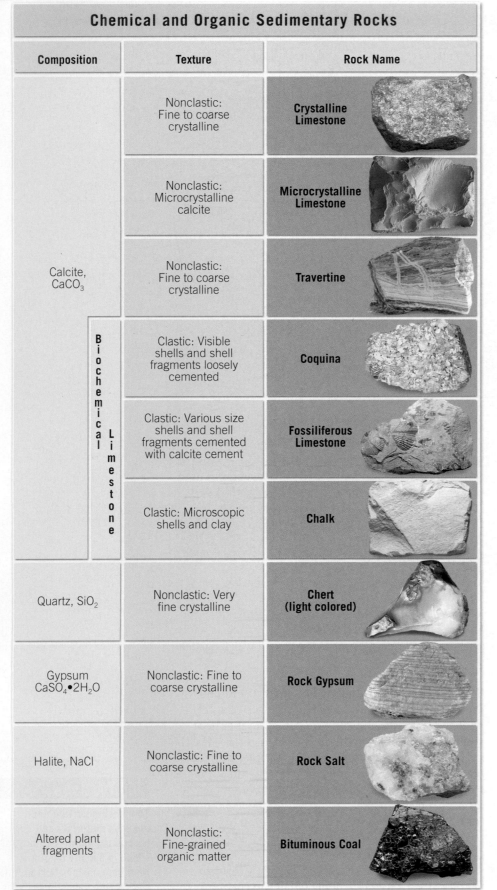

Composition	Texture	Rock Name	
Calcite, CaCO₃	Nonclastic: Fine to coarse crystalline	**Crystalline Limestone**	
	Nonclastic: Microcrystalline calcite	**Microcrystalline Limestone**	
	Nonclastic: Fine to coarse crystalline	**Travertine**	
	Clastic: Visible shells and shell fragments loosely cemented	**Coquina**	
	Clastic: Various size shells and shell fragments cemented with calcite cement	**Fossiliferous Limestone**	
	Clastic: Microscopic shells and clay	**Chalk**	
Quartz, SiO₂	Nonclastic: Very fine crystalline	**Chert (light colored)**	
Gypsum CaSO₄•2H₂O	Nonclastic: Fine to coarse crystalline	**Rock Gypsum**	
Halite, NaCl	Nonclastic: Fine to coarse crystalline	**Rock Salt**	
Altered plant fragments	Nonclastic: Fine-grained organic matter	**Bituminous Coal**	

(Biochemical Limestone spans the Calcite rows from Coquina through Chalk.)

Figure 2.19 Chemical, biochemical, and organic sedimentary rocks

materials called *biochemical sediments*. Many water-dwelling animals and plants extract dissolved mineral matter to form shells and other hard parts. After the organisms die, their skeletons may accumulate on the floor of a lake or an ocean.

Limestone, an abundant sedimentary rock, is composed chiefly of the mineral calcite ($CaCO_3$). Nearly 90 percent of limestone is formed from biochemical sediments secreted by marine organisms, and the remaining amount consists of chemical sediments that precipitated directly from seawater.

One easily identified biochemical limestone is *coquina*, a coarse rock composed of loosely cemented shells and shell fragments (**Figure 2.20**). Another less obvious but familiar example is *chalk*, a soft, porous rock made

← 5 cm →

A.

B.

Figure 2.20 Coquina This variety of limestone consists of shell fragments; therefore, it has a biochemical origin.
(Rock sample photo by E. J. Tarbuck; beach photo by Donald R. Frazier Photolibrary, Inc./Alamy Images)

up almost entirely of the hard parts of microscopic organisms that are no larger than the head of a pin. Among the most famous chalk deposits are the White Chalk Cliffs exposed along the southeast coast of England (**Figure 2.21**).

Inorganic limestone forms when chemical changes or high water temperatures increase the concentration of calcium carbonate to the point that it precipitates. *Travertine*, the type of limestone that decorates caverns, is one example. Groundwater is the source of travertine that is deposited in caves. As water drops reach the air in a cavern, some of the carbon dioxide dissolved in the water escapes, causing calcium carbonate to precipitate.

Dissolved silica (SiO_2) precipitates to form varieties of microcrystalline quartz (**Figure 2.22**). Sedimentary rocks composed of microcrystalline quartz include chert (light color), flint (dark), jasper (red), and agate (banded). These chemical sedimentary rocks may have either an inorganic or biochemical origin, but the mode of origin is usually difficult to determine.

Very often, evaporation causes minerals to precipitate from water. Such minerals include halite, the chief component of *rock salt*, and gypsum, the main ingredient of *rock gypsum*. Both materials have significant commercial importance. Halite is familiar to everyone as the common salt used in cooking and seasoning foods. Of course, it has many other uses and has been considered important enough that people have sought, traded, and fought over it for much of human history. Gypsum is the basic ingredient of plaster of Paris. This material is used most extensively in the construction industry for "drywall" and plaster.

In the geologic past, many areas that are now dry land were covered by shallow arms of the sea that had only narrow connections to the open ocean. Under these

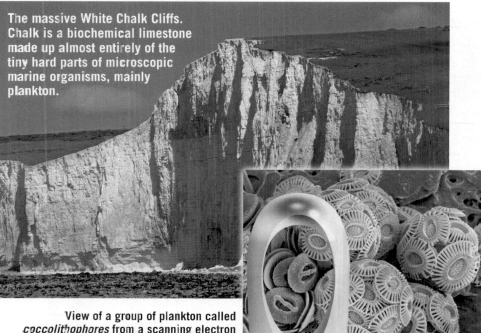

The massive White Chalk Cliffs. Chalk is a biochemical limestone made up almost entirely of the tiny hard parts of microscopic marine organisms, mainly plankton.

View of a group of plankton called *coccolithophores* from a scanning electron microscope. Individual plates shaped like hubcaps are only three one-thousandths of a millimeter in diameter; so tiny they could pass through the eye of a needle.

Figure 2.21 The White Chalk Cliffs This prominent deposit underlies large portions of southern England as well as parts of northern France. (Photos by David Wall/Alamy Images)

Agate

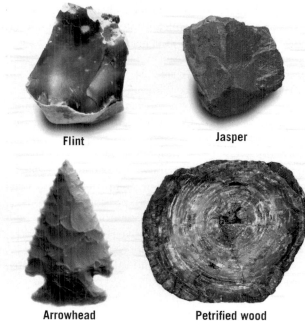

Flint

Jasper

Arrowhead

Petrified wood

Figure 2.22 Colorful chert Chert is the name applied to a number of dense, hard chemical sedimentary rocks made of microcrystalline quartz. (Agate photo by The Natural History Museum/Alamy Images; petrified wood photo by gracious_tiger/Shutterstock; flint and jasper photos by E. J. Tarbuck; arrowhead photo by Daniel Sambraus/Photo Researchers, Inc.)

SmartFigure 2.23
Bonneville salt flats This well-known Utah site was once a large salt lake. (Photo by Stock Connection/Glow Images)

This extensive evaporate deposit is a 30,000 acre expanse of hard white salt that in places is nearly two meters thick.

conditions, water continually moved into the bay to replace water lost by evaporation. Eventually, the waters of the bay became saturated and salt deposition began. Today, these arms of the sea are gone, and the remaining deposits are called **evaporite deposits**.

On a smaller scale, evaporite deposits can be seen in such places as Death Valley, California. Here, following rains or periods of snowmelt in the mountains, streams flow from surrounding mountains into an enclosed basin. As the water evaporates, *salt flats* form from dissolved materials left behind as a white crust on the ground (**Figure 2.23**).

Coal—An Organic Sedimentary Rock *Coal*, in contrast to sedimentary rocks that are calcite or silica rich, consists mostly of **organic matter**. Close examination of a piece of coal under a microscope or magnifying glass often reveals plant structures such as leaves, bark, and wood that have been chemically altered but remain identifiable. This supports the conclusion that coal is the end product of the burial of large amounts of plant material over extended periods (**Figure 2.24**).

The initial stage in coal formation is the accumulation of large quantities of plant remains. However, special conditions are required for such accumulations because dead plants normally decompose when exposed to the atmosphere. An ideal environment that allows for the buildup of plant material is a swamp. Because stagnant swamp water is oxygen deficient, complete decay (oxidation) of the plant material is not possible. At various times during Earth history, such environments have been common. Coal undergoes successive stages of formation. With each successive stage, higher temperatures and pressures drive off impurities and volatiles, as shown in Figure 2.24.

Lignite and bituminous coals are sedimentary rocks, but anthracite is a metamorphic rock. Anthracite forms when sedimentary layers are subjected to the folding and deformation associated with mountain building.

Lithification of Sediment

Lithification refers to the processes by which sediments are transformed into solid sedimentary rocks. One of the most common processes is *compaction*. As sediments accumulate through time, the weight of overlying material compresses the deeper sediments. As the grains are pressed closer and closer, pore space is greatly reduced. For example, when clays are buried beneath several thousand meters of material, the volume of the clay may be reduced as much as 40 percent. Compaction is most significant in fine-grained sedimentary rocks such as shale because sand and other coarse sediments compress little.

Cementation is another important means by which sediments are converted to sedimentary rock. The cementing materials are carried in a solution by water percolating through the pore spaces between particles. Over time, the cement precipitates onto the sediment grains, fills the open spaces, and joins the particles. Calcite, silica, and iron oxide are the most common cements. Identification of the cementing material is simple. Calcite cement will effervesce (fizz) with dilute hydrochloric acid. Silica is the hardest cement and thus produces the hardest sedimentary rocks. When a sedimentary rock has an orange or red color, this usually means iron oxide is present.

Features of Sedimentary Rocks

Sedimentary rocks are particularly important in the study of Earth history. These rocks form at Earth's surface, and as layer upon layer of sediment accumulates, each records the nature of the environment at the time the sediment was deposited. These layers, called **strata**, or **beds**, are the *single most characteristic feature of sedimentary rocks* (see Figure 2.17).

The thickness of beds ranges from microscopically thin to tens of meters thick. Separating the strata are

SWAMP ENVIRONMENT

Burial

PEAT
(Partially altered plant material)

Compaction

Greater burial

LIGNITE
(Soft, brown coal)

Compaction

BITUMINOUS
(Soft, black coal)

Metamorphism

Stress

ANTHRACITE
(Hard, black coal)

SmartFigure 2.24 From plants to coal Successive stages in the formation of coal. (Photos by E. J. Tarbuck)

A.

B.

Figure 2.25 Sedimentary environments A. Ripple marks preserved in sedimentary rocks may indicate a beach or stream channel environment. (Photo by Tim Graham/Alamy Images) **B.** Mud cracks form when wet mud or clay dries and shrinks, perhaps signifying a tidal flat or desert basin. (Photo by Marli Miller)

the end of one episode of sedimentation and the beginning of another.

Sedimentary rocks provide geologists with evidence for deciphering past environments. A conglomerate, for example, indicates a high-energy environment, such as a rushing stream, where only the coarse materials can settle out. By contrast, black shale and coal are associated with a low-energy, organic-rich environment, such as a swamp or lagoon. Other features found in some sedimentary rocks also give clues to past environments (**Figure 2.25**).

Fossils, the traces or remains of prehistoric life, are perhaps the most important inclusions found in some sedimentary rock (**Figure 2.26**). Knowing the nature of the life-forms that existed at a particular time may help answer many questions about the environment. Was it land or ocean, lake or swamp? Was the climate hot or cold, rainy or dry? Was the ocean water shallow or deep, turbid or clear? Furthermore, fossils are

bedding planes, flat surfaces along which rocks tend to separate or break. Generally, each bedding plane marks

Figure 2.26 Fossils—clues to the past Fossils, the remains or traces of prehistoric life, are primarily associated with sediments and sedimentary rocks. A large variety of trilobites are associated with the Paleozoic era. (Photo by Russell Shively/Shutterstock)

important time indicators and play a key role in matching up rocks from different places that are the same age. Fossils are important tools used in interpreting the geologic past and will be examined in some detail in Chapter 11.

2.4 CONCEPT CHECKS

1. Why are sedimentary rocks important?
2. What minerals are most abundant in detrital sedimentary rocks? In which rocks do these sediments predominate?
3. Distinguish between conglomerate and breccia.
4. What are the two categories of chemical sedimentary rock? Give an example of a rock that belongs to each category.
5. How do evaporites form? Give an example.
6. Compaction is an important lithification process with which sediment size?
7. What is the single most characteristic feature of sedimentary rocks?

2.5 Metamorphic Rocks: New Rock from Old

Define *metamorphism,* explain how metamorphic rocks form, and describe the agents of metamorphism.

Recall from the discussion of the rock cycle that metamorphism is the transformation of one rock type into another. **Metamorphic rocks** are produced from preexisting igneous, sedimentary, or even other metamorphic rocks (**Figure 2.27**). Thus, every metamorphic rock has a *parent rock*—the rock from which it was formed.

Metamorphism, which means "to change form," is a process that leads to changes in the mineralogy, texture (for example, grain size), and sometimes the chemical composition of rocks. Metamorphism takes place when preexisting rock is subjected to a physical or chemical environment that is significantly different from that in which it initially formed. In response to these new conditions, the rock gradually changes until a state of equilibrium with the new environment is reached. Most metamorphic changes occur at the elevated temperatures and pressures that exist in the zone beginning

Figure 2.27 Folded and metamorphosed rocks This rock outcrop is in Anza-Borrego Desert State Park, California. (Photo by A. P. Trujillo/APT Photos)

a few kilometers below Earth's surface and extending into the mantle.

Metamorphism often progresses incrementally, from slight changes (*low-grade metamorphism*) to substantial changes (*high-grade metamorphism*) (**Figure 2.28**). For example, under low-grade metamorphism, the common sedimentary rock *shale* becomes the more compact metamorphic rock called *slate* (Figure 2.28A). Hand samples of these rocks are sometimes difficult to distinguish, illustrating that the transition from sedimentary to metamorphic rock is often gradual, and the changes can be subtle.

In more extreme environments, metamorphism causes a transformation so complete that the identity of the parent rock cannot be determined. In high-grade metamorphism, such features as bedding planes, fossils, and vesicles that may have existed in the parent rock are obliterated. Furthermore, when rocks deep in the crust (where temperatures are high) are subjected to directed pressure, the entire mass may deform, producing large-scale structures such as folds (Figure 2.28B).

In the most extreme metamorphic environments, the temperatures approach those at which rocks melt. However, *during metamorphism, the rock must remain essentially solid*, for if complete melting occurs, we have entered the realm of igneous activity.

Most metamorphism occurs in one of two settings:

1. When rock is intruded by magma, **contact metamorphism**, or **thermal metamorphism**, may take place. In such a situation, change is caused by the rise in temperature within the rock surrounding the mass of molten material.

2. During mountain building, great quantities of rock are subjected to pressures and high temperatures associated with large-scale deformation called **regional metamorphism**.

Extensive areas of metamorphic rocks are exposed on every continent. Metamorphic rocks are an important component of many mountain belts, where they make up a large portion of a mountain's crystalline core. Even the

Parent rock (Shale) — Loosely packed clay minerals

Low-grade metamorphism — Low temperatures and pressures

Metamorphic rock (Slate) — Tightly packed chlorite and mica grains

Parent rock (Granodiorite) — Randomly oriented mineral grain

High-grade metamorphism — Strong compressional forces, high temperatures and pressures

Metamorphic rock (Folded gneiss) — Deformed layers of segregated grains

Figure 2.28 Metamorphic grade A. Low-grade metamorphism illustrated by the transformation of the common sedimentary rock shale to the more compact metamorphic rock slate. **B.** High-grade metamorphic environments obliterate the existing texture and often change the mineralogy of the parent rock. High-grade metamorphism occurs at temperatures that approach those at which rocks melt. (Photos by Dennis Tasa)

stable continental interiors, which are generally covered by sedimentary rocks, are underlain by metamorphic basement rocks. In each of these settings, the metamorphic rocks are usually highly deformed and intruded by igneous masses. Consequently, significant parts of Earth's continental crust are composed of metamorphic and associated igneous rocks.

What Drives Metamorphism?

The agents of metamorphism include *heat*, *confining pressure*, *differential stress*, and *chemically active fluids*. During metamorphism, rocks are often subjected to all four metamorphic agents simultaneously. However, the degree of metamorphism and the contribution of each agent vary greatly from one environment to another.

Heat as a Metamorphic Agent Thermal energy (*heat*) is the most important factor driving metamorphism. It triggers chemical reactions that result in the recrystallization of existing minerals and the formation of new minerals. Thermal energy for metamorphism comes mainly from two sources. Rocks experience a rise in temperature when they are intruded by magma rising from below. This is called *contact*, or *thermal*, *metamorphism*. In this situation, the adjacent host rock is "baked" by the emplaced magma.

By contrast, rocks that formed at Earth's surface will experience a gradual increase in temperature as they are taken to greater depths. In the upper crust, this increase in temperature averages about 25°C per kilometer. When buried to a depth of about 8 kilometers (5 miles), where temperatures are between 150° and 200°C, clay minerals tend to become unstable and begin to recrystallize into other minerals, such as chlorite and muscovite, that are stable in this environment. (Chlorite is a mica-like mineral formed by the metamorphism of iron- and magnesian-rich silicates.) However, many silicate minerals, particularly those found in crystalline igneous rocks—quartz and feldspar, for example—remain stable at these temperatures. Thus, metamorphic changes in these minerals require much higher temperatures in order to recrystallize.

SmartFigure 2.29 Confining pressure and differential stress

Confining Pressure and Differential Stress as Metamorphic Agents Pressure, like temperature, increases with depth as the thickness of the overlying rock increases. Buried rocks are subjected to *confining pressure*—similar to water pressure in that the forces are equally applied in all directions (**Figure 2.29A**). The deeper you go in the ocean, the greater the confining pressure. The same is true for buried rock. Confining pressure causes the spaces between mineral grains to close, producing a more compact rock that has greater density. Further, at great depths, confining pressure may cause minerals to recrystallize into new minerals that display more compact crystalline forms.

During episodes of mountain building, large rock bodies become highly crumpled and metamorphosed (**Figure 2.29B**). The forces that generate mountains are unequal in different directions and are called *differential stress*. Unlike confining pressure, which "squeezes" rock equally in all directions, differential stresses are greater in one direction than in others. As shown in Figure 2.29B, rocks subjected to differential stress are shortened in the direction of greatest stress, and they are elongated, or lengthened, in the direction perpendicular to that stress. The deformation caused by differential stresses plays a major role in developing metamorphic textures.

In surface environments where temperatures are relatively low, rocks are *brittle* and tend to fracture when subjected to differential stress. (Think of a heavy boot crushing a piece of fine crystal.) Continued deformation grinds and pulverizes the mineral grains into small fragments. By contrast, in high-temperature, high-pressure environments deep in Earth's

In a depositional environment, as confining pressure increases, rocks deform by decreasing in volume.

A.

During mountain building, rocks subjected to differential stress are shortened in the direction of maximum stress and lengthened in the direction of minimum stress. B.

crust, rocks are *ductile* and tend to flow rather than break. (Think of a heavy boot crushing a soda can.) When rocks exhibit ductile behavior, their mineral grains tend to flatten and elongate when subjected to differential stress. This accounts for their ability to deform by flowing (rather than fracturing) to generate intricate folds (see Figure 2.27).

Chemically Active Fluids Ion-rich fluids composed mainly of water and other volatiles (materials that readily change to gases at surface conditions) are believed to play an important role in some types of metamorphism. Fluids that surround mineral grains act as catalysts to promote recrystallization by enhancing ion migration. In progressively hotter environments, these ion-rich fluids become correspondingly more reactive. Chemically active fluids can produce two types of metamorphism, explained below. The first type changes the arrangement and shape of mineral grains within a rock; the second type changes the rock's chemical composition.

When two mineral grains are squeezed together, the parts of their crystalline structures that touch are the most highly stressed. Atoms at these sites are readily dissolved by the hot fluids and move to the voids between individual grains. Thus, hot fluids aid in the recrystallization of mineral grains by dissolving material from regions of high stress and then precipitating (depositing) this material in areas of low stress. As a result, *minerals tend to recrystallize and grow longer in a direction perpendicular to compressional stresses.*

When hot fluids circulate freely through rocks, ionic exchange may occur between adjacent rock layers, or ions may migrate great distances before they are finally deposited. The latter situation is particularly common when we consider hot fluids that escape during the crystallization of an intrusive mass of magma. If the rocks surrounding the magma differ markedly in composition from the invading fluids, there may be a substantial exchange of ions between the fluids and host rocks. When this occurs, the overall composition of the surrounding rock changes.

Metamorphic Textures

The degree of metamorphism is reflected in a rock's *texture* and *mineralogy*. (Recall that the term *texture* is used to describe the size, shape, and arrangement of grains within a rock.) When rocks are subjected to low-grade metamorphism, they become more compact and thus denser. A common example is the metamorphic rock slate, which forms when shale is subjected to temperatures and pressures only slightly greater than those associated with the compaction that lithifies sediment. In this case, differential stress causes the microscopic clay

minerals in shale to align into the more compact arrangement found in slate.

Under more extreme conditions, stress causes certain minerals to recrystallize. In general, recrystallization encourages the growth of larger crystals. Consequently, many metamorphic rocks consist of visible crystals, much like coarse-grained igneous rocks.

Foliation The term **foliation** refers to any nearly flat arrangement of mineral grains or structural features within a rock (**Figure 2.30**). Although foliation may occur in some sedimentary and even a few types of igneous rocks, it is a fundamental characteristic of regionally metamorphosed rocks—that is, rock units that have been strongly deformed, mainly by folding. In metamorphic environments, foliation is ultimately driven by compressional stresses that shorten rock units, causing mineral grains in preexisting rocks to develop parallel, or nearly parallel, alignments. Examples of foliation include the *parallel*

Platy and elongated mineral grains having random orientation.

When differential stress causes rocks to flatten, the mineral grains rotate and align roughly perpendicular to the direction of maximum differential stress.

Figure 2.30 Rotation of platy and elongated mineral grains Under the pressures of metamorphism, some mineral grains become reoriented and aligned at right angles to the stress. The resulting orientation of mineral grains gives the rock a foliated (layered) texture. If the coarse-grained igneous rock (granite) on the left underwent intense metamorphism, it could end up closely resembling the metamorphic rock on the right (gneiss). (Photos by E. J. Tarbuck) *alignment of platy (flat and disk-like) minerals* such as the micas; the *parallel alignment of flattened pebbles*; *compositional banding*, in which dark and light minerals *separate* generating a layered appearance; and *rock cleavage*, in which rocks can be easily split into tabular slabs.

Nonfoliated Textures Not all metamorphic rocks exhibit a foliated texture. Those that do not are referred to as **nonfoliated** and typically develop in environments where deformation is minimal and the parent rocks are composed of minerals that have a relatively simple chemical composition, such as quartz or calcite. For example, when a fine-grained limestone (made of calcite) is metamorphosed by the intrusion of a hot magma body, the small calcite grains recrystallize to form larger interlocking crystals. The resulting rock, *marble*, exhibits

Figure 2.32 Slate exhibits rock cleavage. Because slate breaks into flat slabs, it has many uses. The larger image shows a quarry near Alta, Norway. (Photo by Fred Bruemmer/ Photolibrary) In the inset photo, slate is used to roof a house in Switzerland. (Photo by E. J. Tarbuck)

large, equidimensional grains that are randomly oriented, similar to those in a coarse-grained igneous rock.

Common Metamorphic Rocks

A chart depicting the common rocks produced by metamorphic processes is found in **Figure 2.31**, and a description of each follows.

Foliated Rocks *Slate* is a very fine-grained foliated rock composed of minute mica flakes that are too small to be visible (see Figure 2.31). A noteworthy characteristic of slate is its excellent rock cleavage, or tendency to break into flat slabs. This property has made slate a useful rock for roof and floor tile, as well as billiard tables (**Figure 2.32**). Slate is usually generated by the low-grade metamorphism of shale. Less frequently, it is produced when volcanic ash is metamorphosed. Slate's color is variable. Black slate contains organic material, red slate gets its color from iron oxide, and green slate is usually composed of chlorite, a greenish micalike mineral.

Phyllite represents a degree of metamorphism between slate and schist. Its constituent platy minerals, mainly muscovite and chlorite, are larger than those in slate but not large enough to be readily identifiable with

Metamorphic Rock	Texture	Comments	Parent Rock
Slate	Foliated	Composed of tiny chlorite and mica flakes, breaks in flat slabs called slaty cleavage, smooth dull surfaces	Shale, mudstone, or siltstone
Phyllite	Foliated	Fine-grained, glossy sheen, breaks along wavy surfaces	Shale, mudstone, or siltstone
Schist	Foliated	Medium- to coarse-grained, scaly foliation, micas dominate	Shale, mudstone, or siltstone
Gneiss	Foliated	Coarse-grained, compositional banding due to segregation of light and dark colored minerals	Shale, granite, or volcanic rocks
Marble	Nonfoliated	Medium- to coarse-grained, relatively soft (3 on the Mohs scale), interlocking calcite or dolomite grains	Limestone, dolostone
Quartzite	Nonfoliated	Medium- to coarse-grained, very hard, massive, fused quartz grains	Quartz sandstone

Figure 2.31 Classification of common metamorphic rocks

A.

B.

Figure 2.33 Marble, because of its workability, is a widely used building stone. A. The white exterior of the Lincoln Memorial in Washington, DC, is constructed mainly of marble that was quarried in Marble, Colorado. Inside, pink Tennessee "marble" was used for the floors, Alabama marble for the ceilings, and Georgia marble for Lincoln's statue. (Photo by Daniel Grill/iStockphoto) **B.** The exterior of India's Taj Mahal is constructed primarily of the metamorphic rock marble. (Photo by Sam DCruz/Shutterstock)

the unaided eye. Although phyllite appears similar to slate, it can be easily distinguished from slate by its glossy sheen and wavy surface (see Figure 2.31).

Schists are moderately to strongly foliated rocks formed by regional metamorphism (see Figure 2.31). They are platy and can be readily split into thin flakes or slabs. Many schists originate from shale parent rock. The term *schist* describes the *texture* of a rock regardless of composition. For example, schists composed primarily of muscovite and biotite are called *mica schists*.

Gneiss (pronounced "nice") is the term applied to banded metamorphic rocks in which elongated and granular (as opposed to platy) minerals predominate (see Figure 2.31). The most common minerals in gneisses are quartz and feldspar, with lesser amounts of muscovite, biotite, and hornblende. Gneisses exhibit strong segregation of light and dark silicates, giving them a characteristic banded texture. While still deep below the surface where temperatures and pressures are great, banded gneisses can be deformed into intricate folds.

Nonfoliated Rocks *Marble* is a coarse, crystalline rock whose parent rock is limestone. Marble is composed of large interlocking calcite crystals, which form from the recrystallization of smaller grains in the parent rock. Because of its color and relative softness (hardness of only 3 on the Mohs scale), marble is a popular building stone.

White marble is particularly prized as a stone from which to carve monuments and statues, such as the Lincoln Memorial in Washington, DC, and the Taj Mahal in India (**Figure 2.33**). The parent rocks from which various marbles form contain impurities that color the stone. Thus, marble can be pink, gray, green, or even black.

Quartzite is a very hard metamorphic rock most often formed from quartz sandstone. Under moderate- to high-grade metamorphism, the quartz grains in sandstone fuse. Pure quartzite is white, but iron oxide may produce reddish or pinkish stains, and dark minerals may impart a gray color.

2.5 CONCEPT CHECKS

1. Metamorphism means "change form." Describe how a rock may change during metamorphism.
2. Briefly describe what is meant by the statement "every metamorphic rock has a parent rock."
3. List the four agents of metamorphism and describe the role of each.
4. Distinguish between regional and contact metamorphism.
5. What feature would easily distinguish schist and gneiss from quartzite and marble?
6. In what ways do metamorphic rocks differ from the igneous and sedimentary rocks from which they formed?

CONCEPTS IN REVIEW: ROCKS: MATERIALS OF THE SOLID EARTH

2.1 Earth as a System: The Rock Cycle

Sketch, label, and explain the rock cycle.

KEY TERMS: rock cycle

- The rock cycle is a good way to think about the transformation of one rock to another due to Earth processes. All igneous rocks are made from molten rock. All sedimentary rocks are made from weathered products of other rocks. All metamorphic rocks are the products of preexisting rocks that are heated or squeezed at high temperatures and/or pressures. Given the right conditions, any kind of rock can be transformed into any other kind of rock.

? Sketch or explain the next stage or possible changes that could occur to the igneous rock shown in this diagram.

Shutterstock   E.J. Tarbuck   USGS   Dennis Tasa

Sediment turns to **sedimentary rock**, which melts to produce **magma**, which then crystallizes into **igneous rock.**

2.2 Igneous Rocks: "Formed by Fire"

Describe the two criteria used to classify igneous rocks and explain how the rate of cooling influences the crystal size of minerals.

KEY TERMS: igneous rocks, magma, lava, extrusive, volcanic, intrusive, plutonic, texture, fine-grained texture, vesicular texture, coarse-grained texture, porphyritic texture, glassy texture, granitic (felsic) composition, basaltic (mafic) composition, andesitic (intermediate) composition, ultramafic, Bowen's reaction series, crystal settling, magmatic differentiation

- Completely or partly molten rock is called magma if it is below Earth's surface and lava if it has erupted onto the surface. It consists of a liquid melt that contains gases (volatiles) such as water vapor, and it may contain solids (mineral crystals).
- Magmas that cool at depth produce intrusive igneous rocks, whereas those that erupt onto Earth's surface produce extrusive igneous rocks.
- To geologists, texture is a description of the size, shape, and arrangement of mineral grains in a rock. A careful observation of the texture of igneous rocks can lead to insights about the conditions under which they formed. The cooling rate is quick for lava on the surface or close

to it and results in a large number of very small crystals. The result is a fine-grained texture. When magma cools at depth, the surrounding rock insulates it, and heat is lost more slowly. This allows sufficient time for the magma's ions to be organized into larger crystals, resulting in a rock with a coarse-grained texture. If crystals begin to form at depth and then the magma moves to a shallow depth or erupts at the surface, it will have a two-stage cooling history. The result is a rock with a porphyritic texture.

- Pioneering experimentation by N. L. Bowen revealed that in a cooling magma, minerals crystallize in a specific order. The dark-colored silicate minerals, such as olivine, crystallize first at the highest temperatures (1250°C [2300°F]), whereas the light silicates, such as quartz, crystallize last at the lowest temperatures (650°C [1200°F]). Separation of minerals by mechanisms such as crystal settling result in igneous rocks having a wide variety of chemical compositions.
- Igneous rocks are divided into broad compositional groups based on the percentage of dark and light silicate minerals they contain. Granitic (or felsic) rocks (such as granite and rhyolite) are composed mostly of the light-colored silicate minerals potassium feldspar and quartz. Rocks of Andesitic (or intermediate) composition (rocks such as andesite) contain plagioclase feldspar and amphibole. Basaltic (or mafic) rocks (such as basalt) contain abundant pyroxene and calcium-rich plagioclase feldspar.

2.3 Weathering of Rocks to Form Sediment

Compare and contrast mechanical and chemical weathering and give examples of each.

KEY TERMS: mechanical weathering, frost wedging, sheeting, exfoliation dome, chemical weathering

- Weathering is the disintegration and decomposition of rocks on Earth's surface. A rock might break down into a large number of smaller pieces through physical processes called mechanical weathering. Rocks can also break down via chemical weathering as their minerals react with oxygen and water to produce new substances that are stable at Earth's surface.
- Mechanical weathering breaks rock into smaller pieces by processes that include the expansion of ice and the growth of plant roots. In addition, rocks that form under lots of pressure deep in Earth will expand when exposed at the surface. This can cause the rock to fracture in onion-like layers, a process called sheeting, which can generate broad, dome-shaped exposures of rock called exfoliation domes.
- Water plays an important role in chemical weathering by facilitating reactions such as rusting, an example of oxidation. Water may also combine with carbon dioxide to produce carbonic acid, which chemically reacts with exposed minerals. The result is new products, mainly clay minerals, which are stable at Earth's surface.

? **How are the two main categories of weathering represented in this image that shows human-made objects?** (Photo by Michael Collier)

2.4 Sedimentary Rocks: Compacted and Cemented Sediment

List and describe the different categories of sedimentary rocks and discuss the processes that change sediment into sedimentary rock.

KEY TERMS: sedimentary rock, sediment, detrital sedimentary rock, chemical sedimentary rock, biochemical sedimentary rock, evaporite deposit, organic matter, lithification, strata (beds), fossil

- Although igneous and metamorphic rocks make up most of Earth's crust by volume, sediment and sedimentary rocks are concentrated near the surface.
- Detrital sedimentary rocks are made of solid particles, mostly quartz grains and microscopic clay minerals. Common detrital sedimentary rocks include shale (the most abundant sedimentary rock), sandstone, and conglomerate.
- Chemical and biochemical sedimentary rocks are derived from mineral matter (ions) that is carried in solution to lakes and seas. Under certain conditions, ions in solution precipitate (settle out) to form chemical sediments as a result of physical processes, such as evaporation. Precipitation may also occur indirectly through life processes of water-dwelling organisms that form materials called biochemical sediments. Many water-dwelling animals and plants extract dissolved mineral matter to form shells and other hard parts. After the organisms die, their skeletons may accumulate on the floor of a lake or an ocean.
- Limestone, an abundant sedimentary rock, is composed chiefly of the mineral calcite ($CaCO_3$). Rock gypsum and rock salt are chemical rocks that form as water evaporates.
- Coal forms from the burial of large amounts of plant matter in low-oxygen depositional environments such as swamps and bogs.
- The transformation of sediment into sedimentary rock is lithification. The two main processes that contribute to lithification are compaction (a reduction in pore space by packing grains more tightly together) and cementation (a reduction in pore space by adding new mineral material that acts as a "glue" to bind the grains to each other).

? **This photo was taken in the Grand Canyon, Arizona. What name(s) do geologists use for the characteristic features found in the sedimentary rocks shown in this image?**

Dennis Tasa

2.5 Metamorphic Rocks: New Rock from Old

Define *metamorphism*, explain how metamorphic rocks form, and describe the agents of metamorphism.

KEY TERMS metamorphic rock, metamorphism, contact (thermal) metamorphism, regional metamorphism, foliation, nonfoliated

- When rocks are subjected to elevated temperatures and pressures, they can change form, producing metamorphic rocks. Every metamorphic rock has a parent rock—the rock it used to be prior to metamorphism. When the minerals in parent rocks are subjected to heat and pressure, new minerals can form.
- Heat, confining pressure, differential stress, and chemically active fluids are four agents that drive metamorphic reactions. Any one alone may trigger metamorphism, or all four may exert influence simultaneously.
- Confining pressure results from burial. The pressure is of the same magnitude in every direction, like the pressure that a swimmer experiences when diving to the bottom of a swimming pool. An increase in confining pressure causes rocks to compact into more dense configurations.
- Differential stress results from tectonic forces. The pressure is greater in some directions and less in other directions. Rocks subjected to

differential stress under ductile conditions deep in the crust tend to shorten in the direction of greatest stress and elongate in the direction(s) of least stress, producing flattened or stretched grains. If the same differential stress is applied to a rock in the shallow crust, it may deform in a brittle fashion instead, breaking into smaller pieces.

- A common kind of texture is foliation, the planar arrangement of mineral grains. Common foliated metamorphic rocks include slate, phyllite, schist, and gneiss. The order listed here is the order of increasing metamorphic grade: Slate is least metamorphosed, and gneiss is the most metamorphosed.
- Common nonfoliated metamorphic rocks include quartzite and marble, recrystallized rocks that form from quartz sandstone and limestone.

? Examine the photograph. Determine whether this rock is foliated or nonfoliated and then determine whether it formed under confining pressure or differential stress. Which of the pairs of arrows shows the direction of maximum stress?

Dennis Tasa

GIVE IT SOME THOUGHT

1. Refer to Figure 2.1. How does the rock cycle diagram—in particular, the labeled arrows—support the fact that sedimentary rocks are the most abundant rock type on Earth's surface?

2. Would you expect all the crystals in an intrusive igneous rock to be the same size? Explain why or why not.

3. Apply your understanding of igneous rock textures to describe the cooling history of each of the igneous rocks pictured on the right.

4. Is it possible for two igneous rocks to have the same mineral composition but be different rocks? Support your answer with an example.

5. Use your understanding of magmatic differentiation to explain how magmas of different composition can be generated in a cooling magma chamber.

6. Dust collecting on furniture is an everyday example of a sedimentary process. Provide another example of a sedimentary process that might be observed in or around where you live.

7. Describe two reasons sedimentary rocks are more likely to contain fossils than igneous rocks.

8. If you hiked to a mountain peak and found limestone at the top, what would that indicate about the likely geologic history of the rock there?

A.

B.

C.

D.

9. The accompanying photos each illustrate either a typical igneous, sedimentary, or metamorphic rock body. Which do you think is a metamorphic rock? Explain why you ruled out the other rock bodies. (Photos by E.J. Tarbuck)

A.   **B.**   **C.**

10. Examine the accompanying photos, which show the geology of the Grand Canyon. Notice that most of the canyon consists of layers of sedimentary rocks, but if you were to hike down into the Inner Gorge, you would encounter the Vishnu Schist, which is metamorphic rock.

    a. What process might have been responsible for the formation of the Vishnu Schist? How does this process differ from the processes that formed the sedimentary rocks that are atop the Vishnu Schist?

    b. What does the Vishnu Schist tell you about the history of the Grand Canyon prior to the formation of the canyon itself?

    c. Why is the Vishnu Schist visible at Earth's surface?

    d. Is it likely that rocks similar to the Vishnu Schist exist elsewhere but are not exposed at Earth's surface? Explain.

A. Inner Gorge of the Grand Canyon

B. Close up of Vishnu Schist (dark color)

MasteringGeology™   www.masteringgeology.com

Looking for additional review and test prep materials? Visit the Study Area in MasteringGeology to enhance your understanding of this chapter's content by accessing a variety of resources, including **Self-Study Quizzes,** **Geoscience Animations, GEODe Tutorials, RSS feeds, flashcards, web links, self-study quizzes,** and an optional **Pearson eText.**

3

Landscapes Fashioned
by Water

Each statement represents the primary learning objective for the corresponding major heading within the chapter. After you complete the chapter, you should be able to:

3.1 List three important external processes and describe where they fit into the rock cycle.

3.2 Explain the role of mass wasting in the development of valleys and discuss the factors that trigger and influence mass-wasting processes.

3.3 List the hydrosphere's major reservoirs and describe the different paths that water takes through the hydrologic cycle.

3.4 Describe the nature of drainage basins and river systems. Sketch four basic drainage patterns.

3.5 Discuss streamflow and the factors that cause it to change.

3.6 Outline the ways in which streams erode, transport, and deposit sediment.

3.7 Contrast bedrock and alluvial stream channels. Distinguish between two types of alluvial channels.

3.8 Contrast narrow V-shaped valleys, broad valleys with floodplains, and valleys that display incised meanders.

3.9 Discuss the formation of deltas and natural levees.

3.10 Discuss the causes of floods and some common flood control measures.

3.11 Discuss the importance of groundwater and describe its distribution and movement.

3.12 Compare and contrast springs, wells, and artesian systems.

3.13 List and discuss three important environmental problems associated with groundwater.

3.14 Explain the formation of caverns and the development of karst topography.

The scenic Little River flows out of Great Smoky Mountains National Park and is a tributary of the Tennessee River. (Photo by Michael Collier)

Earth is a dynamic planet. Volcanic and other internal forces elevate the land, while opposing external processes continually wear it down. The Sun and gravity drive external processes occurring at Earth's surface. Rock is disintegrated and decomposed, moved to lower elevations by gravity, and carried away by water, wind, or ice. Collectively these processes sculpt the physical landscape. This chapter deals with some of these external processes. After a brief examination of mass wasting, we focus on the part of the hydrologic cycle in which water moves from the land back to the sea. Some water travels quickly via a rushing stream, and some moves much more slowly, below the surface. When viewed as part of the Earth system, streams and groundwater represent basic links in the constant cycling of the planet's water.

3.1 Earth's External Processes

List three important external processes and describe where they fit into the rock cycle.

Weathering, mass wasting, and erosion are called **external processes** because they occur at or near Earth's surface and are powered by energy from the Sun. External processes are a basic part of the rock cycle because they are responsible for transforming solid rock into sediment.

To the casual observer, Earth's surface may appear to be without change, unaffected by time. In fact, 200 years ago, most people believed that mountains, lakes, and deserts were permanent features of an Earth that was thought to be no more than a few thousand years old. Today we know Earth is about 4.6 billion years old and that mountains eventually succumb to weathering and erosion, lakes fill with sediment or are drained by streams, and deserts come and go with changes in climate.

Earth is a dynamic body. Some parts of Earth's surface are gradually elevated by mountain building and volcanic activity. These **internal processes** derive their energy from Earth's interior. Meanwhile, opposing external processes are continually breaking rock apart and moving the debris to lower elevations. The latter processes include:

1. **Weathering**—the physical breakdown (disintegration) and chemical alteration (decomposition) of rock at or near Earth's surface.
2. **Mass wasting**—the transfer of rock and soil downslope, under the influence of gravity.
3. **Erosion**—the physical removal of material by a mobile agent such as flowing water, waves, wind, or ice.

Weathering processes were treated in Chapter 2. In this chapter, we will focus on mass wasting and on two important erosional processes—running water and groundwater. Chapter 4 will examine the geologic work of two other significant erosional processes—glacial ice and wind.

3.1 CONCEPT CHECKS

1. Distinguish between internal and external processes.
2. Contrast weathering, mass wasting, and running water.

3.2 Mass Wasting: The Work of Gravity

Explain the role of mass wasting in the development of valleys and discuss the factors that trigger and influence mass-wasting processes.

Earth's surface is never perfectly flat but instead consists of slopes. Some are steep and precipitous; others are moderate or gentle. Some are long and gradual; others are short and abrupt. Some slopes are mantled with soil and covered by vegetation; others consist of barren rock and rubble. Their form and variety are great.

Although most slopes appear to be stable and unchanging, they are not static features because the force of gravity causes material to move downslope. At one extreme, the movement may be gradual and practically imperceptible. At the other extreme, it may consist of a roaring debris flow or a thundering rock avalanche (**Figure 3.1**). Landslides are a worldwide natural hazard. When these natural processes lead to loss of life and property, they become natural disasters.

Figure 3.1 Watch for falling rock! Road signs in mountain areas often warn of this mass-wasting hazard. On August 6, 2006, photographer Herb Dunn witnesses this landslide along the Merced River in California's Yosemite National Park. On the left, the impact of falling rock produces an explosion of dust and debris. On the right, rock debris races down the slope, knocking down trees along the way. (© DunnRight Photograph)

Mass Wasting and Landform Development

Events such as those pictured in **Figures 3.1** and **3.2** are spectacular examples of a common geologic process called mass wasting. *Mass wasting* is the downslope movement of rock and soil under the direct influence of gravity. It is distinct from erosional processes because mass wasting does not require a transporting medium, such as water, wind, or glacial ice. There is a broad array of mass-wasting processes. Four of them are illustrated in **Figure 3.3**.

Figure 3.2 Rock avalanche This rock avalanche occurred along England's White Chalk Cliffs of Dover in March 2012. (Photo by Rex Features via AP images)

Figure 3.3 Types of mass wasting The four processes illustrated here are all considered to be relatively rapid forms of mass wasting. Because the material in slumps and rockslides moves along well-defined surfaces, it is said to move by sliding. By contrast, when material moves like a thick fluid, it is described as a flow. Debris flow and earthflow are examples.

The Role of Mass Wasting

Mass wasting is the step that follows weathering in the evolution of most landforms. Once weathering weakens and breaks rock apart, mass wasting transfers the debris downslope, where a stream, acting as a conveyor belt, usually carries it away (**Figure 3.4**). Although there may be many intermediate stops along the way, most of the sediment is eventually transported to the sea. The combined effects of mass wasting and running water produce stream valleys, which are the most common and conspicuous landforms at Earth's surface and the focus of a later part of this chapter.

If streams alone were responsible for creating the valleys in which they flow, valleys would be very narrow features. However, the fact that most river valleys are much wider than they are deep is a strong indication of the significance of mass-wasting processes in supplying material to streams. The walls of a canyon extend far from the river because of the transfer of weathered debris downslope to the river and its tributaries by mass-wasting processes. In this manner, streams and mass wasting combine to modify and sculpture the surface. Of course, glaciers, groundwater, waves, and wind are also important agents in shaping landforms and developing landscapes.

Slopes Change Through Time

Most rapid and spectacular mass-wasting events occur in areas of rugged, geologically young mountains. Newly formed mountains are rapidly eroded by rivers and glaciers into regions characterized by steep and unstable slopes. It is in such settings that massive destructive landslides occur. As mountain building subsides, mass-wasting and erosional processes lower the land. Over time, steep and rugged mountain slopes give way to gentler, more subdued terrain. Thus, as a landscape ages, massive and rapid mass-wasting processes give way to smaller, less dramatic downslope movements.

Controls and Triggers of Mass Wasting

Gravity is the controlling force of mass wasting, but several factors play important roles in overcoming inertia and creating downslope movements. Long before a landslide occurs, various processes work to weaken slope material, gradually making it more and more susceptible to the pull of gravity. During this span, the slope remains stable but gets closer and closer to being unstable. Eventually, the strength of the slope is weakened to the point that something causes it to cross the threshold from stability to instability. Such an event that initiates downslope movement is called a **trigger**. Remember that the trigger is not the sole cause of a mass-wasting event; it is just the last of many causes. Among the common factors that trigger mass-wasting processes are saturation of material with water, oversteepening of slopes, removal of anchoring vegetation, and ground vibrations from earthquakes.

SmartFigure 3.4
Excavating the Grand Canyon The walls of the Grand Canyon extend far from the channel of the Colorado River. This results primarily from the transfer of weathered debris downslope to the river and its tributaries by mass-wasting processes. (Photo by Bryan Brazil/Shutterstock)

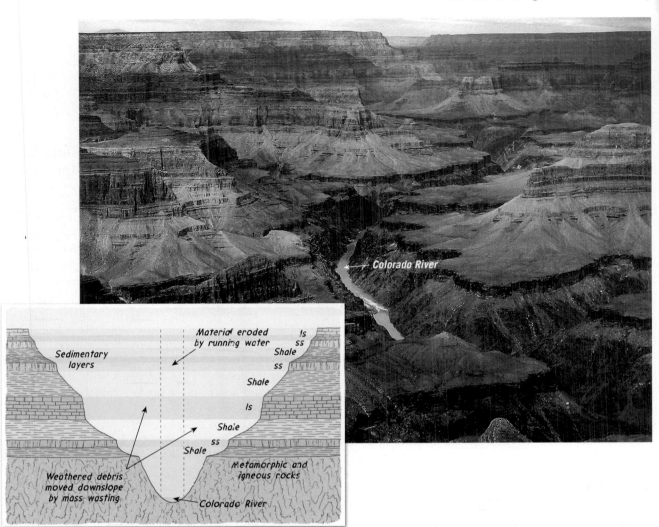

Geologist's Sketch

The Role of Water Mass wasting is sometimes triggered when heavy rains or periods of snowmelt saturate surface materials. Such a situation is shown in **Figure 3.5**.

When the pores in sediment become filled with water, the cohesion among particles is destroyed, allowing the particles to slide past one another with relative ease. For example, when sand is slightly moist, it sticks together quite well. However, if enough water is added to fill the openings between the grains, the sand will ooze out in all directions. Thus, saturation reduces the internal resistance of materials, which are then easily set in motion by the force of gravity. When clay is wetted, it becomes very slick—another example of the "lubricating" effect of water. Water also adds considerable weight to a mass of material. The added weight in itself may be enough to cause the material to slide or flow downslope.

Oversteepened Slopes Oversteepening of slopes is another trigger of many mass movements. This takes place in many situations in nature. A stream undercutting a valley wall and waves pounding against the base of a cliff are two familiar examples. Furthermore, through their activities, people often create oversteepened and unstable slopes that become prime sites for mass wasting.

Unconsolidated, granular (sand-size or coarser) particles assume a stable slope called the **angle of repose**, the steepest angle at which material remains stable (**Figure 3.6**). Depending on the size and shape of the particles, the angle varies from 25 to 40 degrees. The larger, more angular particles maintain the steepest slopes. If the angle is increased, the rock debris will adjust by moving downslope.

Oversteepening is important not only because it triggers movements of unconsolidated granular materials but also because it produces unstable slopes and mass movements in cohesive soils, regolith (weathered rock debris), and bedrock. The response will not be immediate, as with loose, granular material, but sooner or later, one or more mass-wasting processes will eliminate the oversteepening and restore stability to the slope.

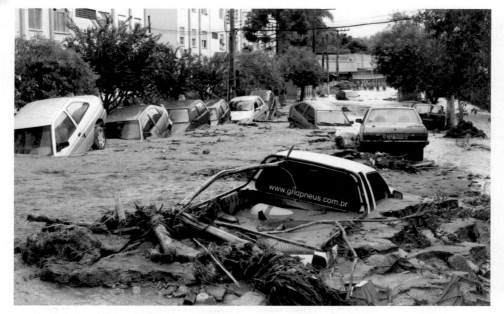

Figure 3.5 Mudflow in Brazil In January 2011, following a period of torrential rains, a thick slurry of mud buried these cars in Nova Friburgo, Brazil. Heavy rains on saturated and unprotected slopes triggers significant mass wasting. (Photo by ZUMA PRESS.com/Newscom)

Removal of Vegetation

Plants protect against erosion and contribute to the stability of slopes because their root systems bind soil and regolith together. Where plants are lacking, mass wasting is enhanced, especially if slopes are steep and water is plentiful. When anchoring vegetation is removed by forest fires or by people (for timber, farming, or development), surface materials frequently move downslope.

Earthquakes as Triggers

Among the most important and dramatic mass-wasting triggers are earthquakes. An earthquake and its aftershocks can dislodge enormous volumes of rock and unconsolidated material. In many areas that are jolted by earthquakes, it is not ground vibrations directly but landslides and ground subsidence triggered by the vibrations that cause the greatest damage.

Landslides Without Triggers?

Do rapid mass-wasting events always require some sort of trigger, such as heavy rains or an earthquake? The answer is "no." Many rapid mass-wasting events occur without discernible triggers. The rockfall shown in Figure 3.1 is an example. Slope materials gradually weaken over time, under the influence of long-term weathering, infiltration of water, and other physical processes. Eventually, if the strength falls below what is necessary to maintain slope stability, a landslide will occur. The timing of such events is random, and accurate prediction is not possible.

3.2 CONCEPT CHECKS

1. How does water affect mass-wasting processes?
2. Describe the significance of the angle of repose.
3. How might a forest fire influence mass wasting?
4. Link earthquakes and landslides.

Figure 3.6 Angle of repose This is the steepest angle at which an accumulation of granular particles remains stable. Larger, more angular particles maintain the steepest slopes. (Photo by G. Leavens/Photo Researchers, Inc.; inset photo by Michael Collier)

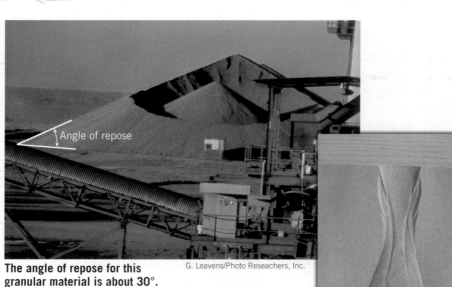

Angle of repose

The angle of repose for this granular material is about 30°.

G. Leavens/Photo Researchers, Inc.

As sand accumulates along a dune crest, the slope gets steeper. Eventually, some sand slides down the oversteepened slope.

Michael Collier

3.3 The Hydrologic Cycle

List the hydrosphere's major reservoirs and describe the different paths that water takes through the hydrologic cycle.

Water is continually on the move, from the ocean to the atmosphere to the land and back again, in an endless cycle called the **hydrologic cycle**. The remainder of this chapter deals with the part of the cycle that returns water to the sea. Some water travels by way of rivers, and some moves more slowly, beneath the surface. We will examine the factors that influence the distribution and movement of water, as well as look at how water sculpts the landscape. The Grand Canyon, Niagara Falls, the Old Faithful geyser, and Mammoth Cave all owe their existence to water on its way to the sea.

Earth's Water Water is almost everywhere on Earth—in the oceans, glaciers, rivers, lakes, air, soil, and living tissue. All of these "reservoirs" constitute Earth's hydrosphere. In all, the water content of the hydrosphere is an estimated 1.36 billion cubic kilometers (326 million cubic miles). The vast bulk of it, about 96.5 percent, is stored in the global ocean. Ice sheets and glaciers account for an additional 1.76 percent, leaving just slightly more than 2 percent to be divided among lakes, streams, groundwater, and the atmosphere (see Figure I.4, p. xx). Although the percentage of Earth's total water found in each of the latter sources is just a small fraction of the total inventory, the absolute quantities are great.

Water's Paths The hydrologic cycle is a gigantic, worldwide system powered by energy from the Sun, in which the atmosphere provides a vital link between the oceans and continents (**Figure 3.7**). **Evaporation**, the process by which liquid water changes into water vapor (gas), is how water enters the atmosphere from the ocean and, to a much lesser extent, from the continents. Winds transport this moisture-laden air, often great distances. Complex processes of cloud formation eventually result in precipitation. The precipitation that falls into the ocean has completed its cycle and

is ready to begin another. The water that falls on the continents, however, must make its way back to the ocean.

What happens to precipitation once it has fallen on land? A portion of the water soaks into the ground (called **infiltration**), slowly moving downward, then moving laterally, finally seeping into lakes, streams, or directly into the ocean. When the rate of rainfall exceeds Earth's ability to absorb it, the surplus water flows over the surface into lakes and streams, a process called **runoff**. Much of the water that infiltrates or runs off eventually returns to the atmosphere because of evaporation from the soil, lakes, and streams. Also, some of the water that soaks into the ground is absorbed by plants, which then release it into the atmosphere. This process is called **transpiration**. Because we cannot clearly distinguish between the amount of water that is evaporated and the amount that is transpired by plants, the term **evapotranspiration** is often used for the combined effect.

Storage in Glaciers When precipitation falls in very cold places—at high elevations or high latitudes—the water may not immediately soak in, run off, or evaporate. Instead, it may become part of a snowfield or a glacier. In this way, glaciers store large quantities of

Hydrologic Cycle

Evaporation 320,000 km³

Precipitation 284,000 km³

Precipitation 96,000 km³

Evaporation/Transpiration 60,000 km³

36,000 km³

Runoff

Infiltration

Oceans

SmartFigure 3.7 The hydrologic cycle The primary movement of water through the cycle is shown by the large arrows. Numbers refer to the annual amount of water taking a particular path.

water on land. If present-day glaciers were to melt and release all their water, sea level would rise by several dozen meters. Such a rise would submerge many heavily populated coastal areas. As you will see in Chapter 4, over the past 2 million years, huge ice sheets have formed and melted on several occasions, each time affecting the balance of the water cycle.

Water Balance   Figure 3.7 also shows Earth's overall *water balance*, or the volume that passes through each part of the cycle annually. The amount of water vapor in the air at any one time is just a tiny fraction of Earth's total water supply. But the *absolute* quantities that are cycled through the atmosphere over a one-year period are immense—some 380,000 cubic kilometers (91,000 cubic miles)—enough to cover Earth's entire surface to a depth of about 1 meter (39 inches).

It is important to know that the hydrologic cycle is *balanced*. Because the total amount of water vapor in the atmosphere remains about the same, the average annual precipitation worldwide must be equal to the quantity of water evaporated. However, for all the continents taken together, precipitation exceeds

evaporation. Conversely, over the oceans, evaporation exceeds precipitation. Because the level of the world ocean is not dropping, the system must be in balance. In Figure 3.7, the 36,000 cubic kilometers (8600 cubic miles) of water that annually makes its way from the land to the ocean causes enormous erosion. In fact, this immense volume of moving water is *the single most important agent sculpturing Earth's land surface*.

In the rest of this chapter, we will observe the work of water running over the surface, including floods, erosion, and the formation of valleys. Then we will look underground at the slow labors of groundwater as it forms springs and caverns and provides drinking water on its long migration to the sea.

3.3 CONCEPT CHECKS

1. Describe or sketch the movement of water through the hydrologic cycle. Once precipitation has fallen on land, what paths might the water take?
2. What is meant by the term *evapotranspiration*?
3. Over the oceans, evaporation exceeds precipitation, yet sea level does not drop. Explain this.

3.4 Running Water

Describe the nature of drainage basins and river systems. Sketch four basic drainage patterns.

Recall that most of the precipitation that falls on land either enters the soil (infiltration) or remains at the surface, moving downslope as runoff. The amount of water that runs off rather than soaking into the ground depends on several factors: (1) intensity and duration of rainfall, (2) amount of water already in the soil, (3) nature of the surface material, (4) slope of the land, and (5) extent and type of vegetation. When the surface material is highly impermeable or when it becomes saturated, runoff is the dominant process. Runoff is also high in urban areas because large areas are covered by impermeable buildings, roads, and parking lots.

Runoff initially flows in broad, thin sheets across hillslopes. This unconfined flow eventually develops threads of current that form tiny channels called *rills*. Rills meet, forming *gullies*, which join to form streams. At first streams are small, but as one intersects another, larger and larger streams form. Eventually rivers develop that carry water from a broad region.

Drainage Basins

The land area that contributes water to a river system is called a **drainage basin** (Figure 3.8). The drainage basin of one stream is separated from the drainage basin of another by an imaginary line called a **divide**. Divides range

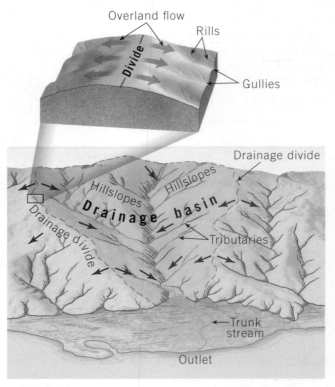

Figure 3.8 Drainage basin and divide A drainage basin is the area drained by a stream and its tributaries. Boundaries between drainage basins are called divides.

SmartFigure 3.9 Mississippi River drainage basin The drainage basin of the Mississippi River, North America's largest, covers about 3 million square kilometers (almost 1.2 million square miles) and consists of many smaller drainage basins. The drainage basin of the Yellowstone River is one of many that contribute water to the Missouri River, which, in turn, is one of many that make up the drainage basin of the Mississippi River.

in scale from a ridge separating two small gullies on a hillside to a *continental divide*, which splits whole continents into enormous drainage basins. The Mississippi River has the largest drainage basin in North America (**Figure 3.9**). Extending between the Rocky Mountains in the West and the Appalachian Mountains in the East, the Mississippi River and its tributaries collect water from more than 3.2 million square kilometers (1.2 million square miles) of the continent.

River Systems

River systems involve not only a network of stream channels but the entire drainage basin. Based on the dominant processes operating within them, river systems can be divided into three zones: *sediment production*—where erosion dominates, *sediment transport*, and *sediment deposition* (**Figure 3.10**). It is important to recognize that sediment is being eroded, transported, and deposited along the entire length of a stream, regardless of which process is dominant within each zone.

The zone of *sediment production*, where most of the water and sediment is derived, is located in the headwaters region of the river system. Much of the sediment carried by streams begins as bedrock that is subsequently broken down by weathering and then transported downslope by mass wasting and overland flow. Bank erosion can also contribute significant amounts of sediment. In addition, scouring of the channel bed deepens the channel and adds to the stream's sediment load.

Sediment acquired by a stream is then transported through the channel network, along sections referred to as *trunk streams*. When trunk streams are in balance, the amount of sediment eroded from their banks equals the amount deposited elsewhere in the channel. Although trunk streams rework their channels over time, they are not a source of sediment, nor do they accumulate or store it.

When a river reaches the ocean, or another large body of water, it slows, and the energy to transport sediment is greatly reduced. Most of the sediments accumulate at the mouth of the river to form a delta, are

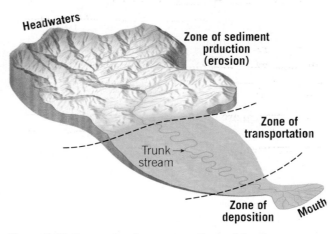

Figure 3.10 Zones of a river system Each of the three zones is based on the dominant process that is operating in that part of the river system.

Figure 3.11 Drainage patterns Networks of streams form a variety of patterns.

Dendritic pattern develops on relatively uniform surface materials

Radial pattern develops on isolated volcanic cones or domes

Volcano

Rectangular pattern develops on highly jointed bedrock

Ridges of resistant rock

Valleys cut in less-resistant rock

Trellis pattern develops in areas of alternating weak and resistant bedrock

reconfigured by wave action to form a variety of coastal features, or are moved far offshore by ocean currents. Because coarse sediments tend to be deposited upstream, it is primarily the fine sediments (clay and fine sand) that eventually reach the ocean. Taken together, erosion, transportation, and deposition are the processes by which rivers move Earth's surface materials and sculpt landscapes.

Drainage Patterns

Drainage systems are networks of streams that together form distinctive patterns. The nature of a drainage pattern can vary greatly from one type of terrain to another, primarily in response to the kinds of rock on which the streams developed and/or the structural pattern of faults and folds. **Figure 3.11** illustrates four drainage patterns.

The most commonly encountered drainage pattern is the **dendritic pattern**. This pattern of irregularly branching tributary streams resembles the branching pattern of a deciduous tree. In fact, the word *dendritic* means "treelike." The dendritic pattern forms where the underlying material is relatively uniform. Because the surface material is essentially uniform in its resistance to erosion, it does not control the pattern of streamflow. Rather, the pattern is determined chiefly by the direction of slope of the land.

When streams diverge from a central area like spokes from the hub of a wheel, the pattern is said to be **radial**. This pattern typically develops on isolated volcanic cones and domal uplifts.

A **rectangular pattern** exhibits many right-angle bends. This pattern develops when the bedrock is crisscrossed by a series of joints and/or faults. Because these structures are eroded more easily than unbroken rock, their geometric pattern guides the directions of valleys.

A **trellis pattern** is a rectangular drainage pattern in which tributary streams are nearly parallel to one another and have the appearance of a garden trellis. This pattern forms in areas underlain by alternating bands of resistant and less-resistant rock.

3.4 CONCEPT CHECKS

1. List several factors that cause infiltration and runoff to vary from place to place and from time to time.
2. Draw a simple sketch of a drainage basin and a divide. Label each.
3. What are the three main parts (zones) of a river system?
4. Prepare a sketch of the four drainage patterns discussed in this section.

3.5 Streamflow

Discuss streamflow and the factors that cause it to change.

Water may flow in one of two ways, either as **laminar flow** or **turbulent flow**. In slow-moving streams, the flow is often laminar, and the water moves in roughly straight-line paths that parallel the stream channel. However, streamflow is usually turbulent, with the water moving in an erratic fashion that can be characterized as a swirling motion. Strong, turbulent flow may be seen in whirlpools and eddies, as well as rolling whitewater rapids (**Figure 3.12**). Even streams that appear smooth on the surface often exhibit turbulent flow near the bottom and sides of the channel. Turbulence contributes to a stream's ability to erode its channel because it acts to lift sediment from the streambed.

An important factor influencing stream turbulence is the water's flow velocity. As the velocity of a stream increases, the flow becomes more turbulent. Flow velocities can vary significantly from place to place along a stream, as well as over time, in response to variations in the amount and intensity of precipitation. If you have ever waded into a stream, you may have noticed that the strength of the current increased as you moved into deeper parts of the channel. This is related to the fact that frictional resistance is greatest near the banks and bed of the stream channel.

Factors Affecting Flow Velocity

The ability of a stream to erode and transport material is directly related to its flow velocity. Even slight variations in flow rate can lead to significant changes in the load of sediment that water can transport. Several factors influence flow velocity and, therefore, control a stream's potential to do "work." These factors include (1) channel slope or gradient, (2) channel size and cross-sectional shape, (3) channel roughness, and (4) the amount of water flowing in the channel.

Gradient The slope of a stream channel expressed as the vertical drop of a stream over a specified distance is the **gradient**. Portions of the lower Mississippi River have very low gradients of 10 centimeters per kilometer or less. By contrast, some mountain

Running the rapids in the Grand Canyon— an extreme example of turbulent flow.

This water is not standing still. It is moving slowly toward the bottom of the image. The flow in the foreground is primarily laminar.

Figure 3.12 Laminar and turbulent flow The water in a sluggish stream may exhibit laminar flow, in which the water moves slowly in straight-line paths. However, streamflow is usually turbulent. (Photos by Michael Collier)

The largest river in North America, the Mississippi, discharges an average of 17,300 cubic meters (611,000 cubic feet) per second. Although this is a huge quantity of water, it is dwarfed by the mighty Amazon in South America, the world's largest river. Fed by a vast rainy region that is nearly three-fourths the size of the conterminous United States, the Amazon discharges 12 times more water than the Mississippi.

The discharges of most rivers are far from constant. This is true because of variables such as rainfall and snowmelt. In areas with seasonal variations in precipitation, streamflow will tend to be highest during the wet season or during spring snowmelt, and it will be lowest during the dry season or during periods when high temperatures increase water losses through evaporation. However, not all channels maintain a continuous flow of water. Streams that exhibit flow only during wet periods are referred to as *intermittent streams*. In arid climates, many streams carry water only occasionally after a heavy rainstorm and are called *ephemeral streams*.

stream channels decrease in elevation at a rate of more than 40 meters per kilometer —a gradient 400 times steeper than the lower Mississippi. Gradient varies not only among different streams but also over a particular stream's length. The steeper the gradient, the more energy available for streamflow. If two streams were identical in every respect except gradient, the stream with the higher gradient would have the greater velocity.

Channel Shape, Size, and Roughness A stream's channel is a conduit that guides the flow of water, but the water encounters friction as it flows. The shape, size, and roughness of the channel affect the amount of friction. Larger channels have more efficient flow because a smaller proportion of water is in contact with the channel. A smooth channel promotes a more uniform flow, whereas an irregular channel filled with boulders creates enough turbulence to slow the stream significantly.

Discharge Streams vary in size from small headwater creeks less than 1 meter wide to large rivers with widths of several kilometers. The size of a stream channel is largely determined by the amount of water supplied from the drainage basin. The measure most often used to compare the sizes of streams is **discharge**—the volume of water flowing past a certain point in a given unit of time. Discharge, usually measured in cubic meters per second or cubic feet per second, is determined by multiplying a stream's cross-sectional area by its velocity.

Changes from Upstream to Downstream

One useful way of studying a stream is to examine its **longitudinal profile**—a cross-sectional view of a stream from its source area (called the *head* or *headwaters*) to its *mouth*, the point downstream where the river empties into another water body—a river, a lake, or an ocean. As shown in **Figure 3.13**, the most obvious feature of a typical profile is a constantly decreasing gradient from the head to the mouth. Although many local irregularities may exist, the overall profile is a relatively smooth concave curve.

The change in slope observed on most stream profiles is usually accompanied by an increase in discharge and channel size, as well as a reduction in sediment particle size. Along most rivers in humid regions, discharge increases toward the mouth because as we move downstream, more and more tributaries contribute water to the main channel. In order to accommodate the growing volume of water, channel size typically increases downstream as well. Recall that flow velocities are higher in large channels than in small channels. Observations also show a general decline in sediment size downstream, making the channel smoother and more efficient.

Although the gradient decreases toward a stream's mouth, the flow velocity generally increases. This

Figure 3.14 The lower Mississippi River Moving downstream toward the mouth of a river, tributaries contribute more and more water. In response, the channel gets larger, and average flow velocity increases. (Photo by Michael Collier)

contradicts our intuitive assumptions of swift, narrow headwater streams and wide, placid rivers flowing across more subtle topography. Increases in channel size and discharge, as well as decreases in channel roughness that occur downstream compensate for the decrease in gradient, thereby making the stream more efficient. Thus the average flow velocity is typically lower in headwater streams than in wide, placid-appearing rivers (**Figure 3.14**).

3.5 CONCEPT CHECKS

1. Contrast laminar flow and turbulent flow.
2. Summarize the factors that influence flow velocity.
3. What is a longitudinal profile?
4. How do gradient, discharge, channel size, and channel roughness typically change from the head to the mouth of a stream?
5. Is flow velocity usually greater at the head or at the mouth of a stream? Explain.

3.6 The Work of Running Water

Outline the ways in which streams erode, transport, and deposit sediment.

Streams are Earth's most important erosional agents. Not only do they have the ability to downcut and widen their channels, but streams also have the capacity to transport the enormous quantities of sediment that are delivered to them by sheet flow, mass wasting, and groundwater. Eventually, much of this material is deposited to create a variety of depositional features.

Stream Erosion

A stream's ability to accumulate and transport soil and weathered rock is aided by the work of raindrops, which knock sediment particles loose. When the ground is saturated, rainwater begins to flow downslope, transporting some of the material it has dislodged. On barren slopes

Figure 3.15 Gully erosion Rainwater can create tiny channels called rills that can join to form larger channels called gullies. (Photo by Carl Purcell/ Photo Researchers, Inc.)

Gully erosion on unprotected soil on a Wisconsin farm. Just one millimeter of soil from a single acre amounts to about 5 tons.

the flow of muddy water, called *sheet flow*, will often erode small channels, or *rills*, which in time may evolve into larger *gullies* (**Figure 3.15**).

Once the surface flow reaches a stream, its ability to erode is greatly enhanced by the increase in water volume. When the flow of water is sufficiently strong, it can dislodge particles from the channel and lift them into the moving water. In this manner, the force of running water swiftly erodes poorly consolidated materials on the bed and sides of a stream channel. On occasion, the banks of the channel may be undercut, dumping even more loose debris into the water to be carried downstream.

In addition to eroding unconsolidated materials, the hydraulic force of streamflow can also cut a channel into solid bedrock. A stream's ability to erode bedrock is greatly enhanced by the particles it carries. These particles vary in size, from large gravel in fast-flowing waters to sand in somewhat slower flows. Just as the particles of grit on sandpaper can wear away a piece of wood, so too can the sand and gravel carried by a stream abrade a bedrock channel. Moreover, pebbles caught in swirling eddies can act like "drills" and bore circular **potholes** into the channel floor (**Figure 3.16**).

Figure 3.16 Potholes The rotational motion of swirling pebbles acts like a drill to create potholes. (Photo by Elmari Joubert/Alamy Images)

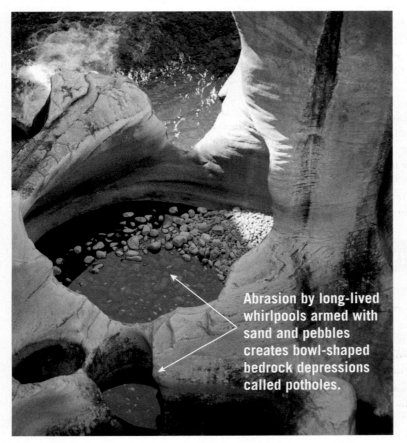

Abrasion by long-lived whirlpools armed with sand and pebbles creates bowl-shaped bedrock depressions called potholes.

Transportation of Sediment by Streams

All streams, regardless of size, transport some rock material. Streams also sort the solid sediment they transport because finer, lighter material is carried more readily than larger, heavier particles. Streams transport their load of sediment in three ways: (1) in solution (**dissolved load**), (2) in suspension (**suspended load**), and (3) sliding or rolling along the bottom (**bed load**).

Dissolved Load Most of the *dissolved load* is brought to a stream by groundwater and is dispersed throughout the flow. When water percolates through the ground, it acquires soluble soil compounds. Then it seeps through cracks and pores in bedrock, dissolving additional mineral matter. Eventually, much of this mineral-rich water finds its way into streams.

Velocity has essentially no effect on a stream's ability to carry its dissolved load; material in solution goes wherever the water flows. Deposition (precipitation) of the dissolved mineral matter occurs when the chemistry of the water changes. In arid regions, the water may enter an inland lake or sea and evaporate, leaving behind the dissolved load.

Suspended Load Most streams carry the largest part of their load in *suspension* (**Figure 3.17**). Indeed, the visible cloud of sediment suspended in the water is the most obvious portion of a stream's load. Usually, only fine particles consisting of silt and clay can be carried this way, but during a flood, larger particles can also be transported in suspension. Also during a flood, the total quantity of material carried in suspension increases dramatically, as can be verified by anyone whose home has been a site for the deposition of this material.

Bed Load A portion of a stream's load consists of sand, gravel, and occasionally large boulders. These coarser particles, which are too large to be carried in suspension, move along the bottom (bed) of the stream channel and constitute the *bed load*. Unlike the suspended and dissolved loads, which are constantly in motion, the bed load is in motion only intermittently, when the force of the water is sufficient to move the larger particles. The smaller particles, mainly sand and gravel are moved by *saltation*, which resembles a series of jumps or skips. Larger particles either roll or slide along the bottom, depending on their shape.

Capacity and Competence A stream's ability to carry solid particles is described using two criteria: *capacity* and *competence*. **Capacity** is the maximum load of solid particles a stream can transport per unit of time. The greater the discharge, the greater the stream's capacity for hauling sediment. Consequently, large rivers with high flow velocities have large capacities.

Competence is a measure of a stream's ability to transport particles based on size rather than quantity. Flow velocity is the key: Swift streams have greater competencies than slow streams, regardless of channel size. A stream's competence increases proportionally to the square of its velocity. Thus, if the velocity doubles, the impact force of the water increases four times; if the velocity triples, the force increases nine times, and so forth. Consequently, large boulders that are often visible during low water and seem immovable can, in fact, be transported during exceptional floods because of the stream's increased competence.

By now it should be clear why the greatest erosion and transportation of sediment occur during floods. The increase in discharge results in a greater capacity, and the increase in velocity results in greater competence. With rising velocity, the water becomes more turbulent, and larger and larger particles are set in motion. In just a few days or perhaps a few hours, a flooding stream can erode and transport more sediment than it does during months of normal flow.

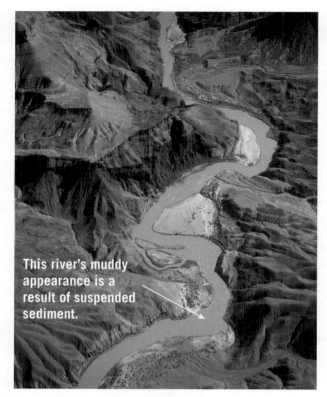

This river's muddy appearance is a result of suspended sediment.

Figure 3.17 Suspended load An aerial view of the Colorado River in the Grand Canyon. Heavy rains washed sediment into the river. (Photo by Michael Collier)

Deposition of Sediment by Streams

When a stream slows down, the situation reverses: As its velocity decreases, the stream's competence is reduced, and sediment begins to drop out, largest particles first. Each particle size has a *critical settling velocity*. As stream flow drops below the critical setting velocity of a certain particle size, sediment in that category begins to settle out. Thus, stream transport provides a mechanism by which solid particles of various sizes are separated. This process, called **sorting**, explains why particles of similar size are deposited together.

The material deposited by a stream is called **alluvium**, the general term for any stream-deposited sediment. Many different depositional features are composed of alluvium. Some occur within stream channels, some occur on the valley floor adjacent to the channel, and some exist at the mouth of the stream. We will consider the nature of these features later in the chapter.

3.6 CONCEPT CHECKS

1. List two ways in which streams erode their channels.
2. In what three ways does a stream transport its load? Which part of the load moves most slowly?
3. What is the difference between capacity and competence?
4. How do streams sort sediment?

3.7 Stream Channels

Contrast bedrock and alluvial stream channels. Distinguish between two types of alluvial channels.

A basic characteristic of streamflow that distinguishes it from overland flow is that it is confined to a channel. A stream channel can be thought of as an open conduit consisting of the streambed and banks that act to confine the flow except during floods.

Although this is somewhat oversimplified, we can divide stream channels into two types. *Bedrock channels* are channels in which the streams are actively cutting into solid rock. In contrast, when the bed and banks are composed mainly of unconsolidated sediment, the channel is called an *alluvial channel*.

Bedrock Channels

In their headwaters, where the gradient is steep, most rivers cut into bedrock. These streams typically transport coarse particles that actively abrade the bedrock channel. Potholes are often visible evidence of the erosional forces at work.

Bedrock channels typically alternate between relatively gently sloping segments where alluvium tends to accumulate and steeper segments where bedrock is exposed. These steeper areas may contain rapids or

Erosion of a cut bank along the Newaukum River in southwestern Washington State.

White River near Vernal, Utah

The cut bank forms on the outside of a meander where flow velocity and turbulence are greatest.

Point bar

As water slows on the inside of a meander, coarser material is deposited as a point bar.

Michael Collier

Deposition of point bar

Erosion of cut bank

Maximum velocity

SmartFigure 3.18 Formation of cut banks and point bars By eroding its outer bank and depositing material on the inside of the bend, a stream is able to shift its channel.

occasionally a waterfall. The channel pattern exhibited by streams cutting into bedrock is controlled by the underlying geologic structure. Even when flowing over rather uniform bedrock, streams tend to exhibit a winding or irregular pattern rather than flow in a straight channel. Anyone who has gone on a white-water rafting trip has observed the steep, winding nature of a stream flowing in a bedrock channel.

Alluvial Channels

Many stream channels are composed of loosely consolidated sediment (alluvium) and therefore can undergo significant changes in shape because the sediments are continually being eroded, transported, and redeposited. The major factors affecting the shapes of these channels are the average size of the sediment being transported and the channel's gradient and discharge.

Cutoff

Green River, WY

Meanders

Neck

Before the formation of a cutoff

Cutoff

After the formation of a cutoff

Oxbow lake

Geologist's Sketch

Figure 3.19 Formation of an oxbow lake Oxbow lakes occupy abandoned meanders. Aerial view of an oxbow lake created by the meandering Green River near Bronx, Wyoming. (Photo by Michael Collier)

Alluvial channel patterns reflect a stream's ability to transport its load at a uniform rate, while expending the least amount of energy. Thus, the size and type of sediment being carried help determine the nature of the stream channel. Two common types of alluvial channels are *meandering channels* and *braided channels*.

Meandering Channels Streams that transport much of their load in suspension generally move in sweeping bends called **meanders**. These streams flow in relatively deep, smooth channels and transport mainly mud (silt and clay). The lower Mississippi River exhibits a channel of this type.

Because of the cohesiveness of consolidated mud, the banks of stream channels carrying fine particles tend to resist erosion. As a consequence, most of the erosion in such channels occurs on the outside of the meander, where velocity and turbulence are greatest. In time, the outside bank is undermined, especially during periods of high water. Because the outside of a meander is a zone of active erosion, it is often referred to as the **cut bank** (**Figure 3.18**). The debris acquired by the stream at a cut bank moves downstream, with the coarser material generally being deposited as **point bars** in zones of decreased velocity on the insides of meanders. In this manner, meanders migrate laterally by eroding the outside of the bends and depositing on the inside.

In addition to migrating laterally, the bends in a channel also migrate down the valley. This occurs because erosion is more effective on the downstream (downslope) side of the meander. The downstream migration of a meander is sometimes slowed when it reaches a more resistant material. This allows the next meander upstream to "catch up" and overtake it, as shown in **Figure 3.19**. The neck of land between the meanders is gradually narrowed. Eventually, the river may erode through the narrow neck of land to the next loop. The new, shorter channel segment is called a **cutoff** and, because of its shape, the abandoned bend is called an **oxbow lake**.

Braided Channels A stream may consist of a complex network of converging and diverging channels that thread their way among numerous islands or gravel bars (**Figure 3.20**). Because these channels have an interwoven appearance, they are said to be **braided channels**. Braided channels form where a large proportion of the stream's load consists of coarse material (sand and gravel) and the stream has a highly variable discharge. Because the bank material is readily erodible, braided channels are wide and shallow.

One setting in which braided streams form is at the end of a glacier, where there is a large seasonal variation in discharge. During summer, large amounts of ice-eroded sediment are dumped into the meltwater streams flowing away from the glacier. However, when flow is sluggish, the stream is unable to move all the sediment and therefore deposits the coarsest material as bars in the channel that force the flow to split and follow several paths. Usually the laterally shifting channels completely

Figure 3.20 Braided stream The Knik River is a classic braided stream with multiple channels separated by migrating gravel bars. The Knik is choked with sediment contributed by the meltwaters of four glaciers in the Chugach Mountains north of Anchorage, Alaska. (Photo by Michael Collier)

rework most of the surface sediments each year, thereby transforming the entire streambed. In some braided streams, however, the bars have built up to form islands that are anchored by vegetation.

In summary, meandering channels develop where the load consists largely of fine-grained particles that are transported as suspended load in a deep, smooth channel. By contrast, wide, shallow braided channels develop where coarse-grained alluvium is transported as bed load.

3.7 CONCEPT CHECKS

1. Are bedrock channels more likely to be found near the head or the mouth of a stream?
2. Describe or sketch the evolution of a meander, including how an oxbow lake forms.
3. Describe a situation that might cause a stream channel to become braided.

3.8 Shaping Stream Valleys

Contrast narrow V-shaped valleys, broad valleys with floodplains, and valleys that display incised meanders.

Streams, with the aid of weathering and mass wasting, shape the landscape through which they flow. As a result, streams continuously modify the valleys that they occupy.

A **stream valley** consists of not only the channel but also the surrounding terrain that directly contributes water to the stream. Thus, it includes the valley bottom, which is the lower, flatter area that is partially or totally occupied by the stream channel, and the

sloping valley walls that rise above the valley bottom on both sides. Most stream valleys are much broader at the top than is the width of their channel at the bottom. This would not be the case if the only agent responsible for eroding valleys were the streams flowing through them. The sides of most valleys are shaped by a combination of weathering, overland flow, and mass wasting. In some arid regions, where weathering is slow and where rock is particularly resistant, narrow valleys that have nearly vertical walls are common.

Stream valleys can be divided into two general types—narrow V-shaped valleys and wide valleys with flat floors—with many gradations between.

Base Level and Stream Erosion

Streams cannot endlessly erode their channels deeper and deeper. There is a lower limit to how deep a stream can erode, and that limit is called **base level**. Most often, a stream's base level occurs where a stream enters the ocean, a lake, or another stream.

Two general types of base level are recognized. Sea level is considered the *ultimate base level* because it is the lowest level to which stream erosion could lower the land. *Temporary*, or *local*, *base levels* include lakes, resistant layers of rock, and main streams that act as base level for their tributaries. For example, when a stream enters a lake, its velocity quickly approaches zero, and its ability to erode ceases. Thus, the lake prevents the stream from eroding below its level at any point upstream from the lake. However, because the outlet of the lake can cut downward and drain the lake, the lake is only a temporary hindrance to the stream's ability to lower its channel. In a similar manner, the layer of resistant rock at the lip of the waterfall in **Figure 3.21** acts as a temporary base level. Until the ledge of hard rock is eliminated, it will limit the amount of downcutting upstream.

Any change in base level will cause a corresponding readjustment of stream activities. When a dam is built along a stream, the reservoir that forms behind it raises the base level of the stream (**Figure 3.22**). Upstream from the dam, the gradient is reduced, lowering the stream's velocity and, hence, its sediment-transporting ability. The stream, now having too little energy to transport its entire load, will deposit sediment. This builds up its channel. Deposition will be the dominant process until the stream's gradient increases sufficiently to transport its load.

A.

Sea — Resistant bed

Stream prior to faulting

B.

Ultimate base level

Sea — Waterfalls — Resistant bed — Fault

Profile of stream if resistant rock did not exist

C.

Ultimate base level

Sea — Rapids — Fault — Resistant bed

Profile of stream if resistant rock did not exist

D.

Ultimate base level

Sea — Graded stream — Resistant bed — Fault

Profile of stream adjusted to ultimate base level

Figure 3.21 Temporary base level This series of diagrams illustrates what would happen if a fault raised a layer of resistant rock across the path of a stream. **A.** Stream course with a smooth profile prior to faulting. **B.** After faulting, the resistant layer acts as a temporary base level and forms a waterfall. Because of the steep gradient, the stream's erosive energy is focused on the resistant bed. **C.** The waterfall evolves to rapids. **D.** Eventually the river reestablishes a smooth profile.

Valley Deepening

When a stream's gradient is steep and the channel is well above base level, downcutting is the dominant activity. Abrasion caused by bed-load sliding and rolling along the bottom, along with the hydraulic power of fast-moving

Profile of stream adjusted to base level

Ultimate base level

Sea

New stream profile formed by deposition of sediment

Sea — Dam — Reservoir — New base level — Original profile

Figure 3.22 Building a dam Base level upstream from the reservoir is raised. This reduces the stream's flow velocity and leads to deposition and a reduced gradient.

Figure 3.23 Yellowstone River The V-shaped valley, rapids, and waterfalls indicate that the river is vigorously downcutting. (Photo by Jorgen Larsson/AGE Fotostock)

Geologist's Sketch

The most prominent features of a V-shaped valley are *rapids* and *waterfalls*. Both occur where a stream's gradient increases significantly, a situation usually caused by variations in the erodibility of the bedrock into which a stream channel is cutting. Resistant beds create rapids by acting as a temporary base level upstream while allowing downcutting to continue downstream. In time, erosion usually eliminates the resistant rock. Waterfalls are places where the stream makes an abrupt vertical drop.

Valley Widening

Once a stream has cut its channel closer to base level, downward erosion becomes less dominant. At this point, the stream's channel takes on a meandering pattern, and more of the stream's energy is directed from side to side. The result is a widening of the valley as the river cuts away first at one bank and then at the other (**Figure 3.24**). The continuous lateral erosion caused by shifting of the stream's meanders produces an increasingly broader, flat valley floor covered with alluvium. This feature, called a **floodplain**, is appropriately named because when a river overflows its banks during flood stage, it inundates the floodplain.

Over time, the floodplain will widen to the point that the stream is actively eroding the valley walls in only a few places. In fact, in large rivers such as the lower Mississippi River valley, the distance from one valley wall to another can exceed 160 kilometers (100 miles).

water slowly lower the streambed. The result is usually a V-shaped valley with steep sides. A classic example of a V-shaped valley is located in the section of Yellowstone River shown in **Figure 3.23**.

Figure 3.24 Formation of an erosional floodplain Continuous side-to-side erosion by shifting meanders gradually produces a broad, flat valley floor covered by alluvium deposited during floods.

Changing Base Level and Incised Meanders

We usually expect a stream with a highly meandering course to be on a floodplain in a wide valley. However, certain rivers exhibit meandering channels that flow in steep, narrow valleys. Such meanders are called **incised** (*incisum* = **to cut into**) **meanders** (**Figure 3.25**). How do such features form?

Originally, the meanders probably developed on the floodplain of a stream that was relatively near base level. Then, a change in base level caused the stream to begin downcutting. One of two events could have occurred.

Figure 3.25 Incised meanders of the Colorado River in Canyonlands National Park, Utah Before uplift of the Colorado Plateau, the river was meandering on a floodplain. During uplift of the plateau, the meanders downcut because of the steepening gradient to create these incised meanders. (Photo by Michael Collier)

Either base level dropped, or the land on which the river flowed was uplifted.

An example of the first circumstance happened during the Ice Age, when large quantities of water were withdrawn from the ocean and locked up in glaciers on land. The result was that sea level (ultimate base level) dropped, causing rivers flowing into the ocean to begin to downcut.

Regional uplift of the land, the second cause of incised meanders, is exemplified by the Colorado Plateau in the southwestern United States. As the plateau was gradually uplifted, numerous meandering rivers adjusted to being higher above base level by downcutting.

3.8 CONCEPT CHECKS

1. Define *base level* and distinguish between ultimate base level and temporary (local) base level.
2. Explain why V-shaped valleys often contain rapids and waterfalls.
3. Describe or sketch how an erosional floodplain develops.
4. Relate the formation of incised meanders to changes in base level.

3.9 Depositional Landforms

Discuss the formation of deltas and natural levees.

Recall that a stream continually picks up sediment in one part of its channel and redeposits it downstream. Such channel deposits are most often composed of sand and gravel, and they are commonly referred to as **bars**. For example, in Figure 3.18, material acquired at the stream's cut bank is carried downstream and deposited as a point bar. Such features, however, are only temporary, as the material will be picked up again and eventually carried to the ocean. In addition to sand and gravel bars, streams also create other depositional features that have somewhat longer life spans. These include deltas and natural levees.

Deltas

A **delta** forms where a sediment-charged stream enters the relatively still waters of an ocean, a lake, or an inland sea (**Figure 3.26**). As the stream's forward motion slows, sediments are deposited by the dying current. As the delta grows outward, the stream's gradient continually lessens. This circumstance eventually causes the channel to become choked with sediment deposited from the slowing water. As a consequence, the river seeks a shorter, higher-gradient route to base level, as

Topset beds are deposited atop the foreset beds during floods.

Distributaries

Foreset beds consist of coarse particles that drop soon after entering the water body. As the delta grows, these beds cover the bottomset beds.

Bottomset beds consist of fine silt and clay particles that settled beyond the mouth of the river.

As the stream extends its channel, the gradient is reduced. During flood stage some of the flow is diverted to a shorter, higher-gradient route forming a new distributary.

Figure 3.26 Formation of a simple delta Structure and growth of a simple delta that forms in relatively quiet water.

illustrated in Figure 3.26. This illustration shows the main channel dividing into several smaller ones, called **distributaries**. Most deltas are characterized by these shifting channels that act in an opposite way to that of tributaries.

Rather than carrying water into the main channel, distributaries carry water *away* from the main channel. After numerous shifts of the channel, a delta may grow into a roughly triangular shape like the Greek letter delta (Δ), for which it is named. Note, however, that many deltas do not exhibit the idealized shape. Differences in the configurations of shorelines and variations in the nature and strength of wave activity result in many shapes. Many large rivers have deltas extending over thousands of square kilometers. The delta of the Mississippi River is one example. It resulted from the accumulation of huge quantities of sediment derived from the vast region drained by the river and its tributaries. Today, New Orleans rests where there was ocean less than 5000 years ago. **Figure 3.27** shows the portion of the Mississippi delta that has been built over the past 6000 years. As you can see, the delta is actually a series of seven coalescing subdeltas. Each formed when the river left its existing channel in favor of a shorter, more direct path to the Gulf of Mexico. The individual subdeltas

interfinger and partially cover one another, producing a very complex structure. The present subdelta, called a *bird-foot* delta because of the configuration of its distributaries, has been built by the Mississippi in the past 500 years.

Natural Levees

Some rivers occupy valleys with broad floodplains and build **natural levees** that parallel their channels on both banks (**Figure 3.28**). Natural levees are built by successive floods over many years. When a stream overflows its banks, its velocity immediately diminishes, leaving coarse sediment deposited in strips bordering the channel. As the water spreads out over the valley, a lesser amount of fine sediment is deposited over the valley floor. This uneven distribution of material produces the very gentle slope of the natural levee.

The natural levees of the lower Mississippi rise 6 meters (20 feet) above the floodplain. The area behind a levee is characteristically poorly drained for the obvious reason that water cannot flow over the levee and into the river. Marshes called **back swamps** result. A tributary stream that cannot enter a river

Figure 3.27 Growth of the Mississippi River delta During the past 6000 years, the river has built a series of seven coalescing subdeltas. The numbers indicate the order in which the subdeltas were deposited. The present bird-foot delta (number 7) represents activity of the past 500 years. (Image courtesy of JPL/Cal Tech/NASA) The left inset shows the point where the Mississippi may sometime break through (arrow) and the shorter path it would take to the Gulf of Mexico.

Figure 3.28 Formation of a natural levee These gently sloping structures that parallel the river channel are created by repeated floods. Because the ground next to the channel is higher than the adjacent floodplain, back swamps and yazoo tributaries may develop.

Coarse sediments deposited   Fine sediments deposited

Floodstage

Developing natural levee

Post flood

Coarse sediments deposited   Fine sediments deposited

Floodstage

Natural levee

Natural levee after numerous floods

because levees block the way often has to flow parallel to the river until it can breach the levee. Such streams are called **yazoo tributaries**, after the Yazoo River, which parallels the Mississippi for more than 300 kilometers (about 190 miles).

3.9 CONCEPT CHECKS

1. What feature may form where a stream enters the relatively still waters of a lake, an inland sea, or an ocean?
2. What are distributaries, and why do they form?
3. Briefly describe the formation of a natural levee. How is this feature related to back swamps and yazoo tributaries?

3.10 Floods and Flood Control

Discuss the causes of floods and some common flood control measures.

When the discharge of a stream becomes so great that it exceeds the capacity of its channel, it overflows its banks as a **flood**. Floods are among the most common and most destructive of all geologic hazards. They are, nevertheless, simply part of the *natural* behavior of streams.

Causes of Floods

Rivers flood because of the weather. Rapid melting of snow in the spring and/or major storms that bring heavy rains over a large region cause most floods. Exceptional rains caused the devastating floods in the upper Mississippi River valley during the summer of 1993 (Figure 3.29).

Unlike the extensive regional floods just mentioned, *flash floods* are more limited in extent. Flash floods occur with little warning and can be deadly because they produce a rapid rise in water levels and can have a devastating flow velocity. Several factors influence flash flooding. Among them are rainfall intensity and duration, surface conditions, and topography. Urban areas are susceptible to flash floods because a high percentage of the surface area is composed of impervious roofs, streets, and parking lots, where runoff is very rapid. Mountainous areas are susceptible because steep slopes can funnel runoff into narrow canyons.

Human interference with the stream system can worsen or even cause floods. A prime example is the failure

Did You Know?

Humans have covered an amazing amount of land with buildings, parking lots, and roads. A recent study indicated that the area of such impervious surfaces in the United States (excluding Alaska and Hawaii) amounts to more than 112,600 sq km (nearly 44,000 sq mi), which is slightly less than the area of the state of Ohio.

Figure 3.29 The great flood of 1993 These satellite images show the St. Louis area, where the Mississippi, Illinois, and Missouri Rivers join. The left image shows the rivers during a drought that occurred in the summer of 1988. The right image shows the peak of the 1993 flood. In all, nearly 14 million acres were inundated, displacing at least 50,000 people. (Images courtesy of Spaceimaging.com)

of a dam or an artificial levee. These structures are built for flood protection. They are designed to contain floods of a certain magnitude. If a larger flood occurs, the dam or levee may be overtopped. If the dam or levee fails or is washed out, the water behind it is released and becomes a flash flood. The bursting of a dam in 1889 on the Little Conemaugh River caused the devastating Johnstown, Pennsylvania, flood that took some 3000 lives. A second dam failure occurred there in 1977, causing 77 fatalities.

Flood Control

Several strategies have been devised to eliminate or lessen the catastrophic effects of floods. Engineering efforts include the construction of artificial levees, the building of flood-control dams, and river channelization.

Artificial Levees *Artificial levees* are earthen mounds built on the banks of a river to increase the volume of water the channel can hold. These most common of stream-containment structures have been used since ancient times and continue to be used today. Artificial levees are usually easy to distinguish from natural levees because their slopes are much steeper. Many artificial levees were not built to withstand periods of extreme flooding. For example, levee failures were numerous in the Midwest during the summer of 1993, when the upper Mississippi and many of its tributaries experienced record floods (**Figure 3.30**).

Flood-Control Dams *Flood-control dams* are built to store floodwater and then let it out slowly. This lowers the flood crest by spreading it out over a longer time

Figure 3.30 Broken artificial levee Water rushes through a break in an artificial levee in Monroe County, Illinois. During the record-breaking 1993 Midwest floods, many artificial levees could not withstand the force of the floodwaters. Sections of many weakened structures were overtopped or simply collapsed. (Photo by James Finley/ AP Wide World Pictures)

span. Since the 1920s, thousands of dams have been built on nearly every major river in the United States. Many dams have significant non-flood-related functions, such as providing water for irrigated agriculture and for hydroelectric power generation. Many reservoirs are also major regional recreational facilities.

Although dams may reduce flooding and provide other benefits, building these structures also has significant costs and consequences. For example, reservoirs created by dams may cover fertile farmland, useful forests, historic sites, and scenic valleys. Of course, dams trap sediment. Therefore, deltas and floodplains downstream erode because they are no longer replenished with silt during floods. Large dams can also cause significant ecologic damage to river environments that took thousands of years to establish.

Building a dam is not a permanent solution to flooding. Sedimentation behind a dam means that the volume of its reservoir will gradually diminish, reducing the effectiveness of this flood-control measure.

Channelization *Channelization* involves altering a stream channel to speed the flow of water to prevent it from reaching flood height. This may simply involve clearing a channel of obstructions or dredging a channel to make it wider and deeper.

A more radical alteration involves straightening a channel by creating *artificial cutoffs*. The idea is that by shortening the stream, the gradient and hence the velocity are increased. By increasing velocity, the larger discharge associated with flooding can be dispersed more rapidly.

Beginning in the early 1930s, the U.S. Army Corps of Engineers created many artificial cutoffs on the Mississippi for the purpose of increasing the efficiency of the channel and reducing the threat of flooding. In all, the river has been shortened more than 240 kilometers (150 miles). The program has been somewhat successful in reducing the height of the river in flood stage. However, because the river's tendency toward meandering still exists, preventing the river from returning to its previous course has been difficult.

A Nonstructural Approach All the flood-control measures described so far have involved structural solutions aimed at "controlling" a river. These solutions are expensive and often give people residing on the floodplain a false sense of security.

Today, many scientists and engineers advocate a nonstructural approach to flood control. They suggest that an alternative to artificial levees, dams, and channelization is sound floodplain management. By identifying high-risk areas, appropriate zoning regulations can be implemented to minimize development and promote more appropriate land use.

3.10 CONCEPT CHECKS

1. Contrast regional floods and flash floods.
2. Describe three basic flood-control strategies.
3. What is meant by a *nonstructural approach* to flood control?

3.11 Groundwater: Water Beneath the Surface

Discuss the importance of groundwater and describe its distribution and movement.

Groundwater is one of our most important and widely available resources. Yet people's perceptions of the subsurface environment from which it comes are often unclear and incorrect. The reason is that groundwater is hidden from view except in caves and mines, and the impressions people gain from these subsurface openings are often misleading. Observations on the land surface give an impression that Earth is "solid." This view is not changed very much when we enter a cave and see water flowing in a channel that appears to have been cut into solid rock.

Because of such observations, many people believe that groundwater occurs only in underground "rivers." But actual rivers underground are extremely rare. In reality, most of the subsurface environment is not "solid" at all. Rather, it includes countless tiny *pore spaces* between grains of soil and sediment plus narrow joints and fractures in bedrock. Together, these spaces add up to an immense volume. Groundwater collects and moves in these tiny openings.

The Importance of Groundwater

Of the entire hydrosphere, or all of Earth's water, only a small percentage occurs underground. Nevertheless, this small percentage, stored in the rocks and sediments beneath Earth's surface, is a vast quantity. When the oceans are excluded and only sources of freshwater are considered, the significance of groundwater becomes more apparent.

Figure 3.31 shows estimates of the distribution of freshwater in the hydrosphere. Clearly, the largest volume occurs as glacial ice. Second in rank is groundwater, with slightly more than 30 percent of the total. However, when ice is excluded and just liquid water is considered, nearly 96 percent is groundwater. Without question, *groundwater represents the largest reservoir*

Figure 3.31 Earth's freshwater Estimated distribution of freshwater in the hydrosphere. Groundwater is the major reservoir of *liquid* freshwater. (U.S. Geological Survey)

of freshwater that is readily available to humans. Its value in terms of economics and human well-being is incalculable.

Worldwide, wells and springs provide water for cities, crops, livestock, and industry. In the United States, groundwater is the source of about 40 percent of the water used for all purposes (except hydroelectric power generation and power plant cooling). Groundwater is the drinking water for more than 50 percent of the population, it is 40 percent of the water for irrigation, and it provides more than 25 percent of industry's needs. In some areas, however, overuse of this basic resource has

caused serious problems, including streamflow depletion, land subsidence, and increased pumping costs. In addition, groundwater contamination resulting from human activities is a real and growing threat in many places.

Groundwater's Geologic Roles

Geologically, groundwater is important as an erosional agent. The dissolving action of groundwater slowly removes rock, allowing surface depressions known as sinkholes to form and creating subterranean caverns (**Figure 3.32**). Groundwater is also an equalizer of streamflow. Much of the water that flows in rivers is not direct runoff from rain and snowmelt. Rather, a large percentage of precipitation soaks in and then moves slowly underground to stream channels. Groundwater is thus a form of storage that sustains streams during periods when rain does not fall. When we see water flowing in a river during a dry period, it is water from rain that fell at some earlier time and was stored underground.

Distribution of Groundwater

When rain falls, some of the water runs off, some returns to the atmosphere through evaporation and transpiration, and the remainder soaks into the ground. This last path is the primary source of practically all groundwater. The amount of water that takes each of these paths, however, varies greatly from time to time and place to place.

A.   **B.**

Figure 3.32 Caverns and sinkholes A. A view of the interior of New Mexico's Carlsbad Caverns. The dissolving action of acidic groundwater created the caverns. Later, groundwater deposited the limestone decorations. (Photo by Clint Farlinger/Alamy Images) **B**. Groundwater was responsible for creating these depressions, called sinkholes, west of Timaru on New Zealand's South Island. The white dots in this photo are grazing sheep. (Photo by David Wall/Alamy Images)

Influential factors include the steepness of the slope, the nature of the surface material, the intensity of the rainfall, and the type and amount of vegetation. Heavy rains falling on steep slopes underlain by impervious materials will obviously result in a high percentage of the water running off. Conversely, if rain falls steadily and gently on more gradual slopes composed of materials that are more easily penetrated by water, a much larger percentage of the water soaks into the ground.

Underground Zones Some of the water that soaks in does not travel far because it is held by molecular attraction as a surface film on soil particles. This near-surface zone is called the *belt of soil moisture*. It is crisscrossed by roots, voids left by decayed roots, and animal and worm burrows that enhance the infiltration of rainwater into the soil. Soil water is used by plants for life functions and transpiration. Some of this water also evaporates directly back into the atmosphere.

Water that is not held as soil moisture penetrates downward until it reaches a zone where all the open spaces in sediment and rock are completely filled with water. This is the **zone of saturation**. Water within it is called **groundwater**. The upper limit of this zone is known as the **water table**. The area above the water table where the soil, sediment, and rock are not saturated is called the **unsaturated zone** (**Figure 3.33**). Although a considerable amount of water can be present in the unsaturated zone, this water cannot be pumped by wells because it clings too tightly to rock and soil particles. By contrast, below the water table, the water pressure is great enough to allow water to enter wells, thus permitting groundwater to be withdrawn for use. We will examine wells more closely later in the chapter.

Water Table The water table is rarely level, as we might expect a table to be. Instead, its shape is usually a subdued replica of the surface, reaching its highest elevations beneath hills and decreasing in height toward valleys (Figure 3.33). The water table of a wetland (swamp) is right at the surface. Lakes and streams generally occupy areas low enough that the water table is above the land surface.

Several factors contribute to the irregular surface of the water table. One important influence is the fact that groundwater moves very slowly. Because of this, water tends to "pile up" beneath high areas between stream valleys. If rainfall were to cease completely, these water "hills" would slowly subside and gradually approach the level of the adjacent valleys. However, new supplies of rainwater are usually added often enough to prevent this. Nevertheless, in times of extended drought, the water table may drop enough to dry up shallow wells. Other causes for the uneven water table are variations in rainfall and permeability of Earth materials from place to place.

Factors Influencing the Storage and Movement of Groundwater

The nature of subsurface materials strongly influences the rate of groundwater movement and the amount of groundwater that can be stored. Two factors are especially important—porosity and permeability.

Porosity Water soaks into the ground because bedrock, sediment, and soil contain countless voids or openings. These openings are similar to those of a

Figure 3.33 Water beneath the surface This diagram illustrates the relative positions of many features associated with subsurface water.

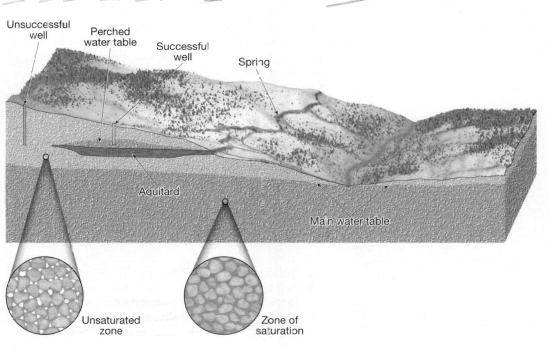

Figure 3.34 Porosity demonstration Porosity is the percentage of the total volume of rock or sediment that consists of pore spaces.

The beaker on the left is filled with 1000 ml of sediment. The beaker on the right is filled with 1000 ml of water.

The sediment-filled beaker now contains 500 ml of water. Pore spaces (porosity) must represent 50 percent of the volume of the sediment.

sponge and are often called *pore spaces*. The quantity of groundwater that can be stored depends on the **porosity** of the material, which is the percentage of the total volume of rock or sediment that consists of pore spaces (**Figure 3.34**). Voids most often are spaces between sedimentary particles, but also common are joints, faults, cavities formed by the dissolving of soluble rock such as limestone, and vesicles (voids left by gases escaping from lava).

Variations in porosity can be great. Sediment is commonly quite porous, and open spaces may occupy 10 percent to 50 percent of the sediment's total volume. Pore space depends on the size and shape of the grains; how they are packed together; the degree of sorting; and, in sedimentary rocks, the amount of cementing material. Most igneous and metamorphic rocks, as well as some sedimentary rocks, are composed of tightly interlocking crystals, so the voids between grains may be negligible. In these rocks, fractures must provide the voids.

Permeability Porosity alone cannot measure a material's capacity to yield groundwater. Rock or sediment may be very porous and still prohibit water from moving through it. The **permeability** of a material indicates its ability to *transmit* a fluid. Groundwater moves by twisting and turning through interconnected small openings. The smaller the pore spaces, the slower the groundwater moves. If the spaces between particles are too small, water cannot move

at all. For example, clay's ability to store water can be great, due to its high porosity, but its pore spaces are so small that water is unable to move through it. Thus, we say that clay is *impermeable*.

Aquitards and Aquifers Impermeable layers such as clay that hinder or prevent water movement are termed **aquitards** (*aqua* = water, *tard* = slow). In contrast, larger particles, such as sand or gravel, have larger pore spaces. Therefore, water moves with relative ease. Permeable rock strata or sediments that transmit groundwater freely are called **aquifers** ("water carriers"). Aquifers are important because they are the water-bearing layers sought after by well drillers.

Groundwater Movement

The movement of most groundwater is exceedingly slow, from pore to pore. A typical rate is a few centimeters per day. The energy that makes the water move is provided by the force of gravity. In response to gravity, water moves from areas where the water table is high to zones where the water table is lower. This means that water usually gravitates toward a stream channel, lake, or spring. Although some water takes the most direct path down the slope of the water table, much of the water follows long, curving paths toward the zone of discharge.

Figure 3.35 shows how water percolates into a stream from all possible directions. Some paths clearly turn upward, apparently against the force of gravity, and enter through the bottom of the channel. This is easily explained: The deeper you go into the zone of saturation, the greater the water pressure. Thus, the looping curves followed by water in the saturated zone may be thought of as a compromise between the downward pull of gravity and the tendency of water to move toward areas of reduced pressure.

Figure 3.35 Groundwater movement Arrows show paths of groundwater movement through uniformly permeable material.

Recharge area
Drainage divide
Recharge area
Discharge area
Discharge area
Water table

The looping flow lines result from the downward pull of gravity and the tendency of groundwater to move toward areas of reduced pressure.

3.11 CONCEPT CHECKS

1. About what percentage of freshwater is groundwater? How does this change if glacial ice is excluded?
2. What are two geologic roles for groundwater?
3. When it rains, what factors influence the amount of water that soaks in?
4. Define *groundwater* and relate it to the water table.
5. Distinguish between porosity and permeability. Contrast aquifer and aquitard.
6. What factors cause water to follow the paths shown in Figure 3.35?

3.12 Springs, Wells, and Artesian Systems

Compare and contrast springs, wells, and artesian systems.

A great deal of groundwater eventually makes its way to the surface. Sometimes this occurs as a naturally flowing spring or as a spectacularly erupting geyser. Much of the groundwater that people use is brought to the surface by pumping it from a well. To understand these phenomena, it is necessary to understand Earth's sometimes complex underground "plumbing."

Springs

Springs have aroused the curiosity and wonder of people for thousands of years. The fact that springs were (and to some people still are) rather mysterious phenomena is not difficult to understand, for here is water flowing freely from the ground in all kinds of weather in seemingly inexhaustible supply but with no obvious source. Today, we know that the source of springs is water from the zone of saturation and that the ultimate source of this water is precipitation.

Whenever the water table intersects Earth's surface, a natural outflow of groundwater results, which we call a **spring**. Springs such as the one pictured in **Figure 3.36** form when an aquitard blocks the downward movement of groundwater and forces it to move laterally. Where the permeable bed (aquifer) outcrops, one or more springs result.

Another situation that can produce a spring is illustrated in Figure 3.33. Here an aquitard is situated above the main water table. As water percolates downward, a portion of it is intercepted by the aquitard, thereby creating a localized zone of saturation and a **perched water table**. Springs, however, are not confined to places where a perched water table creates a flow at the surface. Many geologic situations lead to the formation of springs because subsurface conditions vary greatly from place to place.

Hot Springs There is no universally accepted definition of hot spring. One frequently used definition is that the water in a **hot spring** is 6° to 9°C (10° to 15°F) warmer than the average annual air temperature for the locality where it occurs. In the United States alone, there are well over 1000 such springs.

Temperatures in deep mines and oil wells usually rise with increasing depth, an average of about 2°C per 100 meters (1°F per 100 feet). Therefore, when groundwater circulates at great depths, it becomes heated. If it rises to the surface, the water may emerge as a hot spring. The water of some hot springs in the eastern United States is heated in this manner. The great majority (more than 95 percent) of the hot springs (and geysers) in the United States are found in the West. The reason for such a distribution is that the source of heat for most hot springs is cooling igneous rock, and it is in the West that igneous activity has occurred most recently.

Geysers Intermittent fountains in which columns of hot water and steam are ejected with great force, often rising 30 to 60 meters (100 to 200 feet) into the air, are called **geysers**. After the jet of water ceases, a column of steam rushes out, often with a thunderous roar. Perhaps the most famous geyser in the world is Old Faithful in Yellowstone National Park, which erupts about once each hour (**Figure 3.37**). Geysers are also found in other parts of the world, notably New Zealand and Iceland. In fact, the Icelandic word *geysa*, meaning "to gush," gives us the name *geyser*.

Geysers occur where extensive underground chambers exist within hot igneous rocks. As relatively cool groundwater enters the chambers, it is heated by the surrounding rock. At the bottom of the chamber, the water is under great pressure because of the weight of the overlying water. This great pressure prevents the water from boiling at the normal surface temperature of 100°C (212°F). For example, at the bottom of a 300-meter (1000-foot) water-filled chamber, water must attain a temperature of nearly 230°C (450°F) before it will boil. The heating causes the water to expand, and as a result, some of the water is forced out at the surface. This loss of

Figure 3.36 Thousand Springs These well-known springs are along the Snake River in Hagerman Valley, Idaho. Notice the two people in a canoe (lower right) for scale. (Photo by David R. Frazier/Alamy Images)

Figure 3.37 Old Faithful in Wyoming's Yellowstone National Park (Photo by Art Director & Trip/Alamy Images)

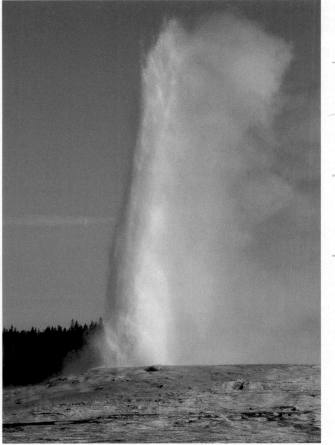

water reduces the pressure on the remaining water in the chamber, which lowers the boiling point. A portion of the water deep within the chamber quickly turns to steam and causes the geyser to erupt. Following the eruption, cool groundwater again seeps into the chamber, and the cycle begins anew.

SmartFigure 3.38
Cone of depression
Drawdown, the lowering of the water table surrounding a well, creates a cone of depression. For most small domestic wells, the cone of depression is negligible. When wells are heavily pumped, the cone of depression can be large and may lower the water table enough that nearby shallower wells may be left dry.

Wells

The most common method for removing groundwater is to use a **well**, a hole bored into the zone of saturation. Wells serve as small reservoirs into which groundwater migrates and from which it can be pumped to the surface. The use of wells dates back many centuries and continues to be an important method of obtaining water. By far the single greatest use of this water in the United States is irrigation for agriculture. More than 65 percent of the groundwater used each year is for this purpose. Industrial uses rank a distant second, followed by the amount used by homes in cities and rural areas.

The water-table level may fluctuate considerably during the course of a year, dropping during dry seasons and rising following periods of precipitation. Therefore, to ensure a continuous supply of water, a well must penetrate below the water table. Whenever a substantial amount of water is withdrawn from a well, the water table around the well is lowered. This effect, termed **drawdown**, decreases with increasing distance from the well. The result is a depression in the water table, roughly conical in shape, known as a **cone of depression** (Figure 3.38). For most small domestic wells, the cone of depression is negligible. However, when wells are used for irrigation or for industrial purposes, the withdrawal of water can be great enough to create a very wide and steep cone of depression that may substantially lower the water table in an area and cause nearby shallow wells to become dry. Figure 3.38 illustrates this situation.

Artesian Systems

In most wells, water cannot rise on its own. If water is first encountered at 30 meters (100 feet) depth, it remains at that level, fluctuating perhaps 1 or 2 meters with seasonal wet and dry periods. However, in some wells, water rises, sometimes overflowing at the surface.

Artesian system refers to a situation in which groundwater rises in a well above the level where it was initially encountered. For such a situation to occur, two conditions must exist (**Figure 3.39**): (1) Water must be confined to an aquifer that is inclined so that one end is exposed at the surface, where it can receive water; and (2) aquitards both above and below the aquifer must be present to prevent the water from escaping. Such an aquifer is called a **confined aquifer**. When such a layer is tapped, the pressure created by the weight of the water above will force the water to rise. If there were no friction, the water in the well would rise to the level of the water at the top of the aquifer. However, friction reduces the height of this pressure surface. The greater the distance from the recharge area (the area where water enters the

SmartFigure 3.39
Artesian systems These groundwater systems occur where an inclined aquifer is surrounded by impermeable beds (aquitards). Such aquifers are called *confined aquifers*. The photo shows a flowing artesian well.
(Photo by James E. Patterson)

inclined aquifer), the greater the friction and the smaller the rise of water.

In Figure 3.39, Well 1 is a *nonflowing artesian well* because at this location, the pressure surface is below ground level. When the pressure surface is above the ground and a well is drilled into the aquifer, a *flowing artesian well* is created (Well 2 in Figure 3.39). Not all artesian systems are wells. *Artesian springs* also exist. In such situations, groundwater may reach the surface by rising along a natural fracture such as a fault rather than through an artificially produced hole. In deserts, artesian springs are sometimes responsible for creating oases.

Artesian systems act as "natural pipelines," transmitting water from remote areas of recharge great distances to the points of discharge. In this manner, water that fell in central Wisconsin years ago is now taken from the ground and used by communities many kilometers to the south, in Illinois. In South Dakota, such a system brings water from the western Black Hills eastward across the state.

On a different scale, city water systems may be considered examples of artificial artesian systems (**Figure 3.40**). A water tower, into which water is pumped, may be considered the area of recharge, the pipes the confined aquifer, and the faucets in homes the flowing artesian wells.

Figure 3.40 City water systems City water systems can be considered to be artificial artesian systems.

3.12 CONCEPT CHECKS

1. Describe the circumstances that created the spring in Figure 3.33.
2. What is the source of heat for most hot springs and geysers?
3. Describe what occurs to cause a geyser to erupt.
4. Relate drawdown to cone of depression.
5. In Figure 3.33, two wells are at the same level. Why was one successful and the other not?
6. Sketch a simple cross section of an artesian system with a flowing artesian well. Label aquitards, the aquifer, and the pressure surface.

3.13 Environmental Problems of Groundwater

List and discuss three important environmental problems associated with groundwater.

As with many of our other valuable natural resources, groundwater is being exploited at an increasing rate. In some areas, overuse threatens the groundwater supply. In other places, groundwater withdrawal has caused the ground and everything resting on it to sink. Still other localities are concerned with the possible contamination of their groundwater supply.

Treating Groundwater as a Nonrenewable Resource

Many natural systems tend to establish a condition of equilibrium. The groundwater system is no exception. The water table's height reflects the rate at which water is added by precipitation and the rate at which water is removed by discharge and withdrawal. An imbalance will either raise or lower the water table. A long-term drop in the water table can occur if there is either a decrease in recharge due to prolonged drought or an increase in groundwater discharge or withdrawal.

For many people, groundwater appears to be an endlessly renewable resource, for it is continually replenished by rainfall and melting snow. But in some regions, groundwater has been and continues to be treated as a *nonrenewable* resource because the amount of water available to recharge the aquifer is significantly less than the amount being withdrawn.

The High Plains, a relatively dry region that extends from South Dakota to western Texas, provides one example of an extensive agricultural economy that is largely dependent on irrigation (**Figure 3.41**). Underlying about 111 million acres (450,000 square kilometers [174,000 square miles]) in parts of eight states, the High Plains aquifer is one of the largest and most agriculturally significant aquifers in the United States. It accounts for about 30 percent of all groundwater withdrawn for irrigation in the country. In the southern part of this region, which includes the Texas panhandle, the natural recharge of the aquifer is very slow, and the problem of declining groundwater levels is acute. In fact, in years of average or below-average precipitation, recharge is negligible because all or nearly all of the meager rainfall is returned to the atmosphere by evaporation and transpiration.

Therefore, where intense irrigation has been practiced for an extended period, depletion of groundwater can be severe. Declines in the water table at rates as great as 1 meter per year have led to an overall drop of between 15 and 60 meters (50 and 200 feet) in some areas. Under these circumstances, it can be said that the groundwater is literally being "mined." Even if pumping were to cease immediately, it would take thousands of years for the groundwater to be fully replenished.

Groundwater depletion has been a concern in the High Plains and other areas of the West for many years, but it is worth pointing out that the problem is not confined to that part of the country. Increased demands on groundwater resources have overstressed aquifers in many areas, not just in arid and semiarid regions.

Land Subsidence Caused by Groundwater Withdrawal

As you will see later in this chapter, surface subsidence can result from natural processes related to groundwater. However, the ground may also sink when water is pumped from wells faster than natural recharge processes

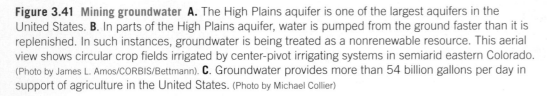

Figure 3.41 Mining groundwater A. The High Plains aquifer is one of the largest aquifers in the United States. **B.** In parts of the High Plains aquifer, water is pumped from the ground faster than it is replenished. In such instances, groundwater is being treated as a nonrenewable resource. This aerial view shows circular crop fields irrigated by center-pivot irrigating systems in semiarid eastern Colorado. (Photo by James L. Amos/CORBIS/Bettmann). **C.** Groundwater provides more than 54 billion gallons per day in support of agriculture in the United States. (Photo by Michael Collier)

Groundwater Contamination

The pollution of groundwater is a serious matter, particularly in areas where aquifers provide a large part of the water supply. One common source of groundwater pollution is sewage emanating from an ever-increasing number of septic tanks. Other sources are inadequate or broken sewer systems and farm wastes.

If sewage water, which is contaminated with bacteria, enters the groundwater system, it may become purified through natural processes. The harmful bacteria may be mechanically filtered by the sediment through which the water percolates, destroyed by chemical oxidation, and/or assimilated by other organisms. For purification to occur, however, the aquifer must be of the correct composition. For example, extremely permeable aquifers (such as highly fractured crystalline rock, coarse gravel, or cavernous limestone) have such large openings that contaminated groundwater may travel long distances without being cleansed. In this case, the water flows too rapidly and is not in contact with the surrounding material long enough for purification to occur. This is the problem at Well 1 in **Figure 3.43A.**

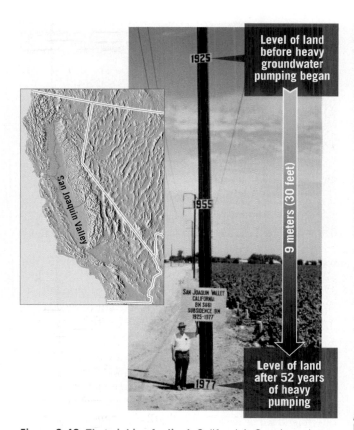

Figure 3.42 That sinking feeling! California's San Joaquin Valley, an important agricultural area, relies heavily on irrigation. Between 1925 and 1977, this part of the valley subsided nearly 9 meters (30 feet) because of the withdrawal of groundwater and the resulting compaction of sediments. (Photo courtesy of U.S. Geological Survey)

can replace it. This effect is particularly pronounced in areas underlain by thick layers of loose sediments. As water is withdrawn, the water pressure drops, and the weight of the overburden is transferred to the sediment. The greater pressure packs the sediment grains more tightly together, and the ground subsides.

Many areas can be used to illustrate such land subsidence. A classic example in the United States occurred in the San Joaquin Valley of California (**Figure 3.42**). Other well-known cases of land subsidence resulting from groundwater pumping in the United States include Las Vegas, Nevada; New Orleans and Baton Rouge, Louisiana; and the Houston–Galveston area of Texas. In the low-lying coastal area between Houston and Galveston, land subsidence ranges from 1.5 to 3 meters (5 to 10 feet). The result is that about 78 square kilometers (30 square miles) are permanently flooded.

Outside the United States, one of the most spectacular examples of subsidence occurred in Mexico City, a portion of which is built on a former lake bed. In the first half of the twentieth century, thousands of wells were sunk into the water-saturated sediments beneath the city. As water was withdrawn, portions of the city subsided by as much as 6 meters (20 feet) or more.

Although the contaminated water has traveled more than 100 meters before reaching Well 1, the water moves too rapidly through the cavernous limestone to be purified.

As the discharge from the septic tank percolates through the permeable sandstone, it moves more slowly and is purified in a relatively short distance.

Figure 3.43 Comparing two aquifers In this example, the limestone aquifer allowed the contamination to reach a well, but the sandstone aquifer did not.

Figure 3.44 Potential sources of groundwater contamination Sometimes leaking gasoline storage tanks and materials leached from landfills contaminate aquifers. (Storage tank photo by Earth Gallery Environment/Alamy Images; landfill photo by Picsfive/Shutterstock)

On the other hand, when an aquifer consists of sand or permeable sandstone, the water can sometimes be purified after traveling only a few dozen meters through it. The openings between sand grains are large enough to permit water movement, yet the movement of the water is slow enough to allow ample time for its purification (Well 2 in **Figure 3.43B**).

Other sources and types of contamination also threaten groundwater supplies (**Figure 3.44**). These include widely used substances such as highway salt, fertilizers that are spread across the land surface, and pesticides. In addition, a wide array of chemicals and industrial materials may leak from pipelines, storage tanks, landfills, and holding ponds. Some of these pollutants are classified as *hazardous*, meaning that they are flammable, corrosive, explosive, or toxic. As rainwater oozes through the refuse, it may dissolve a variety of potential contaminants. If the leached material reaches the water table, it will mix with the groundwater and contaminate the supply.

Because groundwater movement is usually slow, polluted water may go undetected for a long time. In fact, contamination is sometimes discovered only after drinking water has been affected and people become ill. By this time, the volume of polluted water could be quite large, and even if the source of contamination is removed immediately, the problem is not solved. Although the sources of groundwater contamination are numerous, the solutions are relatively few.

Once a source of the problem has been identified and eliminated, the most common practice is simply to abandon the water supply and allow the pollutants to be flushed away gradually. This is the least costly and easiest solution, but the aquifer must remain unused for many years. To accelerate this process, polluted water is sometimes pumped out and treated. Following removal of the tainted water, the aquifer is allowed to recharge naturally or, in some cases, the treated water or other freshwater is pumped back in. This process is costly, time-consuming, and possibly risky because there is no way to be certain that all of the contamination has been removed. Clearly, the most effective solution to groundwater contamination is prevention.

3.13 CONCEPT CHECKS

1. Describe the problem associated with pumping groundwater for irrigation in parts of the High Plains.
2. What happened in the San Joaquin Valley as a result of excessive groundwater pumping?
3. Which aquifer would be most effective in purifying polluted groundwater: one consisting of coarse gravel, sand, or cavernous limestone? Explain.

3.14 The Geologic Work of Groundwater

Explain the formation of caverns and the development of karst topography.

Groundwater dissolves rock. This fact is key to understanding how caverns and sinkholes form. Because soluble rocks, especially limestone, underlie millions of square kilometers of Earth's surface, it is here that groundwater carries on its important role as an erosional agent. Limestone is nearly insoluble in pure water but is quite easily dissolved by water containing small quantities of carbonic acid, and most groundwater contains this acid. It forms because rainwater readily dissolves carbon dioxide from the air and from decaying plants. Therefore, when groundwater comes in contact with limestone, the carbonic acid reacts with calcite (calcium carbonate) in the rocks to form calcium bicarbonate, a soluble material that is then carried away in solution.

Caverns

The most spectacular results of groundwater's erosional handiwork are limestone **caverns**. In the United States alone, about 17,000 caves have been discovered. Although most are relatively small, some have spectacular dimensions. Carlsbad Caverns in southeastern New Mexico and Mammoth Cave in Kentucky are famous examples. One chamber in Carlsbad Caverns has an area equivalent to 14 football fields and enough height to accommodate the U.S. Capitol Building. At Mammoth Cave, the total length of interconnected caverns is over 540 kilometers (340 miles).

Most caverns are created at or just below the water table, in the zone of saturation. Acidic groundwater follows lines of weakness in the rock, such as joints and bedding planes. As time passes, the dissolving process slowly creates cavities and gradually enlarges them into caverns. Material that is dissolved by the groundwater is eventually discharged into streams and carried to the ocean.

Certainly, the features that arouse the greatest curiosity for most cavern visitors are the stone formations that give some caverns a wonderland appearance. These are not erosional features, like the caverns in which they reside, but depositional features. They are created by the seemingly endless dripping of water over great spans of time. The calcium carbonate that is left behind produces the limestone we call travertine. These cave deposits, however, are also commonly called *dripstone*, an obvious reference to their mode of origin.

Although the formation of caverns takes place in the zone of saturation, the deposition of dripstone is not possible until the caverns are above the water table, in the unsaturated

zone. This commonly occurs as nearby streams cut their valleys deeper, lowering the water table as the elevation of the rivers drops. As soon as the chamber is filled with air, the conditions are right for the decoration phase of cavern building to begin.

Of the various dripstone features found in caverns, perhaps the most familiar are **stalactites**. These icicle-like pendants hang from the ceiling of the cavern and form where water seeps through cracks above. When water reaches air in the cave, some of the dissolved carbon dioxide escapes from the drop, and calcite begins to precipitate. Deposition occurs as a ring around the edge of the water drop. As drop after drop follows, each leaves an infinitesimal trace of calcite behind, and a hollow limestone tube is created. Water then moves through the tube, remains suspended momentarily at the end, contributes a tiny ring of calcite, and falls to the cavern floor.

The stalactite just described is appropriately called a *soda straw* (**Figure 3.45A**). Often the hollow tube of the soda straw becomes plugged or its supply of water increases. In either case, the water is forced to flow and deposit along the outside of the tube. As deposition continues, the stalactite takes on the more common conical shape.

A.

B.

Figure 3.45 Cave decorations A. Close-up of a delicate live soda-straw stalactite in Chinn Springs Cave, Independence County, Arkansas. (Photo by Dante Fenolio/Photo Researchers, Inc.) **B.** Stalagmites and stalactites in Carlsbad Caverns National Park, New Mexico. (Photo by glowimages)

Formations that develop on the floor of a cavern and reach upward toward the ceiling are called **stalagmites** (**Figure 3.45B**). The water supplying the calcite for stalagmite growth falls from the ceiling and splatters over the surface. As a result, stalagmites do not have a central tube and are usually more massive in appearance and more rounded on their upper ends than are stalactites. Given enough time, a downward-growing stalactite and an upward-growing stalagmite may join to form a *column*.

Karst Topography

Many areas of the world have landscapes that, to a large extent, have been shaped by the dissolving power of groundwater. Such areas are said to exhibit **karst topography**, named for the *Krs* region in the border area between Slovenia and Italy, where such topography is strikingly developed. In the United States, karst landscapes occur in many areas that are underlain by limestone, including portions of Kentucky, Tennessee, Alabama, southern Indiana, and central and northern Florida (**Figure 3.46**). Generally, arid and semiarid areas do not develop karst topography because there is insufficient groundwater. When solution features exist in such regions, they are likely to be remnants of a time when rainier conditions prevailed.

Sinkholes Karst areas typically have irregular terrain punctuated with many depressions called **sinkholes** or, simply, **sinks** (see Figure 3.32). In the limestone areas of Florida, Kentucky, and southern Indiana, literally tens of thousands of these depressions vary in depth from just 1 to 2 meters to a maximum of more than 50 meters (165 feet).

Sinkholes commonly form in one of two ways. Some develop gradually over many years, without any physical disturbance to the rock. In these situations, the limestone immediately below the soil is dissolved by downward-seeping rainwater that is freshly charged

Figure 3.46 Development of a karst landscape

During early stages, groundwater percolates through limestone along joints and bedding planes. Solution activity creates and enlarges caverns at and below the water table.

Water table

Limestone

Sinkholes

With time, caverns grow larger and the number and size of sinkholes increase. Surface drainage is frequently funneled below ground.

Sinking stream

Sinkholes

Water table

T I M E

Collapse of caverns and coalescence of sinkholes form larger, flat-floored depressions. Eventually solution activity may remove most of the limestone from the area, leaving isolated remnants as in Figure 3.48.

Sinking stream

Sinkholes

Springs

Collapse sink

Sinking stream

Solution valley

Water table

manner are steep sided and deep. When they form in populous areas, they may represent a serious geologic hazard. Such a situation is clearly the case in **Figure 3.47**.

In addition to a surface pockmarked by sinkholes, karst regions characteristically show a striking lack of surface drainage (streams). Following a rainfall, runoff is quickly funneled belowground, through sinks. It then flows through caverns until it finally reaches the water table. Where streams do exist at the surface, their paths are usually short. The names of such streams often give a clue to their fate. The Mammoth Cave area of Kentucky, for example, is home to Sinking Creek, Little Sinking Creek, and Sinking Branch. Some sinkholes become plugged with clay and debris, creating small lakes or ponds.

Tower Karst Landscapes Some regions of karst development exhibit landscapes that look very different from the sinkhole-studded terrain depicted in Figure 3.46. One striking example is an extensive region in southern China that is described as exhibiting *tower karst*. As **Figure 3.48** shows, the term *tower* is appropriate because the landscape consists of a maze of isolated steep-sided hills that rise abruptly from the ground. Each is riddled with interconnected caves and passageways. This type of karst topography forms in wet tropical and subtropical regions having thick beds of highly jointed limestone. In such settings, groundwater dissolves large volumes of limestone, leaving only these residual towers. Karst development is more rapid in tropical climates due to the abundant rainfall and greater availability of carbon dioxide from the decay of lush tropical vegetation. The extra carbon dioxide in the soil means there is more carbonic acid for dissolving limestone. Other tropical areas of advanced karst development include portions of Puerto Rico, western Cuba, and northern Vietnam.

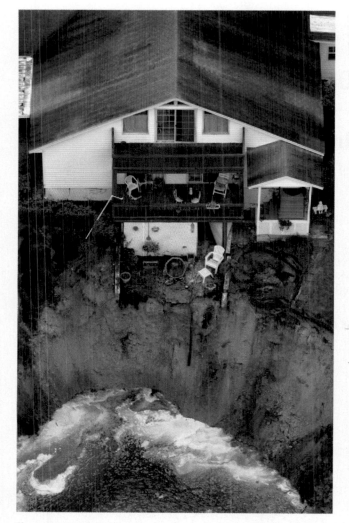

Figure 3.47 Sinkholes can be geologic hazards This sinkhole formed suddenly in the backyard of a home in Lake City, Florida. Sinkholes such as this one form when the roof of a cavern collapses. (AP Photo/*The Florida Times-Union* Jon M. Fletcher)

with carbon dioxide. These depressions are usually not deep and are characterized by relatively gentle slopes. By contrast, sinkholes can also form suddenly and without warning when the roof of a cavern collapses under its own weight. Typically, the depressions created in this

3.14 CONCEPT CHECKS

1. How does groundwater create caverns?
2. How do stalactites and stalagmites form?
3. Describe two ways that sinkholes form.

Figure 3.48 Tower karst landscape in China One of the best-known and most distinctive regions of tower karst development is along the Li River in the Guilin District of southeastern China. (Photo by Philippe Michel)

CONCEPTS IN REVIEW: LANDSCAPES FASHIONED BY WATER

3.1 Earth's External Processes

List three important external processes and describe where they fit into the rock cycle.

KEY TERMS: external process, internal process, weathering, mass wasting, erosion

- Weathering, mass wasting, and erosion are responsible for creating, transporting, and depositing sediment. They are called external processes because they occur at or near Earth's surface and are powered by energy from the Sun.
- Internal processes, such as volcanic activity and mountain building, derive their energy from Earth's interior.

3.2 Mass Wasting: The Work of Gravity

Explain the role of mass wasting in the development of valleys and discuss the factors that trigger and influence mass-wasting processes.

KEY TERMS: trigger, angle of repose

- Mass wasting is the downslope movement of rock and soil under the direct influence of gravity.
- An event that initiates a mass-wasting process is referred to as a trigger. The addition of water, oversteepening of the slope, removal of vegetation, and shaking due to an earthquake are four examples of triggers. Not all landslides are triggered by one of these processes, but many are.

3.3 The Hydrologic Cycle

List the hydrosphere's major reservoirs and describe the different paths that water takes through the hydrologic cycle.

KEY TERMS: hydrologic cycle, evaporation, infiltration, runoff, transpiration, evapotranspiration

- Water moves through the hydrosphere's many reservoirs by evaporating, condensing into clouds, and falling as precipitation. Once it reaches the ground, rain can either soak in, evaporate, be returned to the atmosphere by plant transpiration, or run off. Running water is the most important agent sculpting Earth's varied landscapes.

3.4 Running Water

Describe the nature of drainage basins and river systems. Sketch four basic drainage patterns.

KEY TERMS: drainage basin, divide, dendritic pattern, radial pattern, rectangular pattern, trellis pattern

- The land area that contributes water to a stream is its drainage basin. Drainage basins are separated by imaginary lines called divides.
- As a generalization, river systems tend to erode at the upstream end, transport sediment through the middle section, and deposit sediment at the downstream end.

? Identify each of the drainage patterns depicted in the accompanying sketch.

3.5 Streamflow

Discuss streamflow and the factors that cause it to change.

KEY TERMS: laminar flow, turbulent flow, gradient, discharge, longitudinal profile

- The flow of water in a stream may be laminar or turbulent. A stream's flow velocity is influenced by channel gradient; size, shape, and roughness of the channel; and discharge.
- A cross-sectional view of a stream from head to mouth is a longitudinal profile. Usually the gradient and roughness of the stream channel decrease going downstream, whereas the size of the channel, stream discharge, and flow velocity increase in the downstream direction.

3.6 The Work of Running Water

Outline the ways in which streams erode, transport, and deposit sediment.

KEY TERMS: pothole, dissolved load, suspended load, bed load, capacity, competence, sorting, alluvium

- Streams erode when turbulent water lifts loose particles from the streambed. The focused "drilling" of the stream armed with swirling particles also creates potholes in solid rock.
- Streams transport their load of sediment in solution, in suspension, and along the bottom (bed) of the channel. A stream's ability to transport solid particles is described using two criteria. Capacity refers to how much sediment a stream is transporting, and competence refers to the particle sizes the stream is capable of moving.
- Streams deposit sediment when velocity slows and competence is reduced. This results in sorting, the process by which like-size particles are deposited together.

3.7 Stream Channels

Contrast bedrock and alluvial stream channels. Distinguish between two types of alluvial channels.

KEY TERMS: meander, cut bank, point bar, cutoff, oxbow lake, braided channel

- Bedrock channels are cut into solid rock and are most common in headwaters areas where gradients are steep. Rapids and waterfalls are common features.
- Alluvial channels are dominated by streamflow through alluvium previously deposited by the stream. A floodplain usually covers the valley floor, with the river meandering or moving through braided channels.
- Meanders change shape through erosion at the cut bank (the outer edge of the meander) and deposition of sediment on point bars (the inside of a meander). A meander may become cut off and form an oxbow lake.

3.8 Shaping Stream Valleys

Contrast narrow V-shaped valleys, broad valleys with floodplains, and valleys that display incised meanders.

KEY TERMS: stream valley, base level, floodplain, incised meander

- A stream valley includes the channel itself, the adjacent floodplain, and the relatively steep valley wall. Streams erode downward until they approach base level, the lowest point to which a stream can erode its channel. A river flowing toward the ocean (the ultimate base level) may encounter several local base levels along its route. These could be lakes or resistant rock layers that retard downcutting by the stream.
- A stream valley is widened through the meandering action of the stream, which erodes the valley walls and widens the floodplain. If base level were to drop or if the land were uplifted, a meandering stream might start downcutting and develop incised meanders.

3.9 Depositional Landforms

Discuss the formation of deltas and natural levees.

KEY TERMS: bar, delta, distributary, natural levee, back swamp, yazoo tributary

- Deltas may form where a river deposits sediment in another water body at its mouth. The partitioning of streamflow into multiple distributaries spreads sediment in different directions.
- Natural levees result from sediment deposited along the margins of a stream channel by many flooding events. Because the levees slope gently away from the channel, the adjacent floodplain is poorly drained, resulting in back swamps and yazoo tributaries that flow parallel to the main river.

3.10 Floods and Flood Control

Discuss the causes of floods and some common flood control measures.

KEY TERM: flood

- Floods are triggered by heavy rains and/or snowmelt. Sometimes human interference can worsen or even cause floods. Flood control measures include the building of artificial levees and dams. Channelization may involve creating artificial cutoffs. Many scientists and engineers advocate a nonstructural approach to flood control that involves more appropriate land use.

3.11 Groundwater: Water Beneath the Surface

Discuss the importance of groundwater and describe its distribution and movement.

KEY TERMS: zone of saturation, groundwater, water table, unsaturated zone, porosity, permeability, aquitard, aquifer

- Groundwater represents the largest reservoir of freshwater that is readily available to humans. Geologically, groundwater is an equalizer of streamflow, and the dissolving action of groundwater produces caverns and sinkholes.
- Groundwater is water that occupies the pore spaces in sediment and rock in a zone beneath the surface called the zone of saturation. The upper limit of this zone is called the water table. The zone above the water table where the material is not saturated is called the unsaturated zone.
- The quantity of water that can be stored in the open spaces in rock or sediment is termed porosity. Permeability, the ability of a material to transmit a fluid through interconnected pore spaces, is a key factor affecting the movement of groundwater. Aquifers are permeable materials that transmit groundwater freely, whereas aquitards are impermeable materials.

3.12 Springs, Wells, and Artesian Systems

Compare and contrast springs, wells, and artesian systems.

KEY TERMS: spring, perched water table, hot spring, geyser, well, drawdown, cone of depression, artesian system, confined aquifer

- Springs occur where the water table intersects the land surface and a natural flow of groundwater results. When groundwater circulates deep below the surface, it may become heated and emerge at the surface as a hot spring. Geysers occur when groundwater is heated in underground chambers and expands, with some water quickly changing to steam and causing the geyser to erupt. The source of heat for most hot springs and geysers is hot igneous rock.
- Wells, which are openings bored into the zone of saturation, withdraw groundwater and may create roughly conical depressions in the water table known as cones of depression.

3.13 Environmental Problems of Groundwater

List and discuss three important environmental problems associated with groundwater.

- Some of the current environmental problems involving groundwater include overuse by intense irrigation, land subsidence caused by excessive groundwater withdrawal, and contamination by pollutants.

3.14 The Geologic Work of Groundwater

Explain the formation of caverns and the development of karst topography.

KEY TERMS: cavern, stalactite, stalagmite, karst topography, sinkhole (sink)

- Most caverns form in limestone at or below the water table when acidic groundwater dissolves this soluble rock.
- Karst topography develops in limestone regions and exhibits irregular terrain punctuated with many depressions called sinkholes. Some sinkholes form when the cavern roofs collapse.

? **Identify the cavern deposits that are visible in this photograph.**

Miroslav/AGE Fotostock

GIVE IT SOME THOUGHT

1. During summer, wildfires are relatively common occurrences in the mountains of the western United States. This wildfire occurred near Santa Clarita, California, in July 2004. We have learned that Earth is a system in which various parts of the four major spheres interact in uncountable ways. Let's relate that idea to the situation pictured here.
   a. What atmospheric conditions might have preceded and thus set the stage for this wildfire?
   b. What might have ignited the blaze? Suggest a natural possibility and a human possibility.
   c. Describe at least one way that wildfires, like the one shown here, might influence future mass-wasting processes.

2. Describe at least one situation in which an internal process might cause or contribute to a mass-wasting process.

3. If you collect a jar of water from a stream, what part of its load will settle to the bottom of the jar? What portion will remain in the water indefinitely? What part of the stream's load will probably not be represented in your sample?

Joshua Gates Weisberg/EPA Newscom

4. A river system consists of three zones, based on the dominant process operating in each zone. On the accompanying diagram, match each process with one of the three zones: sediment production (erosion), sediment deposition, and sediment transportation.

5. The meandering White River in Arkansas is a tributary of the Mississippi River.

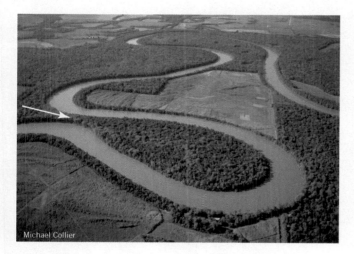

Michael Collier

   a. In this aerial view, the color of the White River is brown. What part of the stream's load gives it this color?
   b. If an artificial channel were created across the narrow neck of land shown by the arrow, how would the river's gradient change?
   c. How would the flow velocity be affected by the change in gradient?

6. During a trip to the grocery store, your friend wants to buy some bottled water. Some brands promote the fact that their product is artesian. Other brands boast that their water comes from a spring. Your friend asks, "Is artesian water or spring water necessarily better than water from other sources?" How would you reply?

7. An acquaintance is considering purchasing a large tract of productive farmland in the Texas Panhandle. His intention is to continue growing crops on the land for years to come. If he asked your opinion about the area he selected and his plans for the future, what advice would you give him?

8. This photo shows the famous Thousand Springs along the Snake River in southern Idaho. The springs are natural outlets for groundwater.

David R. Frazier/Alamy Images

   a. Is there water in this image that can't be seen? If so, where might it be?
   b. Prepare a story that speculates on the answers to the following questions. Suggest several possibilities for each question. What journeys did the water in the image take to get to this place? What paths might the water follow as it makes its way back to the ocean?

9. Which one of the three basic rock types (igneous, sedimentary, or metamorphic) is most likely to be a good aquifer? Explain.

MasteringGeology™

www.masteringgeology.com

Looking for additional review and test prep materials? Visit the Study Area in MasteringGeology to enhance your understanding of this chapter's content by accessing a variety of resources, including **Self-Study Quizzes,**

Geoscience Animations, GEODe Tutorials, RSS feeds, flashcards, web links, self-study quizzes, and an optional **Pearson eText.**

4
Glacial and Arid Landscapes

Each statement represents the primary learning objective for the corresponding major heading within the chapter. After you complete the chapter, you should be able to:

4.1 List the different types of glaciers, summarize their characteristics, and describe the present-day location and extent of glaciers on Earth.

4.2 Describe how glaciers move, the rates at which they move, and the significance of the glacial budget.

4.3 Discuss the processes of glacial erosion and the major features created by these processes.

4.4 Distinguish between the two basic types of glacial deposits and briefly describe the features associated with each type.

4.5 Describe and explain several important effects of Ice Age glaciers other than erosional and depositional landforms.

4.6 Discuss the extent of glaciation and climate variability during the Quaternary Ice Age.

4.7 Describe the general distribution and causes of Earth's dry lands and the role that water plays in modifying desert landscapes.

4.8 Discuss the stages of landscape evolution in the Basin and Range region of the western United States.

4.9 Describe the ways that wind transports sediment and the features created by wind erosion.

4.10 Distinguish between two basic types of wind deposits.

Bear Glacier flows out of the Harding ice field near Seward, Alaska. About 200 years ago, the ice extended to the terminal moraine at the bottom of the image. Like most other glaciers in Alaska, Bear Glacier is retreating into the mountains. (Photo by Michael Collier)

Like the running water and groundwater that were the focus of Chapter 3, glaciers and wind are significant erosional processes. They are responsible for creating many different landforms and are part of an important link in the rock cycle, in which the products of weathering are transported and deposited as sediment.

Climate has a strong influence on the nature and intensity of Earth's external processes. This fact is dramatically illustrated in this chapter. The existence and extent of glaciers are largely controlled by Earth's changing climate. Another excellent example of the strong link between climate and geology is seen when we examine the development of arid landscapes.

Today, glaciers cover nearly 10 percent of Earth's land surface; however, in the recent geologic past, ice sheets were three times more extensive, covering vast areas with ice thousands of meters thick. Many regions still bear the mark of these glaciers. The first part of this chapter will examine glaciers and the erosional and depositional features they create. The second part will explore dry lands and the geologic work of wind. Because desert and near-desert conditions prevail over an area as large as that affected by the massive glaciers of the Ice Age, the nature of such landscapes is indeed worth investigating.

4.1 Glaciers: A Part of Two Basic Cycles

List the different types of glaciers, summarize their characteristics, and describe the present-day location and extent of glaciers on Earth.

Figure 4.1 Kennicott Glacier This 43-kilometer-long (27-mile-long) valley glacier is sculpting the mountains in Alaska's Wrangell–St. Elias National Park. The dark stripes of sediment are called medial moraines. The transport and deposition of sediment make glaciers a basic part of the rock cycle. (Photo by Michael Collier)

Many present-day landscapes were modified by the widespread glaciers of the most recent Ice Age and still strongly reflect the handiwork of ice. The basic features of such diverse places as the Alps, Cape Cod, and Yosemite Valley were fashioned by now-vanished masses of glacial ice. Moreover, Long Island, New York; the Great Lakes; and the fiords of Norway and Alaska all owe their existence to glaciers. Glaciers, of course, are not just a phenomenon of the geologic past. They are still sculpting the landscape and depositing in many regions today.

Glaciers are a part of two of Earth's basic cycles— the hydrologic cycle and the rock cycle. In the hydrologic cycle, when precipitation falls at high elevations or high latitudes, the water may not immediately make its way back toward the sea. Instead, it may become part of a glacier. Although the ice will eventually melt, allowing the water to continue its path to the sea, water can be stored as glacial ice for tens, hundreds, or even thousands of years. During the time that the water is part of a glacier, the moving mass of ice can do enormous amounts of work—scouring the land surface and acquiring, transporting, and depositing great quantities of sediment. This activity is a basic part of the rock cycle (**Figure 4.1**).

A **glacier** is a thick ice mass that forms over hundreds or thousands of years. It originates on land from the accumulation, compaction, and recrystallization of snow. A glacier appears to be motionless, but it is not; glaciers move very slowly. Like running water, groundwater, wind, and waves, glaciers are dynamic erosional agents that accumulate, transport, and deposit sediment. Therefore, glaciers are among the agents that perform a basic function in the rock cycle. Although glaciers are found in many parts of the world today, most are located in remote areas, either near Earth's poles or in high mountains.

Valley (Alpine) Glaciers

Literally thousands of relatively small glaciers exist in lofty mountain areas, where they usually follow valleys originally occupied by streams. Unlike the rivers that previously flowed in these valleys, the glaciers advance slowly, perhaps only a few centimeters per day. Because of their location, these moving ice masses are termed **valley glaciers**, or **alpine glaciers** (see Figure 4.1). Each glacier is a stream of ice, bounded by precipitous rock walls, that flows downvalley from a snow accumulation center near its head. Like rivers, valley glaciers can be long or short, wide or narrow, single or with branching tributaries. The widths of alpine glaciers are generally small compared to their lengths. In length, some extend for just a fraction of a kilometer, whereas others go on for many tens of kilometers. The west branch of the Hubbard Glacier, for example, runs through 112 kilometers (70 miles) of mountainous terrain in Alaska and Canada's Yukon Territory.

Ice Sheets

Ice sheets exist on a much larger scale than valley glaciers. These enormous masses flow out in all directions from one or more snow-accumulation centers and completely obscure all but the highest areas of underlying terrain. The low total annual solar energy reaching the poles makes these regions eligible for such great ice accumulations. Presently each of Earth's polar regions supports an ice sheet—Greenland in the Northern Hemisphere and Antarctica in the Southern Hemisphere (**Figure 4.2**).

Ice Age Ice Sheets About 18,000 years ago, glacial ice not only covered Greenland and Antarctica but also covered large portions of North America, Europe, and Siberia. This date in Earth history is appropriately known as the *Last Glacial Maximum*. The term implies that there were other glacial maximums, and this is indeed the case. Throughout the Quaternary period, which began about 2.6 million years ago and extends to the present, ice sheets formed, advanced over broad areas, and then melted away. These alternating glacial and interglacial periods occurred over and over again.

Greenland and Antarctica Some people mistakenly think that the North Pole is covered by glacial ice, but this is not the case. The ice that covers the Arctic Ocean is **sea ice**—frozen seawater. Sea ice floats because ice is less dense than liquid water. Although sea ice never completely disappears from the Arctic, the area covered expands and contracts with the seasons. The thickness of sea ice ranges from a few centimeters for new ice to 4 meters for sea ice that has survived for years. By contrast, glaciers can be hundreds or thousands of meters thick.

Glaciers form on land, and in the Northern Hemisphere, Greenland supports an ice sheet. Greenland extends between about 60° and 80° north latitude. This largest island on Earth is covered by an imposing ice sheet that occupies 1.7 million square kilometers (more than 660,000 square miles), or about 80 percent of the island. Averaging nearly 1500 meters (5000 feet) thick, the ice extends 3000 meters (10,000 feet) above the island's bedrock floor in some places.

In the Southern Hemisphere, the huge Antarctic Ice Sheet attains a maximum thickness of about 4300 meters (14,000 feet) and covers an area of more than 13.9 million square kilometers (5.4 million square miles)—nearly the entire continent. Because of the proportions of these huge features, they are often called *continental ice sheets*. Indeed, the combined areas of present-day continental ice sheets represent almost 10 percent of Earth's land area.

Ice Shelves Along portions of the Antarctic coast, glacial ice flows into the adjacent ocean, creating features called **ice shelves**. These are large, relatively flat masses of floating ice that extend seaward from the coast but remain attached to the land along one or more sides. The shelves are thickest on their landward sides, and they become thinner seaward. They are sustained by ice from the adjacent ice sheet as well as being nourished by snowfall and the freezing of seawater to their

Greenland's ice sheet occupies 1.7 million square kilometers (663,000 square miles), about 80 percent of the island.

The area of the Antarctic Ice Sheet is almost 14 million square kilometers (5,460,000 square miles). Ice shelves occupy an additional 1.4 million square kilometers (546,000 square miles).

Figure 4.2 Ice sheets The only present-day ice sheets are those covering Greenland and Antarctica. Their combined areas represent nearly 10 percent of Earth's land area.

bases. Antarctica's ice shelves extend over approximately 1.4 million square kilometers (0.6 million square miles). The Ross and Ronne-Filchner Ice Shelves are the largest, with the Ross Ice Shelf alone covering an area approximately the size of Texas (see Figure 4.2).

Other Types of Glaciers

In addition to valley glaciers and ice sheets, other types of glaciers are also recognized. Covering some uplands and plateaus are masses of glacial ice called **ice caps**. Like ice sheets, ice caps completely bury the underlying landscape but are much smaller than the continental-scale features. Ice caps occur in many places, including Iceland and several of the large islands in the Arctic Ocean (**Figure 4.3A**).

Another type, known as **piedmont glaciers**, occupy broad lowlands at the bases of steep mountains and form when one or more valley glaciers emerge from the confining walls of mountain valleys. The advancing ice spreads out to form a broad sheet. The size of individual piedmont glaciers varies greatly. Among the largest is the broad Malaspina Glacier along the coast of southern Alaska. It covers more than 5000 square kilometers (2000 square miles) of the flat coastal plain at the foot of the lofty St. Elias range (**Figure 4.3B**).

Often ice caps and ice sheets feed **outlet glaciers**. These tongues of ice flow down valleys, extending outward from the margins of these larger ice masses. The tongues are essentially valley glaciers that are avenues for ice movement from an ice cap or ice sheet through mountainous terrain to the sea. Where they encounter the ocean, some outlet glaciers spread out as floating ice shelves. Often large numbers of icebergs are produced.

4.1 CONCEPT CHECKS

1. Where are glaciers found on Earth today, and what percentage of Earth's land surface do they cover?
2. Describe how glaciers fit into the hydrologic cycle. What role do they play in the rock cycle?
3. List and briefly distinguish among four types of glaciers.
4. What is the difference between an ice sheet, sea ice, and an ice shelf?

Vatnajökull ice cap

A. Ice caps completely bury the underlying terrain but are much smaller than ice sheets.

Valley glacier

Piedmont glacier

B. When a valley glacier is no longer confined, it spreads out to become a piedmont glacier.

Figure 4.3 Ice caps and piedmont glaciers (Satellite image courtesy of NASA)

4.2 How Glaciers Move

Describe how glaciers move, the rates at which they move, and the significance of the glacial budget.

The way in which ice moves is complex and of two basic types. The first of these, *plastic flow*, involves movement *within* the ice. Ice behaves as a brittle solid until the pressure on it is equivalent to the weight of about 50 meters (165 feet) of ice. Once that load is surpassed, ice behaves as a plastic material, and flow begins. Such flow occurs because of the molecular structure of ice. Glacial ice consists of layers of molecules stacked one

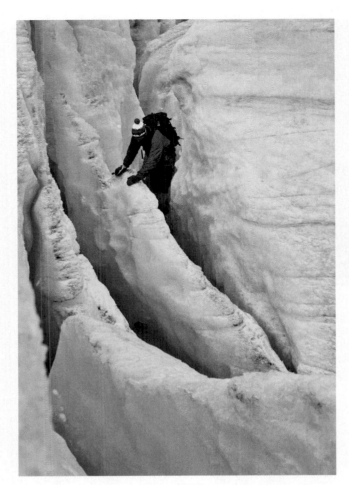

Figure 4.4 Crevasses As a glacier moves, internal stresses cause large cracks to develop in the brittle upper portion of the glacier, called the zone of fracture. Crevasses can extend to depths of 50 meters (165 feet) and can make travel across glaciers dangerous. (Photo by Wave/Glow Images)

upon the other. The bonds between layers are weaker than those within each layer. Therefore, when a stress exceeds the strength of the bonds between the layers, the layers remain intact and slide over one another.

A second and often equally important mechanism of glacial movement occurs when an entire ice mass slips along the ground. The lowest portions of most glaciers probably move by this sliding process.

The uppermost 50 meters (165 feet) of a glacier is appropriately referred to as the *zone of fracture*. Because there is not enough overlying ice to cause plastic flow, this upper part of the glacier consists of brittle ice. Consequently, the ice in this zone is carried along piggyback-style by the ice below. When the glacier moves over irregular terrain, the zone of fracture is subjected to tension, resulting in cracks called **crevasses** (**Figure 4.4**). These gaping cracks, which often make travel across glaciers dangerous, may extend to depths of 50 meters (165 feet). Below this depth, plastic flow seals them off.

Observing and Measuring Movement

Unlike the movement of water in streams, the movement of glacial ice is not obvious. If we could watch a valley glacier move, however, we would see that as with the water in a river, all of the ice does not move downstream at the same rate. Flow is greatest in the center of the glacier because of the drag created by the walls and floor of the valley.

Early in the nineteenth century, the first experiments involving the movement of glaciers were designed and carried out in the Alps. Markers were placed in a straight line across an alpine glacier. The position of the line was marked on the valley walls so that if the ice moved, the change in position could be detected. The positions of the markers were noted periodically, revealing the movement just described. Although most glaciers move too slowly for direct visual detection, the experiments succeeded in demonstrating that movement nevertheless occurs. The experiment illustrated in **Figure 4.5A** was carried out at Switzerland's Rhône Glacier later in the nineteenth century. It not only traced the movement of markers within the ice but also mapped the position of the glacier's terminus.

For many years, time-lapse photography has allowed us to observe glacial movement. Images are taken from the same vantage point on a regular basis (for example, once per day) over an extended span and then played back like a movie. More recently, satellites let us track the movement of glaciers and observe glacial behavior (**Figure 4.5B**). This is especially useful because the remoteness and extreme weather associated with many glacial areas limits on-site study.

How rapidly does glacial ice move? Average rates vary considerably from one glacier to another. Some glaciers move so slowly that trees and other vegetation may become well established in the debris that accumulates on the glacier's surface. Others advance up to several meters each day. Recent satellite radar imaging provided insights into movements within the Antarctic ice sheet. Portions of some outlet glaciers move at rates greater than 800 meters (2600 feet) per year; on the other hand, ice in some interior regions creeps along at less than 2 meters (6.5 feet) per year.

Movement of some glaciers is characterized by occasional periods of extremely rapid advance called *surges*, followed by periods during which movement is much slower.

Budget of a Glacier: Accumulation Versus Wastage

Snow is the raw material from which glacial ice originates. Therefore, glaciers form in areas where more snow falls in winter than can melt during the summer. Glaciers are constantly gaining and losing ice.

Original position
of stakes (1874)

1878 position
of stakes

1882 position
of stakes

Terminus
in 1882

Terminus
in 1878

Terminus of
glacier in 1874

A.

B.

Ice Velocity (m/year)
0 200 400 600 800 1000 1200

Figure 4.5 Measuring the movement of a glacier A. Ice movement and changes in the terminus of Rhône Glacier, Switzerland. In this classic study of a valley glacier, the movement of stakes clearly shows that glacial ice moves and that movement along the sides of the glacier is slower than movement in the center. Also notice that even though the ice front was retreating, the ice within the glacier was advancing. **B.** This satellite image provides detailed information about the movement of Antarctica's Lambert Glacier. Ice velocities are determined from pairs of images obtained 24 days apart using radar data. (NASA)

Glacial Zones Snow accumulation and ice formation occur in the **zone of accumulation** (**Figure 4.6**). The addition of snow thickens the glacier and promotes movement. Beyond this area of ice formation is the **zone of wastage**, where there is a net loss to the glacier as all the

snow from the previous winter melts, as does some of the glacial ice (Figure 4.6).

In addition to melting, glaciers also waste as large pieces of ice break off the front of a glacier, in a process called *calving*. Calving creates **icebergs** when glaciers reach the sea (**Figure 4.7**). Because icebergs are just slightly less dense than seawater, they float very low in the water, with 80 percent or more of their mass submerged. The margins of the Greenland Ice Sheet produce thousands of icebergs each year. Many drift southward and find their way into the North Atlantic, where they are a hazard to navigation.

ZONE OF ACCUMULATION
More snow falls each winter than melts each summer

ZONE OF WASTAGE
All the snow from the previous winter melts along with some glacial ice

Snowline

Crevasses

Braided streams

SmartFigure 4.6 Zones of a glacier The snowline separates the zone of accumulation and the zone of wastage. Whether the ice front advances, retreats, or remains stationary depends on the balance or lack of balance between accumulation and wastage.

Geologist's Sketch

Only about 20 percent or less of an iceberg protrudes above the waterline.

Figure 4.7 Icebergs Icebergs form when large masses of ice break off from the front of a glacier after it reaches a water body, in a process known as calving. (Photo by Radius Images/Photolibrary)

Glacial Budget Whether the margin of a glacier is advancing, retreating, or remaining stationary depends on the *budget of the glacier*. The **glacial budget** refers to the balance or lack of balance between accumulation on the one hand and wastage on the other. If ice accumulation exceeds wastage, the glacial front advances until the two factors balance. At this point, the terminus of the glacier becomes stationary. If a warming trend increases wastage and/or if a drop in snowfall decreases accumulation, the ice front will retreat. As the terminus of the glacier retreats, the extent of the zone of wastage gets smaller. Therefore, in time a new balance will be reached between accumulation and wastage, and the ice front will again become stationary.

Whether the margin of a glacier is advancing, retreating, or stationary, the ice within the glacier continues to move forward. In the case of a receding glacier, the ice simply does not flow forward rapidly enough to offset wastage. This point is illustrated in Figure 4.5A. While the line of stakes within the Rhône Glacier continued to advance downvalley, the terminus of the glacier slowly retreated upvalley.

Glaciers in Retreat Because glaciers are sensitive to changes in temperature and precipitation, they provide clues about changes in climate. With few exceptions, valley glaciers around the world have been retreating at unprecedented rates over the past century. The chapter-opening photo of Bear Glacier is one example. The photos in **Figure 4.8** provide another example. Many valley glaciers have disappeared altogether. For example, 150 years ago, there were 147 glaciers in Montana's Glacier National Park. Today, only 37 remain, and they may vanish by 2030.

1941

2004

Figure 4.8 Retreating glaciers Two images taken 63 years apart from the same vantage point in Alaska's Glacier Bay National Park. Muir Glacier, which is prominent in the 1941 photo, has retreated out of the field of view in the 2004 image. Also, Riggs Glacier (upper right) has thinned and retreated significantly. (Photos courtesy of National Snow and Ice Data Center)

4.2 CONCEPT CHECKS

1. Describe two components of glacial movement.
2. How rapidly does glacial ice move? Provide some examples.
3. What are crevasses, and where do they form?
4. Relate glacial budget to two zones of a glacier.
5. Under what circumstances will the front of a glacier advance? Retreat? Remain stationary?

4.3 Glacial Erosion

Discuss the processes of glacial erosion and the major features created by these processes.

Glaciers erode tremendous volumes of rock. For anyone who has observed the terminus of an alpine glacier, the evidence of its erosive force is plain. You can witness firsthand the release of rock fragments of various sizes from the ice as it melts (**Figure 4.9**). All signs lead to the conclusion that the ice has scraped, scoured, and torn rock debris from the floor and walls of the valley and carried it downvalley. In mountainous regions, mass-wasting processes can also make substantial contributions to the sediment load of a glacier.

Once rock debris is acquired by a glacier, it cannot settle out as does the load carried by a stream or by the wind. Consequently, glaciers can carry huge blocks that

Figure 4.9 Evidence of glacial erosion Close-up view at the terminus of a valley glacier in Alaska. The glacier clearly acquired a significant quantity of rock debris as it advanced across the landscape. (Photo by Michael Collier)

The rock debris dropped by melting ice is a jumbled mixture of different size sediments.

no other erosional agent could possibly budge. Although today's glaciers are of limited importance as erosional agents, many landscapes that were modified by the widespread glaciers of the recent Ice Age still reflect to a high degree the work of ice.

How Glaciers Erode

Glaciers erode land primarily in two ways—by plucking and abrasion. First, as a glacier flows over a fractured bedrock surface, it loosens and lifts blocks of rock and incorporates them into the ice. This process, known as **plucking**, occurs when meltwater penetrates the cracks and joints along the rock floor of the glacier and freezes. As the water freezes, it expands and exerts tremendous leverage that pries rock loose. In this manner, sediment of all sizes becomes part of the glacier's load.

The second major erosional process is **abrasion**. As the ice and its load of rock fragments move over bedrock, they function like sandpaper, smoothing and polishing the surface below. The pulverized rock produced by the glacial "grist mill" is appropriately called **rock flour**. So much rock flour may be produced that meltwater streams leaving a glacier often have the grayish appearance of skim milk—visible evidence of the grinding power of the ice.

When ice at the bottom of a glacier contains larger rock fragments, long scratches and grooves called **glacial striations** may be gouged into the bedrock (**Figure 4.10A**). These linear scratches provide clues to the direction of ice flow. By mapping the striations over large areas, patterns of glacial flow can often be reconstructed.

Not all abrasive action produces striations. The rock surfaces over which a glacier moves may become highly polished by the ice and its load of finer particles. The broad expanses of smoothly polished granite in California's Yosemite National Park provide an excellent example (**Figure 4.10B**).

As is the case with other agents of erosion, the rate of glacial erosion is highly variable. This differential erosion by ice is largely controlled by four factors: (1) rate of glacial movement; (2) thickness of the ice; (3) shape, abundance, and hardness of the rock fragments contained in the ice at the base of the glacier; and (4) the erodibility of the surface beneath the glacier. Variations in any or all of these factors from time to time and/or from place to place mean that the features, effects, and degree of landscape modification in glaciated regions can vary greatly.

Figure 4.10 Glacial abrasion Moving glacial ice, armed with sediment, acts like sandpaper, scratching and polishing rock. (Photos by Michael Collier)

Glacial abrasion created the scratches and grooves in this bedrock.

Glacially polished granite in California's Yosemite National Park.

A.

B.

Landforms Created by Glacial Erosion

Although the erosional accomplishments of ice sheets can be tremendous, landforms carved by these huge ice masses usually do not inspire the same awe as do the erosional features created by valley glaciers. In regions where the erosional effects of ice sheets are significant, glacially scoured surfaces and subdued terrain are the rule. By contrast, in mountainous areas, erosion by valley glaciers produces many truly spectacular features. Much of the rugged mountain scenery so celebrated for its majestic beauty is the product of erosion by valley glaciers.

Take a moment to study **Figure 4.11**, which shows a mountain setting before, during, and after glaciation.

You will refer to this figure often in the following discussion.

Glaciated Valleys Prior to glaciation, alpine valleys are characteristically V-shaped because streams are well above base level and are therefore downcutting (Figure 4.11A). However, in mountainous regions that have been glaciated, the valleys are no longer narrow. As a glacier moves down a valley once occupied by a stream, the ice modifies it in three ways: The glacier widens, deepens, and straightens the valley, so that what was once a narrow V-shaped valley is transformed into a U-shaped **glacial trough** (Figure 4.11C).

The amount of glacial erosion depends in part on the thickness of the ice. Consequently, main glaciers, also called *trunk glaciers*, cut their valleys deeper than

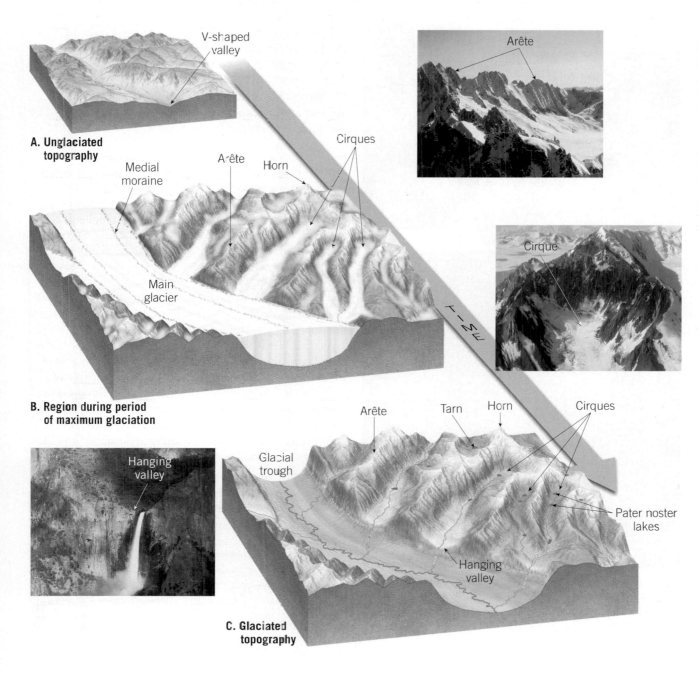

A. Unglaciated topography

B. Region during period of maximum glaciation

C. Glaciated topography

SmartFigure 4.11 Erosional landforms created by alpine glaciers The unglaciated landscape in part A is modified by valley glaciers in part B. After the ice recedes in part C, the terrain looks very different than before glaciation. (Aréte photo by James E. Patterson; cirque photo by Marli Miller; hanging valley photo by John Warden/Superstock)

do their smaller tributary glaciers. Thus, after the ice has receded, the valleys of tributary glaciers are left standing above the main glacial trough and are termed **hanging valleys**. Rivers flowing through hanging valleys may produce spectacular waterfalls, such as those in Yosemite National Park, California (Figure 4.11C).

Cirques   At the head of a glacial valley is a characteristic and often imposing feature associated with an alpine glacier—a **cirque**. As Figure 4.11 illustrates, these hollowed-out, bowl-shaped depressions have precipitous walls on three sides and are open on the downvalley side. The cirque is the focal point of the glacier's growth because it is the area of snow accumulation and ice formation. Cirques begin as irregularities in the mountainside that are subsequently enlarged by frost wedging and plucking

Figure 4.12 Fiords The coast of Norway is known for its many fiords. Frequently these ice-sculpted inlets of the sea are hundreds of meters deep. (Satellite images courtesy of NASA; photo by Inger Helene Boasson/AGE Fotostock)

along the sides and bottom of the glacier. The glacier, in turn, acts as a conveyor belt that carries away the debris. After the glacier has melted away, the cirque basin is often occupied by a small lake called a *tarn*.

Arêtes and Horns   The Alps, Northern Rockies, and many other mountain landscapes carved by valley glaciers reveal more than glacial troughs and cirques. In addition, sinuous, sharp-edged ridges called **arêtes** and sharp, pyramid-like peaks called **horns** project above the surroundings (Figure 4.11C). Both features can originate from the same basic process—the enlargement of cirques produced by plucking and frost action. Cirques around a single high mountain create the spires of rock called horns. As the cirques enlarge and converge, an isolated horn is produced. A famous example is the Matterhorn in the Swiss Alps. This famous peak is on the cover of your book.

Arêtes can form in a similar manner, except that the cirques are not clustered around a point but rather exist on opposite sides of a divide. As the cirques grow, the divide separating them is reduced to a very narrow, knife-like partition. An arête may also be created when the area separating two parallel glacial valleys is narrowed as the glaciers scour and widen their valleys.

Fiords   **Fiords** are deep, often spectacular, steep-sided inlets of the sea that exist in many high-latitude areas of the world where mountains are adjacent to the ocean (**Figure 4.12**). Norway, British Columbia, Greenland, New Zealand, Chile, and Alaska all have coastlines characterized by fiords. They are glacial troughs that became submerged as the ice left the valley and sea level rose following the Ice Age.

The depth of some fiords exceeds 1000 meters (3300 feet). However, the great depths of these flooded troughs are only partly explained by the post–Ice Age rise in sea level. Unlike the situation governing the downward erosional work of rivers, sea level does not act as base level for glaciers. As a consequence, glaciers are capable of eroding their beds far below the surface of the sea. For example, a valley glacier 300 meters (1000 feet) thick can carve its valley floor more than 250 meters (800 feet) below sea level before downward erosion ceases and the ice begins to float.

4.3 CONCEPT CHECKS

1. How do glaciers acquire their load of sediment?
2. How does a glaciated mountain valley differ in appearance from a mountain valley that was not glaciated?
3. Describe the features created by glacial erosion that you might see in an area where valley glaciers recently existed.

4.4 Glacial Deposits

Distinguish between the two basic types of glacial deposits and briefly describe the features associated with each type.

Glaciers pick up and transport a huge load of debris as they slowly advance across the land. These materials are ultimately deposited when the ice melts. Glacial sediment can play a truly significant role in forming the physical landscape in regions where it is deposited. For example, in many areas once covered by the ice sheets of the recent Ice Age, the bedrock is rarely exposed because glacial deposits that are tens or even hundreds of meters thick completely bury the terrain. The general effect of these deposits is to level the topography. Indeed, many of today's familiar rural landscapes—rocky pastures in New England, wheat fields in the Dakotas, rolling farmland in the Midwest—resulted directly from glacial deposition.

Types of Glacial Drift

Long before the theory of an extensive Ice Age was proposed, much of the soil and rock debris covering portions of Europe was recognized as coming from elsewhere. At the time, these foreign materials were believed to have "drifted" into their present positions via floating ice during an ancient flood. As a consequence, the term *drift* was applied to this sediment. Although rooted in a concept that was not correct, this term was so well established by the time the true glacial origin of the debris became widely recognized that it remained in the glacial vocabulary. **Glacial drift** is an all-embracing term for sediments of glacial origin, no matter how, where, or in what form they were deposited.

Glacial drift is divided into two distinct types: (1) materials deposited directly by a glacier, which are known as *till*, and (2) sediments laid down by glacial meltwater, called *stratified drift*. Here is the difference: **Till** is deposited as glacial ice melts and drops its load of rock fragments. Unlike moving water and wind, ice cannot sort the sediment it carries; therefore, deposits of till are characteristically unsorted mixtures of many particle sizes (**Figure 4.13**). **Stratified drift** is sorted according to the size and weight of the fragments. Because ice is not capable of such sorting activity, these sediments are not deposited directly by the glacier. Rather, they reflect the sorting action of glacial meltwater.

Some deposits of stratified drift are made by streams coming directly from a glacier. Other stratified deposits involve sediment that was originally laid down as till and

Glacial till is an unsorted mixture of many different sediment sizes.

A close examination of glacial till often reveals cobbles that have been scratched as they were dragged along by the ice.

Figure 4.13 Glacial till Unlike sediment deposited by running water and wind, material deposited directly by a glacier is not sorted. (Photos by E. J.Tarbuck)

later picked up, transported, and redeposited by meltwater beyond the margin of the ice. Accumulations of stratified drift often consist largely of sand and gravel because the meltwater is not capable of moving larger material and because the finer rock flour remains suspended and is commonly carried far from the glacier. An indication that stratified drift consists primarily of sand and gravel can be seen in many areas where these deposits are actively mined as aggregate for roadwork and other construction projects.

Boulders found in the till or lying free on the surface are called **glacial erratics** if they are different from the

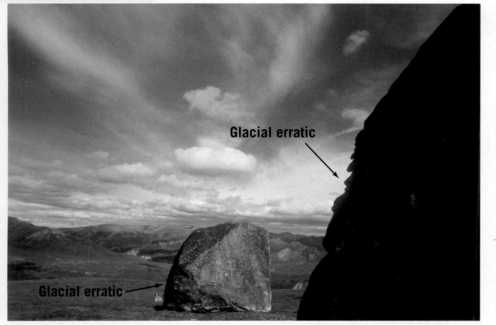

Figure 4.14 Glacial erratics Two large glacially transported boulders in Denali National Park, Alaska. (Photo by Michael Collier)

bedrock below (**Figure 4.14**). Of course, this means that they must have been derived from a source outside the area where they are found. Although the locality of origin for most erratics is unknown, the origin of some can be determined. Therefore, by studying glacial erratics as well as the mineral composition of till, geologists can sometimes trace the path of a lobe of ice. In portions of New England as well as other areas, erratics can be seen dotting pastures and farm fields. In some places, these rocks are cleared from fields and piled to make stone walls and fences.

Moraines, Outwash Plains, and Kettles

Perhaps the most widespread features created by glacial deposition are *moraines*, which are simply layers or ridges of till. Several types of moraines are identified; some are common only to mountain valleys, and others are associated with areas affected by either ice sheets or valley glaciers. Lateral and medial moraines fall in the first category, whereas end moraines and ground moraines are in the second.

Lateral and Medial Moraines The sides of a valley glacier accumulate large quantities of debris from the valley walls. When the glacier wastes away, these materials are left as ridges, called **lateral moraines**, along the sides of the valley. **Medial moraines** form when two advancing valley glaciers come together, forming a

single ice stream. The till that was once carried along the margins of each glacier joins to form a single dark stripe of debris within the newly enlarged glacier. The creation of these dark stripes within the ice stream is one obvious proof that glacial ice moves because the medial moraine could not form if the ice did not flow downvalley. It is common to see several dark debris stripes within a large alpine glacier because one will form each time a tributary glacier joins the main valley. Kennicott Glacier in Figure 4.1 provides an excellent example.

End Moraines and Ground Moraines An **end moraine** is a ridge of till that accumulates at the terminus of a glacier. These relatively common landforms are deposited when a state of equilibrium is attained between wastage and ice accumulation. That is, the end moraine forms when the ice is melting near the end of the glacier at a rate equal to the forward advance of the glacier from its region of nourishment (**Figure 4.15**). Although the terminus of the glacier is stationary, the ice continues to flow forward, delivering a continuous supply of sediment in the same manner a conveyor belt delivers goods to the end of a production line. As the ice melts, the till is dropped and the end moraine grows. The longer the ice front remains stable, the larger the ridge of till will become.

Eventually, wastage exceeds nourishment. At this point, the front of the glacier begins to recede in the direction from which it originally advanced. As the ice front retreats, however, the conveyor-belt action of the glacier continues to provide fresh supplies of sediment to the terminus. In this manner, a large quantity of till is deposited as the ice melts away, creating a rock-strewn, undulating surface. This gently rolling layer of till deposited as the ice front recedes is termed **ground moraine**. Ground moraine has a leveling effect, filling in low spots and clogging old stream channels, often leading to a derangement of the existing drainage system. In areas where this layer of till is still relatively fresh, such as the northern Great Lakes region, poorly drained swampy lands are quite common.

Periodically, a glacier will retreat to a point where wastage and nourishment once again balance. When this happens, the ice front stabilizes, and a new end moraine forms.

The pattern of end moraine formation and ground moraine deposition may be repeated many times before the glacier has completely vanished. Such a pattern is illustrated in **Figure 4.16**. The very first end moraine to form marks the farthest advance of the glacier and is called the *terminal end moraine*. End moraines that form as the ice front occasionally stabilizes during retreat are termed *recessional end moraines*. Terminal and recessional moraines are essentially alike; the only difference between them is their relative positions.

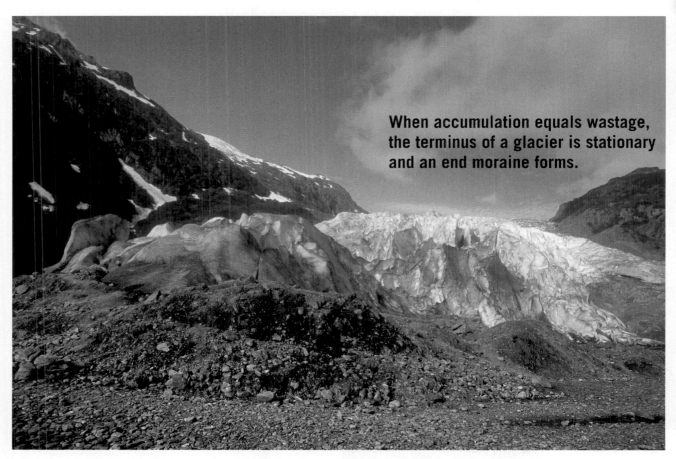

When accumulation equals wastage, the terminus of a glacier is stationary and an end moraine forms.

Figure 4.15 Formation of an end moraine This end moraine is forming at the terminus of Exit Glacier in Kenai Fjords National Park. (Photo by Michael Collier)

End moraines deposited by the most recent major stage of Ice Age glaciation are prominent features in many parts of the Midwest and Northeast. In Wisconsin, the wooded, hilly terrain of the Kettle Moraine near Milwaukee is a particularly picturesque example.

A well-known example in the Northeast is Long Island. This linear strip of glacial sediment that extends northeastward from New York City is part of an end moraine complex that stretches from eastern Pennsylvania to Cape Cod, Massachusetts.

End moraines:
Wisconsinan Age

End moraines:
Illinoian Age

0   150mi

0   150km

Driftless area

Extent of Wisconsinan glaciation

Extent of Illinoian glaciation

Figure 4.16 End moraines of the Great Lakes region End moraines deposited during the most recent stage of glaciation, called the Wisconsinan, are the most prominent.

SmartFigure 4.17
Common depositional land-forms This diagram depicts a hypothetical area that was recently affected by an ice sheet. (Drumlin courtesy of Ward's Natural Science Establishment; kame by John Dankwardt; kettle by Carlyn Iverson/Photo Researchers, Inc.; esker by Richard P. Jacobs; braided stream by Michael Collier)

Figure 4.17 represents a hypothetical area during and following glaciation. It shows the end moraines that were just described as well as the depositional features that are discussed in the sections that follow. This figure depicts landscape features similar to those that might be encountered if you were traveling in the upper Midwest or New England. As you read about other glacial deposits, you will be referred to this figure again.

Outwash Plains and Valley Trains At the same time that an end moraine is forming, meltwater emerges from the ice in rapidly moving streams that are often choked with suspended material and carrying a substantial bed load. As the water leaves the glacier and loses velocity, much of its bed load is dropped, often causing the stream channel to become braided. In this way, a broad, ramplike accumulation of stratified drift is built adjacent to the downstream edge of most end moraines. When the feature is formed in association with an ice sheet, it is termed an **outwash plain**, and when it is confined to a mountain valley, it is commonly referred to as a **valley train** (Figure 4.17).

Kettles Often, end moraines, outwash plains, and valley trains are pockmarked with basins or depressions known as **kettles** (Figure 4.17). Kettles form when blocks of stagnant ice become buried in drift and eventually melt, leaving pits in the glacial sediment. Most

kettles do not exceed 2 kilometers (1.25 miles) in diameter, and the typical depth of most kettles is less than 10 meters (33 feet). Water often fills a kettle depression and forms a pond or lake. One well-known example is Walden Pond near Concord, Massachusetts. It is here that Henry David Thoreau lived alone for two years in the 1840s and about which he wrote *Walden*, his American literature classic.

Drumlins, Eskers, and Kames

Moraines are not the only landforms deposited by glaciers. Some landscapes are characterized by numerous elongate parallel hills made of till. Other areas exhibit conical hills and relatively narrow winding ridges composed mainly of stratified drift.

Drumlins **Drumlins** are streamlined asymmetrical hills composed of till (Figure 4.17). They range in height from 15 to 60 meters (50 to 200 feet) and average 0.4 to 0.8 kilometer (0.25 to 0.50 mile) in length. The steep side of the hill faces the direction from which the ice advanced, while the gentler slope points in the direction the ice moved. Drumlins are not found singly but rather occur in clusters, called *drumlin fields*. One such cluster, east of Rochester, New York, is estimated to contain about 10,000 drumlins. Their streamlined shape indicates that they were molded in the zone of flow within an active glacier. It is thought that drumlins originate when glaciers advance over previously deposited drift and reshape the material.

Eskers and Kames In some areas that were once occupied by glaciers, sinuous ridges composed largely of sand and gravel may be found. These ridges, called **eskers**, are deposits made by streams flowing in tunnels within or beneath the ice, near the terminus of a glacier (Figure 4.17). They may be several meters high and extend for many kilometers. In some areas, they are mined for sand and gravel, and for this reason eskers are disappearing in some localities.

Kames are steep-sided hills that, like eskers, are composed largely of stratified drift (Figure 4.17). Kames originate when glacial meltwater washes sediment into openings and depressions in the stagnant wasting terminus of a glacier. When the ice eventually melts away, the stratified drift is left behind as mounds or hills.

4.4 CONCEPT CHECKS

1. What is the difference between till and stratified drift?
2. Distinguish among terminal end moraine, recessional end moraine, and ground moraine. Relate these moraines to the budget of a glacier.
3. Describe the formation of a medial moraine.
4. List four depositional features other than moraines.

4.5 Other Effects of Ice Age Glaciers

Describe and explain several important effects of Ice Age glaciers other than erosional and depositional landforms.

In addition to the massive erosional and depositional work carried out by Ice Age glaciers, ice sheets had other, sometimes profound, effects on the landscape. For example, as the ice advanced and retreated, animals and plants were forced to migrate. This led to stresses that some organisms could not tolerate. Furthermore, many present-day stream courses bear little resemblance to their preglacial routes. The Missouri River once flowed northward, toward Hudson Bay in Canada. The Mississippi River followed a path through central Illinois, and the head of the Ohio River reached only as far as Indiana (**Figure 4.18**). A comparison of the two parts of Figure 4.18 shows that the Great Lakes were created by glacial erosion during the Ice Age. Prior to the Ice Age, the basins occupied by these huge lakes were lowlands with rivers that ran eastward to the Gulf of St. Lawrence.

In areas that were centers of ice accumulation, such as Scandinavia and northern Canada, the land has been slowly rising for the past several thousand years. The land had downwarped under the tremendous weight of 3-kilometer-thick (almost 2-mile-thick) masses of ice. Following the removal of this immense load, the crust has been adjusting by gradually rebounding upward ever since.

Ice sheets and alpine glaciers can act as dams to create lakes by trapping glacial meltwater and blocking the flow of rivers. Some of these lakes are relatively small and short-lived. Others can be large and exist for hundreds or thousands of years.

Figure 4.19 is a map of Lake Agassiz—the largest lake to form during the Ice Age in North America. With the retreat of the ice sheet came enormous volumes of meltwater. The Great Plains generally slope upward to the

This map shows the Great Lakes and the familiar present-day pattern of rivers. Quaternary ice sheets played a major role in creating this pattern

Reconstruction of drainage systems prior to the Ice Age. The pattern was very different from today, and the Great Lakes did not exist.

Figure 4.18 Changing rivers The advance and retreat of ice sheets caused major changes in the routes followed by rivers in the central United States.

Figure 4.19 Glacial Lake Agassiz This lake was an immense feature—bigger than all of the present-day Great Lakes combined. Modern-day remnants of this proglacial water body are still major landscape features.

west. As the terminus of the ice sheet receded northeastward, meltwater was trapped between the ice on one side and the sloping land on the other, causing Lake Agassiz to deepen and spread across the landscape. It came into existence about 12,000 years ago and lasted for about 4500 years. Such water bodies are termed **proglacial**

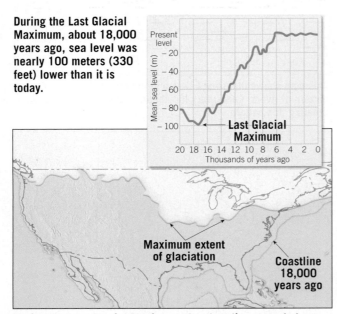

During the Last Glacial Maximum, about 18,000 years ago, sea level was nearly 100 meters (330 feet) lower than it is today.

During the Last Glacial Maximum, the shoreline extended out onto the present day continental shelf.

Figure 4.20 Changing sea level As ice sheets form and then melt away, sea level falls and rises, causing the shoreline to shift.

lakes, referring to their position just beyond the outer limits of a glacier or an ice sheet.

A far-reaching effect of the Ice Age was the worldwide change in sea level that accompanied each advance and retreat of the ice sheets. The snow that nourishes glaciers ultimately comes from moisture evaporated from the oceans. Therefore, when the ice sheets increased in size, sea level fell, and the shoreline moved seaward (**Figure 4.20**). Estimates suggest that sea level was as much as 100 meters (330 feet) lower than it is today. Consequently, the Atlantic coast of the United States was located more than 100 kilometers (60 miles) to the east of New York City. Moreover, France and Britain were joined where the English Channel is today. Alaska and Siberia were connected across the Bering Strait, and Southeast Asia was tied by dry land to the islands of Indonesia.

The formation and growth of ice sheets was an obvious response to significant changes in climate. But the existence of the glaciers themselves triggered climatic changes in the regions beyond their margins. In arid and semiarid areas on all continents, temperatures were lower, which meant evaporation rates were also lower. At the same time, precipitation was moderate. This cooler, wetter climate resulted in the formation of many lakes called **pluvial lakes** (from the Latin term *pluvia*, meaning "rain"). In North America, pluvial lakes were concentrated in the vast Basin and Range region of Nevada and Utah (**Figure 4.21**).

Figure 4.21 Pluvial lakes During the Ice Age, the Basin and Range region experienced a wetter climate than it has today. Many basins were turned into large lakes.

Although most are now gone, a few remnants remain, the largest of which is Utah's Great Salt Lake.

2. Compare the two parts of Figure 4.18 and identify three major changes to the flow of rivers in the central United States during the Ice Age.
3. Examine Figure 4.20 and determine how much sea level has changed since the Last Glacial Maximum.
4. Contrast proglacial lakes and pluvial lakes.

Did You Know?

Taken together, the Great Lakes constitute the largest body of freshwater on Earth. Formed between 10,000 and 12,000 years ago, the lakes currently contain about 19 percent of Earth's *surface* freshwater.

4.6 Extent of Ice Age Glaciation

Discuss the extent of glaciation and climate variability during the Quaternary Ice Age.

During the Ice Age, ice sheets and alpine glaciers were far more extensive than they are today. There was a time when the most popular explanation for what we now know to be glacial deposits was that the material had been drifted in by means of icebergs or perhaps simply swept across the landscape by a catastrophic flood. However, during the nineteenth century, field investigations by many scientists provided convincing proof that an extensive Ice Age was responsible for these deposits and for many other features.

By the beginning of the twentieth century, geologists had largely determined the extent of Ice Age glaciation. Further, they discovered that many glaciated regions did not have just one layer of drift but several. Close examination of these older deposits showed well-developed zones of chemical weathering and soil formation, as well as the remains of plants that require warm temperatures. The evidence was clear: There had not been just one glacial advance but several, each separated by extended periods when climates were as warm as or warmer than at present. The Ice Age was not simply a time when the ice advanced over the land, lingered for a while, and then receded. Rather, it was a complex period characterized by a number of advances and withdrawals of glacial ice.

The glacial record on land is punctuated by many erosional gaps. This makes it difficult to reconstruct the episodes of the Ice Age. But sediment on the ocean floor provides an uninterrupted record of climate cycles for this period. Studies of these seafloor sediments show that glacial/interglacial cycles have occurred about every 100,000 years. About 20 such cycles of cooling and warming were identified for the span we call the Ice Age.

During the Ice Age, ice left its imprint on almost 30 percent of Earth's land area, including about 10 million square kilometers (nearly 4 million square miles) of North America, 5 million square kilometers (2 million square miles) of Europe, and 4 million square kilometers (1.6 million square miles) of Siberia (**Figure 4.22**). The amount of glacial ice in the Northern Hemisphere was roughly twice that in the Southern Hemisphere. The primary reason is that the Southern Hemisphere has little

land in the middle latitudes, and, therefore, the southern polar ice could not spread far beyond the margins of Antarctica. By contrast, North America and Eurasia provided great expanses of land for the spread of ice sheets.

We now know that the Ice Age began between 2 million and 3 million years ago. This means that most of the major glacial episodes occurred during a division of the geologic time scale called the **Quaternary period**. Although the Quaternary is commonly used as a synonym for the Ice Age, this period does not encompass it all. The Antarctic Ice Sheet, for example, formed at least 30 million years ago.

Did You Know?

Antarctica's ice sheet weighs so much that it depresses Earth's crust by an estimated 900 m (3000 ft) or more.

4.6 CONCEPT CHECKS

1. About what percentage of Earth's land surface was affected by glaciers during the Quaternary period?
2. Where were ice sheets more extensive during the Ice Age: the Northern Hemisphere or the Southern Hemisphere? Why?

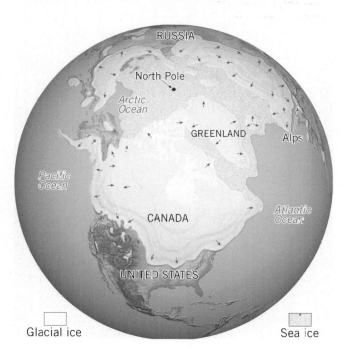

RUSSIA
North Pole
Arctic Ocean
GREENLAND
Alps
Pacific Ocean
Atlantic Ocean
CANADA
UNITED STATES

Glacial ice   Sea ice

Figure 4.22 Where was the ice? This map shows the maximum extent of ice sheets in the Northern Hemisphere during the Ice Age.

4.7 Deserts

Describe the general distribution and causes of Earth's dry lands and the role that water plays in modifying desert landscapes.

The dry regions of the world encompass about 42 million square kilometers (more than 16 million square miles), a surprising 30 percent of Earth's land surface. No other climate group covers so large a land area. The word *desert* literally means "deserted," or "unoccupied." For many dry regions, this is a very appropriate description. Yet where water is available in deserts, plants and animals thrive. Nevertheless, the world's dry regions are among the least familiar land areas on Earth outside of the polar realm.

Desert landscapes frequently appear stark. Their profiles are not softened by a carpet of soil and abundant plant life. Instead, barren rocky outcrops with steep, angular slopes are common. Some rocks are tinted orange and red; others in different locations are gray and brown and streaked with black. For many visitors, desert scenery exhibits a striking beauty; to others, the terrain seems bleak. No matter which feeling is elicited, it is clear that deserts are very different from the more humid places where most people live.

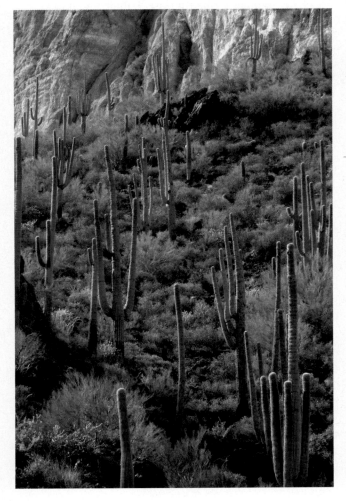

Figure 4.23 Sonoran Desert near Tucson, Arizona Some of these giant saguaro cacti are more than 100 years old. The appearance of desert landscapes varies a great deal from place to place. (Photo by Michael Collier)

As you will see, arid regions are not dominated by a single geologic process. Rather, the effects of tectonic (mountain-building) forces, running water, and wind are all apparent. Because these processes combine in different ways from place to place, the appearance of desert landscapes varies a great deal as well (**Figure 4.23**).

Distribution and Causes of Dry Lands

We all recognize that deserts are dry places, but just what is meant by the word *dry*? That is, how much rain defines the boundary between humid and dry regions?

Sometimes, it is arbitrarily defined by a single rainfall figure, such as 25 centimeters (10 inches) per year of precipitation. However, the concept of dryness is relative; it refers to *any situation in which a water deficiency exists*. Climatologists define **dry climate** as a climate in which yearly precipitation is less than the potential loss of water by evaporation.

Within these water-deficient regions, two climatic types are commonly recognized: **desert**, or arid, and **steppe**, or semiarid. The two categories have many features in common; their differences are primarily a matter of degree. The steppe is a marginal and more humid variant of the desert and represents a transition zone that surrounds the desert and separates it from bordering humid climates. Maps showing the distribution of desert and steppe regions reveal that dry lands are concentrated in the subtropics and in the middle latitudes (**Figure 4.24**).

Deserts in places such as Africa, Arabia, and Australia are primarily the result of the prevailing global distribution of air pressure and winds (**Figure 4.25**). Coinciding with dry regions in the lower latitudes are zones of high air pressure known as the *subtropical highs*. These pressure systems are characterized by subsiding air currents. When air sinks, it is compressed and warmed. Such conditions are just the opposite of what is needed to produce clouds and precipitation. Consequently, these regions are known for their clear skies, sunshine, and ongoing dryness.

Middle-latitude deserts and steppes exist principally because they are sheltered in the deep interiors of large landmasses. They are far removed from the ocean, which is the ultimate source of moisture for cloud formation and precipitation. In addition, the presence of high mountains across the paths of prevailing winds further acts to separate these areas from water-bearing,

SmartFigure 4.24 Dry climates Arid and semi-arid climates cover about 30 percent of Earth's land surface. The dry region of the American West is commonly divided into four deserts, two of which extend into Mexico.

Figure 4.25 Subtropical deserts The distribution of subtropical deserts is closely related to the global distribution of air pressure.

Subtropical deserts and steppes are centered between 20° and 30° north and south latitude in association with the subtropical high-pressure belts. Dry subsiding air inhibits cloud formation and precipitation.

In this view from space, the Sahara Desert, the adjacent Arabian Desert, and the Kalahari and Namib deserts are clearly visible as tan-colored, cloud-free zones. The band of clouds across central Africa and the adjacent oceans coincides with the equatorial low-pressure belt.

When moving air meets a mountain barrier, it is forced to rise. Clouds and precipitation on the windward side often result.

Air descending the leeward side is much drier.

Coast Range

Sierra Nevada Range

Rainshadow

Leeward (dry)

Great Basin

Windward (wet)

Wind

Figure 4.26 Rainshadow deserts Mountains frequently contribute to the aridity of middle-latitude deserts and steppes by creating a rainshadow. The Great Basin Desert is a rain-shadow desert that covers nearly all of Nevada and portions of adjacent states. (Photo on left by Dean Pennala/Shutterstock; photo on right by Dennis Tasa)

maritime air masses, In North America, the Coast Range, Sierra Nevada, and Cascades are the foremost mountain barriers to moisture from the Pacific (**Figure 4.26**). In their rainshadow lies the dry and expansive Basin and Range region of the American West.

Middle-latitude deserts provide an example of how mountain-building processes affect climate. Without mountains, wetter climates would prevail where dry regions exist today.

The Role of Water in Arid Climates

Permanent streams are normal in humid regions, but practically all desert streambeds are dry most of the time. Deserts have **ephemeral streams**, which means that they carry water only in response to specific episodes of rainfall. A typical ephemeral stream may flow only a few days or perhaps just a few hours during the year. In some years, the channel might carry no water at all.

This fact is obvious even to a casual traveler who notices the numerous bridges with no streams beneath them, or numerous dips in the road where dry channels cross. However, when the rare heavy showers occur, so much rain falls in such a short time that it cannot all soak in. Because desert vegetative cover is sparse, runoff is largely unhindered and, therefore, rapid, often creating flash floods along valley floors (**Figure 4.27**). These floods are quite unlike floods in humid regions. A flood on a river like the Mississippi may take days or weeks to reach its crest and then subside. But desert floods arrive suddenly and subside quickly. Because much surface material in a desert is not anchored by vegetation, the amount of erosional work that occurs during a single short-lived rain event is impressive.

Figure 4.27 Ephemeral stream Most of the time these stream channels are dry. This image was taken shortly after a heavy rainstorm. A great deal of erosion can occur in a short time; notice how muddy the water is. Because roads commonly dip into the stream channels, warning signs are common sights. (Stream photo by Universal Images Group/SuperStock; road sign photo by Demetrio Carrasco/Dorling Kindersley)

POTENTIAL FLASH FLOOD AREAS

NEXT 21 MILES

In the dry western United States, different names are used for ephemeral streams, including *wash* and *arroyo* In other parts of the world, a dry desert stream may be a *wadi* (Arabia and North Africa), a *donga* (South America), or a *nullah* (India).

Humid regions are notable for their well-developed drainage systems. But in arid regions, streams usually lack extensive systems of tributaries. In fact, a basic characteristic of desert streams is that they are small and die out before reaching the sea. Because the water table is usually far below the surface, few desert streams can draw upon it as streams do in humid regions. Without a steady supply of water, the combination of evaporation and soaking in soon depletes the stream.

The few permanent streams that do cross arid regions, such as the Colorado River in the American Southwest and the Nile River in North Africa, originate *outside* the desert, often in mountains where precipitation is plentiful. The supply of water must be great to compensate for the losses that occur as the stream crosses the desert. For example, after the Nile leaves its headwaters in the lakes and mountains of central Africa, it crosses almost 3000 kilometers (nearly 1900 miles) of the Sahara Desert *without a single tributary*.

It should be emphasized that *running water, although infrequent, nevertheless is responsible for most of the erosional work in deserts.* This is contrary to a common belief that wind is the most important erosional agent sculpturing desert landscapes. Although wind erosion is indeed more significant in dry areas than elsewhere, most desert landforms are carved by running water. The main role of wind, as you will see later in this chapter, is in the transportation and deposition of sediment, which creates and shapes the ridges and mounds we call dunes.

4.7 CONCEPT CHECKS

1. Define *dry climate*. How extensive are the desert and steppe regions of Earth?
2. Discuss the causes of deserts.
3. What is an ephemeral stream?
4. What is the most important erosional agent in deserts?

4.8 Basin and Range: The Evolution of a Mountainous Desert Landscape

Discuss the stages of landscape evolution in the Basin and Range region of the western United States.

Dry regions typically lack permanent streams and often have **interior drainage**. This means that they exhibit a discontinuous pattern of ephemeral streams that do not flow out of the desert to the ocean. In the United States, the dry Basin and Range region provides an excellent example. The region includes southern Oregon, all of Nevada, western Utah, southeastern California, southern Arizona, and southern New Mexico. The name Basin and Range is an apt description for this almost 800,000-square-kilometer (312,000-square-mile) region, as it is characterized by more than 200 relatively small mountain ranges that rise 900 to 1500 meters (3000 to 5000 feet) above the basins that separate them. The origin of these *fault-block mountains* is examined in Chapter 6. In this discussion we look at how surface processes change the landscape.

In the Basin and Range region, as in others like it around the world, most erosion occurs without reference to the ocean (ultimate base level) because the interior drainage never reaches the sea. Even where permanent streams flow to the ocean, few tributaries exist, and only a narrow strip of land adjacent to the stream has sea level as its ultimate level of land reduction.

The block models in **Figure 4.28** depict how the landscape has evolved in the Basin and Range region. During and following uplift of the mountains, running water carves the elevated masses and deposits large quantities of sediment in the basin. In this early stage, relief (difference in elevation between high and low points in an area) is greatest. As erosion lowers the mountains and sediment fills the basins, elevation differences diminish.

When the occasional torrents of water produced by sporadic rains or periods of snowmelt move down the mountain canyons, they are heavily loaded with sediment. Emerging from the confines of the canyon, the runoff spreads over the gentler slopes at the base of the mountains and quickly loses velocity. Consequently, most of its sediment load is dumped within a short distance. The result is a cone of debris known as an **alluvial fan** at the mouth of a canyon. Over the years, a fan enlarges, eventually coalescing with fans from adjacent canyons to produce an apron of sediment, called a **bajada**, along the mountain front.

On the rare occasions of abundant rainfall, streams may flow across the alluvial fans to the center of the basin, converting the basin floor into a shallow **playa lake**. Playa lakes last only a few days or weeks before evaporation and infiltration remove the water. The dry, flat lake bed that remains is termed a *playa*. Playas

SmartFigure 4.28
Landscape evolution in the Basin and Range region As erosion of the mountains and deposition in the basins continue, relief diminishes.

Geologist's Sketch

Figure 4.29 Death Valley: A classic Basin and Range landscape Shortly before the satellite image (left) was taken in February 2005, heavy rains led to the formation of a small playa lake—the pool of greenish water on the basin floor. By May 2005, the lake had reverted to a salt-covered playa. (NASA) The small photo is a closer view of one of Death Valley's many alluvial fans. (Photo by Michael Collier)

occasionally become encrusted with salts (*salt flats*) that are left behind when the water in which they were dissolved evaporates. **Figure 4.29** includes a satellite image of a portion of California's Death Valley, a classic Basin and Range landscape. Many of the features just described are prominent, including a bajada (left side of valley), alluvial fans, a playa lake, and extensive salt flats.

With the ongoing erosion of the mountain mass and the accompanying sedimentation, the local relief continues to diminish. Eventually, nearly the entire mountain mass is gone. Thus, by the late stages of erosion, the mountain areas are reduced to a few large bedrock knobs (called *inselbergs*) projecting above the sediment-filled basin.

Each of the stages of landscape evolution in an arid climate depicted in Figure 4.29 can be observed in the Basin and Range region. Recently uplifted mountains in an early stage of erosion are found in southern Oregon and northern Nevada. Death Valley, California, and southern Nevada fit into the more advanced middle stage, whereas the late stage, with its inselbergs, can be seen in southern Arizona.

4.8 CONCEPT CHECKS

1. What is meant by *interior drainage*?
2. Describe the features and characteristics associated with each stage in the evolution of a mountainous desert.
3. Where in the United States can each stage of desert landscape evolution be observed?

4.9 Wind Erosion

Describe the ways that wind transports sediment and the features created by wind erosion.

Moving air, like moving water, is turbulent and able to pick up loose debris and transport it to other locations. Just as in a river, the velocity of wind increases with height above the surface. Also like a river, wind transports fine particles in suspension, while it carries heavier ones as bed load. However, the transport of sediment by wind differs from that by running water in two significant ways. First, wind's lower density compared to water renders it less capable of picking up and transporting coarse materials. Second, because wind is not confined to channels, it can spread sediment over large areas, as well as high into the atmosphere.

Compared to running water and glaciers, wind is a relatively insignificant erosional agent. Recall that even in deserts, most erosion is performed by intermittent running water, not by wind. Wind erosion is more effective in arid lands than in humid areas because in humid places, moisture binds particles together and vegetation anchors the soil. For wind to be an effective erosional force, dryness and scant vegetation are important prerequisites. When such circumstances exist, wind may pick up, transport, and deposit great quantities of fine sediment. During the 1930s, parts of the Great Plains experienced vast dust storms (**Figure 4.30**). The plowing under of the natural vegetative cover for farming, followed by severe drought, exposed the land to wind erosion and led to the area being labeled the Dust Bowl.

Deflation, Blowouts, and Desert Pavement

One way that wind erodes is by **deflation**, the lifting and removal of loose material. Wind can suspend only fine sediment such as clay and silt (**Figure 4.31A**). Larger grains of sand are rolled or skipped along the surface, a process called *saltation*, and comprise the bed load (**Figure 4.31B**). Particles larger than sand are usually not transported by

wind. The effects of deflation are sometimes difficult to notice because the entire surface is being lowered at the same time, but it can be significant.

Figure 4.30 1930s Dust Bowl Dust blackens the sky on May 21, 1937, near Elkhart, Kansas. It was because of storms like this one that portions of the Great Plains were called the Dust Bowl in the 1930s. (Photo reproduced from the collection of the Library of Congress)

A.

Saltating sand grains.
B.

Figure 4.31 How wind transports sediment A. This satellite image shows thick plumes of dust from the Sahara Desert blowing across the Red Sea on June 30, 2009. Such dust storms are common in arid North Africa. In fact, this region is the largest dust source in the world. Satellites are an excellent tool for studying the transport of dust on a global scale. They show that dust storms can cover huge areas and that dust can be transported great distances. (NASA) **B.** The bed load carried by wind consists of sand grains, many of which move by bouncing along the surface. Sand never travels far from the surface, even when winds are very strong. (Photo by Berno Zoller/Photolibrary)

Figure 4.32 Blowouts Deflation is especially effective in creating these depressions when the land is dry and largely unprotected by anchoring vegetation. (Photo courtesy of USDA/Natural Resources Conservation Service)

Did You Know?

Deserts do not necessarily consist of mile after mile of drifting sand dunes. Surprisingly, sand accumulations represent only a small percentage of the total desert area. In the Sahara, dunes cover only one-tenth of its area. The sandiest of all deserts, the Arabian, is one-third sand covered.

The man is pointing to where the ground surface was when the grasses began to grow. Wind erosion lowered the land surface to the level of his feet.

Clumps of anchored soil

Unanchored soil

Sand dune

1.2 meters

The most noticeable results of deflation in some places are shallow depressions called **blowouts** (**Figure 4.32**). In the Great Plains region, from Texas north to Montana, thousands of blowouts can be seen. They range from small dimples less than 1 meter (3 feet) deep and 3 meters (10 feet) wide to depressions that are more than 45 meters (150 feet) deep and several kilometers across. The factor that controls the depths of these basins is the local water table. When blowouts are lowered to the water table, damp ground and vegetation usually prevent further deflation.

In portions of many deserts, the surface is characterized by a layer of coarse pebbles and cobbles that are too large to be moved by the wind. This stony veneer, called **desert pavement**, may form as deflation lowers the surface by removing sand and silt from poorly sorted materials.

As **Figure 4.33A** illustrates, the concentration of larger particles at the surface gradually increases as the finer particles are blown away. Eventually, a continuous cover of coarse particles remains.

Studies have shown that the process depicted in Figure 4.33A is not an adequate explanation for all environments in which desert pavement exists. As a result, an alternate explanation was formulated and is illustrated in **Figure 4.33B**. This hypothesis suggests that pavement develops on a surface that initially consists of coarse pebbles. Over time, protruding cobbles trap fine wind-blown grains that settle and sift downward through the spaces between the larger surface stones. The process is aided by infiltrating rainwater.

Once desert pavement becomes established, a process that might take hundreds of years, the surface is effectively protected from further deflation if left undisturbed. However, because the layer is only one or two stones thick, the passage of vehicles or animals can dislodge the pavement and expose the fine-grained material below. If this happens, the surface is no longer protected from deflation.

Wind Abrasion

Like glaciers and streams, wind erodes in part by *abrasion*. In dry regions as well as along some beaches,

SmartFigure 4.33
Formation of desert pavement A. This model shows an area with poorly sorted surface deposits. Over time, deflation lowers the surface, and coarse particles become concentrated.
B. In this model, the surface is initially covered with cobbles and pebbles. Over time, wind-blown dust accumulates at the surface and gradually sifts downward. Infiltrating rainwater aids the process.

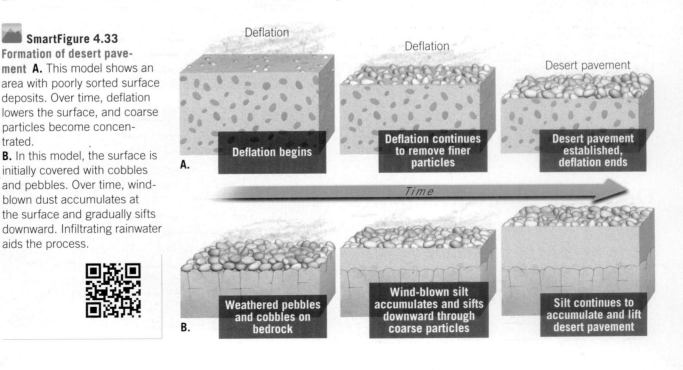

Deflation

Deflation

Desert pavement

Deflation begins

Deflation continues to remove finer particles

Desert pavement established, deflation ends

A.

Time

Weathered pebbles and cobbles on bedrock

Wind-blown silt accumulates and sifts downward through coarse particles

Silt continues to accumulate and lift desert pavement

B.

windblown sand cuts and polishes exposed rock surfaces. Abrasion is often given credit for accomplishments beyond its actual capabilities. Such features as balanced rocks that stand high atop narrow pedestals and intricate detailing on tall pinnacles are not the results of abrasion by windblown sand. Sand is seldom lifted more than a meter above the surface, so the wind's sandblasting effect is obviously limited in vertical extent. But in areas prone to such activity, telephone poles have actually been cut through near their bases. For this reason, collars are often fitted on the poles to protect them from being "sawed" down.

4.10 Wind Deposits

Distinguish between two basic types of wind deposits.

Although wind is relatively unimportant in carving erosional features, significant depositional landforms are created by wind in some regions. Accumulations of windblown sediment are particularly conspicuous in the world's dry lands and along many sandy coasts. Wind deposits are of two distinctive types: (1) extensive blankets of silt, called *loess*, that once were carried in suspension, and (2) mounds and ridges of sand from the wind's bed load, which we call *dunes*.

Loess

In some parts of the world, the surface topography is mantled with deposits of windblown silt, called **loess**. Dust storms deposited this material over thousands of years. When loess is breached by streams or road cuts, it tends to maintain vertical cliffs and lacks any visible layers, as you can see in **Figure 4.34**.

The distribution of loess worldwide indicates two primary sources for this sediment: deserts and glacial deposits of stratified drift. The thickest and most extensive loess deposits occur in western and northern China. They were blown there from the extensive desert basins of central Asia. Accumulations of 30 meters (100 feet) are not uncommon, and thicknesses of more than 100 meters (300 feet) have been measured. It is this fine, buff-colored sediment that gives the Yellow River (Huang He) its name.

In the United States, deposits of loess are significant in many areas, including South Dakota, Nebraska, Iowa, Missouri, and Illinois, as well as portions of the Columbia Plateau in the Pacific Northwest. Unlike the deposits in China, which originated in deserts, the loess in the United States and Europe is an indirect product of glaciation. Its source is deposits of stratified drift. During the retreat of the ice sheets, many river valleys were choked with glacial sediment. Strong winds sweeping across the barren floodplains picked up the finer sediment and dropped it as a blanket on areas adjacent to the valleys.

Sand Dunes

Like running water, wind releases its load of sediment when its velocity falls and the energy available for transport diminishes. Thus, sand begins to accumulate

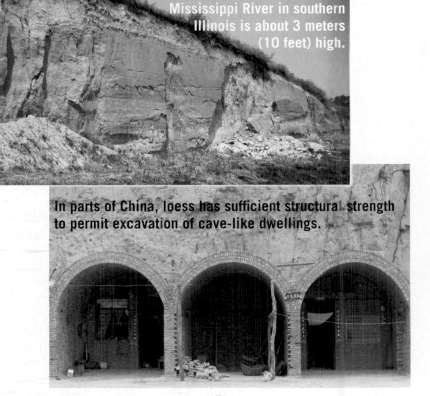

This vertical bluff near the Mississippi River in southern Illinois is about 3 meters (10 feet) high.

In parts of China, loess has sufficient structural strength to permit excavation of cave-like dwellings.

Figure 4.34 Loess In some regions, the surface is mantled with deposits of windblown silt. (Illinois photo by James E. Patterson; China photo by Ashley Cooper/Alamy Images)

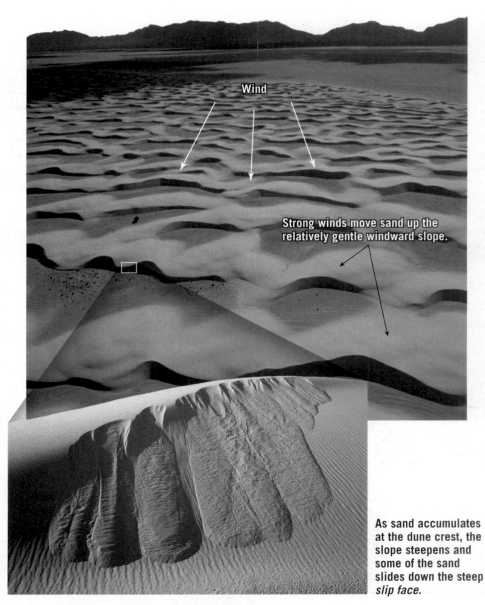

Wind

Strong winds move sand up the relatively gentle windward slope.

As sand accumulates at the dune crest, the slope steepens and some of the sand slides down the steep *slip face.*

Figure 4.35 **White Sands National Monument** The dunes at this landmark in southeastern New Mexico are composed of gypsum. The dunes slowly migrate with the wind. (Photos by Michael Collier)

wherever an obstruction across the path of the wind slows its movement. Unlike deposits of loess, which form blanketlike layers over broad areas, winds commonly deposit sand in mounds or ridges called **dunes** (**Figure 4.35**).

As moving air encounters an object, such as a clump of vegetation or a rock, the wind sweeps around and over it, leaving a shadow of more slowly moving air behind the obstacle as well as a smaller zone of quieter air just in front of the obstacle. Some of the sand grains moving with the wind come to rest in these *wind shadows*. As the accumulation of sand continues, it forms an

increasingly efficient wind barrier to trap even more sand. If there is a sufficient supply of sand and the wind blows steadily long enough, the mound of sand grows into a dune.

Many dunes have an asymmetrical profile, with the leeward (sheltered) slope being steep and the windward slope more gently inclined. Sand is rolled up the gentle slope on the windward side by the force of the wind. Just beyond the crest of the dune, the wind velocity is reduced, and the sand accumulates. As more sand collects, the slope steepens, and eventually some of it slides under the pull of gravity. In this way, the leeward slope

Dunes commonly have an asymmetrical shape and migrate with the wind.

Sand grains deposited on the slip face at the angle of repose create the cross-bedding of dunes.

Cross beds are an obvious characteristic of the Navajo Sandstone in Zion National Park, Utah.

When dunes are buried and become part of the sedimentary rock record, the cross bedding is preserved.

SmartFigure 4.36 Cross bedding As sand is deposited on the slip face, layers form that are inclined in the direction the wind is blowing. With time, complex patterns develop in response to changes in wind direction. (Photo by Dennis Tasa)

of the dune, called the **slip face**, maintains a relatively steep angle. Continued sand accumulation, coupled with periodic slides down the slip face, results in the slow migration of the dune in the direction of air movement (Figure 4.35).

As sand is deposited on the slip face, it forms layers inclined in the direction the wind is blowing. These sloping layers are referred to as **cross bedding**. When the dunes are eventually buried under layers of sediment and become part of the sedimentary rock record, their asymmetrical shape is destroyed, but the cross bedding remains as a testimony to their origin. Nowhere is cross bedding more prominent than in the sandstone walls of Zion Canyon in southern Utah (**Figure 4.36**).

4.10 CONCEPT CHECKS

1. Contrast loess and sand dunes.
2. Describe how sand dunes migrate.
3. What is cross bedding?

CONCEPTS IN REVIEW: GLACIAL AND ARID LANDSCAPES

4.1 Glaciers: A Part of Two Basic Cycles

List the different types of glaciers, summarize their characteristics, and describe the present-day location and extent of glaciers on Earth.

KEY TERMS glacier, valley (alpine) glacier, ice sheet, sea ice, ice shelf, ice cap, piedmont glacier, outlet glacier

- A glacier is a thick mass of ice originating on land from the compaction and recrystallization of snow, and it shows evidence of past or present flow. Glaciers are part of both the hydrologic cycle and the rock cycle. They store and release freshwater, and they transport and deposit large quantities of sediment.
- Valley glaciers flow down mountain valleys, whereas ice sheets are very large masses, such as those that cover Greenland and Antarctica. During the Last Glacial Maximum, around 18,000 years ago, Earth was in an Ice Age that covered large areas with glacial ice.
- When valley glaciers exit confining mountains, they spread out into broad lobes called piedmont glaciers. Similarly, ice shelves form when glaciers flow into the ocean producing a layer of floating ice.
- Ice caps are like small ice sheets. Both ice sheets and ice caps may be drained by outlet glaciers, which often resemble valley glaciers.

4.2 How Glaciers Move

Describe how glaciers move, the rates at which they move, and the significance of the glacial budget.

KEY TERMS crevasse, zone of accumulation, zone of wastage, iceberg, glacial budget

- Glaciers move in part by flowing under pressure. On the surface of a glacier, ice is brittle. Below about 50 meters (165 feet), pressure is great, and ice behaves like a plastic material and flows. Another type of movement consists of the bottom of the glacier sliding along the ground.
- Fast glaciers may move 800 meters (2600 feet) per year, while slow glaciers may move only 2 meters (6.5 feet) per year. Some glaciers experience periodic surges of rapid movement.
- When a glacier has a positive budget, the terminus will advance. This occurs when the glacier gains more snow in its upper zone of accumulation that it loses at its downstream zone of wastage. If wastage exceeds the input of new ice, the glacier's terminus will retreat.

4.3 Glacial Erosion

Discuss the processes of glacial erosion and the major features created by these processes.

KEY TERMS plucking, abrasion, rock flour, glacial striation, glacial trough, hanging valley, cirque, arête, horn, fiord

- Glaciers acquire sediment through plucking from the bedrock beneath the glacier, by abrasion of the bedrock using sediment already in the ice, and when mass-wasting processes drop debris on top of the glacier. Grinding of the bedrock produces grooves and scratches called glacial striations.
- Erosional features produced by valley glaciers include glacial troughs, hanging valleys, cirques, arêtes, horns, and fiords.

? **Examine the illustration of a mountainous landscape after glaciation. Identify the landforms that resulted from glacial erosion.**

4.4 Glacial Deposits

Distinguish between the two basic types of glacial deposits and briefly describe the features associated with each type.

KEY TERMS glacial drift, till, stratified drift, glacial erratic, lateral moraine, medial moraine, end moraine, ground moraine, outwash plain, valley train, kettle, drumlin, esker, kame

- Any sediment of glacial origin is called drift. The two distinct types of glacial drift are (a) till, which is unsorted material deposited directly by the ice, and (b) stratified drift, which is sediment sorted and deposited by meltwater from a glacier.
- The most widespread features created by glacial deposition are layers or ridges of till, called moraines. Associated with valley glaciers are lateral moraines, formed along the sides of the valley, and medial moraines, formed between two valley glaciers that have merged. End moraines, which mark the former position of the front of a glacier, and ground moraines, undulating layers of till deposited as the ice front retreat, are common to both valley glaciers and ice sheets.

? **Examine the illustration of depositional features left behind by a retreating ice sheet. Identify the features and indicate which landforms are composed of till and which are composed of stratified drift.**

4.5 Other Effects of Ice Age Glaciers

Describe and explain several important effects of Ice Age glaciers other than erosional and depositional landforms.

KEY TERMS proglacial lake, pluvial lake

- In addition to erosional and depositional features, other effects of Ice Age glaciers include the forced migration of organisms and adjustments of the crust by rebounding upward after removal or the immense load of ice.
- Advance and retreat of ice sheets caused significant changes to the paths followed by rivers. Proglacial lakes formed when glaciers acted as dams to create lakes by trapping glacial meltwater or blocking rivers. In response to the cooler and wetter glacial climate, pluvial lakes formed in areas such as present-day Nevada.
- Ice sheets are nourished by water that ultimately comes from the ocean, so when ice sheets grow, sea level falls, and when they melt, sea level rises.

4.6 Extent of Ice Age Glaciation

Discuss the extent of glaciation and climate variability during the Quaternary Ice Age.

KEY TERM Quaternary period

- The Ice Age, which began between 2 and 3 million years ago, was a complex period characterized by numerous advances and withdrawals of glacial ice. Most of the major glacial episodes occurred during a span on the geologic time scale called the Quaternary period. Evidence for the occurrence of several glacial advances during the Ice Age is the existence of multiple layers of drift on land and an uninterrupted record of climate cycles preserved in seafloor sediments.

4.7 Deserts

Describe the general distribution and causes of Earth's dry lands and the role that water plays in modifying desert landscapes.

KEY TERMS dry climate, desert, steppe, ephemeral stream

- Dry climates cover about 30 percent of Earth's land area. These regions have yearly precipitation totals that are less than the potential loss of water through evaporation. Deserts are drier than steppes, but both climate types are considered water deficient.
- Dry regions in the lower latitudes coincide with zones of subsiding air and high air pressure known as subtropical highs. Middle-latitude deserts exist because of their positions in the deep interiors of large continents far removed from oceans. Mountains also act to shield these regions from humid marine air masses.
- Practically all desert streams are dry most of the time and are said to be ephemeral. Nevertheless, running water is responsible for most of the erosional work in a desert. Although wind erosion is more significant in dry areas than elsewhere, the main role of wind in a desert is transportation and deposition of sediment.

4.8 Basin and Range: The Evolution of a Mountainous Desert Landscape

Discuss the stages of landscape evolution in the Basin and Range region of the western United States.

KEY TERMS interior drainage, alluvial fan, bajada, playa lake

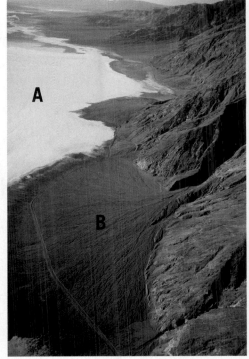

Michael Collier

- The Basin and Range region of the western United States is characterized by interior drainage with streams eroding uplifted mountain blocks and depositing sediment in interior basins. Alluvial fans, bajadas, playas, playa lakes, salt flats, and inselbergs are features often associated with these landscapes.

? Identify the lettered features in this photo. How did they form?

4.9 Wind Erosion

Describe the ways that wind transports sediment and the features created by wind erosion.

KEY TERMS deflation, blowout, desert pavement

- For wind erosion to be effective, dryness and scant vegetation are essential. Deflation, the lifting and removal of loose material, often produces shallow depressions called blowouts and can also lower the surface by removing sand and silt.
- Abrasion, the "sandblasting" effect of wind, is often given too much credit for producing desert features. However, abrasion does cut and polish rock near the surface.

? What term is applied to the layer of coarse pebbles covering this desert surface? How might it have formed?

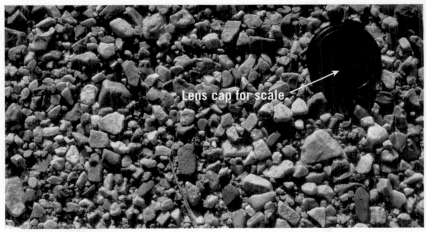

Lens cap for scale

Bobbé Christopherson

4.10 **Wind Deposits**

Distinguish between two basic types of wind deposits.

KEY TERMS loess, dune, slip face, cross bedding

- Wind deposits are of two distinct types: (1) extensive blankets of silt, called loess, that form when wind deposits its suspended load, and (2) mounds and ridges of sand, called dunes, form when sediment that was carried as part of the wind's bed load accumulates.

GIVE IT **SOME THOUGHT**

1. What term is applied to the ice at the North Pole? What term best describes Greenland's ice? Are either or both considered glaciers?

2. The accompanying diagram shows the results of a classic experiment used to determine how glacial ice moves in a mountain valley. The experiment occurred over an 8-year span. Refer to the diagram and answer the following:

   a. What was the average yearly rate at which ice in the center of the glacier advanced?
   b. About how fast was the center of the glacier advancing *per day*?
   c. Calculate the average rate at which ice along the sides of the glacier moved forward.
   d. Why was the rate at the center different than the rate along the sides?

3. Studies have shown that during the Ice Age, the margins of some ice sheets advanced southward from the Hudson Bay region at rates ranging from about 50 to 320 meters per year.
   a. Determine the maximum amount of time required for an ice sheet to move from the southern end of Hudson Bay to the south shore of present-day Lake Erie, a distance of 1600 kilometers.
   b. Calculate the minimum number of years required for an ice sheet to move this distance.

4. The accompanying image shows the surface of a valley glacier in which the ice is fractured.
   a. What term is applied to fractures such as these?
   b. In what vertical zone do these breaks occur?
   c. Do the fractures likely extend to the base of the glacier? Explain.

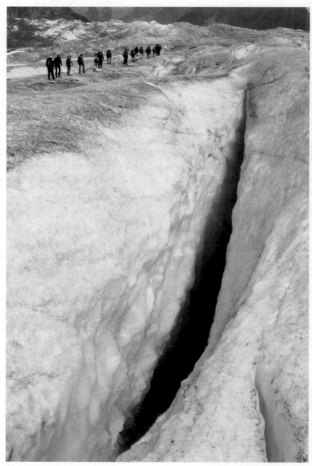

Glow Images

5. If the budget of a valley glacier were balanced for an extended time span, what feature would you expect to find at the terminus of the glacier? Is it composed of till or stratified drift? Now assume that the glacier's budget changes so that wastage exceeds accumulation. How would the terminus of the glacier change? Describe the deposit you would expect to form under these conditions.

6. Albuquerque, New Mexico, receives an average of 20.7 centimeters (8.07 inches) of rainfall annually. Albuquerque is considered a desert when the commonly used Köppen climate classification is applied. Yearly precipitation at the Russian city of Verkhoyansk averages 15.5 centimeters (6.05 inches), about 5 centimeters (2 inches) less than Albuquerque, yet is classified as a humid climate. Explain how this can be the case.

7. Assume that you and a nongeologist friend are visiting Alaska's Hubbard Glacier, shown here. After studying the glacier for quite a long time, your friend asks, "Do these things really move?" How would you convince your friend that this glacier does indeed move, using evidence that is clearly visible in this image?

Michael Collier

8. This photo shows an iceberg floating in the ocean near Greenland. (a) How do icebergs form? What term applies to this process? (b) Using the knowledge you have gained about these features, explain the common phrase "It's only the tip of the iceberg." (c) Is an iceberg the same as sea ice? Explain. (d) If this iceberg were to melt, how would sea level be affected?

Andrzej Gibasiewicz/Shutterstock

9. Bryce Canyon National Park, shown in the accompanying photo, is in dry southern Utah. It is carved into the eastern edge of the Paunsaugunt Plateau. Erosion has sculpted the colorful limestone into bizarre shapes, including spires called "hoodoos." As you and a companion (who has not studied geology) are taking a hike in the national park, your friend says, "It's amazing how wind has created this incredible scenery!" Now that you have studied arid landscapes, how would you respond to your companion's statement?

ozoptimist/Shutterstock

10. Is either of the following statements true? Are they both true? Explain your answer.
   a. Wind is more effective as an agent of erosion in dry places than in humid places.
   b. Wind is the most important agent of erosion in deserts.

11. The accompanying photo is an aerial view of Preston Mesa dunes in northern Arizona.

Michael Collier

   a. Sketch a simple profile (side view) of one of these dunes. Add an arrow to show the prevailing wind direction and label the dune's slip face.
   b. These dunes gradually migrate across the surface. Describe this process.

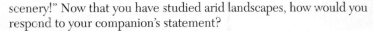

MasteringGeology™

www.masteringgeology.com

Looking for additional review and test prep materials? Visit the Study Area in MasteringGeology to enhance your understanding of this chapter's content by accessing a variety of resources, including **Self-Study Quizzes,**

Geoscience Animations, GEODe Tutorials, RSS feeds, flashcards, web links, self-study quizzes, and an optional **Pearson eText.**

5

Plate Tectonics: A Scientific Revolution Unfolds

Each statement represents the primary learning objective for the corresponding major heading within the chapter. After you complete the chapter, you should be able to:

5.1 Discuss the view that most geologists held prior to the 1960s regarding the geographic positions of the ocean basins and continents.

5.2 List and explain the evidence presented by Wegener to support his continental drift hypothesis.

5.3 Discuss the two main objections to the continental drift hypothesis.

5.4 List the major differences between Earth's lithosphere and asthenosphere and explain the importance of each in the plate tectonics theory.

5.5 Sketch and describe the movement along a divergent plate boundary that results in the formation of new oceanic lithosphere.

5.6 Compare and contrast the three types of convergent plate boundaries and name a location where each type can be found.

5.7 Describe the relative motion along a transform fault boundary and be able to locate several examples on a plate boundary map.

5.8 List the evidence used to support the plate tectonics theory and briefly describe it.

5.9 Summarize what is meant by plate–mantle convection and explain two of the primary driving forces of plate motion.

5.10 Explain why plates such as the African and Antarctic plates are getting larger, while the Pacific plate is getting smaller.

Climber ascending Chang Zheng Peak near Mount Everest. (Photo by Jimmy Chin/Aurora Open/SuperStock)

151

Plate tectonics is the first theory to provide a comprehensive view of the processes that produced Earth's major surface features, including the continents and ocean basins. Within the framework of this theory, geologists have found explanations for the basic causes and distribution of earthquakes, volcanoes, and mountain belts. Further, we are now better able to explain the distribution of plants and animals in the geologic past, as well as the distribution of economically significant mineral deposits.

5.1 From Continental Drift to Plate Tectonics

Discuss the view that most geologists held prior to the 1960s regarding the geographic positions of the ocean basins and continents.

Prior to the late 1960s, most geologists held the view that the ocean basins and continents had fixed geographic positions and were of great antiquity. Researchers came to realize that Earth's continents are not static; instead, they gradually migrate across the globe. Because of these movements, blocks of continental material collide, deforming the intervening crust, thereby creating Earth's great mountain chains (**Figure 5.1**). Furthermore, landmasses occasionally split apart. As continental blocks separate, a new ocean basin emerges between them. Meanwhile, other portions of the seafloor plunge into the mantle. In short, a dramatically different model of Earth's tectonic processes emerged. Tectonic processes are those that deform Earth's crust to create major structural features, such as mountains, continents, and ocean basins.

This profound reversal in scientific thought has been appropriately described as a *scientific revolution*. The revolution began early in the twentieth century as a relatively straightforward proposal called *continental drift*. For more than 50 years, the scientific establishment categorically rejected the idea that continents were capable of movement. Continental drift was particularly distasteful to North American geologists, perhaps because much of the supporting evidence had been gathered from the continents of Africa, South America, and Australia, with which most North American geologists were unfamiliar.

Figure 5.1 Rock pinnacle near Mount Blanc The Alps were created by the collision of the African and Eurasian plates. (Photo by Bildagentur Walhaeus)

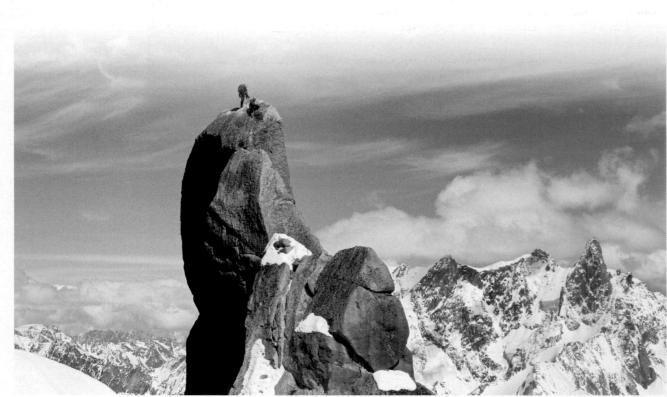

Following World War II, modern instruments replaced rock hammers as the tools of choice for many researchers. Armed with more advanced tools, geologists and a new breed of researchers, including *geophysicists* and *geochemists*, made several surprising discoveries that began to rekindle interest in the drift hypothesis. By 1968 these developments had led to the unfolding of a far more encompassing explanation known as the *theory of plate tectonics*.

In this chapter, we will examine the events that led to this dramatic reversal of scientific opinion. We will also briefly trace the development of the *continental drift hypothesis*, examine why it was initially rejected, and consider the evidence that finally led to the acceptance of its direct descendant—the theory of plate tectonics.

5.1 CONCEPT CHECKS

1. Briefly describe the view held by most geologists regarding the ocean basins and continents prior to the 1960s.
2. What group of geologists were the least receptive to the continental drift hypothesis? Explain.

5.2 Continental Drift: An Idea Before Its Time

List and explain the evidence presented by Wegener to support his continental drift hypothesis.

The idea that continents, particularly South America and Africa, fit together like pieces of a jigsaw puzzle came about during the 1600s, as better world maps became available. However, little significance was given to this notion until 1915, when Alfred Wegener (1880–1930), a German meteorologist and geophysicist, wrote *The Origin of Continents and Oceans*. This book set forth the basic outline of Wegener's hypothesis, called **continental drift**, which dared to challenge the long-held assumption that the continents and ocean basins had fixed geographic positions.

Wegener suggested that a single **supercontinent** consisting of all Earth's landmasses once existed.° He named this giant landmass **Pangaea** (pronounced "Pan-jee-ah," meaning "all lands") (**Figure 5.2**). Wegener further hypothesized that about 200 million years ago, during the early part of the Mesozoic era, this supercontinent began to fragment into smaller landmasses. These continental blocks then "drifted" to their present positions over a span of millions of years.

Wegener and others who advocated the continental drift hypothesis collected substantial evidence to support their point of view. The fit of South America and Africa and the geographic distribution of fossils and ancient climates all seemed to buttress the idea that these now separate landmasses were once joined. Let us examine some of this evidence.

Evidence: The Continental Jigsaw Puzzle

Like a few others before him, Wegener suspected that the continents might once have been joined when he noticed the remarkable similarity between the coastlines on opposite sides of the Atlantic Ocean. However, Wegener's use of present-day shorelines to fit these continents together was challenged immediately by other Earth scientists. These opponents correctly argued that shorelines are continually modified by wave erosion and depositional processes. Even if continental displacement had taken place, a good fit today would be unlikely. Because Wegener's original jigsaw fit of the continents was crude, it is assumed that he was aware of this problem (see Figure 5.2).

Scientists later determined that a much better approximation of the outer boundary of a continent is the seaward edge of its continental shelf, which lies submerged a few hundred meters below sea level. In the early 1960s, Sir Edward Bullard and two associates constructed a map that pieced together the edges of the

SmartFigure 5.2
Reconstructions of Pangaea Pangaea, as it is thought to have appeared 200 million years ago.

Modern reconstruction of Pangaea

Asia

S.E. Asia

Tethys Sea

North America

Africa

P A N G A E A

South America

India

Australia

Antarctica

Wegener's Pangaea, redrawn from his book published in 1912.

North America
Europe
Asia
Africa
South America
Australia
Antarctica

° Wegener was not the first person to conceive of a long-vanished supercontinent. Edward Suess (1831–1914), a distinguished nineteenth-century geologist, pieced together evidence for a giant landmass consisting of the continents of South America, Africa, India, and Australia.

Figure 5.3 Best fit of South America and Africa Drawing that shows the best fit of South America and Africa along the continental slope at a depth of 500 fathoms (about 900 meters [3000 feet]).

Continental shelf

Modern Equator

AFRICA

Modern Equator

SOUTH AMERICA

continental shelves of South America and Africa at a depth of about 900 meters (3000 feet) (**Figure 5.3**). The remarkable fit that was obtained was more precise than even these researchers had expected.

Evidence: Fossils Match Across the Seas

Although the seed for Wegener's hypothesis came from the remarkable similarities of the continental margins on opposite sides of the Atlantic, it was when he learned that identical fossil organisms had been discovered in rocks from both South America and Africa that his pursuit of continental drift became more focused. Through a review

Figure 5.4 Fossil evidence supporting continental drift Fossils of identical organisms have been discovered in rocks of similar age in Australia, Africa, South America, Antarctica, and India—continents that are currently widely separated by ocean barriers. Wegener accounted for these occurrences by placing these continents in their pre-drift locations.

of the literature, Wegener learned that most paleontologists (scientists who study the fossilized remains of ancient organisms) were in agreement that some type of land connection was needed to explain the existence of similar Mesozoic age life-forms on widely separated landmasses. Just as modern life-forms native to North America are quite different from those of Africa and Australia, during the Mesozoic era, organisms on widely separated continents should have been distinctly different.

Mesosaurus To add credibility to his argument, Wegener documented cases of several fossil organisms found on different landmasses, despite the unlikely possibility that their living forms could have crossed the vast ocean presently separating them (**Figure 5.4**). A classic example is *Mesosaurus*, a small aquatic freshwater reptile whose fossil remains are limited to black shales of the Permian period (about 260 million years ago) in eastern South America and southwestern Africa. If *Mesosaurus* had been able to make the long journey across the South Atlantic, its remains would likely be more widely distributed. As this is not the case, Wegener asserted that South America and Africa must have been joined during that period of Earth history.

How did opponents of continental drift explain the existence of identical fossil organisms in places separated by thousands of kilometers of open ocean? Rafting, transoceanic land bridges (isthmian links), and island stepping stones were the most widely invoked explanations for these migrations (**Figure 5.5**). We know, for example, that during the Ice Age that ended about 8,000 years ago, the lowering of sea level allowed mammals (including humans) to cross the narrow Bering Strait that separates Russia and Alaska. Was it possible that land bridges once connected Africa and South America but later subsided below sea level? Modern maps of the seafloor substantiate Wegener's contention that if land bridges of this magnitude once existed, their remnants would still lie below sea level.

Glossopteris Wegener also cited the distribution of the fossil "seed fern" *Glossopteris* as evidence for the existence of Pangaea (see Figure 5.4). This plant, identified by its tongue-shaped leaves and seeds that were too large to be carried by the wind, was known to be widely dispersed among Africa, Australia, India, and South America. Later, fossil remains of *Glossopteris* were also discovered in Antarctica.° Wegener also learned that these seed ferns and associated flora grew only in cool climates—similar to central Alaska. Therefore, he concluded that when these landmasses were joined, they were located much closer to the South Pole.

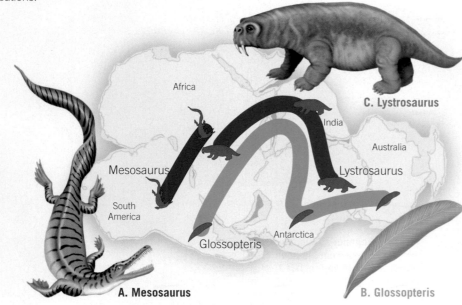

Africa

India

C. **Lystrosaurus**

Australia

Mesosaurus

Lystrosaurus

South America

Antarctica

Glossopteris

A. **Mesosaurus**

B. **Glossopteris**

° In 1912 Captain Robert Scott and two companions froze to death lying beside 35 pounds (16 kilograms) of rock on their return from a failed attempt to be the first to reach the South Pole. These samples, collected on the moraines of Beardmore Glacier, contained fossil remains of *Glossopteris*.

Figure 5.5 How do land animals cross vast oceans? These sketches illustrate various explanations for the occurrence of similar species on landmasses that are presently separated by vast oceans. (Reprinted with permission of John Holden)

Evidence: Rock Types and Geologic Features

Anyone who has worked a jigsaw puzzle knows that its successful completion requires that you fit the pieces together while maintaining the continuity of the picture.

The "picture" that must match in the "continental drift puzzle" is one of rock types and geologic features such as mountain belts. If the continents were together, the rocks found in a particular region on one continent should closely match in age and type those found in adjacent positions on the once adjoining continent. Wegener found evidence of 2.2-billion-year-old igneous rocks in Brazil that closely resembled similarly aged rocks in Africa.

Similar evidence can be found in mountain belts that terminate at one coastline and reappear on landmasses across the ocean. For instance, the mountain belt that includes the Appalachians trends northeastward through the eastern United States and disappears off the coast of Newfoundland (**Figure 5.6A**). Mountains of comparable age and structure are found in the British Isles and Scandinavia. When these landmasses are positioned as they were about 200 million years ago, as shown in **Figure 5.6B**, the mountain chains form a nearly continuous belt.

Wegener described how the similarities in geologic features on both sides of the Atlantic linked these landmasses when he said, "It is just as if we were to refit

Figure 5.6 Matching mountain ranges across the North Atlantic

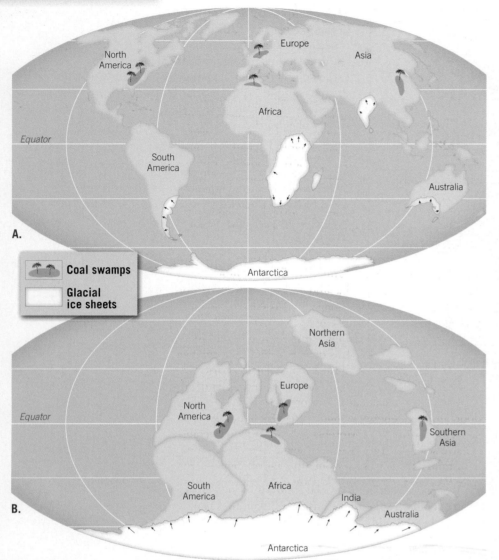

Figure 5.7 Paleoclimatic evidence for continental drift A. About 300 million years ago ice sheets covered extensive areas of the Southern Hemisphere and India. Arrows show the direction of ice movement that can be inferred from the pattern of glacial striations and grooves found in the bedrock. **B.** The continents restored to their pre-drift positions accounts for tropical coal swamps that existed in areas presently located in temperate climates.

the torn pieces of a newspaper by matching their edges and then check whether the lines of print run smoothly across. If they do, there is nothing left but to conclude that the pieces were in fact joined in this way."*

Evidence: Ancient Climates

Because Alfred Wegener was a student of world climates, he suspected that paleoclimatic (*paleo* = ancient, *climatic* = climate) data might also support the idea of mobile continents. His assertion was bolstered when he learned that evidence for a glacial period that dated to the late Paleozoic had been discovered in southern Africa, South America, Australia, and India. This meant that about 300 million years ago, vast ice sheets covered

extensive portions of the Southern Hemisphere as well as India (**Figure 5.7A**). Much of the land area that contains evidence of this period of Paleozoic glaciation presently lies within 30° of the equator in subtropical or tropical climates.

How could extensive ice sheets form near the equator? One proposal suggested that our planet experienced a period of extreme global cooling. Wegener rejected this explanation because during the same span of geologic time, large tropical swamps existed in several locations in the Northern Hemisphere. The lush vegetation in these swamps was eventually buried and converted to coal (**Figure 5.7B**). Today these deposits comprise major coal fields in the eastern United States and Northern Europe. Many of the fossils found in these coal-bearing rocks were produced by tree ferns that possessed large fronds—a feature consistent with warm, moist climates.** The existence of these large tropical swamps, Wegener argued, was inconsistent with the proposals that extreme global cooling caused glaciers to form in what are currently tropical areas.

Wegener suggested that a more plausible explanation for the late Paleozoic glaciation was provided by the supercontinent of Pangaea. In this configuration, the southern continents are joined together and located near the South Pole (see Figure 5.7B). This would account for the conditions necessary to generate extensive expanses of glacial ice over much of these landmasses. At the same time, this geography would place today's northern continents nearer the equator and account for the tropical swamps that generated the vast coal deposits.

How does a glacier develop in hot, arid central Australia? How do land animals migrate across wide expanses of the ocean? As compelling as this evidence may have been, 50 years passed before most of the scientific community accepted the concept of continental drift and the logical conclusions to which it led.

---

* Alfred Wegener, *The Origin of Continents and Oceans*, translated from the 4th revised German ed. of 1929 by J. Birman (London: Methuen, 1966).

** It is important to note that coal can form in a variety of climates, provided that large quantities of plant life are buried.

5.3 The Great Debate

Discuss the two main objections to the continental drift hypothesis.

Wegener's proposal did not attract much open criticism until 1924, when his book was translated into English, French, Spanish, and Russian. From that point until his death in 1930, the drift hypothesis encountered a great deal of hostile criticism. The respected American geologist R. T. Chamberlain stated, "Wegener's hypothesis in general is of the foot-loose type, in that it takes considerable liberty with our globe, and is less bound by restrictions or tied down by awkward, ugly facts than most of its rival theories."

Rejection of the Drift Hypothesis

One of the main objections to Wegener's hypothesis stemmed from his inability to identify a credible mechanism for continental drift. Wegener proposed that gravitational forces of the Moon and Sun that produce Earth's tides were also capable of gradually moving the continents across the globe. However, the prominent physicist Harold Jeffreys correctly countered that tidal forces strong enough to move Earth's continents would have resulted in halting our planet's rotation, which, of course, has not happened.

Wegener also incorrectly suggested that the larger and sturdier continents broke through thinner oceanic crust, much as ice breakers cut through ice. However, no evidence existed to suggest that the ocean floor was weak enough to permit passage of the continents without the continents being appreciably deformed in the process.

In 1930 Wegener made his fourth and final trip to the Greenland Ice Sheet (**Figure 5.8**). Although the primary focus of this expedition was to study this great ice cap and its climate, Wegener continued to test his continental drift hypothesis. While returning from Eismitte, an experimental station located in the center of Greenland, Wegener perished along with his Greenland companion. His intriguing idea, however, did not die.

Why was Wegener unable to overturn the established scientific views of his day? Foremost was the fact that, although the central theme of Wegener's drift hypothesis was correct, it contained some incorrect details. For example, continents do not break through the ocean floor, and tidal energy is much too weak to cause continents to be displaced. Moreover, in order for any comprehensive scientific theory to gain wide acceptance, it must withstand critical testing from all areas of science. Despite Wegener's great contribution to our understanding

Alfred Wegener shown waiting out the 1912–1913 Arctic winter during an expedition to Greenland, where he made a 1200-kilometer traverse across the widest part of the island's ice sheet.

Figure 5.8 Alfred Wegener (Photo courtesy of Archive of Alfred Wegener Institute)

Figure 5.9 Plate movement causes destructive earthquakes Pisco, Peru, following a powerful earthquake on August 16, 2007. (Sergio Erday/epa/Corbis)

of Earth, not *all* of the evidence supported the continental drift hypothesis as he had proposed it.

Although many of Wegener's contemporaries opposed his views, even to the point of open ridicule, some considered his ideas plausible. For those geologists who continued the search, the exciting concept of continents adrift held their interest. Others viewed continental drift as a solution to previously unexplainable observations (**Figure 5.9**). Nevertheless, most of the scientific community, particularly in North America, either categorically rejected continental drift or treated it with considerable skepticism.

5.4 The Theory of Plate Tectonics

List the major differences between Earth's lithosphere and asthenosphere and explain the importance of each in the plate tectonics theory.

Following World War II, oceanographers equipped with new marine tools and ample funding from the U.S. Office of Naval Research embarked on an unprecedented period of oceanographic exploration. Over the next two decades, a much better picture of large expanses of the seafloor slowly and painstakingly began to emerge. From this work came the discovery of a global **oceanic ridge system** that winds through all the major oceans in a manner similar to the seams on a baseball.

In other parts of the ocean, more new discoveries were being made. Studies conducted in the western Pacific demonstrated that earthquakes were occurring at great depths beneath deep-ocean trenches. Of equal importance was the fact that dredging of the seafloor did not bring up any oceanic crust that was older than 180 million years. Further, sediment accumulations in the deep-ocean basins were found to be thin, not the thousands of meters that were predicted. By 1968 these developments, among others, had led to the unfolding of a far more encompassing theory than continental drift, known as the **theory of plate tectonics** (*tekto* = to build).

Rigid Lithosphere Overlies Weak Asthenosphere

According to the plate tectonics model, the crust and the uppermost, and therefore coolest, part of the mantle constitute Earth's strong outer layer, known as the **lithosphere** (*lithos* = stone, *sphere* = ball). The lithosphere varies in both thickness and density, depending on whether it is oceanic lithosphere or continental lithosphere (**Figure 5.10**). Oceanic lithosphere is about 100 kilometers (60 miles) thick in the deep-ocean basins but is considerably thinner along the crest of the oceanic ridge system—a topic we will consider later. By contrast, continental lithosphere averages about 150 kilometers (90 miles) thick but may extend to depths of 200 kilometers (125 miles) or more beneath the stable interiors of the continents. Further, the composition of both the oceanic and continental crusts affects their respective densities. Oceanic crust is composed of rocks that have a mafic (basaltic) composition, and therefore oceanic lithosphere has a greater density than continental lithosphere. Continental crust is composed largely of less dense felsic (granitic) rocks, making continental lithosphere less dense than its oceanic counterpart.

The **asthenosphere** (*asthenos* = weak, *sphere* = ball) is a hotter, weaker region in the mantle that lies

Figure 5.11 Earth's major lithospheric plates

SmartFigure 5.10 Rigid Lithosphere overlies the weak asthenosphere

below the lithosphere (see Figure 5.10). The temperatures and pressures in the upper asthenosphere (100 to 200 kilometers in depth) are such that rocks at this depth are very near their melting temperatures and, hence, respond to forces by *flowing*, similarly to the way a thick liquid would flow. By contrast, the relatively cool and rigid lithosphere tends to respond to forces acting on it by *bending or breaking but not flowing*. Because of these differences, Earth's rigid outer shell is effectively detached from the asthenosphere, which allows these layers to move independently.

Earth's Major Plates

The lithosphere is broken into about two dozen segments of irregular size and shape called **lithospheric plates**, or simply **plates**, that are in constant motion with respect to one another (**Figure 5.11**). Seven major lithospheric plates are recognized and account for 94 percent of Earth's surface area: the *North American*, *South American*, *Pacific*, *African*, *Eurasian*, *Australian-Indian*, and *Antarctic plates*. The largest is the Pacific plate, which encompasses a significant portion of the Pacific basin. The six other large plates include an entire continent plus a significant amount of ocean floor. Notice in **Figure 5.12** that the South American plate encompasses almost all of South America and about one-half of the floor of the South Atlantic. This is a major departure from Wegener's

Figure 5.12 Divergent, convergent, and transform plate boundaries

continental drift hypothesis, which proposed that the continents move through the ocean floor, not with it. Note also that none of the plates are defined entirely by the margins of a single continent.

Intermediate-sized plates include the *Caribbean, Nazca, Philippine, Arabian, Cocos, Scotia,* and *Juan de Fuca plates.* These plates, with the exception of the Arabian plate, are composed mostly of oceanic lithosphere. In addition, several smaller plates (*microplates*) have been identified but are not shown in Figure 5.11.

Plate Boundaries

One of the main tenets of the plate tectonics theory is that plates move as somewhat rigid units relative to all other plates. As plates move, the distance between two locations on different plates, such as New York and London, gradually changes, whereas the distance between sites on the same plate—New York and Denver, for example—remains relatively constant. However, parts of some plates are comparatively "soft," such as southern

China, which is literally being squeezed as the Indian subcontinent rams into Asia proper.

Because plates are in constant motion relative to each other, most major interactions among them (and, therefore, most deformation) occur along their *boundaries*. In fact, plate boundaries were first established by plotting the locations of earthquakes and volcanoes. Plates are bounded by three distinct types of boundaries, which are differentiated by the type of movement they exhibit. These boundaries are depicted in Figure 5.12 and are briefly described here:

1. Divergent plate boundaries (*constructive margins*)—where two plates move apart, resulting in upwelling of hot material from the mantle to create new seafloor (**Figure 5.12A**).

2. Convergent plate boundaries (*destructive margins*)—where two plates move together, resulting in oceanic lithosphere descending beneath an overriding plate, eventually to be reabsorbed into the mantle or possibly in the collision of two continental blocks to create a mountain belt (**Figure 5.12B**).

3. Transform plate boundaries (*conservative margins*)—where two plates grind past each other without the production or destruction of lithosphere (**Figure 5.12C**).

Divergent and convergent plate boundaries each account for about 40 percent of all plate boundaries. Transform faults account for the remaining 20 percent. In the following sections we will summarize the nature of the three types of plate boundaries.

5.5 Divergent Plate Boundaries and Seafloor Spreading

Sketch and describe the movement along a divergent plate boundary that results in the formation of new oceanic lithosphere.

Most **divergent plate boundaries** (*di* = apart, *vergere* = to move) are located along the crests of oceanic ridges and can be thought of as *constructive plate margins* because this is where new ocean floor is generated (**Figure 5.13**). Here, two adjacent plates move away from each other, producing long, narrow fractures in the ocean crust. As a result, hot rock from the mantle below migrates upward to fill the voids left as the crust is being ripped apart. This molten material gradually cools, producing new slivers of seafloor. In a slow and unending manner, adjacent plates spread apart, and new oceanic lithosphere forms between them. For this reason divergent plate boundaries are also referred to as **spreading centers**.

Oceanic Ridges and Seafloor Spreading

The majority of, but not all, divergent plate boundaries are associated with *oceanic ridges*: elevated areas of the seafloor characterized by high heat flow and volcanism.

Figure 5.13 Seafloor spreading Most divergent plate boundaries are situated along the crests of oceanic ridges—the sites of seafloor spreading.

The global ridge system is the longest topographic feature on Earth's surface, exceeding 70,000 kilometers (43,000 miles) in length. As shown in Figure 5.12, various segments of the global ridge system have been named, including the Mid-Atlantic Ridge, East Pacific Rise, and Mid-Indian Ridge.

Representing 20 percent of Earth's surface, the oceanic ridge system winds through all major ocean basins like the seams on a baseball. Although the crest of the oceanic ridge is commonly 2 to 3 kilometers higher than the adjacent ocean basins, the term *ridge* may be misleading because it implies "narrow" when, in fact, ridges vary in width from 1000 kilometers (600 miles) to more than 4000 kilometers (2500 miles). Further, along the crest of some ridge segments is a deep canyonlike structure called a **rift valley** (**Figure 5.14**). This structure is evidence that tensional forces are actively pulling the ocean crust apart at the ridge crest.

The mechanism that operates along the oceanic ridge system to create new seafloor is appropriately called **seafloor spreading**. Typical rates of spreading average around 5 centimeters (2 inches) per year, roughly the same rate at which human fingernails grow. Comparatively slow spreading rates of 2 centimeters per year are found along the Mid-Atlantic Ridge, whereas spreading rates exceeding 15 centimeters (6 inches) per year have been measured along sections of the East Pacific Rise. Although these rates of seafloor production are slow on a human time scale, they are nevertheless rapid enough to have generated all of Earth's ocean basins within the past 200 million years.

The primary reason for the elevated position of the oceanic ridge is that newly created oceanic lithosphere is hot, which means it is less dense than cooler rocks found away from the ridge axis. (Geologists use the term *axis* to refer to a line that follows the general trend of the ridge crest.) As soon as new lithosphere forms, it is slowly yet

Figure 5.14 Rift valley Thingvellir National Park, Iceland, is located on the western margin of a rift valley that is roughly 30 kilometers (20 miles) wide in this region. This rift valley is connected to a similar feature that extends along the crest of the Mid-Atlantic Ridge. The cliff in the left half of the image approximates the eastern edge of the North American plate. (Photo by Ragnar Sigurdsson/Arctic/Alamy)

continually displaced away from the zone of upwelling. Thus, it begins to cool and contract, thereby increasing in density. This thermal contraction accounts for the increase in ocean depths away from the ridge crest. It takes about 80 million years for the temperature of oceanic lithosphere to stabilize and contraction to cease. By this time, rock that was once part of the elevated oceanic ridge system is located in the deep-ocean basin, where it may be buried by substantial accumulations of sediment.

In addition, as the plate moves away from the ridge, cooling of the underlying asthenosphere causes it to become increasingly more rigid. Thus, oceanic lithosphere is generated by cooling of the asthenosphere from the top down. Stated another way, the thickness of oceanic lithosphere is age dependent. The older (cooler) it is, the greater its thickness. Oceanic lithosphere that exceeds 80 million years in age is about 100 kilometers (60 miles) thick—approximately its maximum thickness.

Continental Rifting

Divergent boundaries can develop within a continent, in which case the landmass may split into two or more smaller segments separated by an ocean basin. Continental rifting begins when plate motions produce opposing (tensional) forces that pull and stretch the lithosphere. Stretching, in turn, thins the lithosphere, which promotes mantle upwelling and broad upwarping of the overlying lithosphere (**Figure 5.15A**). During this process, the lithosphere is thinned, while the brittle crustal rocks break into large blocks. As the tectonic forces continue to pull apart the crust, the broken crustal fragments sink, generating an elongated depression called a **continental rift**, which eventually widens to form a narrow sea (**Figure 5.15C**) and then a new ocean basin (**Figure 5.15D**).

A modern example of an active continental rift is the East African Rift (**Figure 5.16**). Whether this rift will

SmartFigure 5.15
Continental rifting: Formation of new ocean basins

A. Continental rifting occurs where plate motions produce opposing tensional forces that thin the lithosphere and promote upwelling in the mantle.

B. Stretching causes the brittle crust to break into large blocks that sink, generating a rift valley.

C. Continued spreading generates a long narrow sea similar to the present-day Red Sea.

D. Eventually, an expansive deep-ocean basin containing a centrally located oceanic ridge is formed by continued seafloor spreading.

Figure 5.16 East African Rift valley

eventually result in the breakup of Africa is a topic of continued research. Nevertheless, the East African Rift is an excellent model of the initial stage in the breakup of a continent. Here, tensional forces have stretched and thinned the lithosphere, allowing molten rock to ascend from the mantle. Evidence for recent volcanic activity includes several large volcanic mountains, including Mount Kilimanjaro and Mount Kenya, the tallest peaks in Africa. Research suggests that if rifting continues, the rift valley will lengthen and deepen (Figure 5.15C). At some point, the rift valley will become a narrow sea with an outlet to the ocean. The Red Sea, which formed when the Arabian Peninsula split from Africa, is a modern example of such a feature and provides us with a view of how the Atlantic Ocean may have looked in its infancy (Figure 5.15D).

5.5 CONCEPT CHECKS

1. Sketch or describe how two plates move in relation to each other along divergent plate boundaries.
2. What is the average rate of seafloor spreading in modern oceans?
3. List four facts that characterize the oceanic ridge system.
4. Briefly describe the process of continental rifting. Where is it occurring today?

5.6 Convergent Plate Boundaries and Subduction

Compare and contrast the three types of convergent plate boundaries and name a location where each type can be found.

New lithosphere is constantly being produced at the oceanic ridges. However, our planet is not growing larger; its total surface area remains constant. A balance is maintained because older, denser portions of oceanic lithosphere descend into the mantle at a rate equal to seafloor production. This activity occurs along **convergent plate boundaries**, where two plates move toward each other and the leading edge of one is bent downward, as it slides beneath the other (**Figure 5.17**).

Convergent boundaries are also called **subduction zones** because they are sites where lithosphere is descending (being subducted) into the mantle. Subduction occurs because the density of the descending lithospheric plate is greater than the density of the underlying asthenosphere. In general, old oceanic lithosphere is about 2 percent more dense than the underlying asthenosphere, which causes it to subduct. Continental lithosphere, in contrast, is less dense and resists subduction. As a consequence, only oceanic lithosphere will subduct to great depths.

Deep-ocean trenches are the surface manifestations produced as oceanic lithosphere descends into the mantle (**Figure 5.18**). These large linear depressions

are remarkably long and deep. The Peru–Chile trench along the west coast of South America is more than 4500 kilometers (3000 miles) long, and its base is as much as 8 kilometers (5 miles) below sea level. The trenches in the western Pacific, including the Mariana and Tonga trenches, tend to be even deeper than those of the eastern Pacific.

Slabs of oceanic lithosphere descend into the mantle at angles that vary from a few degrees to nearly vertical (90 degrees). The angle at which oceanic lithosphere subducts depends largely on its age and therefore its density. For example, when seafloor spreading occurs near a subduction zone, as is the case along the coast of Chile, the subducting lithosphere is young and buoyant, which results in a low angle of descent. As the two plates converge, the overriding plate scrapes over the top of the subducting plate below—a type of forced subduction. Consequently, the region around the Peru–Chile trench experiences great earthquakes, including the 2010 Chilean earthquake—one of the 10 largest on record.

As oceanic lithosphere ages (gets farther from the spreading center), it gradually cools, which causes it to thicken and increase in density. In parts of the western

Pacific, some oceanic lithosphere is 180 million years old—the thickest and densest in today's oceans. The very dense slabs in this region typically plunge into the mantle at angles approaching 90 degrees. This largely explains the fact that most trenches in the western Pacific are deeper than trenches in the eastern Pacific.

Although all convergent zones have the same basic characteristics, they may vary considerably, based on the type of crustal material involved and the tectonic setting. Convergent boundaries can form between *two oceanic plates*, *one oceanic plate and one continental plate*, or *two continental plates*.

Oceanic–Continental Convergence

When the leading edge of a plate capped with continental crust converges with a slab of oceanic lithosphere, the buoyant continental block remains "floating," while the denser oceanic slab sinks into the mantle (see Figure 5.17A). When a descending oceanic slab reaches a depth of about 100 kilometers (60 miles), melting is triggered within the wedge of hot asthenosphere that lies above it. But how does the subduction of a cool slab of oceanic lithosphere cause mantle rock to melt? The answer lies in the fact that water contained in the descending plates acts the way salt does to melt ice. That is, "wet" rock in a high-pressure environment melts at substantially lower temperatures than does "dry" rock of the same composition.

Sediments and oceanic crust contain large amounts of water, which is carried to great depths by a subducting plate. As the plate plunges downward, heat and pressure drive water from the voids in the rock. At a depth of roughly 100 kilometers (60 miles), the wedge of mantle rock is sufficiently hot that the introduction of water from the slab below leads to some melting. This process, called **partial melting**, is thought to generate some molten material, which is mixed with unmelted mantle rock. Being less dense than the surrounding mantle, this hot mobile material gradually rises toward the surface. Depending on the environment, these mantle-derived masses of molten rock may ascend through the crust and give rise to a volcanic eruption. However, much of this material never reaches the surface; rather, it solidifies at depth—a process that thickens the crust.

The volcanoes of the towering Andes are the product of molten rock generated by the subduction of the Nazca plate beneath the South American continent (see Figure 5.12). Mountain systems, such as the Andes, which are produced in part by volcanic activity associated with the subduction of oceanic lithosphere, are called **continental volcanic arcs**. The Cascade Range in Washington, Oregon, and California is another mountain system that consists of several well-known volcanic mountains, including Mount Rainier, Mount Shasta,

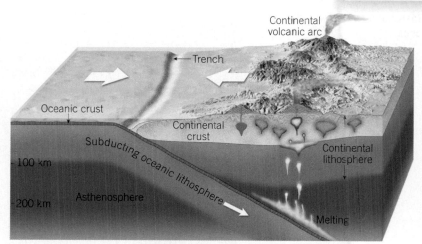

A. Convergent plate boundary where oceanic lithosphere is subducting beneath continental lithosphere.

B. Convergent plate boundary involving two slabs of oceanic lithosphere.

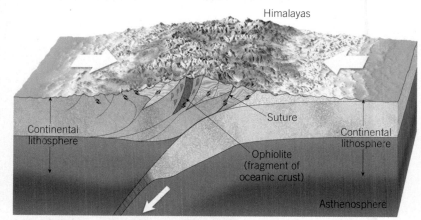

C. Continental collisions occur along convergent plate boundaries when both plates are capped with continental crust.

SmartFigure 5.17
Three types of convergent plate boundaries

Mount St. Helens, and Mount Hood (**Figure 5.19**). This active volcanic arc also extends into Canada, where it includes Mount Garibaldi, Mount Silverthrone, and others.

Figure 5.18 Distribution of the world's oceanic trenches

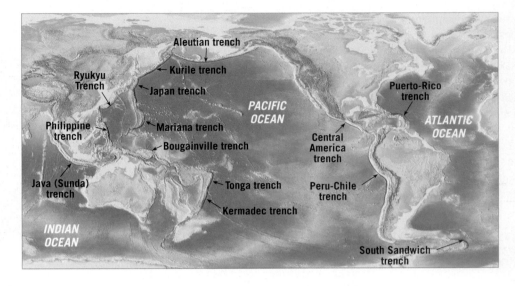

Figure 5.19 Oceanic–continental convergent plate boundary Mount Hood, Oregon, is one of more than a dozen large composite volcanoes in the Cascade Range.

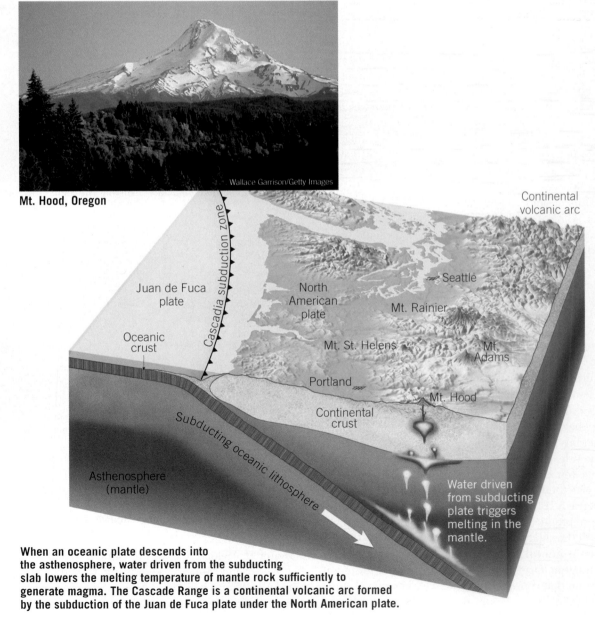

Mt. Hood, Oregon

When an oceanic plate descends into the asthenosphere, water driven from the subducting slab lowers the melting temperature of mantle rock sufficiently to generate magma. The Cascade Range is a continental volcanic arc formed by the subduction of the Juan de Fuca plate under the North American plate.

Oceanic–Oceanic Convergence

An *oceanic–oceanic convergent boundary* has many features in common with oceanic–continental plate margins. Where two oceanic slabs converge, one descends beneath the other, initiating volcanic activity by the same mechanism that operates at all subduction zones (see Figure 5.12). Water squeezed from the subducting slab of oceanic lithosphere triggers melting in the hot wedge of mantle rock above. In this setting, volcanoes grow up from the ocean floor rather than on a continental platform. When subduction is sustained, it will eventually build a chain of volcanic structures large enough to emerge as islands. The newly formed land consisting of an arc-shaped chain of volcanic islands is called a **volcanic island arc**, or simply an **island arc** (**Figure 5.20**).

The Aleutian, Mariana, and Tonga islands are examples of relatively young volcanic island arcs. Island arcs are generally located 100 to 300 kilometers (60 to 200 miles) from a deep-ocean trench. Located adjacent to the island arcs just mentioned are the Aleutian trench, the Mariana trench, and the Tonga trench.

Most volcanic island arcs are located in the western Pacific. Only two are located in the Atlantic—the Lesser Antilles arc, on the eastern margin of the Caribbean Sea, and the Sandwich Islands, located off the tip of South America. The Lesser Antilles are a product of the subduction of the Atlantic seafloor beneath the Caribbean plate. Located within this volcanic arc are the Virgin Islands of the United States and Britain as well as the island of Martinique, where Mount Pelée erupted in 1902, destroying the town of St. Pierre and killing an estimated 28,000 people. This chain of islands also includes Montserrat, where there has been recent volcanic activity. More on these volcanic events is found in Chapter 7.

Island arcs are typically simple structures made of numerous volcanic cones underlain by oceanic crust that is generally less than 20 kilometers (12 miles) thick. By contrast, some island arcs are more complex and are underlain by highly deformed crust that may reach

Figure 5.20 Volcanoes of the Aleutian Islands

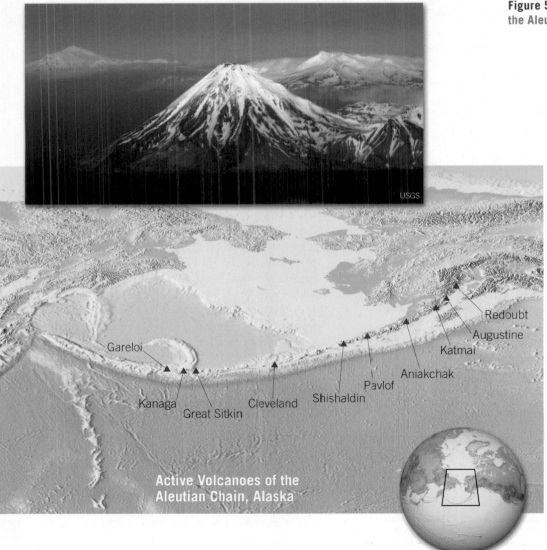

USGS

Gareloi

Redoubt

Augustine

Katmai

Aniakchak

Pavlof

Kanaga   Cleveland   Shishaldin

Great Sitkin

Active Volcanoes of the Aleutian Chain, Alaska

Figure 5.21 Collision of India and Eurasia

The ongoing collision of the subcontinent of India with Eurasia began about 50 million years ago—producing the majestic Himalayas.

35 kilometers (22 miles) in thickness. Examples include Japan, Indonesia, and the Alaskan Peninsula. These island arcs are built on material generated by earlier episodes of subduction or on small slivers of continental crust that have rafted away from the mainland.

Continental–Continental Convergence

The third type of convergent boundary results when one landmass moves toward the margin of another because of subduction of the intervening seafloor (**Figure 5.21A**). Whereas oceanic lithosphere tends to be dense and sink into the mantle, the buoyancy of continental material inhibits it from being subducted. Consequently, a collision between two converging continental fragments ensues (**Figure 5.21B**). This event folds and deforms the accumulation of sediments and sedimentary rocks along the continental margins as if they had been placed in a gigantic vise. The result is the formation of a new mountain belt composed of deformed sedimentary and metamorphic rocks that often contain slivers of oceanic crust.

Such a collision began about 50 million years ago, when the subcontinent of India "rammed" into Asia, producing the Himalayas—the most spectacular mountain range on Earth (see Figure 5.21B). During this collision, the continental crust buckled and fractured and was generally shortened horizontally and thickened vertically. In addition to the Himalayas, several other major mountain systems, including the Alps, Appalachians, and Urals, formed as continental fragments collided. This topic will be considered further in Chapter 6.

5.6 CONCEPT CHECKS

1. Explain why the rate of lithosphere production roughly balances with the rate of lithosphere destruction.
2. Compare a continental volcanic arc and a volcanic island arc.
3. Describe the process that leads to the formation of deep-ocean trenches.
4. Why does oceanic lithosphere subduct, while continental lithosphere does not?
5. Briefly describe how mountain belts such as the Himalayas form.

5.7 Transform Plate Boundaries

Describe the relative motion along a transform fault boundary and be able to locate several examples on a plate boundary map.

Along a **transform plate boundary**, also called a **transform fault**, plates slide horizontally past one another without the production or destruction of lithosphere. The nature of transform faults was discovered in 1965 by Canadian geologist J. Tuzo Wilson, who proposed that these large faults connected two spreading centers (divergent boundaries) or, less commonly, two trenches (convergent boundaries). Most transform faults are found on the ocean floor, where they offset segments of the oceanic ridge system, producing a steplike plate margin (**Figure 5.22A**).

Notice that the zigzag shape of the Mid-Atlantic Ridge in Figure 5.12 roughly reflects the shape of the original rifting that caused the breakup of the supercontinent of Pangaea. (Compare the shapes of the continental margins of the landmasses on both sides of the Atlantic with the shape of the Mid-Atlantic Ridge.)

Typically, transform faults are part of prominent linear breaks in the seafloor known as **fracture zones**, which include both active transform faults as well as their inactive extensions into the plate interior (**Figure 5.22B**).

Fracture zones are long, narrow scar-like features in the seafloor that are roughly perpendicular to the offset ridge segments. They include both the active transform fault and its "fossilized" trace.

Fracture zone

Inactive zone

Transform fault (active)

Inactive zone

Oceanic crust

Most transform faults offset segments of a spreading center, producing a step-like plate margin.

Mid-Atlantic Ridge

South America

Africa

B.

KEY
Spreading centers \
Fracture zones ——
Transform faults ⇄

A. The Mid-Atlantic Ridge, with its zigzag pattern, roughly reflects the shape of the rifting zone which resulted in the break-up of Pangaea.

SmartFigure 5.22 Transform plate boundaries

Figure 5.23 Transform faults facilitate plate motion Seafloor spreading generated along the Juan de Fuca Ridge moves southeastward, past the Pacific plate. Eventually it subducts beneath the North American plate. Thus, this transform fault connects a spreading center (divergent boundary) to a subduction zone (convergent boundary). Also shown is the San Andreas Fault, a transform fault that connects two spreading centers: the Juan de Fuca Ridge and a spreading center located in the Gulf of California.

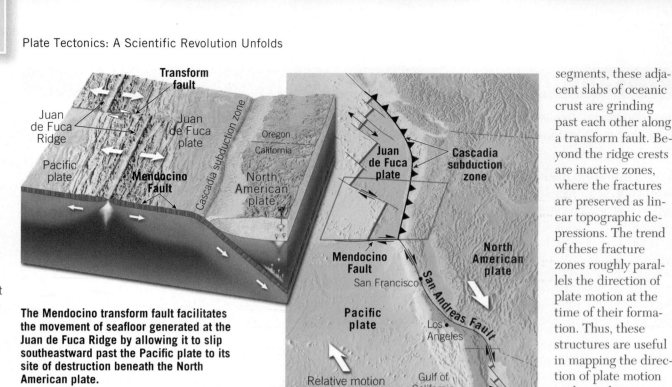

The Mendocino transform fault facilitates the movement of seafloor generated at the Juan de Fuca Ridge by allowing it to slip southeastward past the Pacific plate to its site of destruction beneath the North American plate.

segments, these adjacent slabs of oceanic crust are grinding past each other along a transform fault. Beyond the ridge crests are inactive zones, where the fractures are preserved as linear topographic depressions. The trend of these fracture zones roughly parallels the direction of plate motion at the time of their formation. Thus, these structures are useful in mapping the direction of plate motion in the geologic past.

In another role, transform faults provide the means

Active transform faults lie *only between* the two offset ridge segments and are generally defined by weak, shallow earthquakes. Here seafloor produced at one ridge axis moves in the opposite direction of seafloor produced at an opposing ridge segment. Thus, between the ridge

by which the oceanic crust created at ridge crests can be transported to a site of destruction—the deep-ocean trenches. **Figure 5.23** illustrates this situation. Notice that the Juan de Fuca plate moves in a southeasterly direction, eventually being subducted under

Figure 5.24 Movement along the San Andreas Fault This aerial view shows the offset in the dry channel of Wallace Creek near Taft, California.

the west coast of the United States. The southern end of this plate is bounded by a transform fault called the Mendocino Fault. This transform boundary connects the Juan de Fuca Ridge to the Cascadia subduction zone. Therefore, it facilitates the movement of the crustal material created at the Juan de Fuca Ridge to its destination beneath the North American continent.

Like the Mendocino Fault, most other transform fault boundaries are located within the ocean basins; however, a few cut through continental crust. Two examples are the earthquake-prone San Andreas Fault of California and New Zealand's Alpine Fault. Notice in Figure 5.23 that the San Andreas Fault connects a spreading center located in the Gulf of California to the Cascadia subduction zone and the Mendocino Fault located along the northwest coast of

the United States. Along the San Andreas Fault, the Pacific plate is moving toward the northwest, past the North American plate (**Figure 5.24**). If this movement continues, the part of California west of the fault zone, including the Baja Peninsula of Mexico, will become an island off the west coast of the United States and Canada—with the potential to eventually reach Alaska. However, a more immediate concern is the earthquake activity triggered by movements along this fault system.

5.7 CONCEPT CHECKS

1. Sketch or describe how two plates move in relation to each other along a transform plate boundary.
2. Differentiate between transform faults and the two other types of plate boundaries.

5.8 Testing the Plate Tectonics Model

List the evidence used to support the plate tectonics theory and briefly describe it.

Some of the evidence supporting continental drift has already been presented. With the development of the theory of plate tectonics, researchers began testing this new model of how Earth works. Although new supporting data were obtained, it has often been new interpretations of already existing data that have swayed the tide of opinion.

Evidence: Ocean Drilling

Some of the most convincing evidence for seafloor spreading came from the Deep Sea Drilling Project, which operated from 1968 until 1983. One of the early goals of the project was to gather samples of the ocean floor in order to establish its age. To accomplish this, the *Glomar Challenger*, a drilling ship capable of working in water thousands of meters deep, was built. Hundreds of holes were drilled through the layers of sediments that blanket the ocean crust, as well as into the basaltic rocks below. Rather than radiometrically dating the crustal rocks, researchers used the fossil remains of microorganisms found in the sediments resting directly on the crust to date the seafloor at each site. Radiometric dates of the ocean crust itself are unreliable because of the alteration of basalt by seawater.

When the oldest sediment from each drill site was plotted against its distance from the ridge crest, the plot showed that the sediments increased in age with increasing distance from the ridge. This finding supported the

seafloor-spreading hypothesis, which predicted that the youngest oceanic crust would be found at the ridge crest, the site of seafloor production, and the oldest oceanic crust would be located adjacent to the continents.

The thickness of ocean-floor sediments provided additional verification of seafloor spreading. Drill cores from the *Glomar Challenger* revealed that sediments are almost entirely absent on the ridge crest and that sediment thickness increases with increasing distance from the ridge (**Figure 5.25**). This pattern of sediment distribution should be expected if the seafloor-spreading hypothesis is correct.

The data collected by the Deep Sea Drilling Project also reinforced the idea that the ocean basins are geologically young because no seafloor with an age in excess of 180 million years was found. By comparison, most continental crust exceeds several hundred million years in age, and some has been located that exceeds 4 billion years in age.

In October 2003, the *JOIDES Resolution* became part of a new program, the Integrated Ocean Drilling Program (IODP). This new international effort uses multiple vessels for exploration, including the massive 210-meter-long (nearly 770-foot-long) *Chikyu* (meaning "planet Earth" in Japanese), which began operations in 2007 (see Figure 5.25). One of the goals of the IODP is to recover a complete section of the ocean crust, from top to bottom.

Figure 5.25 Deep-sea drilling Data collected through deep-sea drilling have shown that the ocean floor is indeed youngest at the ridge axis.

Core samples show that the thickness of sediments increase with distance from the ridge crest.

Age of seafloor

Older   Younger   Older

Drilling ship collects core samples of seafloor sediments and basaltic crust

Ocean crust (basalt)

Chikyu is a state-of-the-art drilling ship designed to drill up to 7,000 meters (more than 4 miles) below the seafloor.

AP/Wide World

Evidence: Mantle Plumes and Hot Spots

Mapping volcanic islands and seamounts (submarine volcanoes) in the Pacific Ocean revealed several linear chains of volcanic structures. One of the most-studied chains consists of at least 129 volcanoes that extend from the Hawaiian islands to Midway Island and continue northwestward toward the Aleutian trench (**Figure 5.26**). Radiometric dating of this linear structure, called the *Hawaiian Island–Emperor Seamount chain*, showed that the volcanoes increase in age with increasing distance from the Big Island of Hawaii. The youngest volcanic island in the chain (Hawaii) rose from the ocean floor less than 1 million years ago, whereas Midway Island is 27 million years old, and Detroit Seamount, near the Aleutian trench, is about 80 million years old (see Figure 5.26).

Most researchers agree that a cylindrically shaped upwelling of hot rock, called a **mantle plume**, is located beneath the island of Hawaii. As the hot, rocky plume ascends through the mantle, the confining pressure drops, which triggers partial melting. (This process, called *decompression melting*, is discussed in Chapter 7.) The surface manifestation of this activity is a **hot spot**, an area

of volcanism, high heat flow, and crustal uplifting that is a few hundred kilometers across. As the Pacific plate moved over a hot spot, a chain of volcanic structures known as a **hot-spot track** was built. As shown in Figure 5.26, the age of each volcano indicates how much time has elapsed since it was situated over the mantle plume.

A closer look at the five largest Hawaiian islands reveals a similar pattern of ages, from the volcanically active island of Hawaii to the inactive volcanoes that make up the oldest island, Kauai (see Figure 5.26). Five million years ago, when Kauai was positioned over the hot spot, it was the *only* modern Hawaiian island in existence. Visible evidence of the age of Kauai can be seen by examining its extinct volcanoes, which have been eroded into jagged peaks and vast canyons. By contrast, the relatively young island of Hawaii exhibits many fresh lava flows, and one of its five major volcanoes, Kilauea, remains active today.

Evidence: Paleomagnetism

Anyone who has used a compass to find direction knows that Earth's magnetic field has north and south magnetic poles. Today these magnetic poles roughly align with

Figure 5.26 Hot spots and hot-spot tracks Radiometric dating of the Hawaiian islands shows that volcanic activity increases in age moving away from the Big Island of Hawaii.

Plate motion

Extinct volcano

Hot spot volcanism

Rising mantle plume

Kauai 3.8–5.6

Oahu 2.2–3.3

Molokai 1.3–1.8

Maui less than 1.0

Hawaii 0.7 to present

Detroit: 80 mya

Suiko: 60 mya

Emperor Seamount chain

Koko: 48 mya

Hawaiian chain

Midway Island 27 mya

Hawaii 0.7 mya

Ages given in millions of years

NASA

the geographic poles. (The geographic poles are located where Earth's rotational axis intersects the surface.) Earth's magnetic field is similar to that produced by a simple bar magnet. Invisible lines of force pass through the planet and extend from one magnetic pole to the other (**Figure 5.27**). A compass needle, itself a small magnet free to rotate on an axis, becomes aligned with the magnetic lines of force and points to the magnetic poles.

Earth's magnetic field is less obvious than the pull of gravity because humans cannot feel it. Movement of a compass needle, however, confirms its presence. In addition, some naturally occurring minerals are magnetic and are influenced by Earth's magnetic field. One of the most common is the iron-rich mineral *magnetite*, which is abundant in lava flows of basaltic composition.°

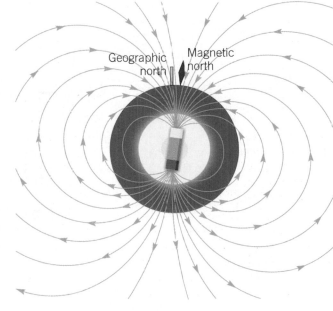

Figure 5.27 Earth's magnetic field Earth's magnetic field consists of lines of force much like those a giant bar magnet would produce if placed at the center of Earth.

Geographic north

Magnetic north

° Some sediments and sedimentary rocks also contain enough iron-bearing mineral grains to acquire a measurable amount of magnetization.

Basaltic lavas erupt at the surface at temperatures greater than 1000°C (1800° F), exceeding a threshold temperature for magnetism known as the **Curie point** (about 585°C [1085° F]). Consequently, the magnetite grains in molten lava are nonmagnetic. However, as the lava cools, these iron-rich grains become magnetized and align themselves in the direction of the existing magnetic lines of force. Once the minerals solidify, the magnetism they possess usually remains "frozen" in this position. Thus, they act like a compass needle because they "point" toward the position of the magnetic poles at the time of their formation. Rocks that formed thousands or millions of years ago and contain a "record" of the direction of the magnetic poles at the time of their formation are said to possess **paleomagnetism**, or **fossil magnetism**.

Apparent Polar Wandering A study of paleomagnetism in ancient lava flows throughout Europe led to an interesting discovery. The magnetic alignment of iron-rich minerals in lava flows of different ages indicated that the position of the paleomagnetic poles had changed through time. A plot of the location of the magnetic north pole, as measured from Europe, revealed that during the past 500 million years, the pole had gradually wandered from a location near Hawaii northeastward to its present location over the Arctic Ocean (**Figure 5.28**). This was strong evidence that either the magnetic north pole had migrated, an idea known as *polar wandering*, or that the poles had remained in

place and the continents had drifted beneath them—in other words, Europe had drifted in relation to the magnetic north pole.

Although the magnetic poles are known to move in a somewhat erratic path, studies of paleomagnetism from numerous locations show that the positions of the magnetic poles, averaged over thousands of years, correspond closely to the positions of the geographic poles. Therefore, a more acceptable explanation for the apparent polar wandering paths was provided by Wegener's hypothesis: If the magnetic poles remain stationary, their *apparent movement* is produced by continental drift.

Further evidence for continental drift came a few years later, when a polar-wandering path was constructed for North America (**Figure 5.28A**). For the first 300 million years or so, the paths for North America and Europe were found to be similar in direction—but separated by about 5000 kilometers (3000 miles). Then, during the middle of the Mesozoic era (180 million years ago), they began to converge on the present North Pole. The explanation for these curves is that North America and Europe were joined until the Mesozoic, when the Atlantic began to open. From this time forward, these continents continuously moved apart. When North America and Europe are moved back to their pre-drift positions, as shown in **Figure 5.28B**, these apparent wandering paths coincide. This is evidence that North America and Europe were once joined and moved relative to the poles as part of the same continent.

Figure 5.28 Apparent polar-wandering paths A. The more westerly path determined from North American data is thought to have been caused by the westward drift of North America by about 24 degrees from Eurasia. **B.** The positions of the wandering paths when the landmasses are reassembled in their predrift locations.

Magnetic Reversals and Seafloor Spreading

Another discovery came when geophysicists learned that over periods of hundreds of thousands of years, Earth's magnetic field periodically reverses polarity. During a **magnetic reversal**, the magnetic north pole becomes the magnetic south pole and vice versa. Lava solidifying during a period of reverse polarity will be magnetized with the polarity opposite that of volcanic rocks being formed today. When rocks exhibit the same magnetism as the present magnetic field, they are said to possess **normal polarity**, whereas rocks exhibiting the opposite magnetism are said to have **reverse polarity**.

Once the concept of magnetic reversals was confirmed, researchers set out to establish a time scale for these occurrences. The task was to measure the magnetic polarity of hundreds of lava flows and use radiometric dating techniques to establish the age of each flow. **Figure 5.29** shows the **magnetic time scale** established using this technique for the past few million years. The major divisions of the magnetic time scale are called *chrons* and last roughly 1 million years. As more measurements became available,

researchers realized that several short-lived reversals (less than 200,000 years long) often occurred during a single chron.

Meanwhile, oceanographers had begun to do magnetic surveys of the ocean floor in conjunction with their efforts to construct detailed maps of seafloor topography. These magnetic surveys were accomplished by towing very sensitive instruments, called **magnetometers**, behind research vessels (**Figure 5.30A**). The goal of these geophysical surveys was to map variations in the strength of Earth's magnetic field that arise from differences in the magnetic properties of the underlying crustal rocks.

The first comprehensive study of this type was carried out off the Pacific coast of North America and had an unexpected outcome. Researchers discovered alternating stripes of high- and low-intensity magnetism, as shown in **Figure 5.30B**. This relatively simple pattern of magnetic variation defied explanation until 1963, when Fred Vine and D. H. Matthews demonstrated that the high- and low-intensity stripes supported the concept of seafloor spreading. Vine and

SmartFigure 5.29 Time scale of magnetic reversals **A.** Time scale of Earth's magnetic reversals for the past 4 million years. **B.** This time scale was developed by establishing the magnetic polarity for lava flows of known age. (Data from Allen Cox and G. B. Dalrymple)

Figure 5.30 Ocean floor as a magnetic recorder A. Magnetic intensities are recorded when a magnetometer is towed across a segment of the ocean floor. **B.** Notice the symmetrical strips of low- and high-intensity magnetism that parallel the axis of the Juan de Fuca Ridge. Vine and Matthews suggested that the strips of high-intensity magnetism occur where normally magnetized oceanic rocks enhanced the existing magnetic field. Conversely, the low-intensity strips are regions where the crust is polarized in the reverse direction, which weakens the existing magnetic field.

Matthews suggested that the stripes of high-intensity magnetism are regions where the paleomagnetism of the ocean crust exhibits normal polarity (see Figure 5.30A). Consequently, these rocks *enhance* (reinforce) Earth's magnetic field. Conversely, the low-intensity stripes are regions where the ocean crust is polarized in the reverse direction and therefore *weaken* the existing magnetic field. But how do parallel stripes of normally and reversely magnetized rock become distributed across the ocean floor?

Vine and Matthews reasoned that as magma solidifies along narrow rifts at the crest of an oceanic ridge, it is magnetized with the polarity of the existing magnetic field (**Figure 5.31**). Because of seafloor spreading, this strip of magnetized crust would gradually increase in width. When Earth's magnetic field reverses polarity, any newly formed seafloor having the opposite polarity would form in the middle of the old strip. Gradually, the two halves of the old strip would be carried in opposite directions, away from the ridge crest. Subsequent reversals would build a pattern of normal and reverse magnetic stripes, as shown in Figure 5.31. Because new rock is added in equal amounts to both trailing edges of the spreading

ocean floor, we should expect the pattern of stripes (size and polarity) found on one side of an oceanic ridge to be a mirror image of those on the other side. In fact, a survey across the Mid-Atlantic Ridge just south of Iceland reveals a pattern of magnetic stripes exhibiting a remarkable degree of symmetry in relation to the ridge axis.

5.8 CONCEPT CHECKS

1. What is the age of the oldest sediments recovered using deep-ocean drilling? How do the ages of these sediments compare to the ages of the oldest continental rocks?
2. Assuming that hot spots remain fixed, in what direction was the Pacific plate moving while the Hawaiian islands were forming? When Suiko Seamount was forming?
3. How do sedimentary cores from the ocean floor support the concept of seafloor spreading?
4. Describe how Fred Vine and D. H. Matthews related the seafloor-spreading hypothesis to magnetic reversals.

A. Normal polarity

Magma

B. Reverse polarity

Magma

C. Normal polarity

Magma

T I M E

Figure 5.31 Magnetic reversal and seafloor spreading When new basaltic rocks form at mid-ocean ridges, they magnetize according to Earth's existing magnetic field. Hence, oceanic crust provides a permanent record of each reversal of our planet's magnetic field over the past 200 million years.

5.9 **What Drives Plate Motions?**

Summarize what is meant by plate–mantle convection and explain two of the primary driving forces of plate motion.

Researchers are in general agreement that some type of convection—where hot mantle rocks rise and cold, dense oceanic lithosphere sinks—is the ultimate driver of plate tectonics. Many of the details of this convective flow, however, remain topics of debate in the scientific community.

Forces That Drive Plate Motion

From geophysical evidence, we have learned that although the mantle consists almost entirely of solid rock, it is hot and weak enough to exhibit a slow, fluid-like convective flow. The simplest type of **convection** is analogous to heating a pot of water on a stove (**Figure 5.32**). Heating the base of a pot causes the water to become less dense (more buoyant), causing it to rise in relatively thin sheets or blobs that spread out at the surface. As the surface layer cools, its density increases, and the cooler

Cooler water sinks

Warm water rises

Convection is a type of heat transfer that involves the actual movement of a substance.

Figure 5.32 Convection in a cooking pot As a stove warms the water in the bottom of a cooking pot, the heated water expands, becomes less dense (more buoyant), and rises. Simultaneously, the cooler, denser water near the top sinks.

Figure 5.33 Forces that act on plates

Friction between the overriding plate and the subducting plate resists plate motion.

Ridge push is a gravity driven force that results from the elevated position of the ridge.

Mid-oceanic ridge

Trench

Slab pull

Mantle drag resists plate subduction.

Slab pull results from the sinking of a cold, dense slab of oceanic lithosphere and is the major driving force of plate motion.

Convection in the mantle enhances plate motion when the velocity of the asthenosphere exceeds that of the overlying plate.

water sinks back to the bottom of the pot, where it is reheated until it achieves enough buoyancy to rise again. Mantle convection is similar to, but considerably more complex than, the model just described.

There is general agreement that subduction of cold, dense slabs of oceanic lithosphere is a major driving force of plate motion (**Figure 5.33**). This phenomenon, called **slab pull**, occurs because cold slabs of oceanic lithosphere are more dense than the underlying warm asthenosphere and hence "sink like a rock" pulled down into the mantle by gravity.

Another important driving force is **ridge push** (see Figure 5.33). This gravity-driven mechanism results from the elevated position of the oceanic ridge, which causes slabs of lithosphere to "slide" down the flanks of the ridge. Ridge push appears to contribute far less to plate motions than slab pull. The primary evidence for this is that fastest-moving plates—the Pacific, Nazca, and Cocos plates—have extensive subduction zones along their margins. By contrast, the spreading rate in the North Atlantic basin, which is nearly devoid of subduction zones, is one of the lowest, at about 2.5 centimeters (1 inch) per year.

Although the subduction of cold, dense lithospheric plates appears to be the dominant force acting on plates, other factors are at work as well. Flow in the mantle, perhaps best described as "mantle drag," is also thought to affect plate motion (see Figure 5.33). When flow in the asthenosphere is moving at a velocity that exceeds that of the plate, mantle drag enhances plate motion. However, if the asthenosphere is moving more slowly than the plate,

or if it is moving in the opposite direction, this force tends to resist plate motion. Another type of resistance to plate motion occurs along some subduction zones, where friction between the overriding plate and the descending slab generates significant earthquake activity.

Models of Plate–Mantle Convection

Although convection in the mantle has yet to be fully understood, researchers generally agree on the following:

- Convective flow in the rocky 2900-kilometer- (1800-mile) thick mantle—in which warm, buoyant rock rises and cooler, denser material sinks—is the underlying driving force for plate movement.
- Mantle convection and plate tectonics are part of the same system. Subducting oceanic plates drive the cold downward-moving portion of convective flow, while shallow upwelling of hot rock along the oceanic ridge and buoyant mantle plumes are the upward-flowing arms of the convective mechanism.
- Convective flow in the mantle is a major mechanism for transporting heat away from Earth's interior to the surface, where it is eventually radiated into space.

What is not known with certainty is the exact structure of this convective flow. Several models have been proposed for plate–mantle convection, and we will look at two of them.

Whole-Mantle Convection Most researchers favor some type of *whole-mantle convection* model, also called the *plume model*, in which cold oceanic lithosphere sinks to great depths and stirs the entire mantle (**Figure 5.34A**). The whole-mantle model suggests that the ultimate burial ground for subducting slabs is the core–mantle boundary. This downward flow is balanced by buoyantly rising mantle plumes that transport hot material toward the surface (Figure 5.34A).

Two kinds of plumes have been proposed—narrow tube-like plumes and giant upwellings. The long, narrow plumes that extend from the core–mantle boundary are thought to produce hot-spot volcanism of the type associated with the Hawaiian islands, Iceland, and Yellowstone. Areas of large mega-plumes, as shown in Figure 5.34A, are thought to occur beneath the Pacific basin and southern Africa. The latter structure is thought to account for the fact that southern Africa has an elevation that is much higher than would be predicted for a stable continental landmass. Heat for both types of plumes is thought to arise mainly from the core, while the deep mantle provides a source for chemically distinct magmas.

Layer Cake Model Some researchers argue that the mantle resembles a "layer cake" divided at a depth of perhaps 660 kilometers (410 miles) but no deeper than 1000 kilometers (620 miles). As shown in **Figure 5.34B**, this layered model has two zones of convection—a thin, dynamic layer in the upper mantle and a thicker, sluggish one located below. As with the whole-mantle model, the

Figure 5.34 Models of mantle convection

A. In the "whole mantle model", sinking slabs of cold oceanic lithosphere are the downward limbs of convection cells, while rising mantle plumes carry hot material from the core-mantle boundary toward the surface.

B. The "layer cake model" has two largely disconnected convective layers. A dynamic upper layer driven by descending slabs of cold oceanic lithosphere and a sluggish lower layer that carries heat upward without appreciably mixing with the layer above.

downward convective flow is driven by the subduction of cold, dense oceanic lithosphere. However, rather than reach the lower mantle, these subducting slabs penetrate to depths of no more than 1000 kilometers (620 miles). Notice in Figure 5.34B that the upper layer in the layer cake model is littered with recycled oceanic lithosphere of various ages. Melting of these fragments is thought to be the source of magma for some of the volcanism that occurs away from plate boundaries, as in the volcanoes of Hawaii.

In contrast to the active upper mantle, the lower mantle is sluggish and does not provide material to support volcanism at the surface. However, very slow convection within this layer likely carries heat upward, but very little mixing between these two layers is thought to occur.

The actual shape of the convective flow in the mantle is a matter of current scientific investigation and is highly debated by geologists. Perhaps a hypothesis that combines features from the layer cake model and the whole-mantle convection model will emerge.

5.9 CONCEPT CHECKS

1. Describe slab pull and ridge push. Which of these forces appears to contribute more to plate motion?
2. What role are mantle plumes thought to play in the convective flow in the mantle?
3. Briefly describe the two models proposed for mantle–plate convection.

5.10 How Do Plates and Plate Boundaries Change?

Explain why plates such as the African and Antarctic plates are getting larger, while the Pacific plate is getting smaller.

Although the total surface area of Earth does not change, the size and shape of individual plates are constantly changing. For example, the African and Antarctic plates, which are mainly bounded by divergent boundaries—sites of seafloor production—are continually growing in size as new lithosphere is added to their margins. By contrast, the Pacific plate is being consumed into the mantle along its northern and western flanks faster that it is growing along the East Pacific Rise and thus is diminishing in size.

Another consequence of plate motion is that boundaries also migrate. For example, the position of the Peru–Chile trench, which is the result of the Nazca plate being bent downward as it descends beneath the South American plate, has changed over time (see Figure 5.12). Because of the westward drift of the South American plate relative to the Nazca plate, the position of the Peru–Chile trench has migrated in a westerly direction as well.

Plate boundaries can also be created or destroyed in response to changes in the forces acting on the lithosphere. Recall that the Red Sea is the site of a relatively new spreading center that came into existence less than 20 million years ago, when the Arabian Peninsula began to split from Africa. At other locations, plates carrying continental crust are presently moving toward one another. Eventually these continental fragments may collide and be sutured together. This could occur, for example, in the South Pacific, where Australia is moving northward toward southern Asia. If Australia continues its northward migration, the boundary separating it from Asia will disappear as these plates become one. The breakup of Pangaea is a classic example of how plate boundaries change through geologic time.

The Breakup of Pangaea

Wegener used evidence from fossils, rock types, and ancient climates to create a jigsaw-puzzle fit of the continents, thereby creating his supercontinent of Pangaea. In a similar manner, but employing modern tools not available to Wegener, geologists have re-created the steps in the breakup of this supercontinent, an event that began about 180 million years ago. From this work, the dates when individual crustal fragments separated from one another and their relative motions have been well established (**Figure 5.35**).

An important consequence of the breakup of Pangaea was the creation of a "new" ocean basin: the Atlantic. As you can see in Figure 5.35, splitting of the supercontinent did not occur simultaneously along the margins of the Atlantic. The first split developed between North America and Africa.

Here, the continental crust was highly fractured, providing pathways for huge quantities of fluid lavas to reach the surface. Today, these lavas are represented by weathered igneous rocks found along the eastern seaboard of the United States—primarily buried beneath the sedimentary rocks that form the continental shelf. Radiometric dating of these solidified lavas indicates that

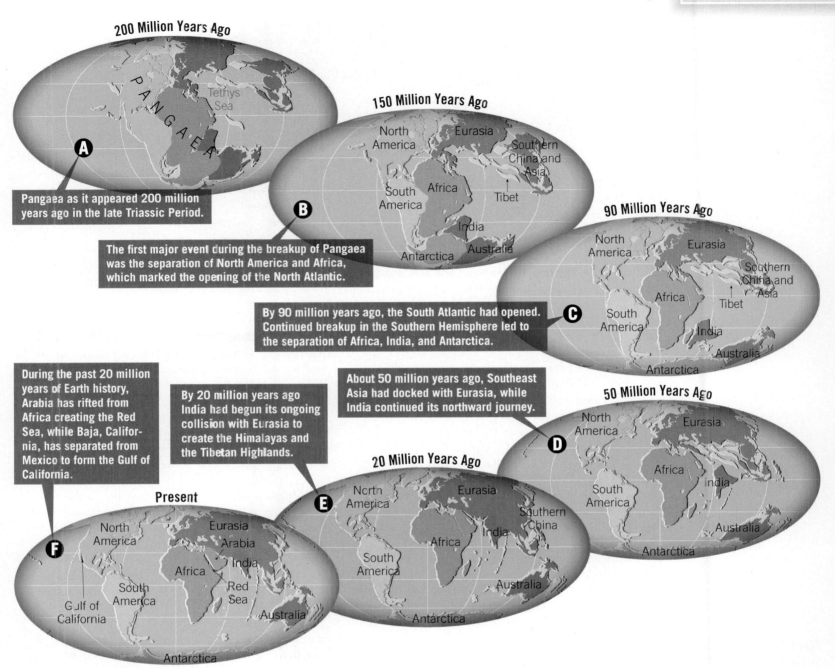

200 Million Years Ago

Pangaea as it appeared 200 million years ago in the late Triassic Period.

150 Million Years Ago

The first major event during the breakup of Pangaea was the separation of North America and Africa, which marked the opening of the North Atlantic.

90 Million Years Ago

By 90 million years ago, the South Atlantic had opened. Continued breakup in the Southern Hemisphere led to the separation of Africa, India, and Antarctica.

During the past 20 million years of Earth history, Arabia has rifted from Africa creating the Red Sea, while Baja, California, has separated from Mexico to form the Gulf of California.

By 20 million years ago India had begun its ongoing collision with Eurasia to create the Himalayas and the Tibetan Highlands.

About 50 million years ago, Southeast Asia had docked with Eurasia, while India continued its northward journey.

50 Million Years Ago

20 Million Years Ago

Present

Figure 5.35 The breakup of Pangaea

rifting began between 180 million and 165 million years ago. This time span represents the "birth date" for this section of the North Atlantic.

By 130 million years ago, the South Atlantic began to open near the tip of what is now South Africa. As this zone of rifting migrated northward, it gradually opened the South Atlantic (compare **Figures 5.35B** and **C**). Continued breakup of the southern landmass led to the separation of Africa and Antarctica and sent India on a northward journey. By the early Cenozoic, about 50 million years ago, Australia had separated from Antarctica, and the South Atlantic had emerged as a full-fledged ocean (**Figure 5.35D**).

A modern map (**Figure 5.35F**) shows that India eventually collided with Asia, an event that began about 50 million years ago and created the Himalayas as well as the Tibetan Highlands. About the same time, the separation of Greenland from Eurasia completed the breakup of the northern landmass. During the past 20 million years or so of Earth's history, Arabia rifted from Africa to form the Red Sea, and Baja California separated from Mexico to form the Gulf of California (**Figure 5.35E**). Meanwhile, the Panama Arc joined North America and South America to produce our globe's familiar modern appearance.

Figure 5.36 The world as it may look 50 million years from now This reconstruction is highly idealized and based on the assumption that the processes that cause the breakup of Pangaea will continue to operate. (Modified after Robert S. Dietz, John C. Holden, C. Scotese, and others)

Plate Tectonics in the Future

Geologists have extrapolated present-day plate movements into the future. **Figure 5.36** illustrates where Earth's landmasses may be 50 million years from now if present plate movements persist during this time span.

In North America we see that the Baja Peninsula and the portion of southern California that lies west of the San Andreas Fault will have slid past the North American plate. If this northward migration takes place, Los Angeles and San Francisco will pass each other in about 10 million years, and in about 60 million years the Baja Peninsula will begin to collide with the Aleutian islands.

If Africa continues on a northward path, it will continue to collide with Eurasia. The result will be the closing of the Mediterranean, the last remnant of a once vast ocean called Tethys Ocean, and the initiation of another major mountain-building episode (see Figure 5.36). In other parts of the world, Australia will be astride the

equator and, along with New Guinea, will be on a collision course with Asia. Meanwhile, North and South America will begin to separate, while the Atlantic and Indian Oceans will continue to grow, at the expense of the Pacific Ocean.

A few geologists have even speculated on the nature of the globe 250 million years into the future. As shown in **Figure 5.37**, the next supercontinent may form as a result of subduction of the floor of the Atlantic Ocean, resulting in the collision of the Americas with the Eurasian–African landmass. Support for the possible closing of the Atlantic comes from a similar event when an ocean that predates the Atlantic closed during the formation of Pangaea. During the next 250 million years, Australia is also projected to collide with Southeast Asia. If this scenario is accurate, the dispersal of Pangaea will end when the continents reorganize into the next supercontinent.

Such projections, although interesting, must be viewed with considerable skepticism because many assumptions must be correct for these events to unfold as just described. Nevertheless, changes in the shapes and positions of continents that are equally profound will undoubtedly occur for many hundreds of millions of years to come. Only after much more of Earth's internal heat has been lost will the engine that drives plate motions cease.

Figure 5.37 Earth as it may appear 250 million years from now

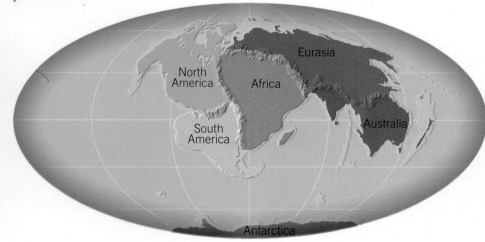

1. List two plates that are growing in size and one that is shrinking in size.
2. What new ocean basin was created by the breakup of Pangaea?
3. Briefly describe some major changes to the globe when we extrapolate present-day plate movements 50 million years into the future.

CONCEPTS IN REVIEW: PLATE TECTONICS: A SCIENTIFIC REVOLUTION UNFOLDS

5.1 From Continental Drift to Plate Tectonics

Discuss the view that most geologists held prior to the 1960s regarding the geographic positions of the ocean basins and continents.

- Fifty years ago, most geologists thought that ocean basins were very old and that continents were fixed in place. Those ideas were discarded with a scientific revolution that revitalized geology: the theory of plate tectonics. Supported by multiple kinds of evidence, plate tectonics is the foundation of modern Earth science.

? Today, we know that the early twentieth-century scientists who rejected the idea of continental drift were wrong. Were they therefore bad scientists? Why or why not?

5.2 Continental Drift: An Idea Before Its Time

List and explain the evidence presented by Wegener to support his continental drift hypothesis.

KEY TERMS: continental drift, supercontinent, Pangaea

- German meteorologist Alfred Wegener formulated the idea of continental drift in 1915. He suggested that Earth's continents were not fixed in place but moved slowly over geologic time.
- Wegener reconstructed a supercontinent called Pangaea that existed about 200 million years ago, during the late Paleozoic and early Mesozoic.
- Wegener's evidence that Pangaea existed but later broke into pieces that drifted apart included (1) the shape of the continents, (2) continental fossil organisms that matched across oceans, (3) matching rock types and modern mountain belts, and (4) sedimentary rocks that recorded ancient climates, including glaciers on the southern portion of Pangaea.

? Why did Wegener choose organisms such as *Glossopteris* and *Mesosaurus* as evidence for continental drift, as opposed to other fossil organisms such as sharks or jellyfish?

Mesosaurus

0   10 cm
0   5 inches

John Cancalosi

5.3 The Great Debate

Discuss the two main objections to the continental drift hypothesis.

- Wegener's hypothesis suffered from two flaws: It proposed tidal forces as the mechanism for the motion of continents, and it implied that the continents would have plowed their way through weaker oceanic crust, like a boat cutting through a thin layer of sea ice. Geologists rejected the idea of continental drift when Wegener proposed it, and it wasn't resurrected for another 50 years.

5.4 The Theory of Plate Tectonics

List the major differences between Earth's lithosphere and asthenosphere and explain the importance of each in the plate tectonics theory.

KEY TERMS: ocean ridge system, theory of plate tectonics, lithosphere, asthenosphere, lithospheric plate (plate)

- Research conducted during World War II led to new insights that helped revive Wegener's hypothesis of continental drift. Exploration of the seafloor revealed previously unknown features, including an extremely long mid-ocean ridge system. Sampling of the oceanic crust revealed that it was quite young relative to the continents.
- The lithosphere is Earth's outermost rocky layer that is broken into plates. It is relatively stiff and deforms by breaking or bending. Beneath the lithosphere is the asthenosphere, a relatively weak layer that deforms by flowing. The lithosphere consists both of crust (either oceanic or continental) and underlying upper mantle.
- There are seven large plates, another seven intermediate-size plates, and numerous relatively small microplates. Plates meet along boundaries that may either be divergent (moving apart from each other), convergent (moving toward each other), or transform (moving laterally past each other).

5.5 Divergent Plate Boundaries and Seafloor Spreading

Sketch and describe the movement along a divergent plate boundary that results in the formation of new oceanic lithosphere.

KEY TERMS: divergent plate boundary, rift valley, seafloor spreading, spreading center, continental rift

- Seafloor spreading leads to the generation of new oceanic lithosphere at mid-ocean ridge systems. As two plates move apart from one another, tensional forces open cracks in the plate, allowing magma to well up and generate new slivers of seafloor. This process generates new oceanic lithosphere at a rate of 2 to 15 centimeters (1 to 6 inches) each year.
- As it ages, oceanic lithosphere cools and becomes denser. As it is transported away from the mid-ocean ridge, it therefore subsides. At the same time, new material is added to its underside, causing the plate to grow thicker.
- Divergent boundaries are not limited to the seafloor. Continents can break apart, too, starting with a continental rift (like modern-day east Africa) and eventually leading to a new ocean basin opening between the two sides of the rift.

5.6 Convergent Plate Boundaries and Subduction

Compare and contrast the three types of convergent plate boundaries and name a location where each type can be found.

KEY TERMS: convergent plate boundary, subduction zone, deep-ocean trench, partial melting, continental volcanic arc, volcanic island arc (island arc)

- When plates move toward one another, oceanic lithosphere is subducted into the mantle, where it is recycled. Subduction manifests itself on the ocean floor as a deep linear trench. The subducting slab of oceanic lithosphere can descend at a variety of angles, from nearly horizontal to nearly vertical.
- Aided by the presence of water, the subducted oceanic lithosphere triggers melting in the mantle, which produces magma. The magma is less dense than the surrounding rock and will rise. It may cool at depth, thickening the crust, or it may make it all the way to Earth's surface, where it erupts as a volcano.
- A line of volcanoes that erupt through continental crust is termed a continental volcanic arc, while a line of volcanoes that erupt through an overriding plate of oceanic lithosphere is a volcanic island arc.
- Continental crust resists subduction due to its relatively low density, and so when an intervening ocean basin is completely destroyed through subduction, the continents on either side collide, generating a new mountain range.

? **Sketch a typical continental volcanic arc and label the key parts. Then repeat the drawing with an overriding plate made of oceanic lithosphere.**

5.7 Transform Plate Boundaries

Describe the relative motion along a transform fault boundary and be able to locate several examples on a plate boundary map.

KEY TERMS: transform plate boundary (transform fault), fracture zone

- At a transform boundary, lithospheric plates slide horizontally past one another. No new lithosphere is generated, and no old lithosphere is consumed. Shallow earthquakes signal the movement of these slabs of rock as they grind past their neighbors.
- The San Andreas Fault in California is an example of a transform boundary in continental crust, while the fracture zones between segments of the Mid-Atlantic Ridge are examples of transform faults in oceanic crust.

? **On the accompanying tectonic map of the Caribbean, find the Enriquillo Fault. (The location of the 2010 Haiti earthquake is shown as a yellow star.) What kind of plate boundary is shown here? Are there any other faults in the area that show the same type of motion?**

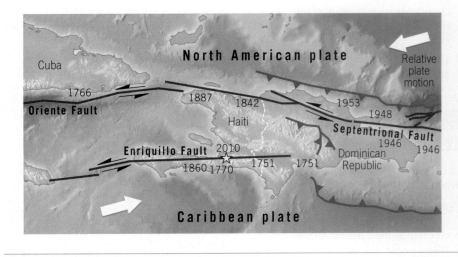

5.8 Testing the Plate Tectonics Model

List the evidence used to support the plate tectonics theory and briefly describe it.

KEY TERMS: mantle plume, hot spot, hot-spot track, Curie point, paleomagnetism (fossil magnetism), magnetic reversal, normal polarity, reverse polarity, magnetic time scale, magnetometer

- The plate tectonics model has been verified by multiple lines of evidence. For instance, the Deep Sea Drilling Project found that the age of the seafloor increases with distance from a mid-ocean ridge. The thickness of sediment atop this seafloor is also proportional to distance from the ridge: Older lithosphere has had more time to accumulate sediment.
- Overall, oceanic lithosphere is quite young, with none older than about 180 million years.
- A hot spot is an area of volcanic activity where a mantle plume reaches the surface. Volcanic rocks generated by hot-spot volcanism provide evidence of both the direction and rate of plate movement over time.
- Magnetic minerals such as magnetite align themselves with Earth's magnetic field as rock forms. These fossil magnets act as recorders of the ancient orientation of Earth's magnetic field. This is useful to geologists in two ways: (1) It allows a given stack of rock layers to be interpreted in terms of changing position relative to the magnetic poles through time, and (2) reversals in the orientation of the magnetic field are preserved as "stripes" of normal and reversed polarity in the oceanic crust. Magnetometers reveal this signature of seafloor spreading as a symmetrical pattern of magnetic stripes parallel to the axis of the mid-ocean ridge.

5.9 What Drives Plate Motions?

Summarize what is meant by plate–mantle convection and explain two of the primary driving forces of plate motion.

KEY TERMS: slab pull, ridge push, convection

- Some kind of convection (upward movement of less dense material and downward movement of more dense material) appears to drive the motion of plates.
- Slabs of oceanic lithosphere sink at subduction zones because the subducted slab is denser than the underlying asthenosphere. In this process, called slab pull, Earth's gravity tugs at the slab, drawing the rest of the plate toward the subduction zone. As lithosphere descends the mid-ocean ridge, it exerts a small additional force, the outward-directed ridge push. Frictional drag exerted on the underside of plates by flowing mantle is another force that acts on plates and influences their direction and rate of movement.

- The exact patterns of mantle convection are not fully understood. Convection may occur throughout the entire mantle, as suggested by the whole-mantle model. Or it may occur in two layers within the mantle—the active upper mantle and the sluggish lower mantle— as proposed in the layer cake model.

(Photo by Steve Bower/ Shutterstock)

? Compare and contrast mantle convection with the operation of a lava lamp.

5.10 How Do Plates and Plate Boundaries Change?

Explain why plates such as the African and Antarctic plates are getting larger, while the Pacific plate is getting smaller.

- Although the total surface area of Earth does not change, the shape and size of individual plates are constantly changing as a result of subduction and seafloor spreading. Plate boundaries can also be created or destroyed in response to changes in the forces acting on the lithosphere.
- The breakup of Pangaea and the collision of India with Eurasia are two examples of how plates change through geologic time.

GIVE IT SOME THOUGHT

1. After referring to the section in the Introduction titled "The Nature of Scientific Inquiry," answer the following:
   a. What observation led Alfred Wegener to develop his continental drift hypothesis?
   b. Why was the continental drift hypothesis rejected by the majority of the scientific community?
   c. Do you think Wegener followed the basic principles of scientific inquiry? Support your answer.

2. Referring to the accompanying diagrams that illustrate the three types of convergent plate boundaries, complete the following:

A. B. C.

   a. Identify each type of convergent boundary.
   b. Volcanic island arcs develop on what type of crust?
   c. Why are volcanoes largely absent where two continental blocks collide?
   d. Describe two ways that oceanic–oceanic convergent boundaries are different from oceanic–continental boundaries. How are they similar?

3. Some predict that California will sink into the ocean. Is this idea consistent with the theory of plate tectonics? Explain.

4. Refer to the accompanying hypothetical plate map to answer the following questions:

✕✕ Oceanic ridge   ▲▲▲ Subduction zone

    a. How many portions of plates are shown?
    b. Are continents A, B, and C moving toward or away from each other? How did you determine your answer?
    c. Explain why active volcanoes are more likely to be found on continents A and B than on continent C.
    d. Provide at least one scenario in which volcanic activity might be triggered on continent C.

5. Volcanoes, such as the Hawaiian chain, that form over mantle plumes are some of the largest shield volcanoes on Earth. However, several shield volcanoes on Mars are gigantic compared to those on Earth. What does this difference tell us about the role of plate motion in shaping the Martian surface?

6. Imagine that you are studying seafloor spreading along two different oceanic ridges. Using data from a magnetometer, you produced the two accompanying maps. From these maps, what can you determine about the relative rates of seafloor spreading along these two ridges? Explain.

Magnetic anomalies

Spreading Center A

Spreading Center B

7. Australian marsupials (kangaroos, koala bears, etc.) have direct fossil links to marsupial opossums found in the Americas. Yet the modern marsupials in Australia are markedly different from their American relatives. How does the breakup of Pangaea help to explain these differences (see Figure 5.35)?

8. Density is a key component in the behavior of Earth materials and is especially important in understanding key aspects of plate tectonics. Describe three different ways that density and/or density differences play a role in plate tectonics.

9. Refer to the accompanying map and the pairs of cities below to complete the following: (Boston, Denver), (London, Boston), (Honolulu, Beijing)
   a. List the pair of cities that is moving apart as a result of plate motion.
   b. List the pair of cities that is moving closer as a result of plate motion.
   c. List the pair of cities that is presently not moving relative to each other.

Plate motion measured in centimeters per year

MasteringGeology™

Restless Earth: Earthquakes, Geologic Structures, and Mountain Building

Each statement represents the primary learning objective for the corresponding major heading within the chapter. After you complete the chapter, you should be able to:

6.1 Sketch and describe the mechanism that generates most earthquakes.

6.2 Compare and contrast the types of seismic waves and describe the principle of the seismograph, which is used to detect them.

6.3 Explain how an earthquake epicenter can be located using seismographs.

6.4 Distinguish between intensity scales and magnitude scales.

6.5 List and describe the major destructive forces that can be triggered by earthquake vibrations.

6.6 Locate Earth's major earthquake belts on a world map and label the regions associated with the largest earthquakes.

6.7 Explain how Earth acquired its layered structure and briefly describe how seismic waves are used to probe Earth's interior.

6.8 List and describe each of Earth's major layers.

6.9 Compare and contrast brittle and ductile deformation.

6.10 List and describe the major types of folds.

6.11 Sketch and describe the relative motion of rock bodies located on opposite sides of normal, reverse, and strike-slip faults.

6.12 Locate and identify Earth's major mountain belts on a world map.

6.13 Sketch an Andean-type mountain belt and describe how each of its major features is generated.

6.14 Summarize the stages in the development of a collisional mountain belt such as the Himalayas.

Tsunami striking the coast of Japan on March 11, 2011. (Photo by Sadatsugu Tomizawa/AFP/Getty Images)

On January 12, 2010, an estimated 316,000 people lost their lives when a 7.0-magnitude earthquake struck the small Caribbean nation of Haiti, the poorest country in the Western Hemisphere. In addition to the staggering death toll, there were more than 300,000 injuries, and more than 280,000 houses were destroyed or damaged.

The quake originated only 25 kilometers (15 miles) from the country's densely populated capital city of Port-au-Prince. It occurred at a depth of just 10 kilometers (6 miles) along a fault similar to California's San Andreas Fault. Because of the quake's shallow depth, ground shaking was exceptional for an event of this magnitude. Other factors that contributed to the Port-au-Prince disaster included the city's geologic setting and the nature of its buildings. The city is built on sediment, which is quite susceptible to ground shaking during an earthquake. More importantly, inadequate or nonexistent building codes meant that buildings collapsed far more readily than they should have. At least 52 aftershocks, measuring magnitude 4.5 or greater, jolted the area and added to the trauma that survivors experienced for days after the original quake.

6.1 What Is an Earthquake?

Sketch and describe the mechanism that generates most earthquakes.

An **earthquake** is ground shaking caused by the sudden and rapid movement of one block of rock slipping past another along fractures in Earth's crust called **faults**. Most faults are locked, except for brief, abrupt movements when sudden slippage produces an earthquake (**Figure 6.1**). Faults are locked because the confining pressure exerted by the overlying crust is enormous, causing these fractures in the crust to be "squeezed shut."

Figure 6.1 Presidential palace damaged during the 2010 Haiti earthquake (Photo by Luis Acosta/AFP/Getty Images)

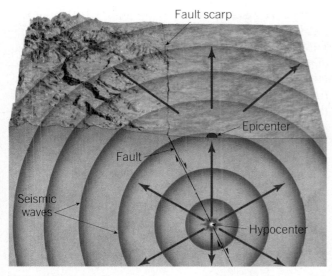

Figure 6.2 Earthquake hypocenter and epicenter The *hypocenter* is the zone at depth where the initial displacement occurs. The *epicenter* is the surface location directly above the hypocenter.

Earthquakes tend to occur along preexisting faults where internal stresses have caused the crustal rocks to rupture or break into two or more units. The location where slippage begins is called the **hypocenter**, or **focus**. Earthquake waves radiate from this spot outward into the surrounding rock. The point on Earth's surface directly above the hypocenter is called the **epicenter** (**Figure 6.2**).

Large earthquakes release huge amounts of stored-up energy as **seismic waves**—a form of energy that travels through the lithosphere and Earth's interior. The energy carried by these waves causes the material that transmits them to shake. Seismic waves are analogous to waves produced when a stone is dropped into a calm pond. Just as the impact of the stone creates a pattern of circular waves, an earthquake generates waves that radiate outward in all directions from the hypocenter. Although seismic energy dissipates rapidly as it moves away from the quake's hypocenter, sensitive instruments can detect earthquakes even when they occur on the opposite side of Earth.

Thousands of earthquakes occur around the world every day. Fortunately, most are small and cannot be detected by people. Of these, only about 15 strong earthquakes (magnitude 7 or greater) are recorded each year, many of them occurring in remote regions. Occasionally, a large earthquake is triggered near a major population center. Such events are among the most destructive natural forces on Earth. The shaking of the ground, coupled with the liquefaction of soils, wreaks havoc on buildings, roadways, and other structures. In addition, when a quake occurs in a populated area, power and gas lines are often ruptured, causing numerous fires. In the famous 1906 San Francisco earthquake, much of the damage was caused by fires that became uncontrollable when broken water mains left firefighters with only trickles of water (**Figure 6.3**).

Did You Know?

Literally thousands of earthquakes occur daily! Fortunately, the majority of them are too small to be felt by people, and the majority of larger ones occur in remote regions. Their existence is known only because of sensitive seismographs.

San Francisco in flames following the 1906 quake. Broken water mains left firefighters without water.

Figure 6.3 Earthquakes trigger fires (Reproduced from the collection of the Library of Congress; inset photo by AFP/Getty Images)

Fire triggered when a gas line ruptured during the Northridge earthquake in southern California in 1994.

Discovering the Causes of Earthquakes

The energy released by volcanic eruptions, massive landslides, and meteorite impacts can generate earthquake-like waves, but these events are usually weak. What mechanism produces a destructive earthquake? As you have learned, Earth is not a static planet. We know that large sections of Earth's crust have been thrust upward because fossils of marine organisms have been discovered thousands of meters above sea level. Other regions, such as California's Death Valley, exhibit evidence of extensive subsidence. In addition to these vertical displacements, offsets in fences, roads, and other structures indicate that horizontal movements between blocks of Earth's crust are also common (**Figure 6.4**).

The actual mechanism of earthquake generation eluded geologists until H. F. Reid of Johns Hopkins University conducted a landmark study following the 1906 San Francisco earthquake. This earthquake was accompanied by horizontal surface displacements of several meters along the northern portion of the San Andreas Fault. Field studies determined that during this single earthquake, the Pacific plate lurched as much as 9.7 meters (32 feet) northward, past the adjacent North American plate. To better visualize this, imagine standing on one side of the fault and watching a person on the other side suddenly slide horizontally 9.7 meters to your right.

What Reid concluded from his investigations is illustrated in **Figure 6.5**. Over tens to hundreds of years, differential stress slowly bends the crustal rocks on both sides of a fault. This is much like a person bending a limber wooden stick, as shown in Figure 6.5. Frictional resistance keeps the fault from rupturing and crustal rocks from slipping. (Friction acts against slippage and is enhanced by irregularities that occur along the fault surface.) At some point, the stress along the fault overcomes the frictional resistance, and slip initiates. Slippage allows the deformed (bent) rock to "snap back" to its original, stress-free, shape; a series of earthquake waves radiate as it slides (see Figure 6.5C,D). Reid termed the "springing back" **elastic rebound** because the rock behaves elastically, much like a stretched rubber band does when it is released.

Faults and Earthquakes

The slippage that occurs along faults can be explained by the plate tectonics theory, which states that large slabs of Earth's lithosphere are continually grinding past one another. These mobile plates interact with neighboring plates, straining and deforming the rocks at their edges. Faults associated with plate boundaries are the source of most large earthquakes.

Transform Fault Boundaries The San Andreas Fault is the transform fault boundary that separates two great sections of Earth's lithosphere: the North American plate and the Pacific plate. Most transform faults, including the San Andreas Fault, are not perfectly straight or continuous; instead, they consist of numerous branches and smaller fractures that display kinks and offsets (**Figure 6.6**). In addition, geologists have learned that displacement along transform faults occurs in discrete

Slippage along a fault produced this offset in an orange grove east of Calexico, California.

This fence was offset 2.5 meters (8.5 feet) during the 1906 San Francisco earthquake.

Figure 6.4 Displacement along a fault (Color photo by John S. Shelton/University of Washington Libraries; inset photo by G.K. Gilbert/USGS)

Deformation of rocks

Deformation of a limber stick

Time — Tens to hundreds of years — Seconds to a few minutes

A. Original position of rocks on opposite sides of a fault.

B. The movement of tectonic plates causes the rocks to bend and store elastic energy.

C. Once the strength of the rocks is exceeded, slippage along the fault produces an earthquake.

D. The rocks return to their original shape, but in a new location.

SmartFigure 6.5 Elastic rebound

segments that often behave differently from one another. Some sections of the San Andreas Fault exhibit slow, gradual displacement known as **fault creep** and produce little seismic shaking. Other segments slip at relatively closely spaced intervals, producing numerous small to moderate earthquakes. Still other segments remain locked and store elastic energy for a few hundred years before they break loose. Rupture on these segments may result in a major earthquake.

Earthquakes that occur along locked segments of the San Andreas Fault tend to be repetitive: As soon as one is over, strain immediately begins accumulating due to the continuous motion of the plates. Decades or centuries later, the fault fails again.

Slippage along large faults does not occur instantaneously. The initial slip begins at the hypocenter and propagates (travels) along the fault surface, at 2 to 4 kilometers per second—faster than a rifle shot. Slippage on one section of the fault adds strain to the adjacent segment, which may also slip. As this zone of slippage advances, it can slow down, speed up, or even jump to a nearby fault segment. The propagation of the rupture zone along a 300-kilometer- (200-mile-) long fault, for example, takes about 1.5 minutes, as compared to

The 1906 San Francisco earthquake caused displacement on the 477-kilometer-long northernmost section of the fault. This event, which had an estimated magnitude of 7.8, likely relieved much of the strain that had been building during the previous 200 years or so.

Located just south of the 1906 rupture is a section of the San Andreas Fault that exhibits fault creep. When plates gradually slide past each other, less strain accumulates than when the fault is locked, diminishing the possibilities of an eventual large quake.

CALIFORNIA

1906 epicenter ★
San Francisco •

★ 1857 epicenter

This 300-kilometer-long section of the San Andreas fault produced the Fort Tejon earthquake of 1857 that had an estimated magnitude of 7.9. Because a portion of the fault has likely accumulated considerable strain since the Fort Tejon quake, the U.S. Geological Survey gives it a 60 percent probability of producing a major earthquake in the next 30 years.

Los Angeles •

The next major quake on the San Andreas may well be on its southernmost 200 kilometers—an area that has not produced a large event in about 300 years.

Figure 6.6 San Andreas Fault California's San Andreas Fault runs diagonally from southeast to northwest for nearly 1300 kilometers (800 miles) through much of the western part of the state. For years, researchers have been trying to predict the location of the next "Big One"—an earthquake with a magnitude of 8 or greater—along this fault system.

about 30 seconds for a 100-kilometer- (60-mile-) long fault. Earthquake waves are generated at every point along the fault as that portion of the fault begins to slip.

Faults Associated with Convergent Plate Boundaries

Strong earthquakes also occur along large faults associated with convergent plate boundaries. Compressional forces associated with continental collisions that result in mountain building generate many of these faults.

In addition, the plate boundary between a subducting slab of oceanic lithosphere and the overlying plate form a fault referred to as a **megathrust fault**. Because these large thrust faults lie beneath the ocean floor and experience vertical motion that can displace the overlying water, they are capable of generating destructive tsunamis. Megathrust faults have produced the majority of Earth's most powerful earthquakes, including the 2011 Japan quake (M 9.0), the 2004 Indian Ocean (Sumatra) quake (M 9.1), the 1964 Alaska quake (M 9.2), and the largest earthquake yet recorded, the 1960 Chile quake (M 9.5).

6.1 CONCEPT CHECKS

1. What is an earthquake? Under what circumstances do most large earthquakes occur?
2. How are faults, hypocenters, and epicenters related?
3. Who was first to explain the mechanism by which most earthquakes are generated?
4. Explain what is meant by *elastic rebound*.
5. What type of faults tend to produce the most destructive earthquakes?
6. During an earthquake, the entire length of a fault slips in a single event that lasts less than 1 second. True or false?
7. Defend or rebut this statement: Faults that do not experience fault creep may be considered safe.

6.2 Seismology: The Study of Earthquake Waves

Compare and contrast the types of seismic waves and describe the principle of the seismograph, which is used to detect them.

Figure 6.7 Ancient Chinese seismograph During an Earth tremor, the dragons located in the direction of the main vibrations would drop one or more balls into the mouth of frogs below.
(Photo by James E. Patterson)

Ball dropping

The study of earthquake waves, **seismology**, dates back to attempts made in China almost 2000 years ago to determine the direction from which these waves originated. The earliest known instrument, invented by Zhang Heng, was a large hollow jar that contained a weight suspended from the top (**Figure 6.7**). The suspended weight (similar to a clock pendulum) was connected to the jaws of several large dragon figurines that encircled the container. The jaws of each dragon held a metal ball. When earthquake waves reached the instrument, the relative motion between the suspended mass and the jar would dislodge some of the metal balls into the waiting mouths of frogs directly below.

Instruments That Record Earthquakes

In principle, modern **seismographs**, or **seismometers**, are similar to the instruments used by early Chinese researchers. A seismograph has a weight freely suspended from a support that is securely attached to bedrock (**Figure 6.8**). When vibrations from an earthquake reach the instrument, the **inertia** of the weight keeps it relatively stationary, while Earth and the support move. Inertia can be simply described by this statement: Objects at rest tend to stay at

Seismograph recording earthquake tremors

Bedrock

Wire

Pivot

Suspended weight

Support

Horizontal ground motion

Rotating drum

Bedrock

B.

A.

Figure 6.8 Principle of the seismograph The inertia of the suspended weight tends to keep it motionless while the recording drum, which is anchored to bedrock, vibrates in response to seismic waves. The stationary weight provides a reference point from which to measure the amount of displacement occurring as a seismic wave passes through the ground. (Photo courtesy of Zephyr/Photo Researchers, Inc.)

rest, and objects in motion tend to remain in motion unless acted upon by an outside force. You probably have experienced inertia when you have tried to stop your automobile quickly and your body has continued to move forward.

To detect very weak earthquakes or a great earthquake that has occurred in another part of the world, most seismographs are designed to amplify ground motion. In earthquake-prone areas, the instruments used can withstand the violent shaking that can occur near the epicenter.

Seismic Waves

The records obtained from seismographs, called **seismograms**, provide useful information about the nature of seismic waves. Seismograms reveal that two main types of seismic waves are generated by the slippage of a rock mass. One of these wave types, called **surface waves**, travels in the rock layers just below Earth's surface (**Figure 6.9**). The other wave types travel through Earth's interior and are called **body waves**.

Body Waves Body waves are further divided into two types—called **primary**, or **P, waves** and **secondary**, or **S, waves**—and are identified by their mode of travel through intervening materials. P waves are "push/pull" waves; they momentarily push (compress) and pull (stretch)

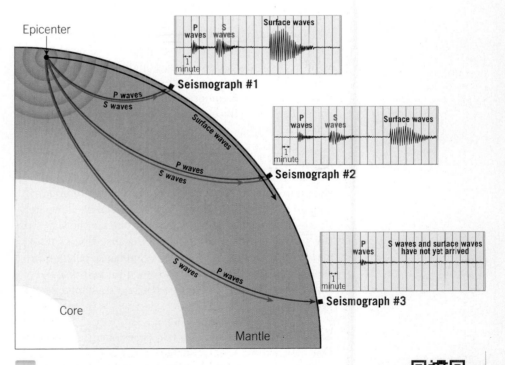

Epicenter

Seismograph #1

Seismograph #2

Seismograph #3

Core

Mantle

SmartFigure 6.9 Body waves (P and S waves) versus surface waves P and S waves travel through Earth's interior, while surface waves travel in the layer directly below the surface. P waves are the first to arrive at a seismic station, followed by S waves, and then surface waves.

Figure 6.10 The characteristic motion of P waves and S waves During a strong earthquake, ground shaking consists of a combination of various kinds of seismic waves.

A. As illustrated by a toy Slinky, P waves alternately compress and expand the material through which they pass.

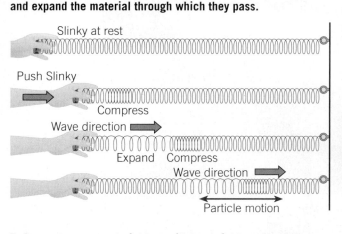

B. S waves cause material to oscillate at right angles to the direction of wave motion.

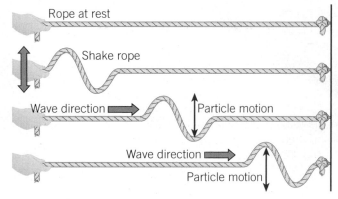

A. One type of surface wave travels along Earth's surface similar to rolling ocean waves. The red arrows show the movement of rock as the wave passes.

B. A second type of surface wave moves the ground from side to side and can be particularly damaging to building foundations.

Figure 6.11 Two types of surface waves

rocks in the direction the wave is traveling (**Figure 6.10A**). This wave motion is similar to that generated by human vocal cords as they move air back and forth to create sound. Solids, liquids, and gases resist stresses that change their volume when compressed and, therefore, elastically spring back once the stress is removed. Therefore, P waves can travel through all these materials.

By contrast, S waves "shake" the particles at right angles to their direction of travel. This can be illustrated by fastening one end of a rope and shaking the other end, as shown in **Figure 6.10B**. Unlike P waves, which temporarily change the *volume* of intervening material by alternately squeezing and stretching it, S waves change the *shape* of the material that transmits them. Because fluids (gases and liquids) do not resist stresses that cause changes in shape—meaning fluids do not return to their original shape once the stress is removed—liquids and gases will not transmit S waves.

Surface Waves There are two types of surface waves. One type causes Earth's surface and anything resting on it to move, much like a ship is tossed by ocean swells (**Figure 6.11A**). The second type of surface wave causes Earth's materials to move side to side. This latter motion is particularly damaging to the foundations of structures (**Figure 6.11B**).

Body Waves Versus Surface Waves By examining the seismogram shown in **Figure 6.12**, you can see that the major difference among seismic waves is their speed of travel. P waves are the first to arrive at a recording station,

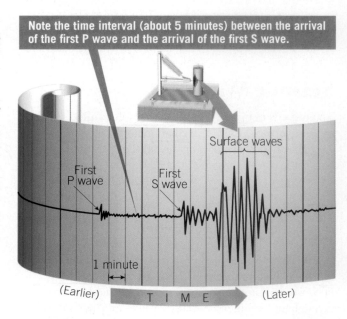

Figure 6.12 Typical seismogram

then S waves, and finally surface waves. Generally, in any solid Earth material, P waves travel about 1.7 times faster than S waves, and S waves are roughly 10 percent faster than surface waves.

In addition to velocity differences, notice in Figure 6.12 that the height, or *amplitude*, of these wave types also varies. S waves have slightly greater amplitudes than P waves, and surface waves exhibit even greater amplitudes. Surface waves also retain their maximum amplitude longer than P and S waves. As a result, surface waves tend to cause greater ground shaking and, hence, greater property damage, than either P or S waves.

6.2 CONCEPT CHECKS

1. Describe the principle of a seismograph.
2. List the major differences between P, S, and surface waves.
3. Which type of seismic waves tend to cause the greatest destruction to buildings?

6.3 Locating the Source of an Earthquake

Explain how an earthquake epicenter can be located using seismographs.

When analyzing an earthquake, the first task seismologists undertake is determining its *epicenter*, the point on Earth's surface directly above the focus (see Figure 6.2). The method used for locating an earthquake's epicenter relies on the fact that P waves travel faster than S waves.

The method is analogous to the results of a race between two automobiles, one faster than the other. The first P wave, like the faster automobile, always wins the race, arriving ahead of the first S wave. The greater the length of the race, the greater the difference in their arrival times at the finish line (the seismic station). Therefore, the greater the interval between the arrival of the first P wave and the arrival of the first S wave, the greater the distance to the epicenter. **Figure 6.13** shows three simplified seismograms for the same earthquake.

Based on the P–S interval, which city—Nagpur, Darwin, or Paris—is farthest from the epicenter?

The system for locating earthquake epicenters was developed by using seismograms from earthquakes whose epicenters could be easily pinpointed from physical evidence. From these seismograms, travel-time graphs were constructed (**Figure 6.14**). Using the sample seismogram for Nagpur, India, in Figure 6.13A and the travel-time curve in Figure 6.14, we can determine the distance separating the recording station from the earthquake in two steps: (1) Using the seismogram, we determine the time interval between the arrival of the first P wave and the arrival of the first S wave, and (2) using the travel-time graph, we find the P–S interval on the

Figure 6.13 Seismograms of the same earthquake recorded at three different locations

Figure 6.14 Travel-time graph A travel-time graph is used to determine the distance to an earthquake's epicenter. The difference in arrival times of the first P and S waves in the example shown is 5 minutes. Thus, the epicenter is roughly 3400 kilometers (2100) miles away.

Figure 6.15 Triangulation to locate an earthquake This method involves using the distance obtained from three or more seismic stations to establish the location of an earthquake.

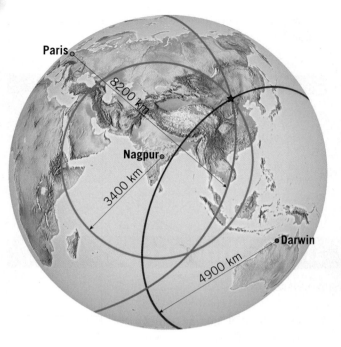

Paris

8200 km

Nagpur

3400 km

Darwin

4900 km

vertical axis and use that information to determine the distance to the epicenter on the horizontal axis. Following this procedure, we can determine that the earthquake occurred 3400 kilometers (2100 miles) from the recording instrument in Nagpur, India.

Now we know the *distance*, but what about *direction*? The epicenter could be in any direction from the seismic station. Using a method called *triangulation*, we can determine the precise location when we know the distance from three or more seismic stations (**Figure 6.15**). On a globe, a circle is drawn around each seismic station. The radius of these circles is equal to the distance from the seismic station to the epicenter. The point where the three circles intersect is the epicenter of the quake.

6.3 CONCEPT CHECKS

1. What information does a travel-time graph provide?
2. Briefly describe the *triangulation* method used to determine the epicenter of an earthquake.

6.4 Determining the Size of an Earthquake

Distinguish between intensity scales and magnitude scales.

Seismologists use a variety of methods to determine two fundamentally different measures that describe the size of an earthquake—*intensity* and *magnitude*. The first of these to be used was **intensity**—a measure of the amount of ground shaking at a particular location, based on observed property damage. Later, with the development of seismographs, it became possible to measure ground motion using instruments. This quantitative measurement, called **magnitude**, relies on data gleaned from seismic records to estimate the amount of energy released at an earthquake's source.

Intensity Scales

Until the mid-1800s, historical records provided the only accounts of the severity of earthquake shaking and destruction. Perhaps the first attempt to scientifically describe the aftermath of an earthquake came following the great Italian earthquake of 1857. By systematically mapping effects of the earthquake, a measure of the intensity of ground shaking was established. The map generated by this study used lines to connect places of equal damage and hence equal ground shaking. Using this technique, zones of intensity were identified, with the zone of highest intensity located near the center of maximum ground shaking and often (but not always) the earthquake epicenter.

In 1902, Giuseppe Mercalli developed a more reliable intensity scale, which in a modified form is still used

today. The **Modified Mercalli Intensity scale**, shown in **Table 6.1**, was developed using California buildings as its standard. For example, on the 12-point Mercalli intensity scale, if an earthquake destroyed some well-built wood structures and most masonry buildings, the affected area would be assigned a Roman numeral X (10). **Figure 6.16** shows the zone of destruction for the 1989 Loma Prieta earthquake, where the intensity of ground shaking was based on the Modified Mercalli Intensity scale.

Magnitude Scales

To more accurately compare earthquakes around the globe, scientists searched for a way to describe the energy released by earthquakes that did not rely on factors such as building practices that vary considerably from one part of the world to another. As a consequence, several magnitude scales were developed.

Richter Magnitude   In 1935 Charles Richter of the California Institute of Technology developed the first magnitude scale to use seismic records. As shown in **Figure 6.17** (top), the **Richter scale** is calculated by measuring the amplitude of the largest seismic wave (usually an S, or surface, wave) recorded on a seismogram. Because seismic waves weaken as the distance between the hypocenter and the seismograph increases, Richter developed a method that accounts for the decrease in wave

TABLE 6.1   Modified Mercalli Intensity Scale

I	Not felt except by a very few under especially favorable circumstances.
II	Felt only by a few persons at rest, especially on upper floors of buildings.
III	Felt quite noticeably indoors, especially on upper floors of buildings, but many people do not recognize it as an earthquake.
IV	During the day felt indoors by many, outdoors by few. Sensation like heavy truck striking building.
V	Felt by nearly everyone, many awakened. Disturbances of trees, poles, and other tall objects sometimes noticed.
VI	Felt by all; many frightened and run outdoors. Some heavy furniture moved; few instances of fallen plaster or damaged chimneys. Damage slight.
VII	Everybody runs outdoors. Damage negligible in buildings of good design and construction; slight to moderate in well-built ordinary structures; considerable in poorly built or badly designed structures.
VIII	Damage slight in specially designed structures; considerable in ordinary substantial buildings with partial collapse; great in poorly built structures. (Fall of chimneys, factory stacks, columns, monuments, walls.)
IX	Damage considerable in specially designed structures. Buildings shifted off foundations. Ground cracked conspicuously.
X	Some well-built wooden structures destroyed. Most masonry and frame structures destroyed. Ground badly cracked.
XI	Few, if any, (masonry) structures remain standing. Bridges destroyed. Broad fissures in ground.
XII	Damage total. Waves seen on ground surfaces. Objects thrown upward into air.

amplitude with increasing distance. Theoretically, as long as equivalent instruments are used, monitoring stations at different locations will obtain the same Richter magnitude for each recorded earthquake. In practice, however, different recording stations often obtain slightly different

Figure 6.16 Seismic intensity map, Loma Prieta earthquake, 1989 Intensity levels are based on the Modified Mercalli Intensity scale, which uses Roman numerals to represent the intensity categories. The zone of maximum intensity during this event roughly corresponded to the epicenter, but this is not always the case.

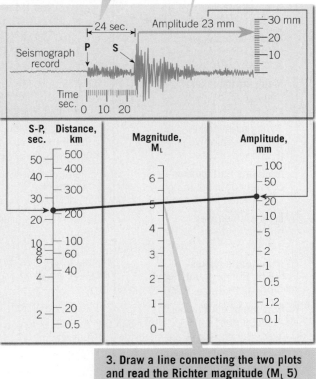

1. Measure the height (amplitude) of the largest wave on the seismogram (23 mm) and plot it on the amplitude scale (right).

2. Determine the distance to the earthquake using the time interval separating the arrival of the first P wave and the arrival of the first S wave (24 seconds) and plot it on the distance scale (left).

3. Draw a line connecting the two plots and read the Richter magnitude (M$_L$ 5) from the magnitude scale (center).

Figure 6.17 Determining the Richter magnitude of an earthquake

Magnitude vs. Ground Motion and Energy

Magnitude Change	Ground Motion Change (amplitude)	Energy Change (approximate)
4.0	10,000 times	1,000,000 times
3.0	1000 times	32,000 times
2.0	100 times	1000 times
1.0	10.0 times	32 times
0.5	3.2 times	5.5 times
0.1	1.3 times	1.4 times

Figure 6.18 Relationship among magnitude, ground motion, and the energy released An earthquake that is one magnitude stronger than another (for example, M 6 versus M 5) produces ground motion that is 10 times greater, and releases about 32 times more energy than the lower magnitude quake.

magnitudes for the same earthquake—a consequence of the variations in rock types through which the waves travel.

Earthquakes vary enormously in strength, and great earthquakes produce wave amplitudes that are thousands of times larger than those generated by weak tremors. To accommodate this wide variation, Richter used a *logarithmic scale* to express magnitude, in which a

tenfold increase in wave amplitude corresponds to an increase of 1 on the magnitude scale. Thus, the intensity of ground shaking for a 5-magnitude earthquake is 10 times greater than that produced by an earthquake having a Richter magnitude of 4 (**Figure 6.18**).

In addition, each unit of increase in Richter magnitude equates to roughly a *32-fold increase in the energy released*. Thus, an earthquake with a magnitude of 6.5 releases 32 times more energy than one with a magnitude of 5.5 and roughly 1000 times (32 × 32) more energy than a 4.5-magnitude quake. A major earthquake with a magnitude of 8.5 releases millions of times more energy than the smallest earthquakes felt by humans (**Figure 6.19**).

The convenience of describing the size of an earthquake by a single number that can be calculated quickly from seismograms makes the Richter scale a powerful tool. Seismologists have since modified Richter's work and developed other Richter-like magnitude scales.

Despite its usefulness, the Richter scale is not adequate for describing very large earthquakes. For example, the 1906 San Francisco earthquake and the 1964 Alaska earthquake have roughly the same Richter magnitudes. However, based on the relative size of the affected areas and the associated tectonic changes, the Alaska

Frequency and Energy Released by Earthquakes of Different Magnitudes

Magnitude (Mw)	Average Per Year	Description	Examples	Energy Release (equivalent kilograms of explosive)
9	<1	**Largest recorded earthquakes—** destruction over vast area massive loss of life possible	Chile, 1960 (M 9.5); Alaska, 1964 (M 9.0); Japan, 2011 (M 9.0)	56,000,000,000,000
8	1	**Great earthquakes—** severe economic impact large loss of life	Sumatra, 2006 (M 8.6); Mexico City, 1980 (M 8.1)	1,800,000,000,000
7	15	**Major earthquakes—** damage ($ billions) loss of life	New Madrid, Missouri 1812 (M 7.7); Turkey, 1999 (M 7.6); Charleston, South Carolina, 1886 (M 7.3)	56,000,000,000
6	134	**Strong earthquakes—** can be destructive in populated areas	Kobe, Japan, 1995 (M 6.9); Loma Prieta, California, 1989 (M 6.9); Northridge, California, 1994 (M 6.7)	1,800,000,000
5	1319	**Moderate earthquakes—** property damage to poorly constructed buildings	Mineral, Virginia, 2011 (M 5.8); Northern New York, 1994 (M 5.8); East of Oklahoma City, Oklahoma, 2011 (M5.6)	56,000,000
4	13,000	**Light earthquakes—** noticeable shaking of items indoors, some property damage	Western Minnesota, 1975 (M 4.6); Arkansas, 2011 (M 4.7)	1,800,000
3	130,000	**Minor earthquakes—** felt by humans, very light property damage, if any	New Jersey, 2009 (M3.0) Maine, 2006 (M3.8)	56,000
2	1,300,000	**Very minor earthquakes—** felt by humans, no property damage		1,800
	Unknown	**Very minor earthquakes—** Generally not felt by humans, but may be recorded		56

Data from USGS

Figure 6.19 Occurrence of earthquakes with various magnitudes (Data from U.S. Geological Survey)

earthquake released considerably more energy than the San Francisco quake. Thus, the Richter scale is considered *saturated* for major earthquakes because it cannot distinguish among them. Despite this shortcoming, Richter-like scales are still used because they can be calculated quickly.

Moment Magnitude For measuring medium and large earthquakes, seismologists have come to favor a newer scale called **moment magnitude** (M_W), which measures the total energy released during an earthquake. Moment magnitude is calculated by determining the average amount of slip on the fault, the area of the fault surface that slipped, and the strength of the faulted rock.

Moment magnitude can also be calculated by modeling data obtained from seismograms. The results are converted to a magnitude number. In addition, as in

the Richter scale, each unit increase in moment magnitude equates to roughly a 32-fold increase in the energy released.

Because moment magnitude estimates the total energy released, it is better for measuring very large earthquakes. In light of this, seismologists have recalculated the magnitudes of older, strong earthquakes using the moment magnitude scale. For example, the 1964 Alaska earthquake, which was originally given a Richter magnitude of 8.3, has since been recalculated using the moment magnitude scale, resulting in an upgrade to 9.2. Conversely, the 1906 San Francisco earthquake that was given a Richter magnitude of 8.3 was downgraded to a moment magnitude of 7.9. The strongest earthquake on record is the 1960 Chilean subduction zone earthquake, with a moment magnitude of 9.5 (**Table 6.2**).

TABLE 6.2   Some Notable Earthquakes

Year	Location	Deaths (est.)	Magnitude*	Comments
856	Iran	200,000		
893	Iran	150,000		
1138	Syria	230,000		
1268	Asia Minor	60,000		
1290	China	100,000		
1556	Shensi, China	830,000		Possibly the greatest natural disaster
1667	Caucasia	80,000		
1727	Iran	77,000		
1755	Lisbon, Portugal	70,000		Tsunami damage extensive
1783	Italy	50,000		
1908	Messina, Italy	120,000		
1920	China	200,000	7.5	Landslide buried a village
1923	Tokyo, Japan	143,000	7.9	Fire caused extensive destruction
1948	Turkmenistan	110,000	7.3	Almost all brick buildings near epicenter collapsed
1960	Southern Chile	5,700	9.5	The largest-magnitude earthquake ever recorded
1964	Alaska	131	9.2	Greatest-magnitude North American earthquake
1970	Peru	70,000	7.9	Great rockslide
1976	Tangshan, China	242,000	7.5	Estimates for the death toll are as high as 655,000
1985	Mexico City	9,500	8.1	Major damage occurred 400 km from epicenter
1988	Armenia	25,000	6.9	Poor construction practices
1990	Iran	50,000	7.4	Landslides and poor construction practices led to great damage
1993	Latur, India	10,000	6.4	Located in stable continental interior
1995	Kobe, Japan	5,472	6.9	Damages estimated to exceed $100 billion
1999	Izmit, Turkey	17,127	7.4	Nearly 44,000 injured and more than 250,000 displaced
2001	Gujarat, India	20,000	7.9	Millions homeless
2003	Bam, Iran	31,000	6.6	Ancient city with poor construction
2004	Indian Ocean (Sumatra)	230,000	9.1	Devastating tsunami damage
2005	Pakistan/Kashmir	86,000	7.6	Many landslides; 4 million homeless
2008	Sichuan, China	87,000	7.9	Millions homeless, some towns will not be rebuilt
2010	Port-au-Prince, Haiti	316,000	7.0	More than 300,000 injured and 1.3 million homeless
2011	Japan	20,000	9.0	Majority of the casualties due to a tsunami

* Widely differing magnitudes have been estimated for some of these earthquakes. When available, moment magnitudes are used.

Source: U.S. Geological Survey.

6.5 Earthquake Destruction

List and describe the major destructive forces that can be triggered by earthquake vibrations.

Did You Know?

It is a commonly held belief that moderate earthquakes decrease the chances of a major earthquake in the same region, but this is not the case. When you compare the amount of energy released by earthquakes of different magnitudes, it turns out that thousands of moderate tremors would be needed to release the huge amount of energy released during one "great" earthquake.

The most violent earthquake ever recorded in North America—the Alaska earthquake—occurred at 5:36 P.M. on March 27, 1964. Felt over most of the state, the earthquake had a moment magnitude (M_W) of 9.2 and lasted 3 to 4 minutes. This event left 128 people dead and thousands homeless, and it badly disrupted the economy of the state. Within 24 hours of the initial shock, 28 aftershocks were recorded, 10 of them exceeding a magnitude of 6. The location of the epicenter and the towns hardest hit by the quake are shown in **Figure 6.20**.

Many factors determine the degree of destruction that accompanies an earthquake. The most obvious is *the magnitude of the earthquake and its proximity to a populated area*. During an earthquake, the region within 20 to 50 kilometers (12 to 30 miles) of the epicenter tends to experience roughly the same degree of ground shaking, and beyond this limit, vibrations usually diminish rapidly. Earthquakes that occur in the stable continental interior, such as the New Madrid earthquake of 1811–1812, are generally felt more over a much larger area than those in earthquake-prone areas such as California.

Destruction from Seismic Vibrations

The 1964 Alaska earthquake provided geologists with insights into the role of ground shaking as a destructive force. As the energy released by an earthquake travels along Earth's surface, it causes the ground to vibrate in a complex manner involving up-and-down as well as side-to-side motion. The amount of damage to human-made structures attributable to the vibrations depends on several factors, including (1) *the intensity* and (2) *the duration of the vibrations*, (3) *the nature of the material on which structures rest*, and (4) *the nature of building materials and construction practices of the region*.

All the multi-story structures in Anchorage were damaged by the vibrations of the 1964 quake. The more flexible wood-frame residential buildings fared best. A striking example of how construction variations affect earthquake damage is shown in **Figure 6.21**. You can see that the steel-frame building on the left withstood the vibrations; whereas the poorly designed JCPenney building was badly damaged. Engineers have learned that buildings built of blocks and bricks that are not reinforced with steel rods are the most serious safety threats in earthquakes. Unfortunately, most of the structures in the developing world are constructed of unreinforced concrete slabs and bricks made of dried mud—a primary reason the death toll in poor countries such as Haiti is usually higher than for earthquakes of similar size in the United States.

The 1964 Alaska earthquake damaged most large structures in Anchorage, even though they were built according to the earthquake provisions of the Uniform Building Code. Perhaps some of that destruction can be attributed to the unusually long duration of the earthquake. Most quakes involve tremors that last less than a minute. For example, the 1994 Northridge earthquake was felt for about 40 seconds, and the strong vibrations of the 1989 Loma Prieta earthquake lasted less than 15 seconds. But the Alaska quake reverberated for 3 to 4 minutes.

Amplification of Seismic Waves Although the region near the epicenter experiences about the same intensity of ground shaking, destruction may vary considerably within this area. These differences are usually attributable to the nature of the ground on which the structures are built. Soft sediments, for example, generally amplify the vibrations more than solid bedrock. Thus, the buildings in Anchorage that were situated on unconsolidated sediments

Figure 6.20 Region most affected by the Alaska earthquake, 1964

Anchorage •   • Valdez
Whittier •   • Chenega
Seward •
Cook Inlet
Kodiak •
Gulf of Alaska

Downtown Anchorage following the 1964 Alaskan earthquake.

Figure 6.22 Ground failure caused this Anchorage street to collapse

Figure 6.21 Comparing damage to structures The extensively damaged five-story JCPenney building in Anchorage, Alaska, was built of concrete. Very little structural camage was incurred by the adjacent steel and glass building. (Courtesy of NOAA)

experienced heavy structural damage (**Figure 6.22**). In contrast, most of the town of Whittier, although much nearer the epicenter, rested on a firm foundation of solid bedrock and suffered much less damage from seismic vibrations.

Liquefaction The intense shaking of an earthquake can cause loosely packed waterlogged materials, such as sandy stream deposits or fill, to be transformed into a substance that acts like fluids (**Figure 6.23**). The phenomenon of transforming a somewhat stable soil into mobile material capable of rising toward Earth's surface is known as **liquefaction**. When liquefaction occurs, the ground may not be capable of supporting buildings, and underground storage tanks and sewer lines may literally float toward the surface (**Figure 6.24**). During the 1989 Loma Prieta earthquake, in San Francisco's Marina District, foundations failed, and geysers of sand and water shot from the ground, evidence that liquefaction had occurred (see Figure 6.23). Liquefaction also contributed to the damage inflicted on San Francisco's water system during the 1906 earthquake. During the 2011 Japan earthquake, liquefaction caused entire buildings to sink several feet.

Landslides and Ground Subsidence

The greatest earthquake-related damage to structures is often caused by landslides and ground subsidence triggered by earthquake vibrations. This was the case during

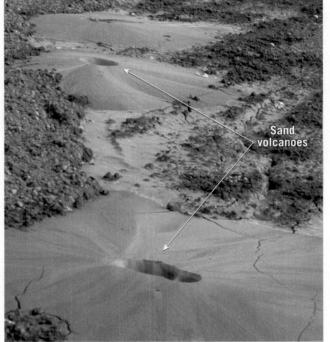

Sand volcanoes

Figure 6.23 Liquefaction These "mud volcanoes" were produced by the Loma Prieta earthquake of 1989. They formed when geysers of sand and water shot from the ground, an indication that liquefaction occurred. (Photo by Richard Hilton, courtesy of Dennis Fox)

Water and sand

Ground settles

Sand volcano

Well packed soil

A. Before earthquake: Loose, saturated, sandy material

B. During earthquake: Water and upper sandy layer mobilize

C. After earthquake: Sandy material becomes more tightly packed

Figure 6.24 Effects of liquefaction (Photo courtesy of USGS)

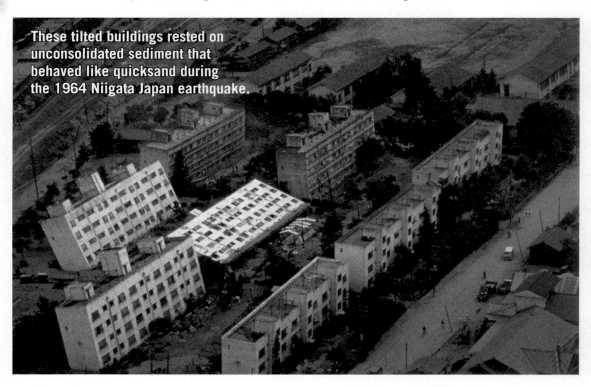

These tilted buildings rested on unconsolidated sediment that behaved like quicksand during the 1964 Niigata Japan earthquake.

 SmartFigure 6.25 The Turnagain Heights slide, caused by the 1964 Alaska earthquake (Photo courtesy of USGS)

Vibrations from the Alaskan earthquake caused cracks to appear near the edge of the Turnagain Heights bluff.

Turnagain Heights

Layers of sand and gravel

Bootlegger Cove clay

Cracks developed

A.

Blocks of land began to slide toward the sea on a weak layer called the Bootlegger Cove clay and in less than 5 minutes, as much as 200 meters of the Turnagain Heights bluff area had been destroyed.

Original profile

200 meters

B.

Photo of a small area of destruction caused by the Turnagain Heights slide.

C.

the 1964 Alaska earthquake in Valdez and Seward, where the violent shaking caused coastal sediments to slump, carrying both waterfronts away. In Valdez, 31 people died when a dock slid into the sea. Because of the threat of recurrence, the entire town of Valdez was relocated to more stable ground about 7 kilometers (4.3 miles) away.

Much of the damage in Anchorage was attributed to landslides. Homes were destroyed in Turnagain Heights when a layer of clay lost its strength and a 200-meter- (660-foot-) wide section of land slid toward the ocean (**Figure 6.25**). A portion of this spectacular landslide was left in its natural condition, as a reminder of this destructive event. The site was appropriately named "Earthquake Park." Downtown Anchorage was also disrupted as sections of the main business district dropped by as much as 3 meters (see Figure 6.22).

Fire

More than 100 years ago, San Francisco was the economic center of the western United States, largely because of gold and silver mining. Then, at dawn on April 18, 1906, a violent earthquake struck, triggering an enormous firestorm (see Figure 6.3). Much of the city was reduced to ashes and ruins. It is estimated that 3000 people died and more than half of the city's 400,000 residents were left homeless.

The historic San Francisco earthquake reminds us of the formidable threat of fire, which started when gas and electrical lines were severed. The initial ground shaking broke the city's water lines into hundreds of disconnected pieces, which made controlling the fires

Figure 6.26 Tsunami generated by displacement of the ocean floor The speed of a wave correlates with ocean depth. Waves moving in deep water advance at speeds in excess of 800 kilometers (500 miles) per hour. Speed gradually slows to 50 kilometers (30 miles) per hour at depths of 20 meters (65 feet). As waves slow in shallow water, they grow in height until they rush onto shore with tremendous force. The size and spacing of these swells are not to scale.

virtually impossible. The fires, which raged out of control for 3 days, were finally contained when buildings were dynamited along a wide boulevard to provide a firebreak, similar to the strategy used in fighting forest fires.

While few deaths were attributed to the San Francisco fires, other earthquake-initiated fires have been more destructive and have claimed many more lives. For example, the 1923 earthquake in Japan triggered an estimated 250 fires, devastating the city of Yokohama and destroying more than half the homes in Tokyo. More than 100,000 deaths were attributed to the fires, which were driven by unusually high winds.

What Is a Tsunami?

Major undersea earthquakes occasionally set in motion a series of large ocean waves that are known by the Japanese name **tsunami** (which means "harbor wave"). Most tsunamis are generated by displacement along a megathrust fault that suddenly lifts a large slab of seafloor (**Figure 6.26**). A tsunami resembles a series of ripples formed when a pebble is dropped into a pond. In contrast to ripples, tsunamis advance across the ocean at amazing speeds, about 800 kilometers (500 miles) per hour— equivalent to the speed of a commercial airliner. Despite this striking characteristic, a tsunami in the open ocean can pass undetected because its height (amplitude) is usually less than 1 meter, and the distance separating wave crests ranges from 100 to 700 kilometers (60 to 435 miles). However, upon entering shallow coastal waters, these destructive waves "feel bottom" and slow, causing the water to pile up (see Figure 6.26). A few exceptional tsunamis have exceeded 30 meters (100 feet) in height. As the crest of a tsunami approaches the shore, it appears as a rapid rise in sea level with a turbulent and chaotic surface; it does not resemble a breaking wave (**Figure 6.27**).

The first warning of an approaching tsunami is often the rapid withdrawal of water from beaches, the result of the trough of the first large wave preceding the crest. Some inhabitants of the Pacific basin have learned to heed this warning and quickly move to higher ground. Approximately 5 to 30 minutes after the retreat of water, a surge capable of extending several kilometers inland occurs. In a successive fashion, each surge is followed by a rapid oceanward retreat of the sea. Therefore, people experiencing a tsunami should not return to the shore once the first surge of water retreats.

Tsunami Damage from the 2004 Indonesia Earthquake A massive undersea earthquake of M_W 9.1 occurred near the island of Sumatra on December 26, 2004, sending waves of water racing across the Indian Ocean and Bay of Bengal (see Figure 6.27). It was one of the deadliest natural disasters of any kind in modern times, claiming more than 230,000 lives. As water surged several kilometers inland, cars and trucks were flung around like toys in a bathtub, and fishing boats were

Figure 6.27 Tsunami generated off the coast of Sumatra, 2004 (AFP/Getty Images Inc.)

Figure 6.28 Japan tsunami, March, 2011 Devastation in Natori City. (Photo by Mike Clark/AFP/ Getty Images)

seismology was the 2011 Tōhoku earthquake (M_W 9.0). This historic earthquake and devastating tsunami resulted in at least 15,861 deaths, more than 3000 people missing, and 6107 injured. Nearly 400,000 buildings, 56 bridges, and 26 railways were destroyed or damaged.

The majority of human casualties and damage were caused by a Pacific-wide tsunami that reached a maximum height of about 40 meters (130 feet) and traveled inland 10 kilometers (6 miles) in the Sendai region (**Figure 6.28**). The chapter-opening photo (pp. 188–189) shows this dramatic event. In addition, meltdowns occurred at three nuclear reactors in Japan's Fukushima Daiichi Nuclear Complex. Across the Pacific in California, Oregon, Peru, and Chile, a few people were killed and several houses, boats, and docks were destroyed. The tsunami was generated when a slab of seafloor located 60 kilometers (37 miles) off the east coast of Japan was suddenly "thrust up" an estimated 5 to 8 meters (16 to 26 feet).

rammed into homes. In some locations, the backwash of water dragged bodies and huge amounts of debris out to sea.

The destruction was indiscriminate, destroying luxury resorts as well as poor fishing hamlets along the Indian Ocean. Damage was reported as far away as the coast of Somalia in Africa, 4100 kilometers (2500 miles) west of the earthquake epicenter.

Japan Tsunami Because of Japan's location along the circum-Pacific belt and its expansive coastline, it is especially vulnerable to tsunami destruction. The most powerful earthquake to strike Japan in the age of modern

Tsunami Warning System In 1946, a large tsunami struck the Hawaiian islands without warning. A wave more than 15 meters (50 feet) high left several coastal villages in shambles. This destruction motivated the U.S. Coast and Geodetic Survey to establish a tsunami warning system for coastal areas of the Pacific that today includes 26 countries. Seismic observatories throughout the region report large earthquakes to the Tsunami Warning Center in Honolulu. Scientists at the center use deep-sea buoys equipped with pressure sensors to detect energy released by an earthquake. In addition, tidal gauges measure the rise and fall in sea level that accompany tsunamis, and warnings are issued within an hour. Although tsunamis travel very rapidly, there is sufficient time to warn all except those in the areas nearest the epicenter. For example, a tsunami generated near the Aleutian islands would take 5 hours to reach Hawaii, and one generated near the coast of Chile would travel 15 hours before reaching the shores of Hawaii (**Figure 6.29**).

Figure 6.29 Tsunami travel times Travel times to Honolulu, Hawaii, from selected locations throughout the Pacific. (Data from NOAA)

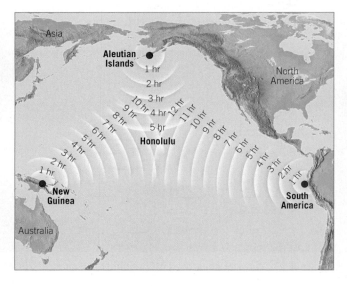

6.5 CONCEPT CHECKS

1. List four factors that affect the amount of destruction that seismic vibrations cause to human-made structures.
2. In addition to the destruction created directly by seismic vibrations, list three other types of destruction associated with earthquakes.
3. What is a *tsunami*? How are tsunamis generated?
4. List at least three reasons an earthquake with a magnitude of 7.0 might inflict more death and destruction than a quake with a magnitude of 8.0.

6.6 Earthquake Belts and Plate Boundaries

Locate Earth's major earthquake belts on a world map and label the regions associated with the largest earthquakes.

About 95 percent of the energy released by earthquakes originates in the few relatively narrow zones, shown in **Figure 6.30**. The zone of greatest seismic activity, called the *circum-Pacific belt*, encompasses the coastal regions of Chile, Central America, Indonesia, Japan, and Alaska, including the Aleutian islands (see Figure 6 30). Most earthquakes in the circum-Pacific belt occur along convergent plate boundaries, where one plate slides at a low angle beneath another. The contacts between the subducting and overlying plates are *megathrust faults*, along which Earth's largest earthquakes are generated.

There are more than 40,000 kilometers (25,000 miles) of subduction boundaries in the circum-Pacific belt where displacement is dominated by thrust faulting. Ruptures occasionally occur along segments that are nearly 1000 kilometers (600 miles) in length, generating catastrophic *megathrust earthquakes* with magnitudes of (M_W) 8 or greater.

Another major concentration of strong seismic activity, referred to as the *Alpine–Himalayan belt*, runs through the mountainous regions that flank the Mediterranean Sea and extends past the Himalaya Mountains (see Figure 6.30). Tectonic activity in this region is mainly attributed to collisions of the African plate with Eurasia and of the Indian plate with Southeast Asia. These plate interactions created many thrust and strike-slip faults that remain active. In addition, numerous faults located a considerable distance from these plate boundaries have been reactivated as India continues its northward advance into Asia. For example, slippage on a complex fault system in 2008 in the Sichuan Province of China killed at least 70,000 people and left 1.5 million others homeless. The "culprit" was the Indian subcontinent continually shoving the Tibetan Plateau eastward against the rocks of the Sichuan basin.

Figure 6.30 shows another continuous earthquake belt that extends for thousands of kilometers through the world's oceans. This zone coincides with the oceanic ridge system, which is an area of frequent but weak seismic activity. As tensional forces pull the plates apart during seafloor spreading, displacement along normal faults generates most of the earthquakes in this zone. The remaining seismic activity is associated with slippage along transform faults located between ridge segments.

Transform faults and smaller strike-slip faults also run through continental crust, where they may generate large earthquakes that tend to occur on a cyclical basis. Examples include California's San Andreas Fault, New Zealand's Alpine Fault, and Turkey's North Anatolian Fault, which produced a deadly earthquake in 1999.

6.6 CONCEPT CHECKS

1. Where does the greatest amount of seismic activity occur?
2. What type of plate boundary is associated with Earth's largest earthquakes?
3. Name another major concentration of strong earthquake activity.

Figure 6.30 Distribution of earthquakes of magnitude 7.5 or greater since 1900 (Data from U.S. Geological Survey)

6.7 Earth's Interior

Explain how Earth acquired its layered structure and briefly describe how seismic waves are used to probe Earth's interior.

If you could slice Earth in half, the first thing you would notice is that it has distinct layers. The heaviest materials (metals) would be in the center. Lighter solids (rocks) would be in the middle, and liquids and gases would be on top. Within Earth we know these layers as the iron core, the rocky mantle and crust, the liquid ocean, and the gaseous atmosphere. More than 95 percent of the variations in composition and temperature within Earth are due to layering. However, this is not the end of the story. If it were, Earth would be a dead, lifeless cinder floating in space.

There are also variations in composition and temperature with depth, which indicate that the interior of our planet is very dynamic. The rocks of the mantle and crust are in constant motion, not only moving about through plate tectonics but also continuously recycling between the surface and the deep interior. Furthermore, it is from Earth's deep interior that the water and air of our oceans and atmosphere are replenished, allowing life to exist at the surface.

Formation of Earth's Layered Structure

As material accumulated to form Earth (and for a short period afterward), the high-velocity impact of nebular debris and the decay of radioactive elements caused the temperature of our planet to steadily increase. During this time of intense heating, Earth became hot enough that iron and nickel began to melt. Melting produced liquid blobs of heavy metal that sank toward the center of the planet. This process occurred rapidly on the scale of geologic time and produced Earth's dense iron-rich core.

The early period of heating resulted in another process of chemical differentiation, whereby melting formed buoyant masses of molten rock that rose toward the surface and solidified to produce a primitive crust. These rocky materials were rich in oxygen and "oxygen-seeking" elements, particularly silicon and aluminum, along with lesser amounts of calcium, sodium, potassium, iron, and magnesium. In addition, some heavy metals such as gold, lead, and uranium, which have low melting points or were highly soluble in the ascending molten masses, were scavenged from Earth's interior and concentrated in the developing crust. This early period of chemical segregation established the three basic divisions of Earth's interior: (1) the iron-rich *core*, (2) the thin *primitive crust*, and (3) Earth's largest layer, called the *mantle*, which is located between the core and crust.

Probing Earth's Interior: "Seeing" Seismic Waves

Discovering the structure and properties of Earth's deep interior has not been easy. Light does not travel through rock, so we must find other ways to "see" into our planet. The best way to learn about Earth's interior is to dig or drill a hole and examine it directly. Unfortunately, this is possible only at shallow depths. The deepest a drilling rig has ever penetrated is only 12.3 kilometers (8 miles), which is about 1/500 of the way to Earth's center! Even this was an extraordinary accomplishment because temperature and pressure increase rapidly with depth.

Many earthquakes are large enough that their seismic waves travel all the way through Earth and can be recorded on the other side (**Figure 6.31**). This means that the seismic waves act like medical x-rays used to take images of a person's insides. There are about 100 to 200 earthquakes each year that are large enough (about $M_w > 6$) to be well recorded by seismographs all around the globe. These large earthquakes provide the means to "see" into our planet and have been the source of most of the data that have allowed us to figure out the nature of Earth's interior.

Interpreting the waves recorded on seismograms in order to identify Earth structure is challenging. Seismic

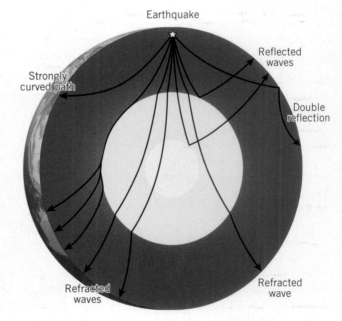

Figure 6.31 Possible paths that earthquake waves follow Notice that in the mantle, the rays follow curved (refracting) paths rather than straight paths because the seismic velocity of rocks increases with depth, as a result of increasing pressure with depth.

waves do not travel along straight paths; instead, seismic waves are *reflected*, *refracted*, and *diffracted* as they pass through our planet. They reflect off boundaries between different layers, they refract (or bend) when passing from one layer to another layer, and they diffract around any obstacles they encounter (see Figure 6.31). These different wave behaviors have been used to identify the boundaries that exist within Earth.

One of the most noticeable behaviors of seismic waves is that they follow strongly curved paths (see Figure 6.31). This occurs because the velocity of seismic waves generally increases with depth. In addition, seismic waves travel faster when rock is stiffer or less compressible. These properties of stiffness and compressibility are then used to interpret the composition and temperature of the rock. For instance, when rock is hotter, it becomes less stiff (imagine heating up a frozen chocolate bar!), and waves travel more slowly. Waves also travel at different speeds through rocks of different compositions. Thus, the speed at which seismic waves travel can help determine both the kind of rock that is inside Earth and how hot it is.

6.7 CONCEPT CHECKS

1. How did Earth acquire its layered structure?
2. Briefly describe how seismic waves are used to probe Earth's interior.

6.8 Earth's Layers

List and describe each of Earth's major layers.

Earth's three compositionally distinct layers—the crust, mantle, and core—can be further subdivided into zones based on physical properties. The physical properties used to define such regions include whether the layer is solid or liquid and how weak or strong it is. Knowledge of both types of layers is essential to our understanding of basic geologic processes, such as volcanism, earthquakes, and mountain building (**Figure 6.32**).

Crust

The **crust** is Earth's relatively thin, rocky outer skin, and there are two types: continental crust and oceanic crust. Both share the word *crust*, but the similarity ends there. The oceanic crust is roughly 7 kilometers (4 miles) thick and composed of the dark igneous rock *basalt*. By contrast, the continental crust averages 35 to 40 kilometers (22 to 25 miles) thick but may exceed 70 kilometers (40 miles) in some mountainous regions such as the Rockies and Himalayas. Unlike the oceanic crust, which has a relatively homogeneous chemical composition, the continental crust consists of many rock types. Although the upper crust has an average composition of a *granitic rock* called *granodiorite*, it varies considerably from place to place.

Continental rocks have an average density of about 2.7 grams per cubic centimeter, and some are 4 billion years old. The rocks of the oceanic crust are younger (180 million years or less) and denser (about 3.0 grams per cubic centimeter) than continental rocks. Liquid water has a density of 1 gram per cubic centimeter; therefore, the density of basalt is three times that of water.

Mantle

Beneath the crust lies the mantle. More than 82 percent of Earth's volume is contained in the **mantle**, a solid, rocky shell that extends to a depth of about 2900 kilometers (1800 miles). The boundary between the crust and mantle represents a marked change in chemical composition. The dominant rock type in the uppermost mantle is *peridotite*, which is richer in the metals magnesium and iron than the minerals found in either the continental or oceanic crust.

The upper mantle extends from the crust–mantle boundary down to a depth of about 660 kilometers (410 miles). The upper mantle can be divided into two different parts. The top portion of the upper mantle is part of the stiff *lithosphere*, and beneath that is the weaker *asthenosphere*. The **lithosphere** ("sphere of rock") consists of the entire crust and uppermost mantle and forms Earth's relatively cool, rigid outer shell. Averaging about 100 kilometers (62 miles) thick, the lithosphere is more than 250 kilometers (155 miles) thick below the oldest portions of the continents (see Figure 6.32). Beneath this stiff layer to a depth of about 350 kilometers (217 miles) lies a soft, comparatively weak layer known as the **asthenosphere** ("weak sphere"). The top portion of the asthenosphere has a temperature/pressure regime that results in a small amount of melting. Within this very weak zone, the asthenosphere and lithosphere are mechanically detached from each other. The result is that the lithosphere is able to move independently of the asthenosphere, a fact we consider in the next chapter.

It is important to emphasize that the strength of various Earth materials is a function of both their composition and the temperature and pressure of their environment. The entire lithosphere does *not* behave like a brittle solid similar to rocks found on the surface. Rather, the rocks of the lithosphere get progressively hotter and weaker (more easily deformed) with increasing depth. At the depth of the uppermost asthenosphere, the rocks are close enough to their melting temperature that they

Figure 6.32 Views of Earth's layered structure The study of seismic waves and other geophysical techniques have shown Earth to be a dynamic planet with many interacting parts. The properties of Earth's layers include the physical state of the material (solid, liquid, or gas) and how stiff the material is (for example, the distinction between the lithosphere and asthenosphere). These studies have shown that Earth's layers are mainly determined by density, with the heaviest materials (iron) at the center and the lightest ones (gases and liquids) on the outside.

are very easily deformed, and some melting may actually occur. Thus, the uppermost asthenosphere is weak because it is near its melting point, just as hot wax is weaker than cold wax.

From 660 kilometers (410 miles) deep to the top of the core, at a depth of 2900 kilometers (1800 miles), is the lower mantle. Because of an increase in pressure (caused by the weight of the rock above), the mantle gradually strengthens with depth. Despite their strength, however, the rocks within the lower mantle are very hot and capable of very gradual flow.

Core

The composition of the **core** is thought to be an iron–nickel alloy, with minor amounts of oxygen, silicon, and sulfur—elements that readily form compounds with iron. At the extreme pressure found in the core, this iron-rich material has an average density of more

than 10 grams per cubic centimeter and is about 13 times the density of water at Earth's center. The core is divided into two regions that exhibit very different mechanical strengths. The **outer core** is a liquid layer 2270 kilometers (1410 miles) thick. It is the movement of metallic iron within this zone that generates Earth's magnetic field. The **inner core** is a sphere with a radius of 1216 kilometers (754 miles). Despite its higher temperature, the iron in the inner core is solid due to the immense pressures that exist in the center of the planet.

6.8 CONCEPT CHECKS

1. How do continental crust and oceanic crust differ?
2. Contrast the physical makeup of the asthenosphere and the lithosphere.
3. How are Earth's inner and outer cores different? How are they similar?

6.9 Rock Deformation

Compare and contrast brittle and ductile deformation.

Earth is a dynamic planet. Shifting lithospheric plates continually change the face of our planet by moving continents across the globe. The results of this tectonic activity are perhaps most strikingly apparent in Earth's major mountain belts. Rocks containing fossils of marine organisms are found thousands of meters above sea level, and massive rock units are bent, contorted, overturned, and sometimes rife with fractures.

Why Rocks Deform

Every body of rock, no matter how strong, has a point at which it will fracture or flow. **Deformation** is a general term that refers to all changes in the shape, position, or orientation of a rock mass. Significant crustal deformation occurs along plate margins. Plate motions and the interactions along plate boundaries generate the tectonic forces that cause rock to deform.

When rocks are subjected to forces (stresses) greater than their own strength, they begin to deform, usually by bending or breaking (**Figure 6.33**). It is easy to visualize how rocks break because we normally think of them

as being brittle. But how can rock masses be bent into intricate folds without being broken during the process? To answer this question, geologists performed laboratory experiments in which rocks were subjected to forces under conditions that simulated those existing at various depths within the crust. Although each rock type deforms somewhat differently, the general characteristics of rock deformation were determined from these experiments.

Types of Rock Deformation

Geologists discovered that when stress is gradually applied, rocks first respond by deforming elastically. Changes that result from **elastic deformation** are recoverable; that is, like a rubber band, the rock will return to nearly its original size and shape when the force is removed. During elastic deformation, the chemical bonds of the minerals within a rock are stretched but do not break. Once the elastic limit (strength) of a rock is surpassed, it either flows (*ductile deformation*) or fractures (*brittle deformation*).

SmartFigure 6.33 Deformed sedimentary strata These deformed strata are exposed in a road cut near Palmdale, California. In addition to the obvious folding, light-colored beds are offset along a fault located on the right side of the photograph. (Photo by E. J. Tarbuck)

Geologist's Sketch

The factors that influence the strength of a rock and how it will deform include temperature, confining pressure, rock type, and time. Rocks near the surface, where temperatures and confining pressures are low, tend to behave like a brittle solid and fracture once their strength is exceeded. This type of deformation is called **brittle deformation**. From our everyday experience, we know that glass objects, wooden pencils, china plates, and even our bones exhibit brittle failure once their strength is surpassed. By contrast, at depth, where temperatures and confining pressures are high, rocks exhibit *ductile* behavior. **Ductile deformation** is a type of solid-state flow that produces a change in the size and shape of an object without fracturing. Ordinary objects that display ductile behavior include modeling clay, bee's wax, caramel candy, and most metals. For example, a copper penny placed on a railroad track will be flattened and deformed (without

breaking) by the force applied by a passing train. Ductile deformation of a rock—strongly aided by high temperature and high confining pressure—is somewhat similar to the deformation of a penny flattened by a train. In rocks, ductile deformation is the result of some chemical bonds breaking, while others are forming, allowing minerals to change shape. Rocks that display evidence of ductile flow were usually deformed at great depth and may exhibit contorted folds that give the impression that the strength of the rock was akin to the strength of soft putty.

6.9 CONCEPT CHECKS

1. Describe elastic deformation.
2. How is brittle deformation different from ductile deformation?

6.10 Folds: Structures Formed by Ductile Deformation

List and describe the major types of folds.

Along convergent plate boundaries, flat-lying sedimentary and volcanic rocks are often bent into a series of wavelike undulations called **folds**. Folds in sedimentary strata are much like those that would form if you were to hold the ends of a sheet of paper and then push them together. In nature, folds come in a wide variety of sizes and configurations. Some folds are broad flexures in which rock units hundreds of meters thick have been slightly warped. Others are very tight microscopic structures found in metamorphic rocks. Size differences notwithstanding, most folds are caused by *compressional forces that result in the shortening and thickening of the crust.*

Anticlines and Synclines

The two most common types of folds are anticlines and synclines (**Figure 6.34**). **Anticlines** usually arise by upfolding, or arching, of sedimentary layers and are sometimes spectacularly displayed along highways that have been cut through deformed strata. Almost always found in association with anticlines are downfolds, or troughs, called **synclines** (see Figure 6.34).* Notice in Figure 6.34 that

the *limb* (or flank) of an anticline is also a limb of the adjacent syncline.

Depending on their orientation, these basic folds are described as *symmetrical* when the limbs are mirror images of each other and *asymmetrical* when they are not. An asymmetrical fold is said to be *overturned* if one or both limbs are tilted beyond the vertical (see Figure 6.34). An overturned fold can also "lie on its side" so a plane extending through the axis of the fold would be horizontal. These *recumbent* folds are common in highly deformed mountainous regions such as the Alps.

Domes and Basins

Broad upwarps in basement rock may deform the overlying cover of sedimentary strata and generate large folds. When this upwarping produces a circular or elongated structure, the feature is called a **dome**. Downwarped structures having a similar shape are termed **basins**.

The Black Hills of western South Dakota is a large domed structure thought to have been generated by upwarping. Erosion has stripped away the highest portions of the upwarped sedimentary beds, exposing older igneous and metamorphic rocks in the center (**Figure 6.35**). Remnants of these once-continuous sedimentary

---

* By strict definition, an *anticline* is a structure in which the oldest strata are found in the center. This most typically occurs when strata are upfolded. A *syncline* is strictly defined as a structure in which the youngest strata are found in the center. This occurs most commonly when strata are downfolded.

SmartFigure 6.34 Common types of folds The upfolded, or arched, structures are anticlines. The downfolds, or troughs, are synclines. Notice that the limb of an anticline is also the limb of the adjacent syncline.

layers are visible, flanking the crystalline core of these mountains.

Several large basins exist in the United States. The basins of Michigan and Illinois have very gently sloping beds similar to saucers. These basins are thought to have resulted from large, heavy accumulations of sediment causing the crust to subside.

Because large basins usually contain sedimentary beds sloping at very low angles, they are usually identified by the age of the rocks comprising them. The youngest rocks are found near the center, and the oldest rocks are at the flanks. This is just the opposite order of a domed structure, such as the Black Hills, where the oldest rocks form the core.

Figure 6.35 The Black Hills of South Dakota, a large structural dome The central core of the Black Hills is composed of resistant igneous and metamorphic rocks that have been exposed by erosion. Map view (above), and cross-sectional view from point X to point Y (below).

6.10 CONCEPT CHECKS

1. Distinguish between anticlines and synclines, domes and basins, anticlines and domes.
2. The Black Hills of South Dakota is a good example of what type of geologic structure?
3. Where do the youngest rocks in an eroded basin outcrop: near the center or near the flanks?

6.11 Faults: Structures Formed by Brittle Deformation

Sketch and describe the relative motion of rock bodies located on opposite sides of normal, reverse, and strike-slip faults.

Faults are fractures in the crust along which appreciable displacement has taken place. Occasionally, small faults can be recognized in road cuts where sedimentary beds have been offset a few meters, as shown in **Figure 6.36**. Faults of this scale usually occur as single discrete breaks. By contrast, large faults, such as the San Andreas Fault in California, have displacements of hundreds of kilometers and consist of many interconnecting fault surfaces. These *fault zones* can be several kilometers wide and are often easier to identify from high-altitude photographs than at ground level.

Dip-Slip Faults

Faults in which movement is primarily parallel to the *dip* (or inclination) of the fault surface are called **dip-slip faults**. It has become common practice to call the rock surface that is immediately above the fault the **hanging wall block** and to call the rock surface below the **foot-wall block** (see Figure 6.36). This nomenclature arose from prospectors and miners who excavated shafts and tunnels along fault zones that were sites of ore deposits. In these tunnels, the miners would walk on the rocks below the mineralized fault zone (the footwall block) and hang their lanterns on the rocks above (the hanging wall block).

Vertical displacements along dip-slip faults may produce long, low cliffs called **fault scarps**. Fault scarps,

Figure 6.37 Fault scarp This scarp was created during the great Alaska earthquake of 1964. (Photo courtesy of USGS)

such as the one shown in **Figure 6.37**, are produced by displacements that generate earthquakes.

Normal Faults Dip-slip faults are classified as **normal faults** when the hanging wall block moves down relative to the footwall block (**Figure 6.38A**). Because of the downward motion of the hanging wall block, normal faults accommodate lengthening, or extension, of the crust.

Normal faults are found in a variety of sizes. Some are small, having displacements of only 1 meter (3 feet) or so, like the one shown in the road cut in Figure 6.36. Others extend for tens of kilometers and may sinuously trace the boundary of a mountain front. Most large normal faults have relatively steep dips that tend to flatten out with depth.

Fault-Block Mountains In the western United States, large normal faults are associated with structures called **fault-block mountains**. Excellent examples of fault-block mountains are found in the Basin and Range Province, a region that encompasses Nevada and portions of the surrounding states (**Figure 6.39**). Here the crust has been elongated and broken to create more than 200 relatively small mountain ranges. Averaging about 80 kilometers (50 miles) in length, the ranges rise 900 to 1500 meters (3000 to 5000 feet) above the adjacent down-faulted basins.

The topography of the Basin and Range Province evolved in association with a system of normal faults that trend roughly north–south. Movements along these faults produced alternating uplifted fault blocks called **horsts** and down-dropped blocks called **grabens**. Horsts generate elevated topography, whereas grabens form basins. As Figure 6.39 illustrates, structures called *half-grabens*, which are titled fault blocks, also contribute to the alternating

Figure 6.36 Faulting caused the displacement of these beds Arrows show the relative motion of rock units. (Photo by Marli Miller)

Tensional stress

Footwal block

Hanging wall block

A. Normal fault

Compressional stress

Hanging wall block

Footwall block

B. Reverse fault

Compressional stress

Hanging wall block

Footwall block

C. Thrust fault

Shear stress

D. Strike-slip fault

Figure 6.38 Block diagrams of four types of faults

Graben

Horst

Fault-block mountains

Extension

Half graben

Horst

Horst

Graben

Detachment fault

Figure 6.39 Normal faulting in the Basin and Range Province Here, tensional stresses have elongated and fractured the crust into numerous blocks. Movement along these fractures has tilted the blocks, producing parallel mountain ranges called *fault-block mountains.* The downfaulted blocks (grabens) form basins, whereas the upfaulted blocks (horsts) are eroded to form rugged mountainous topography. In addition, numerous tilted blocks (half-grabens) form both basins and mountains. (Photo by Michael Collier) topographic highs and lows in the Basin and Range Province. The horsts and higher ends of the tilted fault blocks are the source of sediments that have accumulated in the basins created by the grabens and lower ends of the tilted blocks.

Notice in Figure 6.39 that the slopes of the normal faults decrease with depth and eventually join to form a nearly horizontal fault called a *detachment fault*. These faults form a major boundary between the rocks below, which exhibit ductile deformation, and the rocks above, which exhibit brittle deformation.

Fault motion provides geologists with a method of determining the nature of the tectonic forces at work within Earth. Normal faults are associated with tensional forces that pull the crust apart. This "pulling apart" can be accomplished either by uplifting that causes the surface to stretch and break or by opposing horizontal forces.

Reverse and Thrust Faults Dip-slip faults in which the hanging wall block moves up relative to the footwall block are called **reverse faults** and **thrust faults** (**Figure 6.38B,C**). Thrust faults are reverse faults having dips less than 45 degrees, so the overlying block moves nearly horizontally over the underlying block.

Whereas normal faults occur in tensional environments, reverse faults result from strong compressional stresses. Because the hanging wall block moves up and over the footwall block, reverse and thrust faults accommodate horizontal shortening of the crust.

Most high-angle reverse faults are small and accommodate local displacements in regions dominated by other types of faulting. Thrust faults, on the other hand, exist at all scales, with some large thrust faults having displacements on the order of tens to hundreds of kilometers.

In mountainous regions such as the Alps, Northern Rockies, Himalayas, and Appalachians, thrust faults have displaced strata as far as 100 kilometers (60 miles) over adjacent rock units. The result of this large-scale movement is that older strata end up overlying younger rocks.

Thrust faulting is most pronounced in subduction zones and other convergent boundaries where plates are colliding. Compressional forces generally produce folds as well as faults and result in a thickening and shortening of the material involved.

Strike-Slip Faults

A fault in which the dominant displacement is horizontal and parallel to the *strike* (the compass direction) of the fault surface is called a **strike-slip fault** (Figure 6.38D). The earliest scientific records of strike-slip faulting were made following surface ruptures that produced large earthquakes. One of the most noteworthy of these was the great San Francisco earthquake of 1906. During this strong earthquake, structures such as fences that were built across the San Andreas Fault were displaced as much as 4.7 meters (15 feet). Because movement along the San Andreas causes the crustal block on the opposite side of the fault to move to the right as you face the fault, it is called a *right-lateral* strike-slip fault.

The Great Glen Fault in Scotland is a well-known example of a *left-lateral* strike-slip fault, which exhibits the opposite sense of displacement. The total displacement along the Great Glen Fault is estimated to exceed 100 kilometers (60 miles). Also associated with this fault trace are numerous lakes, including Loch Ness, the home of the legendary monster.

Some strike-slip faults cut through the lithosphere and accommodate motion between two large tectonic plates. Recall that this special kind of strike-slip fault is called a **transform fault**. Numerous transform faults cut the oceanic lithosphere and link spreading oceanic ridges. Others accommodate displacement between continental plates that move horizontally with respect to each other. One of the best-known transform faults is California's San Andreas Fault.

6.11 CONCEPT CHECKS

1. Contrast the movements that occur along normal and reverse faults.
2. What type of faults are associated with fault-block mountains?
3. How are reverse faults different from thrust faults? In what way are they the same?
4. Describe the relative movement along a strike-slip fault.

6.12 Mountain Building

Locate and identify Earth's major mountain belts on a world map.

Mountain building has occurred in the recent geologic past in several locations around the world (**Figure 6.40**). One example is the American Cordillera, which runs along the western margin of the Americas from Cape Horn at the tip of South America to Alaska and includes the Andes and Rocky Mountains. Other examples include the Alpine–Himalaya chain, which extends along the margin of the

Mediterranean, through Iran to northern India, and into Indochina; and the mountainous terrains of the western Pacific, which include volcanic island arcs that comprise Japan, the Philippines, and Sumatra. Most of these young mountain belts have come into existence within the past 100 million years. Some, including the Himalayas, began their growth as recently as 50 million years ago.

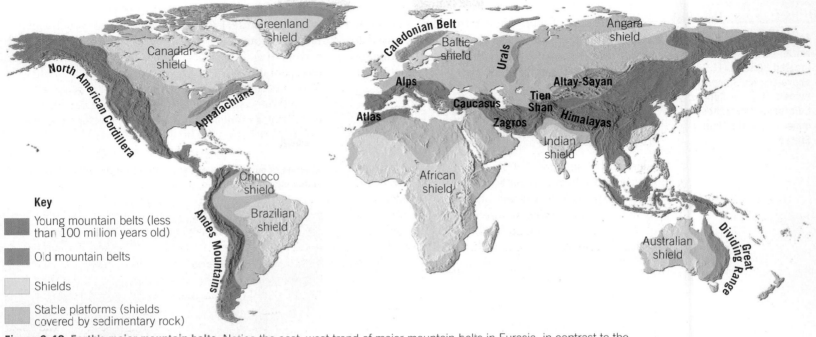

Key

- Young mountain belts (less than 100 million years old)
- Old mountain belts
- Shields
- Stable platforms (shields covered by sedimentary rock)

Figure 6.40 Earth's major mountain belts Notice the east–west trend of major mountain belts in Eurasia, in contrast to the north–south trend of the North and South American Cordillera.

In addition to these young mountain belts, there are several chains of Paleozoic-age mountains on Earth. Although these older structures are deeply eroded and topographically less prominent, they exhibit the same structural features found in younger mountains. The Appalachians in the eastern United States and the Urals in Russia are classic examples of this group of older and well-worn mountain belts.

The term for the processes that collectively produce a mountain belt is **orogenesis** (*oros* = mountain, *genesis* = to come into being). Most major mountain belts display striking visual evidence of great horizontal forces that have shortened and thickened the crust. These **compressional mountains** contain large quantities of preexisting sedimentary and crystalline rocks that have been faulted and contorted into a series of folds. Although folding and thrust faulting are often the most conspicuous signs of orogenesis, varying degrees of metamorphism and igneous activity are always present.

How do mountain belts form? This question has intrigued some of the greatest philosophers and scientists since the time of the ancient Greeks. One early proposal suggested that mountains are simply wrinkles in Earth's crust, produced as the planet cooled from its original semimolten state. According to this idea, Earth contracted and shrank as it lost heat, which caused the crust to deform in a manner similar to how an orange peel wrinkles as the fruit dries out. However, neither this nor any other early hypothesis withstood scientific scrutiny.

With the development of the theory of plate tectonics, a model for orogenesis with excellent explanatory power has emerged. According to this model, the tectonic processes that generate Earth's major mountainous terrains begin at convergent plate boundaries, where oceanic lithosphere descends into the mantle.

6.12 CONCEPT CHECKS

1. Define *orogenesis*.
2. In the plate tectonics model, which type of plate boundary is most directly associated with Earth's major mountain belts?

6.13 Subduction and Mountain Building

Sketch an Andean-type mountain belt and describe how each of its major features is generated.

The subduction of oceanic lithosphere is the driving force of mountain building (orogenesis). Where oceanic lithosphere subducts beneath an oceanic plate, a *volcanic island arc* and related tectonic features develop.

Subduction beneath a continental block, on the other hand, results in the formation of a *continental volcanic arc* and mountainous topography along the margin of a continent. In addition, volcanic island arcs and other crustal fragments "drift" across the ocean basin until they reach a subduction zone, where they collide and become welded to another crustal fragment or larger continental block. If subduction continues long enough, it can ultimately result in the collision of two or more continents.

Island Arc–Type Mountain Building

Island arcs result from the steady subduction of oceanic lithosphere, which may last for 200 million years or more (**Figure 6.41**). Periodic volcanic activity, the emplacement of igneous plutons at depth, and the accumulation of sediment that is scraped from the subducting plate gradually increase the volume of crustal material capping the upper plate. Some large volcanic island arcs, such as Japan, owe their size to having been built upon a preexisting fragment of continental crust or the joining of two or more island arcs.

The continued growth of a volcanic island arc can result in the formation of mountainous topography consisting of belts of igneous and metamorphic rocks. This activity, however, is viewed as just one phase in the development of a major mountain belt.

Andean-Type Mountain Building

Mountain building along continental margins involves the convergence of an oceanic plate and a plate whose leading edge contains continental crust. Exemplified by the Andes Mountains, an *Andean-type convergent zone* results in the formation of a continental volcanic arc and related tectonic features inland of the continental margin.

The first stage in the development of an idealized Andean-type mountain belt occurs prior to the formation of the subduction zone. During this period, the continental margin is a **passive continental margin**; that is, it is not a plate boundary but a part of the same plate as the

adjoining oceanic crust. The East coast of North America provides a present-day example of a passive continental margin. Here, as at other passive continental margins surrounding the Atlantic, deposition of sediment on the continental shelf is producing a thick wedge of shallow-water sandstones, limestones, and shales (**Figure 6.42A**). Beyond the continental shelf, turbidity currents are depositing sediments on the continental slope and rise (see Chapter 9).

A. Passive continental margin with an extensive platform of sediments and sedimentary rocks.

B. Plate convergence generates a subduction zone, and partial melting produces a continental volcanic arc. Compressional forces and igneous activity further deform and thicken the crust, elevating the mountain belt.

C. Subduction ends and is followed by a period of uplift.

Figure 6.42 Mountain building along an Andean-type subduction zone

Figure 6.41 Development of a volcanic island arc A volcanic island arc forms where one slab of oceanic lithosphere is descending beneath another, by the convergence of two oceanic plates. Continuous subduction along these convergent zones results in the development of thick units of continental-type crust.

Continuous subduction of oceanic lithosphere results in the development of thick units of continental-like crust upon the overlying plate.

At some point, the continental margin becomes active. A subduction zone forms, and the deformation process begins (**Figure 6.42B**). A good place to examine an **active continental margin** is the west coast of South America, where the Nazca plate is being subducted beneath the South American plate along the Peru–Chile trench.

In an idealized Andean-type subduction, convergence of the continental block and the subducting oceanic plate leads to deformation and metamorphism of the continental margin. Once the oceanic plate descends to about 100 kilometers (62 miles), partial melting of mantle rock above the subducting slab generates magma that migrates upward (see Figure 6.42B). Thick continental crust greatly impedes the ascent of magma. Consequently, a high percentage of the magma that intrudes the crust never reaches the surface; instead, it crystallizes at depth to form intrusive igneous rocks (plutons). Eventually, uplifting and erosion exhume these igneous bodies and associated metamorphic rocks. Once they are exposed at the surface, these massive structures are called *batholiths* (**Figure 6.42C**). Composed of numerous plutons, batholiths form the core of the Sierra Nevada in California and are prevalent in the Peruvian Andes.

During the development of this continental volcanic arc, sediment derived from the land and scraped from the subducting plate is plastered against the landward side of the trench like piles of dirt in front of a bulldozer. This chaotic accumulation of sedimentary and metamorphic rocks with occasional scraps of ocean crust is called an **accretionary wedge** (see Figure 6.42B). Prolonged subduction can build an accretionary wedge that is large enough to stand above sea level (see Figure 6.42C).

Andean-type mountain belts are composed of two roughly parallel zones. The volcanic arc develops on the continental block. It consists of volcanoes and large intrusive bodies intermixed with high-temperature metamorphic rocks. The seaward segment is the accretionary wedge. It consists of folded and faulted sedimentary and metamorphic rocks (see Figure 6.42C).

Sierra Nevada and Coast Ranges One of the best examples of an inactive Andean-type orogenic belt is found in the western United States. It includes the Sierra Nevada and the Coast Ranges in California. These parallel mountain belts were produced by the subduction of a portion of the Pacific basin under the western edge of the North American plate. The Sierra Nevada batholith is a remnant of a portion of the continental volcanic arc that was produced by several surges of magma over tens of millions of years. Subsequent uplifting and erosion have removed most of the evidence of past volcanic activity and exposed a core of crystalline, igneous, and associated metamorphic rocks.

In the trench region, sediments scraped from the subducting plate, plus those provided by the eroding continental volcanic arc, were intensely folded and faulted into an accretionary wedge. This chaotic mixture of rocks presently constitutes the Franciscan Formation of California's Coast Ranges. Uplifting of the Coast Ranges took place only recently, as evidenced by the young unconsolidated sediments that still mantle portions of these highlands.

6.13 CONCEPT CHECKS

1. Explain why the formation of mountainous topography at a volcanic island arc is considered just one phase in the development of a major mountain belt.
2. Describe and give an example of a passive continental margin.
3. What is an accretionary wedge? Briefly describe its formation.
4. In what ways are the Sierra Nevada and the Andes similar?

6.14 Collisional Mountain Belts

Summarize the stages in the development of a collisional mountain belt such as the Himalayas.

Most major mountain belts are generated when one or more buoyant crustal fragments collide with a continental margin as a result of subduction. Oceanic lithosphere, which is relatively dense, readily subducts, whereas continental lithosphere, which contains significant amounts of low-density crustal rocks, is too buoyant to undergo subduction. Consequently, the arrival of a crustal fragment at a trench results in a collision with the margin of the overlying continental block and usually ends further subduction.

Cordilleran-Type Mountain Building

Cordilleran-type mountain building, named after the North American Cordillera, occurs in Pacific-like ocean basins, where a rapid rate of seafloor spreading is roughly balanced by an equally high rate of subduction. In this setting, it is highly likely that island arcs or small crustal fragments will be carried along until they collide with an

Did You Know?
During the assembly of Pangaea, the landmasses of Europe and Siberia collided to produce the Ural Mountains. Long before the discovery of plate tectonics, this extensively eroded mountain chain was regarded as the boundary between Europe and Asia.

active continental margin. The process of collision and accretion (joining together) of comparatively small crustal fragments to a continental margin has generated many of the mountainous regions that rim the Pacific.

Geologists refer to these accreted crustal blocks as *terranes*. The term **terrane** is used to describe a crustal fragment consisting of a distinct and recognizable series of rock formations that has been transported by plate tectonic processes. By contrast, the term *terrain* is used when describing the shape of the surface topography, or the "lay of the land."

The Nature of Terranes

What is the nature of these crustal fragments, and where did terranes originate? Research suggests that prior to their accretion to a continental block, some of these fragments may have been **microcontinents** similar to the modern-day island of Madagascar, located east of Africa in the Indian Ocean. Many others were island arcs similar to Japan, the Philippines, and the Aleutian islands. Still others may have been submerged oceanic plateaus created by massive outpourings of basaltic lavas associated with mantle plumes. More than 100 of these relatively small crustal fragments, most of which are in the Pacific, are known to exist (**Figure 6.43**).

Accretion and Orogenesis

As oceanic plates move, they carry embedded oceanic plateaus, volcanic island arcs, and microcontinents to an Andean-type subduction zone. When an oceanic plate contains small seamounts, these structures are generally subducted along with the descending oceanic slab. However, large thick units of oceanic crust, such as the Ontong Java Plateau, which is the size of Alaska, or an island arc composed of abundant "light" igneous rocks, render the oceanic lithosphere too buoyant to subduct. In these situations, a collision between the crustal fragment and the continental margin occurs.

The sequence of events that happen when small crustal fragments reach a Cordilleran-type margin is shown in **Figure 6.44**. Rather than subduct, the upper crustal layers of these thickened zones are peeled from the descending plate and thrust in relatively thin sheets upon the adjacent continental block. Convergence does not generally end with the accretion of a crustal fragment. Rather, new subduction zones typically form, and they can carry other island arcs or microcontinents toward a collision with the continental margin. Each collision displaces earlier accreted terranes further inland, adding to the zone of deformation as well as to the thickness and lateral extent of the continental margin.

The North American Cordillera

The correlation between mountain building and the accretion of crustal fragments arose primarily from studies conducted in the North American Cordillera (**Figure 6.45**). Researchers determined that some of the rocks in the orogenic belts of Alaska and British Columbia contained fossil and paleomagnetic evidence which indicated that these strata previously lay much closer to the equator.

It is now known that many of the terranes that make up the North American Cordillera were scattered throughout the eastern Pacific, like the island arcs and oceanic plateaus currently distributed in the western Pacific. During the breakup of Pangaea, the eastern portion of the Pacific basin (Farallon plate) began to subduct under the western margin of North America. This activity resulted in the piecemeal addition of crustal fragments to the entire Pacific margin of the continent—from Mexico's Baja Peninsula to northern Alaska (see Figure 6.45). Geologists expect that many modern microcontinents will likewise be accreted to active continental margins surrounding the Pacific, producing new orogenic belts.

Figure 6.43 Distribution of present-day oceanic plateaus and other submerged crustal fragments (Data from Ben-Avraham and others)

A. A microcontinent and a volcanic island arc are being carried toward a subduction zone.

B. The volcanic island arc is sliced off the subducting plate and thrust onto the continent.

C. A new subduction zone forms seaward of the old subduction zone.

D. The accretion of the microcontinent to the continental margin shoves the remnant island arc further inland and grows the continental margin seaward.

SmartFigure 6.44
Collision and accretion of an island arc and a microcontinent to a continental margin

Figure 6.45 Terranes that have been added to western North America during the past 200 million years (Redrawn after D. R. Hutchinson and others)

ARCTIC OCEAN

ALASKA

Wrangellia

Yukon-Tanana

NORTH AMERICAN CORDILLERA

Stikinia

Wrangellia

Cache Creek

Eastern limit of Cordilleran deformation

CANADA

PACIFIC OCEAN

Franciscan

UNITED STATES

Oceanic terranes

Area deformed by the accretion of terranes

Continental Collisions: Alpine-Type Mountain Building

Alpine-type orogenies, named after the Alps that have been studied intensely for more than 200 years, occur where two continental masses collide. This type of orogeny may also involve the accretion of continental fragments or island arcs that occupied the ocean basin that once separated the two continental blocks. Mountain belts that were formed by the closure of major ocean basins include the Himalayas, Appalachians, Urals, and Alps. Continental collisions result in the development of mountains characterized by shortened and thickened crust achieved through folding and large-scale thrust faulting.

The Himalayas The mountain-building episode that created the Himalayas began roughly 50 million years

ago, when India began to collide with Asia. Prior to the breakup of Pangaea, India was located between Africa and Antarctica in the Southern Hemisphere. As Pangaea fragmented, India moved rapidly, geologically speaking, a few thousand kilometers in a northward direction.

The subduction zone that facilitated India's northward migration was near the southern margin of Asia (**Figure 6.46A**). Continued subduction along Asia's margin created an Andean-type plate margin that contained a well-developed continental volcanic arc and an accretionary wedge. India's northern margin, on the other hand, was a passive continental margin consisting of a thick platform of shallow-water sediments and sedimentary rocks.

Geologists have determined that one or perhaps more small continental fragments were positioned on the subducting plate somewhere between India and Asia. During the closing of the intervening ocean basin, a small crustal fragment, which now forms southern Tibet, reached the trench. This event was followed by the collision of India with Asia. The tectonic forces involved in this collision were immense and caused the more deformable materials located on the seaward edges of these landmasses to become highly folded and faulted (**Figure 6.46B**). The shortening and thickening of the crust elevated great quantities of crustal material, thereby generating the spectacular Himalaya Mountains.

In addition to uplift, crustal shortening caused rocks at the "bottom of the pile" to become deeply buried—an environment where they experienced elevated temperatures and pressures (see Figure 6.46B). Partial melting within the deepest and most deformed region of the developing mountain belt produced magmas that intruded the overlying rocks. It is in these environments that the metamorphic and igneous cores of collisional mountains are generated.

The formation of the Himalayas was followed by a period of uplift that raised the Tibetan Plateau. Seismic evidence suggests that a portion of the Indian subcontinent was thrust beneath Tibet a distance of perhaps 400 kilometers (250 miles). If so, the added crustal thickness would account for the lofty landscape of southern Tibet, which has an average elevation higher than Mount Whitney, the highest point in the contiguous United States.

The collision with Asia slowed but did not stop the northward migration of India, which has since penetrated at least 2000 kilometers (1200 miles) into the mainland of Asia. Crustal shortening accommodated some of this motion. Much of the remaining penetration into Asia caused lateral displacement of large blocks of the Asian crust by a mechanism described as *continental escape*. As shown in **Figure 6.46C**, when India continued its northward trek, parts of Asia were "squeezed" eastward out of the collision zone. These displaced crustal blocks include much of present-day mainland Southeast Asia and sections of China.

A. Prior to the collision of India and Asia, India's northern margin consisted of a thick platform of continental shelf sediments, whereas Asia's was an active continental margin with a well developed accretionary wedge and volcanic arc.

B. The ensuing continental collision folded and faulted the crustal rocks that lay along the margins of these continents to form the Himalayas. This event was followed by the gradual uplift of the Tibetan Plateau as the subcontinent of India was shoved under Asia.

C. Map view showing the southeastward displacement of China and the mainland of Southeast Asia as India plowed into Asia.

SmartFigure 6.46 The collision of India with the Eurasian plate to produce the spectacular Himalayas

Other Continental Collisions A similar, but much earlier, collision is believed to have taken place when the European continent collided with the Asian continent to produce the Ural Mountains, which extend in a north–south direction through Russia. Prior to the discovery of plate tectonics, geologists had difficulty explaining the existence of mountain ranges such as the Urals, which are located deep within continental interiors. How could thousands of meters of marine sediment be deposited and then become highly deformed while situated in the middle of a large, stable landmass?

Other mountain ranges showing evidence of continental collisions are the Alps and the Appalachians. The Appalachians resulted from collisions among North America, Europe, and northern Africa. Although they have since separated, these landmasses were juxtaposed as part of the supercontinent of Pangaea less than 200 million years ago. Detailed studies in the southern

Appalachians indicate that the formation of this mountain belt was more complex than once thought. Rather than forming during a single continental collision, the Appalachians resulted from several distinct episodes of mountain building occurring over a period of nearly 300 million years.

6.14 CONCEPT CHECKS

1. Differentiate between *terrane* and *terrain*.
2. Where might magma be generated in a newly formed collision mountain belt?
3. How does the plate tectonics theory help explain the existence of fossil marine life in rocks atop compressional mountains?
4. List four mountain belts that formed as a result of continental collisions.

CONCEPTS IN REVIEW: RESTLESS EARTH: EARTHQUAKES, GEOLOGIC STRUCTURES, AND MOUNTAIN BUILDING

6.1 What Is an Earthquake?

Sketch and describe the mechanism that generates most earthquakes.

KEY TERMS: earthquake, fault, hypocenter (focus), epicenter, seismic wave, elastic rebound, fault creep, megathrust fault

- Earthquakes are caused by the sudden movement of blocks of rock on opposite sides of faults. The spot where the rock begins to slip is the hypocenter (or focus). Seismic waves radiate from this spot outward into the surrounding rock. The point on Earth's surface directly above the hypocenter is the epicenter.
- Elastic rebound explains why most earthquakes happen: Rock is stressed by movement of Earth's crust. However, frictional resistance keeps the fault locked in place, and the rock bends elastically. Stress builds up until it is greater than the resistance, and the blocks of rock suddenly slip, releasing the pent-up energy, in the form of seismic waves. As elastic rebound occurs, the blocks of rock on either side of the fault return to their original shapes, but they are now in new positions.
- Subduction zones are marked by megathrust faults, large reverse faults that are responsible for the largest earthquakes in recorded history. Megathrust faults are also capable of generating tsunamis.
- Strike-slip faults are important earthquake makers. The San Andreas Fault in California is an example of a large strike-slip fault. Portions of it generate destructive earthquakes. Other parts move slowly and gradually by fault creep; these sections do not produce significant earthquakes because they do not accumulate sufficient strain.

? **Label the blanks on the diagram, showing the relationship between earthquakes and faults.**

6.2 Seismology: The Study of Earthquake Waves

Compare and contrast the types of seismic waves and describe the principle of the seismograph, which is used to detect them.

KEY TERMS: seismology, seismograph (seismometer), inertia, seismogram, surface waves, body waves, P waves (primary waves), S waves (secondary waves)

- Seismology is the study of seismic waves. An instrument called a *seismograph* measures these waves, using the principle of inertia. While the body of the machine moves with the waves, a weight in the center keeps a pen (or an electronic sensor) stationary and records the relative difference between the two. The resulting record is called a *seismogram*.
- Seismograms reveal that there are three main categories of earthquake waves: P waves and S waves (which are both body waves, capable of moving through Earth's interior) and surface waves (which travel along the upper surface of the crust). P waves are the fastest, S waves are intermediate in speed, and surface waves are the slowest. However, surface waves have the greatest amplitude, while S waves are intermediate, and P waves have the least amplitude. Large-amplitude waves produce the most ground shaking, so surface waves are responsible for most damage during earthquakes.
- P waves show compressional motion: They change the volume of rock as they pass through it. In contrast, S waves impart a shaking motion to rock as they pass through it. This changes the shape of the rock but doesn't alter its volume. Because liquids resist a change in volume but not a change in shape, P waves can travel through liquids, but S waves cannot.

? How could you physically demonstrate the difference between P waves and S waves to a friend who hasn't taken a geology course? (*Caution:* Don't hurt your friend!)

6.3 Locating the Source of an Earthquake

Explain how an earthquake epicenter can be located using seismographs.

- The location on Earth's surface directly above the focus of an earthquake is the epicenter. Using the difference in arrival times between P and S waves, the distance separating a recording station from the earthquake epicenter can be determined. When the distances are known from three or more seismic stations, the epicenter can be located using a method called triangulation.

6.4 Determining the Size of an Earthquake

Distinguish between intensity scales and magnitude scales.

KEY TERMS: intensity, magnitude, Modified Mercalli Intensity scale, Richter scale, moment magnitude

- Intensity and magnitude are different measures of earthquake strength. Intensity measures the amount of ground shaking a place experiences due to an earthquake, and magnitude is an estimate of the amount of energy released during an earthquake.
- The Modified Mercalli Intensity scale is based on physical evidence, mainly destruction of human-made structures, to quantify intensity on a 12-point scale. However, it's hard to measure intensity if there are no buildings or people present at a particular location.
- The Richter scale was created to measure an earthquake's magnitude. The scale takes into account both the maximum amplitude of the seismic waves measured at a given seismograph and the distance of that seismograph from the hypocenter. The Richter scale is logarithmic, meaning that each numerical unit up on the scale represents seismic wave amplitudes that are 10 times greater than those of the next number down. Seismic waves with larger amplitudes possess more energy, so each numerical Richter unit represents about 32 times as much energy release as the next number below it.
- The Richter scale isn't good at distinguishing among really large earthquakes, so another scale was devised that can handle data from major quakes. The moment magnitude scale measures the total energy released from an earthquake. It considers the strength of the faulted rock, the amount of slip, and the area of the fault that slipped. Moment magnitude is the modern standard for measuring the size of earthquakes.

? On the Richter scale diagram, first determine the magnitude (M_L) for an earthquake at 400 kilometers distance, with a maximum amplitude of 0.5 millimeters. Second, for this same earthquake (same M_L), determine the amplitude of the biggest waves for a seismograph only 40 kilometers from the earthquake.

S-P, sec.	Distance, km	Magnitude, M_L	Amplitude, mm
50	500		100
40	400	6	50
30	300		20
20	200	5	10
		4	5
10	100		
8	60	3	2
6	40		1
4		2	0.5
	20	1	1.2
2	0.5	0	0.1

6.5 Earthquake Destruction

List and describe the major destructive forces that can be triggered by earthquake vibrations.

KEY TERMS: liquefaction, tsunami

- Several factors influence how much destruction results from an earthquake. The earthquake's magnitude and distance are important, but we must also consider (1) the intensity of the shaking, (2) how long shaking persists, (3) what sort of ground underlies buildings, and (4) building construction standards. Buildings made of unreinforced masonry are more likely than other types of buildings to collapse in a quake.
- Buildings may be constructed on solid bedrock or on deposits of loose sediment. In general, bedrock-grounded buildings fare better in an earthquake, as loose sediments amplify seismic shaking.

- A particular hazard occurs with water-logged sediments or soil: When shaken at certain frequencies, the material will flow like a liquid. Liquefaction can cause sand "volcanoes." Buildings may sink into the ground, or underground tanks or sewer lines may float up to the surface.
- Earthquakes may trigger landslides or ground subsidence, and they may break gas lines, which can initiate devastating fires.
- Tsunamis are large ocean waves that form when water is displaced, usually by a megathrust fault rupturing on the seafloor. Crossing the ocean at about the same speed as a commercial airplane, a tsunami is hardly noticeable in deep water. However, when it gets into shallower coastal waters, the tsunami slows down and piles up, sometimes producing waves more than 30 meters (100 feet) in height. A tsunami looks like a rapid rise in sea level and may be choked with sediment and other debris.

? **Of the secondary earthquake hazards discussed in this section, which is (are) of greatest concern in the region where you live? Why?**

6.6 Earthquake Belts and Plate Boundaries

Locate Earth's major earthquake belts on a world map and label the regions associated with the largest earthquakes.

- Most earthquake energy is released in the circum-Pacific belt, the ring of megathrust faults rimming the Pacific Ocean. Another earthquake belt is the Alpine–Himalayan belt, which runs along the collisional zone between Eurasia and the Indian–Australian and African plates.
- Earth's divergent oceanic ridge system produces another belt of earthquake activity where seafloor spreading generates many frequent quakes of small magnitude. Strike-slip faults in the continental crust, including the San Andreas Fault, can produce large earthquakes.

? **From memory, outline Earth's major earthquake belts on this map.**

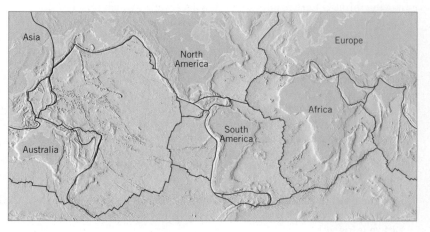

6.7 Earth's Interior

Explain how Earth acquired its layered structure and briefly describe how seismic waves are used to probe Earth's interior.

- The layered internal structure of Earth developed due to gravitational sorting of Earth materials early in the history of the planet. The densest material sunk to the center, while the least dense material rose to the surface.
- Seismic waves allow geoscientists to "look" into Earth's interior, which would otherwise be invisible to scientific investigation. Like x-rays passing through a person's body, seismic waves from large earthquakes reveal details about Earth's layered structure.

6.8 Earth's Layers

List and describe each of Earth's major layers.

KEY TERMS: crust, mantle, lithosphere, asthenosphere, mantle, core, outer core, inner core

- Earth has two distinct kinds of crust: oceanic and continental. Oceanic crust is thinner and denser than continental crust. Oceanic crust also tends to be younger than continental crust, since it readily subducts but continents do not.
- Earth's mantle may be divided by density into upper and lower portions. The uppermost mantle makes up the bulk of rigid lithospheric plates, while a weak layer called the asthenosphere lies beneath it.
- The composition of the core is likely a mix of iron, nickel, and lighter elements. Iron and nickel are common heavy elements in meteorites, the leftover "building blocks" of Earth. The outer core is dense (around 10 g/cm^3). It is known to be liquid, as S waves cannot pass through it. The inner core is solid and very dense (more than 13 g/cm^3).

6.9 Deformation

Compare and contrast brittle and ductile deformation.

KEY TERMS: deformation, elastic deformation, brittle deformation, ductile deformation

- Elastic deformation is a temporary bending of rock, but not past the breaking point. When the stress is released, the rock snaps back to its original shape.
- When stress is greater than the strength of the rock, the rock will deform in either a brittle or ductile fashion. Brittle deformation is the breaking of rocks into smaller pieces, whereas ductile deformation is flow in the manner of modeling clay or warm wax.

6.10 Folds: Rock Structures Formed by Ductile Deformation

List and describe the major types of folds.

KEY TERMS: fold, anticline, syncline, dome, basin

- Folds are wave-like undulations in layered rocks that develop through ductile deformation in rocks undergoing compressional stress.
- Folds that have an arch-like structure are called anticlines, while folds that have a trough-like structure are called synclines. Anticlines and synclines may be symmetrical, asymmetrical, overturned, or recumbent.
- Domes and basins are large folds that produce "bull's-eye"–shaped outcrop patterns. The overall shape of a dome or basin is like a saucer or a bowl, either right side up (basin) or inverted (dome).

? **What name is given to the rock structure shown in the accompanying photo?**

E. J. Tarbuck

6.11 Faults: Structures Formed by Brittle Deformation

Sketch and describe the relative motion of rock bodies located on opposite sides of normal, reverse, and strike-slip faults.

KEY TERMS: fault, dip-slip fault, hanging wall block, footwall block, fault scarp, normal fault, fault-block mountain, horst, graben, reverse fault, thrust fault, strike-slip fault, transform fault

- Faults and joints are fractures in rock that form through brittle deformation. Faults show offset of the rocks on either side of the fracture; joints do not show offset.
- The offset on a fault may be determined by comparing the blocks of rock on either side of the fault surface. If the movement is in the direction of the fault's dip (or inclination), then the rock above the fault plane is the hanging wall block, and the rock below the fault is the footwall block. If the hanging wall moves down relative to the footwall, the fault is a normal fault. If the hanging wall moves up relative to the footwall, the fault is a reverse fault. Reverse faults with low dip angles are thrust faults.
- Faults that intersect Earth's surface may produce a "step" in the land known as a fault scarp. Areas of tectonic extension, such as the Basin and Range Province, produce fault-block mountains: horsts separated by neighboring grabens or half-grabens.
- Strike-slip faults have most of their movement in the direction of the trend of the fault trace. There is minimal up or down movement; instead, the motion is mainly horizontal. Transform faults are strike-slip faults that serve as tectonic boundaries between lithospheric plates.

6.12 Mountain Building

Locate and identify Earth's major mountain belts on a world map.

KEY TERMS: orogenesis, compressional mountains

- Orogenesis is the making of mountains. Most orogenesis occurs along convergent plate boundaries, where compressional forces cause folding and faulting of the rock, thickening the crust vertically while shortening it horizontally. Some mountain belts are very old, while others are actively forming today.

6.13 Subduction and Mountain Building

Sketch an Andean-type mountain belt and describe how each of its major features is generated.

KEY TERMS: passive continental margin, active continental margin, accretionary wedge

- Subduction leads to orogenesis. If a subducted plate is over-ridden by oceanic lithosphere, island arc–type mountain building results—a thick accumulation of erupted volcanic rocks mixed with sediment scraped off the subducted plate.
- Andean-type mountain building occurs where subduction takes place beneath continental lithosphere and results in a continental volcanic arc.
- Sediment scraped off the subducting plate builds up in an accretionary wedge. The geography of central California preserves an accretionary wedge (Coast Ranges) and roots of a continental volcanic arc (Sierra Nevada).

6.14 Collisional Mountain Belts

Summarize the stages in the development of a collisional mountain belt such as the Himalayas.

KEY TERMS: terrane, microcontinent

- Terranes are relatively small crustal fragments (microcontinents, volcanic island arcs, or oceanic plateaus). Terranes may be accreted to continents when subduction brings them to a trench, but they usually do not subduct due to their relatively low density. The terranes are "peeled off" the subducted slab and thrust onto the leading edge of the continent.
- The Himalayas and Appalachians have similar origins in the collision of continents formerly separated by now-subducted ocean basins. The Appalachians are older and were caused by the collision of ancestral North America with ancestral Africa more than 250 million years ago. In contrast, the Himalayas are younger, having been formed by the collision of India and Eurasia starting around 50 million years ago. The Himalayas are still rising.

? Identify modern examples of (1) a microcontinent, (2) a volcanic island arc, and (3) an oceanic plateau on a world map such as Figure 9.14.

GIVE IT SOME THOUGHT

1. Briefly describe the concept of *elastic rebound*. Develop an analogy other than a rubber band to illustrate this concept.

2. The accompanying map shows the locations of the 15 largest earthquakes in the world since 1900. Refer to a map of Earth's plates (see Figure 5.12, page 160) and determine which type of plate boundary is most often associated with these events.

3. Use the accompanying seismogram to answer the following questions:

a. Which of the three types of seismic waves reached the seismograph first?

b. What is the time interval between the arrival of the first P wave and the arrival of the first S wave?

c. Using your answer from Question b and Figure 6.14, determine the distance from the seismic station to the earthquake.

d. Which of the three types of seismic waves had the highest amplitude when the waves reached the seismic station?

4. You go for a jog on a beach and choose to run near the water, where the sand is well packed and solid under your feet. With each step, you notice that your footprint quickly fills with water but not water coming in from the ocean. What is this water's source? For what earthquake-related hazard is this phenomenon a good analogy?

5. Explain, in your own words, why a tsunami often causes a rapid withdrawal of water from beaches before the first surge.

6. Why is it possible to issue a tsunami warning but not provide a warning for an earthquake? Describe a scenario where a tsunami warning would be of little value.

7. Write a brief statement describing each of the accompanying photos, using selected terms from the following list: strike-slip, dip-slip, normal, reverse, right-lateral, left-lateral.

A.

B.

8. Using the Internet, compare and contrast the 2010 earthquakes in Chile and Haiti. Include magnitude, type of plate boundary, and extent of destruction. Explain why the Chile earthquake produced a tsunami, while the Haiti quake did not.

9. Refer to the accompanying diagrams to answer the following:

Diagram 1   Diagram 2   Diagram 3

a. What type of dip-slip fault is shown in Diagram 1?
b. What type of dip-slip fault is shown in Diagram 2?
c. Match the correct pair of arrows in Diagram 3 to the faults in Diagrams 1 and 2.

10. The Ural Mountains exhibit a north–south orientation through Eurasia (see Figure 6.40). How does the theory of plate tectonics explain the existence of this mountain belt in the interior of an expansive landmass?

11. Briefly describe the major differences between the evolution of the Appalachian Mountains and the North American Cordillera.

12. Refer to the accompanying diagram. Which of these features—the *Galapagos Rise* or the *Rio Grande Rise*—will likely end up accreted to a continent? Explain your choice. In the future, how might a geologist determine that this accreted terrane is distinct from the continental crust to which it accreted? (*Hint:* See the plate map in Figure 5.12, page 160.)

Galapagos Rise

Rio Grande Rise

13. Suppose a sliver of oceanic crust were discovered in the interior of a continent. Would this support or refute the theory of plate tectonics? Explain.

14. Which of the accompanying sketches best illustrates an Andean-type orogeny, a Cordilleran-type orogeny, and an Alpine-type orogeny?

A.

B.

C.

7

Volcanoes and Other Igneous Activity

FOCUS ON CONCEPTS

Each statement represents the primary learning objective for the corresponding major heading within the chapter. After you complete the chapter, you should be able to:

7.1 Compare and contrast the 1980 eruption of Mount St. Helen's with the eruption of Kilauea, which began in 1983 and continues today.

7.2 Explain why some volcanic eruptions are explosive and others are quiescent.

7.3 List and describe the three categories of materials extruded during volcanic eruptions.

7.4 Label a diagram that illustrates the basic features of a typical volcanic cone.

7.5 Summarize the characteristics of shield volcanoes and provide one example.

7.6 Describe the formation, size, and composition of cinder cones.

7.7 Explain the formation, distribution, and characteristics of composite volcanoes.

7.8 Discuss the major geologic hazards associated with volcanoes.

7.9 List and describe volcanic landforms other than volcanic cones.

7.10 Compare and contrast these intrusive igneous structures: dikes, sills, batholiths, stocks, and laccoliths.

7.11 Summarize the major processes that generate magma from solid rock.

7.12 Relate the distribution of volcanic activity to plate tectonics.

A recent eruption of Italy's Mount Etna (Stocktrek Images/Superstock)

The significance of igneous activity may not be obvious at first glance. However, because volcanoes extrude molten rock that formed at great depth, they provide our only means of directly observing processes that occur many kilometers below Earth's surface. Furthermore, the atmosphere and oceans have evolved from gases emitted during volcanic eruptions. Either of these facts is reason enough for igneous activity to warrant our attention.

7.1 Mount St. Helens Versus Kilauea

Compare and contrast the 1980 eruption of Mount St. Helen's with the eruption of Kilauea, which began in 1983 and continues today.

On Sunday, May 18, 1980, the largest volcanic eruption to occur in North America in historic times transformed a picturesque volcano into a decapitated remnant (**Figure 7.1**). On that date in southwestern Washington State, Mount St. Helens erupted with tremendous force. The blast blew out the entire north flank of the volcano, leaving a gaping hole. In one brief moment, a prominent

volcano whose summit had been more than 2900 meters (9500 feet) above sea level was lowered by more than 400 meters (1350 feet).

The event devastated a wide swath of timber-rich land on the north side of the mountain (**Figure 7.2**). Trees within a 400-square-kilometer (160-square-mile) area lay intertwined and flattened, stripped of their branches and

1350 feet

The blast blew out the entire north flank of Mount St. Helen's, leaving a gaping hole. In a brief moment, a prominent volcano was lowered by 1350 feet.

Spirit Lake

USGS

Spirit Lake

USGS

Figure 7.1 Before-and-after photographs show the transformation of Mount St. Helens The May 18, 1980, eruption of Mount St. Helens occurred in southwestern Washington.

appearing from the air like toothpicks strewn about. The accompanying mudflows carried ash, trees, and water-saturated rock debris 29 kilometers (18 miles) down the Toutle River. The eruption claimed 59 lives; some died from the intense heat and the suffocating cloud of ash and gases, others from the impact of the blast, and still others from being entrapped in mudflows.

The eruption ejected nearly a cubic kilometer of ash and rock debris. Following the devastating explosion, Mount St. Helens continued to emit great quantities of hot gases and ash. The force of the blast was so strong that some ash was propelled more than 18 kilometers (over 11 miles) into the stratosphere. During the next few days, this very fine-grained material was carried around Earth by strong upper-air winds. Measurable deposits were reported in Oklahoma and Minnesota, and crop damage occurred as far away as central Montana. Meanwhile, ash fallout in the immediate vicinity exceeded 2 meters (6 feet) in depth. The air over Yakima, Washington (130 kilometers [80 miles] to the east), was so filled with ash that residents experienced midnightlike darkness at noon.

Not all volcanic eruptions are as violent as the 1980 Mount St. Helens event. Some volcanoes, such as Hawaii's Kilauea volcano, generate relatively quiet outpourings of fluid lavas. These "gentle" eruptions are not without some fiery displays; occasionally fountains of incandescent lava spray hundreds of meters into the air. Nevertheless, during Kilauea's most recent active phase, which began in 1983, more than 180 homes and a national park visitor center have been destroyed.

Testimony to the "quiet nature" of Kilauea's eruptions is the fact that the Hawaiian Volcanoes Observatory has operated on its summit since 1912, despite the fact that Kilauea has had more than 50 eruptive phases since record keeping began in 1823.

Figure 7.2 Douglas fir trees snapped off or uprooted by the lateral blast of Mount St. Helens (Large photo by Lyn Topinka/AP Photo/ U.S. Geological Survey; inset photo by John M. Burnley/Photo Researchers, Inc.)

7.1 CONCEPT CHECKS

1. Briefly compare the May 18, 1980, eruption of Mount St. Helens to a typical eruption of Hawaii's Kilauea volcano.

7.2 The Nature of Volcanic Eruptions

Explain why some volcanic eruptions are explosive and others are quiescent.

Volcanic activity is commonly perceived as a process that produces a picturesque, cone-shaped structure that periodically erupts in a violent manner. However, many eruptions are not explosive. What determines the manner in which volcanoes erupt?

Factors Affecting Viscosity

The source material for volcanic eruptions is **magma**, molten rock that usually contains some crystals and varying amounts of dissolved gas. Erupted magma is called **lava**. The primary factors that affect the behavior of magma and lava are its *temperature* and *composition*

and, to a lesser extent, the amount of *dissolved gases* it contains. To varying degrees, these factors determine a magma's mobility, or **viscosity** (*viscos* = sticky). The more viscous the material, the greater its resistance to flow. For example, syrup is more viscous and, thus, more resistant to flow, than water.

Temperature The effect of temperature on viscosity is easily seen. Just as heating syrup makes it more fluid (less viscous), temperature also strongly influences the mobility of lava. As lava cools and begins to congeal, its viscosity increases, and eventually the flow halts.

TABLE 7.1   Different Compositions of Magmas Cause Properties to Vary

Composition	Silica Content	Gas Content	Eruptive Temperatures	Viscosity	Tendency to Form Pyroclastics	Volcanic Landform
Basaltic (mafic)	Least (~50%)	Least (1–2%)	1000–1250°C	Least	Least	Shield volcanoes, basalt plateaus, cinder cones
Andesitic (intermediate)	Intermediate (~60%)	Intermediate (3–4%)	800–1050°C	Intermediate	Intermediate	Composite cones
Rhyolitic (felsic)	Most (~70%)	Most (4–6%)	650–900°C	Greatest	Greatest	Pyroclastic flows, lava domes

Composition Another significant factor influencing volcanic behavior is the chemical composition of the magma. Recall that a major difference among various igneous rocks is their silica (SiO_2) content (**Table 7.1**). Magmas that produce mafic rocks such as basalt contain about 50 percent silica, whereas magmas that produce felsic rocks (granite and its extrusive equivalent, rhyolite) contain more than 70 percent silica. Intermediate rock types—andesite and diorite—contain about 60 percent silica.

A magma's viscosity is directly related to its silica content—*the more silica in magma, the greater its viscosity*. Silica impedes the flow of magma because silicate structures start to link together into long chains early in the crystallization process. Consequently, felsic (rhyolitic) lavas are very viscous and tend to form comparatively short, thick flows. By contrast, mafic (basaltic) lavas, which contain less silica, are relatively fluid and have been known to travel 150 kilometers (90 miles) or more before solidifying.

Dissolved Gases The gaseous components in magma, mainly dissolved water, also affect its mobility. Other factors being equal, water dissolved in magma tends to increase fluidity because it reduces formation of long silicate chains by breaking silicon–oxygen bonds. It follows, therefore, that the loss of gases renders magma (lava) more viscous. Gases also give magmas their explosive character.

Quiescent Versus Explosive Eruptions

You learned in Chapter 3 that most magma is generated by partial melting of the rock in the upper mantle and has a basaltic composition. The newly formed magma, which is less dense than the surrounding rock, slowly rises toward the surface. In some settings, high-temperature basaltic magmas reach Earth's surface, where they produce highly fluid lavas. This most commonly occurs on the ocean floor in association with seafloor spreading. In continental settings, however, the density of crustal rocks is less than that of the ascending material, causing the magma to pond at the crust–mantle boundary. Heat from the hot magma is often sufficient to partially melt the overlying crustal rocks, generating a less dense, silica-rich magma, which continues the journey toward Earth's surface.

Quiescent Hawaiian-Type Eruptions Eruptions that involve very fluid basaltic lavas, such as the eruptions of Kilauea on Hawaii's Big Island, are often triggered by the arrival of a new batch of molten rock into a near-surface magma chamber. This event can usually be detected because the summit of the volcano begins to inflate and rise months or even years before an eruption. The injection of a fresh supply of hot molten rock heats and remobilizes the semi-liquid magma chamber. In addition, swelling of the magma chamber fractures the rock above, allowing the fluid magma to move upward along the newly formed openings, often generating outpourings of lava for weeks, months, or possibly years. The eruption of Kilauea that began in 1983 has been ongoing for more than 20 years.

Explosive Eruptions All magmas contain some water vapor and other gases that are kept in solution by the immense pressure of the overlying rock. As magma rises (or the rocks confining the magma fail), the pressure reduces and the dissolved gases begin to separate from the melt, forming tiny bubbles. This is analogous to opening a can of soda and allowing the carbon dioxide bubbles to escape.

When fluid basaltic magmas erupt, the pressurized gases readily escape. At temperatures that often exceed 1100°C (2000°F), these gases can quickly expand to occupy hundreds of times their original volumes. Occasionally, these expanding gases propel incandescent lava hundreds of meters into the air, producing lava fountains (**Figure 7.3**). Although spectacular, these fountains are usually harmless and generally not associated with major explosive events that cause great loss of life and property.

At the other extreme, highly viscous magmas expel particles of fragmented lava and gases at nearly supersonic speeds that evolve into buoyant plumes called **eruption columns**. Eruption columns can rise perhaps 40 kilometers (25 miles) into the atmosphere (**Figure 7.4**). Because silica-rich magmas are sticky (viscous), a significant portion of the gaseous materials remain dissolved until the magma nears Earth's surface, at which time tiny bubbles begin to form and grow. When the pressure of the expanding magma body exceeds the strength of the overlying rock, fracturing occurs. As magma moves up the fractures, a further drop in confining pressure causes more gas bubbles to form and grow. This chain reaction may generate an explosive event in which magma is literally blown into fragments

Figure 7.3 Lava fountain produced by gases escaping fluid basaltic lava Lava erupting from Mount Etna, Italy. (Photo by D. Szczepanski terras/AGE Fotostock)

Gases readily escape hot fluid basaltic flows, producing lava fountains. Although often spectacular, these features generally do not cause great loss of life or property.

Figure 7.4 Eruption columns are generated by viscous, silica-rich magmas Steam and ash eruption column from Mount Augustine, Cook Inlet, Alaska. (Photo by Steve Kaufman/Getty Images)

Eruptions of highly viscous lavas may produce explosive clouds of hot ash and gases called eruption columns.

(ash and pumice) that are carried to great heights by the hot gases. (As exemplified by the 1980 eruption of Mount St. Helens, the collapse of a volcano's flank can result in a reduction in the pressure on the magma below, triggering an explosive eruption.)

When magma in the uppermost portion of the magma chamber is forcefully ejected by the escaping gases, the confining pressure on the molten rock directly below drops suddenly. Thus, rather than a single "bang," volcanic eruptions are really a series of explosions. Following explosive eruptions, degassed lava may slowly ooze out of the vent to form short rhyolitic flows or dome-shaped lava bodies that grow over the vent.

7.2 CONCEPT CHECKS

1. Define *viscosity* and list three factors that influence the viscosity of magma.
2. Explain how the viscosity of magma influences the explosiveness of a volcano.
3. List these three magmas in order from *most* silica rich to *least* silica rich, based on their compositions: mafic (basaltic), felsic (rhyolitic), intermediate (andesitic).
4. The eruption of what type of magma may produce an eruption column?
5. Why is a volcano that is fed by highly viscous magma likely to be a greater threat to life and property than a volcano supplied with very fluid magma?

7.3 Materials Extruded During an Eruption

List and describe the three categories of materials extruded during volcanic eruptions.

Volcanoes extrude lava, large volumes of gas, and pyroclastic materials (broken rock, lava "bombs," fine ash, and dust). In this section, we examine each of these materials.

Lava Flows

The vast majority of lava on Earth—more than 90 percent of the total volume—is estimated to be basaltic in composition. Andesites and other lavas of intermediate composition account for most of the rest, while rhyolitic (felsic) flows make up as little as 1 percent of the total.

Hot basaltic lavas, which are usually very fluid, generally flow in thin, broad sheets or streamlike ribbons. On the island of Hawaii, these lavas have been clocked at 30 kilometers (19 miles) per hour down steep slopes. However, flow rates of 10 to 300 meters (30 to 1000 feet) per hour are more common. By contrast, the movement of silica-rich, rhyolitic lava may be too slow to perceive. Furthermore, most rhyolitic lavas seldom travel more than a few kilometers from their vents. As you might expect, andesitic lavas, which are intermediate in composition, exhibit characteristics that are between the extremes.

Aa and Pahoehoe Flows Fluid basaltic magmas tend to generate two types of lava flows known by Hawaiian names. One of these, called **aa** (pronounced ah-ah) **flows**, have surfaces of rough, jagged blocks with dangerously sharp edges and spiny projections (**Figure 7.5A**). Crossing a hardened aa flow can be a trying and miserable experience. The second type, **pahoehoe** (pronounced "pah-hoy-hoy") **flows**, exhibit smooth surfaces that sometimes resemble the twisted braids of ropes (**Figure 7.5B**). Pahoehoe means "on which one can walk."

Although both lava types can erupt from the same volcano, pahoehoe lavas are hotter and are more fluid than aa flows. In addition, pahoehoe lavas can change into aa lava flows, although the reverse (aa to pahoehoe) does not occur.

Cooling that occurs as the flow moves away from the vent is one factor that facilitates the change from pahoehoe to aa. The reduction in temperature increases viscosity and promotes bubble formation. Escaping gas bubbles produce numerous voids (vesicles) and sharp spines in the surface of the congealing lava. As the molten interior advances, the outer crust is broken, transforming a relatively smooth surface of a pahoehoe flow into an aa flow made up of an advancing mass of rough, sharp, broken lava blocks.

Lava Tubes Pahoehoe flows often develop cave-like tunnels called **lava tubes** that were previously conduits for carrying lava from an active vent to the flow's leading edge (**Figure 7.6**). Lava tubes form in the interior of a lava flow where the temperature remains high long after the surface cools and hardens. Because they serve as insulated pathways that facilitate the advance of lava great distances from its source, lava tubes are important features of fluid lava flows.

Gases

Magmas contain varying amounts of dissolved gases, called **volatiles**, held in the molten rock by confining pressure, just as carbon dioxide is held in cans and bottles of soft drinks. As with soft drinks, as soon as the pressure is reduced, the gases begin to escape. Obtaining gas samples from an erupting volcano is difficult and dangerous, so geologists usually must estimate the amount of gas originally contained in the magma.

The gaseous portion of most magmas makes up 1 to 6 percent of the total weight, with most of this in the form of water vapor. Although the percentage may be small, the actual quantity of emitted gas can exceed

A.

Figure 7.5 Lava flows A. A typical slow-moving, basaltic, aa flow. **B.** A typical fluid pahoehoe (ropy) lava. Both of these lava flows erupted from a rift on the flank of Hawaii's Kilauea Volcano. (Photos courtesy of U.S. Geological Survey)

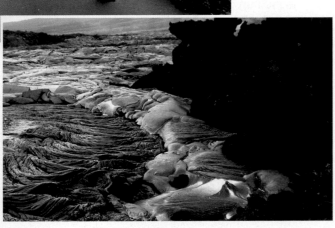

B.

Lava tubes. These cave-like tunnels were once conduits carrying lava from an active vent to the flow's leading edge.

Valentine cave, a lava tube at Lava Beds National Monument, California.

Figure 7.6 Lava tubes A lava flow may develop a solid upper crust while the molten lava below continues to advance in a conduit called a *lava tube*. Some lava tubes exhibit extraordinary dimensions. One such structure, Kazumura Cave, located on the southeastern slope of Hawaii's Mauna Loa Volcano extends more than 60 kilometers (40 miles). (Photo by Dave Bunell)

thousands of tons per day. Occasionally, eruptions emit colossal amounts of volcanic gases that rise high into the atmosphere, where they may reside for several years. Some of these eruptions may have an impact on Earth's climate, a topic we consider later in this chapter.

The composition of volcanic gases is important because these gases contribute significantly to our planet's atmosphere. Analyses of samples taken during Hawaiian eruptions indicate that the gas component is about 70 percent water vapor, 15 percent carbon dioxide, 5 percent nitrogen, and 5 percent sulfur dioxide, with lesser amounts of chlorine, hydrogen, and argon. (The relative proportion of each gas varies significantly from one volcanic region to another.) Sulfur compounds are easily recognized by their pungent odor. Volcanoes are also natural sources of air pollution; some emit large quantities of sulfur dioxide, which readily combines with atmospheric gases to form sulfuric acid and other sulfate compounds.

Pyroclastic Materials

When volcanoes erupt energetically, they eject pulverized rock, lava, and glass fragments from the vent. The particles produced are called **pyroclastic materials** (*pyro* = fire, *clast* = fragment) and are also referred to as *tephra*. These fragments range in size from very fine dust and sand-sized volcanic ash (less than 2 millimeters) to pieces that weigh several tons (**Figure 7.7**).

Ash and *dust* particles are produced when gas-rich viscous magma erupts explosively (see Figure 7.7). As magma moves up in the vent, the gases rapidly expand, generating a melt that resembles the froth that flows from a bottle of champagne. As the hot gases expand explosively, the froth is blown into very fine glassy fragments. When the hot ash falls, the glassy shards often fuse to form a rock called *welded tuff.* Sheets of this material, as well as ash deposits that later consolidate, cover vast portions of the western United States.

Figure 7.7 Types of pyroclastic materials Pyroclastic materials are also commonly referred to as tephra.

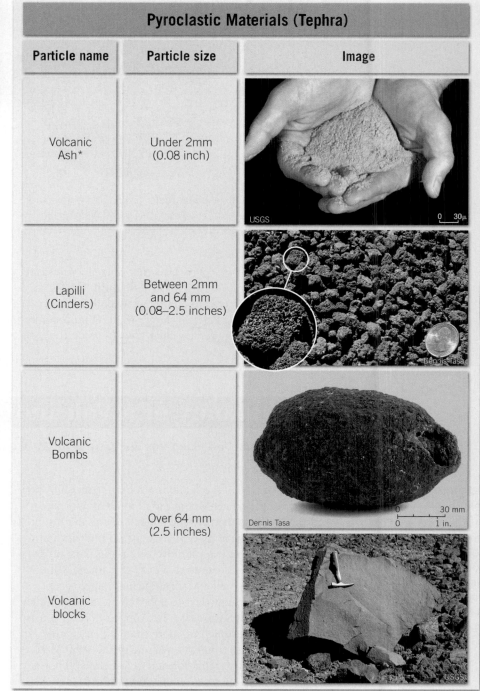

Pyroclastic Materials (Tephra)		
Particle name	**Particle size**	**Image**
Volcanic Ash*	Under 2mm (0.08 inch)	
Lapilli (Cinders)	Between 2mm and 64 mm (0.08–2.5 inches)	
Volcanic Bombs	Over 64 mm (2.5 inches)	
Volcanic blocks		

*The term volcanic dust is used for fine volcanic ash under 0.063 mm (0.0025 inch).

Scoria is a vesicular rock commonly having a basaltic or andesitic composition. Pea-to-basketball size scoria fragments make up a large portion of most cinder cones (also called *scoria cones*).

A.

Pumice is a low density vesicular rock that forms during explosive eruptions of viscous magma having an andesitic to rhyolitic composition.

B.

Figure 7.8 Common vesicular rocks Scoria and pumice are volcanic rocks that exhibit a vesicular texture. Vesicles are small holes left by escaping gas bubbles. (Photos by E. J. Tarbuck)

Somewhat larger pyroclasts that range in size from small beads to walnuts are known as *lapilli* ("little stones"). These ejecta are commonly called *cinders* (2–64 millimeters). Particles larger than 64 millimeters (2.5 inches) in diameter are called *blocks* when they are made of hardened lava and *bombs* when they are ejected as incandescent lava (see Figure 7.7). Because bombs are semimolten upon ejection, they often take on a streamlined shape as they hurtle through the air. Because of their size, bombs and blocks usually fall near the vent; however, they are occasionally propelled great distances. For instance, bombs 6 meters (20 feet) long and weighing about 200 tons were blown 600 meters (2000 feet) from the vent during an eruption of the Japanese volcano Asama.

So far we have distinguished various pyroclastic materials based largely on the sizes of the fragments. Some materials are also identified by their texture and composition. In particular, **scoria** is the name applied to vesicular ejecta that is a product of basaltic magma (**Figure 7.8A**). These black to reddish-brown fragments are generally found in the size range of lapilli and resemble cinders and clinkers produced by furnaces used to smelt iron. When magmas with intermediate (andesitic) or felsic (rhyolitic) compositions erupt explosively, they emit ash and the vesicular rock **pumice** (**Figure 7.8B**). Pumice is usually lighter in color and less dense than scoria, and many pumice fragments have so many vesicles that they are light enough to float.

7.3 CONCEPT CHECKS

1. Describe pahoehoe and aa lava flows.
2. How do lava tubes form?
3. List the main gases released during a volcanic eruption. What role do gases play in eruptions?
4. How do volcanic bombs differ from blocks of pyroclastic debris?
5. What is scoria? How is scoria different from pumice?

7.4 Anatomy of a Volcano

Label a diagram that illustrates the basic features of a typical volcanic cone.

A popular image of a volcano is a solitary, graceful, snow-capped cone, such as Mount Hood in Oregon or Japan's Fujiyama. These picturesque conical mountains are produced by volcanic activity that occurred intermittently over thousands, or even hundreds of thousands, of years. However, many volcanoes do not fit this image. Cinder cones are quite small and form during a single eruptive phase that lasts a few days to a few years. Alaska's Valley of Ten Thousand Smokes is a flat-topped deposit consisting of 15 cubic kilometers (more than 3.5 cubic miles) of ash, more than 20 times the volume of the 1980 Mount St. Helen's eruption, that formed in less than 60 hours and blanketed a river valley to a depth of 200 meters (600 feet).

Volcanic landforms come in a wide variety of shapes and sizes, and each volcano has a unique eruptive history. Nevertheless, volcanologists have been able to classify volcanic landforms and determine their eruptive patterns. In this section, we will consider the general anatomy of an idealized volcanic cone. We will follow this discussion by exploring the three major types of volcanic cones—shield volcanoes, cinder cones, and composite volcanoes—as well as their associated hazards.

Volcanic activity frequently begins when a fissure (crack) develops in Earth's crust, as magma moves forcefully toward the surface. As the gas-rich magma moves up through a fissure, its path is usually localized into a circular **conduit** that terminates at a surface opening called a **vent** (**Figure 7.9**). The cone-shaped structure we call a **volcanic cone** is often created by successive eruptions of lava, pyroclastic material, or frequently a combination of both, often separated by long periods of inactivity.

Located at the summit of most volcanic cones is a somewhat funnel-shaped depression, called a **crater** (*crater* = a bowl). Volcanoes built primarily of pyroclastic materials typically have craters that form by gradual accumulation of volcanic debris on the surrounding rim. Other craters form during explosive eruptions as the rapidly ejected particles erode the crater walls. Craters also form when the summit area of a volcano collapses following an eruption. Some volcanoes have very large circular depressions called *calderas* that have diameters greater than 1 kilometer (0.6 mile) and in some cases can exceed 50 kilometers (30 miles). We will consider the formation of various types of calderas later in this chapter.

During early stages of growth, most volcanic discharges come from a central summit vent. As a volcano matures, material also tends to be emitted from fissures that develop along the flanks or at the base of the volcano. Continued activity from a flank eruption may produce one or more small **parasitic cones** (*parasitus* = one who eats at the table of another). Italy's Mount Etna, for example, has more than 200 secondary vents, some of which have built parasitic cones. Many of these vents, however, emit only gases and are appropriately called **fumaroles** (*fumus* = smoke).

SmartFigure 7.9 Anatomy of a volcano Compare the structure of a "typical" composite cone to that of a shield volcano (Figure 7.10) and a cinder cone (Figure 7.13).

7.4 CONCEPT CHECKS

1. How is a crater different from a caldera?
2. Distinguish among conduit, vent, and crater.
3. What is a parasitic cone, and where does it form?
4. What is emitted from a fumarole?

7.5 Shield Volcanoes

Summarize the characteristics of shield volcanoes and provide one example.

Shield volcanoes are produced by the accumulation of fluid basaltic lavas and exhibit the shape of a broad, slightly domed structure that resembles a warrior's shield (**Figure 7.10**). Most shield volcanoes begin on the ocean floor as seamounts, a few of which grow large enough to form volcanic islands. In fact, with the exception of the volcanic islands that form above subduction zones, most other small oceanic islands are either a single shield volcano or, more often, the coalescence of two or more shields built on massive amounts of pillow lavas. Examples include the Canary islands, the Hawaiian islands, the Galapagos, and Easter Island. In addition, some shield volcanoes form on continental crust. Included in

this group are Nyamuragira, Africa's most active volcano, and Newberry Volcano located in Oregon.

Mauna Loa: Earth's Largest Shield Volcano

Extensive study of the Hawaiian islands revealed that they are constructed of a myriad of thin basaltic lava flows averaging a few meters thick intermixed with relatively minor amounts of pyroclastic ejected material. Mauna Loa is one of five overlapping shield volcanoes that comprise the Big Island of Hawaii (see Figure 7.10).

Did You Know?

According to legend, Pele, the Hawaiian goddess of volcanoes, makes her home at the summit of Kilauea volcano. Evidence for her existence is "Pele's hair"—thin, delicate strands of glass, which are soft and flexible and have a golden-brown color. This threadlike volcanic glass forms when blobs of hot lava are spattered and shredded by escaping gases.

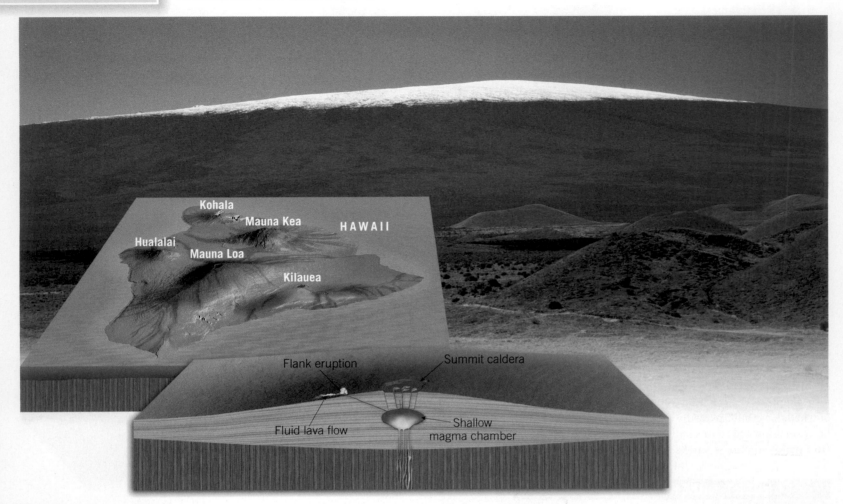

Figure 7.10 Mauna Loa: Earth's largest volcano Mauna Loa is one of five shield volcanoes that collectively make up the Big Island of Hawaii. Shield volcanoes are built primarily of fluid basaltic lava flows and contain only a small percentage of pyroclastic materials. (Photo courtesy of Greg Vaughn/Alamy)

From its base on the floor of the Pacific Ocean to its summit, Mauna Loa is over 9 kilometers (6 miles) high, exceeding the elevation of Mount Everest. The volume of material composing Mauna Loa is roughly 200 times greater than that of a large composite cone such as Mount Rainier (**Figure 7.11**).

The flanks of Mauna Loa have gentle slopes of only a few degrees. This low angle results because very

hot, fluid lava travelled "fast and far" from the vent. In addition, most of the lava (perhaps 80 percent) flowed through a well-developed system of lava tubes. Another feature common to many active shield volcanoes is a large, steep-walled caldera that occupies the summit. A caldera usually forms when the roof of solid rock above a magma chamber collapses. This usually occurs as the magma reservoir empties following a large eruption

Figure 7.11 Profiles comparing scales of different volcanoes A. Profile of Mauna Loa, Hawaii, the largest shield volcano in the Hawaiian chain. Note size comparison with Mount Rainier, Washington, a large composite cone. **B.** Profile of Mount Rainier, Washington. Note how it dwarfs a typical cinder cone. **C.** Profile of Sunset Crater, Arizona, a typical steep-sided cinder cone.

Figure 7.12 Lava "curtain" extruded along the East Rift Zone, Kilauea, Hawaii (Photo by Greg Vaughn/Alamy)

or as magma migrates to the flank of a volcano to feed a fissure eruption.

In the final stage of growth, shield volcanoes erupt more sporadically, and pyroclastic ejections are more common. Further, lavas increase in viscosity, resulting in thicker, shorter flows. These eruptions tend to steepen the slope of the summit area, which often becomes capped with clusters of cinder cones. This may explain why Mauna Kea, which is a more mature volcano that has not erupted in historic times, has a steeper summit than Mauna Loa, which erupted as recently as 1984. Astronomers are so certain that Mauna Kea is "over the hill" that they built an elaborate astronomical observatory on its summit, housing some of the world's most advanced and expensive telescopes.

Kilauea, Hawaii: Eruption of a Shield Volcano

Kilauea, the most active and intensely studied shield volcano in the world, is located on the island of Hawaii in the shadow of Mauna Loa. More than 50 eruptions have been witnessed here since record keeping began in 1823. Several months before each eruptive phase, Kilauea inflates as magma gradually migrates upward and accumulates in a central reservoir located a few kilometers below the summit. For up to 24 hours before an eruption, swarms of small earthquakes warn of the impending activity.

Most of the recent activity on Kilauea has occurred along the flanks of the volcano, in a region called the East Rift Zone. The longest and largest rift eruption ever recorded on Kilauea began in 1983 and continues to this day, with no signs of abating. The first discharge began along a 6-kilometer (4-mile) fissure where a 100-meter- (300-foot-) high "curtain of fire" formed as red-hot lava was ejected skyward (**Figure 7.12**). When the activity became localized, a cinder and spatter cone, given the Hawaiian name *Puu Oo*, was built. Over the next 3 years, the general eruptive pattern consisted of short periods (hours to days) when fountains of gas-rich lava sprayed skyward. Each event was followed by nearly a month of inactivity.

By the summer of 1986, a new vent opened 3 kilometers (nearly 2 miles) downrift. Here smooth-surfaced pahoehoe lava formed a lava lake. Occasionally the lake overflowed, but more often lava escaped through tunnels to feed flows that moved down the southeastern flank of the volcano toward the sea. These flows destroyed nearly 100 rural homes, covered a major roadway, and eventually reached the sea. Lava has been intermittently pouring into the ocean ever since, adding new land to the island of Hawaii.

7.5 CONCEPT CHECKS

1. Describe the composition and viscosity of the lava associated with shield volcanoes.
2. Are pyroclastic materials a significant component of shield volcanoes?
3. Where do most shield volcanoes form—on the ocean floor or on the continents?
4. Relate lava tubes to the extent of lava flows associated with shield volcanoes.
5. Where are the best-known shield volcanoes in the United States? Name some examples in other parts of the world.

7.6 Cinder Cones

Describe the formation, size, and composition of cinder cones.

As the name suggests, **cinder cones** (also called **scoria cones**) are built from ejected lava fragments that begin to harden in flight to produce the vesicular rock *scoria* (**Figure 7.13**). These pyroclastic fragments range in size from fine ash to bombs that may exceed 1 meter in diameter. However, most of the volume of a cinder cone consists of pea- to walnut-sized fragments that are markedly vesicular and have a black to reddish-brown color. In addition, this pyroclastic material tends to have basaltic composition.

Although cinder cones are composed mostly of loose scoria fragments, some produce extensive lava fields. These lava flows generally form in the final stages of the volcano's life span, when the magma body has lost most of its gas content. Because cinder cones are composed of loose fragments rather than solid rock, the lava usually flows out from the unconsolidated base of the cone rather than from the crater.

Cinder cones have very simple, distinct shapes determined by the slope that loose pyroclastic material maintains as it comes to rest (see Figure 7.13). Cinder cones have a high angle of repose (the steepest angle at which material remains stable) and are therefore steep-sided, with slopes between 30 and 40 degrees. In addition, cinder cones have quite large, deep craters in relation to the overall size of the structure. Although relatively symmetrical, some cinder cones are elongated and higher on the side that was downwind during the final eruptive phase.

Most cinder cones are produced by a single, short-lived eruptive event. In one study, half of all cinder cones examined were constructed in less than 1 month, and 95 percent were formed in less than 1 year. Once the event ceases, the magma in the "plumbing" connecting the vent to the magma source solidifies, and the volcano usually does not erupt again. (One exception is Cerro Negro, a cinder cone in Nicaragua, which has erupted more than 20 times since it formed in 1850.) As a consequence of this typically short life span, cinder cones are small, usually between 30 meters (100 feet) and

A young (350 years old) cinder cone, located near Lassen Peak, California.

Ash

Lava flow

Michael Collier

Crater

Pyroclastic material

Central vent filled with rock fragments

Figure 7.13 Cinder cone Cinder cones are built from ejected lava fragments (mostly cinders and bombs) and are relatively small—usually less than 300 meters (1000 feet) in height. This cinder cone is located near Lassen Peak, California. (Photo by Michael Collier)

Parícutin, a cinder cone located in Mexico, erupted for nine years.

Lava flow

Figure 7.14 Parícutin, a well known cinder cone The village of San Juan Parangaricutiro was engulfed by aa lava from Parícutin. Only the church towers remain. (Photos by Michael Collier)

An aa flow emanating from the base of the cone buried much of the village of San Juan Parangaricutiro, leaving only remnants of the village's church.

300 meters (1000 feet). A few rare examples exceed 700 meters (2100 feet) in height.

Cinder cones number in the thousands around the globe. Some occur in groups such as the volcanic field near Flagstaff, Arizona, which consists of about 600 cones. Others are parasitic cones that are found on the flanks or within the calderas of larger volcanic structures.

Parícutin: Life of a Garden-Variety Cinder Cone

One of the very few volcanoes studied by geologists from its very beginning is the cinder cone called Parícutin, located about 320 kilometers (200 miles) west of Mexico City. In 1943 its eruptive phase began, in a cornfield owned by Dionisio Pulido, who witnessed the event as he prepared the field for planting.

For 2 weeks prior to the first eruption, numerous Earth tremors caused apprehension in the nearby village of Parícutin. Then, on February 20, sulfurous gases began billowing from a small depression that had been in the cornfield for as long as local residents could remember. During the night, hot, glowing rock fragments were ejected from the vent, producing a spectacular fireworks display. Explosive discharges continued, throwing hot fragments and ash occasionally as high as 6000 meters

(20,000 feet) into the air. Larger fragments fell near the crater, some remaining incandescent as they rolled down the slope. These built an aesthetically pleasing cone, while finer ash fell over a much larger area, burning and eventually covering the village of Parícutin. In the first day the cone grew to 40 meters (130 feet), and by the fifth day it was more than 100 meters (330 feet) high.

The first lava flow came from a fissure that opened just north of the cone, but after a few months, flows began to emerge from the base of the cone. In June 1944, a clinkery aa flow 10 meters (30 feet) thick moved over much of the village of San Juan Parangaricutiro, leaving only the church steeple exposed (**Figure 7.14**). After 9 years of intermittent pyroclastic explosions and nearly continuous discharge of lava from vents at its base, the activity ceased almost as quickly as it had begun. Today, Parícutin is just one of the scores of cinder cones dotting the landscape in this region of Mexico. Like the others, it will not erupt again.

7.6 CONCEPT CHECKS

1. Describe the composition of cinder cones.
2. How do the size and steepness of slopes of a cinder cone compare with those of a shield volcano?
3. Over what time span does a typical cinder cone form?

7.7 Composite Volcanoes

Explain the formation, distribution, and characteristics of composite volcanoes.

Earth's most picturesque yet potentially dangerous volcanoes are **composite volcanoes**, also known as **stratovolcanoes**. Most are located in a relatively narrow zone that rims the Pacific Ocean, appropriately called the *Ring of Fire* (see Figure 7.33). This active zone consists in part of a chain of continental volcanoes that are distributed along the west coast of the Americas, including the large cones of the Andes in South America and the Cascade Range of the western United States and Canada.

Classic composite cones are large, nearly symmetrical structures consisting of alternating layers of explosively erupted cinders and ash interbedded with lava flows. A few composite cones, notably Italy's Etna and Stromboli, display very persistent eruption activity, and molten lava has been observed in their summit craters for decades. Stromboli is so well known for eruptions that eject incandescent blobs of lava that it has been referred to as the "Lighthouse of the Mediterranean." Mount Etna, on the other hand, has erupted, on average, once every 2 years since 1979 (see the chapter-opening photo).

Just as shield volcanoes owe their shape to fluid basaltic lavas, composite cones reflect the viscous nature of the material from which they are made. In general,

composite cones are the product of silica-rich magma having an andesitic composition. However, many composite cones also emit various amounts of fluid basaltic lava and, occasionally, pyroclastic material having felsic (rhyolitic) composition. The silica-rich magmas typical of composite cones generate thick, viscous lavas that travel less than a few kilometers. In addition, composite cones are noted for generating explosive eruptions that eject huge quantities of pyroclastic material.

A conical shape with a steep summit area and gradually sloping flanks is typical of most large composite cones. This classic profile, which adorns calendars and postcards, is partially a consequence of the way viscous lavas and pyroclastic ejected material contribute to the growth of the cone. Coarse fragments ejected from the summit crater tend to accumulate near their source and contribute to the steep slopes of the summit area. Finer ejected materials, on the other hand, are deposited as a thin layer over a large area that acts to flatten the flank of the cone. In addition, during the early stages of growth, lavas tend to be more abundant and flow greater distances from the vent than later in the volcano's history; this contributes to the cone's broad base. As the volcano matures, the shorter flows that come from the central vent serve to armor and strengthen the summit area. Consequently, steep slopes exceeding 40 degrees are

Figure 7.15 Fujiyama, a classic composite volcano Japan's Fujiyama exhibits the classic form of a composite cone—steep summit and gently sloping flanks. (Photo by Koji Nakano/Getty Images/Sebun)

possible. Two of the most perfect cones—Mount Mayon in the Philippines and Fujiyama in Japan—exhibit the classic form we expect of composite cones, with steep summits and gently sloping flanks (**Figure 7.15**).

Despite the symmetrical forms of many composite cones, most have complex histories. Many composite volcanoes have secondary vents that have produced cinder cones or even much larger volcanic structures on their flanks. Huge mounds of volcanic debris surrounding these structures provide evidence that large sections of these volcanoes slid downslope as massive landslides. Some develop amphitheater-shaped depressions at their summits as a result of explosive lateral eruptions—as occurred during the 1980 eruption of Mount St. Helens.

Often, so much rebuilding has occurred since these eruptions that no trace of the amphitheater-shaped scars remain. Others, such as Crater Lake, have been truncated by the collapse of their summit (see Figure 7.20).

7.7 CONCEPT CHECKS

1. What zone on Earth has the greatest concentration of composite volcanoes?
2. Describe the materials that compose composite volcanoes.
3. How does the composition and viscosity of lava flows differ between composite volcanoes and shield volcanoes?

7.8 Volcanic Hazards

Discuss the major geologic hazards associated with volcanoes.

Roughly 1500 of Earth's known volcanoes have erupted at least once, and some several times, in the past 10,000 years. Based on historical records and studies of active volcanoes, 70 volcanic eruptions can be expected each year and one large-volume eruption every decade. These large-volume eruptions account for the vast majority of human fatalities. For example, the 1902 eruption of Mount Pelée killed 28,000 people, wiping out the entire population of the nearby city of St. Pierre.

Today, an estimated 500 million people from Japan and Indonesia to Italy and Oregon live near active volcanoes. They face a number of volcanic hazards, such as pyroclastic flows, lahars, lava flows, falling volcanic ash, and volcanic gases.

Pyroclastic Flow: A Deadly Force of Nature

One of the most destructive forces of nature is **pyroclastic flows**, which consist of hot gases infused with incandescent ash and large lava fragments. Also referred to as **nuée ardentes** (*glowing avalanches*), these fiery flows are capable of racing down steep volcanic slopes at speeds that can exceed 100 kilometers (60 miles) per hour (**Figure 7.16**). Pyroclastic flows are composed of two parts: a low-density cloud of hot expanding gases containing fine ash particles and a ground-hugging portion that is often composed of pumice and other vesicular pyroclastic material.

Driven by Gravity Pyroclastic flows are propelled by the force of gravity and tend to move in a manner similar to snow avalanches. They are mobilized by expanding volcanic gases released from the lava fragments and by

the expansion of heated air that is overtaken and trapped in the moving front. These gases reduce friction between ash and pumice fragments, which gravity propels downslope in a nearly frictionless environment. This is the reason some pyroclastic flow deposits are found many miles from their source.

Occasionally, powerful hot blasts that carry small amounts of ash separate from the main body of a pyroclastic flow. These low-density clouds, called *surges*, can

Jeep

Figure 7.16 Pyroclastic flows, one of the most destructive volcanic forces This pyroclastic flow occurred on Mount Pinatubo, Philippines, during the 1991 eruption. Pyroclastic flows are composed of hot ash and pumice and/or blocky lava fragments that race down the slope of volcanoes. (Photo by Alberto Garcia/Corbis)

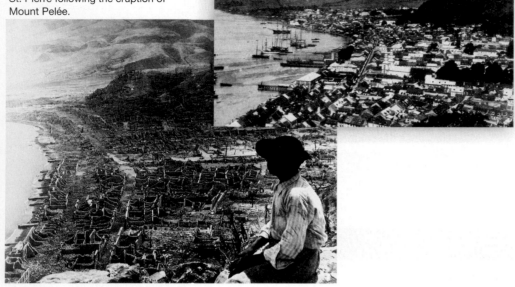

St. Pierre before the 1902 eruption.

St. Pierre following the eruption of Mount Pelée.

Figure 7.17 Destruction of St. Pierre The photo on the left shows St. Pierre as it appeared shortly after the eruption of Mount Pelée in 1902. (Reproduced from the collection of the Library of Congress) The photo on the right shows St. Pierre before the eruption. Many vessels are anchored offshore, as was the case on the day of the eruption. (Photo by Photoshot)

be deadly but seldom have sufficient force to destroy buildings in their paths. Nevertheless, in 1991, a hot ash cloud from Japan's Unzen Volcano engulfed and burned hundreds of homes and moved cars as much as 80 meters (250 feet).

Pyroclastic flows may originate in a variety of volcanic settings. Some occur when a powerful eruption blasts pyroclastic material out of the side of a volcano. More frequently, however, pyroclastic flows are generated by the collapse of tall eruption columns during an explosive event. When gravity eventually overcomes the initial upward thrust provided by the escaping gases, the ejected materials begin to fall, sending massive amounts of incandescent blocks, ash, and pumice cascading downslope.

The Destruction of St. Pierre In 1902 an infamous pyroclastic flow and associated surge from Mount Pelée, a small volcano on the Caribbean island of Martinique, destroyed the port town of St. Pierre. Although the main pyroclastic flow was largely confined to the valley of Riviere Blanche, a low-density fiery surge spread south of the river and quickly engulfed the entire city. The destruction happened in moments and was so devastating that nearly all of St. Pierre's 28,000 inhabitants were killed. Only 1 person on the outskirts of town—a prisoner protected in a dungeon—and a few people on ships in the harbor were spared (**Figure 7.17**).

Within days of this calamitous eruption, scientists arrived on the scene. Although St. Pierre was mantled by only a thin layer of volcanic debris, the scientists discovered that masonry walls nearly 1 meter (3 feet) thick had been knocked over like dominoes, large trees had been uprooted, and cannons had been torn from their mounts.

Lahars: Mudflows on Active and Inactive Cones

In addition to violent eruptions, large composite cones may generate a type of fluid mudflow referred to by its Indonesian name, **lahar**. These destructive flows occur when volcanic ash and debris becomes saturated with water and rapidly moves down steep volcanic slopes, generally following stream valleys. Some lahars are triggered when magma nears the surface of a glacial clad volcano, causing large volumes of ice and snow to melt. Others are generated when heavy rains saturate weathered volcanic deposits. Thus, lahars may occur even when a volcano is *not* erupting.

When Mount St. Helens erupted in 1980, several lahars were generated. These flows and accompanying floodwaters raced down nearby river valleys at speeds exceeding 30 kilometers (20 miles) per hour. These raging rivers of mud destroyed or severely damaged nearly all the homes and bridges along their paths (**Figure 7.18**). Fortunately, the area was not densely populated.

In 1985, deadly lahars were produced during a small eruption of Nevado del Ruiz, a 5300-meter (17,400-foot) volcano in the Andes Mountains of Colombia. Hot pyroclastic material melted ice and snow that capped the mountain (*nevado* means "snow" in Spanish) and sent torrents of ash and debris down three major river valleys that flank the

Figure 7.18 Lahars, mudflows that originate on volcanic slopes This lahar raced down the Muddy River, located southeast of Mount St. Helens, following the May 18, 1980, eruption. Notice the former height of this fluid mudflow, as recorded by the mudflow line on the tree trunks. Note the person (circled) for scale. (Photo by Lyn Topinka/U.S. Geological Survey)

volcano. Reaching speeds of 100 kilometers (60 miles) per hour, these mudflows tragically claimed 25,000 lives.

Many consider Mount Rainier, Washington, to be America's most dangerous volcano because, like Nevado del Ruiz, it has a thick, year-round mantle of snow and glacial ice. Adding to the risk is the fact that more than 100,000 people live in the valleys around Rainier, and many homes are built on deposits left by lahars that flowed down the volcano hundreds or thousands of years ago. A future eruption, or perhaps just a period of heavier-than-average rainfall, may produce lahars that could be similarly destructive.

Other Volcanic Hazards

Volcanoes can be hazardous to human health and property in other various ways. Ash and other pyroclastic material can collapse the roofs of buildings or can be drawn into the lungs of humans and other animals or into aircraft engines (**Figure 7.19**). Volcanoes also emit gases, most notably sulfur dioxide, which affects air quality and, when mixed with rainwater, may destroy vegetation and reduce the quality of groundwater. Despite the known risks, millions of people live in close proximity to active volcanoes.

Volcanic ash can destroy plant life and be drawn into the lungs of humans and other animals.

Ash and other pyroclastic materials can collapse roofs, or completely cover buildings.

Lava flows can destroy homes, roads, and other structures in their paths.

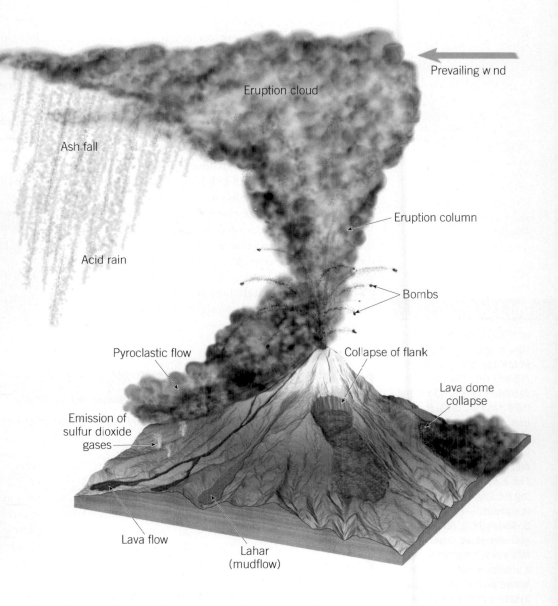

Figure 7.19 Volcanic hazards In addition to generating pyroclastic flows and lahars, volcanoes can be hazardous to human health and property in many other ways.

Volcano-Related Tsunamis One hazard associated with volcanoes is their ability to generate tsunamis. Although they are usually caused by strong earthquakes, these destructive sea waves are sometimes the result of powerful volcanic explosions or the sudden collapse of flanks of volcanoes into the ocean. In 1883, a volcanic eruption on the Indonesian island Krakatoa literally tore the island apart and generated a tsunami that claimed an estimated 36,000 lives.

Volcanic Ash and Aviation During the past 15 years, at least 80 commercial jets have been damaged by inadvertently flying into clouds of volcanic ash. For example, in 1989 a Boeing 747 carrying more than 300 passengers encountered an ash cloud from Alaska's Redoubt Volcano. All four of the engines stalled when they became clogged with ash. Fortunately, the pilots were able to restart the engines and safely land the aircraft in Anchorage.

More recently, in 2010, the eruption of Iceland's Eyjafjallajökull Volcano sent ash high into the atmosphere. The thick plume of ash drifted over Europe, causing airlines to cancel thousands of flights, leaving hundreds of thousands of travelers stranded. Several weeks passed before air travel resumed its normal schedule.

Volcanic Gases and Respiratory Health One of the most destructive volcanic events, called the Laki eruptions, began along a large fissure in southern Iceland in 1783. An estimated 14 cubic kilometers (3.4 cubic miles) of fluid basaltic lavas, in addition to 130 million tons of sulfur dioxide and other poisonous gases, were released. When sulfur dioxide is inhaled, it reacts with moisture in the lungs to produce sulfuric acid. This eruption's sulfur dioxide release caused the death of more than 50 percent of Iceland's livestock. The ensuing famine killed 25 percent of the island's human population. In addition, this huge eruption endangered people and property

across Europe. Crop failure occurred in parts of Western Europe, and thousands of residents perished from lung-related diseases. A recent report estimates that a similar eruption today would cause more than 140,000 cardio-pulmonary fatalities in Europe alone.

Effects of Volcanic Ash and Gases on Weather and Climate Volcanic eruptions can eject dust-sized particles of volcanic ash and sulfur dioxide gas high into the atmosphere. The ash particles reflect solar energy back toward space, producing temporary atmospheric cooling. In the case of the 1783 eruption in Iceland, the atmospheric circulation around the globe appears to have been affected. Drought conditions prevailed in the Nile River valley and in India, while the winter of 1784 brought the longest period of below-zero temperatures in New England's history.

Other eruptions that have produced significant effects on climate worldwide include the eruption of Indonesia's Mount Tambora in 1815, which produced the "year without a summer" (1816), and the eruption of El Chichón in Mexico in 1982. The eruption of El Chichón, although small, emitted an unusually large quantity of sulfur dioxide that reacted with water vapor in the atmosphere to produce a dense cloud of tiny sulfuric-acid droplets. These particles, called **aerosols**, take several years to settle out. Like fine ash, these aerosols lower the mean temperature of the atmosphere by reflecting solar radiation back to space.

7.8 CONCEPT CHECKS

1. Describe pyroclastic flows and explain why they are capable of traveling great distances.
2. What is a lahar?
3. List at least three volcanic hazards other than pyroclastic flows and lahars.

7.9 Other Volcanic Landforms

List and describe volcanic landforms other than volcanic cones.

The most widely recognized volcanic structures are the cone-shaped edifices of composite volcanoes that dot Earth's surface. However, other distinctive and important landforms are also produced by volcanic activity on Earth.

Calderas

Calderas (*caldaria* = a cooking pot) are large steep-sided depressions with diameters that exceed one 1 kilometer (0.6 mile) and have a somewhat circular

form. (Those that are less than one 1 kilometer across are called *collapse pits*, or *craters*.) Most calderas are formed by one of the following processes: (1) the collapse of the summit of a large composite volcano following an explosive eruption of silica-rich pumice and ash fragments (*Crater Lake–type calderas*); (2) the collapse of the top of a shield volcano caused by subterranean drainage from a central magma chamber (*Hawaiian-type calderas*); and (3) the collapse of a large area, caused by the discharge of colossal volumes of silica-rich pumice and ash along ring fractures (*Yellowstone-type calderas*).

Crater Lake–Type Calderas Crater Lake, Oregon, is situated in a caldera approximately 10 kilometers (6 miles) wide and 1175 meters (more than 3800 feet) deep. This caldera formed about 7000 years ago, when a composite cone, named Mount Mazama, violently extruded 50 to 70 cubic kilometers (12 to 17 cubic miles) of pyroclastic material (**Figure 7.20**). With the loss of support, 1500 meters (nearly 1 mile) of the summit of this once-prominent cone collapsed, producing a caldera that eventually filled with rainwater. Later, volcanic activity built a small cinder cone in the caldera. Today this cone, called Wizard Island, provides a mute reminder of past activity (see Figure 7.20).

Hawaiian-Type Calderas Unlike Crater Lake–type calderas, many calderas form gradually because of the loss of lava from a shallow magma chamber underlying the volcano's summit. For example, Hawaii's active shield volcanoes, Mauna Loa and Kilauea, both have large calderas at their summits. Kilauea's measures 3.3 by 4.4 kilometers (about 2 by 3 miles) and is 150 meters (500 feet) deep. The walls of this caldera are almost vertical, and as a result, it looks like a vast, nearly flat-bottomed pit. Kilauea's caldera formed by gradual subsidence as magma slowly drained laterally from the underlying magma chamber, leaving the summit unsupported.

An explosive eruption partially empties a shallow magma chamber.

① Magma chamber

Summit of volcano collapses, enhancing the eruption.

②

Newly formed caldera fills with rain and groundwater.

③

Subsequent eruptions produce the cinder cone called Wizard Island.

④ Wizard Island

Crater Lake

Wizard Island

Michael Collier

Figure 7.20 Formation of Crater Lake–type calderas About 7000 years ago, a violent eruption partly emptied the magma chamber of former Mount Mazama, causing its summit to collapse. Rainfall and groundwater contributed to form Crater Lake, the deepest lake in the United States—594 meters (1949 feet) deep—and the ninth deepest in the world. (After H. Williams, *The Ancient Volcanoes of Oregon.*)

Yellowstone-Type Calderas Historic and destructive eruptions such as Mount St. Helens and Vesuvius pale in comparison to what happened 630,000 years ago in the region now occupied by Yellowstone National Park, when approximately 1000 cubic kilometers (240 cubic miles) of pyroclastic material erupted. This catastrophic eruption sent showers of ash as far as the Gulf of Mexico and resulted in the formation of a caldera 70 kilometers (43 miles) across. Vestiges of this event are the many hot springs and geysers in the Yellowstone region (**Figure 7.21**).

Based on the extraordinary volume of erupted material, researchers have determined that the magma chambers associated with Yellowstone-type calderas must be similarly monstrous. As more and more magma accumulates, the pressure within the magma chamber begins to exceed the pressure exerted by the weight of the overlying rocks. An eruption occurs when the gas-rich magma raises the overlying crust enough to create vertical fractures that extend to the surface. Magma surges upward along these cracks, forming a ring-shaped eruption. With a loss of support, the roof of the magma chamber collapses.

Caldera-forming eruptions are of colossal proportions, ejecting huge volumes of pyroclastic materials, mainly in the form of ash and pumice fragments. Typically, these materials form pyroclastic flows that sweep across the landscape, destroying most living things in their paths. Upon coming to rest, the hot fragments of ash and pumice fuse together, forming a welded tuff that closely resembles a solidified lava flow. Despite the immense size of these calderas, the eruptions that cause them are brief, lasting hours to perhaps a few days.

Large calderas tend to exhibit a complex eruptive history. In the Yellowstone region, for example, three caldera-forming episodes are known to have occurred over the past 2.1 million years. The most recent event (630,000 years ago) was followed by episodic outpourings of degassed rhyolitic and basaltic lavas. Geologic evidence suggests that a magma reservoir still exists beneath Yellowstone; thus, another caldera-forming eruption is likely but not necessarily imminent.

A distinctive characteristic of large caldera-forming eruptions is a slow upheaval of the floor of the caldera that produce a central elevated region called a *resurgent dome* (see Figure 7.21). Unlike calderas associated with shield volcanoes or composite cones, Yellowstone-type depressions are so large and poorly defined that many remained undetected until high-quality aerial and satellite images became available. Other examples of large calderas located in the United States are California's Long Valley Caldera, LaGarita Caldera, located in the San Juan Mountains of southern Colorado, and the Valles Caldera west of Los Alamos, New Mexico. These and similar calderas found elsewhere around the globe are among the largest volcanic structures on Earth, hence the name "supervolcanoes." Volcanologists have compared their destructive force with that of the impact of a small asteroid. Fortunately, no caldera-forming eruption has occurred in historic times.

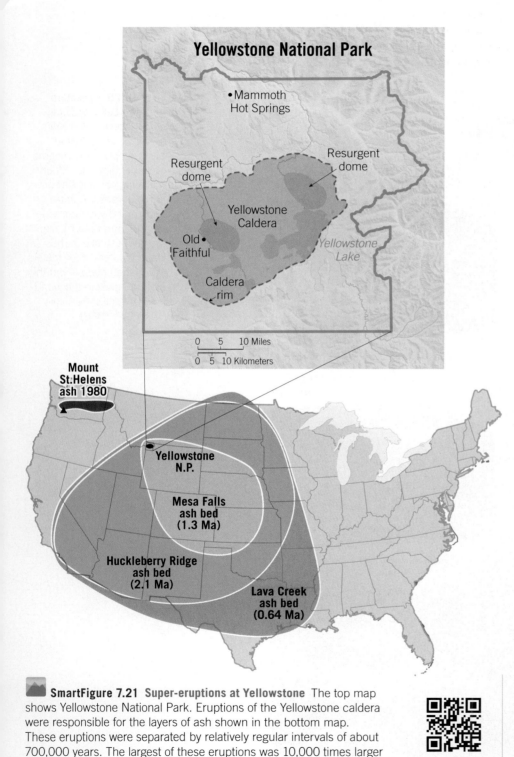

SmartFigure 7.21 Super-eruptions at Yellowstone The top map shows Yellowstone National Park. Eruptions of the Yellowstone caldera were responsible for the layers of ash shown in the bottom map. These eruptions were separated by relatively regular intervals of about 700,000 years. The largest of these eruptions was 10,000 times larger the 1980 eruption of Mount St. Helens.

Fissure Eruptions and Basalt Plateaus

The greatest volume of volcanic material is extruded from fractures in Earth's crust called **fissures** (*fissura =* to split). Rather than building cones, **fissure eruptions** usually emit fluid basaltic lavas that blanket wide areas (**Figure 7.22**). Recall that the recent volcanic activity on Kilauea began along a series of fissures in the East Rift Zone.

In some locations, extraordinary amounts of lava have been extruded along fissures in a relatively short time, geologically speaking. These voluminous accumulations are commonly referred to as **basalt plateaus** because most have a basaltic composition and tend to be rather flat and broad. The Columbia Plateau located in the northwestern United States, which consists of the Columbia River basalts, is the product of this type of activity (**Figure 7.23**). Numerous fissure eruptions have buried the landscape, creating a lava plateau nearly 1 mile thick. Some of the lava remained molten long enough to flow 150 kilometers (90 miles) from its source. The term **flood basalts** appropriately describe these extrusions.

Massive accumulations of basaltic lava, similar to those of the Columbia Plateau, occur elsewhere in the

Figure 7.22 Basaltic fissure eruptions Lava fountaining from a fissure and formation of fluid lava flows called *flood basalts.* The lower photo shows flood basalt flows near Idaho Falls.

USGS

Basaltic lava flows

Lava fountaining   Fissure

Basaltic lava flows

John S. Shelton

Figure 7.23 Columbia River basalts A. The Columbia River basalts cover an area of nearly 164,000 square kilometers (63,000 square miles), commonly called the Columbia Plateau. Activity here began about 17 million years ago, as lava poured out of large fissures, eventually producing a basalt plateau with an average thickness of more than 1 kilometer. **B.** Columbia River basalt flows exposed in the Palouse River Canyon in southwestern Washington State. (Photo by Williamborg)

The Palouse River in Washington State has cut a canyon about 300 meters (1000 feet) deep into the flood basalts of the Columbia Plateau.

KEY

Columbia River Basalts

Other basaltic rocks

▲ Large Cascade volcanoes

world. One of the largest is the Deccan Traps, a thick sequence of flat-lying basalt flows covering nearly 500,000 square kilometers (195,000 square miles) of west-central India. When the Deccan Traps formed about 66 million years ago, nearly 2 million cubic kilometers (480,000 cubic miles) of lava were extruded over a period of approximately 1 million years. Several other huge deposits of flood basalts, including the Ontong Java Plateau, are found on the floor of the ocean.

Volcanic Necks and Pipes

Most volcanic eruptions are fed lava through short conduits that connect shallow magma chambers to vents located at the surface. When a volcano becomes inactive, congealed magma often becomes preserved in the feeding conduit of the volcano as a crudely cylindrical mass. However, all volcanoes will succumb to forces of weathering and erosion. As erosion progresses, the rock occupying the volcanic conduit, which is highly resistant to weathering, may remain standing above the surrounding terrain long after the cone has been worn

away. Shiprock, New Mexico, is a widely recognized and spectacular example of these structures geologists call **volcanic necks** (or **plugs**) **(Figure 7.24)**. More than 420 meters (1380 feet) high, Shiprock is higher than most skyscrapers and one of many such landforms that protrude conspicuously from the red desert landscapes of the American Southwest.

One rare type of conduit, called a **pipe**, carries magma that originated in the mantle at depths that may exceed 150 kilometers (93 miles). The gas-laden magmas that migrate through pipes travel rapidly enough to undergo minimal alteration during their ascent.

The best-known volcanic pipes are the diamond-bearing kimberlite pipes of South Africa. The rocks filling these pipes originated at great depths, where pressure is sufficiently high to generate diamonds and other high-pressure minerals. The process of transporting essentially unaltered magma (along with diamond inclusions) through 150 kilometers (93 miles) of solid rock is exceptional—and accounts for the scarcity of natural diamonds. Geologists consider pipes to be "windows" into Earth that allow us to view rocks that normally form only at great depths.

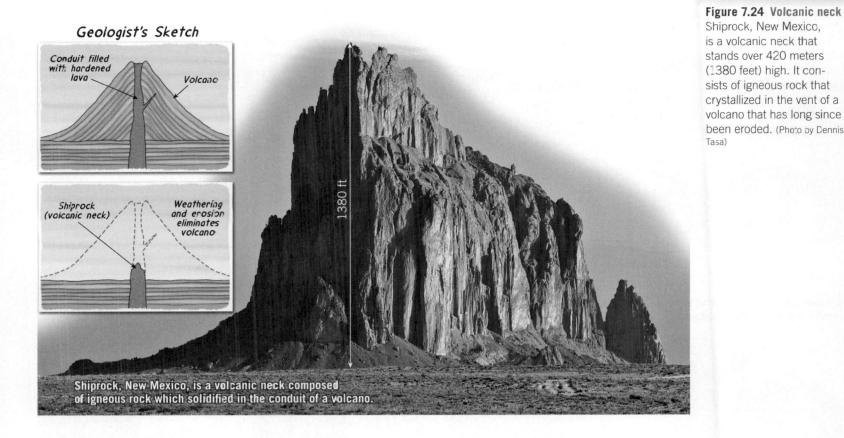

Geologist's Sketch

Conduit filled with hardened lava

Volcano

Shiprock (volcanic neck)

Weathering and erosion eliminates volcano

1380 ft

Shiprock, New Mexico, is a volcanic neck composed of igneous rock which solidified in the conduit of a volcano.

Figure 7.24 Volcanic neck Shiprock, New Mexico, is a volcanic neck that stands over 420 meters (1380 feet) high. It consists of igneous rock that crystallized in the vent of a volcano that has long since been eroded. (Photo by Dennis Tasa)

7.9 CONCEPT CHECKS

1. Describe the formation of Crater Lake. Compare it to the calderas found on shield volcanoes such as Kilauea.
2. Pyroclastic flows are associated with what volcanic structure that is not a cinder cone?
3. How do the eruptions that created the Columbia Plateau differ from eruptions that create large composite volcanoes?
4. Contrast the composition of a typical lava dome and a typical fissure eruption.
5. What type of volcanic structure is Shiprock, New Mexico, and how did it form?

7.10 Intrusive Igneous Activity

Compare and contrast these intrusive igneous structures: dikes, sills, batholiths, stocks, and laccoliths.

Although volcanic eruptions can be violent and spectacular events, most magma is emplaced and crystallizes at depth, without fanfare. Therefore, understanding the igneous processes that occur deep underground is as important to geologists as the study of volcanic events.

Nature of Intrusive Bodies

When magma rises through the crust, it forcefully displaces preexisting crustal rocks referred to as *host*, or *country*, *rock*. The structures that result from the emplacement of magma into preexisting rocks are called **intrusions** or **plutons**. Because all intrusions form far

below Earth's surface, they are studied primarily after uplifting and erosion have exposed them. The challenge lies in reconstructing the events that generated these structures millions of years ago and in vastly different conditions deep underground.

Intrusions are known to occur in a great variety of sizes and shapes. Some of the most common types are illustrated in **Figure 7.25**. Notice that some plutons have a tabular (tablet) shape, whereas others are best described as massive (blob shaped). Also, observe that some of these bodies cut across existing structures, such as sedimentary strata, whereas others form when magma is injected between sedimentary layers. Because of these differences,

Did You Know?

In the eastern United States, some exposed granitic intrusions have dome-shaped, nearly treeless summits, hence the name "summit balds." Examples include Cadillac Mountain in Maine, Chocorua in New Hampshire, Black Mountain in Vermont, and Stone Mountain in Georgia.

A. Interrelationship between volcanism and intrusive igneous activity.

Cinder cones

Composite cones

Laccolith

Fissure eruption

Sills

Pipe

Dikes

Magma chamber

Magma chamber

Sill

B. Basic intrusive structures, some of which have been exposed by erosion.

Laccolith

Volcanic necks

Sills

Dike

Solidified magma bodies (plutons)

Dikes

C. Extensive uplift and erosion exposed a batholith composed of several smaller intrusive bodies (plutons).

Batholith

Plutons

Batholith

Belinda Images/SuperStock

Figure 7.25 Intrusive igneous structures

intrusive igneous bodies are generally classified according to their shape as either **tabular** (*like a tablet*) or **massive** and by their orientation with respect to the host rock. Igneous bodies are said to be **discordant** (*discordare* = to disagree) if they cut across existing structures and **concordant** (*concordare* = to agree) if they inject parallel to features such as sedimentary strata.

Tabular Intrusive Bodies: Dikes and Sills

Tabular intrusive bodies are produced when magma is forcibly injected into a fracture or zone of weakness, such as a bedding surface (see Figure 7.25). **Dikes** are discordant bodies that cut across bedding surfaces or other structures in the country rock. By contrast, **sills** are nearly horizontal, concordant bodies that form when magma exploits weaknesses between sedimentary beds or other structures (**Figure 7.26**). In general, dikes serve as

Sill

E.J. Tarbuck

Figure 7.26 Sill exposed in Salt River Canyon, Arizona
The dark, essentially horizontal band is a sill of basaltic composition that intruded horizontal layers of sedimentary rock.

tabular conduits that transport magma, whereas sills tend to accumulate magma and increase in thickness.

Dikes and sills are typically shallow features, occurring where the country rocks are sufficiently brittle to fracture. They can range in thickness from less than a millimeter to more than a kilometer.

While dikes and sills can occur as solitary bodies, dikes tend to form in roughly parallel groups called *dike swarms*. These multiple structures reflect the tendency for fractures to form in sets when tensional forces stretch brittle country rock. Dikes can also occur radiating from an eroded volcanic neck, like spokes on a wheel. In these situations, the active ascent of magma generated fissures in the volcanic cone out of which lava flowed. Dikes frequently weather more slowly than the surrounding rock. Consequently, when exposed by erosion, dikes tend to have a wall-like appearance, as shown in **Figure 7.27**.

Because dikes and sills are relatively uniform in thickness and can extend for many kilometers, they are assumed to be the product of very fluid and, therefore, mobile magmas. One of the largest and most studied of all sills in the United States is the Palisades Sill. Exposed for 80 kilometers (50 miles) along the west bank of the Hudson River in southeastern New York and northeastern New Jersey, this sill is about 300 meters (1000 feet) thick. Because it is resistant to erosion, the Palisades Sill forms an imposing cliff that can be easily seen from the opposite side of the Hudson.

In many respects, sills closely resemble buried lava flows. Both are tabular and can have a wide aerial extent, and both may exhibit columnar jointing. **Columnar jointing** occurs when igneous rocks cool and develop shrinkage fractures that produce elongated, pillar-like

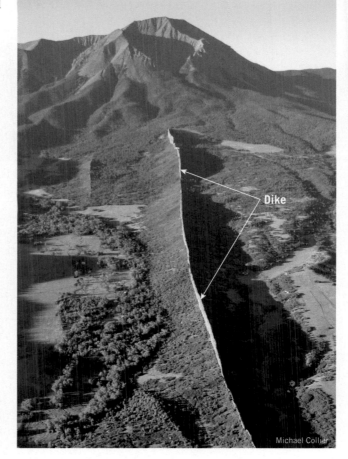

Figure 7.27 Dike exposed in the Spanish Peaks, Colorado This elongated dike is composed of igneous rock that is more resistant to weathering than the surrounding material.

Dike

Michael Collier

columns that most often have six sides (**Figure 7.28**). Further, because sills generally form in near-surface environments and may be only a few meters thick, the emplaced

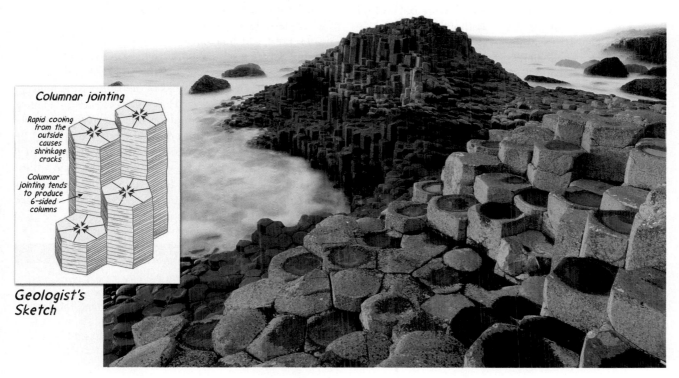

Columnar jointing

Rapid cooling from the outside causes shrinkage cracks

Columnar jointing tends to produce 6-sided columns

Geologist's Sketch

Figure 7.28 Columnar jointing Giant's Causeway in Northern Ireland is an excellent example of columnar jointing. (Photo by John Lawrence/Getty Images)

Figure 7.29 Mount Ellen, one of five peaks that make up Utah's Henry Mountains Although the main intrusions in the Henry Mountains are stocks, numerous laccoliths formed as offshoots of these structures. (Photo by Michael DeFreitas North America/Alamy)

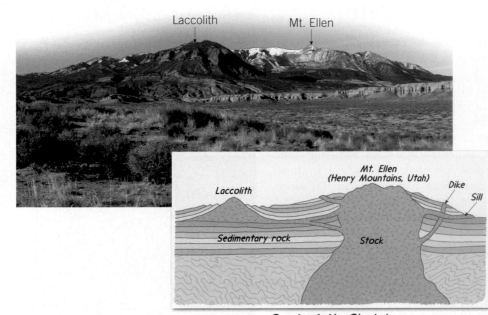

Laccolith   Mt. Ellen

Mt. Ellen (Henry Mountains, Utah)   *Dike*   *Sill*

Laccolith

Sedimentary rock   *Stock*

Geologist's Sketch

magma often cools quickly enough to generate a fine-grained texture. (Recall that most intrusive igneous bodies have a coarse-grained texture.)

Massive Intrusive Bodies: Batholiths, Stocks, and Laccoliths

By far the largest intrusive igneous bodies are **batholiths** (*bathos* = depth, *lithos* = stone). Batholiths occur as mammoth linear structures several hundreds of kilometers long and up to 100 kilometers wide (see Figure 7.25C). The Sierra Nevada batholith, for example, is a continuous granitic structure that forms much of the Sierra Nevada, in California. An even larger batholith extends for over 1800 kilometers (1100 miles) along the Coast Mountains of western Canada and into southern Alaska. Although batholiths can cover a large area, recent gravitational studies indicate that most are less than 10 kilometers (6 miles) thick. Some are even thinner; the Coastal batholith of Peru, for example, is essentially a flat slab with an average thickness of only 2 to 3 kilometers (1 to 2 miles).

Batholiths are typically composed of felsic and intermediate rock types and are often referred to as "granite batholiths." Large granite batholiths consist of hundreds of plutons that intimately crowd against or penetrate one another. These bulbous masses were emplaced over spans of millions of years. The intrusive activity that created the Sierra Nevada batholith, for example, occurred nearly continuously over a 130-million-year period that ended about 80 million years ago.

By definition, a plutonic body must have a surface exposure greater than 100 square kilometers (40 square miles) in order to be considered a batholith. Smaller plutons are termed **stocks**. However, many stocks appear to be portions of much larger intrusive bodies that would be classified as batholiths if they were fully exposed.

Laccoliths A nineteenth-century study by G. K. Gilbert of the U.S. Geological Survey in the Henry Mountains of Utah produced the first clear evidence that igneous intrusions can lift the sedimentary strata they penetrate. Gilbert gave the name **laccoliths** to the igneous intrusions he observed, which he envisioned as igneous rock forcibly injected between sedimentary strata so as to arch the beds above while leaving those below relatively flat. It is now known that the five major peaks of the Henry Mountains are not laccoliths but stocks. However, these central magma bodies are the source material for branching offshoots that are true laccoliths, as Gilbert defined them (**Figure 7.29**).

Numerous other granitic laccoliths have since been identified in Utah. The largest is a part of the Pine Valley Mountains, located north of St. George, Utah. Others are found in the La Sal Mountains near Arches National Park and in the Abajo Mountains directly to the south.

7.10 CONCEPT CHECKS

1. What is meant by the term *country rock*?
2. Describe *dike* and *sill*, using the appropriate terms from the following list: massive, discordant, tabular, concordant.
3. Compare and contrast batholiths, stocks, and laccoliths in terms of size and shape.

7.11 Partial Melting and the Origin of Magma

Summarize the major processes that generate magma from solid rock.

Based on evidence from the study of earthquake waves, *Earth's crust and mantle are composed primarily of solid, not molten, rock.* Although the outer core is liquid, this iron-rich material is very dense and remains deep within Earth. So where does magma come from?

Partial Melting

Recall that igneous rocks are composed of a mixture of minerals. Since these minerals have different melting points, igneous rocks tend to melt over a temperature range of at least 200°C (360°F). As rock begins to melt, the minerals with the lowest melting temperatures are the first to melt. If melting continues, minerals with higher melting points begin to melt, and the composition of the melt steadily approaches the overall composition of the rock from which it was derived.

Most often, however, melting is not complete. The incomplete melting of rocks is known as **partial melting**, a process that produces most magma. Partial melting can be likened to a chocolate chip cookie containing nuts that is left out in the Sun. The chocolate chips represent the minerals with the lowest melting points because they will begin to melt before the other ingredients. Unlike with the chocolate chip cookie, when rock partially melts, the molten material separates from the solid components. Also, because the molten material is less dense than the remaining solids, when enough of the melt collects, it rises toward Earth's surface.

Generating Magma from Solid Rock

Workers in underground mines know that temperatures increase as they descend deeper below Earth's surface. Although the rate of temperature change varies considerably from place to place, it *averages* about 25°C per kilometer in the *upper* crust. This increase in temperature with depth is known as the **geothermal gradient**. As shown in **Figure 7.30**, when a typical geothermal gradient is compared to the melting point curve for the mantle rock peridotite, the temperature at which peridotite melts is higher than the geothermal gradient. Thus, under normal conditions, the mantle is solid. However, tectonic processes exist that trigger melting by reducing the melting point (temperature) of mantle rock.

Decrease in Pressure: Decompression Melting

If temperature were the only factor that determines whether rock melts, our planet would be a molten ball

covered with a thin, solid outer shell. This is not the case because pressure, which also increases with depth, influences the melting temperatures of rocks.

Melting, which is accompanied by an increase in volume, *occurs at progressively higher temperatures with increased depth*. This is the result of the steady increase in confining pressure exerted by the weight of overlying rocks. Conversely, reducing confining pressure lowers a rock's melting temperature. When confining pressure drops sufficiently, **decompression melting** is triggered.

Decompression melting occurs where hot, solid mantle rock ascends in zones of convective upwelling, thereby moving into regions of lower pressure. This process is responsible for generating magma along divergent plate boundaries (oceanic ridges) where plates are rifting apart (**Figure 7.31**). Below the ridge crest, hot mantle rock rises and melts, replacing the material that shifted

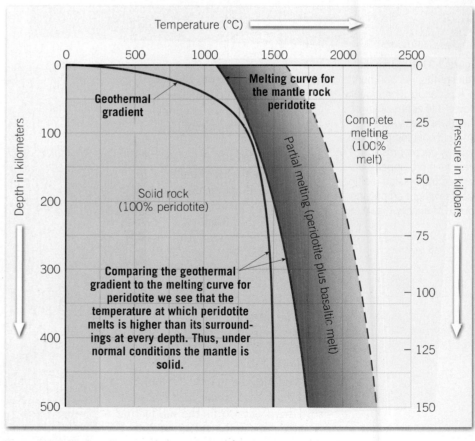

Figure 7.30 Diagram illustrating why the mantle is mainly solid This diagram shows the geothermal gradient (increase in temperature with depth) for the crust and upper mantle. Also illustrated is the melting point curve for the mantle rock peridotite. When we compare the change in temperature with depth (geothermal gradient) to the melting point of peridotite, the temperature at which peridotite melts is higher than the geothermal gradient at every depth. Thus, under normal conditions, the mantle is solid.

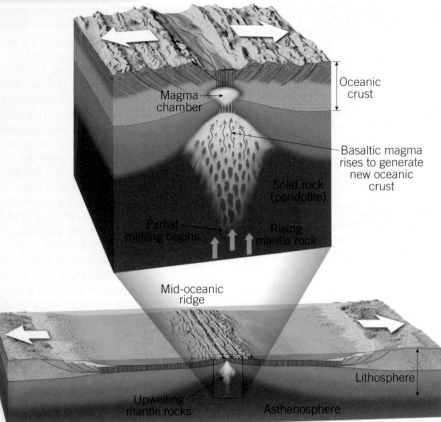

Figure 7.31 Decompression melting As hot mantle rock ascends, it continually moves into zones of lower and lower pressure. This drop in confining pressure initiates *decompression melting* in the upper mantle. Decompression melting occurs without an outside source of energy (heat).

hot basaltic magmas with temperatures that may exceed 1250°C (2300°F).

Temperature Increase: Melting Crustal Rocks When enough mantle-derived basaltic magma forms, it buoyantly rises toward the surface. In continental settings, basaltic magma often "ponds" beneath crustal rocks, which have a lower density and are already near their melting temperature. The hot basaltic magma may heat the overlying crustal rocks sufficiently to generate a secondary, silica-rich magma. If these low-density silica-rich magmas reach the surface, they tend to produce explosive eruptions that we associate with convergent plate boundaries.

Crustal rocks can also melt during continental collisions that result in the formation of a large mountain belt. During such events, the crust is greatly thickened, and some crustal rocks are buried to depths where the temperatures are elevated sufficiently to cause partial melting. The felsic (granitic) magmas produced in this manner usually solidify before reaching the surface, so volcanism in not typically associated with these collision-type mountain belts.

horizontally away from the ridge axis. Decompression melting also occurs when ascending mantle plumes reach the uppermost mantle.

Addition of Water Another important factor affecting the melting temperature of rock is its water content. Water and other volatiles act as salt does to melt ice. Water causes rock to melt at lower temperatures, just as putting rock salt on an icy sidewalk induces melting.

The introduction of water to generate magma occurs mainly at convergent plate boundaries where cool slabs of oceanic lithosphere descend into the mantle (**Figure 7.32**). As an oceanic plate sinks, heat and pressure drive water from the subducting oceanic crust and overlying sediments. These fluids migrate into the wedge of hot mantle that lies directly above. At a depth of about 100 kilometers (60 miles), the addition of water lowers the melting temperature of mantle rock sufficiently to trigger partial melting. Partial melting of the mantle rock peridotite generates

Figure 7.32 The introduction of water can lower the melting temperature of hot mantle rock to trigger melting As an oceanic plate descends into the mantle, water and other volatiles are driven from the subducting crustal rocks into the wedge of mantle above. At a depth of about 100 kilometers (60 miles), mantle rock is hot enough that the addition of water can cause melting.

In summary, magma can be generated three ways: (1) in zones of upwelling, *a decrease in pressure* (without an increase in temperature) can result in *decompression melting*; (2) the *introduction of water* can lower the melting temperature of hot mantle rock sufficiently to generate magma; and (3) crustal rocks can be heated above their melting temperature by hot mantle-derived basaltic magma or during continental collisions.

7.12 Plate Tectonics and Volcanic Activity

Relate the distribution of volcanic activity to plate tectonics.

Geologists have known for decades that the global distribution of most of Earth's volcanoes is not random. Most active volcanoes on land are located along the margins of the ocean basins—notably within the circum-Pacific belt known as the **Ring of Fire** (**Figure 7.33**). These volcanoes consist mainly of composite cones that emit volatile-rich magma having an andesitic (intermediate) composition and occasionally produce awe-inspiring eruptions.

Another group of volcanoes includes the innumerable seamounts that form along the crest of the mid-ocean ridges. At these depths (1–3 kilometers below sea level), the pressures are so intense that the gases emitted quickly dissolve in the seawater and never reach the surface.

There are some volcanic structures, however, that appear to be somewhat randomly distributed around the globe. These volcanic structures comprise most of the islands of the deep-ocean basins, including the Hawaiian islands, the Galapagos islands, and Easter Island.

Prior to the development of the theory of plate tectonics, geologists did not have an acceptable explanation for the distribution of Earth's volcanoes. Recall that most magma originates in rocks from Earth's upper mantle that is essentially solid, *not molten*. The basic connection between plate tectonics and volcanism is that *plate motions provide the mechanisms by which mantle rocks undergo partial melting to generate magma.*

Volcanism at Convergent Plate Boundaries

Recall that along certain convergent plate boundaries, two plates move toward each other, and a slab of oceanic lithosphere descends into the mantle, generating a deep-ocean trench. As the slab sinks deeper into the mantle, the increase in temperature and pressure drives water and carbon dioxide from the oceanic crust. These mobile fluids migrate upward and reduce the melting point of hot mantle rock sufficiently to trigger some melting (**Figure 7.34A**). Recall that the partial melting of mantle rock (peridotite) generates magma having a basaltic composition. After a sufficient quantity of the rock has melted, blobs of buoyant magma slowly migrate upward.

Figure 7.33 Ring of Fire Most of Earth's major volcanoes are located in a zone around the Pacific called the Ring of Fire. Another large group of active volcanoes lays mainly undiscovered along the mid-ocean ridge system.

Pavlof · Hekla · Shishaldin · Bezymianny · Augustine · Laki · Fujiyama · Mt. St. Helens · Surtsey · Vesuvius · Katmai ("Valley of 10,000 Smokes") · Etna · Mt. Unzen · Mariana Is. · Popocatepetl · Pinatubo · Kilauea · Canary Is. · Santorin · Mt. Mayon · Mauna Loa · Parícutin · Pelée · Galapagos Is. · Nevado del Ruiz · Krakatau · Cotopaxi · Misti · Kilimanjaro · Tambora · Tonga Is. · **RING OF FIRE** · Easter Is. · New Zealand · South Sandwich Is. · Deception Is.

A. *Convergent Plate Volcanism* When an oceanic plate subducts, melting in the mantle produces magma that gives rise to a volcanic island arc on the overlying oceanic crust.

Mount Augustine, Alaska (NASA)

C. *Intraplate Volcanism* When an oceanic plate moves over a hot spot, a chain of volcanic structures such as the Hawaiian Islands is created.

Kilauea, Hawaii (USGS)

E. *Convergent Plate Volcanism* When oceanic lithosphere descends beneath a continent, magma generated in the mantle rises to form a continental volcanic arc.

SmartFigure 7.34 Earth's zones of volcanism

B. *Divergent Plate Volcanism* Along the oceanic ridge, where two plates are being pulled apart, upwelling of hot mantle rock creates new seafloor.

Oceanic crust

Magma chamber

Decompression melting

Asthenosphere

Iceland (Wedigo Ferchland)

Mid-Atlantic Rift

Deccan Plateau

Africa

South America

East Africa Rift Valley

Hot spot

Flood basalts

Continental crust

Decompression melting

Rising mantle plume

D. *Intraplate Volcanism* When a large mantle plume ascends beneath continental crust, vast outpourings of fluid basaltic lava like those that formed the Deccan Plateau may be generated.

Rift valley

Continental crust

Decompression melting

Mount Kilimanjaro, Africa (Corbis/Photolibrary)

F. *Divergent Plate Volcanism* When plate motion pulls a continental block apart, stretching and thinning of the lithosphere causes molten rock to ascend from the mantle.

Did You Know?

Iceland, which is the largest volcanic island in the world, has more than 20 active volcanoes and numerous geysers and hot springs. The Icelanders named their first gusher of boiling water *Geysir*, a name that was subsequently used to describe similar features around the world, including "Old Faithful Geyser" in Yellowstone National Park.

Volcanism at a convergent plate margin results in the development of a slightly curved chain of volcanoes called a *volcanic arc*. These volcanic chains develop roughly parallel to the associated trench—at distances of 200 to 300 kilometers (100 to 200 miles). Volcanic arcs can be constructed on oceanic as well as continental lithosphere. Those that develop within the ocean and grow large enough for their tops to rise above the surface are labeled *archipelagos* in most atlases. Geologists prefer the more descriptive term **volcanic island arcs**, or simply **island arcs** (see Figure 7.34A). Several young volcanic island arcs border the western Pacific basin, including the Aleutians, the Tongas, and the Marianas.

Volcanism associated with convergent plate boundaries may also develop where slabs of oceanic lithosphere are subducted under continental lithosphere to produce a **continental volcanic arc** (Figure 7.34E). The mechanisms that generate these mantle-derived magmas are essentially the same as those that create volcanic island arcs. The most significant difference is that continental crust is much thicker and is composed of rocks having higher silica content than oceanic crust. Hence, a mantle-derived magma changes chemically as it rises by assimilating silica-rich crustal rocks. At the same time, extensive magmatic differentiation occurs. (Recall that magmatic differentiation is the formation of secondary magmas from parent magma.) Stated another way, the magma generated in the mantle may change from a fluid basaltic magma to a silica-rich andesitic or rhyolitic magma as it moves up through the continental crust.

The Ring of Fire, a belt of explosive volcanoes, surrounds the Pacific basin. The Ring of Fire exists because oceanic lithosphere is being subducted beneath most of the landmasses that surround the Pacific Ocean. The volcanoes of the Cascade Range in the northwestern United States, including Mount Hood, Mount Rainier, and Mount Shasta, are examples of volcanoes generated at a convergent plate boundary (**Figure 7.35**).

Volcanism at Divergent Plate Boundaries

The greatest volume of magma (perhaps 60 percent of Earth's total yearly output) is produced along the oceanic ridge system in association with seafloor spreading (**Figure 7.34B**). Below the ridge axis where lithospheric

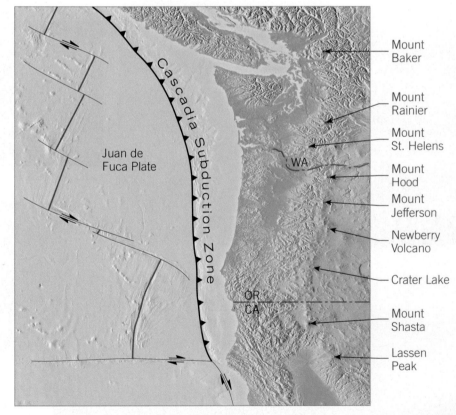

SmartFigure 7.35 Subduction of the Juan de Fuca plate produced the Cascade volcanoes Major volcanic structures of the Cascade Range formed along a convergent plate boundary (Cascadia subduction zone) off the Pacific coast of North America.

A. A rising mantle plume with a large bulbous head is thought to generate Earth's large basalt provinces.

Plate motion
Oceanic lithosphere
Head
Rising mantle plume
Tail

B. Rapid decompression melting of the plume head produces extensive outpourings of flood basalts over a relatively short time span.

Plate motion
Flood basalts
Partial melting
Rising plume tail

C. Because of plate movement, volcanic activity from the rising tail of the plume generates a linear chain of smaller volcanic structures.

Plate motion
Large basaltic province
Volcanic trail
Hot spot volcanic activity
Oceanic lithosphere
Rising plume tail

Figure 7.36 Hot spots and mantle plumes This model of hot-spot volcanism is thought to explain the formation of oceanic plateaus, large basalt provinces on land, and chains of volcanic islands such as the Hawaiian chain.

plates are continually being pulled apart, the solid yet mobile mantle responds by ascending to fill the rift. Recall that as hot rock rises, it experiences a decrease in confining pressure and undergoes melting without the addition of heat. Recall that this process is called *decompression melting*.

Partial melting of mantle rock at spreading centers produces basaltic magma. Because this newly formed magma is less dense than the mantle rock from which it was derived, it rises and collects in reservoirs located just below the ridge crest. This activity continuously adds new basaltic rock to plate margins, temporarily welding them together, only to have them break again as spreading continues. Along some ridge segments, outpourings of pillow lavas build numerous volcanic structures, the largest of which is Iceland.

Although most spreading centers are located along the axis of an oceanic ridge, some are not. In particular, the East African Rift is a site where continental lithosphere is being pulled apart (**Figure 7.34F**). Vast outpourings of fluid basaltic lavas as well as several active composite volcanoes are found in this region of the globe.

Intraplate Volcanism

We know why igneous activity is initiated along plate boundaries, but why do eruptions occur in the interiors of plates? Hawaii's Kilauea, one of the world's most active volcanoes, is situated thousands of kilometers from the nearest plate boundary, in the middle of the vast Pacific plate (**Figure 7.34C**). Other sites of **intraplate volcanism** (meaning "within the plate") include large outpourings of fluid basaltic lavas like those that

compose the Columbia River basalts, the Siberian Traps in Russia, India's Deccan Plateau, and several large oceanic plateaus, including the Ontong Java Plateau located in the western Pacific. These massive structures are estimated to be 10 to 40 kilometers (6 to 25 miles) thick. It is thought that most intraplate volcanism occurs where a mass of hotter-than-normal mantle material called a **mantle plume** ascends toward the surface (**Figure 7.36A**).° Although the depth at which mantle plumes originate is a topic of debate, some appear to form deep within Earth, at the core–mantle boundary. These plumes of solid yet mobile rock rise toward the surface in a manner similar to the blobs that form within a lava lamp. (These are the trendy lamps that contain two immiscible liquids in a glass container. As the base of the lamp is heated, the denser liquid at the bottom becomes buoyant and forms blobs that rise to the top.) Like the blobs in a lava lamp, a mantle plume has a bulbous head that draws out a narrow stalk beneath it as it rises. Once the plume head nears the top of the mantle, decompression melting generates basaltic magma that triggers volcanism at the surface.

Large mantle plumes, dubbed *superplumes*, are thought to be responsible for the vast outpourings of basaltic lava that created the large basalt provinces. When the head of the plume reaches the base of the lithosphere, melting progresses rapidly. This causes the burst of volcanism that emits voluminous outpourings of lava to form a huge basalt province in a matter of a

---
° Recent research on the nature of mantle plumes has caused some geologists to question their role, if any, in the formation of at least some large basalt provinces.

million or so years (**Figure 7.36B**). Due to the extreme nature of the eruptions required to produce the large basalt provinces, some researchers believe the eruptions were responsible for the extinction of many of Earth's life-forms.

The comparatively short initial eruptive phase is often followed by millions of years of less voluminous activity, as the plume tail slowly rises to the surface. Extending away from some large flood basalt provinces is a chain of volcanic structures, similar to the Hawaiian chain (**Figure 7.36C**).

7.12 CONCEPT CHECKS

1. Are volcanoes in the Ring of Fire generally described as quiescent or explosive? Name an example that supports your answer.
2. How is magma generated along convergent plate boundaries?
3. Volcanism at divergent plate boundaries is most often associated with which rock type? What causes rocks to melt in these settings?
4. What is thought to be the source of magma for most intraplate volcanism?
5. At which of the three types of plate boundaries is the greatest quantity of magma generated?

CONCEPTS IN REVIEW: VOLCANOES AND OTHER IGNEOUS ACTIVITY

7.1 Mount St. Helens Versus Kilauea

Compare and contrast the 1980 eruption of Mount St. Helen's with the eruption of Kilauea, which began in 1983 and continues today.

- Volcanic eruptions cover a broad spectrum from explosive eruptions, like Mount St. Helens in 1980, to the quiescent eruptions of Kilauea.

? **Although Kilauea usually erupts in a gentle manner, what risks might you encounter if you lived nearby?**

7.2 The Nature of Volcanic Eruptions

Explain why some volcanic eruptions are explosive and others are quiescent.

KEY TERMS: magma, lava, viscosity, eruption column

- One important characteristic that differentiates various lavas is their viscosity (resistance to flow). In general, the more silica in the lava, the more viscous it is. The lower the silica content, the runnier the lava. Another factor that influences viscosity is temperature. Hot lavas are more fluid, while cool lavas are more viscous.
- High-silica, low-temperature lavas are the most viscous and allow the greatest amount of pressure to build up before they "let go" in an eruption. In contrast, lavas that are hot and low in silica are the most fluid. Because basaltic lavas are less viscous, they produce relatively gentle eruptions, while volcanoes that erupt felsic lavas (rhyolite and andesite) tend to be more explosive.

7.3 Materials Extruded During an Eruption

List and describe the three categories of materials extruded during volcanic eruptions.

KEY TERMS: aa flow, pahoehoe flow, lava tube, volatiles, pyroclastic materials, scoria, pumice

- Volcanoes bring liquid lava, gases, and solid chunks to Earth's surface.
- Because of their low viscosity, basaltic lava flows can extend great distances from a volcano, where they travel over the surface as pahoehoe or aa flows. Sometimes the surface of the flow congeals, but lava continues to flow below in tunnels called lava tubes.

- The gases most commonly emitted by volcanoes are water vapor and carbon dioxide. Upon reaching the surface, these gases rapidly expand, resulting in explosive eruptions that generate an eruptive column and produce a mass of lava fragments called pyroclastic materials.
- Pyroclastic materials come in several sizes. From smallest to largest, they are ash, lapilli, and blocks or bombs, depending on whether the material left the volcano as solid fragments or as liquid blobs.
- If bubbles of gas in lava don't pop before solidification, they are preserved as voids called vesicles. Especially frothy silica-rich lava can cool to make pumice, which may float in water. Basaltic (mafic) lava with lots of bubbles cools to make scoria.

7.4 Anatomy of a Volcano

Label a diagram that illustrates the basic features of a typical volcanic cone.

KEY TERMS: conduit, vent, volcanic cone, crater, parasitic cone, fumarole

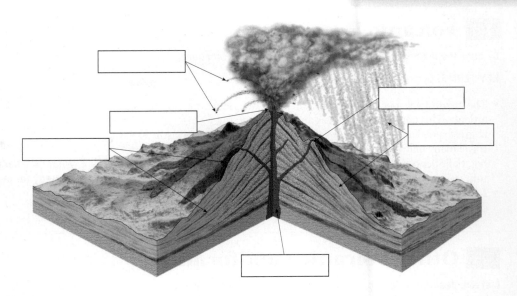

- Volcanoes are varied in form but share a few common features. Most are roughly conical piles of extruded material that collect around a central vent. The vent is usually within a summit crater or caldera. On the flanks of the volcano, there may be smaller vents marked by small parasitic cones, or fumaroles, spots where gas is expelled.

? **Label the diagram with the following terms: conduit, vent, bombs, lava, parasitic cone, pyroclastic material**

7.5 Shield Volcanoes

Summarize the characteristics of shield volcanoes and provide one example.

KEY TERM: shield volcano

- Shield volcanoes consist of many successive layers of low-viscosity basaltic lava and lack significant amounts of pyroclastic debris. Lava tubes help transport lava far from the main vent, resulting in very gentle, shield-like profiles.
- Most large shield volcanoes are associated with hot-spot volcanism. The volcanoes Kilauea, Mauna Loa, and Mauna Kea in Hawaii are classic examples of the low, wide form.

7.6 Cinder Cones

Describe the formation, size, and composition of cinder cones.

KEY TERMS: cinder cone (scoria cone)

- Cinder cones are steep-sided structures composed mainly of pyroclastic debris, typically having a basaltic composition. Lava flows sometimes emerge from the base of a cinder cone but typically do not flow out of the crater.
- Cinder cones are small relative to the other major kinds of volcanoes, reflecting the fact that they form quickly, as single eruptive events. Because they are unconsolidated, cinder cones easily succumb to weathering and erosion.

7.7 Composite Volcanoes

Explain the formation, distribution, and characteristics of composite volcanoes.

KEY TERMS: composite volcano (stratovolcano)

- Composite volcanoes are called "composite" because they consist of both pyroclastic material and lava flows. They typically erupt silica-rich lavas that cool to produce andesite or rhyolite. They are much larger than cinder cones and form from multiple eruptions over a million years or longer.

- Because the andesitic or rhyolitic lava erupted from composite volcanoes is more viscous than basaltic lava, it accumulates at a steeper angle than does the lava from shield volcanoes. Over time, the composite volcano's combination of lava and cinders produces towering volcanoes with a classic symmetrical shape.
- Mount Rainier and the other volcanoes of the Cascade Range in the northwestern United States are good examples of composite volcanoes, as are the other volcanoes of the Pacific Ocean's Ring of Fire.

? **If your family had to live next to a volcano (shield, cinder cone, or composite), which one would you choose? Explain.**

7.8 Volcanic Hazards

Discuss the major geologic hazards associated with volcanoes.

KEY TERMS: pyroclastic flow, nuée ardente, lahar, aerosol

- The greatest volcanic hazard to human life is the pyroclastic flow, or nuée ardente. This dense mix of hot gas and pyroclastic fragments races downhill at great speed and incinerates everything in its path. A pyroclastic flow can travel many miles from its source volcano. Because pyroclastic flows are hot, their deposits frequently "weld" together into a solid rock called welded tuff.
- Lahars are volcanic mudflows. These rapidly moving slurries of ash and debris suspended in water can occur even when a volcano isn't actively erupting. They tend to follow stream valleys and can result in loss of life and/or significant damage to structures in their path.
- Volcanic ash in the atmosphere can be a risk to air travel when it is sucked into airplane engines. Volcanoes at sea level can generate tsunamis when they erupt or their flanks collapse into the ocean. In addition, volcanoes that spew large amounts of gas such as sulfur dioxide can cause human respiratory problems. If volcanic gases reach the stratosphere, they screen out a portion of incoming solar radiation and can trigger short-term cooling at Earth's surface.

? **What do lahars and pyroclastic flows have in common? What is the best strategy for avoiding their worst effects?**

7.9 Other Volcanic Landforms

List and describe volcanic landforms other than volcanic cones.

KEY TERMS: caldera, fissure, fissure eruption, basalt plateau, flood basalt, volcanic neck (plug), pipe

- Calderas are among the largest volcanic structures. They form when the stiff, cold rock above a magma chamber cannot be supported and collapses to create a broad, bowl-like depression. On shield volcanoes, calderas form slowly as lava is drained from the magma chamber beneath the volcano. On composite volcanoes, a caldera collapse often follows an explosive eruption that can result in significant loss of life and destruction of property.
- Fissure eruptions produce massive floods of low-viscosity, silica-poor lava from large cracks in the crust. Layer upon layer of these flood basalts may build up to significant thicknesses, as in the Columbia Plateau or the Deccan Traps. The defining feature of a flood basalt is the broad area it covers.
- An example of a volcanic neck is preserved at Shiprock, New Mexico. The lava in the "throat" of this ancient volcano crystallized to form a "plug" of solid rock, and it weathers more slowly than the conical volcano in which it formed. Now, after the mound of pyroclastic debris has been eroded away, the resistant neck is a distinctive landform.

7.10 Intrusive Igneous Activity

Compare and contrast these intrusive igneous structures: dikes, sills, batholiths, stocks, and laccoliths.

KEY TERMS: intrusion, pluton, tabular, massive, discordant, concordant, dike, sill, columnar jointing, batholith, stock, laccolith

- When magma intrudes other rocks, it may cool and crystallize before reaching the surface, producing intrusions called plutons. Plutons come in many shapes. They may cut across the host rocks without regard for preexisting structures, or the magma may flow along weak zones in the host rock, such as between the horizontal layers of sedimentary bedding.
- Tabular intrusions may be concordant (sills) or discordant (dikes). Massive plutons may be small (stocks) or very large (batholiths). Blister-like intrusions also exist (laccoliths). As solid igneous rock cools, its volume decreases. Contraction can produce columnar jointing, a distinctive pattern consisting mainly of six-sided fractures.

? **Label the accompanying diagram using the following terms; batholith, laccolith, sill, and dike.**

7.11 Partial Melting and the Origin of Magma

Summarize the major processes that generate magma from solid rock.

KEY TERMS: partial melting, geothermal gradient, decompression melting

- Solid rock may melt under three geologic circumstances: when heat is added to the rock, raising its temperature; when already hot rock experiences lower pressures (decompression, as seen at mid-ocean ridges); and when water is added to hot rock that is near its melting point (as occurs at subduction zones).

? **Different processes produce magma in different tectonic settings. Consider situations A, B, and C in the accompanying diagram and describe the processes that would be most likely to trigger melting in each.**

7.12 Plate Tectonics and Volcanic Activity

Relate the distribution of volcanic activity to plate tectonics.

KEY TERMS: Ring of Fire, volcanic island arc, continental volcanic arc, intraplate volcanism, mantle plume

- Volcanoes occur at both convergent and divergent plate boundaries, as well as in intraplate settings.
- Convergent plate boundaries that involve the subduction of oceanic crust are the sites where the explosive volcanoes of the Pacific "Ring of Fire" are most prevalent. Here, release of water from the subducted plate triggers melting in the overlying mantle. The resulting magma interacts with the lower crust of the overlying plate during its ascent and results in the formation of a volcanic arc at the surface.
- At divergent boundaries, decompression melting is the dominant generator of magma. As warm rock rises, it can begin to melt without the addition of heat. The result will be a rift valley if the overlying crust is continental or a mid-ocean ridge if it is oceanic.

- In an intraplate setting, the source of magma is a mantle plume: a column of warm, rising solid mantle rock that begins to melt in the uppermost mantle.

GIVE IT SOME THOUGHT

1. Match each of these volcanic regions with one of the three zones of volcanism (convergent plate boundaries, divergent plate boundaries, or intraplate volcanism):
   a. Crater Lake
   b. Hawaii's Kilauea
   c. Mount St. Helens
   d. East African Rift
   e. Yellowstone
   f. Vesuvius
   g. Deccan Plateau
   h. Mount Etna

2. Examine the accompanying photo and complete the following:
   a. What type of volcano is this? What features helped you make a decision?
   b. What is the eruptive style of such volcanoes? Describe the likely composition and viscosity of the magma.
   c. Which one of the three zones of volcanism is the likely setting for this volcano?
   d. Name a city that is vulnerable to the effects of a volcano of this type.

3. Divergent boundaries, such as the Mid-Atlantic Ridge, are characterized by outpourings of basaltic lava. Answer the following questions about divergent boundaries and their associated lavas:
   a. What is the source of these lavas?
   b. What causes the source rocks to melt?
   c. Describe a divergent boundary that would be associated with lava other than basalt. Why did you choose it, and what type of lava would you expect to erupt there?

4. Explain why volcanic activity occurs in places other than plate boundaries.

5. For each of the accompanying four sketches, identify the geologic setting (zone of volcanism). Which of these settings will most likely generate explosive eruptions? Which will produce outpouring of fluid basaltic lavas?

A.

B.

C.

D.

6. This image shows the Buddhist monastery Taung Kalat, located in central Myanmar (Burma). The monastery sits high on a sheer-sided rock made mainly of magmas that solidified in the conduit of an ancient volcano. The volcano has since been worn away.

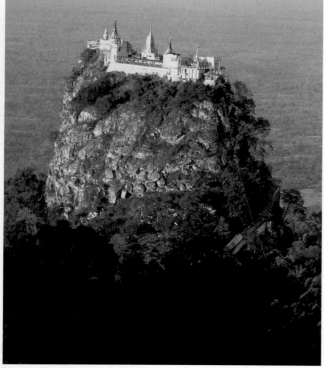

Photo by Dreamstime

   a. Based on this information, what volcanic structure do you think is shown in this photo?
   b. Would this volcanic structure most likely have been associated with a composite volcano or a cinder cone? Explain how you arrived at your answer.

7. Imagine that you are a geologist charged with the task of choosing three sites where state-of-the-art volcano monitoring systems will be deployed. The sites can be anywhere in the world, but the budget and number of experts you can employ to oversee the operations are limited. What criteria would you use to select these sites? List some potential choices and your reasons for considering them.

8. The accompanying image shows an igneous structure (dark color) located in southeastern Utah that was intruded into sedimentary strata.

Photo by Michael Collier

   a. What name is given to this intrusive igneous feature?
   b. The light bands above and below the dark igneous body are metamorphic rock. Identify this type of metamorphism and briefly explain how it alters rock. (*Hint:* Refer to Chapter 2, if necessary.)

9. Explain why an eruption of Mount Rainier would be considerably more destructive than the similar eruption of Mount St. Helens that occurred in 1980.

10. During a field trip with your geology class, you visit an exposure of rock layers similar to the one sketched here. A fellow student suggests that the layer of basalt is a sill. You disagree. Why do you think the other student is incorrect? What is a more likely explanation for the basalt layer?

11. Each of the following descriptions indicates how an intrusive feature appears when exposed at Earth's surface by erosion. Name the feature.
   a. A dome-shaped mountainous structure flanked by upturned layers of sedimentary rocks
   b. A vertical wall-like feature a few meters wide and hundreds of meters long
   c. A huge expanse of granitic rock forming a mountainous terrain tens of kilometers wide
   d. A relatively thin layer of basalt sandwiched between layers of sedimentary rocks exposed on the side of a canyon

12. Mount Whitney, the highest summit (4421 meters [14,500 feet]) in the contiguous United States, is located in the Sierra Nevada batholith. Based on its location, is Mount Whitney likely composed of granitic, andesitic, or basaltic rocks?

Mount Whitney

Photo by Don Smith/Getty Images

MasteringGeology™

Looking for additional review and test prep materials? Visit the Study Area in MasteringGeology to enhance your understanding of this chapter's content by accessing a variety of resources, including **Self-Study Quizzes**, **Geoscience Animations, GEODe Tutorials, RSS feeds, flashcards, web links, self-study quizzes,** and an optional **Pearson eText.**

8

Geologic Time

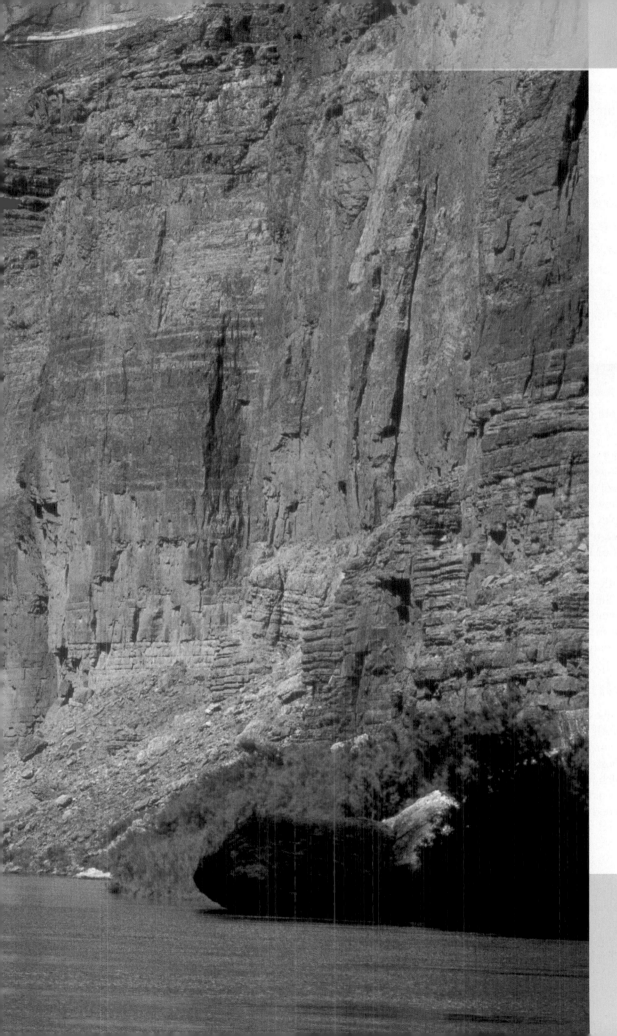

Each statement represents the primary learning objective for the corresponding major heading within the chapter. After you complete the chapter, you should be able to:

8.1 Explain the principle of uniformitarianism and discuss how it differs from catastrophism.

8.2 Distinguish between numerical dates and relative dates and apply relative dating principles to determine a time sequence of geologic events.

8.3 Define *fossil* and discuss the conditions that favor the preservation of organisms as fossils. List and describe various fossil types.

8.4 Explain how rocks of similar age that are in different places can be matched up.

8.5 Discuss three types of radioactive decay and explain how radioactive isotopes are used to determine numerical dates.

8.6 Distinguish among the four basic time units that make up the geologic time scale and explain why the time scale is considered to be a dynamic tool.

8.7 Explain how reliable numerical dates are determined for layers of sedimentary rock.

Rafting the Colorado River in Marble Canyon shortly before entering the Grand Canyon. Millions of years of Earth history are exposed in these walls of rock. (Photo by Michael Collier)

I n the eighteenth century, James Hutton recognized the immensity of Earth history and the importance of time as a component in all geologic processes. In the nineteenth century, others effectively demonstrated that Earth had experienced many episodes of mountain building and erosion, which must have required great spans of geologic time. Although these pioneering scientists understood that Earth was very old, they had no way of knowing its true age. Was it tens of millions, hundreds of millions, or even billions of years old? Because they didn't know exactly how old Earth was, they developed a geologic time scale that showed the sequence of events based on relative dating principles. What were these principles? What part did fossils play? With the discovery of radioactivity and the development of radiometric dating techniques, geologists can now assign fairly accurate dates to many of the events in Earth history. What is radioactivity? Why is it a good "clock" for dating the geologic past? This chapter will answer these questions.

8.1 A Brief History of Geology

Explain the principle of uniformitarianism and discuss how it differs from catastrophism.

The hiker in **Figure 8.1** is perched on the rim of the Grand Canyon. Today we know that the strata beneath the hiker represent hundreds of millions of years of Earth history. We also have the knowledge and skills that are necessary to unravel the complex story contained in these rocks. Such an understanding is a relatively recent accomplishment.

The nature of our Earth—its materials and processes—has been a focus of study for centuries. However, the late 1700s is generally regarded as the beginning of modern geology. It was during this time that James Hutton published his important work *Theory of the Earth*. Prior to that time, a great many explanations about Earth history relied on supernatural events.

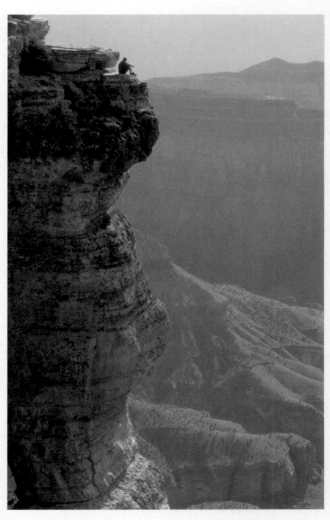

Figure 8.1 Contemplating geologic time This hiker is resting atop the Kaibab Formation, the uppermost layer in the Grand Canyon. (Photo by Michael Collier)

Catastrophism

In the mid-1600s, James Ussher, Anglican Archbishop of Armagh, Primate of All Ireland, published a work that had immediate and profound influence. A respected scholar of the Bible, Ussher constructed a chronology of human and Earth history in which he determined that Earth was only a few thousand years old, having been created in 4004 B.C. Ussher's treatise earned widespread acceptance among Europe's scientific and religious leaders, and his chronology was soon printed in the margins of the Bible itself.

During the 1600s and 1700s, the doctrine of **catastrophism** strongly influenced people's thinking about Earth. Briefly stated, catastrophists believed that Earth's varied landscapes had been fashioned primarily by great catastrophes. Features such as mountains and canyons, which today we know take great periods of time to form, were explained as having been produced by sudden and often worldwide disasters of unknowable causes that no longer operate. This philosophy was an attempt to fit the rate of Earth's processes to the prevailing ideas on Earth's age.

The Birth of Modern Geology

Modern geology began in the late 1700s, when James Hutton, a Scottish physician and gentleman farmer, published his *Theory of the Earth*. In this work, Hutton put forth a fundamental principle that is a pillar of geology today: **uniformitarianism**. It simply states that *the physical, chemical, and biological laws that operate today have also operated in the geologic past*. This means that the forces and processes that we observe presently shaping our planet have been at work for a very long time. Thus, to understand ancient rocks, we must first understand present-day processes and their results. This idea is commonly expressed by saying "The present is the key to the past."

Prior to Hutton's *Theory of the Earth*, no one had effectively demonstrated that geologic processes occur over extremely long periods of time. However, Hutton persuasively argued that processes that appear to be weak and slow acting can, over long spans of time, produce effects that are just as great as those resulting from sudden catastrophic events. Unlike his predecessors, Hutton cited verifiable observations to support his ideas.

For example, when he argued that mountains are sculpted and ultimately destroyed by weathering and the work of running water and that their wastes are carried to the oceans by processes that can be observed, Hutton said, "We have a chain of facts which clearly demonstrates that the materials of the wasted mountains have traveled through the rivers"; and, further, "There is not one step in all this progress that is not to be actually perceived." He went on to summarize these thoughts by asking a question and immediately providing the answer: "What more can we require? Nothing but time."

Geology Today

The basic tenets of uniformitarianism are just as viable today as in Hutton's day. We realize more strongly than ever before that the present gives us insight into the past and

that the physical, chemical, and biological laws that govern geologic processes remain unchanging through time. However, we also understand that the doctrine should not be taken too literally. To say that geologic processes in the past were the same as those occurring today is not to suggest that they always had the same relative importance or that they operated at precisely the same rate. Moreover, some important geologic processes are not currently observable, but evidence that they occur is well established. For example, we know that Earth has experienced impacts from large meteorites even though we have no human witnesses. Such events altered Earth's crust, modified its climate, and strongly influenced life on the planet.

The acceptance of the concept of uniformitarianism, however, meant the acceptance of a very long history for Earth. Although Earth's processes vary in their intensity, they still take a long time to create or destroy major landscape features. For example, geologists have established that mountains once existed in portions of present-day Minnesota, Wisconsin, Michigan, and Manitoba. Today, the region consists of low hills and plains. Erosion (processes that wear away land) gradually destroyed those peaks. The rock record contains evidence that shows Earth has experienced *many cycles* of mountain building and erosion. Concerning the ever-changing nature of Earth through great expanses of geologic time, Hutton made a statement that was to become his most famous. In concluding his classic 1788 paper published in the *Transactions of the Royal Society of Edinburgh*, he stated, "The results, therefore, of our present enquiry is, that we find no vestige of a beginning—no prospect of an end."

It is important to remember that although many features of our physical landscape may seem to be unchanging over the decades we observe them, they are nevertheless changing—but on time scales of hundreds, thousands, or even many millions of years.

8.1 CONCEPT CHECKS

1. Contrast catastrophism and uniformitarianism.
2. How did each philosophy view the age of Earth?

8.2 Creating a Time Scale—Relative Dating Principles

Distinguish between numerical dates and relative dates and apply relative dating principles to determine a time sequence of geologic events.

The Importance of a Time Scale

Like the pages in a long and complicated history book, rocks record the geologic events and changing lifeforms of the past. The book, however, is not complete.

Many pages, especially in the early chapters, are missing. Others are tattered, torn, or smudged. Yet enough of the book remains to allow much of the story to be deciphered.

Interpreting Earth history is a prime goal of the science of geology. Like a modern-day sleuth, a geologist must interpret the clues found preserved in the rocks. By studying rocks and the features they contain, geologists can unravel the complexities of the past.

Geologic events by themselves, however, have little meaning until they are put into a time perspective. Studying history, whether it is the Civil War or the age of dinosaurs, requires a calendar. Among geology's major contributions to human knowledge are the *geologic time scale* and the discovery that Earth history is exceedingly long.

Numerical and Relative Dates

The geologists who developed the geologic time scale revolutionized the way people think about time and perceive our planet. They learned that Earth is much older than anyone had previously imagined and that its surface and interior have been changed over and over again by the same geologic processes that operate today.

Numerical Dates During the late 1800s and early 1900s, attempts were made to determine Earth's age. Although some of the methods appeared promising at

the time, none of these early efforts proved to be reliable. What these scientists were seeking was a **numerical date**. Such dates specify the actual number of years that have passed since an event occurred. Today, our understanding of radiometric dating techniques allows us to accurately determine numerical dates for rocks that represent important events in Earth's distant past. We will study these techniques later in this chapter. Prior to the development of radiometric dating, geologists had no reliable method of numerical dating and had to rely solely on relative dating.

Relative Dates When we place rocks in their proper *sequence of formation*—which formed first, second, third, and so on—we are establishing **relative dates**. Such dates cannot tell us how long ago something took place, only that it followed one event and preceded another. The relative dating techniques that were developed are valuable and still widely used. Numerical dating methods did not replace these techniques; rather, they supplemented them. To establish a relative time scale, a few basic principles or rules had to be discovered and applied. Although they may seem obvious to us today, they were major breakthroughs in thinking at the time, and their discovery was an important scientific achievement.

Principle of Superposition

Nicolas Steno, a Danish anatomist, geologist, and priest (1638–1686), is credited with being the first to recognize a sequence of historical events in an outcrop of sedimentary rock layers. Working in the mountains of western Italy, Steno applied a very simple rule that has come to be the most basic principle of relative dating—the **principle of superposition** (*super* = above; *positum* = to place). The principle simply states that in an undeformed sequence of sedimentary rocks, each bed is older than the one above and younger than the one below. Although it may seem obvious that a rock layer could not be deposited with nothing beneath it for support, it was not until 1669 that Steno clearly stated this principle.

This rule also applies to other surface-deposited materials, such as lava flows and beds of ash from volcanic eruptions. Applying superposition to the beds exposed in the upper portion of the Grand Canyon, we can easily place the layers in their proper order. Among those that are pictured in **Figure 8.2**, the sedimentary rocks in the Supai Group are the oldest, followed in order by the Hermit Shale, Coconino Sandstone, Toroweap Formation, and Kaibab Limestone.

Figure 8.2 Superposition Based on the principle of superposition, among these layers in the upper portion of the Grand Canyon, the Supai Group is oldest, and the Kaibab Formation is youngest. (Photo by E. J. Tarbuck)

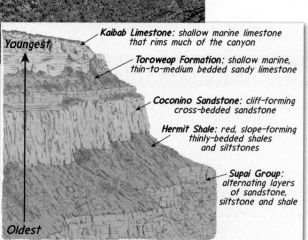

Geologist's Sketch

Youngest

Kaibab Limestone: *shallow marine limestone that rims much of the canyon*

Toroweap Formation: *shallow marine, thin-to-medium bedded sandy limestone*

Coconino Sandstone: *cliff-forming cross-bedded sandstone*

Hermit Shale: *red, slope-forming thinly-bedded shales and siltstones*

Supai Group: *alternating layers of sandstone, siltstone and shale*

Oldest

Principle of Original Horizontality

Steno is also credited with recognizing the importance of another basic principle, called the **principle of original horizontality**. Simply stated, it means that layers of sediment are generally deposited in a horizontal position. Thus, if we observe rock layers that are flat, it means they have not been disturbed and still have their *original* horizontality. The layers in the Grand Canyon illustrate this in Figures 8.1 and 8.2. But if they are folded or inclined at a steep angle, they must have been moved into that position by crustal disturbances sometime *after* their deposition (**Figure 8.3**).

Principle of Cross-Cutting Relationships

Figure 8.4 shows a mass of rock that is offset by a fault, a fracture in rock along which displacement occurs. It is clear that the rocks must be older than the fault that broke them. The **principle of cross-cutting relationships** states that geologic features that cut across rocks must form *after* the rocks they cut through. Igneous

Figure 8.4 Cross-cutting fault The rocks are older than the fault that displaced them. (Photo by Morley Read/Alamy Images)

intrusions provide another example. The dike shown in **Figure 8.5** is a tabular mass of igneous rock that cuts through the surrounding rocks. The magmatic heat from igneous intrusions often creates a narrow "baked" zone of contact metamorphism on the adjacent rock, also indicating that the intrusion occurred after the surrounding rocks were in place.

Figure 8.5 Cross-cutting dike An igneous intrusion is younger than the rocks that are intruded. (Photo by Ken M. Johns/Photo Researchers, Inc.)

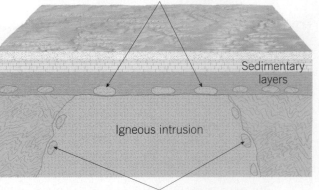

These inclusions of igneous rock contained in the adjacent sedimentary layer indicate the sediments were deposited atop the weathered igneous mass and thus are younger.

Sedimentary layers

Igneous intrusion

Xenoliths are inclusions in an igneous intrusion that form when pieces of surrounding rock are incorporated into magma.

SmartFigure 8.6 Inclusions The rock that contains the inclusions is younger than the inclusions.

Principle of Inclusions

Sometimes inclusions can aid in the relative dating process. **Inclusions** are fragments of one rock unit that have been enclosed within another. The principle of inclusions is logical and straightforward. The rock mass adjacent to the one containing the inclusions must have been there first in order to provide the rock fragments. Therefore, the rock mass that contains the inclusions is the younger of the two. For example, when magma intrudes into surrounding rock, blocks of the surrounding rock may become dislodged and incorporated into the magma. If these pieces do not melt, they remain as inclusions, known as *xenoliths*. In another example, when sediment is deposited atop a weathered mass of bedrock, pieces of the weathered rock become incorporated into the younger sedimentary layer (**Figure 8.6**).

Unconformities

When we observe layers of rock that have been deposited essentially without interruption, we call them **conformable**. Particular sites exhibit conformable beds

representing certain spans of geologic time. However, no place on Earth has a complete set of conformable strata.

Throughout Earth history, the deposition of sediment has been interrupted over and over again. All such breaks in the rock record are termed unconformities. An **unconformity** represents a long period during which deposition ceased, erosion removed previously formed rocks, and then deposition resumed. In each case, uplift and erosion are followed by subsidence and renewed sedimentation. Unconformities are important features because they represent significant geologic events in Earth history. Moreover, their recognition helps us identify what intervals of time are not represented by strata and thus are missing from the geologic record.

There are three basic types of unconformities.

Angular Unconformity Perhaps the most easily recognized unconformity is an **angular unconformity**. It consists of tilted or folded sedimentary rocks that are overlain by younger, more flat-lying strata. An angular unconformity indicates that during the pause in deposition, a period of deformation (folding or tilting) and erosion occurred (**Figure 8.7**).

When James Hutton studied an angular unconformity in Scotland more than 200 years ago, it was clear to him that it represented a major episode of geologic activity (**Figure 8.8**). He and his colleagues also appreciated the immense time span implied by such relationships. When a companion later wrote of their visit to this site, he stated that "the mind seemed to grow giddy by looking so far into the abyss of time."

Disconformity A **disconformity** is a gap in the rock record that represents a period during which erosion rather

SmartFigure 8.7 Formation of an angular unconformity An angular unconformity represents an extended period during which deformation and erosion occurred.

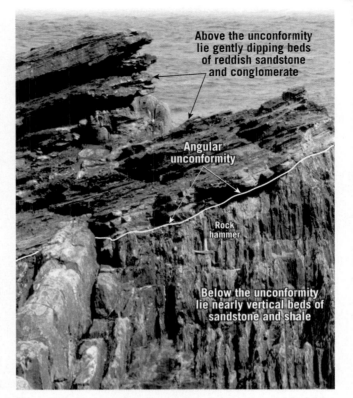

Figure 8.8 Siccar Point, Scotland James Hutton studied this famous unconformity in the late 1700s. (Photo by Marli Miller)

Figure 8.9 Disconformity The layers on either side of this gap in the rock record are essentially parallel.

than deposition occurred. Imagine that a series of sedimentary layers are deposited in a shallow marine setting. Following this period of deposition, sea level falls or the land rises, exposing some the sedimentary layers. During this span, when the sedimentary beds are above sea level, no new sediment accumulates, and some of the existing layers are eroded away. Later sea level rises or the land subsides, submerging the landscape. Now the surface is again below sea level, and a new series of sedimentary beds is deposited. The boundary separating the two sets of beds is a disconformity—a span for which there is no rock record (**Figure 8.9**). Because the layers above and below a disconformity are parallel, these features are sometimes difficult to identify unless you notice evidence of erosion such as a buried stream channel.

Nonconformity The third basic type of unconformity is a **nonconformity**, in which younger sedimentary strata overlie older metamorphic or intrusive igneous rocks (**Figure 8.10**). Just as angular unconformities and some disconformities imply crustal movements, so too do nonconformities. Intrusive igneous masses and metamorphic

rocks originate far below the surface. Thus, for a nonconformity to develop, there must be a period of uplift and the erosion of overlying rocks. Once exposed at the surface, the igneous or metamorphic rocks are subjected to weathering and erosion prior to subsidence and the renewal of sedimentation.

Unconformities in the Grand Canyon The rocks exposed in the Grand Canyon of the Colorado River represent a tremendous span of geologic history. It is a

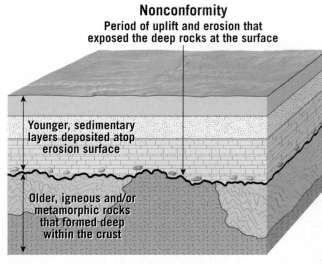

Nonconformity
Period of uplift and erosion that exposed the deep rocks at the surface

Younger, sedimentary layers deposited atop erosion surface

Older, igneous and/or metamorphic rocks that formed deep within the crust

Figure 8.10 Nonconformity Younger sedimentary rocks rest atop older metamorphic or igneous rocks.

wonderful place to take a trip through time. The canyon's colorful strata record a long history of sedimentation in a variety of environments—advancing seas, rivers and deltas, tidal flats and sand dunes. But the record is not continuous. Unconformities represent vast amounts of time that have not been recorded in the canyon's layers. **Figure 8.11** is a geologic cross section of the Grand Canyon. All three types of unconformities can be seen in the canyon walls.

Applying Relative Dating Principles

By applying the principles of relative dating to the hypothetical geologic cross section shown in **Figure 8.12**, the rocks and the events in Earth history they represent can be placed into their proper sequence. The statements within the figure summarize the logic used to interpret the cross section. In this example, we establish a relative time scale for the rocks and events in the area of the cross section. Remember, we do not know how many years of Earth history are represented, nor do we know how this area compares to any other.

8.2 CONCEPT CHECKS

1. Distinguish between numerical dates and relative dates.
2. Sketch and label four simple diagrams that illustrate each of the following: superposition, original horizontality, cross-cuting relationships, and inclusions.
3. What is the significance of an unconformity?
4. Distinguish among angular unconformity, disconformity, and nonconformity.

Figure 8.11 Cross section of the Grand Canyon All three types of unconformities are present. (Center photo by Marli Miller; other photos by E. J. Tarbuck)

Working out the geologic history of a hypothetical region

Interpretation:

1. Beneath the ocean, beds A, B, C, and E were deposited in that order (law of superposition).

2. Uplift and intrusion of a sill (layer D). We know that sill D is younger than beds C and E because of the inclusions in the sill of fragments from beds C and E.

3. Next is the intrusion of dike F. Because the dike cuts through layers A through E, it must be younger (principle of cross-cutting relationships).

4. Layers A through F were tilted and exposed layers were eroded.

5. Next, beds, G, H, I, J, and K were deposited in that order atop the erosion surface to produce an angular unconformity.

6. Finally, the irregular surface and stream valley indicate that another gap in the rock record is being created by erosion.

SmartFigure 8.12 Applying principles Working out the geologic history of a hypothetical region.

8.3 Fossils: Evidence of Past Life

Define *fossil* and discuss the conditions that favor the preservation of organisms as fossils. List and describe various fossil types.

Fossils, the remains or traces of prehistoric life, are important inclusions in sediment and sedimentary rocks. They are basic and important tools for interpreting the geologic past. The scientific study of fossils is called **paleontology**. It is an interdisciplinary science that blends geology and biology, in an attempt to understand all aspects of the succession of life over the vast expanse of geologic time. Knowing the nature of the life-forms that existed at a particular time helps researchers understand past environmental conditions. Further, fossils are important time indicators and play a key role in correlating rocks of similar ages that are from different places.

Types of Fossils

Fossils are of many types. The remains of relatively recent organisms may not have been altered at all. Such objects as teeth, bones, and shells are common examples. Far less common are entire animals, flesh included, that have been preserved because of rather unusual circumstances. Remains of prehistoric elephants called mammoths that were frozen in the Arctic tundra of Siberia and Alaska are examples, as are the mummified remains of sloths preserved in a dry cave in Nevada.

Figure 8.13 Types of fossils The images in this figure are just a sampling of many possibilities. (Photo A by Bernhard Edmaier/Photo Researchers, Inc.; photos B, D, and F by E. J. Tarbuck; photo E by Colin Keates/Dorling Kindersley Media Library)

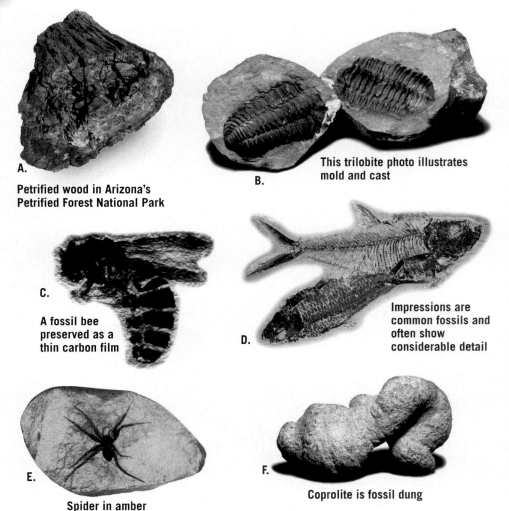

A. Petrified wood in Arizona's Petrified Forest National Park

B. This trilobite photo illustrates mold and cast

C. A fossil bee preserved as a thin carbon film

D. Impressions are common fossils and often show considerable detail

E. Spider in amber

F. Coprolite is fossil dung

Permineralization When mineral-rich groundwater permeates porous tissue such as bone or wood, minerals precipitate out of solution and fill pores and empty spaces, a process called *permineralization*. The formation of *petrified wood* involves permineralization with silica, often from a volcanic source such as a surrounding layer of volcanic ash. The wood is gradually transformed into chert, sometimes with colorful bands from impurities such as iron or carbon (**Figure 8.13A**). The word *petrified* literally means "turned into stone." Sometimes the microscopic details of the petrified structure are faithfully retained.

Molds and Casts Another common class of fossils is *molds* and *casts*. When a shell or another structure is buried in sediment and then dissolved by underground water, a *mold* is created. The mold faithfully reflects only the shape and surface marking of the organism; it does not reveal any information concerning its internal structure. If these hollow spaces are subsequently filled with mineral matter, *casts* are created (**Figure 8.13B**).

Carbonization and Impressions A type of fossilization called *carbonization* is particularly effective in preserving leaves and delicate animal forms. It occurs when fine sediment encases the remains of an organism. As time passes, pressure squeezes out the liquid and gaseous components and leaves behind a thin residue of carbon (**Figure 8.13C**). Black shale deposited as organic-rich mud in oxygen-poor environments often contains abundant carbonized remains. If the film of carbon is lost from a fossil preserved in fine-grained sediment, a replica of the surface, called an *impression*, may still show considerable detail (**Figure 8.13D**).

Amber Delicate organisms, such as insects, are difficult to preserve, and consequently they are relatively rare in the fossil record. Not only must they be protected from decay but they must not be subjected to any pressure that would crush them. One way in which some insects have been preserved is in *amber*, the hardened resin of ancient trees. The spider in **Figure 8.13E** was preserved after being trapped in a drop of sticky resin. Resin sealed off the spider from the atmosphere and protected the remains from damage by water and air. As the resin hardened, a protective pressure-resistant case was formed.

Trace Fossils In addition to the fossils already mentioned, there are numerous other types, many of them only traces of prehistoric life. Examples of such indirect evidence include:

- Tracks—animal footprints made in soft sediment that later turned into sedimentary rock.
- Burrows—tubes in sediment, wood, or rock made by an animal. These holes may later become filled with mineral matter and preserved. Some of the oldest-known fossils are believed to be worm burrows.
- Coprolites—fossil dung and stomach contents that can provide useful information pertaining to the size and food habits of organisms (**Figure 8.13F**).
- Gastroliths—highly polished stomach stones that were used in the grinding of food by some extinct reptiles.

Conditions Favoring Preservation

Only a tiny fraction of the organisms that lived during the geologic past have been preserved as fossils. The remains of an animal or a plant are normally destroyed. Under what circumstances are they preserved? Two special conditions appear to be necessary: rapid burial and the possession of hard parts.

When an organism perishes, its soft parts are usually quickly eaten by scavengers or decomposed by bacteria. Occasionally, however, the remains are buried by sediment. When this occurs, the remains are protected from the environment, where destructive processes operate. Rapid burial, therefore, is an important condition favoring preservation.

In addition, animals and plants have a much better chance of being preserved as part of the fossil record if they have hard parts. Although traces and imprints of soft-bodied animals such as jellyfish, worms, and insects exist, they are not common. Flesh usually decays so rapidly that preservation is exceedingly unlikely. Hard parts such as shells, bones, and teeth predominate in the record of past life.

Because preservation is contingent on special conditions, the record of life in the geologic past is biased. The fossil record of those organisms with hard parts that lived in areas of sedimentation is quite abundant. However, we get only an occasional glimpse of the vast array of other life-forms that did not meet the special conditions favoring preservation.

8.3 CONCEPT CHECKS

1. Describe several ways that an animal or a plant can be preserved as a fossil.
2. List three examples of trace fossils.
3. What conditions favor the preservation of an organism as a fossil?

8.4 Correlation of Rock Layers

Explain how rocks of similar age that are in different places can be matched up.

To develop a geologic time scale that is applicable to the entire Earth, rocks of similar age in different regions must be matched up. Such a task is referred to as **correlation**. Correlating the rocks from one place to another makes possible a more comprehensive view of the geologic history of a region. **Figure 8.14**, for example, shows the correlation of strata at three sites on the Colorado Plateau in southern Utah and northern Arizona. No single locale exhibits the entire sequence, but correlation reveals a more complete picture of the sedimentary rock record.

Correlation Within Limited Areas

Within a limited area, correlating rocks of one locality with those of another may be done simply by walking along the outcropping edges. However, this may not be possible when the rocks are mostly concealed by soil and vegetation. Correlation over short distances is often achieved by noting the position of a bed in a sequence of strata. Or a layer may be identified in another location if it is composed of distinctive or uncommon minerals.

Many geologic studies involve relatively small areas. Although they are important in their own right, their full value is realized only when they are correlated with other regions. Although the methods just described are sufficient to trace a rock formation over relatively short distances, they are not adequate for matching up rocks that are separated by great distances. When correlation between widely separated areas or between continents is the objective, geologists must rely on fossils.

Fossils and Correlation

The existence of fossils had been known for centuries, but it was not until the late 1700s and early 1800s that their significance as geologic tools was made evident. During this period, an English engineer and canal builder, William Smith, discovered that each rock formation in the canals he worked on contained fossils unlike those in the beds either above or below. Further, he noted that sedimentary strata in widely separated areas could be identified—and correlated—by their distinctive fossil content.

Principle of Fossil Succession Based on Smith's classic observations and the findings of many geologists who followed, one of the most important and basic principles in historical geology was formulated: *Fossil organisms succeed one another in a definite and determinable order, and therefore any time period can be recognized by its fossil content.* This has come to be known as the **principle of fossil succession**. In other words, when fossils are arranged according to their age, they do not present a random or haphazard picture. To the contrary, fossils document the evolution of life through time.

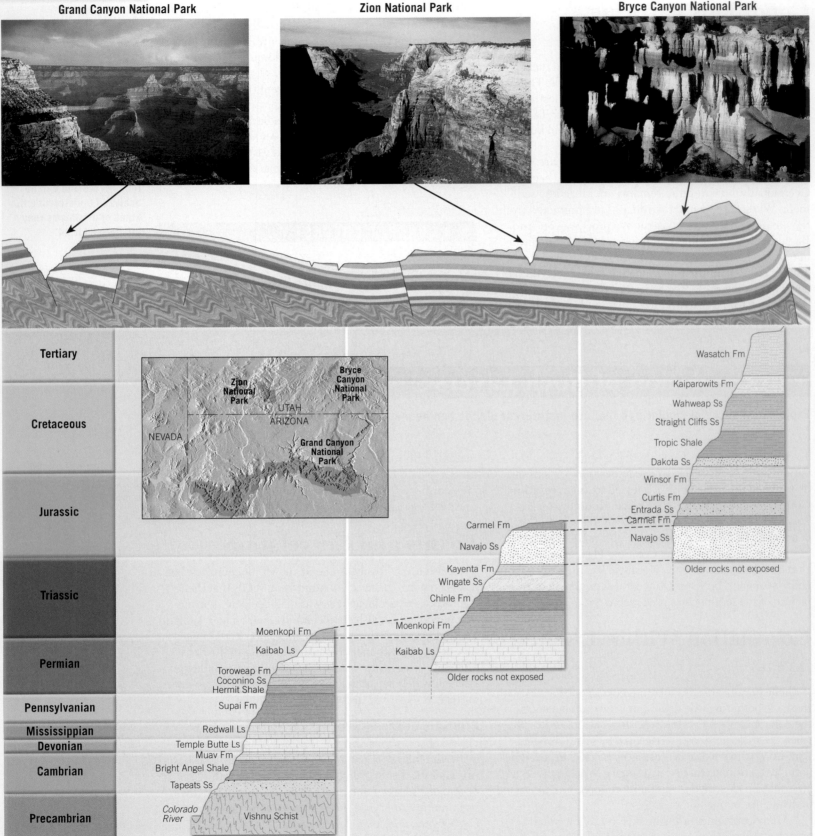

Grand Canyon National Park

Zion National Park

Bryce Canyon National Park

Tertiary

Cretaceous

Jurassic

Triassic

Permian

Pennsylvanian

Mississippian

Devonian

Cambrian

Precambrian

Wasatch Fm

Kaiparowits Fm

Wahweap Ss

Straight Cliffs Ss

Tropic Shale

Dakota Ss

Winsor Fm

Curtis Fm

Entrada Ss

Carmel Fm

Navajo Ss

Older rocks not exposed

Carmel Fm

Navajo Ss

Kayenta Fm

Wingate Ss

Chinle Fm

Moenkopi Fm

Kaibab Ls

Older rocks not exposed

Moenkopi Fm

Kaibab Ls

Toroweap Fm

Coconino Ss

Hermit Shale

Supai Fm

Redwall Ls

Temple Butte Ls

Muav Fm

Bright Angel Shale

Tapeats Ss

Colorado River

Vishnu Schist

Zion National Park

Bryce Canyon National Park

UTAH

ARIZONA

NEVADA

Grand Canyon National Park

Figure 8.14 Correlation Matching strata at three locations on the Colorado Plateau builds a more complete picture of the sedimentary rocks in the region. (Photos by E. J. Tarbuck)

For example, an Age of Trilobites is recognized quite early in the fossil record. Then, in succession, paleontologists recognize an Age of Fishes, an Age of Coal Swamps, an Age of Reptiles, and an Age of Mammals. These "ages" pertain to groups that were especially plentiful and characteristic during particular time periods. Within each of the "ages" are many subdivisions based, for example, on certain species of trilobites and certain types of fish, reptiles, and so on. This same succession of dominant organisms, never out of order, is found on every continent.

Index Fossils and Fossil Assemblages When fossils were found to be time indicators, they became the most useful means of correlating rocks of similar age in different regions. Geologists pay particular attention to certain fossils called **index fossils** (**Figure 8.15**). These fossils are widespread geographically and are limited to a short span of geologic time, so their presence provides an important method of matching rocks of the same age. Rock formations, however, do not always contain a specific index fossil. In such situations a group of fossils, called a **fossil assemblage**, is used to establish the age of the bed. **Figure 8.16** illustrates how an assemblage of fossils may be used to date rocks more precisely than could be accomplished by the use of any one of the fossils.

Environmental Indicators In addition to being important, and often essential, tools for correlation, fossils are important environmental indicators. Although much can be deduced about past environments by studying the nature and characteristics of sedimentary rocks, a close examination of the fossils present can usually provide a great deal more information. For example, when the remains of certain clam shells are found in limestone, a geologist quite reasonably assumes that the region was once covered by a shallow sea. Also, by using what we know of living organisms, we can conclude that fossil animals with thick shells, capable of withstanding pounding and surging waves, inhabited shorelines.

On the other hand, animals with thin, delicate shells probably indicate deep, calm offshore waters. Hence, by looking closely at the types of fossils, the approximate position of an ancient shoreline may be identified. Further, fossils can be used to indicate the former temperature of the water. Certain kinds of present-day corals must live in warm and shallow tropical seas like those around Florida and The Bahamas. When similar types of coral are found in ancient limestones, they indicate the marine environment that must have existed when they were alive. These examples illustrate how fossils can help unravel the complex story of Earth history.

Figure 8.15 Index fossils. Because microfossils are often very abundant, widespread, and quick to appear and become extinct, they constitute ideal index fossils. (Photo by Biophoto Associates/Photo Researchers, Inc.)

SmartFigure 8.16 Fossil assemblage Overlapping ranges of fossils help date rocks more exactly than using a single fossil.

Age ranges of some fossil groups

Rock unit A

Age of rock unit A

Rock unit B

Age of rock unit B

8.4 CONCEPT CHECKS

1. What is the goal of correlation?
2. State the principle of fossil succession in your own words.
3. Contrast *index fossil* and *fossil assemblage*.
4. In addition to being important time indicators, how else are fossils useful to geologists?

8.5 Dating with Radioactivity

Discuss three types of radioactive decay and explain how radioactive isotopes are used to determine numerical dates.

In addition to establishing relative dates by using the principles described in the preceding sections, it is also possible to obtain reliable numerical dates for events in the geologic past. We know that Earth is about 4.6 billion years old and that the dinosaurs became extinct about 65.5 million years ago. Dates that are expressed in millions and billions of years truly stretch our imagination because our personal calendars involve time measured in hours, weeks, and years. Nevertheless, the vast expanse of geologic time is a reality, and radiometric dating allows us to measure it accurately. In this section, you will learn about radioactivity and its application in radiometric dating.

Reviewing Basic Atomic Structure

Recall from Chapter 1 that each atom has a *nucleus* that contains protons and neutrons and that the nucleus is orbited by electrons. An *electron* has a negative electrical charge, and a *proton* has a positive charge. A *neutron* is actually a proton and an electron combined, so it has no charge (it is neutral).

An element's *atomic number* (an element's identifying number) is the number of protons in the nucleus. Every element has a different number of protons in the nucleus and thus a different atomic number (hydrogen = 1, oxygen = 8, uranium = 92, etc.). Atoms of the same element always have the same number of protons, so the atomic number is constant.

Practically all (99.9 percent) of an atom's mass is found in the nucleus, indicating that electrons have practically no mass at all. By adding together the number of protons and neutrons in the nucleus, the *mass number* of the atom is determined. The number of neutrons in the nucleus can vary. These variants, called *isotopes*, have different mass numbers.

We can summarize with an example. Uranium's nucleus always has 92 protons, so its atomic number always is 92. But its neutron population varies, and uranium has three isotopes: uranium-234 (number of protons + neutrons = 234), uranium-235, and uranium-238. All three isotopes are found together in nature. They look the same and behave the same in chemical reactions.

Radioactivity

The forces that bind protons and neutrons together in the nucleus are usually strong. However, in some isotopes, the nuclei are unstable because the forces binding protons and neutrons together are not strong enough.

As a result, the nuclei spontaneously break apart (decay) in a process called **radioactivity**.

Common Examples of Radioactive Decay What happens when unstable nuclei break apart? Three common types of radioactive decay are illustrated in **Figure 8.17** and are summarized as follows:

- *Alpha particles* (α particles) may be emitted from the nucleus. An alpha particle consists of 2 protons and 2 neutrons. Consequently, the emission of an alpha particle means (a) the mass number of the isotope is reduced by 4, and (b) the atomic number is decreased by 2.
- When a *beta particle* (β particle), or electron, is given off from a nucleus, the mass number remains

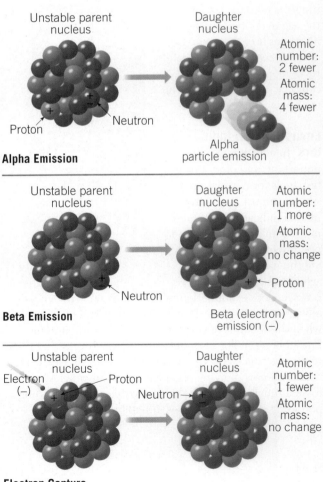

Figure 8.17 Common types of radioactive decay Notice that in each example, the number of protons (atomic number) in the nucleus changes, thus producing a different element.

Figure 8.18 Decay of U-238 Uranium-238 is an example of a radioactive decay series. Before the stable end product (Pb-206) is reached, many different isotopes are produced as intermediate steps.

unchanged, because electrons have practically no mass. However, because the electron has come from a neutron (remember, a neutron is a combination of a proton and an electron), the nucleus contains one more proton than before. Therefore, the atomic number increases by 1.

- Sometimes an electron is captured by the nucleus. The electron combines with a proton and forms an additional neutron. As in the last example, the mass number remains unchanged. However, because the nucleus now contains one less proton, the atomic number decreases by 1.

An unstable (radioactive) isotope is referred to as the *parent*. The isotopes resulting from the decay of the parent are the *daughter products*. **Figure 8.18** provides an example of radioactive decay. When the radioactive parent, uranium-238 (atomic number 92, mass number 238), decays, it follows a number of steps, emitting eight alpha particles and 6 beta particles before finally becoming the stable daughter product lead-206 (atomic number 82, mass number 206).

Radiometric Dating Certainly among the most important properties of radioactivity is that it provides a reliable method of calculating the ages of rocks and minerals that contain particular radioactive isotopes. The procedure is called **radiometric dating**. Why is radiometric dating reliable? The rates of decay for many isotopes have been precisely measured and do not vary under the physical conditions that exist in Earth's outer layers. Therefore, each radioactive isotope used for dating has been decaying at a fixed rate ever since the formation of the rocks in which it

occurs, and the products of decay have been accumulating at a corresponding rate. For example, when uranium is incorporated into a mineral that crystallizes from magma, there is no lead (the stable daughter product) from previous decay. The radiometric "clock" starts at this point. As the uranium in this newly formed mineral disintegrates, atoms of the daughter product are trapped, and measurable amounts of lead eventually accumulate.

Half-Life

The time required for one-half of the nuclei in a sample to decay is called the **half-life** of the isotope. Half-life is a common way of expressing the rate of radioactive disintegration. **Figure 8.19** illustrates what occurs when a radioactive parent decays directly into its stable daughter product. When the quantities of parent and daughter are equal (ratio 1:1), we know that one half-life has transpired. When one-quarter of the original parent atoms remain and three-quarters have decayed to the daughter product, the parent/daughter ratio is 1:3, and we know that two half-lives have passed. After three half-lives, the ratio of parent atoms to daughter atoms is 1:7 (one parent atom for every seven daughter atoms).

If the half-life of a radioactive isotope is known and the parent–daughter ratio can be determined, the age of the sample can be calculated. For example, assume that the half-life of a hypothetical unstable isotope is 1 million years and the parent–daughter ratio in a sample is 1:15. Such a ratio indicates that four half-lives have passed and that the sample must be 4 million years old.

 SmartFigure 8.19 Radioactive decay curve Change is exponential. Half of the radioactive parent remains after one half-life. After a second half-life, one-quarter of the parent remains, and so forth.

Using Various Isotopes

Notice that the *percentage* of radioactive atoms that decay during one half-life is always the same: 50 percent. However, the *actual number* of atoms that decay with the passing of each half-life continually decreases. As the percentage of radioactive parent atoms declines, the proportion of stable daughter atoms rises, with the increase in daughter atoms just matching the drop in parent atoms. This fact is the key to radiometric dating.

Of the many radioactive isotopes that exist in nature, five have proved particularly important in providing radiometric ages for ancient rocks (**Table 8.1**). Rubidium-87, uranium-238, and uranium-235 are used for dating rocks that are millions of years old, but potassium-40 is more versatile. Although the half-life of potassium-40 is 1.3 billion years, analytical techniques make possible the detection of tiny amounts of its stable daughter product, argon-40, in some rocks that are younger than 100,000 years. Another important reason for its frequent use is that potassium is abundant in many common minerals, particularly micas and feldspars.

It is important to realize that an accurate radiometric date can be obtained only if the mineral remained a closed system during the entire period since its formation. A correct date is not possible unless there was neither the addition nor loss of parent or daughter isotopes. This is not always the case. In fact, an important limitation of the potassium–argon method arises from the fact that argon is a gas, and it may leak from minerals, throwing off measurements.

A Complex Process Remember that although the basic principle of radiometric dating is simple, the actual procedure is quite complex. The analysis that determines the quantities of parent and daughter must be painstakingly precise. In addition, some radioactive materials do not decay directly into the stable daughter product. As you saw in Figure 8.18, uranium-238 produces 13 intermediate unstable daughter products before the fourteenth and final daughter product, the stable isotope lead-206, is produced.

Earth's Oldest Rocks Radiometric dating methods have produced literally thousands of dates for events in Earth history. Rocks exceeding 3.5 billion years in age are found on all of the continents. Earth's oldest rocks (so far) may be as old as 4.28 billion years (b.y.). Discovered in northern Quebec, Canada, on the shores of Hudson Bay, these rocks may be remnants of Earth's earliest crust. Rocks from western Greenland have been dated at 3.7 to 3.8 b.y., and rocks nearly as old are found in the Minnesota River valley and northern Michigan (3.5 to 3.7 b.y.), in southern Africa (3.4 to 3.5 b.y.), and in western Australia (3.4 to 3.6 b.y.). Tiny crystals of the mineral zircon with radiometric ages as old as 4.4 b.y. have been found in younger sedimentary rocks in western Australia. The source rocks for these tiny durable grains either no longer exist or have not yet been found.

Dating with Carbon-14

To date very recent events, carbon-14 is used. Carbon-14 is the radioactive isotope of carbon. The process is often called **radiocarbon dating**. Because the half-life of carbon-14 is only 5,730 years, it can be used for dating events from the historic past as well as those from very recent geologic history. In some cases, carbon-14 can be used to date events as far back as 70,000 years.

Carbon-14 is continuously produced in the upper atmosphere as a result of cosmic-ray bombardment. Cosmic rays, which are high-energy particles, shatter the nuclei of gas atoms, releasing neutrons. Some of the neutrons are absorbed by nitrogen atoms (atomic number 7), causing their nuclei to emit a proton. As a result, the atomic number decreases by 1 (to 6), and a different element, carbon-14, is created (**Figure 8.20A**). This isotope of carbon quickly becomes incorporated into carbon dioxide, which circulates in the atmosphere and is absorbed by living matter. As a result, all organisms—including you—contain a small amount of carbon-14.

A. Production of carbon-14

B. Decay of carbon-14

Figure 8.20 Carbon-14 Production and decay of radiocarbon. These sketches represent the nuclei of the respective atoms.

TABLE 8.1	Radioactive isotopes frequently used in radiometric dating	
Radioactive parent	**Stable daughter product**	**Currently accepted half-life values**
Uranium-238	Lead-206	4.5 billion years
Uranium-235	Lead-207	713 million years
Thorium-232	Lead-208	14.1 billion years
Rubidium-87	Strontium-87	47.0 billion years
Potassium-40	Argon-40	1.3 billion years

Figure 8.21 Cave art Chauvet Cave in southern France was discovered in 1994 and contains some of the earliest-known cave paintings. Radiocarbon dating indicates that most of the images were drawn between 30,000 and 32,000 years ago. (Photo by Javier Trueba/MSF/Photo Researchers, Inc.)

While an organism is alive, the decaying radiocarbon is continually replaced, and the proportions of carbon-14 and carbon-12 remain constant. Carbon-12 is the stable and most common isotope of carbon. However, when any plant or animal dies, the amount of carbon-14 gradually decreases as it decays to nitrogen-14 by beta emission (**Figure 8.20B**). By comparing the proportions of carbon-14 and carbon-12 in a sample, radiocarbon dates can be determined.

Although carbon-14 is useful in dating only the last small fraction of geologic time, it has become a valuable tool for anthropologists, archaeologists, and historians, as well as for geologists who study very recent Earth history (**Figure 8.21**). In fact, the development of radiocarbon dating was considered so important that the chemist who discovered this application, Willard F. Libby, received a Nobel Prize.

8.5 CONCEPT CHECKS

1. List three types of radioactive decay. For each type, describe how the atomic number and the atomic mass change.
2. Sketch a simple diagram that explains the idea of half-life.
3. Why is radiometric dating a reliable method for determining numerical dates?
4. For what time span does radiocarbon dating apply?

Did You Know?

Dating with carbon-14 is useful to archeologists and historians as well as geologists. For example, University of Arizona researchers used carbon-14 dating to determine the age of the Dead Sea Scrolls, considered among the great archeological discoveries of the twentieth century. Parchment from the scrolls dates to between 150 B.C. and 5 B.C. Portions of the scrolls contain dates that match those determined by the carbon-14 measurements.

8.6 The Geologic Time Scale

Distinguish among the four basic time units that make up the geologic time scale and explain why the time scale is considered to be a dynamic tool.

Geologists have divided the whole of geologic history into units of varying magnitude. Together, they comprise the **geologic time scale** of Earth history (**Figure 8.22**). The major units of the time scale were delineated during the nineteenth century, principally by scientists in Western Europe and Great Britain. Because radiometric dating was unavailable at that time, the entire time scale was created using methods of relative dating. It was only in the twentieth century that radiometric dating permitted numerical dates to be added.

Structure of the Time Scale

The geologic time scale divides the 4.6-billion-year history of Earth into many different units and provides a meaningful time frame within which the events of the geologic past are arranged. As shown in Figure 8.22, **eons** represent the greatest expanses of time. The eon that began about 542 million years ago is the **Phanerozoic**, a term derived from Greek words meaning "visible life." It is an appropriate description because the rocks and deposits of the Phanerozoic eon contain abundant fossils that document major evolutionary trends.

Another glance at the time scale shows that eons are divided into **eras**. The three eras within the Phanerozoic are the **Paleozoic** ("ancient life"), the **Mesozoic** ("middle life"), and the **Cenozoic** ("recent life"). As the names imply, these eras are bounded by profound worldwide changes in life-forms.

Each era of the Phanerozoic eon is divided into units known as **periods**. The Paleozoic has seven, and the Mesozoic and Cenozoic each have three. Each of these periods is characterized by a somewhat less profound change in life-forms as compared with the eras.

Each of the periods is divided into still smaller units called **epochs**. As you can see in Figure 8.22, seven epochs have been named for the periods of the Cenozoic era. The epochs of other periods, however, are not usually referred to by specific names. Instead, the terms *early*, *middle*, and *late* are generally applied to the epochs of these earlier periods.

Did You Know?

Although movies and cartoons have depicted humans and dinosaurs living side by side, this was never the case. Dinosaurs flourished during the Mesozoic era and became extinct about 65 million years ago. Humans and their close ancestors did not appear on the scene until the late Cenozoic, more than 60 million years *after* the demise of dinosaurs.

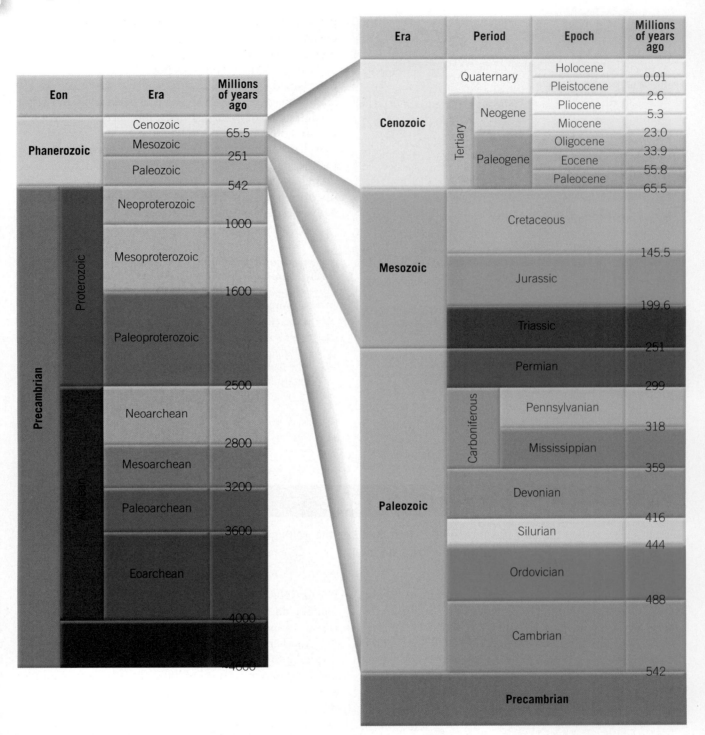

Figure 8.22 Geologic time scale Numbers on the time scale represent time in millions of years before the present. Numerical dates were added long after the time scale was established, using relative dating techniques.

Precambrian Time

Notice that the detail of the geologic time scale does not begin until about 542 million years ago, the date for the beginning of the Cambrian period. The 4 billion years prior to the Cambrian is divided into two eons, the *Archean* and the *Proterozoic*, which are divided into four eras. It is also common for this vast expanse of time to simply be referred to as the **Precambrian**. Although

it represents about 88 percent of Earth history, the Precambrian is not divided into nearly as many smaller time units as is the Phanerozoic eon.

Why is the huge expanse of Precambrian time not divided into numerous eras, periods, and epochs? The reason is that Precambrian history is not known in great enough detail. The quantity of information geologists have deciphered about Earth's past is somewhat analogous to the detail of human history. The farther back we go, the less we know. Certainly, more data and information exist about the past 10 years than for the first decade of the twentieth century; the events of the nineteenth century have been documented much better than the events of the first century A.D., and so on. So it is with Earth history. The more recent past has the freshest, least disturbed, and most observable record. The farther back in time a geologist goes, the more fragmented the record and clues become.

Terminology and the Geologic Time Scale

Some terms are associated with the geologic time scale but are not "officially" recognized as being a part of it. The best-known and the most common example is *Precambrian*—the informal name for the eons that came before the current Phanerozoic eon. Although the term *Precambrian* has no formal status on the geologic time scale, it has been traditionally used as though it does.

Hadean is another informal term that is found on some versions of the geologic time scale and is used by some geologists. It refers to the earliest interval (eon) of Earth history—before the oldest-known rocks. When the term was coined in 1972, the age of Earth's oldest rocks was about 3.8 billion years. Today that number stands at slightly greater than 4 billion, and, of course, is subject to revision. The name *Hadean* derives from *Hades*, Greek for "underworld"—a reference to the "hellish" conditions that prevailed on Earth early in its history.

Effective communication in the geosciences requires that the geologic time scale consist of standardized divisions and dates. So, who determines which names and dates on the geologic time scale are "official"? The organization that is largely responsible for maintaining and updating this important document is the International Committee on Stratigraphy (ICS), a committee of the International Union of Geological Sciences.* Advances in the geosciences

---

*To view the current version of the ICS time scale, go to http://www.stratigraphy.org. *Stratigraphy* is the branch of geology that studies rock layers (strata) and layering (stratification), and thus its primary focus is sedimentary and layered volcanic rocks.

require that the scale be periodically updated to include changes in unit names and boundary age estimates.

For example, the geologic time scale shown in Figure 8.22 was updated as recently as July 2009. After considerable dialogue among geologists who focus on very recent Earth history, the ICS changed the date for the start of the Quaternary period and the Pleistocene epoch from 1.8 million to 2.6 million years ago. Who knows? Perhaps by the time you read this, other changes will have been made.

If you were to examine a geologic time scale from just a few years ago, it is quite possible that you would see the Cenozoic era divided into the Tertiary and Quaternary periods. However, on more recent versions, the space formerly designated as Tertiary is divided into the Paleogene and Neogene periods. As our understanding of this time span has changed, so too has its designation on the geologic time scale. Today, the Tertiary period is considered as a "historic" name and is given no official status on the ICS version of the time scale. Nevertheless, many time scales still contain references to the Tertiary period, including Figure 8.22. One reason for this is that a great deal of past (and some current) geologic literature uses this name.

For those who study historical geology, it is important to realize that the geologic time scale is a dynamic tool that continues to be refined as our knowledge and understanding of Earth history evolves.

8.6 CONCEPT CHECKS

1. What are the four basic units that make up the geologic time scale? List the specific ones that apply to the present day.
2. Why is *zoic* part of so many names on the geologic time scale?
3. What term applies to *all* geologic time prior to the Phanerozoic eon? Why is this span not divided into as many smaller units as the Phanerozoic eon?
4. To what does *Hadean* apply? It is an "official" part of the time scale?

8.7 Determining Numerical Dates for Sedimentary Strata

Explain how reliable numerical dates are determined for layers of sedimentary rock.

Although reasonably accurate numerical dates have been worked out for the periods of the geologic time scale (see Figure 8.22), the task is not without difficulty. The primary problem in assigning numerical dates is the fact that not all rocks can be dated by using radiometric methods. For a

radiometric date to be useful, all minerals in the rock must have formed at about the same time. For this reason, radioactive isotopes can be used to determine when minerals in an igneous rock crystallized and when pressure and heat created new minerals in a metamorphic rock.

Figure 8.23 A problem to date A useful numerical date for this sample of conglomerate is not possible because the gravel that composes it was derived from rocks of diverse ages. (Photo by E. J. Tarbuck)

However, samples of sedimentary rock can only rarely be dated directly by radiometric means. A sedimentary rock may include particles that contain radioactive isotopes, but the rock's age cannot be accurately determined because the grains that make up the rock are not the same age as the rock in which they occur. Rather, the sediments have been weathered from rocks of diverse ages (**Figure 8.23**).

Radiometric dates obtained from metamorphic rocks may also be difficult to interpret because the age of a particular mineral in a metamorphic rock does not necessarily represent the time when the rock initially formed.

Instead, the date may indicate any one of a number of subsequent metamorphic phases.

If samples of sedimentary rocks rarely yield reliable radiometric ages, how can numerical dates be assigned to sedimentary layers? Usually a geologist must relate them to datable igneous masses, as in **Figure 8.24**. In this example, radiometric dating has determined the ages of the volcanic ash bed within the Morrison Formation and the dike cutting the Mancos Shale and Mesaverde Formation. The sedimentary beds below the ash are obviously older than the ash, and all the layers above the ash are younger (based on the principle of superposition). The dike is younger than the Mancos Shale and the Mesaverde Formation but older than the Wasatch Formation because the dike does not intrude the Paleogene rocks (based on the principle of cross-cutting relationships).

From this kind of evidence, geologists estimate that a part of the Morrison Formation was deposited about 160 million years ago, as indicated by the ash bed. Further, they conclude that the Wasatch Formation began after the intrusion of the dike, 66 million years ago. This is one example of literally thousands that illustrate how datable materials are used to "bracket" the various episodes in Earth history within specific time periods. It shows the necessity of combining laboratory dating methods with field observations of rocks.

8.7 CONCEPT CHECKS

1. Briefly explain why it is often difficult to assign a reliable numerical date to a sample of sedimentary rock.
2. How might a numerical date for a layer of sedimentary rock be determined?

Figure 8.24 Dating sedimentary strata Numerical dates for sedimentary layers are usually determined by examining their relationship to igneous rocks.

Rocks of Paleogene age

Rocks of Cretaceous age

Rocks of Jurassic age

Wasatch Formation
Mesaverde Formation
Mancos Shale
Dakota Sandstone
Volcanic ash bed dated at 160 million years
Morrison Formation
Summerville Formation

Igneous dike dated at 66 million years

CONCEPTS IN REVIEW: GEOLOGIC TIME

8.1 A Brief History of Geology

Explain the principle of uniformitarianism and discuss how it differs from catastrophism.

KEY TERMS: catastrophism, uniformitarianism

- Early ideas about the nature of Earth were based on religious traditions and notions of great catastrophes.
- In the late 1700s, James Hutton emphasized that the same slow processes have acted over great spans of time and are responsible for Earth's rocks, mountains, and landforms. This similarity of processes over vast spans of time led to this principle being called uniformitarianism.

8.2 Creating a Time Scale— Relative Dating Principles

Distinguish between numerical and relative dates and apply relative dating principles to determine a time sequence of geologic events.

KEY TERMS: numerical date, relative date, principle of superposition, principle of original horizontality, principle of cross-cutting relationships, inclusion, conformable, unconformity, angular unconformity, disconformity, nonconformity

- The two types of dates that geologists use to interpret Earth history are (1) relative dates, which put events in their proper sequence of formation, and (2) numerical dates, which pinpoint the time in years when an event took place.
- Relative dates can be established using the principles of superposition, original horizontality, cross-cutting relationships, and inclusions. Unconformities, gaps in the geologic record, may be identified during the relative dating process.

? The accompanying diagram is a cross-section of a hypothetical area. Place the lettered features in the proper sequence, from oldest to youngest. Where in the sequence can you identify an unconformity? Which principles did you use to establish the sequence?

Oldest ___ ___ ___ ___ ___ ___ ___ ___ ___ Youngest

8.3 Fossils: Evidence of Past Life

Define fossil and discuss the conditions that favor the preservation of organisms as fossils. List and describe various fossil types.

KEY TERMS: fossil, paleontology

- Fossils are remains or traces of ancient life. Paleontology is the branch of science that studies fossils.
- Fossils can form through many processes. For an organism to be preserved as a fossil, it usually needs to be buried rapidly. Also, an organism's hard parts are most likely to be preserved because soft tissue decomposes rapidly in most circumstances.

8.4 Correlation of Rock Layers

Explain how rocks of similar age that are in different places can be matched up.

KEY TERMS: correlation, principle of fossil succession, index fossil, fossil assemblage

- Matching up exposures of rock that are the same age but are in different places is called correlation. By correlating rocks from around the world, geologists developed the geologic time scale and obtained a fuller perspective on Earth history.
- Fossils can be used to correlate sedimentary rocks in widely separated places by using the rocks' distinctive fossil content and applying the principle of fossil succession. The principle states that fossil organisms succeed one another in a definite and determinable order, and, therefore, a time period can be recognized by examining its fossil content.
- Index fossils are particularly useful in correlation because they are widespread and associated with a relatively narrow time span. The overlapping ranges of fossils in an assemblage may be used to establish an age for a rock layer that contains multiple fossils.
- Fossils may be used to establish ancient environmental conditions that existed when sediment was deposited.

8.5 Dating with Radioactivity

Discuss three types of radioactive decay and explain how radioactive isotopes are used to determine numerical dates.

KEY TERMS: radioactivity, radiometric dating, half-life, radiocarbon dating

- Radioactivity is the spontaneous breaking apart (decay) of certain unstable atomic nuclei. Three common forms of radioactive decay are (1) emission of an alpha particle from the nucleus, (2) emission of a beta particle (electron) from the nucleus, and (3) capture of an electron by the nucleus.
- An unstable radioactive isotope, called a parent, will decay and form daughter products. The length of time for one-half of the nuclei of a radioactive isotope to decay is called the half-life of the isotope. If the half-life of an isotope is known, and the parent–daughter ratio can be measured, the age of the sample can be calculated.

8.6 The Geologic Time Scale

Distinguish among the four basic time units that make up the geologic time scale and explain why the time scale is considered to be a dynamic tool.

KEY TERMS: geologic time scale, eon, Phanerozoic eon, era, Paleozoic era, Mesozoic era, Cenozoic era, period, epoch, Precambrian

- Earth history is divided into units of time on the geologic time scale. Eons are divided into eras, which each contain multiple periods. Periods are divided into epochs.
- Precambrian time includes the Archean and Proterozoic eons. It is followed by the Phanerozoic eon, which is well documented by abundant fossil evidence, resulting in many subdivisions.
- The geologic time scale is a work in progress, continually being refined as new information becomes available.

8.7 Determining Numerical Dates for Sedimentary Strata

Explain how reliable numerical dates are determined for layers of sedimentary rock.

- Sedimentary strata are usually not directly datable using radiometric techniques because they consist of the material produced by the weathering of other rocks. A particle in a sedimentary rock comes from some older source rock. If you were to date the particle using isotopes, you would get the age of the source rock, not the age of the sedimentary deposit.
- One way geologists assign numerical dates to sedimentary rocks is to use relative dating principles to relate them to datable igneous masses, such as dikes and volcanic ash beds. A layer may be older than one igneous feature and younger than another.

? Determine the age of the sandstone layer in the figure as accurately as possible.

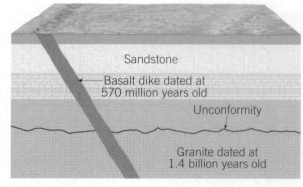

Sandstone

Basalt dike dated at 570 million years old

Unconformity

Granite dated at 1.4 billion years old

GIVE IT SOME THOUGHT

1. The accompanying image shows the metamorphic rock gneiss, a basaltic dike, and a fault. Place these three features in their proper sequence (which came first, second, and third) and explain your logic.

Marli Miller

2. A mass of granite is in contact with a layer of sandstone. Using a principle described in this chapter, explain how you might determine whether the sandstone was deposited on top of the granite or whether the magma that formed the granite was intruded after the sandstone was deposited.

3. The accompanying photo shows two layers of sedimentary rock. The lower layer is shale from the late Mesozoic era. Note the old river channel that was carved into the shale after it was deposited. Above is a younger layer of boulder-rich breccia. Are these layers conformable?

Explain why or why not. What term from relative dating applies to the line separating the two layers?

Callan Bentley

4. Refer to Figure 8.8, which shows the historic angular unconformity at Scotland's Siccar Point that James Hutton studied in the late 1700s. Refer to this photo for the following exercises.
   a. Describe in general what occurred to produce this feature.
   b. Suggest ways in which at least three of Earth's four spheres could have been involved.
   c. The Earth system is powered by energy from two sources. How are both sources represented in the Siccar Point unconformity?

5. These polished stones are called *gastroliths*. Explain how such objects can be considered fossils. What category of fossil are they? Name another example of a fossil in this category.

0 1 2
Centimeters

Francois Gohier/Photo Researchers, Inc.

6. If a radioactive isotope of thorium (atomic number 90, mass number 232) emits 6 alpha particles and 4 beta particles during the course of radioactive decay, what are the atomic number and mass number of the stable daughter product?

7. A hypothetical radioactive isotope has a half-life of 10,000 years. If the ratio of radioactive parent to stable daughter product is 1:3, how old is the rock that contains the radioactive material?

8. Solve the problems below that relate to the magnitude of Earth history. To make calculations easier, round Earth's age to 5 billion years.
  a. What percentage of geologic time is represented by recorded history? (Assume 5000 years for the length of recorded history.)
  b. Humanlike ancestors (hominids) have been around for roughly 5 million years. What percentage of geologic time is represented by these ancestors?
  c. The first abundant fossil evidence does not appear until the beginning of the Cambrian period, about 540 million years ago. What percentage of geologic time is represented by abundant fossil evidence?

9. A portion of a popular college text in historical geology includes 10 chapters (281 pages) in a unit titled "The Story of Earth." Two chapters (49 pages) are devoted to Precambrian time. By contrast, the last two chapters (67 pages) focus on the most recent 23 million years, with 25 of those pages devoted to the Holocene Epoch, which began 10,000 years ago.
  a. Compare the percentage of pages devoted to the Precambrian to the actual percentage of geologic time that this span represents.
  b. How does the number of pages about the Holocene compare to its actual percentage of geologic time?
  c. Suggest some reasons why the text seems to have such an unequal treatment of Earth history.

10. This scene in Montana's Glacier National Park shows layers of Precambrian sedimentary rocks. The darker layer contained within the sedimentary layers is igneous. The narrow, light-colored areas adjacent to the igneous rock were created when the molten material that formed the igneous rock baked the adjacent rock.

Marli Miller

  a. Is the igneous layer more likely a lava flow that was laid down at the surface prior to the deposition of the layers above it or a sill that was intruded after all the sedimentary layers were deposited? Explain.
  b. Is it likely that the igneous layer will exhibit a vesicular texture? Explain.

MasteringGeology™

www.masteringgeology.com

Looking for additional review and test prep materials? Visit the Study Area in MasteringGeology to enhance your understanding of this chapter's content by accessing a variety of resources, including **Self-Study Quizzes**,

Geoscience Animations, GEODe Tutorials, RSS feeds, flashcards, web links, self-study quizzes, and an optional **Pearson eText.**

Oceans: The Last Frontier

FOCUS ON CONCEPTS

Each statement represents the primary learning objective for the corresponding major heading within the chapter. After you complete the chapter, you should be able to:

9.1 Discuss the extent and distribution of oceans and continents on Earth. Identify Earth's four main ocean basins.

9.2 Define *salinity* and list the main elements that contribute to the ocean's salinity. Describe the sources of dissolved substances in seawater and causes of variations in salinity.

9.3 Discuss temperature, salinity, and density changes with depth in the open ocean.

9.4 Define *bathymetry* and summarize the various techniques used to map the ocean floor.

9.5 Compare a passive continental margin with an active continental margin and list the major features of each.

9.6 List and describe the major features associated with deep-ocean basins.

9.7 Summarize the basic characteristics of oceanic ridges.

9.8 Distinguish among three categories of seafloor sediment and explain why some of these sediments can be used to study climate change.

The *Chikyu* (meaning "Earth" in Japanese) is one of the most advanced scientific drillings vessels. It is part of the Integrated Ocean Drilling Program (IODP). (AP Photo/Itsuo Inouye)

How much of Earth is covered by the ocean? What is the difference between freshwater and seawater? How deep is the ocean? What does the seafloor look like? Finding answers to these and other questions about the oceans and the basins they occupy are sometimes elusive and often difficult to determine. Suppose that all of the water were drained from the ocean. What would we see? Plains? Mountains? Canyons? Plateaus? Indeed, the ocean conceals all these features—and more. And what about the carpet of sediment that covers most of the ocean floor? Where did it come from, and what can be learned by studying it? This chapter provides answers to these questions.

9.1 **The Vast World Ocean**

Discuss the extent and distribution of oceans and continents on Earth. Identify Earth's four main ocean basins.

Oceans are a major part of our planet. In fact, Earth is frequently called the *water planet* or the *blue planet*. There is a good reason for these names: Nearly 71 percent of Earth's surface is covered by the global ocean (**Figure 9.1**). The focus of this chapter and Chapter 10 is oceanography. **Oceanography** is an interdisciplinary science that draws on the methods and knowledge of geology, chemistry, physics, and biology to study all aspects of the world ocean.

Geography of the Oceans

The area of Earth is about 510 million square kilometers (197 million square miles). Of this total, approximately 360 million square kilometers (140 million square miles), or 71 percent, is represented by oceans and marginal seas (seas around the ocean's margin, such as the Mediterranean Sea and the Caribbean Sea). Continents and islands comprise the remaining 29 percent, or 150 million square kilometers (58 million square miles).

By studying a globe or world map, it is readily apparent that the continents and oceans are not evenly divided between the Northern and Southern Hemispheres (Figure 9.1). When we compute the percentages of land and water in the Northern Hemisphere, we find that nearly 61 percent of the surface is water, and about 39 percent is land. In the Southern Hemisphere, on the other hand,

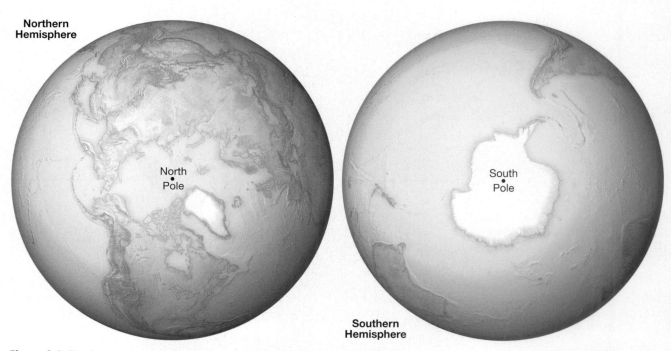

Northern Hemisphere

North Pole

South Pole

Southern Hemisphere

Figure 9.1 North versus south These views of Earth show the uneven distribution of land and water between the Northern and Southern Hemispheres. Almost 81 percent of the Southern Hemisphere is covered by the oceans—20 percent more than the Northern Hemisphere.

SmartFigure 9.2
Distribution of land and water A. The graph shows the amount of land and water in each 5-degree latitude belt. **B.** The world map provides a more familiar view.

almost 81 percent of the surface is water, and only 19 percent is land. It is no wonder then that the Northern Hemisphere is called the *land hemisphere*, and the Southern Hemisphere the *water hemisphere*.

Figure 9.2A shows the distribution of land and water in the Northern and Southern Hemispheres by way of a graph. Between latitudes 45° north and 70° north, there is actually more land than water, whereas between 40° south and 65° south, there is almost no land to interrupt the oceanic and atmospheric circulation.

The world ocean can be divided into four main ocean basins (**Figure 9.2B**):

1. The *Pacific Ocean*, which is the largest ocean (and the largest single geographic feature on the planet), covers more than half of the ocean surface area on Earth. In fact, the Pacific Ocean is so large that all of the continents could fit into the space occupied by it—with room left over! It is also the world's deepest ocean, with an average depth of 3940 meters (12,927 feet).

2. The *Atlantic Ocean* is about half the size of the Pacific Ocean but is not quite as deep. It is a relatively narrow ocean compared to the Pacific and is bounded by almost parallel continental margins.

3. The *Indian Ocean* is slightly smaller than the Atlantic Ocean but has about the same average depth. Unlike the Pacific and Atlantic Oceans, it is largely a Southern Hemisphere water body.

4. The *Arctic Ocean* is about 7 percent of the size of the Pacific Ocean and is only a little more than one-quarter as deep as the rest of the oceans.

Comparing the Oceans to the Continents

A major difference between continents and the ocean basins is their relative levels. The average elevation of the continents above sea level is about 840 meters (2756 feet), whereas the average depth of the oceans is nearly four and a half times this amount—3729 meters (12,234 feet). The volume of ocean water is so large that if Earth's solid mass were perfectly smooth (level) and spherical, the oceans would cover Earth's entire surface to a uniform depth of more than 2000 meters (1.2 miles)!

9.1 CONCEPT CHECKS

1. How does the area of Earth's surface covered by oceans compare with the area covered by continents?
2. Contrast the distribution of land and water in the Northern Hemisphere and the Southern Hemisphere.
3. Excluding the Southern Ocean, name the four main ocean basins. Contrast them in terms of area and depth.
4. How does the average depth of the oceans compare to the average elevation of the continents?

Did You Know?

The Bering Sea is the most northerly marginal sea of the Pacific Ocean and connects to the Arctic Ocean through the Bering Straits. This water body is effectively cut off from the Pacific basin by the Aleutian Islands, which were created by volcanic activity associated with northward subduction of the Pacific plate.

9.2 Composition of Seawater

Define *salinity* and list the main elements that contribute to the ocean's salinity. Describe the sources of dissolved substances in seawater and causes of variations in salinity.

What is the difference between freshwater and seawater? One of the most obvious differences is that seawater contains dissolved substances that give it a distinctly salty taste. These dissolved substances are not simply sodium

chloride (common table salt)—they include various other salts, metals, and even dissolved gases. In fact, every known naturally occurring element is found dissolved in at least trace amounts in seawater. Unfortunately, the salt content of seawater makes it unsuitable for drinking or for irrigating most crops and causes it to be highly corrosive to many materials. Yet, many parts of the ocean are teeming with life that is superbly adapted to the marine environment.

Salinity

Seawater consists of about 3.5 percent (by weight) dissolved mineral substances that are collectively termed *salts.* Although the percentage of dissolved components may seem small, the actual quantity is huge because the ocean is so vast.

Salinity is the total amount of solid material dissolved in water. More specifically, it is the ratio of the mass of dissolved substances to the mass of the water sample. Many common quantities are expressed in percent (%), which is really *parts per hundred.* Because the proportion of dissolved substances in seawater is such a small number, oceanographers typically express salinity *in parts per thousand* (‰). Thus, the average salinity of seawater is 3.5%, or 35‰.

Figure 9.3 shows the principal elements that contribute to the ocean's salinity. Artificial seawater could be approximated by following the recipe in **Table 9.1**. This table shows that most of the salt in seawater is sodium chloride—common table salt. Sodium chloride together with the next four most abundant salts comprise more than 99 percent of all dissolved substances in the sea. Although only eight elements make up these five most abundant salts, seawater contains all of Earth's other naturally occurring elements. Despite their presence in minute quantities, many of these elements are very important in maintaining the necessary chemical environment for life in the sea.

Sources of Sea Salts

What are the primary sources for the vast quantities of dissolved substances in the ocean? Chemical weathering of rocks on the continents is one source. These dissolved

TABLE 9.1 Recipe for Artificial Seawater	
Ingredient	**Amount (grams)**
Sodium chloride (NaCl)	23.48
Magnesium chloride ($MgCl_2$)	4.98
Sodium sulfate (Na_2SO_4)	3.92
Calcium chloride ($CaCl_2$)	1.10
Potassium chloride (KCl)	0.66
Sodium bicarbonate ($NaHCO_3$)	0.192
Potassium bromide (KBr)	0.096
Hydrogen borate (H_3BO_3)	0.026
Strontium chloride ($SrCl_2$)	0.024
Sodium fluoride (NaF)	0.003
Then add:	
Pure water (H_2O) to form 1000 grams of solution.	

materials are delivered to the oceans by streams at an estimated rate of more than 2.5 billion tons annually. The second major source of elements found in ocean water is Earth's interior. Through volcanic eruptions, large quantities of water and dissolved gases have been emitted during much of geologic time. This process, called *outgassing,* is the principal source of water in the oceans and in the atmosphere. Certain elements—notably chlorine, bromine, sulfur, and boron—were outgassed along with water and exist in the ocean in much greater abundance than could be explained by weathering of rocks alone.

Although rivers and volcanic activity continually contribute salts to the oceans, the salinity of seawater is not increasing. In fact, evidence suggests that the composition of seawater has been relatively stable for millions of years. Why doesn't the sea get saltier? The answer is that material is being removed just as rapidly as it is added. For example, some dissolved components are withdrawn from seawater by plants and animals as they build hard parts. Other components are removed when they chemically precipitate from the water as sediment. Still others are exchanged at oceanic ridges by hydrothermal activity. The net effect is that the overall makeup of seawater remains relatively constant through time.

Processes Affecting Seawater Salinity

Because the ocean is well mixed, the relative abundances of the major components in seawater are essentially constant, no matter where the ocean is sampled. Variations in salinity, therefore, are primarily a result of changes in the water content of the solution.

Various surface processes alter the amount of water in seawater, thereby affecting salinity. Processes that add large amounts of freshwater to seawater—and thereby decrease salinity—include precipitation, runoff from land, iceberg melting, and sea ice melting. Processes that remove large amounts of freshwater from seawater—and

Figure 9.3 Composition of seawater Relative proportions of water and dissolved components in seawater. Dissolved components shown by chemical symbol are chlorine (Cl^-), sodium (Na^+), sulfate (SO_4^{2-}), magnesium (Mg^{2+}), calcium (Ca^{2+}), potassium (K^+), strontium (Sr^{2+}), bromine (Br^-), and carbon (C).

SmartFigure 9.4 **Variations in surface temperature and salinity with latitude** Average temperatures are highest near the equator and get colder toward the poles. Important factors influencing salinity are variations in rainfall and rates of evaporation. For example, in the dry subtropics in the vicinity of the Tropics of Cancer and Capricorn, high evaporation rates remove more water than is replaced by the meager rainfall, resulting in high surface salinities. In the wet equatorial region, abundant rainfall reduces surface salinities.

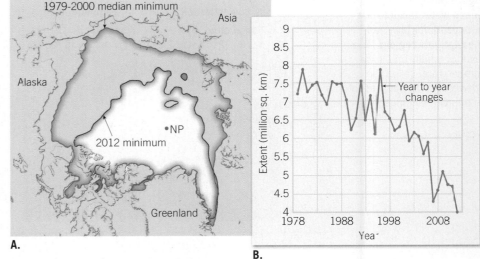

A.

B.

Figure 9.5 **Tracking sea ice changes A.** Sea ice is frozen seawater. The formation and melting of sea ice influence the surface salinity of the ocean. In winter the Arctic Ocean is completely ice covered. In summer a portion of the ice melts. This map shows the extent of sea ice in early September 2012 compared to the average extent for the period 1979 to 2000. The extent in 2012 was the lowest on record. **B.** The graph clearly depicts the trend in the area covered by Arctic sea ice at the end of the summer melt period. The trend is likely related to global climate change. (NASA)

thereby increase seawater salinity—include evaporation and the formation of sea ice. High salinities, for example, are found where evaporation rates are high, as is the case in the dry subtropical regions (roughly between 25° and 35° north or south latitude). Conversely, where large amounts of precipitation dilute ocean waters, as in the midlatitudes (between 35° and 60° north or south latitude) and near the equator, lower salinities prevail (**Figure 9.4**).

Surface salinity in polar regions varies seasonally due to the formation and melting of sea ice. When seawater freezes in winter, sea salts do not become part of the ice. Therefore, the salinity of the remaining seawater increases. In summer, when sea ice melts, the addition of the relatively freshwater dilutes the solution, and salinity decreases (**Figure 9.5**).

Surface salinity variation in the open ocean normally ranges from 33‰ to 38‰. Some marginal seas, however, demonstrate extraordinary extremes. For example, in the

restricted waters of the Middle East's Persian Gulf and Red Sea—where evaporation far exceeds precipitation—salinity may exceed 42‰. Conversely, very low salinities occur where large quantities of freshwater are supplied by rivers and precipitation. Such is the case for northern Europe's Baltic Sea, where salinity is often below 10‰.

9.2 CONCEPT CHECKS

1. What is salinity, and how is it usually expressed? What is the average salinity of the ocean?
2. What are the six most abundant elements dissolved in seawater? What is produced when the two most abundant elements combine?
3. What are the two primary sources for the elements that comprise the dissolved components in seawater?
4. List several factors that cause salinity to vary from place to place and from time to time.

9.3 Variations in Temperature and Density with Depth

Discuss temperature, salinity, and density changes with depth in the open ocean.

Temperature and density are basic ocean water properties that influence such things as deep-ocean circulation and the distribution and types of life-forms. A sampling of the open ocean from the surface to the seafloor shows that these basic properties change with depth and that the changes are not the same everywhere. How and why temperature and density change with depth is the focus of this section.

Figure 9.6 Variations in ocean water temperature with depth for low- and high-latitude regions The layer of rapidly changing temperature, called the *thermocline*, is not present in the high latitudes.

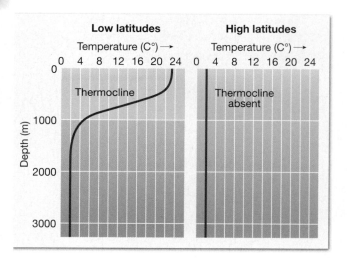

Temperature Variations

If a thermometer were lowered from the surface of the ocean into deeper water, what temperature pattern would be found? Surface waters are warmed by the Sun, so they generally have higher temperatures than deeper waters. Surface water temperatures are also higher in the tropics and get colder toward the poles.

Figure 9.6 shows two graphs of temperature versus depth: one for high-latitude regions and one for low-latitude regions. The low-latitude curve begins at the surface with high temperature, but the temperature decreases rapidly with depth because of the inability of the Sun's rays to penetrate very far into the ocean. At a depth of about 1000 meters (3300 feet), the temperature remains just a few degrees above freezing and is relatively constant from this level down to the ocean floor. The layer of ocean water between about 300 meters (980 feet) and 1000 meters (3300 feet), where there is a rapid change of temperature with depth, is called the **thermocline** (*thermo* = heat, *cline* = slope). The thermocline is a very important zone in the ocean because it creates a vertical barrier to many types of marine life.

The high-latitude curve in Figure 9.6 displays a pattern quite different from the low-latitude curve. Surface-water temperatures in high latitudes are much cooler than in low latitudes, so the curve begins at the surface with low temperature. Deeper in the ocean, the temperature of the water is similar to that at the surface (just a few degrees above freezing), so the curve remains vertical and there is no rapid change of temperature with depth. A thermocline is not present in high latitudes; instead, the water column is *isothermal* (*iso* = same, *thermal* = heat).

Some high-latitude waters can experience minor warming during the summer months. Thus, certain high-latitude regions experience an extremely weak seasonal thermocline. Midlatitude waters, on the other hand, experience a more dramatic seasonal thermocline and exhibit characteristics intermediate between high- and low-latitude regions.

Density Variations

Density is defined as mass per unit volume but can be thought of as a measure of *how heavy something is for its size*. For instance, an object that has low density is lightweight for its size—for example, a dry sponge, foam packing, or a surfboard. Conversely, an object that has high density is heavy for its size—for example, cement and many metals.

Density is an important property of ocean water because it determines the water's vertical position in the ocean. Furthermore, density differences cause large areas of ocean water to sink or float. For example, when high-density seawater is added to low-density freshwater, the denser seawater sinks below the freshwater.

Factors Affecting Seawater Density Seawater density is influenced by two main factors: *salinity* and *temperature*. An increase in salinity adds dissolved substances and results in an increase in seawater density (**Figure 9.7**). An increase in temperature, on the other hand, causes water to expand and results in a decrease in seawater density. Such a relationship, where one variable decreases as a result of another variable's increase, is known as an *inverse relationship*, where one variable is *inversely proportional* with the other.

Temperature has the greatest influence on surface seawater density because variations in surface seawater temperature are greater than salinity variations. In fact, only in the extreme polar areas of the ocean, where temperatures are low and remain relatively constant, does salinity significantly affect density. Cold water that also has high salinity is some of the highest-density water in the world.

Density Variation with Depth By extensively sampling ocean waters, oceanographers have learned that temperature and salinity—and the water's resulting density—vary with depth. **Figure 9.8** shows two graphs of density

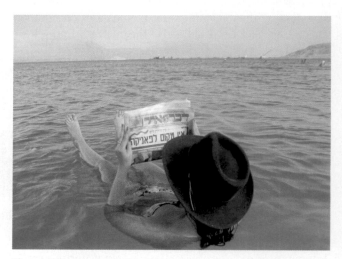

Figure 9.7 The Dead Sea This water body, which has a salinity of 330‰ (almost 10 times the average salinity of seawater), has high density. As a result, it also has high buoyancy that allows swimmers to float easily. (Photo by Peter Guttman/Corbis/Bettmann)

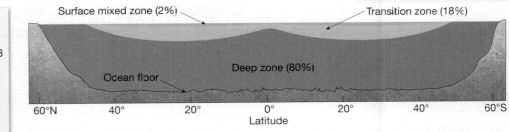

Figure 9.9 The Ocean's layers Oceanographers recognize three main layers in the ocean, based on water density, which varies with temperature and salinity. The warm *surface mixed layer* accounts for only 2 percent of ocean water; the *transition zone* includes the thermocline and the pycnocline and accounts for 18 percent of ocean water; and the *deep zone* contains cold, high-density water that accounts for 80 percent of ocean water.

SmartFigure 9.8 Variations in ocean-water density with depth for low- and high-latitude regions The layer of rapidly changing density, called the *pycnocline*, is present in the low latitudes but absent in the high latitudes.

versus depth: one for high-latitude regions and one for low-latitude regions.

The low-latitude curve in Figure 9.8 begins at the surface with low density (related to high surface-water temperatures). However, density increases rapidly with depth because the water temperature is getting colder. At a depth of about 1000 meters (3300 feet), seawater density reaches a maximum value related to the water's low temperature. From this depth to the ocean floor, density remains constant and high. The layer of ocean water between about 300 meters (980 feet) and 1000 meters (3300 feet), where there is a rapid change of density with depth, is called the **pycnocline** (*pycno* = density, *cline* = slope). A pycnocline has a high gravitational stability and presents a significant barrier to mixing between low-density water above and high-density water below.

The high-latitude curve in Figure 9.8 is also related to the temperature curve for high latitudes shown in Figure 9.6. Figure 9.8 shows that in high latitudes, there is high-density (cold) water at the surface and high-density (cold) water below. Thus, the high-latitude density curve remains vertical, and there is no rapid change of density with depth. A pycnocline is not present in high latitudes; instead, the water column is *isopycnal* (*iso* = same, *pycno* = density).

Ocean Layering

The ocean, like Earth's interior, is layered according to density. Low-density water exists near the surface, and higher-density water occurs below. Except for some shallow inland seas with a high rate of evaporation, the highest-density water is found at the greatest ocean depths. Oceanographers generally recognize a three-layered structure in most parts of the open ocean: a shallow surface mixed zone, a transition zone, and a deep zone (**Figure 9.9**).

Because solar energy is received at the ocean surface, it is here that water temperatures are warmest.

The mixing of these waters by waves as well as turbulence from currents and tides distributes heat gained at the surface through a shallow layer. Consequently, this *surface mixed zone* has nearly uniform temperatures. The thickness and temperature of this layer vary, depending on latitude and season. The zone usually extends to about 300 meters (980 feet) but may attain a thickness of 450 meters (1500 feet). The surface mixed zone accounts for only about 2 percent of ocean water.

Below the sun-warmed zone of mixing, the temperature falls abruptly with depth (see Figure 9.6). Here, a distinct layer called the *transition zone* exists between the warm surface layer above and the deep zone of cold water below. The transition zone includes a prominent thermocline and associated pycnocline and accounts for about 18 percent of ocean water.

Below the transition zone is the *deep zone*, where sunlight never reaches and water temperatures are just a few degrees above freezing. As a result, water density remains constant and high. Remarkably, the deep zone includes about 80 percent of ocean water, indicating the immense depth of the ocean. (The average depth of the ocean is more than 3700 meters [12,200 feet].)

In high latitudes, the three-layer structure does not exist because the water column is isothermal and isopycnal, which means that there is no rapid change in temperature or density with depth. Consequently, good vertical mixing between surface and deep waters can occur in high-latitude regions. Here, cold high-density water forms at the surface, sinks, and initiates deep-ocean currents, which are discussed in Chapter 10.

9.3 CONCEPT CHECKS

1. Contrast temperature variations with depth in the high and low latitudes. Why do high-latitude waters generally lack a thermocline?
2. What two factors influence seawater density? Which one has the greater influence on surface seawater density?
3. Contrast density variations with depth in the high and low latitudes. Why do high-latitude waters generally lack a pycnocline?
4. Describe the ocean's layered structure. Why does the three-layer structure not exist in high latitudes?

9.4 An Emerging Picture of the Ocean Floor

Define *bathymetry* and summarize the various techniques used to map the ocean floor.

If all the water were removed from the ocean basins, a great variety of features would be seen, including broad volcanic peaks, deep trenches, extensive plains, linear mountain chains, and large plateaus. In fact, the scenery would be nearly as diverse as that on the continents.

Mapping the Seafloor

The complex nature of ocean-floor topography did not begin to unfold until the historic three-and-a-half-year voyage of the HMS *Challenger* (**Figure 9.10**). From December 1872 to May 1876, the *Challenger* expedition made the first comprehensive study of the global ocean ever attempted. During the 127,500-kilometer (79,200-mile) voyage, the ship and its crew of scientists traveled to every ocean except the Arctic. Throughout the voyage, they sampled a multitude of ocean properties, including water depth, which was accomplished by lowering a long weighted line overboard and then retrieving it. Using this laborious process, the *Challenger* made the first recording of the deepest-known point on the ocean floor in 1875. This spot, on the floor of the western Pacific, was later named the *Challenger Deep*.

Modern Bathymetric Techniques The measurement of ocean depths and the charting of the shape (topography) of the ocean floor is known as **bathymetry** (*bathos* = depth, *metry* = measurement). Today, sound energy is used to measure water depths. The basic approach employs **sonar**, an acronym for *so*und *na*vigation and *r*anging.

The first devices that used sound to measure water depth, called **echo sounders**, were developed early in the twentieth century. Echo sounders work by transmitting a sound wave (called a *ping*) into the water in order to produce an echo when it bounces off any object, such as a large marine organism or the ocean floor (**Figure 9.11**). A sensitive receiver intercepts the echo reflected from the bottom, and a clock precisely measures the travel time to fractions of a second. By knowing the speed of sound waves in water—about 1500 meters (4900 feet) per second—and the time required for the energy pulse to reach the ocean floor and return, depth can be calculated. Depths determined from continuous monitoring of these echoes are plotted to create a profile of the ocean floor. By laboriously combining profiles from several adjacent traverses, a chart of the seafloor was produced.

Following World War II, the U.S. Navy developed *sidescan sonar* to look for explosive devices that had been deployed in shipping lanes (**Figure 9.12A**). These torpedo-shaped instruments can be towed behind a ship, where they send out a fan of sound extending to either side of the ship's track. By combining swaths of sidescan sonar data, researchers produced the first photograph-like images of the seafloor. Although sidescan sonar provides valuable views of the seafloor, it does not provide bathymetric (water depth) data.

This drawback was resolved with the development of *high-resolution multibeam* instruments (Figure 9.12A). These systems use hull-mounted sound sources that

Figure 9.10 HMS *Challenger* The first systematic bathymetric measurements of the ocean were made aboard the HMS *Challenger*, which departed England in December 1872 and returned in May 1876. (Image courtesy of the Library of Congress)

Figure 9.11 **Echo sounder** An echo sounder determines water depth by measuring the time interval required for an acoustic wave to travel from a ship to the seafloor and back. The speed of sound in water is 1500 m/sec. Therefore, depth = 1/2 (1500 m/sec × Echo travel time).

A. Sidescan sonar and multibeam sonar operating from the same research vessel.

B. Color enhanced perspective map of the seafloor and coastal landforms in the Los Angeles area of California.

Figure 9.12 **Sidescan and multibeam sonar**

send out a fan of sound and then record reflections from the seafloor through a set of narrowly focused receivers aimed at different angles. Rather than obtain the depth of a single point every few seconds, this technique makes it possible for a survey ship to map the features of the ocean floor along a strip tens of kilometers wide. These systems can collect bathymetric data of such high resolution that they can distinguish depths that differ by less than a meter (**Figure 9.12B**). When multibeam sonar is used to make a map of a section of seafloor, the ship travels through the area in a regularly spaced back-and-forth pattern known as "mowing the lawn."

Despite their greater efficiency and enhanced detail, research vessels equipped with multibeam sonar travel at a mere 10 to 20 kilometers (6 to 12 miles) per hour. It would take at least 100 vessels outfitted with this equipment hundreds of years to map the entire seafloor. This explains why only about 5 percent of the seafloor has been mapped in detail—and why large areas of the seafloor have not been mapped with sonar at all.

Mapping the Ocean Floor from Space

Another technological breakthrough that has led to an enhanced understanding of the seafloor involves measuring the shape of the ocean surface from space. After compensating for waves, tides, currents, and atmospheric effects, it was discovered that the ocean surface is not perfectly "flat." Because massive seafloor features exert stronger-than-average gravitational attraction, they produce elevated areas on the ocean surface. Conversely, canyons and trenches create slight depressions.

Satellites equipped with *radar altimeters* are able to measure these subtle differences by bouncing microwaves off the sea surface (**Figure 9.13**). These devices can measure variations as small as a few centimeters. Such data have added greatly to the knowledge of ocean-floor topography. Combined with traditional sonar depth measurements, the data are used to produce detailed ocean-floor maps, such as the one shown in **Figure 9.14**.

SmartFigure 9.13 Satellite altimeter A satellite altimeter measures the variation in sea-surface elevation, which is caused by gravitational attraction and mimics the shape of the seafloor. The sea-surface anomaly is the difference between the measured and theoretical ocean surface.

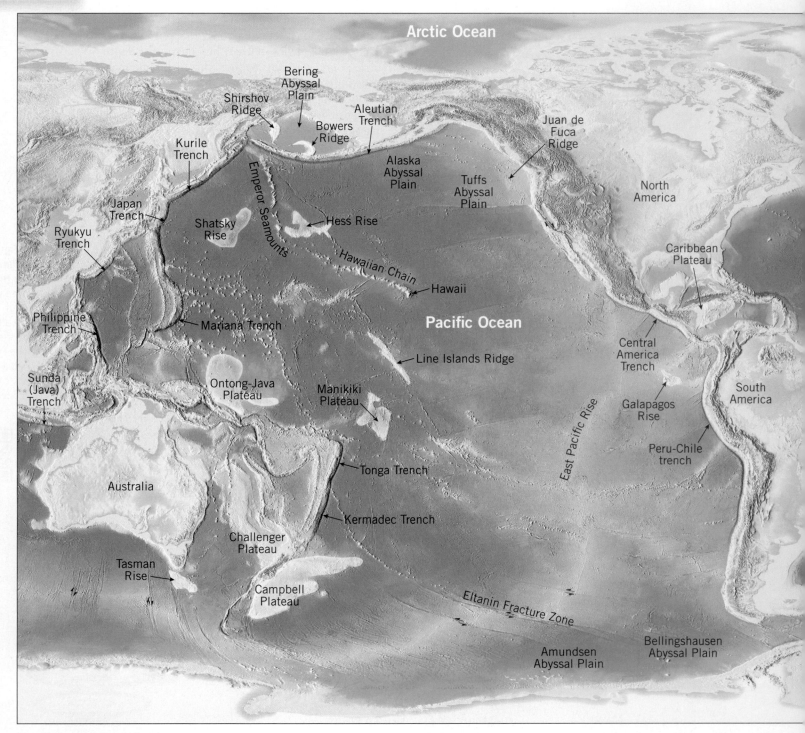

Figure 9.14 Major features of the seafloor

1. Continental shelf
2. Continental slope
3. Continental rise
4. Seamount
5. Abyssal plain
6. Rift valley
7. Abyssal plain
8. Seamount
9. Continental rise
10. Continental slope
11. Continental shelf

Figure 9.15 Major topographic divisions of the North Atlantic Map view (*top*) and corresponding profile view (*bottom*). On the profile, the vertical scale has been exaggerated to make topographic features more conspicuous.

Provinces of the Ocean Floor

Oceanographers studying the topography of the ocean floor have delineated three major areas: *continental margins*, the *deep-ocean basins*, and the *oceanic (mid-ocean) ridges*. The map in **Figure 9.15** outlines these provinces for the North Atlantic Ocean, and the profile at the bottom of the illustration shows the varied topography. Such profiles usually have their vertical dimension exaggerated many times—40 times in this case—to make topographic features more conspicuous. Vertical exaggeration, however, makes slopes shown in seafloor profiles appear to be *much* steeper than they actually are.

9.4 CONCEPT CHECKS

1. Define *bathymetry*.
2. Describe how satellites orbiting Earth can determine features on the seafloor without being able to directly observe them beneath several kilometers of seawater.
3. List the three major provinces of the ocean floor.

9.5 Continental Margins

Compare a passive continental margin with an active continental margin and list the major features of each.

As the name implies, the **continental margins** are the outer margins of the continents where continental crust transitions to oceanic crust. Two types of continental margin have been identified: *passive* and *active*. Nearly the entire Atlantic Ocean and a large portion of the Indian Ocean are surrounded by passive continental margins (see Figure 9.14). By contrast, most of the Pacific Ocean is bordered by active continental margins that are represented by subduction zones in **Figure 9.16**. Notice that many of those active subduction zones lie far beyond the margins of the continents.

Passive Continental Margins

Passive continental margins are geologically inactive regions located some distance from plate boundaries. As a result, they are not associated with strong earthquakes or volcanic activity. Passive continental margins develop when continental blocks rift apart and are separated by continued seafloor spreading. As a result, the continental blocks are firmly attached to the adjacent oceanic crust.

Most passive margins are relatively wide and are sites where large quantities of sediments are deposited.

Figure 9.16 Distribution of Earth's subduction zones Most of the active subduction zones surround the Pacific basin.

The features comprising passive continental margins include the continental shelf, the continental slope, and the continental rise (**Figure 9.17**).

Continental Shelf
The **continental shelf** is a gently sloping, submerged surface extending from the shoreline toward the deep-ocean basin. It consists mainly of continental crust capped with sedimentary rocks and sediments eroded from adjacent landmasses.

The continental shelf varies greatly in width. Although almost nonexistent along portions of some continents, the shelf extends seaward more than 1500 kilometers (930 miles) along others. The average inclination of the continental shelf is only about one-tenth of 1 degree, a slope so slight that it would appear to an observer to be a horizontal surface.

The continental shelf tends to be relatively featureless; however, some areas are mantled by extensive glacial deposits and thus are quite rugged. In addition, some continental shelves are dissected by large valleys running from the coastline into deeper waters. Many of these *shelf valleys* are the seaward extensions of river valleys on the adjacent landmass. They were eroded during the last Ice Age (Quaternary period), when enormous quantities of water were stored in vast ice sheets on the continents, causing sea levels to drop at least 100 meters (330 feet). Because of this sea-level drop, rivers extended their courses, and land-dwelling plants and animals migrated to the newly exposed portions of the continents. Dredging off the coast of North America has retrieved the ancient remains of numerous land dwellers, including mammoths, mastodons, and horses, providing further evidence that portions of the continental shelves were once above sea level.

Although continental shelves represent only 7.5 percent of the total ocean area, they have economic and political significance because they contain important reservoirs of oil and natural gas, and they support important fishing grounds.

Continental Slope
Marking the seaward edge of the continental shelf is the **continental slope**, a relatively steep structure that marks the boundary between continental crust and oceanic crust. Although the inclination of the continental slope varies greatly from place to place, it averages about 5 degrees and in places exceeds 25 degrees.

Figure 9.17 Passive continental margin Note that the slopes shown for the continental shelf and continental slope are greatly exaggerated. The continental shelf has an average slope of one-tenth of 1 degree, whereas the continental slope has an average slope of about 5 degrees.

Continental Rise The continental slope merges into a more gradual incline known as the **continental rise** that may extend seaward for hundreds of kilometers. The continental rise consists of a thick accumulation of sediment that has moved down the continental slope and onto deep-ocean floor. Most of the sediments are delivered to the seafloor by *turbidity currents* that periodically flow down *submarine canyons.* (We will discuss these shortly.) When these muddy slurries emerge from the mouth of a canyon onto the relatively flat ocean floor, they deposit sediment that forms a **deep-sea fan**. As fans from adjacent submarine canyons grow, they merge to produce a continuous wedge of sediment at the base of the continental slope, forming the continental rise.

Submarine Canyons and Turbidity Currents Deep, steep-sided valleys known as **submarine canyons** are cut into the continental slope and may extend across the entire continental rise to the deep-ocean basin (**Figure 9.18**). Although some of these canyons appear to be the seaward extensions of river valleys, many others do not line up in this manner. Furthermore, submarine canyons extend to depths far below the maximum lowering of sea level during the Ice Age, so we cannot attribute their formation to stream erosion.

These submarine canyons have probably been excavated by turbidity currents. **Turbidity currents** are downslope movements of dense, sediment-laden water. They are created when sand and mud on the continental shelf and slope are dislodged and thrown into suspension. Because the mud-choked water is denser than normal seawater, it moves downslope as a mass, eroding and

accumulating more sediment as it goes. The erosional work repeatedly carried on by these muddy torrents is thought to be the major force in the excavation of most submarine canyons.

Turbidity currents usually originate along the continental slope and continue across the continental rise, still cutting channels. Eventually, they lose momentum and come to rest along the floor of the deep-ocean basin. As these currents slow, suspended sediments begin to settle out. First, the coarser sand is dropped, followed by successively finer accumulations of silt and then clay. These deposits, called *turbidites*, display a decrease in sediment grain size from bottom to top, a feature known as *graded bedding.* Turbidity currents are an important mechanism of sediment transport in the ocean. By the action of turbidity currents, submarine canyons are excavated, and sediments are carried to the deep-ocean floor.

Active Continental Margins

Active continental margins are located along convergent plate boundaries where oceanic lithosphere is being subducted beneath the leading edge of a continent (**Figure 9.19**). Deep-ocean trenches are the major topographic expression at convergent plate boundaries. These deep, narrow furrows surround most of the Pacific Rim.

Along some subduction zones, sediments from the ocean floor and pieces of oceanic crust are scraped from the descending oceanic plate and plastered against the edge of the overriding plate. This chaotic accumulation of deformed sediment and scraps of oceanic crust is

Figure 9.18 Turbidity currents and submarine canyons Turbidity currents are an important factor in the formation of submarine canyons. (Photo by Marli Miller)

Beds deposited by turbidity currents are called turbidites. Each event produces a single bed characterized by a decrease in sediment size from bottom to top, a feature known as graded bedding.

Turbidity current

Turbidite deposits

Turbidity currents are downslope movements of dense, sediment-laden water. They are created when sand and mud on the continental shelf and/or slope are dislodged and thrown into suspension. Because the mud-choked water is denser than normal seawater, it flows downslope, eroding and accumulating more sediment.

Fine particles settle out last

Graded beds

Coarse particles settle out first

Turbidity current

Deep-sea fans

Submarine canyons

Active continental margins are located along convergent plate boundaries where oceanic lithosphere is being subducted beneath the leading edge of a continent.

SmartFigure 9.19 Active continental margins

Continental volcanic arc

Trench

Oceanic crust

Continental crust

Continental lithosphere

- 100 km

Asthenosphere

Subducting oceanic lithosphere

Melting

- 200 km

Shallow trench

Accretionary wedge

Continental crust

Subducting oceanic lithosphere

Deep trench

Continental crust

Erosion

Subducting oceanic lithosphere

A. Accretionary wedges develop along subduction zones where sediments from the ocean floor are scraped from the descending oceanic plate and pressed against the edge of the overriding plate.

B. Subduction erosion occurs where sediment and rock scraped off the bottom of the overriding plate is carried into the mantle by the subducting plate.

called an **accretionary wedge** (*ad* = toward, *crescere* = to grow). Prolonged plate subduction can produce massive accumulations of sediment along active continental margins.

The opposite process, known as **subduction erosion**, characterizes many other active continental margins. Rather than sediment accumulating along the front of the overriding plate, sediment and rock are scraped off the bottom of the overriding plate and transported into the mantle by the subducting plate. Subduction erosion is particularly effective when the angle of descent is steep. Sharp bending of the subducting plate causes faulting in the ocean crust and a rough surface, as shown in Figure 9.19.

9.5 CONCEPT CHECKS

1. List the three major features of a passive continental margin. Which of these features is considered a flooded extension of the continent? Which one has the steepest slope?

2. Describe the differences between active and passive continental margins. Where is each type found?

3. Discuss the process that is responsible for creating most submarine canyons.

4. How are active continental margins related to plate tectonics?

5. Briefly explain how an accretionary wedge forms. What is meant by *subduction erosion*?

9.6 Features of Deep-Ocean Basins

List and describe the major features associated with deep-ocean basins.

Between the continental margin and the oceanic ridge lies the **deep-ocean basin** (see Figure 9.15). The size of this region—almost 30 percent of Earth's surface—is roughly comparable to the percentage of the surface that presently projects above sea level as land. This region includes *deep-ocean trenches*, which are extremely deep linear depressions in the ocean floor; remarkably flat regions, known as *abyssal plains*; tall volcanic peaks, called *seamounts* and *guyots*; and extensive areas of lava flows piled one atop the other, called *oceanic plateaus*.

Deep-Ocean Trenches

Deep-ocean trenches are long, relatively narrow troughs that are the deepest parts of the ocean. Most trenches are located along the margins of the Pacific Ocean, where many exceed 10,000 meters (33,000 feet) in depth (see Figure 9.14). A portion of one—the Challenger Deep in the Mariana trench—has been measured at 11,022 meters (36,163 feet) below sea level, making it the deepest known part of the world ocean. Only two trenches are located in the Atlantic—the Puerto Rico trench and the South Sandwich trench.

Although deep-ocean trenches represent only a very small portion of the area of the ocean floor, they are nevertheless significant geologic features. Trenches are sites of plate convergence where slabs of oceanic lithosphere subduct and plunge back into the mantle. In addition to earthquakes being created as one plate "scrapes" beneath another, volcanic activity is also associated with these regions. Thus, trenches are often paralleled by an arc-shaped row of active volcanoes called a **volcanic island arc**. Furthermore, **continental volcanic arcs**, such as those making up portions of the Andes and Cascades, are located parallel to trenches that lie adjacent to continental margins (see Figure 9.19). The volcanic activity associated with the trenches that surround the Pacific Ocean explains why the region is called the *Ring of Fire*.

Abyssal Plains

Abyssal plains are deep and incredibly flat features; in fact, these regions are likely the most level places on Earth. The abyssal plain found off the coast of Argentina, for example, has less than 3 meters (10 feet) of relief over a distance exceeding 1300 kilometers (800 miles). The monotonous topography of abyssal plains is occasionally interrupted by the protruding summit of a buried volcanic peak (seamount).

Using *seismic reflection profilers*, instruments whose signals penetrate far below the ocean floor, researchers have determined that abyssal plains consist of thick accumulations of sediment that have buried an otherwise rugged ocean floor (**Figure 9.20**). The nature of the sediment indicates that these plains consist primarily of three types of sediment: (1) fine sediments transported far out to sea by turbidity currents, (2) mineral matter that has precipitated out of seawater, and (3) shells and skeletons of microscopic marine organisms.

Abyssal plains are found in all oceans. However, the floor of the Atlantic has the most extensive abyssal plains because it has few trenches to act as traps for sediment carried down the continental slope.

Seismic reflection profile

Geologist's Sketch

Figure 9.20 Seismic profile This seismic cross section and matching sketch across a portion of the Madeira Abyssal Plain in the eastern Atlantic show the irregular oceanic crust buried by sediments. (Image courtesy of Charles Hollister, Woods Hole Oceanographic Institution)

Volcanic Structures on the Ocean Floor

Dotting the seafloor are numerous volcanic structures of various sizes. Many occur as isolated features that resemble volcanoes on land. Others occur as long, narrow chains that stretch for thousands of kilometers, while still others are massive structures that cover an area the size of Texas.

Seamounts and Volcanic Islands Submarine volcanoes, called **seamounts**, may rise hundreds of meters above the surrounding topography. It is estimated that more than 1 million seamounts exist. Some grow large enough to become oceanic islands, but most do not have a sufficiently long eruptive history to build a structure above sea level. Although seamounts are found on the floors of all the oceans, they are most common in the Pacific.

Some, like the Hawaiian Island–Emperor seamount chain, which stretches from the Hawaiian Islands to the Aleutian trench, form over volcanic hot spots (see Figure 9.14 and Figure 5.26, page 173). Others are born near oceanic ridges.

If a volcano grows large enough before it is carried from its magma source by plate motion, the structure may emerge as a *volcanic island*. Examples of volcanic islands include Easter Island, Tahiti, Bora Bora, Galapagos, and the Canary Islands.

Guyots During their existence, inactive volcanic islands are gradually but inevitably lowered to near sea level by the forces of weathering and erosion. As a moving plate slowly carries inactive volcanic islands away from the elevated oceanic ridge or hot spot over which they formed, they gradually sink and disappear below the water surface. Submerged, flat-topped seamounts that formed in this manner are called **guyots**.*

Oceanic Plateaus The ocean floor contains several massive **oceanic plateaus**, which resemble lava plateaus composed of flood basalts found on the continents. Oceanic plateaus, which can be more than 30 kilometers (20 miles) thick, are generated from vast outpourings of fluid basaltic lavas.

---

*The term *guyot* is named after Arnold Guyot, Princeton University's first geology professor. It is pronounced "GEE-oh" with a hard *g* as in "give."

Some oceanic plateaus appear to have formed quickly in geologic terms. Examples include the Ontong Java Plateau, which formed in less than 3 million years, and the Kerguelen Plateau, which formed in 4.5 million years (see Figure 9.14). Like basalt plateaus on land, oceanic plateaus are thought to form when the bulbous head of a rising mantle plume melts and produces a vast outpouring of basaltic lava.

9.6 CONCEPT CHECKS

1. Explain how deep-ocean trenches are related to plate boundaries.
2. Why are abyssal plains more extensive on the floor of the Atlantic than on the floor of the Pacific?
3. How does a flat-topped seamount, called a *guyot*, form?
4. What features on the ocean floor most resemble basalt plateaus on the continents?

9.7 The Oceanic Ridge

Summarize the basic characteristics of oceanic ridges.

Along well-developed divergent plate boundaries, the seafloor is elevated, forming a broad linear swell called the **oceanic ridge** or **rise**, or **mid-ocean ridge**. Our knowledge of the oceanic ridge system comes from soundings of the ocean floor, core samples from deep-sea drilling, visual inspection using deep-diving submersibles, and firsthand study of slices of ocean floor that have been thrust onto dry land during continental collisions. At oceanic ridges we find extensive faulting, earthquakes, high heat flow, and volcanism.

Anatomy of the Oceanic Ridge

The oceanic ridge system winds through all major oceans in a manner similar to the seam on a baseball and is the longest topographic feature on Earth, exceeding 70,000 kilometers (43,000 miles) in length (**Figure 9.21**). The crest of the ridge typically stands 2 to 3 kilometers above the adjacent deep-ocean basins and marks the divergent plate boundary where new oceanic crust is created.

Notice in Figure 9.21 that large sections of the oceanic ridge system have been named based on their locations in the various ocean basins. Some ridges run through the middle of ocean basins, where they are appropriately called *mid-ocean* ridges. The Mid-Atlantic Ridge and the Mid-Indian Ridge are examples. By contrast, the East Pacific Rise is *not* a "mid-ocean" feature.

Rather, as its name implies, it is located in the eastern Pacific, far from the center of the ocean.

The term *ridge* is somewhat misleading because these features are not narrow and steep, as the term implies, but have widths of from 1000 to 4000 kilometers (about 620 to 2500 miles) and the appearance of broad, elongated swells that exhibit varying degrees of ruggedness. Furthermore, the ridge system is broken into segments that range from a few tens to hundreds of kilometers in length. Each segment is offset from an adjacent segment by a transform fault.

Oceanic ridges are as high as some mountains on the continents, but the similarities end there. Whereas most mountain ranges on land form when the compressional

Figure 9.21 Distribution of the oceanic ridge system The map shows ridge segments that exhibit slow, intermediate, and fast spreading rates.

Figure 9.22 Rift valleys
The axis of some segments of the oceanic ridge system contain deep down-faulted structures called *rift valleys*. Some may exceed 50 kilometers (31 miles) in width and 2000 meters (6600 feet) in depth.

Why Is the Oceanic Ridge Elevated?

The primary reason for the elevated position of the ridge system is that newly created oceanic lithosphere is hot and therefore less dense than cooler rocks of the deep-ocean basin. As the newly formed basaltic crust travels away from the ridge crest, it is cooled from above as seawater circulates through the pore spaces and fractures in the rock. In addition, it cools because it gets farther and farther from the zone of hot mantle upwelling. As a result, the lithosphere gradually cools, contracts, and becomes denser. This thermal contraction accounts for the greater ocean depths that occur away from the ridge. After about 80 million years of cooling and contraction, rock that was once part of an elevated ocean-ridge system becomes part of the deep-ocean basin.

As lithosphere is displaced away from the ridge crest, cooling also causes a gradual increase in lithospheric thickness. This happens because the boundary between the lithosphere and asthenosphere is a thermal (temperature) boundary. Recall that the lithosphere is Earth's cool, stiff outer layer, whereas the asthenosphere is a comparatively hot and weak layer. As material in the uppermost asthenosphere ages (cools), it becomes stiff and rigid. Thus, the upper portion of the asthenosphere is gradually converted to lithosphere simply by cooling. Oceanic lithosphere continues to thicken until it is about 80 to 100 kilometers (50 to 60 miles) thick. Thereafter, its thickness remains relatively unchanged until it is subducted.

forces associated with continental collisions fold and metamorphose thick sequences of sedimentary rocks, oceanic ridges form where upwelling from the mantle generates new oceanic crust. Oceanic ridges consist of layers and piles of newly formed basaltic rocks that are buoyantly uplifted by the hot mantle rocks from which they formed.

Along the axis of some segments of the oceanic ridge system are deep, down-faulted structures that are called **rift valleys** because of their striking similarity to the continental rift valleys found in East Africa (**Figure 9.22**). Some rift valleys, including those along the rugged Mid-Atlantic Ridge, are typically 30 to 50 kilometers (20 to 30 miles) wide and have walls that tower 500 to 2500 meters (1640 to 8200 feet) above the valley floor. This makes them comparable to the deepest and widest part of Arizona's Grand Canyon.

9.7 CONCEPT CHECKS

1. Write a brief summary describing oceanic ridges.
2. Although oceanic ridges can be as tall as some mountains on the continents, list some ways that oceanic ridges are different.
3. Where do rift valleys form along the oceanic ridge system?
4. What is the primary reason for the elevated position of the oceanic ridge system?

9.8 Seafloor Sediments

Distinguish among three categories of seafloor sediment and explain why some of these sediments can be used to study climate change.

Except for steep areas of the continental slope and areas near the crest of the mid-ocean ridge, the ocean floor is covered with sediment. Part of this material has been deposited by turbidity currents, and the rest has slowly settled to the seafloor from above. The thickness of this carpet of debris varies greatly. In some trenches, which act as traps for sediments originating on the continental

margin, accumulations may approach 10 kilometers (6 miles). In general, however, sediment accumulations are considerably less. In the Pacific Ocean, for example, uncompacted sediment measures about 600 meters (2000 feet) or less, whereas on the floor of the Atlantic, the thickness varies from 500 to 1000 meters (1600 to 3300 feet).

Types of Seafloor Sediments

Seafloor sediments can be classified according to their origin into three broad categories: (1) **terrigenous** ("derived from land"), (2) **biogenous** ("derived from organisms"), and (3) **hydrogenous** ("derived from water"). Although each category is discussed separately, remember that all seafloor sediments are mixtures. No body of sediment comes entirely from a single source.

Terrigenous Sediment Terrigenous sediment consists primarily of mineral grains that were weathered from continental rocks and transported to the ocean. Larger particles (sand and gravel) usually settle rapidly near shore, whereas the very smallest particles take years to settle to the ocean floor and may be carried thousands of kilometers by ocean currents. As a result, virtually every area of the ocean receives some terrigenous sediment. The rate at which this sediment accumulates on the deep-ocean floor, though, is very slow. Forming a 1-centimeter (0.4-inch) abyssal clay layer, for example, requires as much as 50,000 years. Conversely, on the continental margins near the mouths of large rivers, terrigenous sediment accumulates rapidly and forms thick deposits.

Biogenous Sediment Biogenous sediment consists of shells and skeletons of marine animals and algae (**Figure 9.23**). This debris is produced mostly by microscopic organisms living in the sunlit waters near the ocean surface. Once these organisms die, their hard *tests* (*testa* = shell) continually "rain" down and accumulate on the seafloor.

The most common biogenous sediment is *calcareous* (CaCO$_3$) *ooze*, which, as the name implies, has the consistency of thick mud. This sediment is produced from the tests of organisms that inhabit warm surface waters. When calcareous hard parts slowly sink through a cool layer of water, they begin to dissolve. This results because

Figure 9.23 Marine microfossils: An example of biogenous sediment These tiny, single-celled organisms are sensitive to even small fluctuations in temperature. Seafloor sediments containing these fossils are useful sources of data on climate change.

the deeper cold seawater is richer in carbon dioxide and is thus more acidic than warm water. In seawater deeper than about 4500 meters (15,000 feet), calcareous tests completely dissolve before they reach bottom. Consequently, calcareous ooze does not accumulate at these greater depths.

Other biogenous sediments include *siliceous* (SiO$_2$) *ooze* and phosphate-rich material. The former is composed primarily of tests of diatoms (single-celled algae) and radiolaria (single-celled animals), whereas the latter is derived from the bones, teeth, and scales of fish and other marine organisms.

Hydrogenous Sediment Hydrogenous sediment consists of minerals that crystallize directly from seawater through various chemical reactions. For example, some limestones are formed when calcium carbonate precipitates directly from the water; however, most limestone is composed of biogenous sediment.

Some of the most common types of hydrogenous sediment include the following:

- *Manganese nodules* are rounded, hard lumps of manganese, iron, and other metals that precipitate in concentric layers around a central object (such as a volcanic pebble or a grain of sand). The nodules can be up to 20 centimeters (8 inches) in diameter and are often littered across large areas of the deep seafloor (**Figure 9.24A**).
- *Calcium carbonates* form by precipitating directly from seawater in warm climates. If this material is buried and hardened, it forms limestone. Most limestone, however, is composed of biogenous sediment.
- *Metal sulfides* are usually precipitated as coatings on rocks near black smokers associated with the crest of a mid-ocean ridge (**Figure 9.24B**). These deposits contain iron, nickel, copper, zinc, silver, and other metals in varying proportions.
- *Evaporites* form where evaporation rates are high and there is restricted open-ocean circulation. As water evaporates from such areas, the remaining seawater becomes saturated with dissolved minerals, which then begin to precipitate. Heavier than seawater, they sink to the bottom or form a characteristic white crust of evaporite minerals around the edges of these areas. Collectively termed *salts*, some evaporite minerals taste salty, such as *halite* (common table salt, NaCl), and some do not, such as the calcium sulfate minerals *anhydrite* (CaSO$_4$) and *gypsum* (CaSO$_4$ • 2H$_2$O).

Seafloor Sediment—A Storehouse of Climate Data

Reliable climate records go back only a couple hundred years, at best. How do scientists learn about climates and climate change prior to that time? The answer is that they must reconstruct past climates from *indirect*

Did You Know?

Communities of organisms reside around hydrothermal vents (black smokers) in dark, hot sulfur-rich environments where photosynthesis cannot occur (see Figure 9.24B). The base of the food web is provided by bacterialike organisms that use a process called *chemosynthesis* and heat from the vents to produce sugars and other foods that allow them and many other organisms to live in this extreme environment.

A.   **B.**

🔺 **SmartFigure 9.24 Examples of hydrogenous sediment A.** Manganese nodules. (Photo by Charles A. Winters/ Photo Researchers, Inc.) **B.** This black smoker is spewing hot, mineral-rich water. When the heated solutions meet cold seawater, metal sulfides precipitate and form mounds of minerals around these hydrothermal vents. (Photo by Fisheries and Oceans Canada/Uvic—Verena Tunnicliffe/Newscom)

evidence; that is, they must analyze phenomena that respond to and reflect changing atmospheric conditions. An interesting and important technique for analyzing Earth's climate history is the study of biogenous seafloor sediments.

We know that the parts of the Earth system are linked so that a change in one part can produce changes in any or all of the other parts. In this example,

you will see how changes in atmospheric and oceanic temperatures are reflected in the nature of life in the sea.

Most seafloor sediments contain the remains of organisms that once lived near the sea surface (the ocean–atmosphere interface). When such near-surface organisms die, their shells slowly settle to the floor of the ocean, where they become part of the sedimentary record. These seafloor sediments are useful recorders of worldwide climate change because the numbers and types of organisms living near the sea surface change with the climate.

Thus, in seeking to understand climate change as well as other environmental transformations, scientists are tapping the huge reservoir of data in seafloor sediments (**Figure 9.25**). The sediment cores gathered by drilling ships such as the one in the chapter-opening photo and other research vessels have provided invaluable data that have significantly expanded our knowledge and understanding of past climates.

One notable example of the importance of seafloor sediments to our understanding of climate change relates to unraveling the fluctuating atmospheric conditions of the Ice Age. The records of temperature changes contained in cores of sediment from the ocean floor have proven critical to our present understanding of this recent span of Earth history.

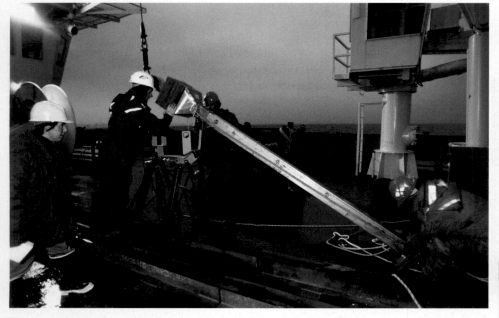

Figure 9.25 Data from the seafloor Scientists aboard the National Science Foundation's research vessel *Nathaniel B. Palmer* drilling for sediment in the South Atlantic. Cores of sediment provide data that allow for a more complete understanding of past climates.

9.8 CONCEPT CHECKS

1. List and describe the three basic types of seafloor sediments. Give at least one example of each.
2. Why are seafloor sediments useful in studying past climates?

CONCEPTS IN REVIEW: OCEANS: THE LAST FRONTIER

9.1 The Vast World Ocean

Discuss the extent and distribution of oceans and continents on Earth. Identify Earth's four main ocean basins.

KEY TERM: oceanography

- Oceanography is an interdisciplinary science that draws on the methods and knowledge of biology, chemistry, physics, and geology to study all aspects of the world ocean.
- Earth's surface is dominated by oceans. Nearly 71 percent of the planet's surface area is oceans and marginal seas. In the Southern Hemisphere, about 81 percent of the surface is water. Of the three major oceans—the Pacific, Atlantic, and Indian—the Pacific Ocean is the largest, contains slightly more than half of the water in the world ocean, and has the greatest average depth—3940 meters (12,927 feet).

9.2 Composition of Seawater

Define *salinity* and list the main elements that contribute to the ocean's salinity. Describe the sources of dissolved substances in seawater and causes of variations in salinity.

KEY TERM: salinity

- Salinity is the proportion of dissolved salts to pure water, usually expressed in parts per thousand (‰). The average salinity in the open ocean ranges from 35‰ to 37‰. The principal elements that contribute to the ocean's salinity are chlorine (55%) and sodium (31%). The primary sources for the elements in sea salt are chemical weathering of rocks on the continents and volcanic outgassing on the ocean floor.
- Variations in seawater salinity are primarily caused by changing the water content. Natural processes that add large amounts of freshwater to seawater and decrease salinity include precipitation, runoff from land, iceberg melting, and sea ice melting. Processes that remove large amounts of freshwater from seawater and increase salinity include the formation of sea ice and evaporation.

9.3 Variations in Temperature and Density with Depth

Discuss temperature, salinity, and density changes with depth in the open ocean.

KEY TERMS: thermocline, density, pycnocline

- The ocean's surface temperature is related to the amount of solar energy received and varies as a function of latitude. Low-latitude regions have relatively warm surface water and distinctly colder water at depth, creating a thermocline, which is a layer of rapid temperature change. No thermocline exists in high-latitude regions because there is little temperature difference between the top and bottom of the water column. The water column is isothermal.
- Seawater density is mostly affected by water temperature but also by salinity. Cold, high-salinity water is densest. Low-latitude regions have distinctly denser (colder) water at depth than at the surface, creating a pycnocline, which is a layer of rapidly changing density. No pycnocline exists in high-latitude regions because the water column is isopycnal.
- Most open-ocean regions exhibit a three-layered structure based on water density. The shallow surface mixed zone has warm and nearly uniform temperatures. The transition zone includes a prominent thermocline and associated pycnocline. The deep zone is continually dark and cold and accounts for 80 percent of the water in the ocean. In high latitudes, the three-layered structure does not exist.

9.4 An Emerging Picture of the Ocean Floor

Define *bathymetry* and summarize the various techniques used to map the ocean floor.

KEY TERMS: bathymetry, sonar, echo sounder

- Seafloor mapping is done with sonar—shipboard instruments that emit pulses of sound that "echo" off the bottom. Satellites are also used to map the ocean floor. Their instruments measure slight variations in sea level that result from differences in the gravitational pull of features on the seafloor. Accurate maps of seafloor topography can be made using these data.
- Mapping efforts have revealed three major areas of the ocean floor: continental margins, deep-ocean basins, and oceanic ridges.

? **Do the math: How many seconds would it take an echo sounder's ping to make the trip from a ship to the Challenger Deep (10,994 meters [36,069 feet]) and back? Recall that depth = ½ (1500 m/sec × Echo travel time).**

9.5 Continental Margins

Compare a passive continental margin with an active continental margin and list the major features of each.

KEY TERMS: continental margin, passive continental margin, continental shelf, continental slope, continental rise, deep-sea fan, submarine canyon, turbidity current, active continental margin, accretionary wedge, subduction erosion

- Continental margins are transition zones between continental and oceanic crust. Active continental margins occur where a plate boundary and the edge of a continent coincide, usually on the leading edge of a plate. Passive continental margins are on the trailing edges of continents, far from plate boundaries.
- Heading offshore from the shoreline of a passive margin, a submarine traveler would first encounter the gently sloping continental shelf and then the steeper continental slope, which marks the end of the continental crust and the beginning of the oceanic crust. Beyond the continental slope is another gently sloping section, the continental rise: It is made of sediment transported by turbidity currents through submarine canyons and piled up in deep-sea fans atop the oceanic crust.
- Submarine canyons are deep, steep-sided valleys that originate on the continental slope and may extend to the deep-ocean basin. Many submarine canyons have been excavated by turbidity currents (downslope movements of dense, sediment-laden water).
- At an active continental margin, material may be added to the leading edge of a continent in the form of an accretionary wedge (common at shallow-angle subduction zones), or material may be scraped off the edge of a continent by subduction erosion (common at steeply dipping subduction zones).

? **The image shows a perspective view of the continental margin looking southwest toward the Palos Verdes Peninsula near Los Angeles. Match the features labeled 1 through 4 with the following terms: continental shelf, continental slope, continental rise, and submarine canyon. Based on the features in this image, what type of continental margin is shown here?**

9.6 Features of Deep-Ocean Basins

List and describe the major features associated with deep-ocean basins.

KEY TERMS: deep-ocean basin, deep-ocean trench, volcanic island arc, continental volcanic arc, abyssal plain, seamount, guyot, oceanic plateau

- The deep-ocean basin makes up about half of the ocean floor's area. Much of it is abyssal plain (deep, featureless sediment-draped crust). Subduction zones and deep-ocean trenches also occur in deep-ocean basins. Paralleling trenches are volcanic island arcs (if the subduction goes underneath oceanic lithosphere) or continental volcanic arcs (if the overriding plate has continental lithosphere on its leading edge).
- There are a variety of volcanic structures on the deep-ocean floor. Seamounts are submarine volcanoes; if they pierce the surface of the ocean, we call them volcanic islands. Guyots are old volcanic islands that have had their tops eroded off before sinking below sea level. Oceanic plateaus are unusually thick sections of oceanic crust formed by massive underwater eruptions of lava.

? **On the cross-sectional view of a deep-ocean basin, label the following features: a seamount, a guyot, a volcanic island, an oceanic plateau, and an abyssal plain.**

9.7 The Oceanic Ridge

Summarize the basic characteristics of oceanic ridges.

KEY TERMS: oceanic ridge or rise (mid-ocean ridge), rift valley

- The oceanic ridge system is the longest topographic feature on Earth, wrapping around the world through all major ocean basins. It is a few kilometers tall, a few thousand kilometers wide, and a few tens of thousands of kilometers long. The summit is the place where new oceanic crust is generated, often marked by a rift valley.
- Oceanic ridges are elevated features because they are warm and therefore less dense than older, colder oceanic lithosphere. As oceanic crust moves away from the ridge crest, heat loss causes the oceanic crust to become denser and subside. After 80 million years, crust that was once part of an oceanic ridge is in the deep ocean basin, far from the ridge.

9.8 Seafloor Sediments

Distinguish among three categories of seafloor sediment and explain why some of these sediments can be used to study climate change.

KEY TERMS: terrigenous sediment, biogenous sediment, hydrogenous sediment

- There are three broad categories of seafloor sediments. Terrigenous sediment consists primarily of mineral grains that were weathered from continental rocks and transported to the ocean; biogenous sediment consists of shells and skeletons of marine animals and plants; and hydrogenous sediment includes minerals that crystallize directly from seawater through various chemical reactions.
- Seafloor sediments are helpful in studying worldwide climate change because they often contain the remains of organisms that once lived near the sea surface. The numbers and types of these organisms change as the climate changes, and their remains in seafloor sediments record these changes.

GIVE IT SOME THOUGHT

1. Refer to Figure 9.2 to answer the following questions:
   a. Water dominates Earth's surface, but not everywhere. In what Northern Hemisphere latitude belt is there more land than water?
   b. In what latitude belt is there no land at all?

2. Assuming that the average speed of sound waves in water is 1500 meters per second, determine the water depth if a signal sent out by an echo sounder on a research vessel requires 6 seconds to strike bottom and return to the recorder aboard the ship.

3. The accompanying photo is of sea ice. How do seasonal changes in the amount of sea ice influence the salinity of the remaining surface water? Is water density greater before or after sea ice forms? Explain.

Michael Collier

4. Assume that someone brings several water samples to your laboratory. His problem is that the labels are incomplete. He knows that samples A and B are from the Atlantic Ocean; he also knows that one came from near the equator, and the other came from near the Tropic of Cancer. But he does not know which one is which. He has a similar problem with samples C and D. One is from the Red Sea, and the other is from the Baltic Sea. Applying your knowledge of ocean salinity, how would you identify the location of each sample? How were you able to figure this out?

5. The accompanying graph depicts variations in ocean-water density and temperature with depth for a location near the equator. Which line represents temperature, and which line represents density? Explain.

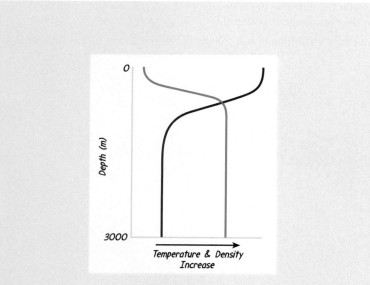

6. Refer to the accompanying map and diagram showing the eastern seaboard of the United States to complete the following:

a. Which letter is associated with each of the following: continental shelf, continental slope, shelf break?
b. How does the size of the continental shelf surrounding Florida compare to the size of the Florida peninsula?
c. Why are there no deep-ocean trenches on this map?

7. Are the continental margins surrounding the Atlantic Ocean primarily active or passive? How about the margins surrounding the Pacific Ocean? Based on your response to the previous questions, indicate whether each ocean basin is getting larger, shrinking, or staying the same size. Explain your answer.

8. Examine the accompanying sketch, which shows three sediment layers on the ocean floor. What term is applied to such layers? What was your clue? What process was responsible for creating these layers? Are these layers more likely part of a deep-sea fan or an accretionary wedge?

9. Imagine that while you and a passenger are in a deep-diving submersible in the North Pacific near Alaska's Aleutian Islands, you encounter a long, narrow depression on the ocean floor. Your passenger asks whether you think it is a submarine canyon, a rift valley, or a deep-ocean trench. How would you respond? Explain your response.

MasteringGeology™

www.masteringgeology.com

Looking for additional review and test prep materials? Visit the Study Area in MasteringGeology to enhance your understanding of this chapter's content by accessing a variety of resources, including **Self-Study Quizzes,** **Geoscience Animations, GEODe Tutorials, RSS feeds, flashcards, web links, self-study quizzes,** and an optional **Pearson eText.**

The Restless Ocean

Each statement represents the primary learning objective for the corresponding major heading within the chapter. After you complete the chapter, you should be able to:

10.1 Discuss the factors that create and influence ocean currents and describe the affect ocean currents have on climate.

10.2 Explain the processes that produce coastal upwelling and the ocean's deep circulation.

10.3 Explain why the shoreline is considered a dynamic interface.

10.4 List and discuss the factors that influence the height, length, and period of a wave and describe the motion of water within a wave.

10.5 Describe how waves erode and move sediment along the shore.

10.6 Describe the features typically created by wave erosion and those resulting from sediment deposited by longshore transport processes.

10.7 Summarize the ways in which people deal with shoreline erosion problems.

10.8 Contrast the erosion problems faced along different parts of America's coasts. Distinguish between emergent and submergent coasts.

10.9 Explain the cause of tides, their monthly cycles, and patterns. Describe the horizontal flow of water that accompanies the rise and fall of tides.

North Carolina's Outer Banks. Oregon Inlet Bridge connects Bodie Island (foreground) and Hatteras Island. These narrow wisps of sand are part of an extensive barrier island system. The beach and dunes on the left face the Atlantic Ocean. On the right are the quieter waters of Pamlico Sound. (Photo by Michael Collier)

The restless waters of the ocean are constantly in motion, powered by many different forces. Winds, for example, generate surface currents, which influence coastal climate and provide nutrients that affect the abundance of algae and other marine life in surface waters. Winds also produce waves that carry energy from storms to distant shores, where their impact erodes the land. In some regions, density differences create deep-ocean circulation, which is important for ocean mixing and nutrient recycling. In addition, the Moon and the Sun produce tides, which periodically raise and lower sea level. This chapter will examine these movements of ocean waters and their effects on coastal regions.

10.1 The Ocean's Surface Circulation

Discuss the factors that create and influence ocean currents and describe the affect ocean currents have on climate.

You may have heard of the Gulf Stream, an important surface current in the Atlantic Ocean that flows northward along the East Coast of the United States (**Figure 10.1**). Surface currents like this one are set in motion by the wind. At the water surface, where the atmosphere and ocean meet, energy is passed from moving air to the water through friction. The drag exerted by winds blowing steadily across the ocean causes the surface layer of water to move. Thus, major horizontal movements of surface waters are closely related to the global pattern of prevailing winds.° As an example, the small map in **Figure 10.2** shows how the wind belts known as the *trade winds* and the *westerlies* create large, circular-moving loops of water in the Atlantic Ocean. The same wind belts influence the other oceans as well, so that a similar pattern of currents can also be seen in the Pacific and Indian Oceans. Essentially, the pattern of surface-ocean circulation closely matches the pattern of global winds but is also strongly influenced by the distribution of major landmasses and by the spinning of Earth on its axis.

The Pattern of Ocean Currents

Huge circular-moving current systems dominate the surfaces of the oceans. These large whirls of water within an ocean basin are called **gyres** (*gyros* = circle). The large map in Figure 10.2 shows the world's five main gyres: the *North Pacific Gyre*, the *South Pacific Gyre*, the *North Atlantic Gyre*, the *South Atlantic Gyre*, and the *Indian Ocean Gyre* (which exists mostly within the Southern Hemisphere). The center of each gyre coincides with the subtropics at about 30° north or south latitude, so they are often called *subtropical gyres*.

Coriolis Effect As shown in Figure 10.2, subtropical gyres rotate clockwise in the Northern Hemisphere and counterclockwise in the Southern Hemisphere. Why do the gyres flow in different directions in the two hemispheres? Although wind is the force that generates surface currents, other factors also influence the movement of ocean waters. The most significant of these is the

Figure 10.1 The Gulf Stream In this satellite image off the east coast of the United States, orange and yellow represent higher water temperatures, and blue indicates cooler water temperatures. The current transports heat from the subtropics far into the North Atlantic. (Courtesy of the NOAA.)

°Details about the global pattern of winds appear in Chapter 13, The Atmosphere in Motion.

SmartFigure 10.2
Major surface-ocean currents The ocean's surface circulation is organized into five major gyres. Poleward-moving currents are warm, and equatorward-moving currents are cold. Ocean currents play an important role in redistributing heat around the globe. Note that cities mentioned in the text discussion are shown on this map. In the smaller inset map, broad arrows show the idealized surface circulation for the Atlantic, and the thin arrows show prevailing winds. Winds provide the energy that drives the ocean's surface circulation.

Coriolis effect. Because of Earth's rotation, currents are deflected to the *right* in the Northern Hemisphere and to the *left* in the Southern Hemisphere. You can see the influence of the Coriolis effect by comparing the wind arrows and the arrows representing the ocean currents on the small map in Figure 10.2. (The Coriolis effect is more fully explained in Chapter 13.) As a result, gyres flow in opposite directions in the two different hemispheres.

North Pacific Currents Four main currents generally exist within each gyre (see Figure 10.2). The North Pacific Gyre, for example, consists of the North Equatorial Current, the Kuroshio Current, the North Pacific Current, and the California Current. The tracking of floating objects that are released into the ocean intentionally or accidentally reveals that it takes about 6 years for the objects to go all the way around the loop.

North Atlantic Currents Like the North Pacific, the North Atlantic Ocean has four main currents (see Figure 10.2). Beginning near the equator, the North Equatorial Current is deflected northward through the Caribbean, where it becomes the Gulf Stream. As the Gulf

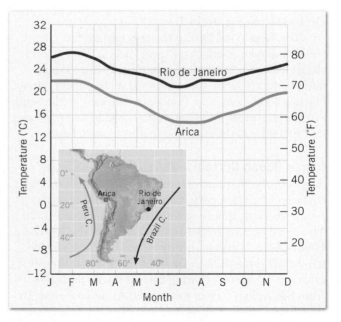

Figure 10.3 The chilling effect of a cold current Monthly mean temperatures for Rio de Janeiro, Brazil, and Arica, Chile, both of which are coastal cities near sea level. Even though Arica is closer to the equator, its temperatures are cooler than Rio de Janeiro's. Arica is influenced by the cold Peru Current, whereas Rio de Janeiro is adjacent to the warm Brazil Current.

Stream moves along the east coast of the United States, it is strengthened by the prevailing westerly winds and is deflected to the east (to the right) offshore of North Carolina into the North Atlantic. As it continues northeastward, it gradually widens and slows until it becomes a vast, slowly moving current known as the North Atlantic Current, which because of its sluggish nature is also known as the North Atlantic Drift.

As the North Atlantic Current approaches Western Europe, it splits, part of it moving northward past Great Britain, Norway, and Iceland, carrying heat to these otherwise chilly areas. The other part is deflected southward as the cool Canary Current. As the Canary Current moves southward, it eventually merges into the North Equatorial Current, completing the gyre. Because the North Atlantic Ocean basin is about half the size of the North Pacific basin, it takes floating objects about 3 years to go completely around this gyre.

The circular motion of gyres leaves a large central area that has no well-defined currents. In the North Atlantic, this zone of calmer waters is known as the Sargasso Sea, named for the large quantities of *Sargassum*, a type of floating seaweed encountered there.

Southern Hemisphere Currents

The ocean basins in the Southern Hemisphere exhibit a similar pattern of flow as in the Northern Hemisphere basins, with surface currents influenced by wind belts, the position of continents, and the Coriolis effect. In the South Atlantic and South Pacific, for example, surface-ocean circulation is very much the same as in their Northern Hemisphere counterparts, except that the direction of flow is counterclockwise (see Figure 10.2).

The West Wind Drift is the only current that completely encircles Earth (see Figure 10.2). It flows around the ice-covered continent of Antarctica, where no large landmasses are in the way, so its cold surface waters circulate in a continuous loop. It moves in response to the Southern Hemisphere's prevailing westerly winds, and portions of it split off into the adjoining southern ocean basins.

Indian Ocean Currents

The Indian Ocean exists mostly in the Southern Hemisphere, and it follows a surface circulation pattern similar to those in other Southern Hemisphere ocean basins (see Figure 10.2). The small portion of the Indian Ocean in the Northern Hemisphere, however, is influenced by the seasonal wind shifts known as the summer and winter *monsoons* (*mausim* = season). During the summer, winds blow from the Indian Ocean toward the Asian landmass. In the winter, the winds reverse and blow out from Asia over the Indian Ocean. You can see this reversal when you compare the January and July winds in Figure 13.18. When the winds change direction, the surface currents also reverse direction.

Ocean Currents Influence Climate

Surface-ocean currents have an important effect on climate. It is known that for Earth as a whole, the gains in solar energy equal the losses to space of heat radiated from the surface. When most latitudes are considered individually, however, this is not the case. There is a net gain of energy in lower latitudes and a net loss at higher latitudes. Because the tropics are not becoming progressively warmer, nor are the polar regions becoming colder, there must be a large-scale transfer of heat from areas of excess to areas of deficit. This is indeed the case. *The transfer of heat by winds and ocean currents equalizes these latitudinal energy imbalances.* Ocean water movements account for about one-quarter of this total heat transport, and winds account for the remaining three-quarters.

The Effect of Warm Currents The moderating effect of poleward-moving warm ocean currents is well known. The North Atlantic Drift, an extension of the warm Gulf Stream, keeps wintertime temperatures in Great Britain and much of Western Europe warmer than would be expected for their latitudes. London is farther north than St. John's, Newfoundland, yet is not nearly so frigid in winter. (Cities mentioned in this section are shown in Figure 10.2). Because of the prevailing westerly winds, the moderating effects are carried far inland. For example, Berlin (52° north latitude) has a mean January temperature similar to that experienced at New York City, which lies 12° latitude farther south. The January mean at London (51° north latitude) is 4.5°C (8.1°F) higher than at New York City.

Cold Currents Chill the Air In contrast to warm ocean currents like the Gulf Stream, the effects of which are felt most during the winter, cold currents exert their greatest influence in the tropics or during the summer months in the middle latitudes. For example, the cool Benguela Current off the western coast of southern Africa moderates the tropical heat along this coast. Walvis Bay (23° south latitude), a town adjacent to the Benguela Current, is 5°C (9°F) cooler in summer than Durban, which is 6° latitude farther poleward but on the eastern side of South Africa, away from the influence of the cold current. The east and west coasts of South America provide another example. **Figure 10.3** shows monthly mean temperatures for Rio de Janeiro, Brazil, which is influenced by the warm Brazil Current, and Arica, Chile, which is adjacent to the cold Peru Current. Closer to home, because of the cold California Current, summer temperatures in subtropical coastal southern California are lower by 6°C (10.8°F) or more compared to East Coast stations.

Cold Currents Increase Aridity In addition to influencing temperatures of adjacent land areas, cold currents have other climatic influences. For example, where tropical deserts exist along the west coasts of continents, cold ocean currents have a dramatic impact. The principal west coast deserts are the Atacama in Peru and Chile and the Namib in southwestern Africa (**Figure 10.4**). The aridity along these coasts is intensified because the lower atmosphere is chilled by cold offshore waters. When this occurs, the air becomes very stable and resists the upward movement necessary to create precipitation-producing clouds. In addition, the presence of cold currents causes temperatures to approach and often reach the dew point, the temperature at which water vapor condenses. As a result, these areas are characterized by high relative humidities and much fog. Thus, not all tropical deserts are hot with low humidities and clear skies. Rather, the presence of cold currents transforms some tropical deserts into relatively cool, damp places that are often shrouded in fog.

10.1 CONCEPT CHECKS

1. What is the primary driving force of surface-ocean currents?
2. How does the Coriolis effect influence ocean currents?
3. Name the five subtropical gyres and identify the main surface currents in each.
4. How do ocean currents influence climate? Provide at least three examples.

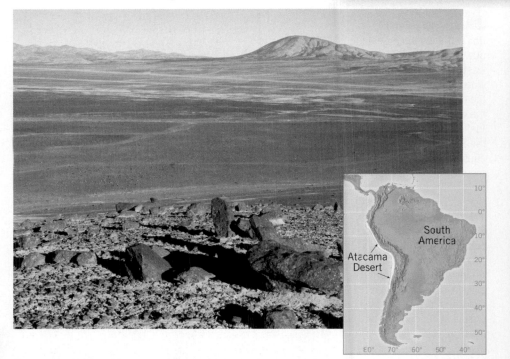

Figure 10.4 Chile's Atacama Desert This is the driest desert on Earth. Average rainfall at the wettest locations is not more than 3 millimeters (0.12 inch) per year. Stretching nearly 1000 kilometers (600 miles), the Atacama is situated between the Pacific Ocean and the towering Andes Mountains. The cold Peru Current makes this slender zone cooler and drier than it would otherwise be. (Photo by Jaques Jangoux/Photo Researchers, Inc.)

10.2 Upwelling and Deep-Ocean Circulation

Explain the processes that produce coastal upwelling and the ocean's deep circulation.

The preceding discussion focused mainly on the horizontal movements of the ocean's surface waters. In this section, you will learn that the ocean also exhibits significant vertical movements and a slow-moving, multi-layered deep-ocean circulation. Some vertical movements are related to wind-driven surface currents, whereas deep-ocean circulation is strongly influenced by density differences.

Coastal Upwelling

In addition to producing surface currents, winds can also cause *vertical* water movements. **Upwelling**, the rising of cold water from deeper layers to replace warmer surface water, is a common wind-induced vertical movement. One type of upwelling, called *coastal upwelling*, is most characteristic along the west coasts of continents, most notably along California, western South America, and West Africa.

Coastal upwelling occurs in these areas when winds blow toward the equator and parallel to the coast (**Figure 10.5**). Coastal winds combined with the Coriolis effect cause surface water to move away from shore. As the surface layer moves away from the coast, it is replaced by water that "upwells" from below the surface. This slow upward movement of water from depths of 50 to 300 meters (165 to 1000 feet) brings water that is cooler than the original surface water and results in lower surface-water temperatures near the shore.

For swimmers who are accustomed to the warm waters along the mid-Atlantic shore of the United States, a swim in the Pacific off the coast of central California can be a chilling surprise. In August, when temperatures

Did You Know?

About half of the world's human population lives within 100 km (60 mi) of a coast. The proportion of the U.S. population residing within 75 km (45 mi) of a coast in 2010 is in excess of 50 percent. The concentration of such large numbers of people so near the shore means that hurricanes and tsunamis place millions at risk.

SmartFigure 10.5
Coastal upwelling Coastal upwelling occurs along the west coasts of continents, where winds blow toward the equator and parallel to the coast. The Coriolis effect (deflection to the left in the Southern Hemisphere) causes surface water to move away from the shore, which brings cold, nutrient-rich water to the surface. This satellite image shows chlorophyll concentration along the southwest coast of Africa (February 21, 2001). An instrument aboard the satellite detects changes in seawater color caused by changing concentrations of chlorophyll. High chlorophyll concentrations indicate high amounts of photosynthesis, which is linked to the upwelling nutrients. Red indicates high concentrations, and blue indicates low concentrations.
(Provided by the SeaWiFS Project, NASA/Goddard Space Flight Center and ORBIMAGE)

in the Atlantic are 21°C (70°F) or higher, central California's surf is only about 15°C (60°F).

Upwelling brings greater concentrations of dissolved nutrients, such as nitrates and phosphates, to the ocean surface. These nutrient-enriched waters from below promote the growth of microscopic plankton, which in turn support extensive populations of fish and other marine organisms. Figure 10.5 includes a satellite image that shows high productivity due to coastal upwelling off the southwest coast of Africa.

Deep-Ocean Circulation

Deep-ocean circulation has a significant vertical component and accounts for the thorough mixing of deep-water masses. This component of ocean circulation is a response to density differences among water masses that cause denser water to sink and slowly spread out beneath the surface. Because the density variations that cause deep-ocean circulation are caused by differences in temperature and salinity, deep-ocean circulation is also referred to as **thermohaline** (*thermo* = heat, *haline* = salt) **circulation**.

An increase in seawater density can be caused by either a decrease in temperature or an increase in salinity. Density changes due to salinity variations are important in very high latitudes, where water temperature remains low and relatively constant.

Most water involved in deep-ocean currents (thermohaline circulation) begins in high latitudes at the surface. In these regions, where surface waters are cold, salinity increases when sea ice forms (**Figure 10.6**). When seawater freezes to form sea ice, salts do not become

Figure 10.6 Sea ice near Antarctica When seawater freezes, sea salts do not become part of the ice. Consequently, the surface salinity of the remaining seawater increases, which makes it denser and prone to sink. (Photo by John Higdon/agefotostock)

Figure 10.7 The ocean conveyor belt Source areas for dense water masses exist in high-latitude regions where cold, high-salinity water sinks and flows into all the oceans. This water eventually ascends and completes the conveyor by returning to the source areas as warm surface currents.

part of the ice. As a result, the salinity (and therefore the density) of the remaining seawater increases. When this surface water becomes dense enough, it sinks, initiating deep-ocean currents. Once this water sinks, it is removed from the physical processes that increased its density in the first place, and so its temperature and salinity remain largely unchanged for the duration of the time it spends in the deep ocean.

Near Antarctica, surface conditions create the highest-density water in the world. This cold saline brine slowly sinks to the seafloor, where it moves throughout the ocean basins in sluggish currents. After sinking from the surface of the ocean, deep waters will not reappear at the surface for an average of 500 to 2000 years.

A simplified model of ocean circulation is similar to a conveyor belt that travels from the Atlantic Ocean through the Indian and Pacific Oceans and back again (**Figure 10.7**). In this model, warm water in the ocean's upper layers flows poleward, converts to dense water, and returns toward the equator as cold deep water that eventually upwells to complete the circuit. As this "conveyor belt" moves around the globe, it influences global climate by converting warm water to cold and liberating heat to the atmosphere.

10.2 CONCEPT CHECKS

1. Describe the process of coastal upwelling. Why is an abundance of marine life associated with these areas?
2. Why is deep-ocean circulation referred to as *thermohaline circulation*?
3. Beginning in the far North Atlantic, describe or make a simple sketch of the ocean's conveyor-belt circulation.

10.3 The Shoreline: A Dynamic Interface

Explain why the shoreline is considered a dynamic interface.

Shorelines are dynamic environments. Their topography, geologic makeup, and climate vary greatly from place to place. Continental and oceanic processes converge along coasts to create landscapes that frequently undergo rapid change. When it comes to the deposition of sediment, they are transition zones between marine and continental environments.

Nowhere is the restless nature of the ocean's water more noticeable than along the shore—the dynamic interface among air, land, and sea. An **interface** is a

Figure 10.8 Hurricane Sandy A portion of the New Jersey shoreline shortly after this huge storm struck in late October 2012. The extraordinary storm surge caused much of the damage pictured here. Many shoreline areas are intensively developed. Often the shifting shoreline sands and the desire of people to occupy these areas are in conflict. There is more about hurricanes and hurricane damage in Chapter 14. (AP Photo by Mike Groll)

common boundary where different parts of a system interact. This is certainly an appropriate designation for the coastal zone, where we can see the rhythmic rise and fall of tides and observe waves rolling in and breaking. Sometimes, the waves are low and gentle. At other times, they pound the shore with awesome fury.

Although it may not be obvious, the shoreline is constantly being modified by waves. Crashing surf erodes the land. Wave activity also moves sediment toward and away from the shore, as well as along it. Such activity sometimes produces narrow sandbars and fragile offshore islands that frequently change size and shape as storm waves come and go.

The nature of present-day shorelines is not just the result of the relentless attack of the land by the sea. Indeed, the shore has a complex character that results from multiple geologic processes. For example, practically all coastal areas were affected by the worldwide rise in sea level that accompanied the melting of glaciers following the Last Glacial Maximum (see Figure 4.20). As the sea encroached landward, the shoreline retreated, becoming superimposed upon existing landscapes that had resulted from such diverse processes as stream erosion, glaciation, volcanic activity, and the forces of mountain building.

Today, the coastal zone is experiencing intensive human activity. Unfortunately, people often treat the shoreline as if it were a stable platform on which structures can safely be built. This attitude inevitably leads to conflicts between people and nature. In October 2012, this fact was tragically reinforced when the storm surge from Hurricane Sandy struck parts of New York City and the narrow barrier islands along the coast of New Jersey (**Figure 10.8**). Many coastal landforms, especially beaches and barrier islands, are relatively fragile, short-lived features that are frequently inappropriate sites for development.

10.3 CONCEPT CHECKS

1. Define *interface*.
2. Aside from ocean waves, what other factors influence the nature of present-day shorelines?

10.4 Ocean Waves

List and discuss the factors that influence the height, length, and period of a wave and describe the motion of water within a wave.

Ocean waves are energy traveling along the interface between ocean and atmosphere, often transferring energy from a storm far out at sea over distances of several thousand kilometers. That's why even on calm days, the ocean still has waves that travel across its surface. When observing waves, always remember that you are watching *energy* travel through a medium (water). If you make waves by tossing a pebble into a pond, or by splashing in a pool, or by blowing across the surface of a cup of coffee, you are imparting *energy* to the water, and the waves you see are the visible evidence of the energy passing through.

Wind-generated waves provide most of the energy that shapes and modifies shorelines. Where the land and sea meet, waves that may have traveled unimpeded for hundreds or thousands of kilometers suddenly encounter a barrier that will not allow them to advance farther and must absorb their energy. Stated another way, the shore is the location where a practically irresistible force confronts an almost immovable object. The conflict that results is never ending and sometimes dramatic.

Wave Characteristics

Most ocean waves derive their energy and motion from the wind. When a breeze is less than 3 kilometers (2 miles) per hour, only small wavelets appear. At greater wind speeds, more stable waves gradually form and advance with the wind.

Characteristics of ocean waves are illustrated in **Figure 10.9**, which shows a simple, nonbreaking waveform. The tops of the waves are the *crests*, which are separated by *troughs*. Halfway between the crests and troughs is the *still water level*, which is the level that the water would occupy if there were no waves. The vertical distance between trough and crest is called the **wave height**, and the horizontal distance between successive crests (or troughs) is the **wavelength**. The time it takes one full wave—one wavelength—to pass a fixed position is the **wave period**.

The height, length, and period that are eventually achieved by a wave depend on three factors: (1) wind speed, (2) length of time the wind has blown, and (3) **fetch**, the distance that wind has traveled across open water. As the quantity of energy transferred from the wind to the water increases, both the height and steepness of the waves increase. Eventually, a critical point is reached where waves grow so tall that they topple over, forming ocean breakers called *whitecaps*.

For a particular wind speed, there is a maximum fetch and duration of wind beyond which waves will

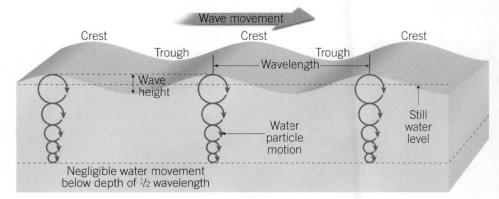

Figure 10.9 Wave basics An idealized nonbreaking wave showing its basic parts and the movement of water with increasing depth.

no longer increase in size. When the maximum fetch and duration are reached for a given wind velocity, the waves are said to be "fully developed." The reason that waves can grow no further is that they are losing as much energy through the breaking of whitecaps as they are receiving from the wind.

When the wind stops or changes direction, or the waves leave the storm area where they were created, they continue on without relation to local winds. The waves also undergo a gradual change to *swells*, which are lower in height and longer in length and may carry a storm's energy to distant shores. Because many independent wave systems exist at the same time, the sea surface acquires a complex and irregular pattern, sometimes producing very large waves. The sea waves that are seen from shore are usually a mixture of swells from faraway storms and waves created by local winds.

Circular Orbital Motion

Waves can travel great distances across ocean basins. In one study, waves generated near Antarctica were tracked as they traveled through the Pacific Ocean basin. After more than 10,000 kilometers (more than 6000 miles), the waves finally expended their energy a week later, along the shoreline of the Aleutian Islands of Alaska. The water itself does not travel the entire distance, but the waveform does. As the wave travels, the water passes the energy along by moving in a circle. This movement is called *circular orbital motion*.

Observation of an object floating in waves reveals that it moves not only up and down but also slightly forward and backward with each successive wave. **Figure 10.10** shows that a floating object moves up and backward as the crest approaches, up and forward as the crest

Wave movement

Toy boat

SmartFigure 10.10
Passage of a wave The movements of the toy boat show that the waveform advances, but the water does not advance appreciably from the original position. In this sequence, the wave moves from left to right as the boat (and the water in which it is floating) rotates in an imaginary circle.

passes, down and forward after the crest, down and backward as the trough approaches, and then it rises and moves backward again as the next crest advances. When the movement of the toy boat shown in Figure 10.10 is traced as a wave passes, it can be seen that the boat moves in a circle, and it returns to essentially the same place. Circular orbital motion allows a waveform (the wave's shape) to move forward *through the water*, while the individual water particles that transmit the wave move around in a circle. Wind moving across a field of wheat causes a similar phenomenon: The wheat itself does not travel across the field, but the waves do.

The energy contributed by the wind to the water is transmitted not only along the surface of the sea but also downward. However, beneath the surface, the circular motion rapidly diminishes until, at a depth equal to one-half the wavelength measured from the still water level, the movement of water particles becomes negligible. This depth is known as the *wave base*. The dramatic decrease of wave energy with depth is shown by the rapidly diminishing diameters of water-particle orbits in Figure 10.9.

Waves in the Surf Zone

As long as a wave is in deep water, it is unaffected by water depth (**Figure 10.11**, *left*). However, when a wave approaches the shore, the water becomes shallower and influences wave behavior. The wave begins to "feel bottom" at a water depth equal to its wave base. Such depths interfere with water movement at the base of the wave and slow its advance (Figure 10.11, *center*).

As a wave advances toward the shore, the slightly faster waves farther out to sea catch up, decreasing the wavelength. As the speed and length of the wave diminish, the wave steadily grows higher. Finally, a critical point is reached when the wave is too steep to support itself and the wave front collapses, or *breaks* (Figure 10.11, *right*), causing water to advance up the shore.

The turbulent water created by breaking waves is called **surf**. On the landward margin of the surf zone, the turbulent sheet of water from collapsing breakers, called *swash*, moves up the slope of the beach. When the energy of the swash has been expended, the water flows back down the beach toward the surf zone as *backwash*.

10.4 CONCEPT CHECKS

1. List three factors that determine the height, length, and period of a wave.
2. Describe the motion of a floating object as a wave passes.
3. How do a wave's speed, wavelength, and height change as it moves into shallow water and breaks?

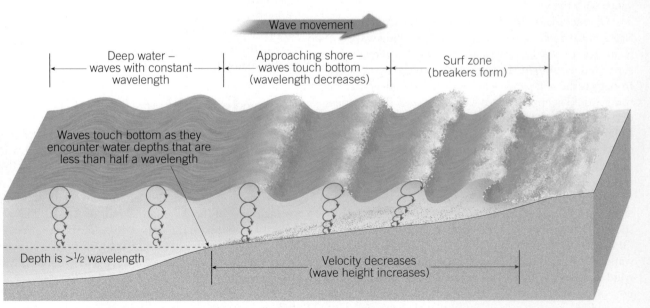

Wave movement

Deep water – waves with constant wavelength

Approaching shore – waves touch bottom (wavelength decreases)

Surf zone (breakers form)

Waves touch bottom as they encounter water depths that are less than half a wavelength

Depth is >½ wavelength

Velocity decreases (wave height increases)

Figure 10.11 Waves approaching the shore Waves touch bottom as they encounter water depths that are less than half a wavelength. As a result, the wave speed decreases, and the faster-moving waves farther from shore begin to catch up, which causes the distance between waves (the wavelength) to decrease. This causes an increase in wave height to the point where the waves finally pitch forward and break in the surf zone.

10.5 Beaches and Shoreline Processes

Describe how waves erode and move sediment along the shore.

For many, a beach is a sandy area where people lie in the sun and walk along the water's edge. Technically, a **beach** is an accumulation of sediment found along the landward margins of an ocean or a lake. Along straight coasts, beaches may extend for tens or hundreds of kilometers. Where coasts are irregular, beach formation may be confined to the relatively quiet waters of bays.

Beaches are composed of whatever material is locally abundant. The sediment for some beaches is derived from the erosion of adjacent cliffs or nearby coastal mountains. Other beaches are built from sediment delivered to the coast by rivers. Although the mineral makeup of many beaches is dominated by durable quartz grains, other minerals may be dominant. For example, in areas such as southern Florida, where there are no mountains or other sources of rock-forming minerals nearby, most beaches are composed largely of shell fragments and the remains of

organisms that live in coastal waters (**Figure 10.12A**). Some beaches on volcanic islands in the open ocean are composed of weathered grains of the basaltic lava that comprise the islands or of coarse debris eroded from coral reefs that develop around many tropical islands (**Figure 10.12B**).

Regardless of the composition, the material that comprises a beach does not stay in one place. Instead, the waves that crash along the shoreline are constantly moving it. Thus, beaches can be thought of as material in transit along the shoreline.

Wave Erosion

During calm weather, wave action is minimal. During storms, however, waves can cause much erosion. The impact of large, high-energy waves against the shore can

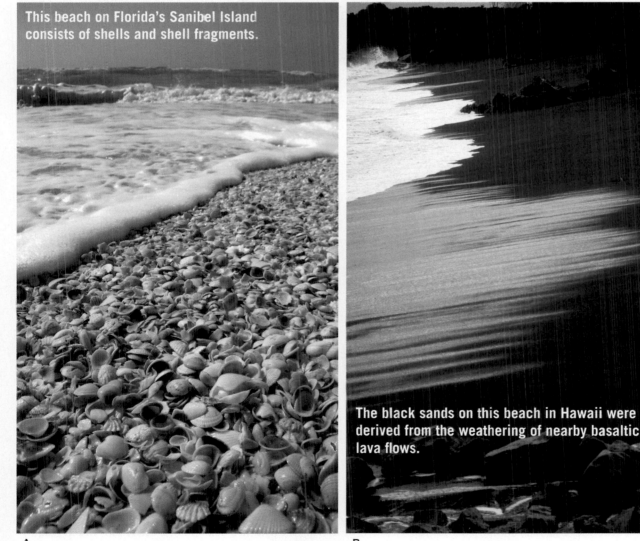

This beach on Florida's Sanibel Island consists of shells and shell fragments.

The black sands on this beach in Hawaii were derived from the weathering of nearby basaltic lava flows.

A.

B.

Figure 10.12 Beaches A beach is an accumulation of sediment on the landward margin of an ocean or a lake and can be thought of as material in transit along the shore. Beaches are composed of whatever material is locally available. (Photo A by David R. Frazier/Photo Library/Alamy Images; photo B by E. J. Tarbuck)

Figure 10.13 Storm waves When large waves break against the shore, the force of the water can be powerful, and the erosional work that is accomplished can be great. These storm waves are breaking along the coast of Wales. (The Photo Library/Alamy Images)

be awesome in its violence. Each breaking wave may hurl thousands of tons of water against the land, sometimes causing the ground literally to tremble. The pressures exerted by Atlantic waves in winter, for example, average nearly 10,000 kilograms per square meter (more than 2000 pounds per square foot). The force during storms is even greater.

It is no wonder that cracks and crevices are quickly opened in cliffs, coastal structures, and anything else that is subjected to these enormous shocks (**Figure 10.13**). Water is forced into every opening, causing air in the cracks to become highly compressed by the thrust of crashing waves. When a wave subsides, the air expands rapidly, dislodging rock fragments and enlarging and extending fractures.

In addition to the erosion caused by wave impact and pressure, **abrasion**, the sawing and grinding action of the water armed with rock fragments, is also important. In fact, abrasion is probably more intense in the surf zone than in any other environment. Smooth, rounded stones and pebbles along the shore are obvious reminders of the relentless grinding action of rock against rock in the surf zone (**Figure 10.14A**). Rock fragments are also used as "tools" by the waves as they cut horizontally into the land (**Figure 10.14B**).

Sand Movement on the Beach

Beaches are sometimes called "rivers of sand." The reason is that the energy from breaking waves often causes large quantities of sand to move along the beach face and in the surf zone roughly parallel to the shoreline. Wave energy also causes sand to move perpendicular to (toward and away from) the shoreline.

Movement Perpendicular to the Shoreline If you stand ankle-deep in water at the beach, you will see that swash and backwash move sand toward and away from the shoreline. Whether there is a net loss or addition of sand depends on the level of wave activity. When wave activity is relatively light (less energetic waves), much of the swash soaks into the beach, which reduces the backwash. Consequently, the swash dominates and causes a net movement of sand up the beach face.

When high-energy waves prevail, the beach is saturated from previous waves, so much less of the swash soaks in. As a result, the beach erodes because backwash is strong and causes a net movement of sand toward open water.

Along many beaches, light wave activity is the rule during the summer. Therefore, a wide sandy beach gradually develops. During winter, when storms are frequent and

A.

B.

Smooth, rounded rocks along the shore are an obvious reminder that abrasion can be intense in the surf zone.

This sandstone cliff at Gabriola Island, British Columbia was undercut by wave action.

Figure 10.14 Abrasion—sawing and grinding Breaking waves armed with rock debris can do a great deal of erosional work. (Photo A by Michael Collier; photo B by Fletcher and Baylis/Photo Researchers, Inc.)

more powerful, strong wave activity erodes and narrows the beach. A wide beach that may have taken months to build can be dramatically narrowed in just a few hours by the high-energy waves created by a strong winter storm.

Wave Refraction The bending of waves, called **wave refraction**, plays an important part in shoreline processes (**Figure 10.15**). It affects the distribution of energy along the shore and thus strongly influences where and to

As these waves approach nearly straight on, refraction causes the wave energy to be concentrated at headlands (resulting in erosion) and dispersed in bays (resulting in deposition).

Beach deposits

Headland

Waves travel at original speed in deep water

Waves "feel bottom" and slow down in surf zone

Shoreline

Result: waves bend so that they strike the shore more directly

Wave refraction at Rincon Point, California

SmartFigure 10.15 Wave refraction As waves first touch bottom in the shallows along an irregular coast, they are slowed, causing them to bend (refract) and align nearly parallel to the shoreline. (Photo by Rich Reid/National Geographic/Getty Images)

Path of sand particles

Beach drift

Net movement of sand grains

Longshore current

Beach drift occurs as incoming waves carry sand at an angle up the beach, while the water from spent waves carries it directly down the slope of the beach. Similar movements occur offshore in the surf zone to create the longshore current.

Longshore current

These waves approaching the beach at a slight angle near Oceanside, California, produce a longshore current moving from left to right.

SmartFigure 10.16 The longshore transport system The two components of this sediment-moving system, beach drift and longshore currents, are created by breaking waves that approach the shoreline at an angle. These processes move large quantities of material along the beach and in the surf zone. (Photo by John S. Shelton/University of Washington Libraries)

and deposition in the bays will straighten an irregular shoreline.

Longshore Transport Although waves are refracted, most still reach the shore at a slight angle. Consequently, the uprush of water from each breaking wave (the swash) is at an oblique angle to the shoreline. However, the backwash is straight down the slope of the beach. The effect of this pattern of water movement is to transport sediment in a zigzag pattern along the beach face (**Figure 10.16**). This movement is called **beach drift**, and it can transport sand and pebbles hundreds or even thousands of meters daily. However, a more typical rate is 5 to 10 meters (16 to 33 feet) per day.

Waves that approach the shore at an angle also produce currents within the surf zone that flow parallel to the shore and move substantially more sediment than beach drift (see Figure 10.16). Because the water here is turbulent, these **longshore currents** easily move the fine suspended sand and roll larger sand and gravel along the bottom. When the sediment transported by longshore currents is added to the quantity moved by beach drift, the total amount can be very large. At Sandy Hook, New Jersey, for example, the quantity of sand transported along the shore over a 48-year period averaged almost 680,000 metric tons (750,000 short tons) annually. For a 10-year period at Oxnard, California, more than 1.4 million metric tons (1.5 million short tons) of sediment moved along the shore each year.

Both rivers and coastal zones move water and sediment from one area (*upstream*) to another (*downstream*). As a result, the beach is often characterized as a "river of sand." Beach drift and longshore currents, however, move in a zigzag pattern, whereas rivers flow mostly in a turbulent, swirling fashion. In addition, the direction of flow of longshore currents along a shoreline can change, whereas rivers flow in the same direction (downhill). Longshore currents change direction because waves approach the beach in different directions each season. Nevertheless, longshore currents generally flow southward along both the Atlantic and Pacific shores of the United States.

what degree erosion, sediment transport, and deposition will take place.

Waves seldom approach the shore straight on. Rather, most waves move toward the shore at a slight angle. However, when they reach the shallow water of a smoothly sloping bottom, the wave crests are refracted (bent) and tend to become parallel to the shore. Such bending occurs because the part of the wave nearest the shore touches bottom and slows first, whereas the part of the wave that is still in deeper water continues forward at its full speed. The net result is a wave that approaches nearly parallel to the shore, regardless of its original orientation.

Because of refraction, wave energy is concentrated against the sides and ends of headlands that project into the water, whereas wave attack is weakened in bays. This differential wave attack along irregular coastlines is illustrated in Figure 10.15. Because the waves reach the shallow water in front of the headland sooner than they do in adjacent bays, they are bent more nearly parallel to the protruding land and strike it from all three sides. By contrast, refraction in the bays causes waves to diverge and expend less energy. In these zones of weakened wave activity, sediments can accumulate and form sandy beaches. Over a long period, erosion of the headlands

10.5 CONCEPT CHECKS

1. What is a beach?
2. Why do waves approaching the shoreline often bend?
3. What is the effect of wave refraction along an irregular coastline?
4. Describe the two processes that contribute to longshore transport.

10.6 Shoreline Features

Describe the features typically created by wave erosion and those resulting from sediment deposited by longshore transport processes.

A fascinating assortment of shoreline features can be observed along the world's coastal regions. These shoreline features vary, depending on the type of rocks exposed along the shore, the intensity of waves, and the nature of coastal currents, as well as whether the coast is stable, sinking, or rising. Features that owe their origin primarily to the work of erosion are called *erosional features*, whereas deposits of sediment produce *depositional features*.

Erosional Features

Many coastal landforms owe their origin to erosional processes. Such erosional features are common along the rugged and irregular New England coast and along the steep shorelines of the west coast of the United States.

Wave-Cut Cliffs, Wave-Cut Platforms, and Marine Terraces As the name implies, **wave-cut cliffs** originate in the cutting action of the surf against the base of coastal land. As erosion progresses, rocks overhanging the notch at the base of the cliff crumble into the surf and the cliff retreats. A relatively flat, benchlike surface, called a **wave-cut platform**, is left behind by the receding cliff (**Figure 10.17**, *left*). The platform broadens as wave attack continues. Some debris produced by the breaking waves remains along the water's edge as sediment on the beach, and the remainder is transported farther seaward. If a wave-cut platform is uplifted above sea level by tectonic forces, it becomes a **marine terrace** (Figure 10.17, *right*). Marine terraces are easily recognized by their gentle seaward-sloping shape and are often perceived as desirable sites for coastal development.

Sea Arches and Sea Stacks Headlands that extend into the sea are vigorously attacked by waves because of refraction. The surf erodes the rock selectively, wearing away the softer or more highly fractured rock at the fastest rate. At first, sea caves may form. When two caves on opposite sides of a headland unite, a **sea arch** results (**Figure 10.18**). Eventually, the arch falls in, leaving an isolated remnant, or **sea stack**, on the wave-cut platform (see Figure 10.18). In time, it, too, will be consumed by the action of the waves.

Depositional Features

Sediment from a beach is transported along the shore and deposited in areas where wave energy is low. Such processes produce a variety of depositional features.

Did You Know?

Along shorelines composed of unconsolidated material rather than hard rock, the rate of erosion by breaking waves can be extraordinary. In parts of Britain, where waves have the easy task of eroding glacial deposits of sand, gravel, and clay, the coast has been worn back 3 to 5 km (2 to 3 mi) since Roman times (2000 years ago), sweeping away many villages and ancient landmarks.

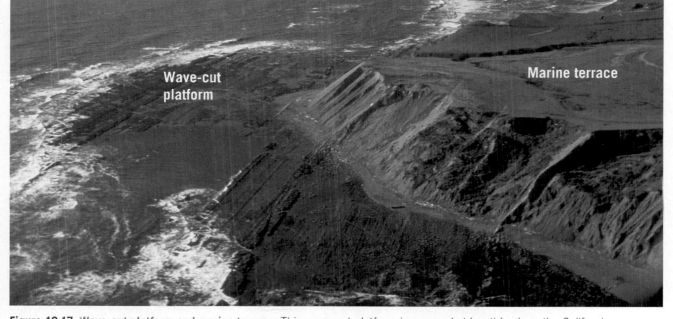

Figure 10.17 Wave-cut platform and marine terrace This wave-cut platform is exposed at low tide along the California coast at Bolinas Point near San Francisco. A wave-cut platform was uplifted to create the marine terrace. (Photo by John S. Shelton/University of Washington Libraries)

Figure 10.18 Sea arch and sea stack These features at the tip of Mexico's Baja Peninsula resulted from the vigorous wave attack of a headland. (Photo by Lew Robertson/Getty Images)

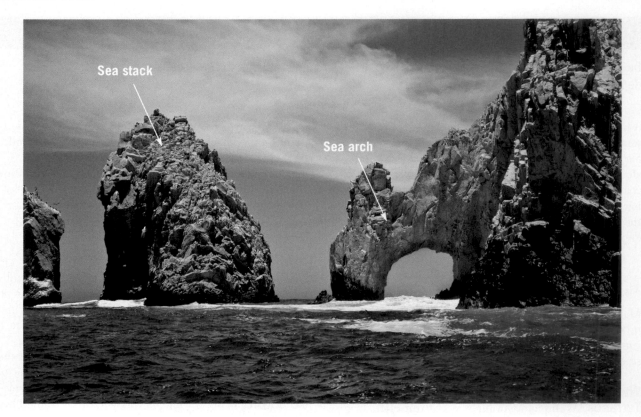

Figure 10.19 Some depositional features A. High-altitude image of a well-developed spit and baymouth bar along the coast of Martha's Vineyard, Massachusetts. (Image courtesy of USDA-ASCS) **B.** This photograph, taken from the International Space Station, shows Provincetown Spit at the tip of Cape Cod. (NASA image)

A.

B.

Spits, Bars, and Tombolos Where beach drift and longshore currents are active, several features related to the movement of sediment along the shore may develop. A **spit** is an elongated ridge of sand that projects from the land into the mouth of an adjacent bay. Often the end in the water hooks landward in response to the dominant direction of the longshore current. Both images in **Figure 10.19** show spits. The term **baymouth bar** is applied to a sandbar that completely crosses a bay, sealing it off from the open ocean. Such a feature tends to form across bays where currents are weak, allowing a spit to extend to the other side (Figure 10.19A). A **tombolo**, a ridge of sand that connects an island to the mainland or to another island, forms in much the same manner as a spit.

Barrier Island The Atlantic and Gulf coastal plains are relatively flat and slope gently seaward. The shore zone is characterized by **barrier islands**. These low ridges of sand parallel the coast at distances from 3 to 30 kilometers (2 to 20 miles) offshore. From Cape Cod, Massachusetts, to Padre Island, Texas, nearly 300 barrier islands rim the coast (**Figure 10.20**).

Most barrier islands are from 1 to 5 kilometers (0.6 to 3 miles) wide and 15 to 30 kilometers (9 to 19 miles) long. The tallest features are sand dunes, which usually reach heights of 5 to 10 meters (16 to 33 feet). The lagoons that separate these narrow islands from the shore are zones of relatively quiet

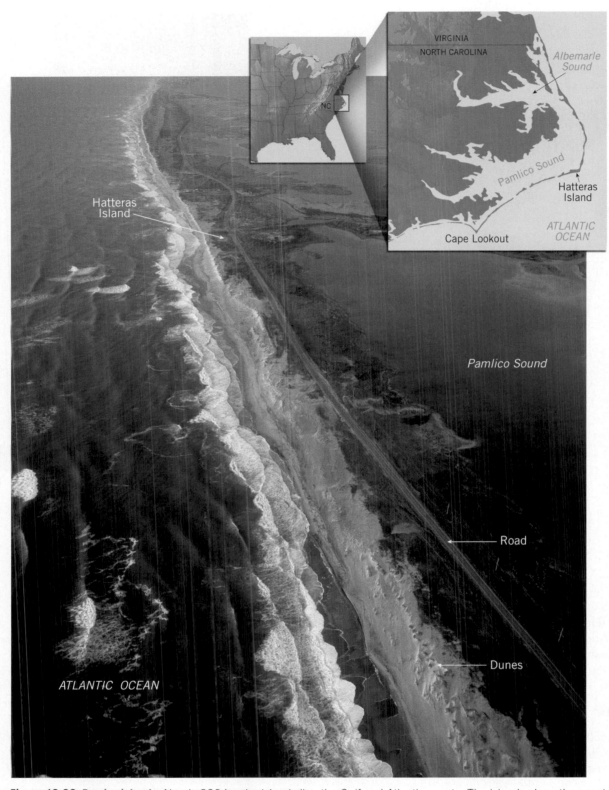

Figure 10.20 Barrier islands Nearly 300 barrier islands line the Gulf and Atlantic coasts. The islands along the coast of North Carolina are excellent examples. (Photo by Michael Collier)

water that allow small craft traveling between New York and northern Florida to avoid the rough waters of the North Atlantic.

Barrier islands probably form in several ways. Some originate as spits that were subsequently severed from the mainland by wave erosion or by the general rise in sea level following the last episode of glaciation. Others were created when turbulent waters in the line of breakers heaped up sand that had been scoured from the bottom. Finally, some barrier islands may be former

Figure 10.21 The evolving shore These diagrams illustrate changes that can take place through time along an initially irregular coastline that remains tectonically stable. The diagrams also serve to illustrate many of the features described in the section on shoreline features. (Top and bottom photos by E. J. Tarbuck; middle photo by Michael Collier)

sand-dune ridges that originated along the shore during the last glacial period, when sea level was lower. As the ice sheets melted, sea level rose and flooded the area behind the beach–dune complex.

The Evolving Shore

A shoreline continually undergoes modification, regardless of its initial configuration. At first, most coastlines are irregular, although the degree of and reason for the irregularity may differ considerably from place to place. Along a coastline that is characterized by varied geology, the pounding surf may initially increase its irregularity because the waves will erode the weaker rocks more easily than the stronger ones. However, if a shoreline remains stable, marine erosion and deposition will eventually produce a straighter, more regular coast.

Figure 10.21 illustrates the evolution of an initially irregular coast that remains relatively stable and shows many of the coastal features discussed in the previous section. As headlands are eroded and erosional features such as wave-cut cliffs and wave-cut platforms are created, sediment is produced that is carried along the shore by beach drift and longshore currents. Some material is deposited in the bays, whereas other debris is formed into depositional features such as spits and baymouth bars. At the same time, rivers fill the bays with sediment. Ultimately, a generally straight, smooth coast results.

10.6 CONCEPT CHECKS

1. Explain how the marine terrace in Figure 10.17 formed.

2. Describe the formation of each labeled feature in Figure 10.21.

3. Along which of America's coastal areas are barrier islands common? How do these islands form?

10.7 Stabilizing the Shore

Summarize the ways in which people deal with shoreline erosion problems.

The coastal zone teems with human activity. Unfortunately, people often treat the shoreline as if it were a stable platform on which structures can be built safely. This approach jeopardizes both people and the shoreline because many coastal landforms are relatively fragile, short-lived features that are easily damaged by development. As anyone who has endured a strong coastal storm knows, the shoreline is not always a safe place to live. A glance back at Figure 10.8 reminds us of that fact.

Compared with other natural hazards, such as earthquakes, volcanic eruptions, and landslides, shoreline erosion appears to be a more continuous and predictable process that causes relatively modest damage to limited areas. In reality, the shoreline is one of Earth's most dynamic places, changing rapidly in response to natural forces. Exceptional storms are capable of eroding beaches and cliffs at rates that far exceed the long-term average. Such bursts of accelerated erosion not only have a significant impact on the natural evolution of a coast but can also have a profound impact on people who reside in the coastal zone. Erosion along the coast causes significant property damage. Huge sums are spent annually not only to repair damage but also to prevent or control erosion. Already a problem at many sites, shoreline erosion is certain to become increasingly serious as extensive coastal development continues.

Although the same processes cause change along every coast, not all coasts respond in the same way. Interactions among different processes and the relative importance of each process depend on local factors. The factors include (1) the proximity of a coast to sediment-laden rivers, (2) the degree of tectonic activity, (3) the topography and composition of the land, (4) prevailing winds and weather patterns, and (5) the configuration of the coastline and near-shore areas.

During the past 100 years, growing affluence and increasing demands for recreation have brought unprecedented development to many coastal areas. As both the number and the value of buildings have increased, so, too, have efforts to protect property from storm waves by stabilizing the shore. Also, controlling the natural migration of sand is an ongoing struggle in many coastal areas. Such interference can result in unwanted changes that are difficult and expensive to correct.

Hard Stabilization

Structures built to protect a coast from erosion or to prevent the movement of sand along a beach are known as **hard stabilization**. Hard stabilization can take many forms and often results in predictable yet unwanted outcomes. Hard stabilization includes groins, breakwaters, and seawalls.

Groins To maintain or widen beaches that are losing sand, groins are sometimes constructed. A **groin** is a barrier built at a right angle to the beach to trap sand that is moving parallel to the shore. Groins are usually constructed of large rocks but may also be composed of wood. These structures often do their job so effectively that the longshore current beyond the groin becomes sand-starved. As a result, the current erodes sand from the beach on the downstream side of the groin.

To offset this effect, property owners downstream from the structure may erect a groin on their property. In this manner, the number of groins multiplies, resulting in a *groin field* (**Figure 10.22**). An example of such proliferation is the shoreline of New Jersey, where hundreds of these structures have been built. Because it has been shown that groins often do not provide a satisfactory solution, using them is no longer the preferred method of keeping beach erosion in check.

Breakwaters and Seawalls Hard stabilization can be built parallel to the shoreline. One such structure is a **breakwater**, which is designed to protect boats from the force of large breaking waves by creating a quiet water zone near the shore. However, when this is done, the reduced wave activity along the shore behind the structure

Figure 10.22 Groins These wall-like structures trap sand that is moving parallel to the shore. This series of groins is along the shoreline near Chichester, Sussex, England. (Photo by Sandy Stockwell/London Aerial Photo Library/CORBIS)

Figure 10.23 Breakwater
Aerial view of a breakwater at Santa Monica, California. The structure appears as a line in the water behind which many boats are anchored. The construction of the breakwater disrupted longshore transport and caused the seaward growth of the beach. (Photo by John S. Shelton/University of Washington Libraries)

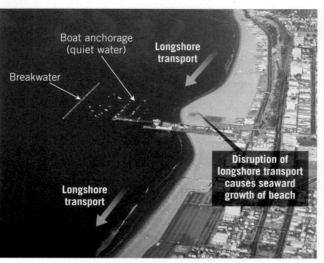

Boat anchorage (quiet water)

Longshore transport

Breakwater

Disruption of longshore transport causes seaward growth of beach

Longshore transport

farther downstream where longshore currents continue to move the sand down the coast (**Figure 10.23**).

Another type of hard stabilization built parallel to the shore is a **seawall**, which is designed to armor the coast and defend property from the force of breaking waves. Waves expend much of their energy as they move across an open beach. Seawalls cut this process short by reflecting the force of unspent waves seaward. As a consequence, the beach to the seaward side of the seawall experiences significant erosion and may, in some instances, be eliminated entirely (**Figure 10.24**). Once the width of the beach is reduced, the seawall is subjected to even greater pounding by the waves. Eventually, this battering takes its toll on the seawall such that the seawall will fail and a larger, more expensive structure must be built to take its place.

The wisdom of building temporary protective structures along shorelines is increasingly questioned. The opinions of many coastal scientists and engineers is that halting an eroding shoreline with protective structures benefits only a few and seriously degrades or destroys the natural beach and the value it holds for the majority. Protective structures divert the ocean's energy temporarily from private properties, but usually refocus that energy

may allow sand to accumulate. If this happens, the boat anchorage will eventually fill with sand, while the downstream beach erodes and retreats. At Santa Monica, California, where the building of a breakwater has created such a problem, the city uses a dredge to remove sand from the protected quiet water zone and deposit it

Figure 10.24 Seawall
Seabright in northern New Jersey once had a broad, sandy beach. A seawall 5 to 6 meters (16 to 18 feet) high and 8 kilometers (5 miles) long was built to protect the town and the railroad that brought tourists to the beach. After the wall was built, the beach narrowed dramatically. (Photo by Rafael Macia/Photo Researchers, Inc.)

Seawall

on the adjacent beaches. Many structures interrupt the natural sand flow in coastal currents, robbing affected beaches of vital sand replacement.

Alternatives to Hard Stabilization

Armoring the coast with hard stabilization has several potential drawbacks, including the cost of the structure and the loss of sand on the beach. Alternatives to hard stabilization include beach nourishment and relocation.

Beach Nourishment One approach to stabilizing shoreline sands without hard stabilization is **beach nourishment**. As the term implies, this practice involves adding large quantities of sand to the beach system (**Figure 10.25**). Extending beaches seaward causes buildings along the shoreline to become less vulnerable to destruction by storm waves and enhances recreational uses.

The process of beach nourishment is straightforward. Sand is pumped by dredges from offshore or trucked from inland locations. The "new" beach, however, will not be the same as the former beach. Because replenishment sand is from somewhere else, typically not another beach, it is new to the beach environment. The new sand is usually different in size, shape, sorting, and composition. These differences pose problems in terms of erodibility and the kinds of life the new sand will support.

Beach nourishment is not a permanent solution to the problem of shrinking beaches. The same processes that removed the sand in the first place will eventually remove the replacement sand as well. Nevertheless, the number of nourishment projects has increased in recent years and many beaches, especially along the Atlantic coast, have had their sand replenished many times. Virginia Beach, Virginia, has been nourished more than 50 times.

Beach nourishment is costly. For example, a modest project might involve 50,000 cubic yards of sand distributed across a half mile of shoreline. A good-sized dump truck holds about 10 cubic yards of sand. So this small project would require about 5000 dump-truck loads. Many projects extend for many miles. Nourishing beaches typically costs millions of dollars per mile.

Relocation Instead of building structures such as groins and seawalls to hold the shoreline in place or adding sand

Did You Know?

A significant impact of global climate change is a rise in sea level. As this occurs, coastal cities, wetlands, and low-lying islands could be threatened with more frequent flooding, increased shoreline erosion, and saltwater encroachment into coastal rivers and aquifers. There's more about global climate change in Chapter 11.

Figure 10.25 Beach nourishment If you visit a beach along the Atlantic coast, it is more and more likely that you will walk into the surf zone atop an artificial beach. In this image, an offshore dredge is pumping sand to a beach. (Photo by Michael Weber/imagebroker/Alamy Images)

to replenish eroding beaches, another option is available. Many coastal scientists and planners are calling for a policy shift from defending and rebuilding beaches and coastal property in high-hazard areas to *relocating* storm-damaged or at-risk buildings and letting nature reclaim the beach. This approach is similar to an approach the federal government adopted for river floodplains following the devastating 1993 Mississippi River floods, in which vulnerable structures were abandoned and relocated on higher, safer ground.

Such proposals, of course, are controversial. People with significant shoreline investments want to rebuild and defend coastal developments from the erosional wrath of the sea. Others, however, argue that with sea level rising, the impact of coastal storms will get worse in the decades to come, and oft-damaged structures should be relocated or abandoned to improve personal safety and to reduce costs. Such ideas will no doubt be the focus of much study and debate as states and communities evaluate and revise coastal land-use policies.

10.7 CONCEPT CHECKS

1. List two examples of hard stabilization and describe what each is intended to do. How does each affect the distribution of sand on a beach?
2. What are two alternatives to hard stabilization, and what potential problems are associated with each?

10.8 Contrasting America's Coasts

Contrast the erosion problems faced along different parts of America's coasts. Distinguish between emergent and submergent coasts.

The shoreline along the Pacific coast of the United States is strikingly different from that of the Atlantic and Gulf coast regions. Some of the differences are related to plate tectonics. The west coast represents the leading edge of the North American plate; therefore, it experiences active uplift and deformation. By contrast, the east coast is a tectonically quiet region that is far from any active plate margin. Because of this basic geologic difference, the nature of shoreline erosion problems along America's opposite coasts is different.

Atlantic and Gulf Coasts

Much of the coastal development along the Atlantic and Gulf coasts has occurred on barrier islands. Typically, a barrier island consists of a wide beach that is backed by dunes and separated from the mainland by marshy lagoons. The broad expanses of sand and exposure to the ocean have made barrier islands exceedingly attractive sites for development. Unfortunately, development has taken place more rapidly than has our understanding of barrier island dynamics.

Because barrier islands face the open ocean, they receive the full force of major storms that strike the coast. When a storm occurs, the barriers absorb the energy of the waves primarily through the movement of sand. **Figure 10.26**, which shows changes at Cape Hatteras National Seashore, reinforces this point. This process and the dilemma that results have been described as follows:

> Waves may move sand from the beach to offshore areas or, conversely, into the dunes; they may erode the dunes, depositing sand onto the beach or carrying it out to sea; or they may carry sand from the beach and the dunes into the marshes behind the barrier, a process known as *overwash*. The common factor is movement. Just as a flexible reed may survive a wind that destroys an oak tree, so the barriers survive hurricanes and nor'easters not through unyielding strength but by giving before the storm.

This picture changes when a barrier is developed for homes or a resort. Storm waves that previously rushed harmlessly through gaps between the dunes now encounter buildings and roadways. Moreover, since the dynamic nature of the barriers is readily perceived only during storms, homeowners tend to attribute damage to a particular storm, rather than to the basic mobility of coastal barriers. With their homes or investments at stake, local residents are more likely to seek to hold the sand in place and the waves at bay than to admit that development was improperly placed to begin with.[*]

Pacific Coast

In contrast to the broad, gently sloping coastal plains of the Atlantic and Gulf coasts, much of the Pacific coast is characterized by relatively narrow beaches that are backed by steep cliffs and mountain ranges. Recall that America's western margin is a more rugged and tectonically active region than the eastern margin. Because uplift continues, the rise in sea level in the West is not so readily apparent. Nevertheless, like the shoreline erosion

---

[*]Frank Lowenstein, "Benches or Bedrooms—The Choice as Sea Level Rises," *Oceanus* 28 (No. 3, Fall 1985): 22.

Figure 10.26 **Relocating the Cape Hatteras Lighthouse** Despite a number of efforts to protect this 21-story lighthouse, the nation's tallest, from being destroyed due to a receding shoreline, the structure finally had to be moved. (Photos by U.S. Geological Survey and Drew Wilson, *Virginia-Pilot*, © 1999)

Various attempts to protect the lighthouse failed. They included building groins and beach nourishment. By 1999, when this photo was taken, the lighthouse was only 36 meters (120 ft.) from the water.

Former location of lighthouse

884 meters (2900 ft.)

To save the famous candy-striped landmark, the National Park Service authorized moving the structure. After the $12 million move, it is expected to be safe for 50 years or more.

problems facing the East's barrier islands, west coast difficulties also stem largely from the alteration of natural systems by people.

A major problem facing the Pacific shoreline—particularly along southern California—is a significant narrowing of many beaches. The bulk of the sand on many of these beaches is supplied by rivers that transport it from the mountainous regions to the coast. Over the years, this natural flow of material to the coast has been interrupted by dams built for irrigation and flood control. The reservoirs effectively trap the sand that would otherwise nourish the beach environment. When the beaches were wider, they protected the cliffs behind them from the force of storm waves. Now, however, the waves move across the narrowed beaches without losing much energy and cause more rapid erosion of the sea cliffs.

Although the retreat of the cliffs provides material to replace some of the sand impounded behind dams, it also endangers homes and roads built on the bluffs.

In addition, development atop the cliffs aggravates the problem. Urbanization increases runoff, which, if not carefully controlled, can result in serious bluff erosion. Watering lawns and gardens adds significant quantities of water to the slope. This water percolates downward toward the base of the cliff, where it may emerge in small seeps. This action reduces the slope's stability and facilitates mass wasting.

Shoreline erosion along the Pacific coast varies considerably from one year to the next, largely because of the sporadic occurrence of storms. As a result, when the infrequent but serious episodes of erosion occur, the damage is often blamed on the unusual storms and not on coastal development or the sediment-trapping dams that may be great distances away. If, as predicted, sea level experiences a significant rise in the years to come because of global climate change, increased shoreline erosion and sea-cliff retreat should be expected along many parts of the Pacific coast.

Coastal Classification

The great variety of shorelines demonstrates their complexity. Indeed, to understand any particular coastal area, many factors must be considered, including rock types, size and direction of waves, frequency of storms, tidal range, and offshore topography. In addition, practically all coastal areas were affected by the worldwide rise in sea level that accompanied the melting of Ice Age glaciers following the Last Glacial Maximum. Finally, tectonic events that uplift or downdrop the land or change the volume of ocean basins must be taken into account. The large number of factors that influence coastal areas makes shoreline classification difficult.

Many geologists classify coasts based on changes that have occurred with respect to sea level. This commonly used classification system divides coasts into two general categories: emergent and submergent. **Emergent coasts** develop either because an area experiences uplift or as a result of a drop in sea level. Conversely, **submergent coasts** are created when sea level rises or the land adjacent to the sea subsides.

Emergent Coasts In some areas, the coast is clearly emergent because rising land or falling water levels expose wave-cut cliffs and marine terraces above sea level. Excellent examples include portions of coastal California where uplift has occurred in the recent geologic past. The elevated marine terrace in Figure 10.17 illustrates this situation. In the case of the Palos Verdes Hills, south of Los Angeles, California, seven different terrace levels exist, indicating at least seven episodes of uplift. The ever-persistent sea is now cutting a new platform at the base of the cliff. If uplift follows, it, too, will become an elevated marine terrace.

Other examples of emergent shores include regions that were once buried beneath great ice sheets. When glaciers were present, their weight depressed the crust, and when the ice melted, the crust began gradually to spring back. Consequently, prehistoric shoreline features today are found high above sea level. The Hudson Bay region of Canada is one such area; portions of it are still rising at a rate of more than 1 centimeter (0.4 inch) per year.

Submergent Coasts In contrast to the preceding examples, other coastal areas show definite signs of submergence. Shorelines that have been submerged in the relatively recent past are often highly irregular because the sea typically floods the lower reaches of river valleys flowing into the ocean. The ridges separating the valleys, however, remain above sea level and project into the sea as headlands. These drowned river mouths, which are called **estuaries**, characterize many coasts today. Along the Atlantic coastline, Chesapeake and Delaware bays are examples of estuaries created by submergence (**Figure 10.27**). The picturesque coast of Maine, particularly in the vicinity of Acadia National Park, is another excellent example of an area that was flooded by the

SmartFigure 10.27 East coast estuaries The lower portions of many river valleys were flooded by the rise in sea level that followed the end of the Quaternary Ice Age, creating large estuaries such as the Chesapeake and Delaware Bays.

postglacial rise in sea level and transformed into a highly irregular submerged coastline.

Keep in mind that most coasts have a complicated geologic history. With respect to sea level, many coasts have at various times emerged and then submerged again. Each time, they retain some of the features created during the previous event.

10.8 CONCEPT CHECKS

1. Briefly describe what happens when storm waves strike an undeveloped barrier island.
2. How might building a dam on a river that flows to the sea affect a beach?
3. What is an observable feature that would lead you to classify a coastal area as emergent?
4. Are estuaries associated with submergent or emergent coasts? Explain.

10.9 Tides

Explain the cause of tides, their monthly cycles, and patterns. Describe the horizontal flow of water that accompanies the rise and fall of tides.

Tides are daily changes in the elevation of the ocean surface. Their rhythmic rise and fall along coastlines have been known since antiquity. Other than waves, they are the easiest ocean movements to observe (**Figure 10.28**).

Although known for centuries, tides were not explained satisfactorily until Sir Isaac Newton applied the law of gravitation to them. Newton showed that there is a mutual attractive force between two bodies, as between Earth and the Moon. Because both the atmosphere and the ocean are fluids and are free to move, both are deformed by this force. Hence, ocean tides result from the gravitational attraction exerted upon Earth by the Moon and, to a lesser extent, by the Sun.

Causes of Tides

It is easy to see how the Moon's gravitational force can cause the water to bulge on the side of Earth nearest the Moon. In addition, however, an equally large tidal bulge is produced on the side of Earth directly opposite the Moon (**Figure 10.29**).

Both tidal bulges are caused, as Newton discovered, by the pull of gravity. Gravity is inversely proportional to the square of the distance between two objects, meaning simply that it quickly weakens with distance. In this case, the two objects are the Moon and Earth. Because the force of gravity decreases with distance, the Moon's gravitational pull on Earth is slightly greater on the near side of Earth than on the far side. The result of this differential pulling is to stretch (elongate) the "solid" Earth very slightly. In contrast, the world ocean, which is mobile, is deformed quite dramatically by this effect to produce the two opposing tidal bulges.

Because the position of the Moon changes only moderately in a single day, the tidal bulges remain in place while Earth rotates "through" them. If you stand on the seashore for 24 hours, Earth will rotate you through alternating areas of higher and lower water. As you are carried into each tidal bulge, the tide rises, and as you are carried into the intervening troughs between the tidal bulges, the tide falls. Therefore, most places on Earth experience two high tides and two low tides each day.

In addition, the tidal bulges migrate as the Moon revolves around Earth about every 29 days. As a result, the tides, like the time of moonrise, shift about 50 minutes

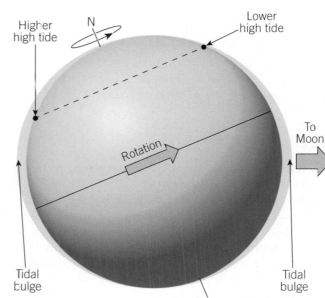

Figure 10.28 Bay of Fundy tides High tide and low tide on Nova Scotia's Minas Basin in the Bay of Fundy. Tidal flats are exposed during low tide. (Photos courtesy of Nova Scotia Department of tourism)

Figure 10.29 Idealized tidal bulges caused by the Moon If Earth were covered to a uniform depth with water, there would be two tidal bulges: one on the side of Earth facing the Moon (right) and the other on the opposite side of Earth (left). Depending on the Moon's position, tidal bulges may be inclined relative to Earth's equator. In this situation, Earth's rotation causes an observer to experience two unequal high tides in a day.

Did You Know?

Tides can be used to generate electrical power. Using tides in this way involves constructing a dam across the mouth of a bay or an estuary in a coastal area that has a large tidal range. The narrow opening between the bay and the open ocean magnifies the variations in water level as tides rise and fall, and the strong in-and-out flow is used to drive turbines. The largest tidal electricity plant yet constructed is along the coast of France.

later each day. After 29 days the cycle is complete and a new one begins.

Many locations may show an inequality between the high tides during a given day. Depending on the Moon's position, the tidal bulges may be inclined to the equator as in Figure 10.29. This figure illustrates that one high tide experienced by an observer in the Northern Hemisphere is considerably higher than the high tide half a day later. In contrast, a Southern Hemisphere observer would experience the opposite effect.

Monthly Tidal Cycle

The primary body that influences the tides is the Moon, which makes one complete revolution around Earth every 29 days. The Sun, however, also influences the tides. It is far larger than the Moon, but because it is much farther away, its effect is considerably less. In fact, the Sun's tide-generating effect is only about 46 percent that of the Moon.

Near the times of new and full moons, the Sun and Moon are aligned, and their forces on tides are added together (**Figure 10.30A**). The combined gravity of these two tide-producing bodies causes larger tidal bulges (higher high tides) and larger tidal troughs (lower low tides), producing a large tidal range. These are called the **spring tides**, which have no connection with the spring season but occur twice a month, during the times when the Earth–Moon–Sun system is aligned. Conversely, at about the time of the first and third quarters of the Moon, the gravitational forces of the Moon and Sun act on Earth at right angles, and each partially offsets the influence of the other (**Figure 10.30B**). As a result, the daily tidal range is less. These are called **neap tides**, and they also occur twice each month. Each month, then, there are two spring tides and two neap tides, each about 1 week apart.

Tidal Patterns

The basic causes and types of tides have been explained. Keep in mind, however, that these theoretical considerations cannot be used to predict either the height or the time of actual tides at a particular place. Many factors—including the shape of the coastline, the configuration of ocean basins, and water depth—greatly influence the tides. Consequently, tides at various locations respond differently to tide-producing forces. Thus, the nature of the tide at any coastal location can be determined most accurately by actual observation. The predictions in tidal tables and tidal data on nautical charts are based on such observations.

Three main tidal patterns exist worldwide. A **diurnal** (*diurnal* = daily) **tidal pattern** is characterized by a single high tide and a single low tide each tidal day (**Figure 10.31**). Tides of this type occur along the northern shore of the Gulf of Mexico, among other locations. A **semidiurnal** (*semi* = twice, *diurnal* = daily) **tidal**

A. Spring Tide When the Moon is in the full or new position, the tidal bulges created by the Sun and Moon are aligned and there is a large tidal range.

B. Neap Tide When the Moon is in the first-or third-quarter position, the tidal bulges produced by the Moon are at right angles to the bulges created by the Sun and the tidal range is smaller.

Figure 10.30 Spring and neap tides Earth–Moon–Sun positions influence the tides.

pattern exhibits two high tides and two low tides each tidal day, with the two highs about the same height and the two lows about the same height (see Figure 10.31). This type of tidal pattern is common along the Atlantic coast of the United States. A **mixed tidal pattern** is similar to a semidiurnal pattern except that it is characterized by a large inequality in high-water heights, low-water heights, or both (see Figure 10.31). In this case, there are usually two high tides and two low tides each day, with high tides of different heights and low tide of different heights. Such tides are prevalent along the Pacific coast of the United States and in many other parts of the world.

Tidal Currents

Tidal current is the term used to describe the *horizontal* flow of water accompanying the rise and fall of the tides. These water movements induced by tidal forces can

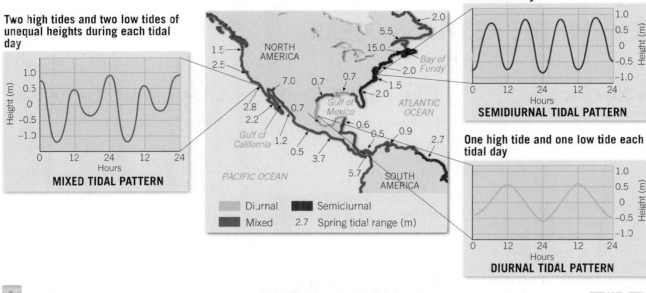

Two high tides and two low tides of unequal heights during each tidal day

MIXED TIDAL PATTERN

Two high tides and two low tides of approximately equal heights during each tidal day

SEMIDIURNAL TIDAL PATTERN

One high tide and one low tide each tidal day

DIURNAL TIDAL PATTERN

Diurnal Semidiurnal
Mixed 2.7 Spring tidal range (m)

 SmartFigure 10.31 Tidal patterns A diurnal tidal pattern (lower right) shows one high tide and one low tide each tidal day. A semidiurnal pattern (upper right) shows two high tides and two low tides of approximately equal heights during each tidal day. A mixed tidal pattern (left) shows two high tides and two low tides of unequal heights during each tidal day.

be important in some coastal areas. Tidal currents that advance into the coastal zone as the tide rises are called *flood currents*. As the tide falls, seaward-moving water generates *ebb currents*. Periods of little or no current, called *slack water*, separate flood and ebb. The areas affected by these alternating tidal currents are called **tidal flats** (see Figure 10.28). Depending on the nature of the coastal zone, tidal flats vary from narrow strips seaward of the beach to zones that may extend for several kilometers.

Although tidal currents are not important in the open sea, they can be rapid in bays, river estuaries, straits, and other narrow places. Off the coast of Brittany in France, for example, tidal currents that accompany a high tide of 12 meters (40 feet) may attain a speed of 20 kilometers (12 miles) per hour. Tidal currents are not generally considered to be major agents of erosion and sediment transport. Notable exceptions occur where tides move through narrow inlets, where they scour the narrow entrances to many harbors that would otherwise be blocked.

Sometimes deposits called **tidal deltas** are created by tidal currents (**Figure 10.32**). They may develop either as *flood deltas* landward of an inlet or as *ebb deltas* on the seaward side of an inlet. Because wave activity and longshore currents are reduced on the sheltered landward side, flood deltas are more common and actually more prominent (see Figure 10.19A). They form after the tidal current moves rapidly through an inlet. As the current emerges into more open waters from the narrow passage, it slows and deposits its load of sediment.

Figure 10.32 Tidal deltas As a rapidly moving tidal current (flood current) moves through a barrier island's inlet into the quiet waters of the lagoon, the current slows and deposits sediment, creating a tidal delta. Because this tidal delta has developed on the landward side of the inlet, it is called a *flood delta*. Such a tidal delta is shown in Figure 10.19A.

10.9 CONCEPT CHECKS

1. Explain why an observer can experience two *unequal* high tides during one day.
2. Distinguish between neap tides and ebb tides.
3. How is a mixed tidal pattern different from a semidiurnal tidal pattern?
4. Contrast flood current and ebb current.

CONCEPTS IN REVIEW: THE RESTLESS OCEAN

10.1 The Ocean's Surface Circulation

Discuss the factors that create and influence ocean currents and describe the affect ocean currents have on climate.

KEY TERMS: gyre, Coriolis effect

- The ocean's surface currents follow the general pattern of the world's major wind belts. Surface currents are parts of huge, slowly moving loops of water called *gyres* that are centered in the subtropics of each ocean basin. The positions of the continents and the Coriolis effect also influence the movement of ocean water within gyres. Because of the Coriolis effect, subtropical gyres move clockwise in the Northern Hemisphere and counterclockwise in the Southern Hemisphere. Generally, four main currents comprise each subtropical gyre.
- Ocean currents can have a significant effect on climate. Poleward-moving *warm* ocean currents moderate winter temperatures in the middle latitudes. Cold currents exert their greatest influence during summer in middle latitudes and year-round in the tropics. In addition to cooler temperatures, cold currents are associated with greater fog frequency and drought.

? Assume that arrow A represents prevailing winds in a Northern Hemisphere ocean. Which arrow, A, B, or C, best represents the surface-ocean current in this region? Explain.

10.2 Upwelling and Deep-Ocean Circulation

Explain the processes that produce coastal upwelling and the ocean's deep circulation.

KEY TERMS: upwelling, thermohaline circulation

- Upwelling, the rising of colder water from deeper layers, is a wind-induced movement that brings cold, nutrient-rich water to the surface. Coastal upwelling is most characteristic along the west coasts of continents.
- In contrast to surface currents, deep-ocean circulation is governed by gravity and driven by density differences. The two factors that are most significant in creating a dense mass of water are temperature and salinity, so the movement of deep-ocean water is often termed thermohaline circulation. Most water involved in thermohaline circulation begins in high latitudes at the surface, when the salinity of the cold water increases as a result of sea ice formation. This dense water sinks, initiating deep-ocean currents.

10.3 The Shoreline: A Dynamic Interface

Explain why the shoreline is considered a dynamic interface.

KEY TERM: interface

- The shoreline is a transition zone between marine and continental environments. It is a dynamic interface, a boundary where land, sea, and air meet and interact.
- Energy from waves plays an important role in shaping the shoreline, but many factors contribute to the character of particular shorelines.

10.4 Ocean Waves

List and discuss the factors that influence the height, length, and period of a wave and describe the motion of water within a wave.

KEY TERMS: wave height, wavelength, wave period, fetch, surf

- Waves are moving energy, and most ocean waves are initiated by wind. The three factors that influence the height, wavelength, and period of a wave are (1) wind speed, (2) length of time the wind has blown, and (3) fetch, the distance that the wind has traveled across open water. Once waves leave a storm area, they are termed *swells*, which are symmetrical, longer-wavelength waves.
- As waves travel, water particles transmit energy by circular orbital motion, which extends to a depth equal to one-half the wavelength (the wave base). When a wave enters water that is shallower than the wave base, it slows down, which allows waves farther from shore to catch up. As a result, wavelength decreases and wave height increases. Eventually the wave breaks, creating turbulent surf in which water rushes toward the shore.

10.5 Beaches and Shoreline Processes

Describe how waves erode and move sediment along the shore.

KEY TERMS: beach, abrasion, wave refraction, beach drift, longshore current

- A beach is composed of any locally derived material that is in transit along the shore. Wave erosion is caused by wave impact pressure and abrasion (the sawing and grinding action of water armed with rock fragments). The bending of waves, called wave refraction, causes wave impact to be concentrated against the sides and ends of headlands and dispersed in bays.
- Most waves reach the shore at an angle. The uprush (swash) and backwash of water from each breaking wave move the sediment in a zigzag pattern along the beach. This movement is called beach drift. Oblique waves also produce longshore currents within the surf zone that flow parallel to the shore and transport more sediment than does beach drift.

10.6 Shoreline Features

Describe the features typically created by wave erosion and those resulting from sediment deposited by longshore transport processes.

KEY TERMS: wave-cut cliff, wave-cut platform, marine terrace, sea arch, sea stack, spit, baymouth bar, tombolo, barrier island

- Erosional features include wave-cut cliffs (which originate from the cutting action of the surf against the base of coastal land), wave-cut platforms (relatively flat, bench-like surfaces left behind by receding cliffs), and marine terraces (uplifted wave-cut platforms). Erosional features also include sea arches (formed when a headland is eroded and two sea caves from opposite sides unite) and sea stacks (formed when the roof of a sea arch collapses).
- Some of the depositional features that form when sediment is moved by beach drift and longshore currents are spits (elongated ridges of sand that project from the land into the mouth of an adjacent bay), baymouth bars (sandbars that completely cross a bay), and tombolos (ridges of sand that connect an island to the mainland or to another island). Along the Atlantic and Gulf coastal plains, the coastal region is characterized by offshore barrier islands, which are low ridges of sand that parallel the coast.

? **Identify the lettered features in this diagram.**

10.7 Stabilizing the Shore

Summarize the ways in which people deal with shoreline erosion problems.

KEY TERMS: hard stabilization, groin, breakwater, seawall, beach nourishment

- Local *factors that influence shoreline erosion* are (1) the proximity of a coast to sediment-laden rivers, (2) the degree of tectonic activity, (3) the topography and composition of the land, (4) prevailing winds and weather patterns, and (5) the configuration of the coastline and near-shore areas.
- Hard stabilization is a term that pertains to any structures built along the coastline to prevent movement of sand. Groins are oriented perpendicular to the coast, with the goal of slowing beach erosion by longshore currents. Breakwaters are parallel to the coast but located some distance offshore; their goal is to blunt the force of incoming ocean waves, often to protect boats. Like breakwaters, seawalls are parallel to the coast, but they are built on the shoreline itself. Often the installation of hard stabilization actually leads to increased erosion.
- Beach nourishment is an expensive alternative to hard stabilization. Sand is pumped onto the beach from some other area, temporarily replenishing the sediment supply. An alternative to hard stabilization and beach nourishment is relocating buildings away from high-risk areas and leaving the beach to be shaped by natural processes.

? **Based on their position and orientation, identify the three types of hard stabilization illustrated in this diagram.**

10.8 Contrasting America's Coasts

Contrast the erosion problems faced along different parts of America's coasts. Distinguish between emergent and submergent coasts.

KEY TERMS: emergent coast, submergent coast, estuary

- The Atlantic and Gulf coasts of the United States are markedly different from the Pacific coast. The Atlantic and Gulf coasts are lined in many places by barrier islands—dynamic expanses of sand that see a lot of change during storm events. Many of these low and narrow islands have also been prime sites for real estate development.
- The Pacific coast's big issue is the thinning of beaches due to sediment starvation. Rivers that drain to the coast (bringing it sand) have been dammed, resulting in reservoirs that trap sand before it can make it to the coast. Thinner beaches offer less resistance to incoming waves, often leading to erosion of bluffs behind the beach.
- Coasts may be classified by their changes relative to sea level. Emergent coasts are sites of either land uplift or sea-level fall. Marine terraces are features of emergent coasts. Submergent coasts are sites of land subsidence or sea-level rise. One characteristic of submergent coasts is drowned river valleys called estuaries.

? **What term is applied to the masses of rock protruding from the water in this photo? How did they form? Is the location more likely along the Gulf Coast or the coast of California? Explain.**

Michael Collier

10.9 Tides

Explain the cause of tides, their monthly cycles, and patterns. Describe the horizontal flow of water that accompanies the rise and fall of tides.

KEY TERMS: tide, spring tide, neap tide, diurnal tidal pattern, semidiurnal tidal pattern, mixed tidal pattern, tidal current, tidal flat, tidal delta

- Tides are daily changes in ocean surface elevation. They are caused by gravitational pull on ocean water by the Moon and, to a lesser extent, the Sun. When the Sun, Earth, and Moon all line up about every 2 weeks (full moon or new moon), the tides are most exaggerated. When a quarter moon is in the sky, the Moon is pulling on Earth's water at a right angle relative to the Sun, and the daily tidal range is minimized as the two forces partially counteract one another.

- Tides are strongly influenced by local conditions, including the shape of the local coastline and the depth of the ocean basin. Tidal patterns may be diurnal (one high tide per day), semidiurnal (two high tides per day), or mixed (similar to semidiurnal but with significant inequality between high tides).

- A flood current is the landward movement of water during the shift between low tide and high tide. When high tide transitions to low tide again, the movement of water away from the land is an ebb current. Ebb currents may expose tidal flats to the air. If a tide passes through an inlet, the current may carry sediment that gets deposited as a tidal delta.

GIVE IT SOME THOUGHT

1. In this chapter you learned that global winds are the force that drives surface-ocean currents. A glance at the accompanying map, however, shows a surface current that does not exactly coincide with the prevailing wind. Provide an explanation.

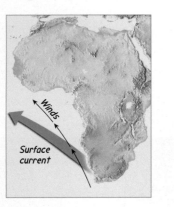

2. If the North Atlantic Drift were to cease, how might the climate of Western Europe change?

3. During a visit to the beach, you and a friend get in a rubber raft and paddle out into deep water *beyond* the surf zone. Tiring, you stop and take a rest. Describe the movement of the raft during your rest. How does this movement differ, if at all, from what you would have experienced if you had stopped paddling while *in* the surf zone?

4. Answer these questions that relate to the photo of a surfer enjoying a ride on a large wave.
   a. What is the likely source of energy that created this wave?
   b. How was the wavelength changing just prior to the time this photo was taken?
   c. Why was the wavelength changing?
   d. Many waves exhibit circular orbital motion. Is that true of the wave in this photo? Explain.

Ron Dahlquist/Getty Images

5. Examine the accompanying aerial photo that shows a portion of the New Jersey coast. What term is applied to the wall-like structures that extend into the water? What is their purpose? In what direction are beach drift and longshore currents moving sand—toward the top or toward the bottom of the photo?

John S. Shelton/University of Washington Libraries

6. You and a friend set up an umbrella and chairs at a beach. Your friend then goes into the surf zone to play Frisbee with another person. Several minutes later, your friend looks back toward the beach and is surprised to see that she is no longer near where the umbrella and chairs are set up. Although she is still in the surf zone, she is 30 yards away from where she started. How would you explain to your friend why she moved along the shore?

7. A friend wants to purchase a vacation home on a barrier island. If consulted, what advice would you give your friend?

8. This photo shows a small portion of a coastal area in the United States. Is this an emergent coast or a submergent coast? How did you figure this out? Is the location more likely the coast of North Carolina or California? Explain.

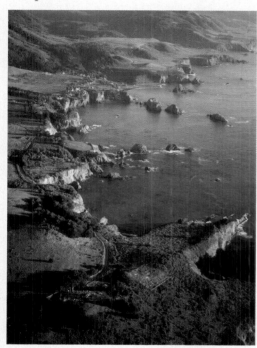

Michael Collier

9. The force of gravity plays a critical role in creating ocean tides. The more massive an object, the stronger its pull of gravity. Explain why the Sun's influence is only about half that of the Moon, even though the Sun is much more massive than the Moon.

10. This photo shows a portion of the Maine coast. The brown muddy area in the foreground is influenced by tidal currents. What term is applied to this muddy area? Name the type of tidal current this area will experience in the hours to come.

Marli Miller

MasteringGeology™   www.masteringgeology.com

Looking for additional review and test prep materials? Visit the Study Area in MasteringGeology to enhance your understanding of this chapter's content by accessing a variety of resources, including **Self-Study Quizzes,**

Geoscience Animations, GEODe Tutorials, RSS feeds, flashcards, web links, self-study quizzes, and an optional **Pearson eText.**

11

Heating the Atmosphere

FOCUS ON CONCEPTS

Each statement represents the primary learning objective for the corresponding major heading within the chapter. After you complete the chapter, you should be able to:

11.1 Distinguish between weather and climate and name the basic elements of weather and climate.

11.2 List the major gases composing Earth's atmosphere and identify the components that are most important to understanding weather and climate.

11.3 Interpret a graph that shows changes in air pressure from Earth's surface to the top of the atmosphere. Sketch and label a graph that shows atmospheric layers based on temperature.

11.4 Explain what causes the Sun angle and length of daylight to change during the year and describe how these changes produce the seasons.

11.5 Distinguish between heat and temperature. List and describe the three mechanisms of heat transfer.

11.6 Sketch and label a diagram that shows the paths taken by incoming solar radiation. Summarize the greenhouse effect.

11.7 Summarize the nature and cause of the atmosphere's changing composition since about 1750. Describe the atmosphere's response and some possible future consequences.

11.8 Calculate five commonly used types of temperature data and interpret a map that depicts temperature data using isotherms.

11.9 Discuss the principal controls of temperature and use examples to describe their effects.

11.10 Interpret the patterns depicted on world maps of January and July temperatures.

Mauna Loa Observatory is an important atmospheric research facility on Hawaii's Big Island. It has been collecting data and monitoring atmospheric change since the 1950s. (Photo by Forrest M. Mims III)

Earth's atmosphere is unique. No other planet in our solar system has an atmosphere with the exact mixture of gases or the heat and moisture conditions necessary to sustain life as we know it. The gases that make up Earth's atmosphere and the controls to which they are subject are vital to our existence. In this chapter, we will begin our examination of the ocean of air in which we all must live. What is the composition of the atmosphere? Where does the atmosphere end, and where does outer space begin? What causes the seasons? How is air heated? What factors control temperature variations around the globe?

11.1 Focus on the Atmosphere

Distinguish between weather and climate and name the basic elements of weather and climate.

Weather influences our everyday activities, our jobs, and our health and comfort. Many of us pay little attention to the weather unless we are inconvenienced by it or it adds to our enjoyment outdoors. Nevertheless, few other aspects of our physical environment affect our lives more than the phenomena we collectively call the weather.

Weather in the United States

The United States occupies an area that stretches from the tropics to the Arctic Circle. It has thousands of miles of coastline and extensive regions that are far from the influence of the ocean. Some landscapes are mountainous and others are dominated by plains. It is a place where Pacific storms strike the West coast, and the East is sometimes influenced by events in the Atlantic and the Gulf of Mexico. For those in the center of the country, it is common to experience weather events triggered when frigid southward-bound Canadian air masses clash with northward-moving ones.

Stories about weather are a routine part of the daily news. Articles and items about the effects of heat, cold, floods, drought, fog, snow, ice, and strong winds are commonplace. Of course, storms of all kinds are frequently front-page news (**Figure 11.1**). Beyond its direct impact on the lives of individuals, the weather has a strong effect on the world economy, influencing agriculture, energy use, water resources, transportation, and industry.

Weather clearly influences our lives a great deal. Yet, it is important to realize that people influence the atmosphere and its behavior as well (**Figure 11.2**). There are, and will continue to be, significant political and scientific decisions that must be made involving these impacts. Important examples are air pollution control and the effects of human activities on global climate and the atmosphere's protective ozone layer. So there is a need for increased awareness and understanding of our atmosphere and its behavior.

Weather and Climate

Acted on by the combined effects of Earth's motions and energy from the Sun, our planet's formless and invisible envelope of air reacts by producing an infinite variety of weather, which in turn creates the basic patterns of global climates. Although not identical, weather and climate have much in common.

Weather is constantly changing, sometimes from hour to hour and at other times from day to day. It is a term that refers to the state of the atmosphere at a given time and place. Whereas changes in the weather are continuous and sometimes seemingly erratic, it is nevertheless possible to arrive at a generalization of these variations. Such a description of aggregate weather conditions is termed **climate**. It is based on observations that have been accumulated over many years. Climate is often defined simply as "average weather," but this is an inadequate definition. To more accurately portray the character of an area, variations and extremes must also be included, as well as the probabilities that such departures will take place. For example, it is not only necessary for farmers to know the average rainfall during the growing season but also important to know the frequency of extremely wet and extremely dry years. Thus, climate is the sum of all statistical weather information that helps describe a place or region.

Suppose you were planning a vacation trip to an unfamiliar place. You would probably want to know what kind of weather to expect. Such information would help as you selected clothes to pack and could influence decisions regarding activities you might engage in during your stay. Unfortunately, weather forecasts that go beyond a few days are not very dependable. Therefore, you might ask someone who is familiar with the area about what kind of weather to expect. "Are thunderstorms common?" "Does it get cold at night?" "Are the afternoons sunny?" What you are seeking is information about the climate, the conditions that are typical for that place.

Figure 11.1 Stormy weather Few aspects of our physical environment influence our daily lives more than the weather. A winter blizzard of historic proportions struck Chicago, Illinois, in early February 2011. (AP Photo/Kiichiro Sato) Severe thunderstorms are relatively common occurrence in many parts of the United States during warmer months. (Photo by Mark Newman/SuperStock)

Figure 11.2 People influence the atmosphere An air pollution episode in Shanghai, China. (Photo by Doable/Amana Images/Glow Images)

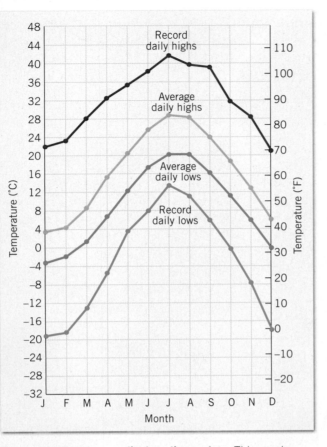

Figure 11.3 Graphs can display climate data This graph shows daily temperature data for New York City. In addition to the average daily maximum and minimum temperatures for each month, extremes are also shown. As this graph shows, there can be significant departures from the average.

Another useful source of information is the great variety of climate tables, maps, and graphs that are available. For example, the graph in **Figure 11.3** shows average daily high and low temperatures for each month, as well as extremes, for New York City.

Such information could no doubt help as you planned your trip. But it is important to realize that *climate data cannot predict the weather.* Although a place may usually (climatically) be warm, sunny, and dry during the time of your planned vacation, you may actually experience cool, overcast, and rainy weather. There is a well-known saying that summarizes this idea: "Climate is what you expect, but weather is what you get."

The nature of both weather and climate is expressed in terms of the same basic **elements**—quantities or properties that are measured regularly. The most important elements are (1) air temperature, (2) humidity, (3) type and amount of cloudiness, (4) type and amount of precipitation, (5) air pressure, and (6) the speed and direction of the wind. These elements are the major variables by which weather patterns and climate types are depicted. Although you will study these elements separately at first, keep in mind that they are very much interrelated. A change in one of the elements often produces changes in the others.

11.1 CONCEPT CHECKS

1. Distinguish between weather and climate.
2. Write two brief statements about your current location: one that relates to weather and one that relates to climate.
3. What is an *element*?
4. List the basic elements of weather and climate.

11.2 Composition of the Atmosphere

List the major gases composing Earth's atmosphere and identify the components that are most important to understanding weather and climate.

Sometimes the term *air* is used as if it were a specific gas, but it is not. Rather, **air** is a *mixture* of many discrete gases, each with its own physical properties, in which varying quantities of tiny solid and liquid particles are suspended.

Major Components

The composition of air is not constant; it varies from time to time and from place to place. If the water vapor, dust, and other variable components were removed from the atmosphere, we would find that its makeup is very stable worldwide, up to an altitude of about 80 kilometers (50 miles).

As you can see in **Figure 11.4**, two gases—nitrogen and oxygen—make up 99 percent of the volume of clean, dry air. Although these gases are the most plentiful components of the atmosphere and are of great significance to life on Earth, they are of little or no importance in affecting weather phenomena. The remaining 1 percent of dry air is mostly the inert gas argon (0.93 percent) plus tiny quantities of a number of other gases.

Carbon Dioxide (CO_2)

Carbon dioxide, although present in only minute amounts (0.0391 percent, or 395 parts per million [ppm]), is nevertheless an important constituent of air. Carbon dioxide

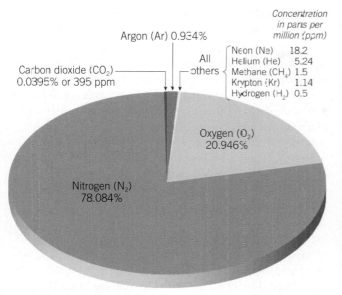

Concentration
in parts per
million (ppm)

Argon (Ar) 0.934%

Carbon dioxide (CO₂)
0.0395% or 395 ppm

All
others

Neon (Ne)	18.2
Helium (He)	5.24
Methane (CH₄)	1.5
Krypton (Kr)	1.14
Hydrogen (H₂)	0.5

Oxygen (O₂)
20.946%

Nitrogen (N₂)
78.084%

Figure 11.4 Composition of the atmosphere This graph shows the proportional volume of gases composing dry air. Nitrogen and oxygen clearly dominate.

is of great interest to meteorologists because it is an efficient absorber of energy emitted by Earth and thus influences the heating of the atmosphere. Although the proportion of carbon dioxide in the atmosphere is relatively uniform, its percentage has been rising steadily for 200 years. **Figure 11.5** is a graph that shows the growth in atmospheric CO_2 since 1958. Much of this rise is

attributed to the burning of ever-increasing quantities of fossil fuels, such as coal and oil. Some of this additional carbon dioxide is absorbed by the ocean or is used by plants, but more than 40 percent remains in the air. Estimates project that by sometime in the second half of the twenty-first century, CO_2 levels will be twice as high as pre-industrial level.

Most atmospheric scientists agree that increased carbon dioxide concentrations have contributed to a warming of Earth's atmosphere over the past several decades and will continue to do so in the decades to come. The magnitude of such temperature changes is uncertain and depends partly on the quantities of CO_2 contributed by human activities in the years ahead. The role of carbon dioxide in the atmosphere and its possible effects on climate are examined later in the chapter.

Variable Components

Air includes many gases and particles that vary significantly from time to time and from place to place. Important examples include water vapor, dust particles, and ozone. Although usually present in small percentages, they can have significant effects on weather and climate.

Water Vapor You are probably familiar with the term *humidity* from watching weather reports on TV. Humidity is a reference to water vapor in the air. As you will learn in Chapter 12, there are several ways to express humidity. The amount of water vapor in the air varies considerably, from practically none at all up to about 4 percent by volume. Why is such a small fraction of the atmosphere so significant? The fact that water vapor is the source of all clouds and precipitation would be enough to explain its importance. However, water vapor has other roles. Like carbon dioxide, water vapor absorbs heat energy emitted by Earth as well as some solar energy. It is therefore important when we examine the heating of the atmosphere.

When water changes from one state to another (see Figure 12.2, page 389), it absorbs or releases heat. This energy is termed *latent heat*, which means "hidden" heat. As we will see in later

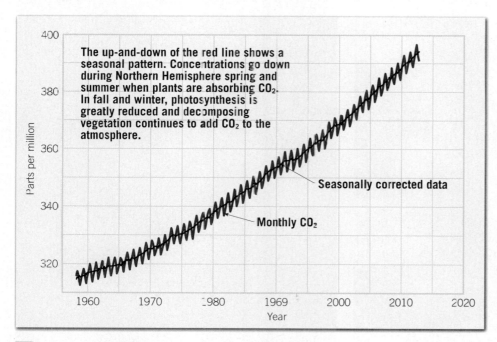

The up-and-down of the red line shows a seasonal pattern. Concentrations go down during Northern Hemisphere spring and summer when plants are absorbing CO₂. In fall and winter, photosynthesis is greatly reduced and decomposing vegetation continues to add CO₂ to the atmosphere.

Seasonally corrected data

Monthly CO₂

Parts per million

400
380
360
340
320

1960   1970   1980   1969   2000   2010   2020

Year

SmartFigure 11.5 Monthly CO₂ concentrations Atmospheric CO_2 has been measured at Mauna Loa Observatory, Hawaii, since 1958 (see the chapter-opening photo). There has been a consistent increase since monitoring began. This graphic portrayal is known as the *Keeling Curve* in honor of the scientist who originated the measurements. (NOAA)

chapters, water vapor in the atmosphere transports this latent heat from one region to another, and it is the energy source that helps drive many storms.

Aerosols The movements of the atmosphere are sufficient to keep a large quantity of solid and liquid particles suspended within it. Although visible dust sometimes clouds the sky, these relatively large particles are too heavy to stay in the air very long. Many other particles are microscopic and remain suspended for considerable periods of time. They may originate from many sources, both natural and human made, and include sea salts from breaking waves; fine soil blown into the air; smoke and soot from fires; pollen and microorganisms lifted by the wind, ash, and dust from volcanic eruptions; and more (**Figure 11.6A**). Collectively, these tiny solid and liquid particles are called **aerosols**.

From a meteorological standpoint, these tiny, often invisible particles can be significant. First, many act as surfaces on which water vapor can condense, an important function in the formation of clouds and fog. Second, aerosols can absorb or reflect incoming solar radiation. Thus, when an air pollution episode is occurring or when ash fills the sky following a volcanic eruption, the amount of sunlight reaching Earth's surface can be measurably reduced. Finally, aerosols contribute to an optical phenomenon we have all observed—the varied hues of red and orange at sunrise and sunset (**Figure 11.6B**).

Ozone Another important component of the atmosphere is **ozone**. It is a form of oxygen that combines three oxygen atoms into each molecule (O_3). Ozone is not the same as the oxygen we breathe, which has two atoms per molecule (O_2). There is very little ozone in the atmosphere, and its distribution is not uniform. It is concentrated between 10 and 50 kilometers (6 and 31 miles) above the surface, in a layer called the *stratosphere*.

In this altitude range, oxygen molecules (O_2) are split into single atoms of oxygen (O) when they absorb ultraviolet radiation emitted by the Sun. Ozone is then created when a single atom of oxygen (O) and a molecule of oxygen (O_2) collide. This must happen in the presence of a third, neutral molecule that acts as a *catalyst* by allowing the reaction to take place without itself being consumed in the process. Ozone is concentrated in the 10- to 50-kilometer (6- to 31-mile) height range because a crucial balance exists there: The ultraviolet radiation from the Sun is sufficient to produce single atoms of oxygen, and there are enough gas molecules to bring about the required collisions.

The presence of the ozone layer in our atmosphere is crucial to those who dwell on Earth. The reason is that

A.

B.

Figure 11.6 Aerosols A. This satellite image from November 11, 2002, shows two examples of aerosols. First, a large dust storm is blowing across northeastern China toward the Korean Peninsula. Second, a dense haze toward the south (bottom center) is human-generated air pollution. (NASA) **B.** Dust in the air can cause sunsets to be especially colorful. (Photo by elwynn/Shutterstock)

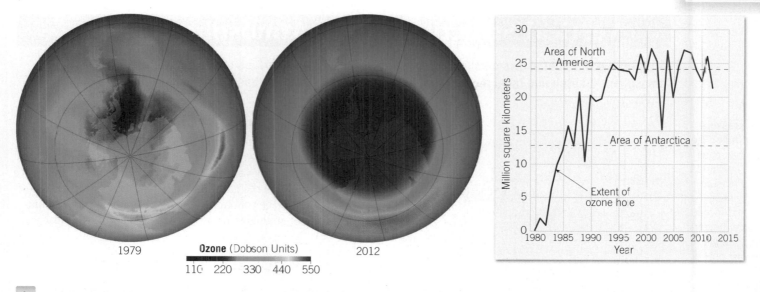

1979   Ozone (Dobson Units)   2012
110   220   330   440   550

SmartFigure 11.7 Antarctic ozone hole The two satellite images show ozone distribution in the Southern Hemisphere on the days in September 1979 and 2012 when the ozone hole was largest. The dark blue shades over Antarctica correspond to the region with the sparsest ozone. The ozone hole is not technically a "hole" where no ozone is present but is actually a region of exceptionally depleted ozone in the stratosphere over the Antarctic that occurs in the spring. The small graph traces changes in the maximum size of the ozone hole, 1980 to 2012. (NASA)

ozone absorbs the potentially harmful ultraviolet (UV) radiation from the Sun. If ozone did not filter a great deal of the ultraviolet radiation, and if the Sun's UV rays reached the surface of Earth undiminished, our planet would be uninhabitable for most life as we know it. Thus, anything that reduces the amount of ozone in the atmosphere could affect the well-being of life on Earth. Just such a problem exists and is described in the next section.

Ozone Depletion—A Global Issue

Although stratospheric ozone is 10 to 50 kilometers (6 to 31 miles) above Earth's surface, it is vulnerable to human activities. Chemicals we produce are breaking up ozone molecules in the stratosphere, weakening our shield against UV rays. This loss of ozone is a serious global-scale environmental problem. Measurements since the late 1970s confirm that ozone depletion is occurring worldwide and is especially pronounced near Earth's poles. You can see this effect over the South Pole in **Figure 11.7**.

Over the past half century, people have unintentionally placed the ozone layer in jeopardy by polluting the atmosphere. The most significant of the offending chemicals are known as *chlorofluorocarbons* (*CFCs* for short). Over the decades, CFCs were used as coolants for air-conditioning and refrigeration equipment, as cleaning solvents for electronic components, and as propellants for aerosol sprays, and they were used in the production of certain plastic foams.

Because CFCs are practically inert (that is, not chemically active) in the lower atmosphere, some of these gases gradually make their way to the ozone layer, where

sunlight separates the chemicals into their constituent atoms. The chlorine atoms released this way break up some of the ozone molecules.

Because ozone filters out most of the UV radiation from the Sun, a decrease in its concentration permits more of these harmful wavelengths to reach Earth's surface. The most serious threat to human health is an increased risk of skin cancer. An increase in damaging UV radiation also can impair the human immune system as well as promote cataracts, a clouding of the eye lens that reduces vision and may cause blindness if not treated.

In response to this problem, an international agreement known as the *Montreal Protocol* was developed under the sponsorship of the United Nations to eliminate the production and use of CFCs. Although relatively strong action was taken, CFC levels in the atmosphere will not drop rapidly. Once in the atmosphere, CFC molecules can take many years to reach the ozone layer, and once there, they can remain active for decades. This does not promise a near-term reprieve for the ozone layer. Between 2060 and 2075, the abundance of ozone-depleting gases is projected to fall to values that existed before the ozone hole began to form in the 1980s.

11.2 CONCEPT CHECKS

1. Is *air* a specific gas? Explain.
2. What are the two major components of clean, dry air? What proportion does each represent?
3. Why are water vapor and aerosols important constituents of Earth's atmosphere?
4. What is ozone? Why is ozone important to life on Earth?
5. What are CFCs, and what is their connection to the ozone problem?

Did You Know?

Although naturally occurring ozone in the stratosphere is critical to life on Earth, it is regarded as a pollutant when produced at ground level because it can damage vegetation and can be harmful to human health. Ozone is a major component in a noxious mixture of gases and particles called *photochemical smog*. It forms as a result of reactions triggered by sunlight that occur among pollutants emitted by motor vehicles and industrial sources.

11.3 Vertical Structure of the Atmosphere

Interpret a graph that shows changes in air pressure from Earth's surface to the top of the atmosphere. Sketch and label a graph that shows atmospheric layers based on temperature.

Figure 11.8 Air pressure changes with altitude The rate of pressure decrease with an increase in altitude is not constant. Pressure decreases rapidly near Earth's surface and more gradually at greater heights. Put another way, the graph shows that the vast bulk of the gases making up the atmosphere is very near Earth's surface and that the gases gradually merge with the emptiness of space.

One-half of the atmosphere lies below an altitude of 5.6 kilometers (3.5 miles). At about 16 kilometers (10 miles), 90 percent of the atmosphere has been traversed, and above 100 kilometers (62 miles), only 0.00003 percent of all the gases making up the atmosphere remain. Even so, traces of our atmosphere extend far beyond this altitude, gradually merging with the emptiness of space.

Temperature Changes

By the early twentieth century, much had been learned about the lower atmosphere. The upper atmosphere was partly known from indirect methods. Data from balloons and kites showed that near Earth's surface air, temperature drops with increasing height. This phenomenon is felt by anyone who has climbed a high mountain and is obvious in pictures of snowcapped mountaintops rising above snow-free lowlands (**Figure 11.9**). We divide the atmosphere vertically into four layers, on the basis of temperature (**Figure 11.10**).

Troposphere The lowermost layer in which we live, where temperature decreases with an increase in altitude, is the **troposphere**. The term literally means the region where air "turns over," a reference to the turbulent weather in this lowermost zone. The troposphere is the chief focus of meteorologists, because it is in this layer that essentially all important weather phenomena occur.

To say that the atmosphere begins at Earth's surface and extends upward is obvious. However, where does the atmosphere end, and where does outer space begin? There is no sharp boundary; the atmosphere rapidly thins as you travel away from Earth, until there are too few gas molecules to detect.

Pressure Changes

To understand the vertical extent of the atmosphere, let us examine changes in atmospheric pressure with increasing height. Atmospheric pressure is simply the weight of the air above. At sea level, the average pressure is slightly more than 1000 millibars (mb). This corresponds to a weight of slightly more than 1 kilogram per square centimeter (14.7 pounds per square inch). The pressure at higher altitudes is less (**Figure 11.8**).

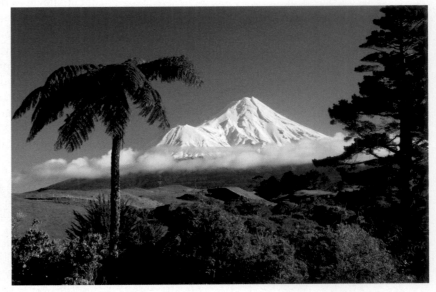

Figure 11.9 Temperatures drop with an increase in altitude in the troposphere Snow-capped mountains and snow-free lowlands are a reminder that temperatures decrease as we go higher in the troposphere. (Photo by David Wall/Alamy)

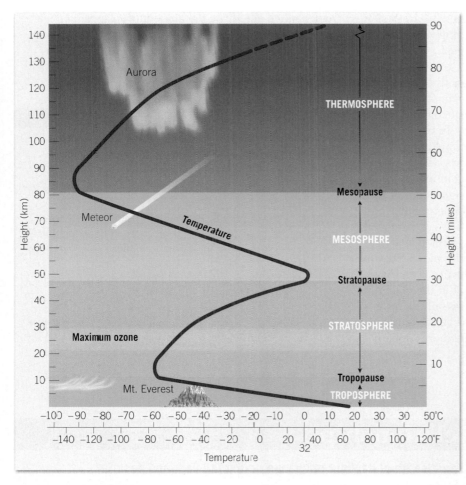

Figure 11.10 **Thermal structure of the atmosphere** Earth's atmosphere is traditionally divided into four layers, based on temperature.

Figure 11.11 **Radiosonde** A radiosonde is a lightweight package of instruments that is carried aloft by a small weather balloon. It transmits data on vertical changes in temperature, pressure, and humidity in the troposphere. The troposphere is where practically all weather phenomena occur, so it is very important to have frequent measurements (Photo by David R. Frazier/DanitaDelimont.com/Newscom)

The temperature decrease in the troposphere is called the **environmental lapse rate**. Although its average value is 6.5°C per kilometer (3.5°F per 1000 feet), a figure known as the *normal lapse rate*, its value is variable. To determine the actual environmental lapse rate for any particular time and place, as well as to gather information about vertical changes in pressure, wind, and humidity, radiosondes are used. A **radiosonde** is an instrument package that is attached to a balloon and transmits data by radio as it ascends through the atmosphere (**Figure 11.11**).

The thickness of the troposphere is not the same everywhere; it varies with latitude and the season. On average, the temperature drop continues to a height of about 12 kilometers (7.4 miles). The outer boundary of the troposphere is the *tropopause*.

Stratosphere Beyond the tropopause is the **stratosphere**. In the stratosphere, the temperature remains constant to a height of about 20 kilometers

(12 miles) and then begins an increase that continues until the *stratopause*, at a height of about 50 kilometers (31 miles) above Earth's surface. Below the tropopause, atmospheric properties such as temperature and humidity are readily transferred by large-scale turbulence and mixing. Above the tropopause, in the stratosphere, they are not. The reason for the increased temperatures in the stratosphere is that the atmosphere's ozone is concentrated in this layer. Recall that ozone absorbs ultraviolet radiation from the Sun. As a consequence, the stratosphere is heated.

Mesosphere In the third layer, the **mesosphere**, temperatures again decrease with height until, at the *mesopause*—approximately 80 kilometers (50 miles) above the surface—the temperature approaches −90°C (−130°F). The coldest temperatures anywhere in the atmosphere occur at the mesopause. Because accessibility is difficult, the mesosphere is one of the least-explored regions of the atmosphere. It cannot be reached by the highest research balloons, nor is it accessible to the lowest orbiting satellites. Recent technical developments are just beginning to fill this knowledge gap.

Thermosphere The fourth layer extends outward from the mesopause and has no well-defined upper limit. This is the **thermosphere**, a layer that contains only a tiny fraction of the atmosphere's mass. In the extremely rarefied air of this outermost layer, temperatures again increase, due to the absorption of short-wave high-energy solar radiation by atoms of oxygen and nitrogen.

Temperatures rise to extremely high values of more than 1000°C (1832°F) in the thermosphere. But such temperatures are not comparable to those experienced near Earth's surface. Temperature is defined in terms of the average speed at which molecules move. Because the gases of the thermosphere are moving at very high speeds, the temperature is very high. But the gases are so sparse that, collectively, they possess only an insignificant quantity of heat. For this reason, the temperature of a satellite orbiting Earth in the thermosphere is determined chiefly by the amount of solar radiation it absorbs and not by the high temperature of the almost nonexistent surrounding air. If an astronaut inside a satellite were to expose his or her hand, the atmosphere would not feel hot.

11.3 CONCEPT CHECKS

1. Does air pressure increase or decrease with an increase in altitude? Is the rate of change constant or variable? Explain.
2. Is the outer edge of the atmosphere clearly defined? Explain.
3. The atmosphere is divided vertically into four layers, on the basis of temperature. List and describe these layers in order, from lowest to highest. In which layer does practically all of our weather occur?
4. What is the environmental lapse rate, and how is it determined?
5. Why are temperatures in the thermosphere not strictly comparable to those experienced near Earth's surface?

11.4 Earth–Sun Relationships

Explain what causes the Sun angle and length of daylight to change during the year and describe how these changes produce the seasons.

Nearly all the energy that drives Earth's variable weather and climate comes from the Sun. Earth intercepts only a minute percentage of the energy given off by the Sun—less than 1 two-billionth. This may seem to be an insignificant amount, until we realize that it is several hundred thousand times the electrical-generating capacity of the United States.

Solar energy is not distributed evenly over Earth's land–sea surface. The amount of energy received varies with latitude, time of day, and season of the year. Contrasting images of polar bears on ice rafts and palm trees along a remote tropical beach serve to illustrate the extremes. It is the unequal heating of Earth that ultimately creates winds and drives ocean currents. These movements, in turn, transport heat from the tropics toward the poles in an unending attempt to balance energy inequalities. The consequences of these processes are the phenomena we call weather.

If the Sun were to be "turned off," global winds would quickly subside. Yet, as long as the Sun shines, the winds will blow and the phenomena we know as weather will persist. So to understand how the atmosphere's dynamic weather machine works, we must first know why different latitudes receive varying quantities of solar energy and why the amount of solar energy changes to produce the seasons. As we will see, the variations in solar heating are caused by the motions of Earth relative to the Sun and by variations in Earth's land–sea surface.

Earth's Motions

Earth has two principal motions—rotation and revolution. **Rotation** is the spinning of Earth about its axis. The axis is an imaginary line that runs through the poles. Our planet rotates once every 24 hours, producing the daily cycle of daylight and darkness. At any moment, half of Earth is experiencing daylight and the other half darkness. The line separating the dark half of Earth from the lighted half is called the **circle of illumination**.

Revolution refers to the movement of Earth in a slightly elliptical orbit around the Sun. The distance between Earth and Sun averages about 150 million kilometers (93 million miles). Because Earth's orbit is not perfectly circular, however, the distance varies during the course of a year. Each year, on about January 3, our planet is about 147.3 million kilometers (91.5 million miles) from the Sun, closer than at any other time—a position known as *perihelion*. About 6 months later, on July 4, Earth is about 152 million kilometers

A. Summer solstice at 40°N latitude   **B. Spring or fall equinox at 40°N latitude**   **C. Winter solstice at 40°N latitude**

SmartFigure 11.12 The changing Sun angle Daily paths of the Sun for a place located at 40° north latitude for **A.** summer solstice, **B.** spring or fall equinox, and **C.** winter solstice As we move from summer to winter, the angle of the noon Sun decreases from 73½, to 26½ degrees—a difference of 47 degrees. Notice also how the location of sunrise (east) and sunset (west) changes during a year.

(94.5 million miles) from the Sun, farther away than at any other time—a position called *aphelion*. Although Earth is closest to the Sun and receives up to 7 percent more energy in January than in July, this difference plays only a minor role in producing seasonal temperature variations, as evidenced by the fact that Earth is closest to the Sun during the Northern Hemisphere winter.

What Causes the Seasons?

If variations in the distance between the Sun and Earth do not cause seasonal temperature changes, what does? The gradual but significant change in the length of daylight certainly accounts for some of the difference we notice between summer and winter. Furthermore, a gradual change in the angle (altitude) of the Sun

above the horizon is also a major contributing factor (**Figure 11.12**). For example, someone living in Chicago, Illinois, experiences the noon Sun highest in the sky in late June. But as summer gives way to autumn, the noon Sun appears lower in the sky, and sunset occurs earlier each evening.

The seasonal variation in the angle of the Sun above the horizon affects the amount of energy received at Earth's surface in two ways. First, when the Sun is directly overhead (at a 90-degree angle), the solar rays are most concentrated and thus most intense (**Figure 11.13A**). The lower the angle, the more spread out and less intense is the solar radiation that reaches the surface (**Figures 11.13B, C**). To illustrate this principle, hold a flashlight at a right angle to a surface and then change the angle.

Figure 11.13 The Sun angle influences the intensity of solar radiation at Earth's surface Changes in the Sun's angle cause variations in the amount of solar energy reaching Earth's surface. The nearer the angle to 90 degrees, the more intense the solar radiation.

Figure 11.14 Sun angle affects the path of sunlight through the atmosphere Notice that rays striking Earth at a low angle (toward the poles) must travel through more of the atmosphere than rays striking at a high angle (around the equator) and are thus subject to greater depletion by reflection, scattering, and absorption.

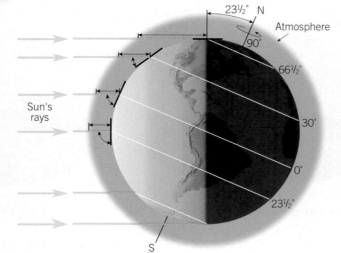

cannot look directly at the midday Sun, but we can enjoy gazing at a sunset.

It is important to remember that Earth's shape is spherical. On any given day, the only places that will receive vertical (90-degree) rays from the Sun are located along one particular line of latitude. As we move either north or south of that location, the Sun's rays strike at ever-decreasing angles. The nearer a place is to the latitude receiving vertical rays of the Sun, the higher will be its noon Sun and the more concentrated will be the radiation it receives (see Figure 11.14).

Earth's Orientation

What causes fluctuations in Sun angle and length of daylight during the course of a year? Variations occur because Earth's orientation to the Sun continually changes as it travels along its orbit. Earth's axis (the imaginary line through the poles around which Earth rotates) is not perpendicular to the plane of its orbit around the Sun. Instead, it is tilted 23½° from the perpendicular, as shown in Figure 11.14. This is termed the **inclination of the axis**. If the axis were not inclined, Earth would lack seasons. Because the axis remains pointed in the same direction (toward the North Star), the orientation of Earth's axis to the Sun's rays is constantly changing (**Figure 11.15**).

For example, on one day in June each year, Earth's position in orbit is such that the Northern Hemisphere is "leaning" 23½° *toward* the Sun (see Figure 11.15, left). Six months later, in December, when Earth has moved to

Second, but of lesser importance, the angle of the Sun determines the path solar rays take as they pass through the atmosphere (**Figure 11.14**). When the Sun is directly overhead, the rays strike the atmosphere at a 90-degree angle and travel the shortest possible route to the surface. This distance is referred to as *1 atmosphere*. However, rays entering at a 30-degree angle must travel twice this distance before reaching the surface, while rays at a 5-degree angle travel through a distance roughly equal to the thickness of 11 atmospheres. The longer the path, the greater the chance that sunlight will be dispersed by the atmosphere, which reduces the intensity at the surface. These conditions account for the fact that we

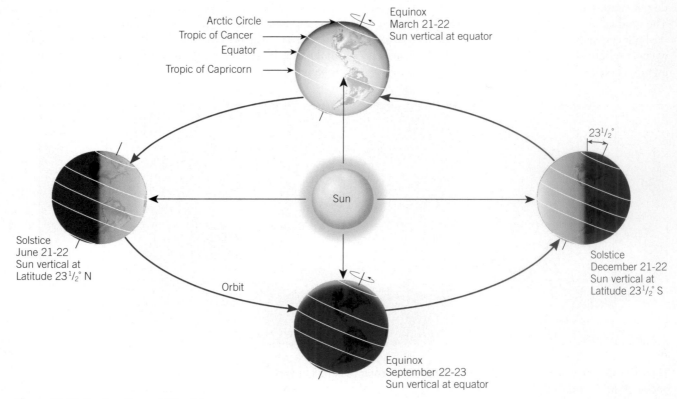

Figure 11.15 Earth—Sun relationships

the opposite side of its orbit, the Northern Hemisphere leans 23½° *away* from the Sun (see Figure 11.15, right). On days between these extremes, Earth's axis is leaning at amounts less than 23½° to the rays of the Sun. This change in orientation causes the spot where the Sun's rays are vertical to make a yearly migration from 23½° north of the equator to 23½° south of the equator.

In turn, this migration causes the angle of the noon Sun to vary by as much as 47° (23½ + 23½) during the year at places located poleward of latitude 23½°. For example, a midlatitude city such as New York (about 40° north latitude) has a maximum noon Sun angle of 73½° when the Sun's vertical rays reach their farthest northward location in June and a minimum noon Sun angle of 26½° 6 months later.

Solstices and Equinoxes

Historically, 4 days each year have been given special significance, based on the annual migration of the direct rays of the Sun and its importance to the yearly weather cycle. On June 21 or 22, Earth is in a position such that the north end of its axis is tilted 23½° *toward* the Sun (**Figure 11.16A**). At this time, the vertical rays of the Sun strike 23½° north latitude (23½° north of the equator), a latitude known as the **Tropic of Cancer**. For people in the Northern Hemisphere, June 21 or 22 is known as the **summer solstice**, the first "official" day of summer.

Six months later, on about December 21 or 22, Earth is in the opposite position, with the Sun's vertical rays striking at 23½° south latitude (**Figure 11.16B**). This parallel is known as the **Tropic of Capricorn**. For those in the Northern Hemisphere, December 21 and 22 is the **winter solstice**. However, at the same time in the Southern Hemisphere, people are experiencing just the opposite—the summer solstice.

Midway between the solstices are the equinoxes. September 22 or 23 is the date of the **autumnal equinox** in the Northern Hemisphere, and March 21 or 22 is the date of the **spring equinox**. On these dates, the vertical rays of the Sun strike the equator (0° latitude) because Earth is in such a position in its orbit that the axis is tilted neither toward nor away from the Sun (**Figure 11.16C**).

The length of daylight versus darkness is also determined by Earth's position in orbit. The length of daylight on June 21, the summer solstice in the Northern Hemisphere, is greater than the length of night. This fact can be established from Figure 11.16A by comparing the fraction of a given latitude that is on the "day" side of the circle of illumination with the fraction on the "night" side. The opposite is true for the winter solstice, when the nights are longer than the days. Again for comparison, let us consider New York City, which has about 15 hours of daylight on June 21 and only about 9 hours on December 21 (you can see this in Figure 11.16 and

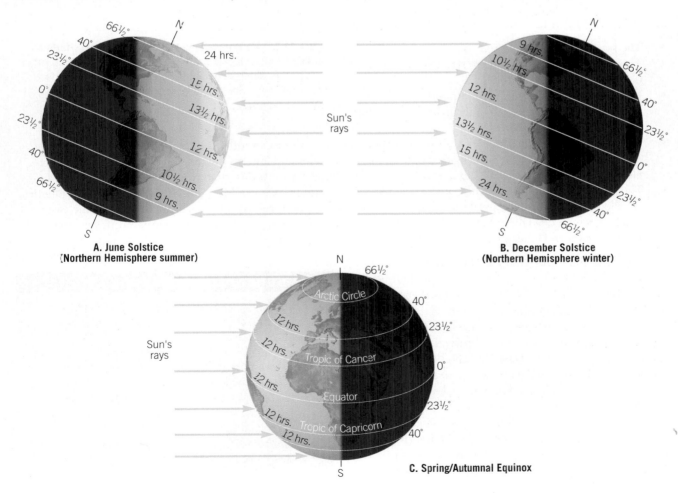

A. June Solstice (Northern Hemisphere summer)

B. December Solstice (Northern Hemisphere winter)

C. Spring/Autumnal Equinox

SmartFigure 11.16
Characteristics of the solstices and equinoxes

Latitude (degrees)	Summer Solstice	Winter Solstice	Equinoxes
0	12 h	12 h	12 h
10	12 h 35 min	11 h 25 min	12
20	13 h 12	10 h 48	12
30	13 h 56	10 h 04	12
40	14 h 52	9 h 08	12
50	16 h 18	7 h 42	12
60	18 h 27	5 h 33	12
70	24 h (for 2 mo)	0 00	12
80	24 h (for 4 mo)	0 00	12
90	24 h (for 6 mo)	0 00	12

TABLE 11.1 Length of Daylight

TABLE 11.1). Also note from Table 11.1 that the farther north of the equator you are on June 21, the longer the period of daylight. When you reach the Arctic Circle (66½° north latitude), the length of daylight is 24 hours. This is the land of the "midnight Sun," which does not set for about 6 months at the North Pole.

During an equinox (meaning "equal night"), the length of daylight is 12 hours *everywhere* on Earth, because the circle of illumination passes directly through the poles, dividing the latitudes in half (see Figure 11.16C).

As a review of the characteristics of the summer solstice for the Northern Hemisphere, examine Figure 11.16A and Table 11.1 and consider the following facts:

- The solstice occurs on June 21 or 22.
- The vertical rays of the Sun are striking the Tropic of Cancer (23½° north latitude).
- Locations in the Northern Hemisphere are experiencing their greatest length of daylight (opposite for the Southern Hemisphere).
- Locations north of the Tropic of Cancer are experiencing their highest noon Sun angles (opposite for places south of the Tropic of Capricorn).
- The farther you are north of the equator, the longer the period of daylight, until the Arctic Circle is reached, where daylight lasts for 24 hours (opposite for the Southern Hemisphere).

The facts about the winter solstice are just the opposite. It should now be apparent why a midlatitude location is warmest in the summer, for it is then that days are longest and Sun's altitude is highest.

These seasonal changes, in turn, cause the month-to-month variations in temperature observed at most locations outside the tropics. **Figure 11.17** shows mean monthly temperatures for selected cities at different latitudes. Notice that the cities located at more poleward latitudes experience larger temperature differences from summer to winter than do cities located nearer the equator. Also notice that temperature minimums for Southern Hemisphere locations occur in July, whereas they occur in January for most places in the Northern Hemisphere.

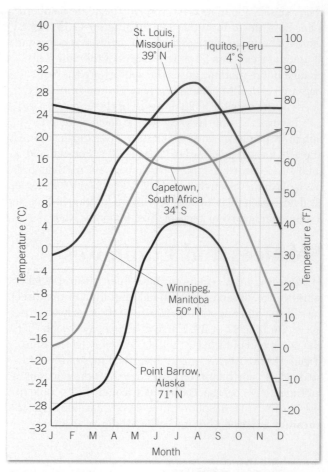

Figure 11.17 Monthly temperatures for cities at different latitudes Places located at higher latitudes experience larger temperature differences between summer and winter. Note that Cape Town, South Africa, experiences winter in June, July, and August.

All places at the same latitude have identical Sun angles and lengths of daylight. If the Earth–Sun relationships just described were the only controls of temperature, we would expect these places to have identical temperatures as well. Obviously, this is not the case. Other factors that influence temperature are discussed later in the chapter.

11.4 CONCEPT CHECKS

1. Do the annual variations in Earth–Sun distance adequately account for seasonal temperature changes? Explain.
2. Use a simple sketch to show why the intensity of solar radiation striking Earth's surface changes when the Sun angle changes.
3. Briefly explain the primary cause of the seasons.
4. What is the significance of the Tropic of Cancer and the Tropic of Capricorn?
5. After examining Table 11.1, write a general statement that relates the season, latitude, and the length of daylight.

11.5 Energy, Heat, and Temperature

Distinguish between heat and temperature. List and describe the three mechanisms of heat transfer.

The universe is made up of a combination of matter and energy. The concept of matter is easy to grasp because it is the "stuff" we can see, smell, and touch. Energy, on the other hand, is abstract and therefore more difficult to describe. For our purposes, we will define energy simply as *the capacity to do work.* We can think of work as being accomplished whenever matter is moved. You are likely familiar with some of the common forms of energy, such as thermal, chemical, nuclear, radiant (light), and gravitational energy. One type of energy is described as *kinetic energy,* which is energy of motion. Recall that matter is composed of atoms or molecules that are constantly in motion and therefore possesses kinetic energy.

Heat is a term that is commonly used synonymously with *thermal energy.* In this usage, heat is energy possessed by a material arising from the internal motions of its atoms or molecules. Whenever a substance is heated, its atoms move faster and faster, which leads to an increase in its heat content. **Temperature**, on the other hand, is related to the average kinetic energy of a material's atoms or molecules. Stated another way, the term *heat* generally refers to the quantity of energy present, whereas the word *temperature* refers to the intensity—that is, the degree of "hotness."

Heat and temperature are closely related concepts. Heat is the energy that flows because of temperature differences. In all situations, *heat is transferred from warmer to cooler objects.* Thus, if two objects of different temperature are in contact, the warmer object will become cooler and the cooler object will become warmer until they both reach the same temperature.

The three mechanisms of heat transfer are conduction, convection, and radiation. Although we present them separately, all three processes go on simultaneously in the atmosphere. In addition, these mechanisms operate to transfer heat between Earth's surface (both land and water) and the atmosphere.

Mechanisms of Heat Transfer: Conduction

Conduction is familiar to all of us. Anyone who has touched a metal spoon that was left in a hot pan has discovered that heat was conducted through the spoon. **Conduction** *is the transfer of heat through matter by molecular activity.* The energy of molecules is transferred through collisions between one molecule and another, with the heat flowing from the higher temperature to the lower temperature.

SmartFigure 11.18
The three mechanisms of heat transfer

The ability of substances to conduct heat varies considerably. Metals are good conductors, as those of us who have touched hot metal have quickly learned (**Figure 11.18**). Air, on the other hand, is a very poor conductor of heat. Consequently, conduction is important only between Earth's surface and the air directly in contact with the surface. As a means of heat transfer for the atmosphere as a whole, conduction is the least significant.

Mechanisms of Heat Transfer: Convection

Much of the heat transport in Earth's atmosphere and oceans is carried on by convection. **Convection** *is the transfer of heat by mass movement or circulation within a substance.* It takes place in fluids (for example, liquids such as the ocean and gases such as air) where the atoms and molecules are free to move about.

The pan of water in Figure 11.18 illustrates the nature of simple convective circulation. Radiation from the fire warms the bottom of the pan, which conducts heat to the water near the bottom of the container. As the water is heated, it expands and becomes less dense than the water above. Because of this new buoyancy, the warmer water rises. At the same time, cooler, denser water near the top of the pan sinks to the bottom, where it becomes heated. As long as the water is heated unequally—that is, from the bottom up—the water will continue to "turn over," producing a *convective circulation.* In a similar manner, most of the heat acquired in the lowest layer of the atmosphere by way of radiation and conduction is transferred by convective flow.

On a global scale, convection in the atmosphere creates a huge, worldwide circulation of air. This is responsible for the redistribution of heat between hot equatorial regions and the frigid poles. This important process is discussed in detail in Chapter 13.

Mechanisms of Heat Transfer: Radiation

The third mechanism of heat transfer is **radiation**. As shown in Figure 11.18, radiation travels out in all directions from its source. Unlike conduction and convection, which need a medium to travel through, radiant energy readily travels through the vacuum of space. Thus, radiation is the heat-transfer mechanism by which solar energy reaches our planet.

Solar Radiation From our everyday experience, we know that the Sun emits light and heat as well as the ultraviolet rays that cause suntan. Although these forms of energy comprise a major portion of the total energy that radiates from the Sun, they are only part of a large array

of energy called radiation or **electromagnetic radiation**. This array, or spectrum, of electromagnetic energy is shown in **Figure 11.19**. All radiation, whether x-rays, microwaves, or radio waves, transmit energy through the vacuum of space at 300,000 kilometers (186,000 miles) per second and only slightly slower through our atmosphere.

Nineteenth-century physicists were so puzzled by the seemingly impossible phenomenon of energy traveling through the vacuum of space without a medium to transmit it that they assumed that a material, which they named *ether*, existed between the Sun and Earth. This medium was thought to transmit radiant energy in much the same way that air transmits sound waves. Of course, this was incorrect. We now know that, like gravity, radiation requires no material for transmission.

In some respects, the transmission of radiant energy parallels the motion of the gentle swells in the open ocean. Like ocean swells, electromagnetic waves come in various sizes. For our purpose, the most important characteristic is their *wavelength*, or the distance from one crest to the next. Radio waves have the longest wavelengths, ranging to tens of kilometers, whereas gamma waves are the shortest, being less than one-billionth of a centimeter long.

Visible light, as the name implies, is the only portion of the spectrum we can see. We often refer to visible light as "white" light since it appears "white" in color. However, it is easy to show that white light is really a mixture of colors, each corresponding to a particular wavelength. Using a prism, we can divide white light into a color array rainbow. Figure 11.19 shows that violet has the shortest wavelength—0.4 micrometer—and red has the longest—0.7 micrometer.

Located adjacent to red and having a longer wavelength is **infrared** radiation, which we cannot see but which we can detect as heat. The closest invisible waves to violet are called **ultraviolet (UV)** rays and are responsible for sunburn after an intense exposure to the Sun. Although we divide radiant energy into groups based on our ability to perceive them, all forms of radiation are basically the same. When any form of radiant energy is absorbed by an object, the result is an increase in molecular motion, which causes a corresponding increase in temperature.

Laws of Radiation To obtain a better understanding of how the Sun's radiant energy interacts with Earth's atmosphere and land–sea surface, it is helpful to have a general understanding of the basic laws governing radiation:

1. *All objects, at whatever temperature, emit radiant energy.* Hence, not only hot objects such as the Sun but also Earth, including its polar ice caps, continually emit energy.

2. *Hotter objects radiate more total energy per unit area than do colder objects.* The Sun, which has a surface temperature of nearly 6000°C (10,000°F), emits about 160,000 times more energy per unit area than does Earth, which has an average surface temperature of about 15°C (59°F).

Figure 11.19 The electromagnetic spectrum This diagram illustrates the wavelengths and names of various types of radiation. Visible light consists of an array of colors we commonly call the "colors of the rainbow." (Photo by Michael Giannechini/Photo researchers, Inc.)

3. *Hotter objects radiate more energy in the form of short-wavelength radiation than do cooler objects.* We can visualize this law by imagining a piece of metal that, when heated sufficiently (as occurs in a blacksmith shop), produces a white glow. As the metal cools, it emits more of its energy in longer wavelengths and glows a reddish color. Eventually, no light is given off, but if you place your hand near the metal, the still-longer infrared radiation will be detected as heat. The Sun radiates maximum energy at 0.5 micrometer, which is in the visible range. The maximum radiation for Earth occurs at a wavelength of 10 micrometers, well within the infrared (heat) range. Because the maximum Earth radiation is roughly 20 times longer than the maximum solar radiation, Earth radiation is often called *long-wave radiation*, and solar radiation is called *short-wave radiation*.

4. *Objects that are good absorbers of radiation are good emitters as well.* Earth's surface and the Sun approach being perfect radiators because they absorb and radiate with nearly 100 percent efficiency for their respective temperatures. On the other hand, *gases are selective absorbers and radiators.* Thus, the atmosphere, which is nearly transparent (does not absorb) to certain wavelengths of radiation, is nearly opaque (a good absorber) to others. Our experience tells us that the atmosphere is transparent to visible light; hence, it readily reaches Earth's surface. This is not the case for the longer-wavelength radiation emitted by Earth.

Figure 11.18 summarizes the various mechanisms of heat transfer. A portion of the radiant energy generated by the campfire is absorbed by the pan. This energy is readily transferred through the metal container by the process of conduction. Conduction also increases the temperature of the water at the bottom. Once warmed, this layer of water moves upward and is replaced by cool water descending from above. Thus, convection currents that redistribute the newly acquired energy throughout the pan are established. Meanwhile, the camper is warmed by radiation emitted by the fire and the pan. Furthermore, because metals are good conductors, the camper's hand is likely to be burned if he or she does not use a potholder. As in this example, the heating of Earth's atmosphere involves the processes of conduction, convection, and radiation, all of which occur simultaneously.

11.5 CONCEPT CHECKS

1. Distinguish between heat and temperature.
2. Describe the three mechanisms of heat transfer. Which mechanism is *least* important as a means of heat transfer in the atmosphere?
3. In what part of the electromagnetic spectrum does the Sun radiate maximum energy? How does this compare to Earth?
4. Describe the relationship between the temperature of a radiating body and the wavelengths it emits.

11.6 Heating the Atmosphere

Sketch and label a diagram that shows the paths taken by incoming solar radiation. Summarize the greenhouse effect.

The goal of this section is to describe how energy from the Sun heats Earth's surface and atmosphere. It is important to know the paths taken by incoming solar radiation and the factors that cause the amount of solar radiation taking each path to vary.

What Happens to Incoming Solar Radiation?

When radiation strikes an object, three different results usually occur. First, some of the energy is *absorbed* by the object. Recall that when radiant energy is absorbed, it is converted to heat, which causes an increase in temperature. Second, substances such as water and air are transparent to certain wavelengths of radiation. Such materials simply *transmit* this energy. Radiation that is transmitted does not contribute energy to the object. Third, some radiation may "bounce off" the object without being absorbed or transmitted. *Reflection* and *scattering* are responsible for redirecting incoming solar radiation. In summary, *radiation may be absorbed, transmitted, or redirected (reflected or scattered)*.

Figure 11.20 shows the fate of incoming solar radiation averaged for the entire globe. Notice that the atmosphere is quite transparent to incoming solar radiation. On average, about 50 percent of the solar energy that reaches the top of the atmosphere is absorbed at Earth's surface. Another 30 percent is reflected back to space by the atmosphere, clouds, and reflective surfaces. The remaining 20 percent is absorbed by clouds and the atmosphere's gases. What determines whether solar radiation will be transmitted to the surface, scattered, reflected outward, or absorbed by the atmosphere? As you will see, it depends greatly on the wavelength of the energy being transmitted, as well as on the nature of the intervening material.

SmartFigure 11.20 Paths taken by solar radiation This diagram shows the average distribution of incoming solar radiation, by percentage. More solar radiation is absorbed by Earth's surface than by the atmosphere.

Reflection and Scattering

Reflection is the process whereby light bounces back from an object at the same angle at which it encounters a surface and with the same intensity (**Figure 11.21A**). By contrast, **scattering** produces a larger number of weaker rays that travel in different directions. Although scattering disperses light both forward and backward (*backscattering*), more energy is dispersed in the forward direction (**Figure 11.21B**).

Reflection and Earth's Albedo Energy is returned to space from Earth in two ways: reflection and emission of radiant energy. The portion of solar energy that is reflected back to space leaves in the same short wavelengths in which it came to Earth. About 30 percent of the solar energy that reaches the outer atmosphere is reflected back to space. Included in this figure is the amount sent skyward by backscattering. This energy is lost to Earth and does not play a role in heating the atmosphere.

The fraction of the total radiation that is reflected by a surface is called its **albedo**. Thus, the albedo for Earth as a whole (the *planetary albedo*) is 30 percent. However, the albedo from place to place as well as from time to time in the same locale varies considerably, depending on the amount of cloud cover and particulate matter in the air, as well as on the angle of the Sun's rays and the nature of the surface. A lower Sun angle means that more atmosphere must be penetrated, thus making the "obstacle course" longer and the loss of solar radiation greater (see Figure 11.14). **Figure 11.22** shows the albedo for various surfaces. Note that the angle at which the Sun's rays strike a water surface greatly affects its albedo.

A. Reflected light bounces back from a surface at the same angle at which it strikes that surface and with the same intensity.

B. When a beam of light is scattered, it results in a larger number of weaker rays, traveling in all directions. Usually more energy is scattered in the forward direction than is backscattered.

Figure 11.21 Reflection and scattering

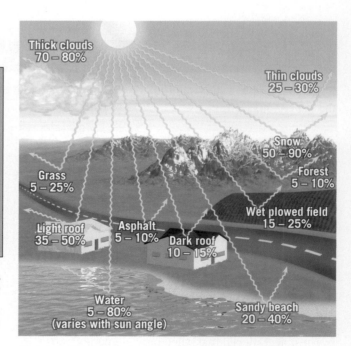

Figure 11.22 Albedo (reflectivity) of various surfaces In general, light-colored surfaces tend to reflect more sunlight than dark-colored surfaces and thus have higher albedos.

Scattering Although incoming solar radiation travels in a straight line, small dust particles and gas molecules in the atmosphere scatter some of this energy in all directions. The result, called **diffused light**, explains how light reaches into the area beneath a shade tree and how a room is lit in the absence of direct sunlight. Further, scattering accounts for the brightness and even the blue color of the daytime sky. In contrast, bodies such as the Moon and Mercury, which are without atmospheres, have dark skies and "pitch-black" shadows, even during daylight hours. Overall, about half of the solar radiation that is absorbed at Earth's surface arrives as diffused (scattered) light.

Absorption

As stated earlier, gases are selective absorbers, meaning that they absorb strongly in some wavelengths, moderately in others, and only slightly in still others. When a gas molecule absorbs radiation, the energy is transformed into internal molecular motion, which is detectable as a rise in temperature.

Nitrogen, the most abundant constituent in the atmosphere, is a poor absorber of all types of incoming solar radiation. Oxygen and ozone are efficient absorbers of ultraviolet radiation. Oxygen removes most of the shorter ultraviolet radiation high in the atmosphere, and ozone absorbs most of the remaining UV rays in the stratosphere. The absorption of UV radiation in the stratosphere accounts for the high temperatures experienced there. The only other significant absorber of incoming solar radiation is water vapor, which, along with oxygen and ozone, accounts for most of the solar radiation absorbed directly by the atmosphere.

For the atmosphere as a whole, none of the gases are effective absorbers of visible radiation. This explains why most visible radiation reaches Earth's surface and why we say that the atmosphere is *transparent* to incoming solar radiation. Thus, the atmosphere does not acquire the bulk of its energy directly from the Sun. Rather, it is heated chiefly by energy that is first absorbed by Earth's surface and then reradiated to the sky.

Heating the Atmosphere: The Greenhouse Effect

Approximately 50 percent of the solar energy that strikes the top of the atmosphere reaches Earth's surface and is absorbed. Most of this energy is then reradiated skyward. Because Earth has a much lower surface temperature than the Sun, the radiation that it emits has longer wavelengths than solar radiation.

The atmosphere as a whole is an efficient absorber of the longer wavelengths emitted by Earth (*terrestrial radiation*). Water vapor and carbon dioxide are the principal absorbing gases. Water vapor absorbs roughly five times more terrestrial radiation than do all the other gases combined and accounts for the warm temperatures found in the lower troposphere, where it is most highly concentrated. Because the atmosphere is quite transparent to shorter-wavelength solar radiation and more readily absorbs longer-wavelength radiation emitted by Earth, the atmosphere is heated from the ground up rather than vice versa. This explains the general drop in temperature with increasing altitude experienced in the troposphere. The farther from the "radiator," the colder it becomes.

When the gases in the atmosphere absorb terrestrial radiation, they warm; but they eventually radiate this energy away. Some energy travels skyward, where it may be reabsorbed by other gas molecules, a possibility that is less likely with increasing height because the concentration of water vapor decreases with altitude. The remainder travels Earthward and is again absorbed by Earth. For this reason, Earth's surface is continually being supplied with heat from the atmosphere as well as from the Sun. Without these absorptive gases in our atmosphere, Earth would not be a suitable habitat for humans and numerous other life-forms. This very important phenomenon has been termed the **greenhouse effect** because it was once thought that greenhouses were heated in a similar manner (**Figure 11.23**).

The gases of our atmosphere, especially water vapor and carbon dioxide, act very much like the glass in the greenhouse. They allow shorter-wavelength solar radiation to enter, where it is absorbed by the objects inside. These objects in turn radiate energy, but at longer wavelengths, to which glass is nearly opaque. The heat therefore is "trapped" in the greenhouse. However, a more important factor in keeping a greenhouse warm is the fact that the greenhouse itself prevents mixing of air inside with cooler air outside. Nevertheless, the term *greenhouse effect* is still used.

11.6 CONCEPT CHECKS

1. What are the three paths taken by incoming solar radiation?
2. What factors cause albedo to vary from time to time and from place to place?
3. Explain why the atmosphere is heated chiefly by radiation emitted from Earth's surface rather than by direct solar radiation.
4. Prepare a sketch with labels that explains the greenhouse effect.

Airless bodies like the Moon All incoming solar radiation reaches the surface. Some is reflected back to space. The rest is absorbed by the surface and radiated directly back to space. As a result the lunar surface has a much lower average surface temperature than Earth.

Bodies with modest amounts of greenhouse gases like Earth The atmosphere absorbs some of the longwave radiation emitted by the surface. A portion of this energy is radiated back to the surface and is responsible for keeping Earth's surface 33°C (59°F) warmer than it would otherwise be.

Bodies with abundant greenhouse gases like Venus Venus experiences extraordinary greenhouse warming, which is estimated to raise its surface temperature by 523°C (941°F).

All outgoing longwave energy is reradiated directly back to space

Incoming shortwave solar radiation

David Cole/Alamy

Some outgoing longwave radiation absorbed by greenhouse gases

Greenhouse gases reradiate some energy Earthward

Incoming shortwave solar radiation

NASA

Most outgoing longwave radiation absorbed by greenhouse gases

Greenhouse gases reradiate considerable energy toward the Venusian surface

Incoming shortwave solar radiation

NASA

SmartFigure 11.23 The greenhouse effect Earth's greenhouse effect is compared with two of our close solar system neighbors.

11.7 Human Impact on Global Climate

Summarize the nature and cause of the atmosphere's changing composition since about 1750. Describe the atmosphere's response and some possible future consequences.

Climate not only varies from place to place but is naturally variable over time. During the great expanse of Earth history, and long before humans were roaming the planet, there were many shifts—from warm to cold and from wet to dry and back again. Today scientists understand that, in addition to the natural forces that cause variations in climate, humans are playing a significant role as well. One important example results from the addition of carbon dioxide and other trace gases to the atmosphere.

In the preceding section, you learned that carbon dioxide (CO_2) absorbs some of the radiation emitted by Earth and thus contributes to the greenhouse effect. Because CO_2 is an important heat absorber, it follows that a change in the atmosphere's CO_2 content could influence air temperature.

CO_2 Levels Are Rising

Earth's tremendous industrialization of the past two centuries has been fueled—and still is fueled—by burning fossil fuels: coal, natural gas, and petroleum (**Figure 11.24**). Combustion of these fuels has added great quantities of carbon dioxide to the atmosphere. Figure 11.5 shows changes in CO_2 concentrations at Hawaii's Mauna Loa Observatory, where measurements have been made since 1958 (see the chapter-opening photo).

The use of coal and other fuels is the most prominent means by which humans add CO_2 to the atmosphere, but it is not the only way. The clearing of forests also contributes substantially because CO_2 is released as vegetation is burned or decays. Deforestation is particularly

SOURCES OF ENERGY

This number is the amount of energy produced (quadrillion Btu) —

This is the percentage of the total for this category. —

Petroleum
35.3
(36.2%)

Natural gas
24.9
(25.5%)

Coal
19.9
(20.4%)

Renewable energy
9.1 (9.3%)

Nuclear electric power
8.3 (8.5%)

 SmartFigure 11.24 U.S. energy consumption The graph shows energy consumption in 2011. The total was 97.5 quadrillion Btu. A quadrillion is 10 raised to the 12th power, or a million million. The burning of fossil fuels represents slightly more than 83 percent of the total. (Data from U.S. Energy Information Administration)

Figure 11.25 Tropical deforestation Clearing the tropical rain forest is a serious environmental issue. In addition to causing a loss of biodiversity, it is a significant source of carbon dioxide. Fires are frequently used to clear the land. This scene is in Brazil's Amazon basin. (Photo by Pete Oxford/Nature Picture Library)

Figure 11.26 CO_2 concentrations over the past 400,000 years Most of these data come from the analysis of air bubbles trapped in ice cores. The record since 1958 comes from direct measurements at Mauna Loa Observatory, Hawaii. The rapid increase in CO_2 concentrations since the onset of the Industrial Revolution is obvious. (NOAA)

pronounced in the tropics, where vast tracts are cleared for ranching and agriculture or subjected to inefficient commercial logging operations (**Figure 11.25**).

Some of the excess CO_2 is taken up by plants or is dissolved in the ocean. It is estimated that about 45 percent remains in the atmosphere. **Figure 11.26** is a graphic record of changes in atmospheric CO_2 extending back more than 400,000 years. Over this long span, natural fluctuations varied from about 180 to 300 parts per million (ppm). As a result of human activities, the present CO_2 level is about 30 percent higher than its highest level over at least the past 650,000 years. The rapid increase in CO_2 concentrations since the onset of industrialization is obvious. The annual rate at which atmospheric CO_2 concentrations are growing has been increasing over the past several decades. **Figure 11.27** shows the yearly "carbon footprint" of an average American.

The Atmosphere's Response

Given the increase in the atmosphere's carbon dioxide content, have global temperatures actually increased? The answer is "yes." According to a 2007 report by the Intergovernmental Panel on Climate Change (IPCC), "Warming of the climate system is unequivocal, as is now evident from observations of increases in global average air and ocean temperatures, widespread melting of snow and ice, and rising global sea level." Most of the observed increase in global average temperatures since the mid-twentieth

Figure 11.27 Yearly CO$_2$ contribution of an average American Americans are responsible for nearly 20 percent of the world's greenhouse gas emissions. That amounts to about 24,300 kilograms (54,000 pounds) of CO$_2$ each year for an average American. This diagram represents some of the ways people in the United States contribute. Your carbon footprint may be very different, depending on your lifestyle.

17,000 pounds of CO$_2$ by using 1,100 kilowatt-hours of electricity per month

8,800 pounds of CO$_2$ by using 6,300 cubic feet of natural gas per month

1,000 pounds of CO$_2$ by creating 4.5 pounds of trash per day

8,900 pounds of CO$_2$ by driving 160 miles per week

1,000 pounds of CO$_2$ by flying 1,900 miles per year

century is *very likely* due to the observed increase in human-generated greenhouse gas concentrations. (As used by the IPCC, *very likely* indicates a probability of 90–99 percent.) Global warming since the mid-1970s is now about 0.6°C (1°F), and total warming in the past century is about 0.8°C (1.4°F). The upward trend in surface temperatures is shown in **Figure 11.28A**. The world map in **Figure 11.28B** compares surface temperatures for 2011 to the base period (1951–1980). You can see that the greatest warming has been in the Arctic and neighboring high-latitude regions. Here are some related facts:

- When we consider the time span for which there are instrumental records (since 1850), the warmest 16 years have all been in the 17-year span 1995–2011.
- Global mean temperature is now higher than at any time in at least the past 500 to 1000 years.
- The average temperature of the global ocean has increased to depths of at least 3000 meters (10,000 feet).

Are these temperature trends caused by human activities, or would they have occurred anyway? The scientific consensus of the IPCC is that human activities were *very likely* responsible for most of the temperature increase since 1950.

What about the future? Projections for the years ahead depend in part on the quantities of greenhouse gases that are emitted. **Figure 11.29** shows the best estimates of global

Figure 11.28 Global temperatures The year 2011 was one of the 10 warmest years on record. **A.** The graph depicts global temperature change in degrees Celsius since the year 1880. **B.** The world map shows how temperatures in 2011 deviated from the mean for the 1951–1980 base period. The high latitudes in the Northern Hemisphere clearly stand out. (NASA/Goddard Institute for Space Studies)

A.

Temperature difference (°C) compared to 1951-80 mean

-4.1 -4   -2   -1 -.5 -.2   .2   .5   1   2   4   4.1
Temperature difference (°C) compared to 1951-80 mean

B.

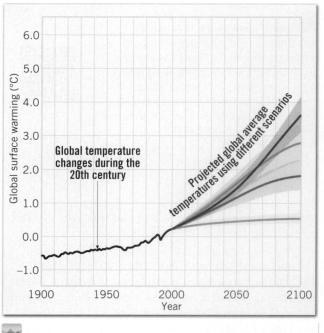

SmartFigure 11.29 Temperature projections to 2100

Global surface warming (°C)

Global temperature changes during the 20th century

Projected global average temperatures using different scenarios

Year

 SmartFigure 11.29 Temperature projections to 2100 The right half of the graph shows projected global warming, based on different emissions scenarios. The shaded zone adjacent to each colored line shows the uncertainty range for each scenario. The basis for comparison (0.0 on the vertical axis) is the global average for the period 1980 to 1999. The orange line represents the scenario in which CO$_2$ concentrations were held constant at values for the year 2000. (NOAA)

warming for several different scenarios. The 2007 IPCC report also states that if there is a doubling of the pre-industrial level of carbon dioxide (280 ppm) to 560 ppm, the "likely" temperature increase will be in the range of 2 to 4.5°C (3.5 to 8.1°F). The increase is "very unlikely" (1 to 10 percent probability) to be less than 1.5°C (2.7°F), and values higher than 4.5°C (8.1°F) cannot be excluded.

Some Possible Consequences

What consequences can be expected if the carbon dioxide content of the atmosphere reaches a level that is twice what it was early in the twentieth century? Because the climate system is so complex, predicting the distribution of particular regional changes is speculative. It is not yet possible to pinpoint specifics, such as where or when it will become drier or wetter. Nevertheless, plausible scenarios can be given for larger scales of space and time.

One important impact of human-induced global warming is a rise in sea level. Potential weather changes include shifts in the paths of large-scale storms, which in turn would affect the distribution of precipitation and the occurrence of severe weather. Other possibilities include stronger tropical storms and increases in the frequency and intensity of heat waves and droughts (**Table 11.2**).

The changes that occur will probably take the form of gradual environmental shifts that will be imperceptible to

TABLE 11.2	Global Climate Change in the Twenty-First Century: Projected Changes and Estimated Probabilities*
Higher maximum temperatures; more hot days and heat waves over nearly all land areas *(virtually certain)*	
Higher minimum temperatures; fewer cold days, frost days, and cold waves over nearly all land areas *(virtually certain)*	
Frequency of heavy precipitation events increases over most areas *(very likely)*	
Area affected by drought increases *(likely)*	
Intense tropical cyclone activity *(likely)*	

Virtually certain indicates a probability greater than 99 percent, *very likely* indicates a probability of 90–99 percent, and *likely* indicates a probability of 67–90 percent.

most people from year to year. Although the changes may seem gradual, the effects will clearly have powerful economic, social, and political consequences.

11.7 CONCEPT CHECKS

1. Why has the CO_2 level of the atmosphere been increasing for the past 200 years?
2. How has the atmosphere responded to the growing CO_2 levels?
3. How are temperatures in the lower atmosphere likely to change as CO_2 levels continue to increase?

11.8 For the Record: Air Temperature Data

Calculate five commonly used types of temperature data and interpret a map that depicts temperature data using isotherms.

Changes in air temperature are probably noticed by people more often than changes in any other element of weather. At a weather station, the temperature is monitored on a regular basis from instruments mounted in an instrument shelter (**Figure 11.30**). The shelter protects the instruments from direct sunlight and allows a free flow of air.

The daily maximum and minimum temperatures are the bases for much of the basic temperature data compiled by meteorologists:

1. By adding the maximum and minimum temperatures and then dividing by two, the **daily mean temperature** is calculated.
2. The **daily range** of temperature is computed by finding the difference between the maximum and minimum temperatures for a given day.
3. The **monthly mean** is calculated by adding together the daily means for each day of the month and dividing by the number of days in the month.
4. The **annual mean** is an average of the 12 monthly means.

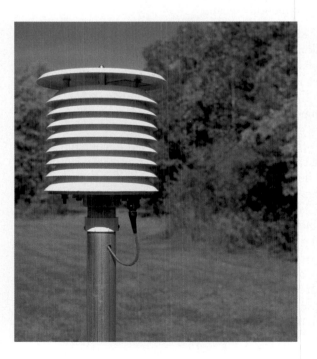

Figure 11.30 Measuring temperature This modern shelter contains an electrical thermometer called a *thermistor*. A shelter protects instruments from direct sunlight and allows for the free flow of air. (Photo by Bobbé Christopherson)

SmartFigure 11.31 Isotherms Map showing high temperatures for a spring day. Isotherms are lines that connect points of equal temperature. The temperatures on this map are in degrees Fahrenheit. Showing temperature distribution in this way makes patterns easier to see. Notice that most isotherms do not pass directly through the observing stations. It is usually necessary to draw isotherms by estimating their proper position between stations. On television and in many newspapers, temperature maps are in color. Rather than labeling isotherms, the area *between* isotherms is labeled. For example, the zone between the 60° and 70° isotherms is labeled "60s."

5. The **annual temperature range** is computed by finding the difference between the highest and lowest monthly means.

Mean temperatures are particularly useful for making comparisons, whether on a daily, monthly, or annual basis. It is quite common to hear a weather reporter state, "Last month was the hottest July on record," or "Today, Chicago was 10 degrees warmer than Miami." Temperature ranges are also useful statistics, because they give an indication of extremes.

To examine the distribution of air temperatures over large areas, isotherms are commonly used. An **isotherm** is a line that connects points on a map that have the same temperature (*iso* = equal, *therm* = temperature). Therefore, all points through which an isotherm passes have identical temperatures for the time period indicated. Generally, isotherms representing 5° or 10° temperature differences are used, but any interval may be chosen. **Figure 11.31** illustrates how isotherms are drawn on a map. Notice that most isotherms do not pass directly through the observing stations because the station readings may not coincide with the values chosen for the isotherms. Only an occasional station temperature will be exactly the same as the

value of the isotherm, so it is usually necessary to draw the lines by estimating the proper position between stations.

Maps with isotherms are valuable tools because they clearly make temperature distribution visible at a glance. Areas of low and high temperatures are easy to pick out. In addition, the amount of temperature change per unit of distance, called the **temperature gradient**, is easy to visualize. Closely spaced isotherms indicate a rapid rate of temperature change, whereas more widely spaced lines indicate a more gradual rate of change. You can see this in Figure 11.31. The isotherms are closer in Colorado and Utah (steeper temperature gradient), whereas the isotherms are spread farther in Texas (gentler temperature gradient). Without isotherms, a map would be covered with numbers representing temperatures at dozens or hundreds of places, which would make patterns difficult to see.

11.8 CONCEPT CHECKS

1. How are the following temperature data calculated: daily mean, daily range, monthly mean, annual mean, annual range?

2. What are isotherms, and what is their purpose?

11.9 Why Temperatures Vary: The Controls of Temperature

Discuss the principal controls of temperature and use examples to describe their effects.

A **temperature control** is any factor that causes temperature to vary from place to place and from time to time. Earlier in this chapter, we examined the single greatest cause for temperature variations—differences in the receipt of solar radiation. Because variations in Sun angle and length of daylight are a function of latitude, they are responsible for warm temperatures in the tropics and colder temperatures at more poleward locations. Of course, seasonal temperature changes at a given latitude occur as the Sun's vertical rays migrate toward and away from a place during the year.

But latitude is not the only control of temperature; if it were, we would expect that all places along the same parallel of latitude would have identical temperatures. This is clearly not the case. For example, Eureka, California, and New York City are both coastal cities at about the same latitude, and both have an average mean temperature of 11°C (52°F). However, New York City is 9°C (16°F) warmer than Eureka in July and 10°C cooler in January. In another example, two cities in Ecuador—Quito and Guayaquil—are relatively close to one another, but the mean annual temperatures at these two cities differ by 12°C (21°F). To explain these situations and countless others, we must understand that factors other than latitude exert a strong influence on temperature. Among the most important are the differential heating of land and water, altitude, geographic position, and ocean currents.*

Why do land and water heat and cool differently? Several factors are responsible:

- The *specific heat* (amount of energy needed to raise 1 gram of a substance 1°C) is far greater for water than for land. Thus, water requires a great deal more heat to raise its temperature the same amount as an equal quantity of land.
- Land surfaces are opaque, so heat is absorbed only at the surface. Water, being more transparent, allows heat to penetrate to a depth of many meters.
- The water that is heated often mixes with water below, thus distributing the heat through an even larger mass.
- Evaporation (a cooling process) from water bodies is greater than that from land surfaces.

All these factors collectively cause water to warm more slowly, store greater quantities of heat, and cool more slowly than land.

Monthly temperature data for two cities will demonstrate the moderating influence of a large water body and the extremes associated with land (**Figure 11.32**). Vancouver, British Columbia, is located along a windward coast, whereas Winnipeg, Manitoba, is in a continental position

Land and Water

The heating of Earth's surface directly influences the heating of the air above. Therefore, to understand variations in air temperature, we must understand the variations in heating properties of the different surfaces that Earth presents to the Sun—soil, water, trees, ice, and so on. Different land surfaces absorb varying amounts of incoming solar energy, which in turn cause variations in the temperature of the air above. The greatest contrast, however, is not between different land surfaces but between land and water. *Land heats more rapidly and to higher temperatures than water and cools more rapidly and to lower temperatures than water.* Temperature variations, therefore, are considerably greater over land than over water.

SmartFigure 11.32
Monthly mean temperatures for Vancouver, British Columbia, and Winnipeg, Manitoba Vancouver has a much smaller annual temperature range because of the strong marine influence of the Pacific Ocean. The curve for Winnipeg illustrates the greater extremes associated with an interior location.

---

*For a discussion of the effects of ocean currents on temperatures, see Chapter 10.

| TABLE 11.3 | Variation in Mean Annual Temperature Range (°C) with Latitude | |
Latitude	Northern Hemisphere	Southern Hemisphere
0	0	0
15	3	4
30	13	7
45	23	6
60	30	11
75	32	26
90	40	31

Figure 11.33 Monthly mean temperatures for Concepción and La Paz, Bolivia Both cities have nearly the same latitude (about 16° south). However, because La Paz is high in the Andes, at 4103 meters (13,461 feet), it experiences much cooler temperatures than Concepción, which is at an elevation of 490 meters (1608 feet).

far from the influence of water. Both cities are at about the same latitude and thus experience similar Sun angles and lengths of daylight. Winnipeg, however, has a mean January temperature that is 20°C (36°F) lower than Vancouver's. Winnipeg's July mean is 2°C (3.6°F) higher than Vancouver's. Although their latitudes are nearly the same, Winnipeg, which has no water influence, experiences much greater temperature extremes than Vancouver, which does. The key to Vancouver's moderate year-round climate is the Pacific Ocean.

On a different scale, the moderating influence of water may also be demonstrated when temperature variations in the Northern and Southern Hemispheres are compared. In the Northern Hemisphere, 61 percent is covered by water, and land accounts for the remaining 39 percent. However, in the Southern Hemisphere, 81 percent is covered by water and 19 percent by land. The Southern Hemisphere is therefore correctly called the *water hemisphere* (see Figure 9.1, page 296). **Table 11.3** portrays the considerably smaller annual temperature variations in the water-dominated Southern Hemisphere as compared with the Northern Hemisphere.

Altitude

The two cities in Ecuador mentioned earlier—Quito and Guayaquil—demonstrate the influence of altitude on mean temperatures. Although both cities are near the equator and not far apart, the annual mean at Guayaquil is 25°C (77°F), as compared to Quito's mean of 13°C (55°F). The difference is explained largely by the difference in the cities' elevations: Guayaquil is only 12 meters (40 feet) above sea level, whereas Quito is high in the Andes Mountains, at 2800 meters (9200 feet). **Figure 11.33** provides another example.

Recall that temperatures drop an average of 6.5°C per kilometer (3.5°F per 1000 feet) in the troposphere; thus, cooler temperatures are to be expected at greater heights. Yet, the magnitude of the difference is not totally explained by the normal lapse rate. If this figure were used, we would expect Quito to be about 18°C (nearly 33°F) cooler than Guayaquil; the difference, however,

is only 12°C. The fact that high-altitude places, such as Quito, are warmer than the value calculated using the normal lapse rate results from the absorption and reradiation of solar energy by the ground surface.

Geographic Position

The geographic setting can greatly influence the temperatures experienced at a specific location. A coastal location where prevailing winds blow from the ocean onto the shore (a *windward* coast) experiences considerably different temperatures than does a coastal location where prevailing winds blow from the land toward the ocean (a *leeward* coast). The windward coast will experience the full moderating influence of the ocean—cool summers and mild winters—compared to an inland location at the same latitude.

A leeward coast, however, will have a more continental temperature regime because the winds do not carry the ocean's influence onshore. Eureka, California, and New York City, the two cities mentioned earlier, illustrate this aspect of geographic position. The annual temperature range at New York City is 19°C (34°F) greater than Eureka's (**Figure 11.34**).

Seattle and Spokane, both in Washington State, illustrate a second aspect of geographic position—mountains that act as barriers. Although Spokane is only about 360 kilometers (220 miles) east of Seattle,

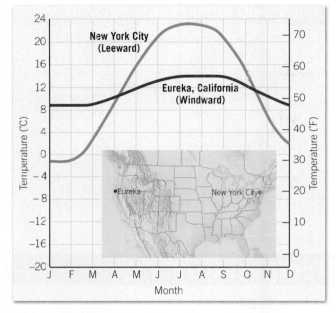

Figure 11.34 Monthly mean temperatures for Eureka, California, and New York City Both cities are coastal and located at about the same latitude. Because Eureka is strongly influenced by prevailing winds from the ocean and New York City is not, the annual temperature range at Eureka is much smaller.

the towering Cascade Range separates the cities. Consequently, Seattle's temperatures show a marked marine influence, but Spokane's are more typically continental (**Figure 11.35**). Spokane is 7°C (13°F) cooler than Seattle in January and 4°C (7°F) warmer than Seattle in July. The annual range at Spokane is 11°C (20°F) greater than at

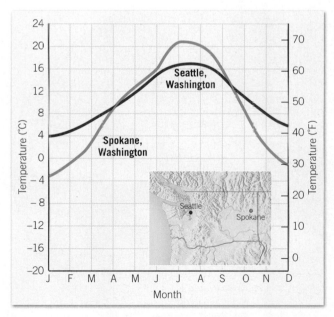

Figure 11.35 Monthly mean temperatures for Seattle and Spokane, Washington Because the Cascade Mountains cut off Spokane from the moderating influence of the Pacific Ocean, its annual temperature range is greater than Seattle's.

Seattle. The Cascade Range effectively cuts off Spokane from the moderating influence of the Pacific Ocean.

Cloud Cover and Albedo

You may have noticed that clear days are often warmer than cloudy ones and that clear nights are usually cooler than cloudy ones. This demonstrates that cloud cover is another factor that influences temperature in the lower atmosphere. Studies using satellite images show that at any particular time, about half of our planet is covered by clouds. Cloud cover is important because many clouds have a high albedo; therefore, clouds reflect a significant portion of the sunlight that strikes them back into space (see Figure 11.22). By reducing the amount of incoming solar radiation, daytime temperatures will be lower than if the clouds were not present and the sky were clear.

At night, clouds have the opposite effect as during daylight. They act as a blanket by absorbing radiation emitted by Earth's surface and reradiating a portion of it back to the surface. Consequently, some of the heat that otherwise would have been lost remains near the ground. Thus, nighttime air temperatures do not drop as low as they would on a clear night. The effect of cloud cover is to reduce the daily temperature range by lowering the daytime maximum and raising the nighttime minimum (**Figure 11.36**).

SmartFigure 11.36 The daily cycle of temperature at Peoria, Illinois, for two July days Clouds reduce daily temperature range. During daylight hours, clouds reflect solar radiation back to space. Therefore, the maximum temperature is lower than if the sky were clear. At night, the minimum temperature will not fall as low because clouds retard the loss of heat.

Clouds are not the only phenomenon that increases albedo and thereby reduces air temperatures. We also recognize that snow- and ice-covered surfaces have high albedos. This is one reason why mountain glaciers do not melt away in the summer and why snow may still be present on a mild spring day. In addition, during the winter, when snow covers the ground, daytime maximums on a sunny day are lower than they otherwise would be because energy that the land would have absorbed and used to heat the air is reflected and lost.

11.10 World Distribution of Temperature

Interpret the patterns depicted on world maps of January and July temperatures.

Take a moment to study the two world maps in **Figures 11.37** and **11.38**. From hot colors near the equator to cool colors toward the poles, these maps portray sea-level temperatures in the seasonally extreme months of January and July. Temperature distribution is shown by using isotherms. On these maps, you can study global temperature patterns and the effects of the controlling factors of temperature, especially latitude, the distribution of land and water, and ocean currents. Like most other isothermal maps of large regions, all temperatures on these world maps have been reduced to sea level to eliminate the complications caused by differences in altitude.

On both maps, the isotherms generally trend east–west and show a decrease in temperatures poleward from the tropics. They illustrate one of the most fundamental aspects of world temperature distribution: that the effectiveness of incoming solar radiation in heating Earth's surface and the atmosphere above is largely a function of latitude.

Figure 11.37 World mean sea-level temperatures in January, in Celsius (°C) and Fahrenheit (°F)

SmartFigure 11.38 World mean sea-level temperatures in July, in Celsius (°C) and Fahrenheit (°F)

Moreover, there is a latitudinal shifting of temperatures caused by the seasonal migration of the Sun's vertical rays. To see this, compare the color bands by latitude on the two maps. On the January map, the "hot spots" of 30°C are *south* of the equator, but in July they have shifted *north* of the equator.

If latitude were the only control of temperature distribution, our analysis could end here, but this is not the case. The added effect of the differential heating of land and water is also reflected on the January and July temperature maps. The warmest and coldest temperatures are found over land. Because temperatures do not fluctuate as much over water as over land, the seasonal north–south migration of isotherms is greater over the continents than over the oceans.

In addition, it is clear that the isotherms in the Southern Hemisphere, where there is little land and the oceans predominate, are much straighter and more stable than in the Northern Hemisphere, where they bend sharply northward in July and southward in January over the continents.

Isotherms also reveal the presence of ocean currents. Warm currents cause isotherms to be deflected poleward, whereas cold currents cause an equatorward bending. The horizontal transport of water poleward warms the overlying air and results in air temperatures that are higher than otherwise would be expected for the latitude. Conversely, currents moving toward the equator produce air temperatures cooler than expected.

Because Figures 11.37 and 11.38 show the seasonal extremes of temperature, they can be used to evaluate variations in the annual range of temperature from place to place. A comparison of the two maps shows that a station near the equator will record a very small annual temperature range. This is true because such a station experiences little variation in the length of daylight and always has a relatively high Sun angle. By contrast, a site in the middle latitudes experiences much wider variations in Sun angle and length of daylight and thus larger temperature variations. Therefore, we can state that the annual temperature range increases with an increase in latitude.

Land and water also affect seasonal temperature variations, especially outside the tropics. A continental location must endure hotter summers and colder winters than a coastal location. Consequently, the annual range will increase with an increase in continentality.

11.10 CONCEPT CHECKS

1. Why do isotherms generally trend east–west?
2. Why do isotherms shift north and south from season to season?
3. Where do isotherms shift most—over land or over water? Explain.
4. Which area on Earth experiences the highest annual temperature range?

CONCEPTS IN REVIEW: HEATING THE ATMOSPHERE

11.1 Focus on the Atmosphere

Distinguish between weather and climate and name the basic elements of weather and climate.

KEY TERMS: weather, climate, elements (of weather and climate)

- Weather is the state of the atmosphere at a particular place for a short period of time. Climate, on the other hand, is a generalization of the weather conditions of a place over a long period of time.
- The most important elements—quantities or properties that are measured regularly—of weather and climate are (1) air temperature, (2) humidity, (3) type and amount of cloudiness, (4) type and amount of precipitation, (5) air pressure, and (6) the speed and direction of the wind.

? **This is a scene on a day in late April in southern Arizona's Organ Pipe Cactus National Monument. Write two brief statements about the locale in this image—one that relates to weather and one that relates to climate.**

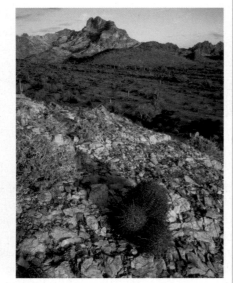

Michael Collier

11.2 Composition of the Atmosphere

List the major gases composing Earth's atmosphere and identify the components that are most important to understanding weather and climate.

KEY TERMS: air, aerosols, ozone

- Air is a mixture of many discrete gases, and its composition varies from time to time and from place to place. If water vapor, dust, and other variable components of the atmosphere are removed, clean, dry air is composed almost entirely of nitrogen (N_2) and oxygen (O_2). Carbon dioxide (CO_2), although present only in minute amounts, is important because it has the ability to absorb heat radiated by Earth and thus helps keep the atmosphere warm.
- Among the variable components of air, water vapor is important because it is the source of all clouds and precipitation. Like carbon dioxide, water vapor can absorb heat emitted by Earth. When water changes from one state to another, it absorbs or releases heat. In the atmosphere, water vapor transports this latent ("hidden") heat from place to place and is the energy that helps to drive many storms.
- Aerosols are tiny solid and liquid particles that are important because they may act as surfaces on which water vapor can condense and are also absorb+ers and reflectors of incoming solar radiation.
- Ozone, a form of oxygen that combines three oxygen atoms into each molecule (O_3), is a gas concentrated in the 10- to 50-kilometer (6- to 31-mile) height range in the atmosphere and is important to life because of its ability to absorb potentially harmful ultraviolet radiation from the Sun.

11.3 Vertical Structure of the Atmosphere

Interpret a graph that shows changes in air pressure from Earth's surface to the top of the atmosphere. Sketch and label a graph that shows atmospheric layers based on temperature.

KEY TERMS: troposphere, environmental lapse rate, radiosonde, stratosphere, mesosphere, thermosphere

- Because the atmosphere gradually thins with increasing altitude, it has no sharp upper boundary but simply blends into outer space.
- Based on temperature, the atmosphere is divided vertically into four layers. The *troposphere* is the lowermost layer. In the troposphere, temperature usually decreases with increasing altitude. This *environmental lapse rate* is variable but averages about 6.5°C per kilometer (3.5°F per 1000 feet). Essentially, all important weather phenomena occur in the troposphere.
- Beyond the troposphere is the *stratosphere*, which exhibits warming because of absorption of UV radiation by ozone. In the *mesosphere*, temperatures again decrease. Upward from the mesosphere is the

thermosphere, a layer with only a tiny fraction of the atmosphere's mass and no well-defined upper limit.

? **When the weather balloon in this photo was launched, the surface temperature was 17°C. The balloon is now at an altitude of 1 kilometer. What term is applied to the instrument package being carried aloft by the balloon? In what layer of the atmosphere is the balloon? If average conditions prevail, what is the air temperature at this altitude? How did you figure this out?**

David R. Frazier/Photo Researchers, Inc.

11.4 Earth—Sun Relationships

Explain what causes the Sun angle and length of daylight to change during the year and describe how these changes produce the seasons.

KEY TERMS: rotation, circle of illumination, revolution, inclination of the axis, Tropic of Cancer, summer solstice, Tropic of Capricorn, winter solstice, autumnal equinox, spring equinox

- The two principal motions of Earth are (1) rotation, the spinning about its axis that produces the daily cycle of daylight and darkness; and (2) revolution, the movement in its orbit around the Sun.
- The seasons are caused by changes in the angle at which the Sun's rays strike the surface and the changes in the length of daylight at each latitude. These seasonal changes are the result of the tilt of Earth's axis as it revolves around the Sun.

11.5 Energy, Heat, and Temperature

Distinguish between heat and temperature. List and describe the three mechanisms of heat transfer.

KEY TERMS: heat, temperature, conduction, convection, radiation, electromagnetic radiation, visible light, infrared radiation, ultraviolet (UV) radiation

- Heat refers to the quantity of energy present in a material, whereas temperature refers to intensity, the degree of "hotness."
- The three mechanisms of heat transfer are (1) conduction, the transfer of heat through matter by molecular activity (2) convection, the transfer of heat by the movement of a mass or substance from one place to another; and (3) radiation, the transfer of heat by electromagnetic waves.
- Electromagnetic radiation is energy emitted in the form of rays, or waves, called electromagnetic waves. All radiation is capable of transmitting energy through the vacuum of space. One of the most important differences between electromagnetic waves is their wavelengths, which range from very long radio waves to very short gamma rays. Visible light is the only portion of the electromagnetic spectrum we can see.
- Some basic laws that relate to radiation are (1) all objects emit radiant energy; (2) hotter objects radiate more total energy than do colder objects; (3) the hotter the radiating body, the shorter the wave lengths of maximum radiation; and (4) objects that are good absorbers of radiation are good emitters as well.

11.6 Heating the Atmosphere

Sketch and label a diagram that shows the paths taken by incoming solar radiation. Summarize the greenhouse effect.

KEY TERMS: reflection, scattering, albedo, diffused light, greenhouse effect

- About 50 percent of the solar radiation that strikes the atmosphere reaches Earth's surface. About 30 percent is reflected back to space. The fraction of radiation reflected by a surface is called its albedo. The remaining 20 percent of the energy is absorbed by clouds and the atmosphere's gases.
- Radiant energy absorbed at Earth's surface is eventually radiated skyward. Because Earth has a much lower surface temperature than the Sun, its radiation is in the form of long-wave infrared radiation. Because atmospheric gases, primarily water vapor and carbon dioxide, are more efficient absorbers of long-wave radiation, the atmosphere is heated from the ground up.
- The selective absorption of Earth radiation by water vapor and carbon dioxide that results in Earth's average temperature being warmer than it would be otherwise is referred to as the greenhouse effect.

11.7 Human Impact on Global Climate

Summarize the nature and cause of the atmosphere's changing composition since about 1750. Describe the atmosphere's response and some possible future consequences.

- By adding carbon dioxide to the atmosphere, primarily by burning fossil fuels and through deforestation, humans are very likely contributing significantly to global warming.
- Some consequences of global warming include (1) shifts in temperature and rainfall patterns, (2) a gradual rise in sea level, (3) changing storm tracks and a higher frequency and greater intensity of tropical storms, and (4) an increase in the frequency and intensity of heat waves and droughts.

11.8 For the Record: Air Temperature Data

Calculate five commonly used types of temperature data and interpret a map that depicts temperature data using isotherms.

KEY TERMS: daily mean temperature, daily range, monthly mean, annual mean, annual temperature range, isotherm, temperature gradient

- Daily mean temperature is an average of the daily maximum and daily minimum temperatures, whereas the daily range is the difference between the daily maximum and daily minimum temperatures. The monthly mean is determined by averaging the daily means for a particular month. The annual mean is an average of the 12 monthly means, whereas the annual temperature range is the difference between the highest and lowest monthly means.
- Temperature distribution is shown on a map by using isotherms, which are lines of equal temperature. Temperature gradient is the amount of temperature change per unit of distance. Closely spaced isotherms indicate a rapid rate of change.

11.9 Why Temperatures Vary: The Controls of Temperature

Discuss the principal controls of temperature and use examples to describe their effects.

KEY TERM: temperature control

- Controls of temperature are factors that cause temperature to vary from place to place and from time to time. Latitude (Earth–Sun relationships) is one example. Ocean currents (discussed in Chapter 10) is another example.
- Unequal heating of land and water is a temperature control. Because land and water heat and cool differently, land areas experience greater temperature extremes that water-dominated areas.
- Altitude is an easy-to-visualize control: The higher up you go, the colder it gets; therefore, mountains are cooler than adjacent lowlands.
- Geographic position as a temperature control involves such factors as mountains acting as barriers to marine influence or whether a place is on a windward or a leeward coast.

? **The graph shows monthly high temperatures for Urbana, Illinois, and San Francisco, California. Although both cities are located at about the same latitude, the temperatures they experience are quite different. Which line on the graph represents Urbana, and which represents San Francisco? How did you figure this out?**

11.10 World Distribution of Temperature

Interpret the patterns depicted on world maps of January and July temperatures.

- On world maps showing January and July mean temperatures, isotherms generally trend east–west and show a decrease in temperature moving poleward from the equator. When the two maps are compared, a latitudinal shifting of temperatures is seen. Bending isotherms reveal the locations of ocean currents.
- Annual temperature range is small near the equator and increases with an increase in latitude. Outside the tropics, annual temperature range also increases as marine influence diminishes.

GIVE IT SOME THOUGHT

1. Determine which statements refer to weather and which refer to climate. (*Note*: One statement includes aspects of both weather and climate.)
   a. The baseball game was rained out today.
   b. January is Omaha's coldest month.
   c. North Africa is a desert.
   d. The high this afternoon was 25°C.
   e. Last evening a tornado ripped through central Oklahoma.
   f. I am moving to southern Arizona because it is warm and sunny.
   g. Thursday's low of –20°C is the coldest temperature ever recorded for that city.
   h. It is partly cloudy.

2. Refer to the graph in Figure 11.3 to answer the following questions about temperatures in New York City:
   a. What is the approximate average daily high temperature in January? In July?
   b. Approximately what are the highest and lowest temperatures ever recorded?

3. The jet in this photo is at an altitude of 10 kilometers (6.2 miles). Refer to the graph in Figure 11.8. What is the approximate air pressure where the jet is flying? About what percentage of the atmosphere is below the jet (assuming that the pressure at the surface is 1000 millibars)?

interlight/Shutterstock

4. If you were to see the noon Sun *directly overhead* (90-degree angle) in New York City (40° north latitude) on the summer solstice, at what angle would Earth's axis be tilted? If Earth were inclined at this angle, what might the seasons be like?

5. The Sun shines continually at the North Pole for 6 months, from the spring equinox until the autumnal equinox, yet temperatures never get very warm. Explain why this is the case.

6. Imagine being at the beach in this photo on a sunny summer afternoon.
   a. Describe the temperatures you would expect if you measured the surface of the beach and at a depth of 12 inches in the sand.
   b. If you stood waist deep in the water and measured the water's surface temperature and its temperature at a depth of 12 inches, how would these measurements compare to those taken on the beach?

Tequilab/Shutterstock

7. On which summer day would you expect the greatest temperature range? Which would have the smallest range in temperature? Explain your choices.
   a. Cloudy skies during the day and clear skies at night
   b. Clear skies during the day and cloudy skies at night
   c. Clear skies during the day and clear skies at night
   d. Cloudy skies during the day and cloudy skies at night

8. The accompanying sketch map represents a hypothetical continent in the Northern Hemisphere. One isotherm has been placed on the map.
   a. Is the temperature higher at City A or City B?
   b. Is the season winter or summer? How are you able to determine this?

c. Describe (or sketch) the position of this isotherm 6 months later.

9. This photo shows a snow-covered area in the middle latitudes on a sunny day in late winter. Assume that 1 week after this photo was taken, conditions were essentially identical except that the snow was gone. Would you expect the air temperatures to be different on the two days? If so, which day would be warmer? Suggest an explanation.

CoolR/Shutterstock

MasteringGeology™

www.masteringgeology.com

Looking for additional review and test prep materials? Visit the Study Area in MasteringGeology to enhance your understanding of this chapter's content by accessing a variety of resources, including **Self-Study Quizzes,**

Geoscience Animations, GEODe Tutorials, RSS feeds, flashcards, web links, self-study quizzes, and an optional **Pearson eText.**

12

Moisture, Clouds, and Precipitation

Each statement represents the primary learning objective for the corresponding major heading within the chapter. After you complete the chapter, you should be able to:

12.1 List and describe the processes that cause water to change from one state of matter to another. Define *latent heat* and explain why it is important.

12.2 Distinguish between relative humidity and dew point. Write a generalization relating how temperature changes affect relative humidity.

12.3 Explain how adiabatic cooling results in cloud formation.

12.4 List and describe the four mechanisms that cause air to rise.

12.5 Describe how atmospheric stability is determined and compare conditional instability with absolute instability.

12.6 List the necessary conditions for condensation and briefly describe the two criteria used for cloud classification.

12.7 Define *fog* and explain how the various types of fog form.

12.8 Describe the two mechanisms that produce precipitation.

12.9 List the different types of precipitation and explain how each type forms.

12.10 Explain how precipitation is measured.

Cumulonimbus clouds are often associated with thunderstorms and severe weather.
(Photo by Cusp/SuperStock)

Water vapor is an odorless, colorless gas that mixes freely with the other gases of the atmosphere. Unlike oxygen and nitrogen—the two most abundant components of the atmosphere—water can change from one state of matter to another (solid, liquid, or gas) at the temperatures and pressures experienced on Earth. Because of this unique property, water freely leaves the oceans as a gas and returns again as a liquid or solid.

As you observe day-to-day weather changes, you might ask: Why is it generally more humid in the summer than in the winter? Why do clouds form on some occasions but not on others? Why do some clouds look thin and harmless, whereas others form gray and ominous towers? Answers to these questions involve the role of water vapor in the atmosphere, the central theme of this chapter.

12.1 Water's Changes of State

List and describe the processes that cause water to change from one state of matter to another. Define *latent heat* and explain why it is important.

Water is the only substance that exists in the atmosphere as a solid, liquid, and gas (**Figure 12.1**). It is made of hydrogen and oxygen atoms that are bonded together to form water molecules (H_2O). In all three states of matter (even ice), these molecules are in constant motion; the higher the temperature, the more vigorous the movement. The chief difference among liquid water, ice, and water vapor is the arrangement of the water molecules.

Figure 12.1 Caught in a downpour (Photo by AP Photo/Keystone, Marcel Bier)

Ice, Liquid Water, and Water Vapor

Ice is composed of water molecules that are held together by mutual molecular attractions. The molecules form a tight, orderly network, as shown in **Figure 12.2**. As a consequence, the water molecules in ice are not free to move relative to each other but rather vibrate about fixed sites. When ice is heated, the molecules oscillate more rapidly. When the rate of molecular movement increases sufficiently, the bonds between some of the water molecules are broken, resulting in melting.

In the liquid state, water molecules are still tightly packed but are moving fast enough that they are able to slide past one another. As a result, liquid water is fluid and will take the shape of its container.

As liquid water gains heat from its environment, some of the molecules will acquire enough energy to break the remaining molecular attractions and escape from the surface, becoming water vapor. Water-vapor molecules are widely spaced compared to liquid water and exhibit very energetic random motion. What distinguishes a gas from a liquid is its compressibility (and expandability). For example, you can easily put more and more air into a tire and increase its volume only slightly. However, you can't put 10 gallons of gasoline into a 5-gallon can.

To summarize, when water changes state, it does not turn into a different substance; only the distances and interactions among the water molecules change.

Latent Heat

Whenever water changes state, heat is exchanged between water and its surroundings. When water evaporates, heat

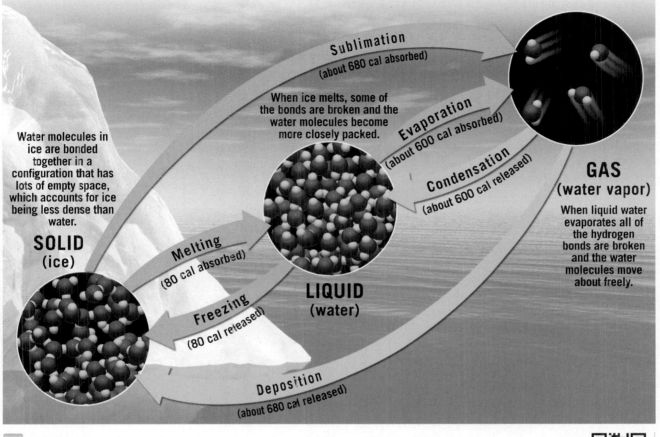

Sublimation
(about 680 cal absorbed)

When ice melts, some of
the bonds are broken and the
water molecules become
more closely packed.

Evaporation
(about 600 cal absorbed)

Condensation
(about 600 cal released)

GAS
(water vapor)

When liquid water
evaporates all of
the hydrogen
bonds are broken
and the water
molecules move
about freely.

Water molecules in
ice are bonded
together in a
configuration that has
lots of empty space,
which accounts for ice
being less dense than
water.

SOLID
(ice)

Melting
(80 cal absorbed)

LIQUID
(water)

Freezing
(80 cal released)

Deposition
(about 680 cal released)

 SmartFigure 12.2 Changes of state involve an exchange of heat The numbers shown here are the approximate numbers of calories either absorbed or released when 1 gram of water changes from one state of matter to another.

is absorbed (see Figure 12.2). Meteorologists often measure heat energy in calories. One **calorie** is the amount of heat required to raise the temperature of 1 gram of water 1°C (1.8°F). Thus, when 10 calories of heat are absorbed by 1 gram of water, the molecules vibrate faster, and a 10°C (18°F) temperature rise occurs.

Under certain conditions, heat may be added to a substance without an accompanying rise in temperature. For example, when a glass of ice water is warmed, the temperature of the ice–water mixture remains a constant 0°C (32°F) until all the ice has melted. If adding heat does not raise the temperature, where does this energy go? In this case, the added energy goes into breaking the molecular attractions between the water molecules in the ice cubes.

Because the heat used to melt ice does not produce a temperature change, it is referred to as **latent heat**. (*Latent* means "hidden," like the latent fingerprints hidden at a crime scene.) This energy can be thought of as being stored in liquid water, and it is not released to its surroundings as heat until the liquid returns to the solid state.

Melting 1 gram of ice requires 80 calories, an amount referred to as *latent heat of melting*. Freezing, the reverse process, releases these 80 calories per gram to the environment as *latent heat of fusion*.

Evaporation and Condensation We saw that heat is absorbed when ice is converted to liquid water. Heat is also absorbed during **evaporation**, the process of converting a liquid to a gas (vapor). The energy absorbed by water molecules during evaporation is used to give them the motion needed to escape the surface of the liquid and become a gas. This energy is referred to as the *latent heat of vaporization*. During the process of evaporation, it is the higher-temperature (faster-moving) molecules that escape the surface. As a result, the average molecular motion (temperature) of the remaining water is reduced—hence the common expression "evaporation is a cooling process." You have undoubtedly experienced this cooling effect when stepping dripping wet from a swimming pool or bathtub. In this situation, the energy used to evaporate water comes from your skin—hence, you feel cool.

The reverse process, **condensation**, occurs when water vapor changes to the liquid state. During condensation, water-vapor molecules release energy (*latent heat of condensation*) in an amount equivalent to what was absorbed during evaporation. When condensation occurs in the atmosphere, it results in the formation of such phenomena as fog and clouds.

As you will see, latent heat plays an important role in many atmospheric processes. In particular, when water

Figure 12.3 Examples of condensation and deposition (Top: Photo by NaturePL/Superstock; Bottom: Photo by Stockxpert/Thinkstock)

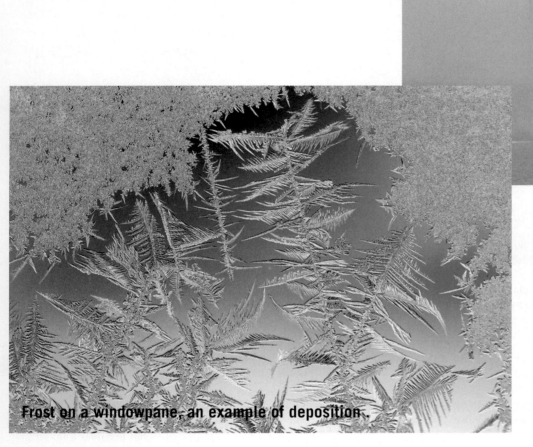

Condensation of water vapor generates phenomena such as dew, clouds, and fog.

Frost on a windowpane, an example of deposition.

B.

vapor condenses to form cloud droplets, latent heat of condensation is released, warming the surrounding air and giving it buoyancy. When the moisture content of air is high, this process can spur the growth of towering storm clouds.

Sublimation and Deposition   You are probably least familiar with the last two processes illustrated in Figure 12.2—sublimation and deposition. **Sublimation** is the conversion of a solid directly to a gas, without passing through the liquid state. Examples you may have observed include the gradual shrinking of unused ice cubes in the freezer and the rapid conversion of dry ice (frozen carbon dioxide) to wispy clouds that quickly disappear.

   Deposition refers to the reverse process, the conversion of a vapor directly to a solid. This change occurs, for example, when water vapor is deposited as ice on

solid objects such as grass or windows (**Figure 12.3**). These deposits are called *white frost* or *hoar frost* and are frequently referred to simply as *frost*. A household example of the process of deposition is the "frost" that accumulates in a freezer. As shown in Figure 12.2, deposition releases an amount of energy equal to the total amount released by condensation and freezing.

12.1 CONCEPT CHECKS

1. Summarize the processes by which water changes from one state of matter to another. Indicate whether energy is absorbed or released.
2. What is *latent heat*?
3. What is a common example of sublimation?
4. How does frost form?

12.2 Humidity: Water Vapor in the Air

Distinguish between relative humidity and dew point. Write a generalization relating how temperature changes affect relative humidity.

Water vapor constitutes only a small fraction of the atmosphere, varying from as little as one-tenth of 1 percent up to about 4 percent by volume. But the importance of water in the air is far greater than these small percentages would indicate. Indeed, scientists agree that *water vapor* is the most important gas in the atmosphere when it comes to understanding atmospheric processes.

Humidity is the general term for the amount of water vapor in air. Meteorologists employ several methods to express the water-vapor content of the air; we examine three: mixing ratio, relative humidity, and dew-point temperature.

Saturation

Before we consider these humidity measures further, it is important to understand the concept of **saturation**. Imagine a closed jar that contains water overlain by dry air, both at the same temperature. As the water begins to evaporate from the water surface, a small increase in pressure can be detected in the air above. This increase is the result of the motion of the water-vapor molecules that were added to the air through evaporation. In the open atmosphere, this pressure is termed **vapor pressure** and is defined as the part of the total atmospheric pressure that can be attributed to the water-vapor content.

In the closed container, as more and more molecules escape from the water surface, the steadily increasing vapor pressure in the air above forces more and more of these molecules to return to the liquid. Eventually the number of vapor molecules returning to the surface will balance the number leaving. At that point, the air is *saturated*. If we add heat to the container, thereby increasing the temperature of the water and air, more water will evaporate before a balance is reached. Consequently, at higher temperatures, more moisture is required for saturation. The amount of water vapor required for saturation at various temperatures is shown in **Table 12.1**.

Mixing Ratio

Not all air is saturated, of course. Thus, we need ways to express how humid a parcel of air is. One method specifies the amount of water vapor contained in a unit of air. The **mixing ratio** is the mass of water vapor in a unit of air compared to the remaining mass of dry air:

$$\text{mixing ratio} = \frac{\text{mass of water vapor (grams)}}{\text{mass of dry air (kilograms)}}$$

TABLE 12.1 Amount of Water Vapor Needed to Saturate 1 Kilogram of Air at Various Temperatures

Temperature °C (°F)	Water-Vapor Content at Saturation (grams)
−40 (−40)	0.1
−30 (−22)	0.3
−20 (−4)	0.75
−10 (14)	2
0 (32)	3.5
5 (41)	5
10 (50)	7
15 (59)	10
20 (68)	14
25 (77)	20
30 (86)	26.5
35 (95)	35
40 (104)	47

Table 12.1 shows the mixing ratios of saturated air at various temperatures. For example, at 25°C (77°F) a saturated parcel of air (1 kilogram) would contain 20 grams of water vapor.

Because the mixing ratio is expressed in units of mass (usually in grams per kilogram), it is not affected by changes in pressure or temperature. However, measuring the mixing ratio by direct sampling is time-consuming. Thus, other methods are employed to express the moisture content of the air. These include relative humidity and dew-point temperature.

Relative Humidity

The most familiar and, unfortunately, the most misunderstood term used to describe the moisture content of air is relative humidity. **Relative humidity** *is a ratio of the air's actual water-vapor content compared with the amount of water vapor required for saturation at that temperature (and pressure).* Thus, relative humidity indicates how near the air is to saturation rather than the actual quantity of water vapor in the air.

To illustrate, we see from Table 12.1 that at 25°C (77°F), air is saturated when it contains 20 grams of water vapor per kilogram of air. Thus, if the air contains 10 grams of water vapor per kilogram on a 25°C day, the relative humidity is expressed as 10/20, or 50 percent. If air with a temperature of 25°C had a

A. Initial condition: 5 grams of water vapor

Temperature 25°C

1 kg air

5 grams H$_2$O vapor

1. Saturation mixing ratio at 25°C = 20 grams*
2. H$_2$O vapor content = 5 grams
3. Relative humidity = $^5/_{20}$ = 25%

*See Table 12.1

B. Addition of 5 grams of water vapor = 10 grams

25°C

1 kg air

10 grams H$_2$O vapor

Evaporation

1. Saturation mixing ratio at 25°C = 20 grams*
2. H$_2$O vapor content = 10 grams
3. Relative humidity = $^{10}/_{20}$ = 50%

C. Addition of 10 grams of water vapor = 20 grams

25°C

1 kg air

20 grams H$_2$O vapor

Evaporation

1. Saturation mixing ratio at 25°C = 20 grams*
2. H$_2$O vapor content = 20 grams
3. Relative humidity = $^{20}/_{20}$ = 100%

Figure 12.4 The relationship between relative humidity and water vapor content At a constant temperature, the capacity (saturation mixing ratio) remains unchanged at 20 grams per kilogram, while the relative humidity rises from 25 to 100 percent as the water-vapor content increases.

water-vapor content of 20 grams per kilogram, the relative humidity would be expressed as 20/20, or 100 percent. When the relative humidity reaches 100 percent, the air is saturated.

Figure 12.5 How relative humidity varies with temperature When the water-vapor content (mixing ratio) is constant, a decrease in air temperature causes an increase in relative humidity. In this example, when the temperature of the air in the flask was lowered from 20° to 10°C, the relative humidity increased from 50 to 100 percent. Further cooling (from 10° to 0°C) causes one-half of the water vapor to condense. In nature, cooling of air below its saturation mixing ratio generally causes condensation in the form of clouds, dew, or fog.

Because relative humidity is based on the air's water-vapor content, as well as the amount of moisture required for saturation, it can be changed in either of two ways. First, relative humidity can be changed by the addition or removal of water vapor. Second, because the amount of moisture required for saturation is a function of air temperature, relative humidity varies with temperature. (Recall that the amount of water vapor required for saturation is temperature dependent, and at higher temperatures, it takes more water vapor to saturate air than at lower temperatures.)

Adding or Subtracting Moisture Notice in **Figure 12.4** that when water vapor is added to a parcel of air, its relative humidity increases until saturation occurs (100 percent relative humidity). What if even more moisture is added to this parcel of saturated air? Does the relative humidity exceed 100 percent? Normally, this situation does not occur. Instead, the excess water vapor condenses to form liquid water.

In nature, moisture is added to the air mainly via evaporation from the oceans. However, plants, soil, and smaller bodies of water also make substantial contributions.

Changes with Temperature The second condition that affects relative humidity is air temperature. Examine **Figure 12.5** carefully. Note in Figure 12.5A that when air at 20°C contains 7 grams of water vapor per kilogram, it has a relative humidity of 50 percent. This can be verified by referring to Table 12.1. Here we can see that at 20°C, air is saturated when it contains 14 grams of water vapor per kilogram of air. Because the air in Figure 12.5A

A. Initial condition: 20°C

Temperature 20°C

1 kg air

7 grams H$_2$O vapor

1. Saturation mixing ratio at 20°C = 14 grams*
2. H$_2$O vapor content = 7 grams
3. Relative humidity = $^7/_{14}$ = 50%

*See Table 12.1

B. Cooled to 10°C

10°C

1 kg air

7 grams H$_2$O vapor

1. Saturation mixing ratio at 10°C = 7 grams*
2. H$_2$O vapor content = 7 grams
3. Relative humidity = $^7/_7$ = 100%

C. Cooled to 0°C

0°C

3.5 grams H$_2$O liquid

1 kg air

3.5 grams H$_2$O vapor

1. Saturation mixing ratio at 0°C = 3.5 grams*
2. H$_2$O vapor content = 3.5 grams
3. Relative humidity = $^{3.5}/_{3.5}$ = 100%

contains 7 grams of water vapor, its relative humidity is 7/14, or 50 percent.

When the flask is cooled from 20°to 10°C, as shown in Figure 12.5B, the relative humidity increases from 50 to 100 percent. We can conclude from this that when the water-vapor content remains constant, *a decrease in temperature results in an increase in relative humidity.*

What happens when the air is cooled below the temperature at which saturation occurs? Figure 12.5C illustrates this situation. Notice from Table 12.1 that when the flask is cooled to 0°C, the air is saturated at 3.5 grams of water vapor per kilogram of air. Because this flask originally contained 7 grams of water vapor, 3.5 grams of water vapor will condense to form liquid droplets that collect on the walls of the container. The relative humidity of the air inside remains at 100 percent. This brings up an important concept. When air aloft is cooled below its saturation level, some of the water vapor condenses to form clouds. As clouds are made of liquid droplets (or ice crystals), they are no longer part of the *water-vapor* content of the air. (Clouds are not water vapor; they are composed of liquid water droplets or ice crystals too tiny to fall to Earth.)

We can summarize the effects of temperature on relative humidity as follows: When the water-vapor content of air remains at a constant level, a decrease in air temperature results in an increase in relative humidity, and an increase in temperature causes a decrease in relative humidity. In **Figure 12.6** the variations in temperature and relative humidity during a typical day demonstrate the relationship just described.

Dew-Point Temperature

Another important measure of humidity is the dew-point temperature. The **dew-point temperature**, or simply the **dew point**, is the temperature to which a parcel of air would need to be cooled to reach saturation. For example, in Figure 12.5, the unsaturated air in the flask had to be cooled to 10°C before saturation occurred. Therefore, 10°C is the dew-point temperature for this air. In nature, cooling below the dew point causes water vapor to condense, typically as dew, fog, or clouds. The term *dew point* stems from the fact that during nighttime hours, objects near the ground often cool below the dew-point temperature and become coated with dew (**Figure 12.7**).

Unlike relative humidity, which is a measure of how near the air is to being saturated, dew-point temperature is a measure of its *actual moisture* content. Because the dew-point temperature is directly related to the amount of water vapor in the air, and because it is easy to determine, it is one of the most useful measures of humidity.

The amount of water vapor needed for saturation is temperature dependent: For every 10°C (18°F) increase

Figure 12.6 Typical daily variations in temperature and relative humidity This graph shows the daily variations in temperature and relative humidity during a spring day at Washington, DC. When temperature increases, relative humidity drops (see mid-afternoon) and vice versa.

Figure 12.7 Condensation, or "dew," on a cold drinking glass The cold glass chills the surrounding layer of air below its dew-point temperature, causing condensation. (amana images inc./ Aamy)

SmartFigure 12.8 Map showing dew-point temperatures on a September day Dew-point temperatures above 60°F dominate the southeastern United States, indicating that this region is blanketed with humid air, while much drier conditions prevail over the American Southwest.

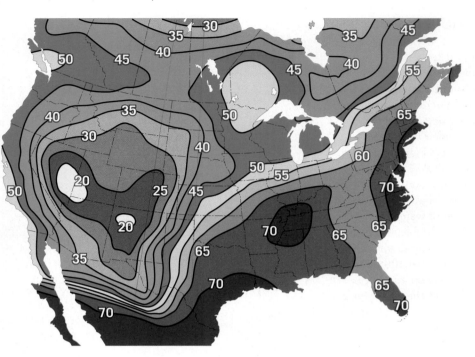

in temperature, the amount of water vapor needed for saturation doubles (see Table 12.1). Therefore, relatively cold *saturated air* (0°C [32°F]) contains about half the water vapor of *saturated air* having a temperature of

10°C (50°F) and roughly one-fourth that of warm *saturated air* with a temperature of 20°C (68°F). Because the dew point is the temperature at which saturation occurs, we can conclude that high dew-point temperatures indicate moist air, and low dew-point temperatures indicate dry air. More precisely, we know that air over Fort Myers, Florida, with a dew point of 25°C (77°F), contains about twice the water vapor of air situated over St. Louis, Missouri, with a dew point of 15°C (59°F), and four times that of Tucson, Arizona, with a dew point of 5°C (41°F).

Because the dew-point temperature is a good measure of the amount of water vapor in the air, it is the measure of atmospheric moisture that appears on a variety of weather maps. Notice on the map in **Figure 12.8** that most of the places located near the warm Gulf of Mexico have dew-point temperatures that exceed 70°F (21°C). When the dew point exceeds 65°F (18°C), most people consider the air to be humid, and air with a dew point of 75°F (24°C) or higher is oppressive. Also notice in Figure 12.8 that although the Southeast is dominated by humid conditions (dew points above 65°F), most of the remainder of the country is experiencing drier air.

Measuring Humidity

Relative humidity is commonly measured using a **hygrometer** (*hygro* = moisture, *metron* = measuring instrument). One type of hygrometer, called a **psychrometer**, consists of two identical thermometers mounted side by side (**Figure 12.9**). One thermometer, the *dry-bulb*

Figure 12.9 Sling psychrometer used to determine both relative humidity and dew point (Photo by E. J. Tarbuck)

A. The dry-bulb thermometer gives the current air temperature.

Wet bulb — Dry bulb

Room temperature water

B. The wet-bulb thermometer is covered with a cloth wick that is dipped in water.

C. The thermometers are spun until the temperature of the wet-bulb thermometer stops declining. Then the thermometers are read and the data is interpreted using the tables in Appendix C.

thermometer, gives the current air temperature. The other, called the *wet-bulb* thermometer, has a thin muslin wick tied around the end (see the ends of the thermometers in Figure 12.9).

To use the psychrometer, the cloth sleeve is saturated with water, and a continuous current of air is passed over the wick (see Figure 12.9). This is done either by swinging the instrument freely in the air or by fanning air past it. As a consequence, water evaporates from the wick, and the heat absorbed by the evaporating water makes the temperature of the wet bulb drop. The loss of heat that was required to evaporate water from the wet bulb lowers the thermometer reading.

The amount of cooling that takes place is directly proportional to the dryness of the air. The drier the air, the greater the amount of moisture that evaporates. The more heat the evaporating water absorbs, the greater the cooling. Therefore, the larger the difference that is observed between the thermometer readings, the lower the relative humidity; the smaller the difference, the higher the relative humidity. If the air is saturated, no evaporation will occur, and the two thermometers will have identical readings.

To determine the precise relative humidity from the thermometer readings, a standard table is used (refer to Appendix C, Table C.1). With the same information but using a different table (Table C.2), the dew-point temperature can also be calculated.

12.2 CONCEPT CHECKS

1. Describe how air temperature and saturation are related.
2. List three measures that are used to express humidity.
3. How do relative humidity and mixing ratio differ?
4. If the temperature remains unchanged and the amount of water vapor in the air decreases, how does relative humidity change?
5. On a warm summer day when the relative humidity is high, it may seem even warmer than the thermometer indicates. Why do we feel so uncomfortable on these muggy days?

12.3 The Basis of Cloud Formation: Adiabatic Cooling

Explain how adiabatic cooling results in cloud formation.

We have considered basic properties of water vapor and how its variability is measured. This section examines some of the important roles that water vapor plays in weather, especially the formation of clouds.

Fog and Dew Versus Cloud Formation

Recall that condensation occurs when water vapor changes to a liquid. Condensation may form dew, fog, or clouds. Although these three forms are different, all require that air reach saturation. As indicated earlier, saturation occurs either when sufficient water vapor is added to the air or, more commonly, when the air is cooled to its dew point.

Near Earth's surface, heat is readily exchanged between the ground and the air above. During evening hours, the surface radiates heat away, causing the surface and adjacent air to cool rapidly. This radiation cooling accounts for the formation of dew and some types of fog. Thus, surface cooling that occurs after sunset accounts for some condensation. However, cloud formation often takes place during the warmest part of the day. Clearly some other mechanism must operate aloft that cools air sufficiently to generate clouds.

Adiabatic Temperature Changes

The process that is responsible for most cloud formation is easily demonstrated if you have ever pumped up a bicycle tire and noticed that the pump barrel became quite warm. The heat you felt was a result of the work you did on the air to compress it. When energy is used to compress air, the motion of the gas molecules increases, and therefore the temperature of the air rises. Conversely, air that is allowed to escape from a bicycle tire *expands and cools*. This results because the expanding air pushes

Figure 12.10 Dry verses wet adiabatic rates of cooling Rising air cools at the dry adiabatic rate of 10° per 1000 meters, until the air reaches the dew point and condensation (cloud formation) begins. As air continues to rise, the latent heat released by condensation reduces the rate of cooling. The wet adiabatic rate is therefore always less than the dry adiabatic rate.

(does work on) the surrounding air and must cool by an amount equivalent to the energy expended.

You have probably felt the cooling effect of the propellant gas expanding as you have applied hairspray or spray deodorant. As the compressed gas in an aerosol spray can is released, it quickly expands and cools. This drop in temperature occurs *even though heat is neither added nor subtracted*. Such variations are known as **adiabatic temperature changes** and result when air is compressed or allowed to expand. In summary, *when air is allowed to expand, it cools, and when it is compressed, it warms*.

Adiabatic Cooling and Condensation

To simplify the following discussion, it helps to imagine a volume of air enclosed in a thin elastic cover. Meteorologists call this imaginary volume of air a **parcel**. Typically, we consider a parcel to be a few hundred cubic meters in volume, and we assume that it acts independently of the surrounding air. It is also assumed that no heat is transferred into, or out of, the parcel. Although this image is highly idealized, over short time spans, a parcel of air behaves in a manner much like an actual volume of air moving vertically in the atmosphere.

Dry Adiabatic Rate As you travel from Earth's surface upward through the atmosphere, the atmospheric pressure rapidly diminishes because there are fewer and fewer gas molecules. Thus, any time a parcel of air moves upward, it passes through regions of successively lower pressure, and

the ascending air expands. As it expands, it cools adiabatically. Unsaturated air cools at a constant rate of 10°C for every 1000 meters of ascent (5.5°F per 1000 feet).

Conversely, descending air comes under increasingly higher pressures, compresses, and is heated 10°C for every 1000 meters of descent. This rate of cooling or heating applies only to *unsaturated air* and is known as the **dry adiabatic rate**.

Wet Adiabatic Rate If a parcel of air rises high enough, it will eventually cool to its dew point, where the process of condensation begins. From this point on along its ascent, *latent heat of condensation* stored in the water vapor will be liberated. Although the air will continue to cool after condensation begins, the released latent heat works against the adiabatic process, thereby reducing the rate at which the air cools. This slower rate of cooling caused by the addition of latent heat is called the **wet adiabatic rate** of cooling. Because the amount of latent heat released depends on the quantity of moisture present in the air, the wet adiabatic rate varies from 5°C per 1000 meters for air with a high moisture content to 9°C per 1000 meters for dry air.

Figure 12.10 illustrates the role of adiabatic cooling in the formation of clouds. Note that from the surface up to the condensation level, the air cools at the dry adiabatic rate. The wet adiabatic rate begins at the condensation level.

12.3 CONCEPT CHECKS

1. How is the formation of dew different from the formation of a cloud? How are they similar?
2. Why does air cool when it rises through the atmosphere?
3. What do meteorologists mean by the word *parcel*?
4. Why does the adiabatic rate of cooling change when condensation begins? Why is the wet adiabatic rate not a constant figure?
5. The contents of an aerosol spray can are under very high pressure. When you push the nozzle on such a can, the spray feels cold. Explain.

12.4 Processes That Lift Air

List and describe the four mechanisms that cause air to rise.

To review, when air rises, it expands and cools adiabatically. If air is lifted sufficiently, it will eventually cool to its dew-point temperature, saturation will occur, and clouds will develop. Why does air rise on some occasions but not on others?

Generally, air tends to resist vertical movement; air located near the surface tends to stay near the surface, and air aloft tends to remain aloft. Exceptions to these patterns include conditions in the atmosphere that give air sufficient buoyancy to rise without the aid of outside forces. One example, called **convective lifting**, occurs when a mass of air is warmer and therefore less dense than the surrounding air. In most situations, however, when you see clouds forming, there is some mechanical phenomenon at work that forces the air to rise (at least initially).

Here we will look at four mechanisms that cause air to rise:

1. *Orographic lifting*—Air is forced to rise over a mountainous barrier.
2. *Frontal wedging*—Warmer, less dense air is forced over cooler, denser air.
3. *Convergence*—A pileup of horizontal airflow results in upward movement.
4. *Localized convective lifting*—Unequal surface heating causes localized pockets of air to rise because of their buoyancy.

A. Orographic lifting leads to precipitation on windward slopes.

B. By the time air reaches the leeward side of the mountains, much of the moisture has been lost, resulting in a *rain shadow desert*.

Figure 12.11 Orographic lifting and rainshadow deserts (Photo A by Dean Pennala/Shutterstock; photo B by Dennis Tasa)

Orographic Lifting

Orographic lifting occurs when elevated terrains, such as mountains, act as barriers to the flow of air (**Figure 12.11**). As air ascends a mountain slope, adiabatic cooling often generates clouds and copious precipitation. In fact, many of the rainiest places in the world are located on windward mountain slopes.

By the time air reaches the leeward side of a mountain, much of its moisture has been lost. If the air descends, it warms adiabatically, making condensation and precipitation even less likely. As shown in Figure 12.11, the result can be a **rainshadow desert**. The Great Basin Desert of the western United States lies only a few hundred kilometers from the Pacific Ocean, but it is effectively cut off from the ocean's moisture by the imposing Sierra Nevada (see Figure 12.11). The Gobi Desert of Mongolia, the Takla Makan of China, and the Patagonia Desert of Argentina are other examples of deserts that exist because they are on the leeward sides of mountains. (For a map showing deserts, see Figure 4.24, page 137.)

Frontal Wedging

If orographic lifting were the only mechanism that forced air aloft, the relatively flat central portion of North America would be an expansive desert instead of the nation's breadbasket. Fortunately, this is not the case.

In central North America, masses of warm and cold air collide, producing a **front**. Here the cooler, denser air acts as a barrier over which the warmer, less dense air rises. This process, called **frontal wedging**, is illustrated in **Figure 12.12**.

It should be noted that weather-producing fronts are associated with storm systems called *middle-latitude cyclones*. Because these storms are responsible for producing a high percentage of the precipitation in the middle latitudes, we examine them closely in Chapter 14.

Convergence

We saw that the collision of contrasting air masses forces air to rise. In a more general sense, whenever air in the lower atmosphere flows together, lifting results. This phenomenon is called **convergence**. When air flows

Did You Know?

Many of the rainiest places in the world are located on windward mountain slopes. A station at Mount Waialeale, Hawaii, records the highest average annual rainfall, some 1234 cm (486 in). The greatest recorded rainfall for a single 12-month period occurred at Cherrapunji, India, where an astounding 2647 cm (1042 in [over 86 ft]) fell. Much of this rainfall occurred in the month of July—a record 930 cm (366 in [more than 30 ft]). This is 10 times more rain than Chicago receives in an average year.

Figure 12.12 Frontal wedging Colder, denser air acts as a barrier over which warmer, less dense air rises.

in from more than one direction (converges), it must go somewhere. Because it cannot go down, it goes up. This leads to adiabatic cooling and possible cloud formation.

Convergence can also occur whenever an obstacle slows or restricts horizontal airflow (wind). When air moves from a relatively smooth surface, such as the ocean, onto an irregular landscape, its speed is reduced. The result is a pileup of air (convergence). This is similar to what happens when people leave a well-attended sporting event and a pileup results at the exits. When air converges, the air molecules do not simply squeeze closer together (like people); rather, there is a net upward flow.

The Florida peninsula provides an excellent example of the role that convergence can play in initiating cloud development and precipitation (**Figure 12.13**). On warm days, the airflow is from the ocean to the land along both coasts of Florida. This leads to a pileup of air along the coasts and general convergence over the peninsula. This pattern of air movement and the uplift that results is aided by intense solar heating of the land. This is why the peninsula of Florida experiences the greatest frequency of mid-afternoon thunderstorms in the United States.

Convergence as a mechanism of forced lifting is a major contributor to the weather associated with middle-latitude cyclones and hurricanes. The low-level horizontal airflow associated with these systems is inward and upward around their centers. These important weather producers are covered in more detail later, but for now remember that convergence near the surface results in a general upward flow.

Localized Convective Lifting

On warm summer days, unequal heating of Earth's surface may cause pockets of air to be warmed more than the surrounding air. For instance, air above a paved parking lot will be warmed more than the air above an adjacent wooded park. Consequently, the parcel of air above the parking lot, which is warmer (less dense) than the surrounding air, will be buoyed upward (**Figure 12.14**).

These rising parcels of warmer air are called *thermals*. Birds such as hawks and eagles use these thermals to carry them to great heights, from which they can gaze down on unsuspecting prey. People have learned to employ these rising parcels using hang gliders as a way to "fly."

The phenomenon that produces rising thermals is called **localized convective lifting**. When these warm parcels of air rise above the lifting condensation level, clouds form, which can produce mid-afternoon rain showers. The accompanying rains, although occasionally heavy, are of short duration and widely scattered.

SmartFigure 12.13 Surface convergence enhances cloud development When surface air converges, the column of air increases in height because the air is crammed into a smaller and smaller space. The result is forced lifting and adiabatic cooling that generates clouds. This photo shows southern Florida viewed from the Space Shuttle. On warm days, airflow from the Atlantic Ocean and Gulf of Mexico onto the Florida peninsula generates many mid-afternoon thunderstorms. (Photo by NASA/Media Services)

TIME

Figure 12.14 **Localized convective lifting** Unequal heating of Earth's surface causes pockets of air to be warmed more than the surrounding air. These buoyant parcels of hot air (thermals) rise, and if they reach the condensation level, clouds form.

Although localized convective lifting by itself is not a major producer of precipitation, the added buoyancy that results from surface heating contributes significantly to the lifting initiated by the other mechanisms. In addition, even though the other mechanisms force air to rise, convective lifting occurs because the air is warmer (less dense) than the surrounding air and rises for the same reasons as does a hot-air balloon.

12.4 CONCEPT CHECKS

1. List four mechanisms that cause air to rise.
2. How do orographic lifting and frontal wedging force air to rise?
3. Explain why the Great Basin area of the western United States is dry. What term is applied to this situation?
4. What causes the Florida peninsula to experience the greatest frequency of mid-afternoon thunderstorms in the United States?

12.5 The Weathermaker: Atmospheric Stability

Describe how atmospheric stability is determined and compare conditional instability with absolute instability.

When air rises, it cools and usually produces clouds. Why do clouds vary so much in size, and why does the resulting precipitation vary so much? The answers are closely related to the *stability* of the air.

Recall that a parcel of air can be thought of as having a thin, flexible cover that allows it to expand but prevents it from mixing with the surrounding air (picture a hot-air balloon). If this parcel were forced to rise, *its temperature would decrease because of expansion*. By comparing the parcel's temperature to that of the surrounding air, we can determine the stability of the bubble. If the parcel's temperature is lower than that of the surrounding environment, it will be denser; if it is allowed to move freely, it will sink to its original position. Air of this type, called **stable air**, resists vertical movement.

If, however, our imaginary rising parcel is *warmer* and hence less dense than the surrounding air, it will continue to rise until it reaches an altitude where its temperature equals that of its surroundings. This is exactly how a hot-air balloon works, rising as long as it is warmer and less dense than the surrounding air (**Figure 12.15**). This type of air is classified as **unstable air**. In summary,

Figure 12.15 **Air that is warmer and therefore less dense than its surroundings will rise** Hot-air balloons rise up through the atmosphere for this reason. (Photo by Steve Vidler/SuperStock)

Figure 12.16 When an unsaturated parcel of air is lifted, it expands and cools at the dry adiabatic rate (10°C per 1000 meters) Because the temperature of the rising parcel of air is lower than the surrounding environment, it will be heavier and, it will sink to its original position if it is allowed to do so. Air of this type is referred to as *stable*.

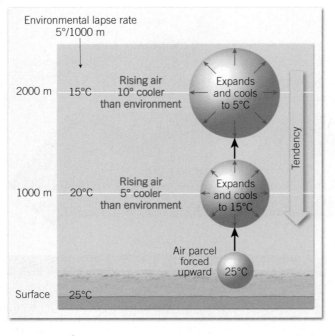

stability is a property of air that describes its tendency to remain in its original position (stable) or to rise (unstable).

Types of Stability

The stability of air is determined by measuring the temperature of the atmosphere at various heights. Recall from Chapter 11 that this measure is called the *environmental lapse rate*. The environmental lapse rate is the rate of temperature change with altitude, as determined from observations made by radiosondes and aircraft. It is important not to confuse this with *adiabatic temperature changes*,

which are changes in temperature caused by expansion or compression as a parcel of air rises or descends.

To illustrate, we examine a situation in which the environmental lapse rate is 5°C per 1000 meters (**Figure 12.16**). Under this condition, when air at the surface has a temperature of 25°C, the air at 1000 meters will be 5° cooler, or 20°C, the air at 2000 meters will have a temperature of 15°C, and so forth. At first glance, it appears that the air at the surface is less dense than the air at 1000 meters because it is 5° warmer. However, if the air near the surface were unsaturated and were to rise to 1000 meters, it would expand and cool at the dry adiabatic rate of 10°C per 1000 meters. Therefore, upon reaching 1000 meters, its temperature would have dropped 10°C. Being 5° cooler than its environment, it would be denser and tend to sink to its original position. Hence, we say that the air near the surface is potentially cooler than the air aloft and therefore will not rise on its own. The air just described is *stable* and resists vertical movement.

Absolute Stability Stated quantitatively, **absolute stability** prevails when the environmental lapse rate is less than the wet adiabatic rate. **Figure 12.17** depicts this situation using an environmental lapse rate of 5°C per 1000 meters and a wet adiabatic rate of 6°C per 1000 meters. Note that at 1000 meters the temperature of the surrounding air is 15°C, while the rising parcel of air has cooled to 10°C and is therefore the denser air. Even if this stable air were to be forced above the condensation level, it would remain cooler and denser than its environment, and thus it would tend to return to the surface.

The most stable conditions occur when the temperature in a layer of air actually increases with altitude, rather than decrease. When such a reversal occurs, a

SmartFigure 12.17 *Absolute stability* prevails when the environmental lapse rate is less than the wet adiabatic rate **A.** The rising parcel of air is always cooler and heavier than the surrounding air and is therefore stable. **B.** Graphic representation of the conditions shown in part A.

Environmental lapse rate 12°C/1000 m

ABSOLUTE INSTABILITY

4000 m	−8°C		Rising air 16° warmer than environment	8°C
3000 m	4°C	Wet rate 6°C/1000 m	Rising air 10° warmer than environment	14°C
2000 m	16°C		Rising air 4° warmer than environment	20°C
1000 m	28°C	Dry rate 10°C/1000 m	Rising air 2° warmer than environment	30°C
Surface	40°C			40°C

Tendency

Solar heating

B. Temperature −10 0 10 20 30 40(°C)

Wet adiabatic rate

Condensation level

Dry adiabatic rate

Environmental lapse rate

A.

Figure 12.18 Atmospheric conditions that result in *absolute instability* **A.** Absolute instability develops when solar heating causes the lowermost layer of the atmosphere to be warmed to a higher temperature than the air aloft. The result is a steep environmental lapse rate that renders the atmosphere unstable. **B.** Graphic representation of the conditions shown in part A.

temperature inversion is said to exist. Temperature inversions frequently occur on clear nights as a result of radiation cooling of Earth's surface. Under these conditions, an inversion is created because the ground and the air immediately above will cool more rapidly than the air aloft. When warm air overlies cooler air, it acts as a lid and prevents appreciable vertical mixing. Because of this, temperature inversions are responsible for trapping pollutants in a narrow zone near Earth's surface.

Absolute Instability At the other extreme from absolute stability, air is said to exhibit **absolute instability** when the environmental lapse rate is greater than the dry adiabatic rate. As shown in **Figure 12.18**, the ascending parcel of air is always warmer than its environment and will continue to rise because of its own buoyancy. However, absolute instability is generally found near Earth's surface. On hot, sunny days the air above some surfaces, such as shopping center parking lots, is heated more than the air over adjacent surfaces. These invisible pockets of more intensely heated air, being less dense than the air aloft, will rise like a hot-air balloon. This phenomenon produces the small, fluffy clouds we associate with fair weather. Occasionally, when the surface air is considerably warmer than the air aloft, clouds with considerable vertical development can form.

Conditional Instability A more common type of atmospheric instability is called **conditional instability**. This occurs when moist air has an environmental lapse rate between the dry and wet adiabatic rates (between 5 and 10°C per 1000 meters). Simply, the atmosphere is said to be conditionally unstable when it is *stable* for an *unsaturated* parcel of air but *unstable* for a *saturated* parcel of

air. Notice in **Figure 12.19** that the rising parcel of air is cooler than the surrounding air for nearly 3000 meters. With the addition of latent heat above the lifting condensation level, the parcel becomes warmer than the surrounding air. From this point along its ascent, the parcel will continue to rise because of its own buoyancy, without an outside force. Thus, conditional instability depends on whether the rising air is saturated. The word *conditional* is used because the air must be forced upward, such as over mountainous terrain, before it becomes unstable and rises because of its own buoyancy.

In summary, the stability of air is determined by measuring the temperature of the atmosphere at various heights. In simple terms, a column of air is deemed unstable when the air near the bottom of this layer is significantly warmer (less dense) than the air aloft, indicating a steep environmental lapse rate. Under these conditions, the air actually turns over, as the warm air below rises and displaces the colder air aloft. Conversely, the air is considered to be stable when the temperature drops gradually with increasing altitude. The most stable conditions occur during a temperature inversion, when the temperature actually increases with height. Under these conditions, there is very little vertical air movement.

Stability and Daily Weather

From the previous discussion, we can conclude that stable air resists vertical movement, whereas unstable air ascends freely because of its own buoyancy. But how do these facts manifest themselves in our daily weather?

Because stable air resists upward movement, we might conclude that clouds will not form when stable

Figure 12.19 *Conditional instability* may result when warm air is forced to rise along a frontal boundary **A.** Note that the environmental lapse rate of 9°C per 1000 meters lies between the dry and wet adiabatic rates. The parcel of air is cooler than the surrounding air up to nearly 3000 meters, where its tendency is to sink toward the surface (stable). Above this level, however, the parcel is warmer than its environment and will rise because of its own buoyancy (unstable). Thus, when conditionally unstable air is forced to rise, the result can be towering cumulus clouds. **B.** Graphic representation of the conditions shown in part A.

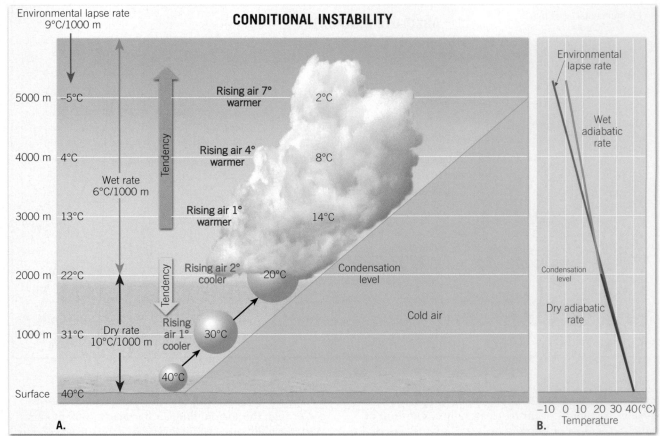

Figure 12.19 *Conditional instability* may result when warm air is forced to rise along a frontal boundary **A.** Note that the environmental lapse rate of 9°C per 1000 meters lies between the dry and wet adiabatic rates. The parcel of air is cooler than the surrounding air up to nearly 3000 meters, where its tendency is to sink toward the surface (stable). Above this level, however, the parcel is warmer than its environment and will rise because of its own buoyancy (unstable). Thus, when conditionally unstable air is forced to rise, the result can be towering cumulus clouds. **B.** Graphic representation of the conditions shown in part A.

conditions prevail in the atmosphere. Although this seems reasonable, recall that processes exist that *force* air aloft. These processes include orographic lifting, frontal wedging, and convergence. When stable air is forced aloft, the clouds that form are widespread and have little vertical thickness when compared to their horizontal dimension, and precipitation, if any, is light to moderate.

By contrast, clouds associated with the lifting of unstable air are towering and often generate thunderstorms and occasionally even a tornado. For this reason, we can conclude that on a dreary, overcast day with light drizzle, stable air has been forced aloft. On the other hand, during a day when cauliflower-shaped clouds appear to be growing as if bubbles of hot air are surging upward, we can be fairly certain that the ascending air is unstable.

In summary, stability plays an important role in determining our daily weather. To a large degree, stability determines the type of clouds that develop and whether precipitation will come as a gentle shower or a heavy downpour.

12.6 Condensation and Cloud Formation

List the necessary conditions for condensation and briefly describe the two criteria used for cloud classification.

To review briefly, condensation occurs when water vapor in the air changes to a liquid. The result of this process may be dew, fog, or clouds. For any of these forms of condensation to occur, the air must be saturated. Saturation occurs most commonly when air is cooled to its dew point, or less often when water vapor is added to the air.

Generally, there must be a surface on which the water vapor can condense. When dew occurs, objects at or near the ground, such as grass and car windows, serve this purpose. But when condensation occurs in the air above the ground, tiny bits of particulate matter, known as **condensation nuclei**, serve as surfaces for water-vapor condensation. These nuclei are very important, for in their absence a relative humidity well in excess of 100 percent is needed to produce clouds.

Condensation nuclei such as microscopic dust, smoke, and salt particles (from the ocean) are profuse in the lower atmosphere. Because of this abundance of particles, relative humidity rarely exceeds 101 percent. Some particles, such as ocean salt, are particularly good nuclei because they absorb water. These particles are termed **hygroscopic** (*hygro* = moisture, *scopic* = to seek) **nuclei**. When condensation takes place, the initial growth rate of cloud droplets is rapid. It diminishes quickly because the excess water vapor is quickly absorbed by the numerous competing particles. This results in the formation of a cloud consisting of millions upon millions of tiny water droplets, all so fine that they remain suspended in air. When cloud formation occurs at below-freezing temperatures, tiny ice crystals form. Thus, a cloud might consist of water droplets, ice crystals, or both.

The slow growth of cloud droplets by additional condensation and the immense size difference between cloud droplets and raindrops suggest that condensation alone is not responsible for the formation of drops large enough to fall as rain. We first examine clouds and then return to the question of how precipitation forms.

Types of Clouds

Clouds are among the most conspicuous and observable aspects of the atmosphere and its weather. **Clouds** are a form of condensation best described as *visible aggregates of minute droplets of water or tiny crystals of ice*. In addition to being prominent and sometimes spectacular features in the sky, clouds are of continual interest to meteorologists because they provide a visible indication of what is going on in the atmosphere. Anyone who observes clouds with the hope of recognizing different types often finds that there is a bewildering variety of these familiar white and gray masses streaming across the sky. Once one comes to know the basic classification scheme for clouds, most of the confusion vanishes.

Clouds are classified on the basis of their *form* and *height* (**Figure 12.20**). Three basic forms are recognized:

- **Cirrus** (*cirrus* = a curl of hair) clouds are high, white, and thin. They can occur as patches or as

SmartFigure 12.20
Classification of clouds, based on height and form

delicate veil-like sheets or extended wispy fibers that often have a feathery appearance.

- **Cumulus** (*cummulus* = a pile) clouds consist of individual globular cloud masses. They normally exhibit a flat base and have the appearance of rising domes or towers. Such clouds are frequently described as having a cauliflower structure.
- **Stratus** (*stratum* = a layer) clouds are best described as sheets or layers that cover much or all of the sky. While there may be minor breaks, there are no distinct individual cloud units.

All other clouds reflect one of these three basic forms or are combinations or modifications of them.

Four levels of cloud heights are recognized: high, middle, low, and clouds of vertical displacement (see Figure 12.20). **High clouds** normally have bases above 6000 meters (20,000 feet), **middle clouds** generally occupy heights from 2000 to 6000 meters (6500 to 20,000 feet), and **low clouds** form below 2000 meters (6500 feet). The altitudes listed for each height category are not hard and fast. There is some seasonal as well as latitudinal variation. For example, at high latitudes or during cold winter months in the midlatitudes, high clouds are often found at lower altitudes.

High Clouds Three cloud types make up the family of high clouds (above 6000 meters [20,000 feet]): cirrus, cirrostratus, and cirrocumulus. *Cirrus* clouds are thin and delicate and sometimes appear as hooked filaments called "mares' tails" (**Figure 12.21A**). As the names suggest, *cirrocumulus* clouds consist of fluffy masses (**Figure 12.21B**), whereas *cirrostratus* clouds are flat layers (**Figure 12.21C**). Because of the low temperatures and small quantities of water vapor present at high altitudes, all high clouds are thin and white and are

Figure 12.21 These photos depict common forms of several different cloud types. (Photos A, B, D, E, F, and G by E. J. Tarbuck; photo C by Jung-Pang/Getty Images; photo H by Doug Millar/Science Source/ Photo Researchers, Inc.)

A. Cirrus

B. Cirrocumulus

C. Cirrostratus

D. Altocumulus

made up of ice crystals. Furthermore, these clouds are not considered precipitation makers. However, when cirrus clouds are followed by cirrocumulus clouds and increased sky coverage, they may warn of impending stormy weather.

Middle Clouds Clouds that appear in the middle range (2000–6000 meters [6500–20,000 feet]) have the prefix *alto* as part of their name. *Altocumulus* clouds are composed of globular masses that differ from cirrocumulus clouds in that they are larger and denser (**Figure 12.21D**). *Altostratus* clouds create a uniform white to grayish sheet covering the sky, with the Sun or Moon visible as a bright spot (**Figure 12.21E**). Infrequent light snow or drizzle may accompany these clouds.

Low Clouds There are three members in the family of low clouds: stratus, stratocumulus, and nimbostratus.

Stratus are a uniform foglike layer of clouds that frequently covers much of the sky. On occasion these clouds may produce light precipitation. When stratus clouds develop a scalloped bottom that appears as long parallel rolls or broken globular patches, they are called *stratocumulus* clouds.

Nimbostratus clouds derive their name from the Latin *nimbus*, which means rainy cloud, and *stratus*, which means to cover with a layer (**Figure 12.21F**). As the name suggests, nimbostratus clouds are one of the chief precipitation producers. Nimbostratus clouds form in association with stable conditions. We might not expect clouds to grow or persist in stable air, but cloud growth of this type is common when air is forced to rise, as occurs along a mountain range, along a front, or near the center of a cyclone where converging winds cause air to ascend. Such forced ascent of stable air leads to the formation of a stratified cloud layer that is large horizontally compared to its depth.

Figure 12.21 *Continued*

E. Altostratus

F. Nimbostratus

G. Cumulus

H. Cumulonimbus

TABLE 12.2 Cloud Types and Characteristics

Cloud Family and Height	Cloud Type	Characteristics
High clouds—above 6000 meters (20,000 feet)	Cirrus	Thin, delicate, fibrous, ice-crystal clouds. Sometimes appear as hooked filaments called "mares' tails." (Figure 12.21A)
	Cirrocumulus	Thin, white, ice-crystal clouds in the form of ripples, waves, or globular masses all in a row. May produce a "mackerel sky." Least common of the high clouds. (Figure 12.21B)
	Cirrostratus	Thin sheet of white, ice-crystal clouds that may give the sky a milky look. Sometimes produce halos around the Sun or Moon. (Figure 12.21C)
Middle clouds—2000–6000 meters (6500–20,000 feet)	Altocumulus	White to gray clouds often composed of separate globules; "sheep-back" clouds. (Figure 12.21D)
	Altostratus	Stratified veil of clouds that are generally thin and may produce very light precipitation. When thin, the Sun or Moon may be visible as a bright spot, but no halos are produced. (Figure 12.21E)
Low clouds—below 2000 meters (6500 feet)	Stratocumulus	Soft gray clouds in globular patches or rolls. Rolls may join together to make a continuous cloud.
	Stratus	Low uniform layer resembling fog but not resting on the ground. May produce drizzle.
	Nimbostratus	Amorphous layer of dark gray clouds. One of the chief precipitation-producing clouds. (Figure 12.21F)
Clouds of vertical development— 500–18,000 meters (1600–60,000 feet)	Cumulus	Dense, billowy clouds often characterized by flat bases. May occur as isolated clouds or closely packed. (Figure 12.21G)
	Cumulonimbus	Towering cloud sometimes spreading out on top to form an "anvil head." Associated with heavy rainfall, thunder, lightning, hail, and tornadoes. (Figure 12.21H)

Clouds of Vertical Development Some clouds do not fit into any one of the three height categories just mentioned. Such clouds have their bases in the low height range but often extend upward into the middle or high altitudes. Consequently, clouds in this category are called **clouds of vertical development**. They are all related to one another and are associated with unstable air. Although *cumulus* clouds are often connected with fair weather (**Figure 12.21G**), they may grow dramatically under the proper circumstances. Once upward movement is triggered, acceleration is powerful, and clouds with great vertical extent form. The end result is often a towering cloud, called a *cumulonimbus*, that usually produces rain showers or a thunderstorm (**Figure 12.21H**).

Definite weather patterns can often be associated with particular clouds or certain combinations of cloud types, so it is important to become familiar with cloud descriptions and characteristics. **Table 12.2** lists the 10 basic cloud types that are recognized internationally and gives some characteristics of each.

12.6 CONCEPT CHECKS

1. As you drink an ice-cold beverage on a warm, humid day, the outside of the glass or bottle becomes wet. Explain.
2. What is the function of condensation nuclei in cloud formation?
3. What is the basis for the classification of clouds?
4. Why are high clouds always thin?
5. Which cloud types are associated with the following characteristics: thunder, halos, precipitation, hail, lightning, mares' tails?

12.7 Fog

Define *fog* and explain how the various types of fog form.

Fog is generally considered to be an atmospheric hazard. When it is light, visibility is reduced to 2 or 3 kilometers (1 or 2 miles). However, when it is dense, visibility may be cut to a few dozen meters or less, making travel by any mode not only difficult but often dangerous. Officially, visibility must be reduced to 1 kilometer (0.6 mile) or less before fog is reported. While this figure is arbitrary, it does provide an objective criterion for comparing fog frequencies at different locations.

Fog is defined as *a cloud with its base at or very near the ground.* There is no basic physical difference between fog and a cloud; their appearance and structure are the same. The essential difference is the method and place of formation. Whereas clouds result when air rises and cools adiabatically, most fogs result from radiation cooling or the movement of air over a cold surface. (The exception is upslope fog.) In other circumstances, fogs form when enough water vapor is added to the air to bring about saturation (evaporation fogs). We examine both of these types.

Fogs Caused by Cooling

Three common fogs form when air at Earth's surface is chilled below its dew point. Often a fog is a combination of these types.

Advection Fog   When warm, moist air moves over a cool surface, the result might be a blanket of fog called **advection fog** (**Figure 12.22**). Examples of such fogs are very common. The foggiest location in the United States, and perhaps in the world, is Cape Disappointment, Washington. The name is indeed appropriate because this station averages 2552 hours of fog each year (there are about 8760 hours in a year). The fog experienced at Cape Disappointment, and at other West Coast locations, is produced when warm, moist air from the Pacific Ocean moves over the cold California Current and is then carried onshore by the prevailing winds. Advection fogs are also relatively common in the winter season, when warm air from the Gulf of Mexico moves across cold, often snow-covered surfaces of the Midwest and East.

Radiation Fog   **Radiation fog** forms on cool, clear, calm nights, when Earth's surface cools rapidly by radiation. As the night progresses, a thin layer of air in contact with the ground is cooled below its dew point. As the air cools and becomes denser, it drains into low areas, resulting in "pockets" of fog. The largest pockets

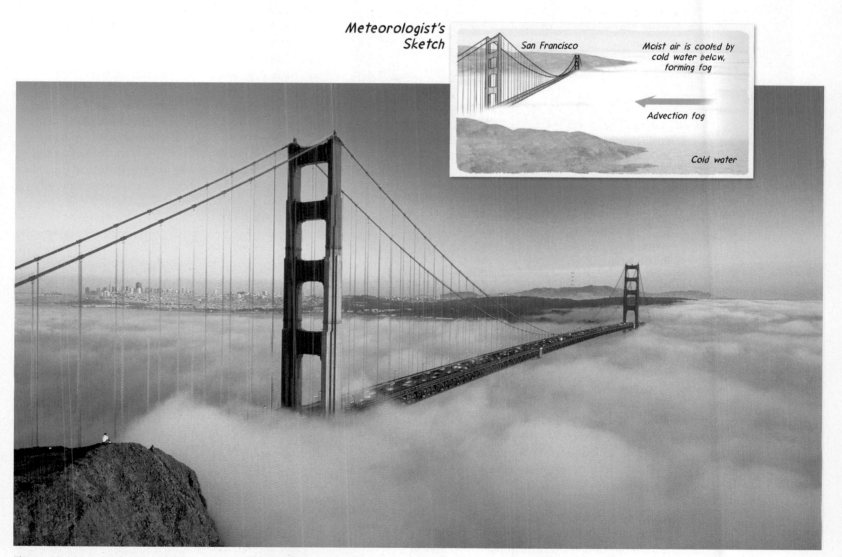

Figure 12.22 Advection fog rolling into San Francisco Bay Advection fog in coastal areas often occurs when warm, moist air flows over the cold waters of an ocean current and is cooled below its dew-point temperature. (Photo by Ed Pritchard/Getty Images, Inc.–Stone Allstock)

Figure 12.23 Radiation fog A. Satellite image of dense fog in California's San Joaquin Valley on November 20, 2002. This early morning fog was caused by radiation cooling that occurred on a clear, cool evening. It was responsible for several accidents in the region, including a 14-car pileup. The white areas to the east of the fog are the snowcapped Sierra Nevada. (NASA image)
B. Radiation fog can make the morning commute hazardous. (Photo by Tim Gainey/Alamy)

upslope toward the Rocky Mountains, the air gradually rises, resulting in an adiabatic temperature decrease of about 13°C. When the difference between the air temperature and dew point of westward-moving air is less than 13°C, an extensive fog can result in the western plains.

Evaporation Fogs

When the saturation of air occurs primarily because of the addition of water vapor, the resulting fogs are called *evaporation fogs.* Two types of evaporation fogs are recognized: steam fog and frontal, or precipitation, fog.

Steam Fog When cool air moves over warm water, enough moisture may evaporate from the water surface to produce saturation. As the rising water vapor meets the cold air, it immediately recondenses and rises with the air that is being warmed from below. Because the water has a steaming appearance, the phenomenon is called **steam fog**. Steam fog is fairly common over lakes and rivers in the fall and early winter, when the water may still be relatively warm and the air is rather crisp (**Figure 12.24**). Steam fog is often shallow because as the steam rises, it evaporates in the unsaturated air above.

Frontal Fog When frontal wedging occurs, warm air is lifted over colder air. If the resulting clouds yield rain, and the cold air below is near the dew point, enough rain will evaporate to produce fog. A fog formed in this manner is called **frontal fog**, or **precipitation fog**. The result is a more-or-less continuous zone of condensed

are often river valleys, where thick accumulations may occur (**Figure 12.23**).

Radiation fog may also form when the skies clear after a rainfall. In these situations, the air near the surface is close to saturation and only a small amount of radiation cooling is needed to promote condensation. Radiation fog of this type often occurs around sunset, and it can make driving hazardous.

Upslope Fog As its name implies, **upslope fog** is created when relatively humid air moves up a gradually sloping plain or up the steep slopes of a mountain. As a result of this upward movement, air expands and cools adiabatically. If the dew point is reached, an extensive layer of fog may form. In the United States, the Great Plains offers an excellent example. When humid easterly or southeasterly winds blow westward from the Mississippi River

Figure 12.24 Steam fog rising from Sierra Lake, Blanca, Arizona (Photo by Michael Collier)

SmartFigure 12.25
Map showing the average number of days per year with heavy fog Notice that the frequency of dense fog varies considerably from place to place. Coastal areas, particularly the Pacific Northwest and New England, where cold currents prevail, have high occurrences of dense fog.

water droplets reaching from the ground up through the clouds.

Both steam fog and frontal fog result from the addition of moisture to a layer of air. The air is usually cool or cold and already near saturation. Because air's capacity for water vapor at low temperatures is small, only a relatively modest amount of evaporation is necessary to produce saturated conditions and fog.

The frequency of dense fog varies considerably from place to place (**Figure 12.25**). As might be expected, fog incidence is highest in coastal areas, especially where cold currents prevail, as along the Pacific and New England

coasts. Relatively high frequencies are also found in the Great Lakes region and in the humid Appalachian Mountains of the East. By contrast, fogs are rare in the interior of the continent, especially in the arid and semiarid areas of the West.

12.7 CONCEPT CHECKS

1. Define *fog*.
2. List five main types of fog and describe how each type forms.

12.8 How Precipitation Forms

Describe the two mechanisms that produce precipitation.

All clouds contain water. But why do some clouds produce precipitation while others drift placidly overhead? This simple question perplexed meteorologists for many years. First, cloud droplets are very small, averaging less than 10 micrometers in diameter; for comparison, a human hair is about 75 micrometers in diameter (**Figure 12.26**). Because of their small size, cloud droplets fall incredibly slowly. In addition, clouds are made up of many billions of these droplets, all competing for the available water vapor; thus, their continued growth via condensation is extremely slow. So, what causes precipitation?

A raindrop large enough to reach the ground without completely evaporating contains roughly 1 million times more water than a single cloud droplet. Therefore, for precipitation to form, millions of cloud droplets must somehow coalesce (join together) into drops large enough to sustain themselves during their descent to the surface.

Figure 12.26 Comparative diameters of particles involved in condensation and precipitation processes

Two mechanisms have been proposed to explain this phenomenon: the *Bergeron process* in **Figure 12.27** and the *collision-coalescence process* in **Figure 12.28**.

Precipitation from Cold Clouds: The Bergeron Process

You have probably watched a TV documentary in which mountain climbers have braved intense cold and ferocious snow storms to scale ice-covered peaks. Although it is hard to imagine, similar conditions exist in the upper portions of towering cumulonimbus clouds, even on sweltering summer days. It is within these cold clouds that a mechanism called the **Bergeron process** generates much of the precipitation that occurs in the middle latitudes.

The Bergeron process is based on the fact that cloud droplets remain liquid at temperatures as low as –40°C (–40°F). Liquid water at temperatures below freezing is referred to as **supercooled**, and it becomes solid, or freezes, upon impact with objects. This explains why airplanes collect ice when they pass through a cloud composed of subzero droplets, a condition called *icing*. Supercooled water droplets also freeze upon contact with particles in the atmosphere known as **freezing nuclei**. Because freezing nuclei are relatively sparse, cold clouds primarily consist of supercooled droplets intermixed with a lesser amount of ice crystals.

When ice crystals and supercooled water droplets coexist in a cloud, the conditions are ideal for generating precipitation. Because ice crystals have a greater affinity for water vapor, they collect the available water vapor at a much faster rate than does liquid water. In turn, the water droplets evaporate to replenish the diminishing water vapor, thereby providing a continual source of moisture for the growth of ice crystals. As shown in Figure 12.27, the result is that ice crystals

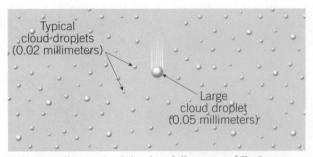

A. Because large cloud droplets fall more rapidly than smaller droplets, they are able to sweep up the smaller ones in their path and grow.

B. As drops increase in size, their fall velocity increases, resulting in increased air resistance, which causes the raindrop to flatten.

C. As the raindrop approaches 4 millimeters in size, it develops a depression in the bottom.

D. Finally, when the diameter exceeds about 5 millimeters, the depression grows upward almost explosively, forming a donut–like ring of water that immediately breaks into smaller drops.

Figure 12.28 The collision–coalescence process Most cloud droplets are so small that the motion of the air keeps them suspended. The drops are not drawn to scale; a typical raindrop has a volume equal to roughly 1 million cloud droplets.

Figure 12.27 The Bergeron process Ice crystals grow at the expense of cloud droplets until they are large enough to fall. The size of these particles has been greatly exaggerated.

grow larger, at the expense of the water droplets, which shrink in size.

Eventually, this process generates ice crystals large enough to fall as snowflakes. During their descent, these ice crystals become larger as they intercept supercooled cloud droplets that freeze on them. When the surface temperature is about 4°C (39°F) or higher, snowflakes usually melt before they reach the ground and continue their descent as rain.

Precipitation from Warm Clouds: The Collision–Coalescence Process

A few decades ago, meteorologists believed that the Bergeron process was responsible for the formation of most precipitation. However, it was discovered that copious rainfall may be produced within clouds located well below the freezing level (*warm clouds*), particularly in the tropics. This led to the proposal of a second mechanism thought to produce precipitation—the **collision–coalescence process**.

Research has shown that clouds composed entirely of liquid droplets must contain some droplets larger than 20 micrometers (0.02 millimeters) for precipitation to form. These large droplets usually form when *hygroscopic particles* (particles that attract water), such as sea salt, are abundant in the atmosphere. Hygroscopic

particles begin to remove water vapor from the air when the relative humidity is under 100 percent and the cloud droplets that form on them can grow quite large. Because the rate at which drops fall is size dependent, these "giant" droplets fall most rapidly. As they plummet, they collide with smaller, slower droplets (Figure 12.28). After many such collisions, these droplets may grow large enough to fall to the surface without evaporating. Updrafts also aid this process because they cause the droplets to traverse the cloud repeatedly.

Raindrops can grow to a maximum size of 5 millimeters, at which point they fall at a rate of 33 kilometers (20 miles) per hour. At this size and speed, the water's surface tension, which holds the drop together, is overcome by the drag imposed by the air, causing the drops to break apart. The resulting breakup of a large raindrop produces numerous smaller drops that begin anew the task of sweeping up cloud droplets. Drops that are less than 0.5 millimeter (0.02 inch) upon reaching the ground are termed *drizzle* and require about 10 minutes to fall from a cloud 1000 meters (3300 feet) overhead.

12.8 CONCEPT CHECKS

1. What is the difference between precipitation and condensation?
2. Describe the Bergeron process in your own words.
3. What conditions favor the collision–coalescence process?

12.9 Forms of Precipitation

List the different types of precipitation and explain how each type forms.

Much of the world's precipitation begins as snow crystals or other solid forms, such as hail or graupel (**Table 12.3**). Entering the warmer air below the cloud, these ice particles often melt and reach the ground as raindrops. In some parts of the world, particularly the subtropics, precipitation often forms in clouds that are warmer than 0°C (32°F). These rains frequently occur over the ocean, where cloud condensation nuclei are not plentiful and those that exist vary in size. Under such conditions, cloud droplets can grow rapidly by the collision–coalescence process to produce copious amounts of rain.

Because atmospheric conditions vary greatly from place to place as well as seasonally, several different forms of precipitation are possible. Rain and snow are the most common and familiar, but the other forms of precipitation listed in Table 12.3 are important as well. The occurrence of sleet, glaze, or hail is often associated with important weather events. Although limited in occurrence and sporadic in both time and space, these forms, especially glaze and hail, can cause considerable damage.

Rain

In meteorology, the term **rain** is restricted to drops of water that fall from a cloud and have a diameter of at least 0.5 millimeter (0.02 inch). Most rain originates either in nimbostratus clouds or in towering cumulonimbus clouds that are capable of producing unusually heavy rainfalls known as *cloudbursts*. Raindrops rarely exceed about 5 millimeters (0.2 inch) in diameter. Larger drops do not survive because surface tension, which holds the drops together, is exceeded by the frictional drag of the air. Consequently, large raindrops regularly break apart into smaller ones.

Fine, uniform drops of water having a diameter less than 0.5 millimeter (0.02 inch) are called *drizzle*. Drizzle can be so fine that the tiny drops appear to float, and their impact is almost imperceptible. Drizzle and small raindrops generally are produced in stratus or nimbostratus clouds, where precipitation can be continuous for several hours or, on rare occasions, for days.

TABLE 12.3	**Forms of Precipitation**		
Type	**Appropriate Size**	**State of Matter**	**Description**
Mist	0.005–0.05 mm	Liquid	Droplets large enough to be felt on the face when air is moving 1 meter/second. Associated with stratus clouds.
Drizzle	0.05–0.5 mm	Liquid	Small uniform drops that fall from stratus clouds, generally for several hours.
Rain	0.5–5 mm	Liquid	Generally produced by nimbostratus or cumulonimbus clouds. When heavy, it can show high variability from one place to another.
Sleet	0.5–5 mm	Solid	Small, spherical to lumpy ice particles that form when raindrops freeze while falling through a layer of subfreezing air. Because the ice particles are small, damage, if any, is generally minor. Sleet can make travel hazardous.
Glaze	Layers 1 mm–2 cm thick	Solid	Produced when supercooled raindrops freeze on contact with solid objects. Glaze can form a thick coating of ice having sufficient weight to seriously damage trees and power lines.
Rime	Variable accumulations	Solid	Deposits usually consisting of ice feathers that point into the wind. These delicate, frostlike accumulations form as supercooled cloud or fog droplets encounter objects and freeze on contact.
Snow	1 mm–2 cm	Solid	The crystalline nature of snow allows it to assume many shapes, including six-sided crystals, plates, and needles. Produced in supercooled clouds, where water vapor is deposited as ice crystals that remain frozen during their descent.
Hail	5 mm–10 cm or larger	Solid	Precipitation in the form of hard, rounded pellets or irregular lumps of ice. Produced in large cumulonimbus clouds, where frozen ice particles and supercooled water coexist.
Graupel	2–5 mm	Solid	Sometimes called soft hail, graupel forms when rime collects on snow crystals to produce irregular masses of "soft" ice. Because these particles are softer than hailstones, they normally flatten out upon impact.

Snow

Snow is precipitation in the form of ice crystals (snowflakes) or, more often, aggregates of crystals. The size, shape, and concentration of snowflakes depend to a great extent on the temperature at which they form.

Recall that at very low temperatures, the moisture content of air is small. The result is the formation of very light, fluffy snow made up of individual six-sided ice crystals. This is the "powder" that downhill skiers talk so much about. By contrast, at temperatures warmer than about −5°C (23°F) the ice crystals join together into larger clumps consisting of tangled aggregates of crystals. Snowfalls composed of these composite snowflakes are generally heavy and have a high moisture content, which makes them ideal for making snowballs.

Sleet and Glaze

Sleet is a wintertime phenomenon and refers to the fall of small particles of ice that are clear to translucent. For sleet to be produced, a layer of air with temperatures above freezing must overlie a subfreezing layer near the ground. When raindrops, which are often melted snow, leave the warmer air and encounter the colder air below, they freeze and reach the ground as small pellets of ice the size of the raindrops from which they formed.

On some occasions, when the vertical distribution of temperatures is similar to that associated with the formation of sleet, **glaze** (freezing rain) results instead. In such situations, the subfreezing air near the ground is not thick enough to allow the raindrops to freeze. The raindrops, however, do become supercooled as they fall through the cold air and turn to ice upon colliding with solid objects. The result can be a thick coating of ice having sufficient weight to break tree limbs, down power lines, and make walking or driving extremely hazardous (**Figure 12.29**).

Figure 12.29 Glaze forms when supercooled raindrops freeze on contact with objects In January 1998, an ice storm of historic proportions caused enormous damage in New England and southeastern Canada. Nearly 5 days of freezing rain (glaze) left millions without electricity—some for as long as a month. (Photo by Dick Blume/Syracuse Newspapers/The Image Works)

A. Hailstones begin as small ice pellets that grow by adding supercooled water droplets as they move through a cloud. Strong updrafts may carry stones upward in several cycles, increasing the size of the hail by adding a new layer with each cycle. Eventually, the hailstones encounter a downdraft or grow too large to be supported by the updraft.

B. This cut hailstone fell at Coffeyville, Kansas, in 1970 and weighed 0.75 kilogram (1.67 pounds). Notice its layered structure.

Smartfigure 12.30 Formation of hail-stones (Photo courtesy of University Corporation for Atmospheric Research/National Science Foundation/NCAR).

Hail

Hail is precipitation in the form of hard, rounded pellets or irregular lumps of ice. Moreover, large hailstones often consist of a series of nearly concentric shells of differing densities and degrees of opaqueness (**Figure 12.30**). Most hailstones have diameters between 1 centimeter (0.4 inch, pea size) and 5 centimeters (2 inches, golf ball size), although some can be as big as an orange or larger.

Occasionally, hailstones weighing a pound or more have been reported. Many of these were probably composites of several stones frozen together.

Hail is produced only in large cumulonimbus clouds where updrafts can sometimes reach speeds approaching 160 kilometers (100 miles) per hour and where there is an abundant supply of supercooled water. Hailstones begin as small embryonic ice pellets that grow by collecting supercooled water droplets as they fall through the cloud (see Figure 12.30A). If they encounter a strong updraft, they may be carried upward again and begin the downward journey anew. Each trip through the supercooled portion of the cloud results in an additional layer of ice. Hailstones can also form from a single descent through an updraft. Either way, the process continues until the hailstone encounters a downdraft or grows too heavy to remain suspended by the thunderstorm's updraft.

The record for the largest hailstone ever found in the United States was set on July 23, 2010, in Vivian, South Dakota. The stone was over 20 centimeters (8 inches) in diameter and weighed nearly 900 grams (2 pounds). The stone that held the previous record of 0.75 kilograms (1.67 pounds) fell in Coffeyville, Kansas, in 1970 (see Figure 12.30B). The diameter of the stone found in South Dakota also surpassed the previous record of a 17.8-centimeter (7-inch) stone that fell in Aurora, Nebraska, in 2003. Even larger hailstones have reportedly been recorded in Bangladesh, where a 1987 hailstorm killed more than 90 people. It is estimated that large hailstones hit the ground at speeds exceeding 160 kilometers (100 miles) per hour.

The destructive effects of large hailstones are well known, especially to farmers whose crops have been devastated in a few minutes and to people whose windows and roofs have been damaged (**Figure 12.31**). In the

Figure 12.31 Hail damage to auto windshield. (Photo by Jim Reed/RGB Ventures LLC dba SuperStock/Alamy)

United States, hail damage each year is in the hundreds of millions of dollars.

Rime

Rime is a deposit of ice crystals formed by the freezing of supercooled fog or cloud droplets on objects whose surface temperature is below freezing. When rime forms on trees, it adorns them with its characteristic ice feathers, which can be spectacular to behold (**Figure 12.32**). In these situations, objects such as pine needles act as freezing nuclei, causing the supercooled droplets to freeze on contact. When the wind is blowing, only the windward surfaces of objects will accumulate the layer of rime.

12.9 CONCEPT CHECKS

1. List the forms of precipitation and the circumstances of their formation.
2. Explain why snow can sometimes reach the ground as rain, but the reverse does not occur.
3. How is sleet different from glaze? Which is usually more hazardous?

Figure 12.32 Rime Rime consists of delicate ice crystals that form when supercooled fog or cloud droplets freeze on contact with objects. (Photo by Siepman/Photolibrary)

12.10 Measuring Precipitation

Explain how precipitation is measured.

The most common form of precipitation—rain—is the easiest to measure. Any open container having a consistent cross section throughout can be a rain gauge. In general practice, however, more sophisticated devices are used so that small amounts of rainfall can be measured accurately and losses from evaporation can be reduced. The *standard rain gauge* (**Figure 12.33**) has a diameter of about 20 centimeters (8 inches) at the top. When the water is caught, a funnel conducts the rain into a cylindrical measuring tube that has a cross-sectional area only one-tenth as large as the receiver. Consequently, rainfall depth is magnified 10 times, which allows for accurate measurements to the nearest 0.025 centimeter (0.01 inch). The narrow opening also minimizes evaporation. When the amount of rain is less than 0.025 centimeter, it is reported as a *trace* of precipitation.

Measuring Snowfall

Snow records are typically kept with two measurements: depth and water equivalent. The depth of snow is usually measured with a calibrated stick. The actual

Figure 12.33 Precipitation measurement using a standard rain gauge A standard rain gauge allows for accurate rainfall measurement to the nearest 0.025 centimeter (0.01 inch). Because the cross-sectional area of the measuring tube is only one-tenth as large as the collector, rainfall is magnified 10 times.

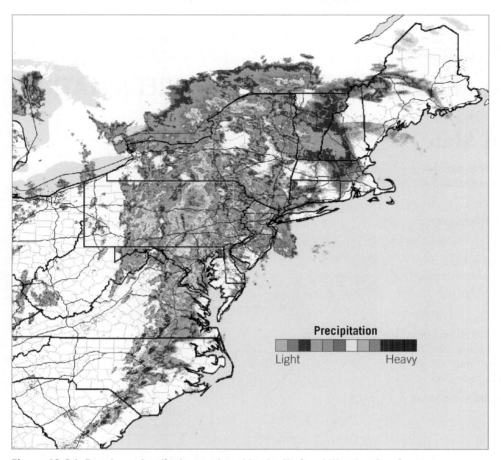

Figure 12.34 Doppler radar display produced by the National Weather Service Colors indicate different intensities of precipitation. Note the band of heavy precipitation along the eastern seaboard. (Courtesy of NOAA)

measurement is simple, but choosing a *representative* spot often poses a dilemma. Even when winds are light or moderate, snow drifts freely, making measurement of snowfall difficult. As a rule, it is best to take several measurements in an open place away from trees and obstructions and then average them. To obtain the water equivalent, samples are melted and then weighed or measured as rain.

The quantity of water in a given volume of snow is not constant. A general ratio of 10 units of snow to 1 unit of water is often used when exact information is not available, but the actual water content of snow may deviate widely from this figure. It may take as much as 30 centimeters of light, fluffy dry snow or as little as 4 centimeters of wet snow to produce 1 centimeter of water.

meteorologists an important tool to probe storm systems as far as a few hundred kilometers away.

Weather radar units have transmitters that send out short pulses of radio waves at wavelengths that can penetrate clouds consisting of small droplets but are reflected by larger raindrops, ice crystals, or hailstones. The reflected signal, called an *echo*, is received and displayed on a monitor. Because the echo is "brighter" when the precipitation is more intense, weather radar is able to depict not only the regional extent of the precipitation but also the rate of rainfall. Figure 12.34 is a typical radar display in which precipitation intensity is shown using colors. Weather radar is also an important tool for determining the rate and direction of storm movement.

Precipitation Measurement by Weather Radar

Weather forecasts utilize maps like the one in **Figure 12.34** to depict precipitation patterns. The instrument that produces these images, called *weather radar*, has given

CONCEPTS IN REVIEW: MOISTURE, CLOUDS, AND PRECIPITATION

12.1 Water's Changes of State

List and describe the processes that cause water to change from one state of matter to another. Define *latent heat* and explain why it is important.

KEY TERMS: calorie, latent heat, evaporation, condensation, sublimation, deposition

- Water vapor, an odorless, colorless gas, changes from one state of matter (solid, liquid, or gas) to another at the temperatures and pressures experienced near Earth's surface.
- The processes that result in a change of state are evaporation, condensation, melting, freezing, sublimation, and deposition. Evaporation, melting, and sublimation require heat, whereas condensation, freezing, and deposition result in latent (hidden) heat being released.

? Label the accompanying diagram with the appropriate terms for the changes of state that are shown.

SOLID (ice) LIQUID (water) GAS (water vapor)

12.2 Humidity: Water Vapor in the Air

Distinguish between relative humidity and dew point. Write a generalization relating how temperature changes affect relative humidity.

KEY TERMS: humidity, saturation, vapor pressure, mixing ratio, relative humidity, dew-point temperature (dew point), hygrometer, psychrometer

- Humidity is a general term to describe the amount of water vapor in the air. The methods used to express humidity quantitatively include (1) mixing ratio, the mass of water vapor in a unit of air compared to the remaining mass of dry air; (2) vapor pressure, the part of the total atmospheric pressure attributable to its water-vapor content; (3) relative humidity, the ratio of the air's actual water-vapor content to the amount of water vapor required for saturation at that temperature; and (4) dew point, the temperature to which a parcel of air would need to be cooled to reach saturation.
- When air is saturated, the pressure exerted by the water vapor is called saturation vapor pressure. At saturation, a balance is reached between the number of water molecules leaving the surface of the water and the number returning. At higher temperatures, more water vapor is required to reach saturation because the saturation vapor pressure is temperature dependent.
- Relative humidity can be changed in two ways: by adding or subtracting water vapor or by changing the air's temperature. For example, when air is cooled, the relative humidity increases.

? Refer to the accompanying photo and explain how the relative humidity inside the house compares to the relative humidity outside the house on this particular day.

(Photo by Clynt Garnham Housing/Alamy)

12.3 The Basis of Cloud Formation: Adiabatic Cooling

Explain how adiabatic cooling results in cloud formation.

KEY TERMS: adiabatic temperature change, parcel, dry adiabatic rate, wet adiabatic rate

- Cooling of air as it rises and expands, due to successively lower air pressure, is the basic cloud-forming process. Temperature changes that result when air is compressed or when air expands are called adiabatic temperature changes.
- Unsaturated air warms by compression and cools by expansion at the rather constant rate of 10°C per 1000 meters (5.5°F per 1000 feet) of altitude change, a quantity called the dry adiabatic rate. When air rises high enough, it will cool sufficiently to cause condensation and form clouds. Air that continues to rise above the condensation level will cool at the wet adiabatic rate, which varies from 5°C to 9°C per 1000 meters of ascent. The difference in the wet and dry adiabatic rates results because condensation releases latent heat, thereby reducing the rate at which the air cools.

12.4 Processes That Lift Air

List and describe the four mechanisms that cause air to rise.

KEY TERMS: orographic lifting, rainshadow desert, front, frontal wedging, convergence, localized convective lifting

- Four mechanisms that can initiate the vertical movement of air are (1) orographic lifting, which occurs when elevated terrains, such as mountains, act as barriers to the flow of air; (2) frontal wedging, when cool air acts as a barrier over which warmer, less dense air rises; (3) convergence, which happens when air flows together, causing an upward movement of air; and (4) localized convective lifting, caused by unequal surface heating which results in pockets of air rising because of their lower density (buoyancy).

12.5 The Weathermaker: Atmospheric Stability

Describe how atmospheric stability is determined and compare conditional instability with absolute instability.

KEY TERMS: stable air, unstable air, absolute stability, absolute instability, conditional instability

- The stability of air is determined by comparing the environmental lapse rate (change in temperature with altitude) to the wet and dry adiabatic rates.
- Absolute stability prevails when the environmental lapse rate is less than the wet adiabatic rate. Under these conditions a parcel of air that is forced to rise will always be heavier than the surrounding air, producing stability.
- Absolute instability results when the environmental lapse rate is greater than the dry adiabatic rate. Specifically, a column of air exhibits absolute instability when the air near the ground is significantly warmer (less dense) than the air aloft.
- Conditional instability occurs when moist air has an environmental lapse rate between the dry and wet adiabatic rates. Simply, the air is stable for an unsaturated parcel or air, but becomes unstable if the parcel of air be forced high enough for it be become saturated.

? **Describe the atmospheric conditions that were likely associated with the development of the towering cloud shown in the accompanying photo.**

(Photo by Rolf Nussbaumer Photography/Alamy)

12.6 Condensation and Cloud Formation

List the necessary conditions for condensation and briefly describe the two criteria used for cloud classification.

KEY TERMS: condensation nuclei, hygroscopic nuclei, cloud, cirrus, cumulus, stratus, high clouds, middle clouds, low clouds, clouds of vertical development

- For condensation to occur, air must reach saturation. Saturation occurs most commonly when air is cooled to its dew point, or when water vapor is added to the air. Condensation normally requires a surface on which water vapor can condense. In cloud and fog formation, tiny particles called condensation nuclei serve this purpose.
- Clouds are a form of condensation described as visible aggregates of minute droplets of water or tiny crystals of ice.
- Clouds are classified on the basis of their form and height. The three basic forms are cirrus (high, white, thin, wispy fibers), cumulus (globular, individual cloud masses), and stratus (sheets or layers that cover much or all of the sky). The four categories based on height are high clouds (bases normally above 6000 meters [20,000 feet]), middle clouds (from 2000 to 6000 meters [6500 to 20,000 feet]), low clouds (below 2000 meters [6500 feet]), and clouds of vertical development.

12.7 Fog

Define *fog* and explain how the various types of fog form.

KEY TERMS: fog, advection fog, radiation fog, upslope fog, steam fog, frontal fog (precipitation fog)

- Fog is a cloud with its base at or very near the ground. Fogs form when air is cooled below its dew point or when enough water vapor is added to the air to cause saturation.
- Advection fog forms when warm, moist air moves over a cool surface. Radiation fog forms on cool, clear, calm nights, when Earth's surface cools rapidly by the process of radiation cooling.
- When air becomes saturated through the addition of water vapor, the resulting fogs are called evaporation fogs. During a gentle rain when the air near the surface is cool and near saturation, enough rain may evaporate to produce fog. A fog formed in this manner is called frontal fog, or precipitation fog.

? Identify the type of fog shown in the accompanying photo.

(Photo by Pat and Chuck Blackley/Alamy)

12.8 How Precipitation Forms

Describe the two mechanisms that produce precipitation.

KEY TERMS: Bergeron process, supercooled, freezing nuclei, collision–coalescence process

- For precipitation to form, millions of cloud droplets must join together into drops that are large enough to reach the ground before evaporating. Two mechanisms have been proposed to explain this phenomenon: the Bergeron process and the collision–coalescence process.
- The Bergeron process operates in cold clouds where ice crystals and supercooled droplets coexist. The ice crystals, which have a greater affinity for water vapor, grow larger at the expense of supercooled cloud droplets. When the ice crystals grow large enough, they usually fall as snowflakes in winter, whereas in the summer they melt to form rain.
- The collision–coalescence process operates in warm clouds that contain large hygroscopic (water-seeking) nuclei, such as salt particles. The hydroscopic particles form large droplets, which collide and join with smaller water droplets as they descend. After many collisions, the droplets grow large enough to fall to the ground as rain.

12.9 Forms of Precipitation

List the different types of precipitation and explain how each type forms.

KEY TERMS: rain, snow, sleet, glaze (freezing rain), hail, rime

- The forms of precipitation include rain, snow, sleet, glaze, hail, and rime. The term rain is restricted to drops of water that fall from a cloud and have a diameter of at least 0.5 millimeter (0.02 inch). Snow is precipitation in the form of ice crystals (snowflakes) or, more often, aggregates of crystals. Sleet is a wintertime phenomenon and refers to the fall of small particles of ice that are clear to translucent, while glaze forms when supercooled raindrops turn to ice upon colliding with solid objects. Hail is precipitation in the form of hard, rounded pellets or irregular lumps of ice that usually have diameters between 1 centimeter (0.4 inch, pea size) and 5 centimeters (2 inches, golf ball size). Rime is a deposit of feathery-looking ice crystals formed by the freezing of supercooled fog droplets on objects.

12.10 Measuring Precipitation

Explain how precipitation is measured.

- Any open container having a consistent cross section throughout can be used as a rain gauge.
- Weather radar units have transmitters that emit short pulses of radio waves at wavelengths that can penetrate clouds consisting of small droplets but are reflected by larger raindrops, ice crystals, or hailstones. Because the echo is "brighter" when the precipitation is more intense, weather radar is able to depict not only the regional extent of the precipitation but also the rate of rainfall.

GIVE IT SOME THOUGHT

1. Refer to Figure 12.2 to complete the following:
   a. In which state of matter is water densest?
   b. In which state of matter are water molecules most energetic?
   c. In which state of matter is water compressible?

2. The accompanying photo shows a cup of hot coffee. What state of matter is the "steam" rising from the liquid? Explain your answer.

3. The primary mechanism by which the human body cools itself is perspiration.
   a. Explain how perspiring cools the skin.

(Photo by Dmitry Kolmakov/Shutterstock)

b. Refer to the data for Phoenix, Arizona, and Tampa, Florida, in Table A.

c. In which city would it be easier to stay cool by perspiring? Explain your choice.

Table A

City	Temperature	Dew-Point Temperature
Phoenix, AZ	101°F	47°F
Tampa, FL	101°F	77°F

4. During hot summer weather, many people put "koozies" around their beverages to keep the drinks cold. In addition to preventing a warm hand from heating the container through conduction, what other mechanism(s) slow the process of warming beverages?

5. Refer to Table 12.1 to answer this question. How much more water is contained in saturated air at a tropical location with a temperature of 40°C compared to a polar location with a temperature of −10°C?

6. Refer to the data for Phoenix, Arizona, and Bismarck, North Dakota, in Table B, to complete the following:

Table B

City	Temperature	Dew-Point Temperature
Phoenix, AZ	101°F	47°F
Bismark, ND	39°F	38°F

a. Which city has a higher relative humidity?

b. Which city has the greatest quantity of water vapor in the air?

c. In which city is the air closest to its saturation point with respect to water vapor?

d. In which city does the air have the greatest holding capacity for water vapor?

7. The accompanying graph shows how air temperature and relative humidity change on a typical summer day in the Midwest. Assuming that the dew-point temperature remained constant, what would be the best time of day to water a lawn to minimize evaporation of the water spread on the grass?

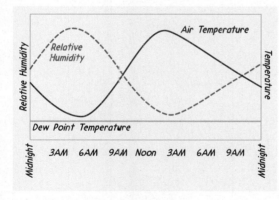

8. The diagram in the next column shows air flowing from the ocean over a coastal mountain range. Assume that the dew-point temperature remains constant in dry air (that is, relative humidity less than 100 percent). If the air parcel becomes saturated the dew-point temperature will cool at the wet adiabatic rate as it ascends, but it will not change as the air parcel descends. Use this information to complete the following:

a. Determine the air temperature and dew-point temperature for the air parcel at each location (B–G) shown on the diagram.

b. At what elevation will clouds begin to form (with relative humidity = 100 percent)?

c. Compare the air temperatures at points A and G. Why are they different?

d. How did the water vapor content of the air change as the parcel of air traversed the mountain? (*Hint:* Compare dew-point temperatures.)

e. On which side of the mountain might you expect lush vegetation and on which side would you expect desert-like conditions?

f. Where in the United States might you find a situation like what is pictured here?

Temperature at Point A = 27°C
Dew point temperature at Point A = 17°C
Dry adiabatic rate = 10°C/km
Wet adiabatic rate = 5°C/km

9. The cumulonimbus cloud pictured in Figure 12.21H are roughly 12 kilometers tall, 8 kilometers wide, and 8 kilometers long. Assume that the droplets in each cubic meter of the cloud total 0.5 cubic centimeter. How much liquid does the cloud contain? How many gallons is this (3785 cm³ = 1 gallon)?

10. Cloud droplets form and grow as water vapor condenses onto a hygroscopic condensation nuclei. Research has shown that the maximum radius for cloud droplets is about 0.05 millimeter. However, typical raindrops have volumes thousands of times greater. How do cloud droplets become raindrops?

11. Why does radiation fog form mainly on clear nights rather than on cloudy nights?

12. Which winter storm is likely to produce deeper snowfall: a low-pressure system that passes through the midwestern states of Nebraska, Iowa, and Illinois (26°F average temperature at the time of the storm) *or* exactly the same system passing through North Dakota, Minnesota, and Wisconsin (16°F average temperature at the time of the storm).

13. Weather radar provides information on the intensity of precipitation in addition to the total amount of precipitation that falls over a given time period. Table C shows the relationship between radar reflectivity values and rainfall rates. If radar measured a reflectivity value of 47 dBZ for 2½ hours over a location, how much rain will have fallen there?

14. What are the advantages and disadvantages of rain gauges compared to weather radar in measuring rainfall?

Table C
Conversion of Radar Reflectivity to Rainfall Rate

Radar Reflectivity (dBZ)	Rainfall Rate (inches/hr)
65	16+
60	8.0
55	4.0
52	2.5
47	1.3
41	0.5
36	0.3
30	0.1
20	trace

MasteringGeology™

13

The Atmosphere in Motion

FOCUS ON CONCEPTS

Each statement represents the primary learning objective for the corresponding major heading within the chapter. After you complete the chapter, you should be able to:

13.1 Define *air pressure* and describe the instruments used to measure this weather element.

.13.2 Discuss the three forces that act on the atmosphere to either create or alter winds.

13.3 Contrast the weather associated with low-pressure centers (cyclones) and high-pressure centers (anticyclones).

13.4 Summarize Earth's idealized global circulation. Describe how continents and seasonal temperature changes complicate the idealized pattern.

13.5 List three types of local winds and describe their formation.

13.6 Describe the instruments used to measure wind. Explain how wind direction is expressed using compass directions.

13.7 Discuss the major factors that influence the global distribution of precipitation.

Wind turbines, Tehachapi, California.
(Photo by T. J. Florian/Rainbow/Age Fotostock/Robert Harding)

O f the various elements of weather and climate, changes in air pressure are the least noticeable. When listening to a weather report, we are generally interested in moisture conditions (humidity and precipitation), temperature, and perhaps wind. Rarely to people wonder about air pressure. Although the hour-to-hour and day-to-day variations in air pressure are generally not noticed by people, they are very important in producing changes in our weather. Variations in air pressure from place to place cause the movement of air we call wind and are a significant factor in weather forecasting. As we will see, air pressure is closely tied to the other elements of weather in a cause-and-effect relationship.

13.1 Understanding Air Pressure

Define *air pressure* and describe the instruments used to measure this weather element.

In Chapter 11 we noted that **air pressure** is simply the pressure exerted by the weight of air above. Average air pressure at sea level is about 1 kilogram per square centimeter, or 14.7 pounds per square inch. Specifically, a column of air 1 square inch in cross section, measured from sea level to the top of the atmosphere, would weigh about 14.7 pounds (**Figure 13.1**). This is roughly the same pressure that is produced by a 1-square-inch column of water 10 meters (33 feet) in height. With some simple arithmetic, you can calculate that the air pressure exerted on the top of a small (50 centimeter-by-100 centimeter [20 inch-by-40 inch]) school desk exceeds 5000 kilograms (11,000 pounds), or about the weight of a 50-passenger school bus. Why doesn't the desk collapse under the weight of the ocean of air above? Simply, air pressure is exerted in all directions—down, up, and sideways. Thus, the air pressure pushing down on the desk exactly balances the air pressure pushing up on the desk.

Visualizing Air Pressure

Imagine a tall aquarium that has the same dimensions as the small desk mentioned in the preceding paragraph. When this aquarium is filled to a height of 10 meters (33 feet), the water pressure at the bottom equals 1 atmosphere (14.7 pounds per square inch). Now, imagine what will happen if this aquarium is placed on top of our student desk so that all the force is directed downward. Compare this to what results when the desk is placed inside the aquarium and allowed to sink to the bottom. In the latter example, the desk survives because the water pressure is exerted in all directions, not just downward, as in our earlier example. The desk, like your body, is "built" to withstand the pressure of 1 atmosphere. It is important to note that although we do not generally notice the pressure exerted by the ocean of air around us, except when ascending or descending in an elevator or airplane, it is nonetheless substantial. The pressurized suits that astronauts use on space walks are designed to duplicate the atmospheric pressure experienced at Earth's surface. Without these protective suits to keep body fluids from boiling away, astronauts would perish in minutes.

The concept of air pressure can also be understood if we examine the behavior of gas molecules. Gas

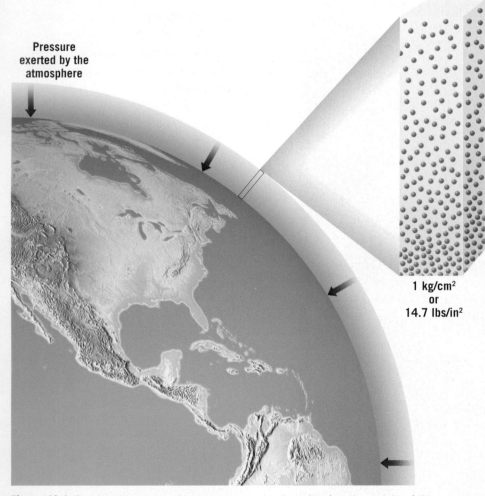

Pressure exerted by the atmosphere

1 kg/cm² or 14.7 lbs/in²

Figure 13.1 Sea-level pressure Air pressure can be thought of as the weight of the atmosphere above. A column of air 1-square inch in cross section extending from sea level to the top of the atmosphere would weigh about 14.7 pounds.

molecules, unlike those of the liquid and solid phases, are not bound to one another but freely move about, filling all the space available to them. When two gas molecules collide, which happens frequently under normal conditions, they bounce off each other like elastic balls. If a gas is confined to a container, this motion is restricted by its sides, much as the walls of a handball court redirect the motion of the handball. The continuous bombardment of gas molecules against the sides of the container exerts an outward push that we call air pressure. Although the atmosphere is without walls, it is confined from below by Earth's surface and effectively from above because the force of gravity prevents its escape. Here, we can define *air pressure* as the force exerted against a surface by the continuous collision of gas molecules.

Measuring Air Pressure

When meteorologists measure atmospheric pressure, they use a unit called the *millibar*. Standard sea-level pressure is 1013.2 millibars. Although the millibar has been the unit of measure on all U.S. weather maps since January 1940, the media use "inches of mercury" to describe atmospheric pressure. In the United States, the National Weather Service converts millibar values to inches of mercury for public and aviation use (**Figure 13.2**).

Figure 13.3 Simple mercury barometer The weight of the column of mercury is balanced by the pressure exerted on the dish of mercury by the air above. If the pressure drops, the column of mercury falls; if the pressure increases, the column rises.

Inches of mercury is easy to understand. The use of mercury for measuring air pressure dates from 1643, when Torricelli, a student of the famous Italian scientist Galileo, invented the **mercury barometer**. Torricelli correctly described the atmosphere as a vast ocean of air that exerts pressure on us and all objects about us. To measure this force, he filled a glass tube, which was closed at one end, with mercury. The tube was then inverted into a dish of mercury (**Figure 13.3**). Torricelli found that the mercury flowed out of the tube until the weight of the column was balanced by the pressure that the atmosphere exerted on the surface of the mercury in the dish. In other words, the weight of mercury in the column equaled the weight of the same diameter column of air that extended from the ground to the top of the atmosphere.

When air pressure increases, the mercury in the tube rises. Conversely, when air pressure decreases, so does the height of the mercury column. With some refinements, the mercury barometer invented by Torricelli is still the standard pressure-measuring instrument used today. Standard atmospheric pressure at sea level equals 29.92 inches of mercury.

The need for a smaller and more portable instrument for measuring air pressure led to the development of the **aneroid barometer** (*aneroid* means "without liquid"). Instead of having a mercury column held up by air pressure, an aneroid barometer uses a partially evacuated metal chamber (**Figure 13.4**). The chamber is extremely sensitive to variations in air pressure and changes shape, compressing as the pressure increases and expanding as the pressure decreases. A series of levers transmits the movements of the chamber to a pointer on a dial that is

Inches of Hg	Milli- bars	
32.48	1100	
32.19	1090	
31.89	1080	Highest recorded sea level pressure: 1084 mb (32.01 in.) Agata, Siberia (December, 1968)
31.60	1070	
31.30	1060	Highest recorded sea level pressure in United States: 1064 mb (31.42 in.) Miles City, Montana (December, 1983)
31.00	1050	
30.71	1040	Strong high pressure system (anticyclone)
30.42	1030	
30.12	1020	**Average sea level pressure: 1013.25 mb (29.92 in.)**
29.82	1010	
29.53	1000	
29.24	990	Strong low pressure system (Midlatitude cyclone)
28.94	980	
28.64	970	
28.35	960	
28.05	950	
27.76	940	
27.46	930	
27.17	920	
26.87	910	
26.58	900	Hurricane Katrina (August, 2005): 902 mb (26.71 in.)
26.28	890	Hurricane Wilma (October, 2005): 382 mb (26.12 in.)
25.99	880	Lowest pressure recorded for an Atlantic hurricane
25.69	870	Lowest recorded sea level pressure: 870 mb (25.70 in.) Typhoon Tip (October, 1979)
25.40	860	
25.10	850	

SmartFigure 13.2 Inches and millibars A comparison of two units commonly used to express air pressure.

Figure 13.4 Aneroid barometer The black pointer shows the current air pressure. When the barometer is read, the observer moves the other pointer to coincide with the current air pressure. Later, when the barometer is checked, the observer can see whether the air pressure has been risen, fallen, or remained steady. The bottom diagram is a cross section. An aneroid barometer has a partially evacuated chamber that changes shape, compressing as air pressure increases and expanding as pressure decreases.

Figure 13.5 Aneroid barograph This instrument makes a continuous record of air pressure. The cylinder with the graph is a clock that turns once per day or once per week. (Photo courtesy of Qualimetrics, Inc., Sacramento, California)

Did You Know?

A barometer could be constructed using any number of different liquids, including water. The problem with a water-filled barometer is size. Because water is 13.6 times less dense than mercury, the height of a water column at standard sea-level pressure would be 13.6 times taller than that of a mercury barometer. Consequently, a water-filled barometer would need to be nearly 34 ft (10.2 meters) tall.

calibrated to read in inches of mercury and/or millibars.

As shown in Figure 13.4, the face of an aneroid barometer intended for home use is inscribed with words such as *fair*, *change*, *rain*, and *stormy*. Notice that "fair" corresponds with high-pressure readings, whereas "rain" is associated with low pressures. Barometric readings, however, may not always indicate the weather. The dial may point to "fair" on a rainy day or to "rain" on a fair day. To "predict" the local weather, the change in air pressure over the past few hours is more important than the current pressure reading. Falling pressure is often associated with increasing cloudiness and the possibility of precipitation, whereas rising air pressure generally

indicates clearing conditions. It is useful to remember, however, that particular barometer readings or trends do not always correspond to specific types of weather.

Another advantage of an aneroid barometer is that it can easily be connected to a recording mechanism. The resulting instrument is a **barograph**, which provides a continuous record of pressure changes with the passage of time (**Figure 13.5**). Another important adaptation of the aneroid barometer is its use to indicate altitude for aircraft, mountain climbers, and mapmakers.

13.1 CONCEPT CHECKS

1. Describe air pressure in your own words.
2. What is standard sea-level pressure in millibars, in inches, and in pounds per square inch?
3. Describe the operating principles of the mercury barometer and the aneroid barometer.
4. List two advantages of the aneroid barometer.

13.2 Factors Affecting Wind

Discuss the three forces that act on the atmosphere to either create or alter winds.

In Chapter 12, we examined the upward movement of air and its role in cloud formation. As important as vertical motion is, far more air moves horizontally, the phenomenon we call **wind**. What causes wind?

Simply stated, wind is the result of horizontal differences in air pressure. *Air flows from areas of higher pressure to areas of lower pressure.* You may have experienced

this when opening something that is vacuum-packed. The noise you hear is caused by air rushing from the higher pressure outside the can or jar to the lower pressure inside. Wind is nature's attempt to balance such inequalities in air pressure. Because unequal heating of Earth's surface generates these pressure differences, *solar radiation is the ultimate energy source for most wind.*

If Earth did not rotate, and if there were no friction between moving air and Earth's surface, air would flow in a straight line from areas of higher pressure to areas of lower pressure. But because Earth does rotate and friction does exist, wind is controlled by the following combination of forces: pressure gradient force, Coriolis effect, and friction.

Pressure Gradient Force

If an object experiences an unbalanced force in one direction, it will accelerate (experience a change in velocity). The force that generates winds results from horizontal pressure differences. When air is subjected to greater pressure on one side than on another, the imbalance produces a force that is directed from the region of higher pressure toward the area of lower pressure. Thus, pressure differences cause the wind to blow, and the greater these differences, the greater the wind speed.

Variations in air pressure over Earth's surface are determined from barometric readings taken at hundreds of weather stations. These pressure measurements are shown on surface weather maps using **isobars** (*iso* = "equal," *bar* = "pressure"), or lines connecting places of equal air pressure (**Figure 13.6**). The *spacing* of the isobars indicates the amount of pressure change occurring over a given distance and is called the **pressure gradient force.** Pressure gradient is analogous to gravity acting on a ball rolling down a hill. A steep pressure gradient, like a steep hill, causes greater acceleration of a parcel of air than does a weak pressure gradient (a gentle hill). Thus, the relationship between wind speed and the pressure gradient is straightforward: *Closely spaced isobars indicate a steep*

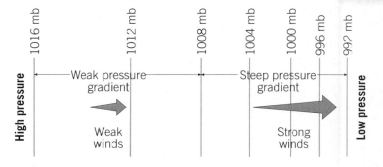

Figure 13.6 Isobars show the pressure gradient Isobars are lines connecting places of equal air pressure. The spacing of isobars indicates the amount of pressure change occurring over a given distance—called the pressure gradient. Closely spaced isobars indicate a strong pressure gradient and high wind speeds, whereas widely spaced isobars indicate a weak pressure gradient and low wind speeds.

pressure gradient and strong winds; widely spaced isobars indicate a weak pressure gradient and light winds. Figure 13.6 illustrates the relationship between the spacing of isobars and wind speed. Note also that the pressure gradient force is always directed at *right angles* to the isobars.

In order to draw isobars on a weather map to show air pressure patterns, meteorologists must compensate for the *elevation* of each station. Otherwise, all high-elevation locations, such as Denver, Colorado, would always be mapped as having low pressure. This compensation is accomplished by converting all pressure measurements to sea-level equivalents.

Figure 13.7 is a surface weather map that shows isobars (representing corrected sea-level air pressure) and winds. Wind *direction* is shown as wind arrow shafts and

Wind speed symbols	Miles per hour
◎	Calm
⎯	1–2
⟍	3–8
⟍	9–14
⟍	15–20
⟍	21–25
⟍	26–31
⟍	32–37
⟍	38–43
⟍	44–49
⟍	50–54
⟍	55–60
⟍	61–66
⟍	67–71
⟍	72–77
⟍	78–83
⟍	84–89
⟍	119–123

Isobars (4 mb interval)

SmartFigure 13.7 Isobars on a weather map Isobars are used to show the distribution of pressure on daily weather maps. Isobars are seldom straight but usually form broad curves. Concentric isobars indicate cells of high and low pressure. The "wind flags" indicate the expected airflow surrounding pressure cells and are plotted as "flying" with the wind (that is, the wind blows toward the station circle). Notice on this map that the isobars are more closely spaced and the wind speed is faster around the low-pressure center than around the high.

speed as wind bars (see the accompanying key). Isobars, used to depict pressure patterns, are rarely straight or evenly spaced on surface maps. Consequently, wind generated by the pressure gradient force typically changes speed and direction as it flows.

The area of somewhat circular closed isobars in eastern North America represented by the red letter *L* is a *low-pressure system.* In western Canada, a *high-pressure system,* denoted by the blue letter *H,* can also be seen. We will discuss *highs* and *lows* in the next section.

In summary, the *horizontal pressure gradient is the driving force of wind.* The magnitude of the pressure gradient force is shown by the spacing of isobars. The direction of force is always from areas of higher pressure toward areas of lower pressure and at right angles to the isobars.

Coriolis Effect

Figure 13.7 shows the typical air movements associated with high- and low-pressure systems. As expected, the air moves out of the regions of higher pressure and into the regions of lower pressure. However, the wind does not cross the isobars at right angles as the pressure gradient force directs it to do. The direction deviates as a result of Earth's rotation. This has been named the **Coriolis effect**, after the French scientist who first thoroughly described it.

All free-moving objects or fluids, including the wind, are deflected to the *right* of their path of motion in the Northern Hemisphere and to the *left* in the Southern Hemisphere. The reason for this deflection can be illustrated by imagining the path of a rocket launched from the North Pole toward a target located on the equator (**Figure 13.8**). If the rocket took an hour to reach its target, during its flight, Earth would have rotated 15° to the east. To someone standing on Earth, it would look as if the rocket had veered off its path and hit Earth 15° west of its target. The true path of the rocket is straight and would appear so to someone out in space looking at Earth. It is Earth turning under the rocket that gives it its *apparent* deflection.

Note that the rocket is deflected to the right of its path of motion because of the counterclockwise rotation of the Northern Hemisphere. In the Southern Hemisphere, the effect is reversed. Clockwise rotation produces a similar deflection, but to the *left* of the path of motion. The same deflection is experienced by wind regardless of the direction it is moving.

We attribute the apparent shift in wind direction to the Coriolis effect. This deflection (1) is always directed at right angles to the direction of airflow; (2) affects only wind direction, not wind speed; (3) is affected by wind speed (the stronger the wind, the greater the deflection); and (4) is strongest at the poles and weakens equatorward, becoming nonexistent at the equator.

Note that any free-moving object will experience a deflection caused by the Coriolis effect. The U.S. Navy dramatically discovered this fact in World War II. During target practice, long-range guns on battleships continually missed their targets by as much as several hundred yards until ballistic corrections were made for the changing position of a seemingly stationary target. Over a short distance, however, the Coriolis effect is relatively small.

Friction with Earth's Surface

The effect of friction on wind is important only within a few kilometers of Earth's surface. Friction acts to slow air movement and, as a consequence, alters wind direction. To illustrate friction's effect on wind direction, let us look at a situation in which friction has no role. Above

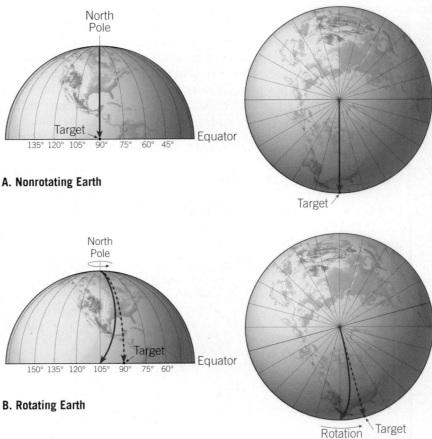

A. Nonrotating Earth

B. Rotating Earth

SmartFigure 13.8 Coriolis effect The Coriolis effect, illustrated using the 1-hour flight of a rocket traveling from the North Pole to a location on the equator. **A.** On a nonrotating Earth, the rocket would travel straight to its target. **B.** However, Earth rotates 15° each hour. Thus, although the rocket travels in a straight line, when we plot the path of the rocket on Earth's surface, it follows a curved path that veers to the right of the target.

the friction layer, the pressure gradient force and Coriolis effect work together to direct the flow of air. Under these conditions, the pressure gradient force causes air to start moving across the isobars. As soon as the air starts to move, the Coriolis effect acts at right angles to this motion. The faster the wind speed, the greater the deflection.

Eventually, the Coriolis effect will balance the pressure gradient force, and the wind will blow parallel to the isobars (**Figure 13.9**). Upper-air winds generally take this path and are called **geostrophic winds**. The lack of friction with Earth's surface allows geostrophic winds to travel at higher speeds than do surface winds. This can be observed in **Figure 13.10** by noting the wind flags, many of which indicate winds of 50 to 100 miles per hour.

The most prominent features of upper-level flow are **jet streams**. First encountered by high-flying bombers during World War II, these fast-moving "rivers" of air travel between 120 and 240 kilometers (75 and 150 miles) per hour in a west-to-east direction. One such stream is situated over the polar front, which is the zone separating cool polar air from warm subtropical air.

Below 600 meters (2000 feet), friction complicates the airflow just described. Recall that the Coriolis effect is proportional to wind speed. Friction lowers the wind

Figure 13.9 The geostrophic wind The only force acting on a stationary parcel of air is the pressure gradient force. Once the air begins to accelerate, the Coriolis effect deflects it to the right in the Northern Hemisphere. Greater wind speeds result in a stronger Coriolis effect (deflection) until the flow is parallel to the isobars. At this point, the pressure gradient force and Coriolis effect are in balance, and the flow is called a *geostrophic wind*. It is important to note that in the "real" atmosphere, airflow is continually adjusting for variations in the pressure field. As a result, the adjustment to geostrophic equilibrium is much more irregular than shown.

Upper-level weather chart

Representation of upper-level chart

Wind speed symbols	Miles per hour
◎	Calm
—	1–2
⌐	3–8
⌐	9–14
⌐	15–20
⌐	21–25
⌐	26–31
⌐	32–37
⌐	38–43
⌐	44–49
⌐	50–54
⌐	55–60
⌐	61–66
⌐	67–71
⌐	72–77
⌐	78–83
⌐	84–89
⌐	119–123

Figure 13.10 Upper-air weather chart This simplified weather chart shows the direction and speed of the upper-air winds. Note from the flags that the airflow is almost parallel to the contours. Like most other upper-air charts, this one shows variations in the height (in meters) at which a selected pressure (500 millibars) is found instead of showing variations in pressure at a fixed height, like surface maps. Do not let this confuse you because there is a simple relationship between height contours and pressure. Places experiencing 500-millibar pressure at higher altitudes (toward the south on this map) are experiencing higher pressures than are places where the height contours indicate lower altitudes. Thus, *higher-elevation* contours indicate *higher* pressures, and *lower-elevation* contours indicate *lower* pressures.

speed, so it reduces the Coriolis effect. Because the pressure gradient force is not affected by wind speed, it wins the tug of war shown in **Figure 13.11**. The result is a movement of air at an angle across the isobars, toward the area of lower pressure.

The roughness of the terrain determines the angle of airflow across the isobars. Over the smooth ocean surface, friction is low, and the angle is small. Over rugged terrain, where friction is higher, the angle that air makes as it flows across the isobars can be as great as 45 degrees.

In summary, upper airflow is nearly parallel to the isobars, whereas the effect of friction causes the surface winds to move more slowly and cross the isobars at an angle.

A. Upper-level wind (no friction)

B. Surface wind (effect of friction)

Figure 13.11 The effects of friction on wind Friction has little effect on winds aloft, so airflow is parallel to the isobars. In contrast, friction slows surface winds, which weakens the Coriolis effect, causing winds to cross the isobars and move toward the lower pressure.

13.2 CONCEPT CHECKS

1. List three factors that combine to direct horizontal airflow (wind).
2. What force is responsible for generating wind?
3. Write a generalization relating the spacing of isobars to wind speed.
4. Briefly describe how the Coriolis effect influences air movement.
5. Unlike winds aloft, which blow nearly parallel to the isobars, surface winds generally cross the isobars. Explain what causes this difference.

13.3 Highs and Lows

Contrast the weather associated with low-pressure centers (cyclones) and high-pressure centers (anticyclones).

Among the most common features on a weather map are areas designated as pressure centers. **Cyclones**, or **lows**, are centers of low pressure, and **anticyclones**, or **highs**, are high-pressure centers. As **Figure 13.12** illustrates, the pressure decreases from the outer isobars toward the center in a cyclone. In an anticyclone, just the opposite is the case—the values of the isobars increase from the outside toward the center. Knowing just a few basic facts about centers of high and low pressure greatly increases your understanding of current and forthcoming weather.

Cyclonic and Anticyclonic Winds

In the preceding section, you learned that the two most significant factors that affect wind are the pressure gradient force and the Coriolis effect. Winds move from higher pressure toward lower

Figure 13.12 Cyclonic and anticyclonic winds in the Northern Hemisphere Arrows show the winds blowing inward and counterclockwise around a low and outward and clockwise around a high.

Figure 13.13 Cyclonic circulation in the Northern and Southern Hemispheres The cloud patterns in these images allow us to "see" the circulation pattern in the lower atmosphere. (NASA images)

A. This satellite image shows a large low-pressure center in the Gulf of Alaska. The cloud pattern clearly shows an inward and counterclockwise spiral.

B. This satellite image shows a strong cyclonic storm in the South Atlantic near the coast of Brazil. The cloud pattern shows an inward and clockwise circulation.

pressure and are deflected to the right or left by Earth's rotation. When these controls of airflow are applied to pressure centers in the Northern Hemisphere, the result is that winds blow inward and counterclockwise around a low (**Figure 13.13A**). Around a high, they blow outward and clockwise (see Figure 13.12).

In the Southern Hemisphere, the Coriolis effect deflects the winds to the left; therefore, winds around a low blow clockwise, and winds around a high move counterclockwise (**Figure 13.13B**). In either hemisphere, friction causes a net inflow (**convergence**) around a cyclone and a net outflow (**divergence**) around an anticyclone.

Weather Generalizations About Highs and Lows

Rising air is associated with cloud formation and precipitation, whereas subsidence produces clear skies. In this section we will discuss how the movement of air can itself create pressure change and generate winds. After doing so, we will examine the relationship between horizontal and vertical flow, and their effects on weather.

Let us first consider a surface low-pressure system where the air is spiraling inward. The net inward transport of air causes a shrinking of the area occupied by the air mass, a process that is termed *horizontal convergence*. Whenever air converges horizontally, it must pile up—that is, increase in height to allow for the decreased area it now occupies. This generates a "taller" and

therefore heavier air column. Yet a surface low can exist only as long as the column of air exerts less pressure than that occurring in surrounding regions. We seem to have encountered a paradox—a low-pressure center causes a net accumulation of air, which increases its pressure. Consequently, a surface cyclone should quickly eradicate itself in a manner not unlike what happens when a vacuum-packed can is opened.

For a surface low to exist for very long, compensation must occur at some layer aloft. For example, surface convergence could be maintained if divergence (spreading out) aloft occurred at a rate equal to the inflow below. **Figure 13.14** shows the relationship between surface convergence and divergence aloft that is needed to maintain a low-pressure center.

Divergence aloft may even exceed surface convergence, thereby resulting in intensified surface inflow and accelerated vertical motion. Thus, divergence aloft can intensify storm centers as well as maintain them. On the other hand, inadequate divergence aloft permits surface flow to "fill" and weaken the accompanying cyclone.

Note that surface convergence about a cyclone causes a net upward movement. The rate of this vertical movement is slow, generally less than 1 kilometer (0.6 mile) per day. Nevertheless, because rising air often results in cloud formation and precipitation, a low-pressure center is generally related to unstable conditions and stormy weather (**Figure 13.15A**).

As often as not, it is divergence aloft that creates a surface low. Spreading out aloft initiates upflow in the

Figure 13.14 Airflow associated with surface cyclones (L) and anticyclones (H) A low, or cyclone, has converging surface winds and rising air, resulting in cloudy conditions. A high, or anticyclone, has diverging surface winds and descending air, which leads to clear skies and fair weather.

atmosphere directly below, eventually working its way to the surface, where inflow is encouraged.

Like their cyclonic counterparts, anticyclones must be maintained from above. Outflow near the surface is accompanied by convergence aloft and general subsidence of the air column (see Figure 13.14). Because descending air is compressed and warmed, cloud formation and precipitation are unlikely in an anticyclone. Thus, fair weather can usually be expected with the approach of a high-pressure center (**Figure 13.15B**).

For reasons that should now be obvious, it has been common practice to print on household barometers the words "stormy" at the low-pressure end and "fair" on the high-pressure end. By noting whether the pressure is rising, falling, or steady, we have a good indication of what the forthcoming weather will be. Such a determination, called the **pressure,** or **barometric, tendency**, is a useful aid in short-range weather prediction.

You should now be better able to understand why television weather reporters emphasize the positions and projected paths of cyclones and anticyclones. The "villain" on these weather programs is always the low-pressure center, which produces "bad" weather in any season. Lows move in roughly a west-to-east direction across the United States and require a few days to more than a week for the journey. Because their paths can be somewhat erratic, accurate prediction of their migration is difficult, although essential, for short-range forecasting.

Meteorologists must also determine whether the flow aloft will intensify an embryo storm or act to suppress its development. Because of the close tie between conditions at the surface and those aloft, a great deal of

(A)

(B)

Figure 13.15 Weather generalizations related to pressure centers **A.** A rainy day in London. Low-pressure systems are frequently associated with cloudy conditions and precipitation. (Photo by Lourens Smak/Alamy) **B.** Clear skies and "fair" weather may be expected when an area is under the influence of high pressure. (Photo by Prisma Bildagentur/Alamy)

emphasis has been placed on the importance and understanding of the total atmospheric circulation, particularly in the midlatitudes. We will now examine the workings of Earth's general atmospheric circulation and then again consider the structure of the cyclone in light of this knowledge.

13.3 CONCEPT CHECKS

1. Prepare a diagram with isobars and wind arrows that shows the winds associated with surface cyclones and anticyclones in both the Northern and Southern Hemispheres.
2. For a surface low-pressure center to exist for an extended period, what condition must exist aloft?
3. What general weather conditions are to be expected when the pressure tendency is rising? When the pressure tendency is falling?

13.4 General Circulation of the Atmosphere

Summarize Earth's idealized global circulation. Describe how continents and seasonal temperature changes complicate the idealized pattern.

The underlying cause of wind is unequal heating of Earth's surface. In tropical regions, more solar radiation is received than is radiated back to space. In polar regions, the opposite is true—less solar energy is received than is lost. Attempting to balance these differences, the atmosphere acts as a giant heat-transfer system, moving warm air poleward and cool air equatorward. On a smaller scale, but for the same reason, ocean currents also contribute to this global heat transfer. The general circulation is complex, and a great deal has yet to be explained. We can, however, develop a general understanding by first considering the circulation that would occur on a nonrotating Earth having a uniform surface. We will then modify this system to fit observed patterns.

Circulation on a Nonrotating Earth

On a hypothetical nonrotating planet with a smooth surface of either all land or all water, two large thermally produced cells would form (**Figure 13.16**). The heated equatorial air would rise until it reached the tropopause, which acts like a lid and deflects the air poleward. Eventually, this upper-level airflow would reach the poles, sink, spread out in all directions at the surface, and move back toward the equator. Once there, it would be

reheated and start its journey over again. This hypothetical circulation system has upper-level air flowing poleward and surface air flowing equatorward.

If we add the effect of rotation, this simple convection system will break down into smaller cells. **Figure 13.17** illustrates the three pairs of cells proposed to carry on the task of heat redistribution on a rotating planet. The polar and tropical cells retain the characteristics of the thermally generated convection described earlier. The nature of the midlatitude circulation is more complex and will be discussed in more detail in a later section.

Idealized Global Circulation

Near the equator, the rising air is associated with the pressure zone known as the **equatorial low**. This region of ascending moist, hot air is marked by abundant precipitation. Because this region of low pressure is a zone where winds converge, it is also referred to as the **intertropical convergence zone (ITCZ)**. As the upper-level flow from the equatorial low reaches 20° to 30° latitude, north or

Figure 13.16 Global circulation on a nonrotating Earth A simple convection system is produced by unequal heating of the atmosphere.

SmartFigure 13.17 Idealized global circulation proposed for the three-cell circulation model of a rotating Earth

south, it sinks back toward the surface. This subsidence and associated adiabatic heating produce hot, arid conditions. The center of this zone of subsiding dry air is the **subtropical high**, which encircles the globe near 30° latitude, north and south (see Figure 13.17). The great deserts of Australia, Arabia, and Africa exist because of the stable, dry condition caused by the subtropical highs.

At the surface, airflow is outward from the center of the subtropical high. Some of the air travels equatorward and is deflected by the Coriolis effect, producing the reliable **trade winds**. The remainder travels poleward and is also deflected, generating the prevailing **westerlies** of the midlatitudes. As the westerlies move poleward,

they encounter the cool **polar easterlies** in the region of the **subpolar low**. The interaction of these warm and cool winds produces the stormy belt known as the **polar front**. The source region for the variable polar easterlies is the **polar high**. Here, cold polar air is subsiding and spreading equatorward.

In summary, this simplified global circulation is dominated by four pressure zones. The subtropical and polar highs are areas of dry subsiding air that flows outward at the surface, producing the prevailing winds. The low-pressure zones of the equatorial and subpolar regions are associated with inward and upward airflow accompanied by clouds and precipitation.

Figure 13.18 Average surface air pressure These maps show the average surface air pressure, in millibars, for **A.** January and **B.** July, with associated winds.

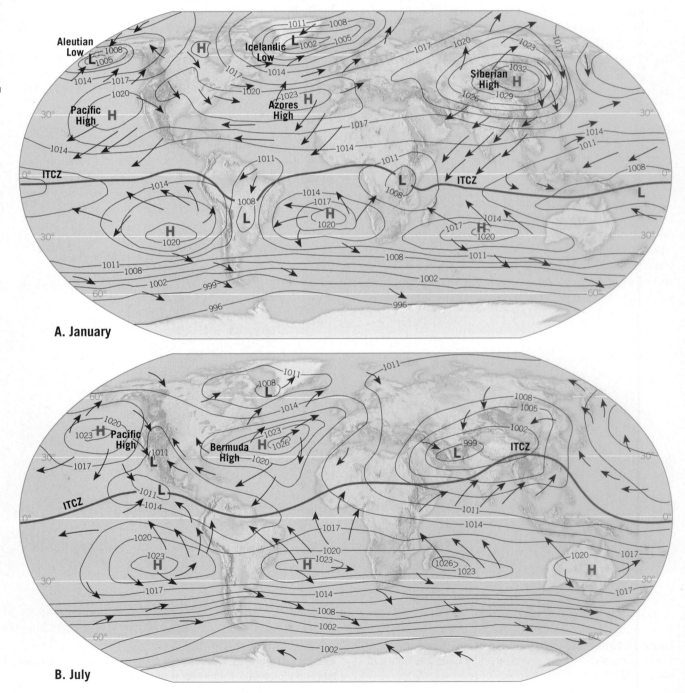

A. January

B. July

Influence of Continents

Up to this point, we have described the surface pressure and associated winds as continuous belts around Earth. However, the only truly continuous pressure belt is the subpolar low in the Southern Hemisphere, where the ocean is uninterrupted by landmasses. At other latitudes, particularly in the Northern Hemisphere, where landmasses break up the ocean surface, large seasonal temperature differences disrupt the pattern. **Figure 13.18** shows the resulting pressure and wind patterns for January and July. The circulation over the oceans is dominated by semipermanent cells of high pressure in the subtropics and cells of low pressure over the subpolar regions. The subtropical highs are responsible for the trade winds and westerlies, as mentioned earlier.

The large landmasses, on the other hand, particularly Asia, become cold in the winter and develop a seasonal high-pressure system from which surface flow is directed off the land (see Figure 13.18). In the summer, the opposite occurs: The landmasses are heated and develop a low-pressure cell, which permits air to flow onto the land. These seasonal changes in wind direction are known as the **monsoons**. During warm months, areas such as India experience a flow of warm, water-laden air from the Indian Ocean, which produces the rainy summer monsoon. The winter monsoon is dominated by dry continental air. A similar situation exists, but to a lesser extent, over North America.

In summary, the general circulation is produced by semipermanent cells of high and low pressure over the oceans and is complicated by seasonal pressure changes over land.

The Westerlies

The circulation in the midlatitudes, the zone of the westerlies, is complex and does not fit the convection system proposed for the tropics. Between 30° and 60° latitude, the general west-to-east flow is interrupted by the migration of cyclones and anticyclones. In the Northern Hemisphere, these cells move from west to east around the globe, creating an anticyclonic

(clockwise) flow or a cyclonic (counterclockwise) flow in their area of influence. A close correlation exists between the paths taken by these surface pressure systems and the position of the upper-level airflow, indicating that the upper air steers the movement of cyclonic and anticyclonic systems.

Among the most obvious features of the flow aloft are the seasonal changes. The steep temperature gradient across the middle latitudes in the winter months corresponds to a stronger flow aloft. In addition, the polar jet stream fluctuates seasonally such that its average position migrates southward with the approach of winter and northward as summer nears. By midwinter, the jet core may penetrate as far south as central Florida.

Because the paths of low-pressure centers are guided by the flow aloft, we can expect the southern tier of states to experience more of their stormy weather in the winter season. During the hot summer months, the storm track is across the northern states, and some cyclones never leave Canada. The northerly storm track associated with summer also applies to Pacific storms, which move toward Alaska during the warm months, thus producing an extended dry season for much of the West coast. The number of cyclones generated is seasonal as well, with the largest number occurring in the cooler months, when the temperature gradients are greatest. This fact is in agreement with the role of cyclonic storms in the distribution of heat across the midlatitudes.

13.4 CONCEPT CHECKS

1. Referring to the idealized model of atmospheric circulation, in which belt of prevailing winds is most of the United States?
2. The trade winds diverge from which pressure belt?
3. Which prevailing wind belts converge in the stormy region known as the polar front?
4. Which pressure belt is associated with the equator?
5. Explain the seasonal change in winds associated with India. What term is applied to this seasonal wind shift?

13.5 Local Winds

List three types of local winds and describe their formation.

Now that we have examined Earth's large-scale circulation, let us turn briefly to winds that influence much smaller areas. Remember that all winds are produced for the same reason: pressure differences that arise because of temperature differences caused by unequal heating

of Earth's surface. **Local winds** are small-scale winds produced by a locally generated pressure gradient. Those described here are caused either by topographic effects or by variations in surface composition in the immediate area.

Land and Sea Breezes

In coastal areas during the warm summer months, the land is heated more intensely during the daylight hours than is the adjacent body of water. As a result, the air above the land surface heats, expands, and rises, creating an area of lower pressure. A **sea breeze** then develops because cooler air over the water (higher pressure) moves toward the warmer land (lower pressure) (**Figure 13.19A**). The sea breeze begins to develop shortly before noon and generally reaches its greatest intensity during the mid- to late afternoon. These relatively cool winds can be a significant moderating influence on afternoon temperatures in coastal areas. Small-scale sea breezes can also develop along the shores of large lakes. People who live in a city near the Great Lakes, such as Chicago, recognize this lake effect, especially in the summer. They are reminded daily by weather reports of the cooler temperatures near the lake as compared to warmer outlying areas.

At night, the reverse may take place. The land cools more rapidly than the sea, and the **land breeze** develops (**Figure 13.19B**).

Mountain and Valley Breezes

A daily wind similar to land and sea breezes occurs in many mountainous regions. During daylight hours, the air along the slopes of the mountains is heated more intensely than the air at the same elevation over the valley floor. Because this warmer air is less dense, it glides up along the slope and generates a **valley breeze** (**Figure 13.20A**). The occurrence of these daytime upslope breezes can often be identified by the cumulus clouds that develop on adjacent mountain peaks.

After sunset, the pattern may reverse. Rapid radiation cooling along the mountain slopes produces a layer of cooler air next to the ground. Because cool air is denser than warm air, it drains downslope into the valley. This movement of air is called a **mountain breeze** (**Figure 13.20B**). The same type of cool air drainage can occur in places that have very modest slopes. The result is that the coldest pockets of air are usually found in the lowest spots. Like many other winds, mountain and valley breezes have seasonal preferences. Although valley breezes are most common during the warm season, when solar heating is most intense, mountain breezes tend to be more dominant in the cold season.

SmartFigure 13.19
Sea and land breezes

A. During daylight hours, cooler and denser air over the water moves onto the land, generating a sea breeze.

B. At night the land cools more rapidly than the sea, generating an offshore flow called a land breeze.

A. Valley breeze

B. Mountain breeze

Figure 13.20 Valley and mountain breezes

Chinook and Santa Ana Winds

Warm, dry winds sometimes move down the east slopes of the Rockies, where they are called **chinooks**. Such winds are often created when a strong pressure gradient develops in a mountainous region. As the air descends the leeward slopes of the mountains, it is heated adiabatically (by compression). Because condensation may have occurred as the air ascended the windward side, releasing latent heat, the air descending the leeward slope will be warmer and drier than it was at a similar elevation on the windward side. Although the temperature of these winds is generally less than 10°C (50°F), which is not particularly warm, the winds occur mostly in the winter and spring, when the affected areas may be experiencing below-freezing temperatures. Thus, by comparison, these dry, warm winds often bring a drastic change. When the ground has a snow cover, these winds are known to melt it in short order.

A chinooklike wind that occurs in southern California is the **Santa Ana**. This hot, desiccating wind greatly increases the threat of fire in this already dry area (**Figure 13.21**).

13.5 CONCEPT CHECKS

1. What is a local wind?
2. Describe the formation of a sea breeze.
3. Does a land breeze blow toward or away from the shore?
4. During what time of day would you expect to experience a well-developed valley breeze—midnight, late morning, or late afternoon?

Figure 13.21 Santa Ana winds Wildfires fanned by Santa Ana winds raged in southern California in October 2003. These fires scorched more than 740,000 acres and destroyed more than 3000 homes. The inset shows an idealized high-pressure center composed of cool, dry air that drives Santa Ana winds. Adiabatic heating causes the air temperature to increase and the relative humidity to decrease. (NASA)

13.6 Measuring Wind

Describe the instruments used to measure wind. Explain how wind direction is expressed using compass directions.

Two basic wind measurements, direction and speed, are particularly significant to weather observers. One simple device for determining both measurements is a *wind sock*, which is a common sight at small airports and landing strips (**Figure 13.22A**). The cone-shaped bag is open at both ends and is free to change position with shifts in wind direction. The degree to which the sock is inflated is an indication of wind speed.

Winds are always labeled by the direction from which they blow. A north wind blows *from* the north *toward* the south, an east wind *from* the east *toward* the west. The instrument most commonly used to determine wind direction is the **wind vane** (**Figure 13.22B**, upper right). This instrument, a common sight on many buildings, always points *into* the wind. The wind direction is often shown on a dial that is connected to the wind vane. The dial indicates wind direction, either by points of the compass (N, NE, E, SE, etc.) or by a scale of 0 to 360 degrees. On the latter scale, 0 degrees and 360 degrees are both north, 90 degrees is east, 180 degrees is south, and 270 degrees is west.

Wind speed is commonly measured using a **cup anemometer** (see Figure 13.22B, upper left). The wind

Did You Know?

"Snow eaters" is a local term for chinooks, the warm, dry winds that descend the eastern slopes of the Rockies. These winds have been known to melt more than a foot of snow in a single day. A chinook that moved through Granville, North Dakota, on February 21, 1918, caused temperatures to rise from −33°F to 50°F, an increase of 83°F!

A.

B.

Cup anemometer

Wind vane

Figure 13.22 Measuring the wind The two basic wind measurements are speed and direction. (Photo A by Lourens Smak/Alamy; photo B by Belfort Instrument Company, Baltimore, MD)

Wind frequency for winter in the northeastern United States.

Wind frequency for winter in northeastern Australia. Note the reliability of the southeast trade winds in Australia as compared to the westerlies in the northeastern United States.

Figure 13.23 Wind roses These graphs show the percentage of time winds come from various directions.

speed is read from a dial much like the speedometer of an automobile. Places where winds are steady and speeds are relatively high are potential sites for tapping wind energy.

When the wind consistently blows more often from one direction than from any other, it is called a **prevailing wind**. You may be familiar with the prevailing westerlies that dominate the circulation in the midlatitudes.

In the United States, for example, these winds consistently move the "weather" from west to east across the continent. Embedded within this general eastward flow are cells of high and low pressure, with their characteristic clockwise and counterclockwise flow. As a result, the winds associated with the westerlies, as measured at the surface, often vary considerably from day to day and from place to place. By contrast, the direction of airflow associated with the belt of trade winds is much more consistent, as can be seen in **Figure 13.23**.

By knowing the locations of cyclones and anticyclones in relation to where you are, you can predict the changes in wind direction that will occur as a pressure center moves past. Because changes in wind direction often bring changes in temperature and moisture conditions, the ability to predict the winds can be very useful. In the Midwest, for example, a north wind may bring cool, dry air from Canada, whereas a south wind may bring warm, humid air from the Gulf of Mexico. Sir Francis Bacon summed it up nicely when he wrote, "Every wind has its weather."

13.6 CONCEPT CHECKS

1. What are the two basic wind measurements? What instruments are used to make these measurements?

2. *From* what direction does a northeast wind blow? *Toward* what direction does a south wind blow?

13.7 Global Distribution of Precipitation

Discuss the major factors that influence the global distribution of precipitation.

A casual glance at **Figure 13.24** shows a relatively complex pattern for the distribution of precipitation. Although the map appears to be complicated, the general features of the map can be explained by applying our knowledge of global winds and pressure systems.

In general, regions influenced by high pressure, with its associated subsidence and diverging winds, experience relatively dry conditions. On the other hand, regions under the influence of low pressure and its converging winds and ascending air receive ample precipitation. This pattern is illustrated by noting that the tropical regions dominated by the equatorial low are the rainiest regions on Earth. It is here that we find the rain forests of the Amazon basin in South America and the Congo basin in Africa. The warm, humid trade winds converge to yield abundant rainfall throughout the year. By contrast, areas dominated by the subtropical high-pressure cells clearly receive much smaller amounts of precipitation. These are the regions of extensive subtropical deserts. In the Northern Hemisphere, the largest desert is the Sahara. Examples in the Southern Hemisphere include the Kalahari in southern Africa and the dry lands of Australia.

If Earth's pressure and wind belts were the only factors controlling precipitation distribution, the pattern shown in Figure 13.24 would be simpler. The inherent nature of the air is also an important factor in determining precipitation potential. Because cold air has a low capacity for moisture compared with warm air, we would expect a latitudinal variation in precipitation, with low latitudes receiving the greatest amounts of precipitation and high latitudes receiving the smallest amounts. Figure 13.24 indeed shows heavy rainfall in equatorial regions and meager precipitation in high-latitude areas. Recall that the dry region in the warm subtropics is explained by the presence of the subtropical high.

The distribution of land and water also complicates the precipitation pattern. Large landmasses in the middle latitudes commonly experience decreased precipitation toward their interiors. For example, central North America and central Eurasia receive considerably less precipitation than do coastal regions at the same latitude. Mountain barriers also alter precipitation patterns. Windward mountain slopes receive abundant precipitation, whereas leeward slopes and adjacent lowlands are often deficient in moisture.

13.7 CONCEPT CHECKS

1. With which global pressure belt are the rain forests of Africa's Congo basin associated? Which pressure system is linked to the Sahara Desert?
2. What factors, in addition to the distribution of wind and pressure, influence the global distribution of precipitation?

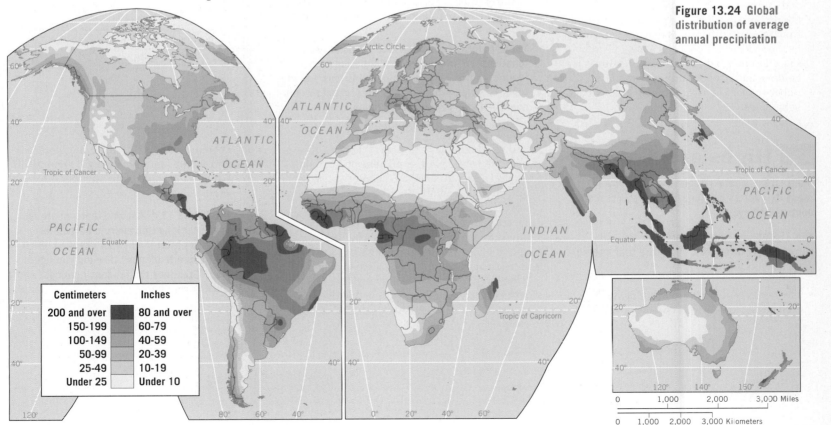

Figure 13.24 Global distribution of average annual precipitation

Centimeters	Inches
200 and over	80 and over
150-199	60-79
100-149	40-59
50-99	20-39
25-49	10-19
Under 25	Under 10

CONCEPTS IN REVIEW: THE ATMOSPHERE IN MOTION

13.1 Understanding Air Pressure

Define *air pressure* and describe the instruments used to measure this weather element.

KEY TERMS: air pressure, mercury barometer, aneroid barometer, barograph

- Air has weight: At sea level, it exerts a pressure of 1 kilogram per square centimeter (14.7 pounds per square inch).

- Air pressure is the force exerted by the weight of air above. With increasing altitude, there is less air above to exert a force, and thus air pressure decreases with altitude—rapidly at first and then much more slowly.
- The unit meteorologists use to measure atmospheric pressure is the millibar. Standard sea-level pressure is expressed as 1013.2 millibars. Isobars are lines on a weather map that connect places of equal air pressures.
- A mercury barometer measures air pressure using a column of mercury in a glass tube sealed at one end and inverted in a dish of

mercury. It measures atmospheric pressure in inches of mercury, the height of the column of mercury in the barometer. Standard atmospheric pressure at sea level equals 29.92 inches of mercury. As air pressure increases, the mercury in the tube rises, and when air pressure decreases, so does the height of the column of mercury.
- Aneroid ("without liquid") barometers consist of partially evacuated metal chambers that compress as air pressure increases and expand as pressure decreases.

13.2 Factors Affecting Wind

Discuss the three factors that act on the atmosphere to either create or alter winds.

KEY TERMS: wind, isobar, pressure gradient force, Coriolis effect, geostrophic wind, jet stream

- Wind is controlled by a combination of (1) the pressure gradient force, (2) the Coriolis effect, and (3) friction. The pressure gradient force is the primary driving force of wind that results from pressure differences that are depicted by the spacing of isobars on a map. Closely spaced isobars indicate a steep pressure gradient and strong winds; widely spaced isobars indicate a weak pressure gradient and light winds.

- The Coriolis effect produces deviation in the path of wind due to Earth's rotation (to the right in the Northern Hemisphere and to the left in the Southern Hemisphere). Friction, which significantly influences airflow near Earth's surface, is negligible above a height of a few kilometers.
- Above a height of a few kilometers, the Coriolis effect is equal to and opposite the pressure gradient force, which results in geostrophic winds. Geostrophic winds flow in a nearly straight path, parallel to the isobars, with velocities proportional to the pressure gradient force.

? These diagrams show surface winds at two locations. All factors in both situations are identical except that one surface is land and the other is water. Which diagram represents winds over the land? Explain your choice.

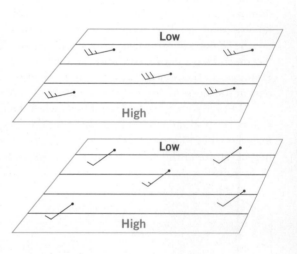

13.3 Highs and Lows

Contrast the weather associated with low-pressure centers (cyclones) and high-pressure centers (anticyclones).

KEY TERMS: cyclone, low, anticyclone, high, convergence, divergence, pressure (barometric) tendency

- The two types of pressure centers are (1) cyclones, or lows (centers of low pressure),

and (2) anticyclones, or highs (high-pressure centers). In the Northern Hemisphere, winds around a low (cyclone) are counterclockwise and inward. Around a high (anticyclone), they are clockwise and outward. In the Southern Hemisphere, the Coriolis effect causes winds to be clockwise around a low and counterclockwise around a high.
- Because air rises and cools adiabatically in a low-pressure center, cloudy conditions and precipitation are often associated with their passage. In a high-pressure center, descending

air is compressed and warmed; therefore, cloud formation and precipitation are unlikely in an anticyclone, and "fair" weather is usually expected.

? Refer to Figure 13.4. Assume that you are an observer checking this aneroid barometer several hours after it was last checked. What is the pressure tendency? How did you figure this out? What does the tendency indicate about forthcoming weather?

13.4 General Circulation of the Atmosphere

Summarize Earth's idealized global circulation. Describe how continents and seasonal temperature changes complicate the idealized pattern.

KEY TERMS: equatorial low, intertropical convergence zone (ITCZ), subtropical high, trade winds, westerlies, polar easterlies, subpolar low, polar front, polar high, monsoon

- If Earth's surface were uniform, four belts of pressure oriented east to west would exist in each hemisphere. Beginning at the equator, the four belts would be the (1) equatorial low, also referred to as the intertropical convergence zone (ITCZ), (2) subtropical high at about 25° to 35° on either side of the equator, (3) subpolar low, situated at about 50° to 60° latitude, and (4) polar high, near Earth's poles.

- Particularly in the Northern Hemisphere, large seasonal temperature differences over continents disrupt the idealized, or zonal, global patterns of pressure and wind. In winter, large, cold landmasses develop a seasonal high-pressure system from which surface airflow is directed off the land. In summer, landmasses are heated, and a low-pressure system develops over them, which permits air to flow onto the land. These seasonal changes in wind direction are known as monsoons.

- In the middle latitudes, between 30° and 60° latitude, the general west-to-east flow of the westerlies is interrupted by the migration of cyclones and anticyclones. The paths taken by these cyclonic and anticyclonic systems is closely correlated to upper-level airflow and the polar jet stream. The average position of the polar jet stream, and hence the paths followed by cyclones, migrates southward with the approach of winter and northward as summer nears.

13.5 Local Winds

List three types of local winds and describe their formation.

KEY TERMS: local wind, sea breeze, land breeze, valley breeze, mountain breeze, chinook, Santa Ana

- Local winds are small-scale winds produced by a locally generated pressure gradient. Sea and land breezes form along coasts and are brought about by temperature contrasts between land and water. Valley and mountain breezes occur in mountainous areas where the air along slopes heats differently than does the air at the same elevation over the valley floor. Chinook and Santa Ana winds are warm, dry winds created when air descends the leeward side of a mountain and warms by compression.

? With which local wind are the clouds in this photo most likely associated? Explain.

Herbert Koeppel/Alamy

13.6 Measuring Wind

Describe the instruments used to measure wind. Explain how wind direction is expressed using compass directions.

KEY TERMS: wind vane, cup anemometer, prevailing wind

- The two basic wind measurements are direction and speed. Winds are always labeled by the direction from which they blow. Wind direction is measured with a wind vane, and wind speed is measured using a cup anemometer.

? When designing an airport, it is important to have planes take off into the wind. Refer to the accompanying wind rose and describe the orientation of the runway and the direction planes would usually travel when they took off. Where on Earth might you find a wind rose like this?

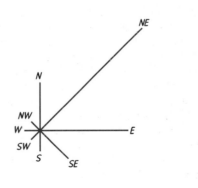

13.7 Global Distribution of Precipitation

Discuss the major factors that influence the global distribution of precipitation.

- The general features of the global distribution of precipitation can be explained by global winds and pressure systems. In general, regions influenced by high pressure, with its associated subsidence and divergent winds, experience dry conditions. Regions under the influence of low pressure and its converging winds and ascending air receive ample precipitation.

- Air temperature, the distribution of continents and oceans, and the location of mountains also influence the distribution of precipitation.

GIVE IT SOME THOUGHT

1. Mercury is 13.5 times denser (heavier) than water. If you built a barometer using water rather than mercury, how tall, in inches, would it have to be to record standard sea-level pressure?

2. If divergence in the jet stream above a surface low-pressure center exceeds convergence at the surface, will surface winds likely get stronger or weaker? Explain.

3. The accompanying map is a simplified surface weather map for April 2, 2011, on which the centers of three pressure cells are numbered.
   a. Identify which of the pressure cells are anticyclones (highs) and which are cyclones (lows).
   b. Which pressure center has the steepest pressure gradient and therefore the strongest winds?
   c. Refer to Figure 13.2 to determine whether pressure center 3 should be considered strong or weak.

5. If you live in the Northern Hemisphere and are directly west of the center of a cyclone, what is the probable wind direction at your location? What if you were west of an anticyclone?

6. If Earth did not rotate on its axis and if its surface were completely covered with water, what direction would a boat drift if it started its journey in the middle latitudes of the Northern Hemisphere? (*Hint:* What would the global circulation pattern be like for a nonrotating Earth?)

7. The accompanying sketch shows a cross section of the idealized circulation in the Northern Hemisphere. Match the appropriate number on the sketch to each of the following features:
   a. equatorial low
   b. polar front
   c. subtropical high
   d. polar high

4. You and a friend are watching TV on a rainy day, when the weather reporter says, "The barometric pressure is 28.8 inches and rising." Hearing this, you say, "It looks like fair weather is on its way." Your friend responds with the following questions. "I thought air pressure had something to do with the weight of air. How does inches relate to weight? And why do you think the weather is going to improve?" How would you respond to your friend's queries?

8. It is late afternoon on a warm summer day and you are enjoying some time at the beach. Until the last hour or two, winds were calm. Then a breeze began to develop. Is it more likely a cool breeze from the water or a warm breeze from the adjacent land area? Explain.

9. The accompanying maps of Africa show the distribution of precipitation for July and January. Which map represents July, and which represents January? How were you able to figure this out?

	< 25 mm
	25–50 mm
	50–150 mm
	> 150 mm

MasteringGeology™

14

Weather Patterns and Severe Weather

Each statement represents the primary learning objective for the corresponding major heading within the chapter. After you complete the chapter, you should be able to:

14.1 Discuss air masses, their classification, and associated weather.

14.2 Compare and contrast typical weather associated with a warm front and a cold front. Describe an occluded front and a stationary front.

14.3 Summarize the weather associated with the passage of a mature midlatitude cyclone. Describe how airflow aloft is related to cyclones and anticyclones at the surface.

14.4 List the basic requirements for thunderstorm formation and locate places on a map that exhibit frequent thunderstorm activity. Describe the stages in the development of a thunderstorm.

14.5 Summarize the atmospheric conditions and locations that are favorable to the formation of tornadoes. Discuss tornado destruction and tornado forecasting.

14.6 Identify areas of hurricane formation on a world map and discuss the conditions that promote hurricane formation. List the three broad categories of hurricane destruction.

This satellite image shows Hurricane Sandy, called Superstorm Sandy in the media, battering the East Coast on October 30, 2012. This view of the storm is looking south from Canada. Florida is near the top of the image. (NASA)

Tornadoes and hurricanes rank among nature's most destructive forces. Each spring, newspapers report the death and destruction left in the wake of "bands" of tornadoes. During late summer and fall, we hear occasional news reports about hurricanes. Storms with names such as Katrina, Rita, Sandy, and Ike make front-page headlines. Thunderstorms, although less intense and far more common than tornadoes and hurricanes, are also part of our discussion of severe weather in this chapter. Before looking at violent weather, however, we will study the atmospheric phenomena that most often affect our day-to-day weather: air masses, fronts, and traveling middle-latitude cyclones. We will see the interplay of the elements of weather discussed in Chapters 11, 12, and 13.

14.1 Air Masses

Discuss air masses, their classification, and associated weather.

For many people who live in the middle latitudes, which include much of the United States, summer heat waves and winter cold spells are familiar experiences. In the first instance, several days of high temperatures and oppressive humidities may finally end when a series of thunderstorms pass through the area, followed by a few days of relatively cool relief. By contrast, the clear skies that often accompany a span of frigid subzero days may be replaced by thick gray clouds and a period of snow as temperatures rise to levels that seem mild compared to those that existed just a day earlier. In both examples, what was experienced was a period of generally constant weather conditions followed by a relatively short period of change and then the reestablishment of a new set of weather conditions that remained for perhaps several days before changing again.

What Is an Air Mass?

The weather patterns just described result from movements of large bodies of air, called air masses. An **air mass** is an immense body of air, typically 1600 kilometers (1000 miles) or more across and perhaps several kilometers thick, that is characterized by a similarity of temperature and moisture at any given altitude. When this air moves out of its region of origin, it carries those temperatures and moisture conditions elsewhere, eventually affecting a large portion of a continent (**Figure 14.1**).

Figure 14.1 Lake-effect snow storm This satellite image shows a cold, dry air mass moving from its source region in Canada across Lake Superior. It illustrates the process that leads to lake-effect snow storms. (NASA)

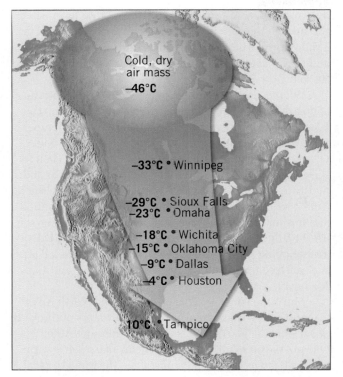

Figure 14.2 An invasion of frigid air As this very cold air mass moved southward from Canada, it brought some of the coldest weather of the winter to the areas in its path. As it advanced into the United States, the air mass slowly got warmer. Thus, the air mass was gradually modified at the same time that it modified the weather in the areas over which it moved. (From *Physical Geography: A Landscape Appreciation*, 9th edition, by Tom L. McKnight and Darrell Hess, ©2008. Reprinted and electronically reproduced by permission of Pearson Education, Inc., Upper Saddle River, NJ.)

An excellent example of the influence of an air mass is illustrated in **Figure 14.2**, which shows a cold, dry mass from northern Canada moving southward. With a beginning temperature of −46°C (−51°F), the air mass warms to −33°C (−27°F) by the time it reaches Winnipeg. It continues to warm as it moves southward through the Great Plains and into Mexico. Throughout its southward journey, the air mass becomes warmer. But it also brings some of the coldest weather of the winter to the places in its path. Thus, the air mass is modified, but it also modifies the weather in the areas over which it moves.

The horizontal uniformity of an air mass is not perfect, of course. Because air masses extend over large areas, small differences occur in temperature and humidity from place to place. Still, the differences observed within an air mass are small compared to the rapid changes experienced across air-mass boundaries.

Since it may take several days for an air mass to move across an area, the region under its influence will probably experience fairly constant weather, a situation called **air-mass weather**. Certainly, there may be some day-to-day variations, but the events will be very unlike those in an adjacent air mass.

The air-mass concept is an important one because it is closely related to the study of atmospheric disturbances. Most disturbances in the middle latitudes originate along the boundary zones that separate different air masses.

Source Regions

When a portion of the lower atmosphere moves slowly or stagnates over a relatively uniform surface, the air will assume the distinguishing features of that area, particularly with regard to temperature and moisture conditions.

The area where an air mass acquires its characteristic properties of temperature and moisture is called its **source region**. The source regions that produce air masses that influence North America are shown in **Figure 14.3**.

Air masses are classified according to their source region. **Polar (P)** and **arctic (A) air masses** originate in high latitudes toward Earth's poles, whereas those that form in low latitudes are called **tropical (T) air masses**. The designation *polar, arctic,* or *tropical* gives an indication of the temperature characteristics of an air mass. *Polar* and *arctic* indicate cold, and *tropical* indicates warm.

Figure 14.3 Air-mass source regions for North America Source regions are largely confined to subtropical and subpolar locations. The fact that the middle latitudes are where cold and warm air masses clash, often because the converging winds of a traveling cyclone draw them together, means that this zone lacks the conditions necessary to be a source region. The differences between polar and arctic are relatively small and serve to indicate the degree of coldness of the respective air masses. By comparing the winter (A) and summer (B) maps, it is clear that the extent and temperature characteristics fluctuate.

In addition, air masses are classified according to the nature of the surface in the source region. **Continental (c) air masses** form over land, and **maritime (m) air masses** originate over water. The designation *continental* or *maritime* suggests the moisture characteristics of the air mass. Continental air is likely to be dry, and maritime air is likely to be humid.

The basic types of air masses according to this scheme of classification are continental polar (cP), continental arctic (cA), continental tropical (cT), maritime polar (mP), and maritime tropical (mT).

Weather Associated with Air Masses

Continental polar and maritime tropical air masses influence the weather of North America most, especially east of the Rocky Mountains. Continental polar air masses originate in northern Canada, interior Alaska, and the Arctic—areas that are uniformly cold and dry in winter and cool and dry in summer. In winter, an invasion of continental polar air brings the clear skies and cold temperatures we associate with a cold wave as it moves southward from Canada into the United States. In summer, this air mass may bring a few days of cooling relief.

Although cP air masses are not normally associated with heavy precipitation, those that cross the Great Lakes in late autumn and winter sometimes bring snow to the leeward shores. These localized storms often form when the surface weather map indicates no apparent cause for a snowstorm. These are known as **lake-effect snows**, and they make Buffalo and Rochester, New York, among the snowiest cities in the United States (**Figure 14.4**).

What causes lake-effect snow? During late autumn and early winter, the temperature contrast between the lakes and adjacent land areas can be large.° The temperature contrast can be especially great when a very cold cP air mass pushes southward across the lakes. When this occurs, the air acquires large quantities of heat and moisture from the relatively warm lake surface. By the time it reaches the opposite shore, the air mass is humid and unstable, and heavy snow showers are likely. Figure 14.1 illustrates this process. In this image, cold cP air is moving south from Canada. Notice that the cP air is cloud free until it moves over Lake Superior and Lake Michigan.

Maritime tropical air masses affecting North America most often originate over the warm waters of the Gulf of Mexico, the Caribbean Sea, or the adjacent Atlantic Ocean. As you might expect, these air masses are warm, moisture laden, and usually unstable. Maritime tropical air is the source of much, if not most, of the precipitation in the eastern two-thirds of the United States. In

---

° Recall that land cools more rapidly and to lower temperatures than water. See the discussion "Land and Water" in the section "Why Temperatures Vary: The Controls of Temperature" in Chapter 11.

SmartFigure 14.4 Snowfall map The snowbelts of the Great Lakes are easy to pick out on this snowfall map. (Data from NOAA) The photo was taken following a 6-day lake-effect snowstorm in November 1996 that dropped 175 centimeters (nearly 69 inches) of snow on Chardon, Ohio, setting a new state record. (Photo by Tony Dejak/AP/Wide World Photos)

AVERAGE ANNUAL SNOWFALL	
cm	in.
>330	>130
250–329	100–129
150–249	60–99
90–149	36–59
60–89	24–35
<60	<24

summer, when an mT air mass invades the central and eastern United States, and occasionally southern Canada, it brings the high temperatures and high humidity typically associated with its source region.

Of the two remaining air masses, maritime polar and continental tropical, the latter has the least influence on the weather of North America. Hot, dry continental tropical air, originating in the Southwest and Mexico during the summer, only occasionally affects the weather outside its source region.

During winter, maritime polar air masses coming from the North Pacific often originate as continental polar air masses in Siberia. The cold, dry cP air is transformed into relatively mild, humid, unstable mP air during its long journey across the North Pacific (**Figure 14.5**). As this mP air arrives at the western shore of North America, it is often accompanied by low clouds and shower activity. When this air advances inland against the western mountains, orographic uplift produces heavy rain or snow on the windward slopes of the mountains. Maritime polar air also originates in the North Atlantic, off the coast of eastern Canada, and occasionally influences the weather of the northeastern United States. In winter, when New England is on the northern or northwestern side of a passing low, the counterclockwise cyclonic winds draw in maritime polar air. The result is a storm characterized by snow and cold temperatures, known locally as a *nor'easter*.

Figure 14.5 Air-mass modification During winter, maritime polar (mP) air masses in the North Pacific usually begin as continental polar (cP) air masses in Siberia. The cP air is modified to mP as it slowly crosses the ocean.

14.1 CONCEPT CHECKS

1. Define *air mass*. What is air-mass weather?
2. On what basis are air masses classified?
3. Compare the temperature and moisture characteristics of the following air masses: cP, mP, mT, and cT.
4. Which air mass is associated with lake-effect snow? What causes lake-effect snow?

14.2 Fronts

Compare and contrast typical weather associated with a warm front and a cold front. Describe an occluded front and a stationary front.

Fronts are boundaries that separate different air masses, one warmer than the other and often higher in moisture content. A front can form between any two contrasting air masses. Considering the vast size of the air masses involved, fronts are relatively narrow, being 15- to 200-kilometer-wide (9- to 120-mile-wide) bands of discontinuity. On the scale of a weather map, they are generally narrow enough to be represented by a broad line.

Above Earth's surface, a front slopes at a low angle so that warmer air overlies cooler air (**Figure 14.6**). In the ideal case, the air masses on both sides of the front move in the same direction and at the same speed. Under this condition, the front acts simply as a barrier, moving along between the two contrasting air masses. Generally, however, an air mass on one side of a front moves faster relative to the frontal boundary than the air mass on the other side. Thus, one air mass actively advances into another and "clashes" with it. In fact, during World War I, Norwegian meteorologists called these boundaries *fronts*, analogous to battle lines between two armies. Along

Figure 14.6 Warm front This diagram shows the idealized clouds and weather associated with a warm front. During most of the year, warm fronts produce light to moderate precipitation over a wide area.

these "battlegrounds," centers of low pressure develop and generate much of the precipitation and severe weather in the middle latitudes.

As one air mass advances into another, limited mixing occurs along the frontal surface, but for the most part, the air masses retain their distinct identities as one is displaced upward over the other. No matter which air mass is advancing, *it is always the warmer (less dense) air that is forced aloft*, whereas *the cooler (denser) air acts as the wedge on which lifting takes place.* The term **overrunning** is generally applied to warm air gliding up along a cold air mass. We will now take a look at different types of fronts.

Warm Fronts

When the surface position of a front moves so that warm air occupies territory formerly covered by cooler air, it is called a **warm front** (see Figure 14.6). On a weather map, the surface position of a warm front is shown by a red line with red semicircles protruding into the cooler air.

East of the Rockies, warm tropical air often enters the United States from the Gulf of Mexico and overruns receding cool air. As the cold air retreats, friction with the ground slows the advance of the surface position of the front more so than its position aloft. Stated another way, less dense, warm air has a hard time displacing denser, cold air. For this reason, the boundary separating these air masses acquires a very gradual slope. The average slope of a warm front is about 1:200, which means that if you are 200 kilometers (120 miles) ahead of the surface location of a warm front, you will find the frontal surface at a height of 1 kilometer (0.6 mile).

As warm air ascends the retreating wedge of cold air, it expands and cools adiabatically to produce clouds and, frequently, precipitation. The sequence of clouds shown in Figure 14.6 typically precedes a warm front. The first sign of the approach of a warm front is the appearance of cirrus clouds overhead. These high clouds form 1000 kilometers (600 miles) or more ahead of the surface front, where the overrunning warm air has ascended high up the wedge of cold air.

As the front nears, cirrus clouds grade into cirrostratus, which blend into denser sheets of altostratus. About 300 kilometers (180 miles) ahead of the front, thicker stratus and nimbostratus clouds appear, and rain or snow begins. Because of their slow rate of advance and very low slope, warm fronts usually produce light to moderate precipitation over a large area for an extended period (**Figure 14.7**). Warm fronts, however, are occasionally associated with cumulonimbus clouds and thunderstorms. This occurs when the overrunning air is unstable and the temperatures on opposite sides of the front contrast sharply. At the other extreme, a warm front associated with a dry air mass could pass unnoticed at the surface.

Figure 14.7 Rain associated with a warm front As the surface position of a warm front approaches, it is often preceded by light to moderate precipitation from nimbostratus clouds. (Photo by David Grossman/Alamy Images)

A gradual increase in temperature occurs with the passage of a warm front. The increase is most noticeable when there is a large temperature difference between the adjacent air masses. The moisture content and stability of the encroaching warm air mass largely determine when clear skies will return. During the summer, cumulus, and occasionally cumulonimbus, clouds are embedded in the warm unstable air mass that follows the front. Precipitation from these clouds can be heavy but is usually scattered and of short duration.

Cold Fronts

When dense cold air is actively advancing into a region occupied by warmer air, the boundary is called a **cold front** (**Figure 14.8**). As with warm fronts, friction tends to slow the surface position of a cold front more so than its position aloft. However, because of the relative positions of the adjacent air masses, the cold front steepens as it moves. On average, cold fronts are about twice as steep as warm fronts, having a slope of perhaps 1:100. In addition, cold fronts advance at speeds around 35 to 50 kilometers (20 to 35 miles) per hour compared to 25 to 35 kilometers (15 to 20 miles) per hour for warm fronts. These two differences—rate of movement and steepness of slope—largely account for the more violent nature of cold-front weather compared to the weather generally accompanying a warm front (**Figure 14.9**).

As a cold front approaches, generally from the west or northwest, towering clouds can often be seen in the distance. Near the front, a dark band of ominous clouds foretells the ensuing weather. The forceful lifting of air along a cold front is often so rapid that the latent heat released when water vapor condenses increases the air's buoyancy appreciably. The heavy downpours and vigorous wind gusts associated with mature cumulonimbus clouds frequently result. A cold front produces roughly the same amount of lifting as a warm front, but over a shorter distance. As a result, the intensity of precipitation is greater, but the duration is shorter. In addition, a marked temperature drop and a wind shift from the south to west or northwest accompany the passage of the front. The sometimes violent weather and sharp temperature contrast along the cold front are symbolized on a weather map by a blue line with blue triangle-shaped points that extend into the warmer air mass (see Figure 14.8).

The weather behind a cold front is dominated by a subsiding and relatively cold air mass. Thus, clearing usually begins soon after the front passes. Although the compression of air due to subsidence causes some adiabatic heating, the effect on surface temperatures is minor. In winter, the long, cloudless nights that often follow the passage of a cold front allow for abundant radiation cooling that reduces surface temperatures. When a cold front moves over a relatively warm area, surface heating can produce shallow convection. This, in turn, may generate low cumulus or stratocumulus clouds behind the front.

SmartFigure 14.8 Cold front Fast-moving cold front and cumulonimbus clouds. Thunderstorms may occur if the warm air is unstable.

Figure 14.9 Splashing hailstones Cumulonimbus clouds along a cold front produced hail and heavy rain at this ballpark in Wichita, Kansas. Dozens of cars in the parking lot were damaged. (AP Photo/*The Wichita Eagle*, Fernando Salazar)

Stationary Fronts and Occluded Fronts

Occasionally, the flow on both sides of a front is neither toward the cold air mass nor toward the warm air mass but almost parallel to the line of the front. Thus,

Figure 14.10 Stages in the formation of an occluded front

In this example, the air behind the cold front is colder and denser than the air ahead of the warm front.

The surface cold front moves faster than the surface warm front and overtakes it to form an occluded front.

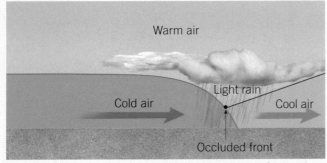

The denser cold air lifts the warm air and advances into and displaces the cool air.

the surface position of the front does not move. This condition is called a **stationary front**. On a weather map, stationary fronts are shown with blue triangular points on one side of the front and red semicircles on the other. At times, some overrunning occurs along a stationary front, most likely causing gentle to moderate precipitation.

The fourth type of front is an **occluded front**, an active cold front that overtakes a warm front, as shown in **Figure 14.10**. As the advancing cold air wedges the warm front upward, a new front emerges between the advancing cold air and the air over which the warm front is gliding. The weather of an occluded front is generally complex. Most precipitation is associated with the warm air being forced aloft. When conditions are suitable, however, the newly formed front is capable of initiating precipitation of its own.

A word of caution is in order concerning the weather associated with various fronts. Although the preceding discussion will help you recognize the weather patterns associated with fronts, remember that these descriptions are generalizations. The weather generated along any individual front may or may not conform fully to this idealized picture. Fronts, like all other aspects of nature, do not lend themselves to classification as easily as we would like.

14.2 CONCEPT CHECKS

1. Compare the weather of a typical warm front with that of a typical cold front.
2. Why is cold-front weather usually more severe than warm-front weather?
3. Describe a stationary front and an occluded front.

14.3 Midlatitude Cyclones

Summarize the weather associated with the passage of a mature midlatitude cyclone. Describe how airflow aloft is related to cyclones and anticyclones at the surface.

So far, we have examined the basic elements of weather as well as the dynamics of atmospheric motions. We are now ready to apply our knowledge of these diverse phenomena to an understanding of day-to-day weather patterns in the middle latitudes. For our purposes, *middle latitudes* refers to the region between southern Florida and Alaska. The primary weather producers here are **midlatitude**, or **middle-latitude, cyclones**. On weather maps they are shown by an *L*, meaning *low-pressure system*. **Figure 14.11**

shows two views of a large idealized midlatitude cyclone with probable air masses, fronts, and surface wind patterns.

Midlatitude cyclones are large centers of low pressure that generally travel from west to east. Lasting from a few days to more than a week, these weather systems have a counterclockwise circulation, with an airflow inward toward their centers. Most midlatitude cyclones also have a cold front extending from the central area of low pressure, and frequently a warm

A. Map view

Key
━━● Warm front
━━▲ Cold front
━━▼● Stationary front
━━●● Occluded front

B. Three-dimentional view from point A to point B

SmartFigure 14.11
Idealized structure of a large, mature midlatitude cyclone A. This map view shows fronts, air masses, and surface winds. **B**. The three-dimensional view is a cross section through warm and cold fronts along a line from point A to point B.

front as well. Convergence and forceful lifting initiate cloud development and frequently cause abundant precipitation.

As early as the 1800s, it was known that middle-latitude cyclones were the bearers of precipitation and severe weather. But it was not until the early part of the 1900s that a model was developed to explain how cyclones form. A group of Norwegian scientists formulated and published this model in 1918. The model was created primarily from near-surface observations.

Years later, as data from the middle and upper troposphere and from satellite images became available, modifications were necessary. However, this model is still a useful working tool for interpreting the weather. If you keep this model in mind when you observe changes in the weather, the changes will no longer come as a surprise. You should begin to see some order in what once appeared to be disorder, and you might even occasionally "predict" the impending weather.

Idealized Weather of a Midlatitude Cyclone

The midlatitude cyclone model provides a useful tool for examining the weather patterns of the middle latitudes. **Figure 14.12** illustrates the distribution of clouds and thus the regions of possible precipitation associated with a mature system. Compare this drawing to the satellite image shown in **Figure 14.13**. It is easy to see why we often refer to the cloud pattern of a midlatitude cyclone as having a "comma" shape.

Guided by the westerlies aloft, cyclones generally move eastward across the United States, so we can expect the first signs of their arrival in the west. However, often in the region of the Mississippi River valley, cyclones begin a more northeasterly path and occasionally move directly northward. A midlatitude cyclone typically requires 2 to 4 days to move completely across a region. During that brief period, abrupt changes in atmospheric

Figure 14.12 Cloud patterns typically associated with a mature middle-latitude cyclone The middle section is a map view. Note the cross-sectional lines (F–G, A–E). Above the map is a vertical cross section along line F–G. Below the map is a section along A–E. For cloud abbreviations, refer to Figures 14.6 and 14.8.

Figure 14.13 Satellite view of a mature midlatitude cyclone This storm swept across the central United States and produced strong wind gusts (up to 125 kilometers [78 miles] per hour), rain, hail, and snow. It also spawned 61 tornadoes on October 26, 2010. This cyclone set a record for the lowest pressure not associated with a hurricane ever recorded over land in the continental United States: 28.21 inches of mercury. It is easy to see why we often refer to the cloud pattern of a cyclone as having a "comma" shape. (NASA)

conditions may be experienced. This is particularly true in the winter and spring, when the largest temperature contrasts occur across the middle latitudes.

Using Figure 14.12 as a guide, we will now consider these weather producers and what we should expect from them as they move over an area. To facilitate our discussion, Figure 14.12 includes two profiles along lines A–E and F–G:

- Imagine the change in weather as you move along profile A–E. At point A, the sighting of high cirrus clouds would be the first sign of the approaching cyclone. These high clouds can precede the surface front by 1000 kilometers (600 miles) or more, and they generally are accompanied by falling pressure. As the warm front advances, a lowering and thickening of the cloud deck is noticed.

- Usually within 12 to 24 hours after the first sighting of cirrus clouds, light precipitation begins (point B). As the front nears, the rate of precipitation increases, a rise in temperature is noticed, and winds begin to change from east or southeast to south or southwest.

- With the passage of the warm front, an area is under the influence of a maritime tropical air mass (point C). Generally, the region affected by this sector of the cyclone experiences warm to hot temperatures, southwesterly winds, fairly high humidity, and clear to partly cloudy skies containing cumulus clouds.

- The relatively warm, humid weather of the warm sector passes quickly and is replaced by gusty winds and precipitation generated along the cold front. The approach of a rapidly advancing cold front is marked by a wall of dark clouds (point D). Severe weather accompanied by heavy precipitation, hail, and an occasional tornado is a definite possibility, especially during spring and summer. The passage of the cold front is easily detected by a wind shift; the southwest winds are replaced by winds from the west to northwest and by a pronounced drop in temperature. Also, the rising pressure hints of the subsiding cool, dry air behind the front.

- Once the front passes, skies clear as cooler air invades the region (point E). Often a day or two of almost cloudless deep blue skies occurs unless another cyclone is edging into the region.

A very different set of weather conditions will prevail in the regions north of the storm's center along profile F–G of Figure 14.12. In this part of the storm, temperatures remain cool. The first hints of the approaching low-pressure center are a continual drop in air pressure and increasingly overcast conditions that bring varying amounts of precipitation. This section of the cyclone most often generates snow during the winter months.

Once the formation of an occluded front begins, the character of the storm changes. Because occluded fronts tend to move more slowly than other fronts, the entire wishbone-shaped frontal structure of the storm rotates counterclockwise. As a result, the occluded front appears to "bend over backward." This effect adds to the misery of the region influenced by the occluded front because it lingers over the area longer than the other fronts.

The Role of Airflow Aloft

When the earliest studies of midlatitude cyclones were made, little was known about the nature of the airflow in the middle and upper troposphere. Since then, a close relationship has been established between surface disturbances and the flow aloft. Airflow aloft plays an important role in maintaining cyclonic and anticyclonic circulation. In fact, more often than not, these rotating surface wind systems are actually generated by upper-level flow.

Recall that the airflow around a cyclone (low-pressure system) is inward, a fact that leads to mass convergence, or coming together (**Figure 14.14**). The resulting accumulation of air must be accompanied by a corresponding increase in surface pressure. Consequently, we might expect a low-pressure system to "fill" rapidly and be eliminated. However, this does not occur. On the contrary, cyclones often exist for a week or longer. For this to happen, surface convergence must be offset by a mass outflow at some level aloft (see Figure 14.14). As long as divergence (spreading out) aloft is equal to or greater than surface inflow, the low pressure and its accompanying convergence can be sustained.

Because cyclones are bearers of stormy weather, they have received far more attention than anticyclones. Nevertheless, a close relationship exists, which makes it difficult to separate any discussion of these two types of pressure systems. The surface air that feeds a cyclone, for example, generally originates as air flowing out of an anticyclone. Consequently, cyclones and anticyclones typically are found adjacent to each other. Like a cyclone, an anticyclone depends on the flow far above to maintain its circulation. Divergence at the surface is balanced by convergence aloft and general subsidence of the air column (see Figure 14.14).

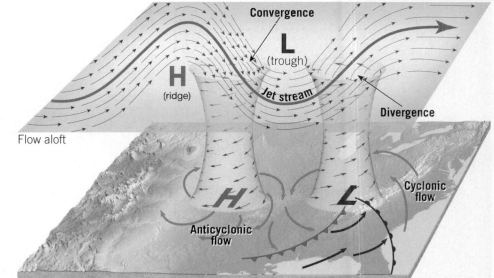

Figure 14.14 Flow aloft influences surface winds and pressure Idealized depiction showing the support that divergence and convergence aloft provide to cyclonic and anticyclonic circulation at the surface. Divergence aloft initiates upward air movement, reduces surface pressure, and cyclonic flow. On the other hand, convergence along the jet stream results in general subsidence of the air column, increased surface pressure, and anticyclonic surface winds.

14.3 CONCEPT CHECKS

1. Briefly describe the weather associated with the passage of a mature midlatitude cyclone when the center of low pressure is about 200 to 300 kilometers (125 to 200 miles) north of your location.
2. If the midlatitude cyclone described in Question 1 took 3 days to pass your location, on which day would temperatures likely be warmest? On which day would they likely be coldest?
3. What winter weather might be expected with the passage of a mature midlatitude cyclone when the center of low pressure is located about 100 to 200 kilometers (60 to 125 miles) south of your location?
4. Briefly explain how flow aloft aids the formation of cyclones at the surface.

Did You Know?

The term *extratropical cyclone* is another name that meteorologists use for a midlatitude cyclone. *Extratropical* means "outside the tropics." By contrast, hurricanes and tropical storms, which form in the lower latitudes, are examples of *tropical cyclones*.

14.4 Thunderstorms

List the basic requirements for thunderstorm formation and locate places on a map that exhibit frequent thunderstorm activity. Describe the stages in the development of a thunderstorm.

Thunderstorms are the first of three severe weather types we will examine in this chapter. Sections on tornadoes and hurricanes follow. All these phenomena can be related to low-pressure systems (cyclones).

Severe weather is more fascinating than everyday weather phenomena. The lightning display and booming thunder generated by a severe thunderstorm can be a spectacular event that elicits both awe and fear (**Figure 14.15**). Of course, hurricanes and tornadoes also attract a great deal of much-deserved attention. A single tornado outbreak or hurricane can cause many deaths as well as billions of dollars in property damage. In a typical year, the United States experiences thousands of violent thunderstorms, hundreds of tornadoes, and several hurricanes.

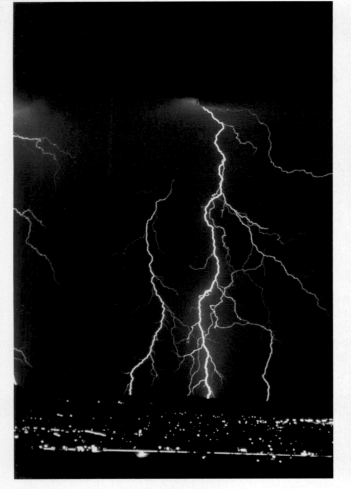

Figure 14.15 Summertime lightning display A storm is classified as a thunderstorm only after thunder is heard. Because thunder is produced by lightning, lightning must also occur. (Photo by AGE Fotostock/SuperStock)

In southern Asia and Australia, the term *cyclone* is applied to storms that are called *hurricanes* in the United States. This image shows Cyclone Yasi, which struck eastern Australia in February 2011.

In parts of the Great Plains, *cyclone* is a synonym for *tornado*. The nickname for the athletic teams at Iowa State University is the *Cyclones**

Figure 14.16 The term *cyclone* Sometimes the use of the term *cyclone* can be confusing. (Satellite image courtesy of NASA; logo courtesy of Iowa State University)

What's in a Name?

Up to now we have examined midlatitude cyclones, which play an important role in causing day-to-day weather changes. Yet the use of the term *cyclone* is often confusing. To many people, the term implies only an intense storm, such as a tornado or a hurricane. When a hurricane unleashes its fury on India or Bangladesh, for example, it is usually reported in the media as a *cyclone* (the term denoting a hurricane in that part of the world).

Similarly, tornadoes are referred to as *cyclones* in some places. This custom is particularly common in portions of the Great Plains of the United States. Recall that in *The Wizard of Oz*, Dorothy's house was carried from her Kansas farm to the land of Oz by a cyclone. Indeed, the nickname for the athletic teams at Iowa State University is the *Cyclones* (**Figure 14.16**). Although hurricanes and tornadoes are, in fact, cyclones, the vast majority of cyclones are *not* hurricanes or tornadoes. The term *cyclone* simply refers to the

circulation around any low-pressure center, no matter how large or intense it is.

Tornadoes and hurricanes are both smaller and more violent than middle-latitude cyclones. Middle-latitude cyclones can have a diameter of 1600 kilometers (1000 miles) or more. By contrast, hurricanes average only 600 kilometers (375 miles) across, and tornadoes, with a typical diameter of just 0.25 kilometer (0.16 mile), are much too small to show up on a weather map.

The thunderstorm, a much more familiar weather event, hardly needs to be distinguished from tornadoes, hurricanes, and midlatitude cyclones. Unlike the flow of air about these latter storms, the circulation associated with thunderstorms is characterized by strong up-and-down movements. Winds in the vicinity of a

---

* Iowa State University is the only Division I school to use Cyclones as its team name. The pictured logo was created to better communicate the school's image by combining the mascot, a cardinal bird named Cy, and the Cyclone team name, a swirling twister in the bird's tail.

thunderstorm do not follow the inward spiral of a cyclone, but they are typically variable and gusty.

Although thunderstorms form "on their own," away from cyclonic storms, they also form in conjunction with cyclones. For instance, thunderstorms are frequently spawned along the cold front of a midlatitude cyclone, where on rare occasions a tornado may descend from the thunderstorm's cumulonimbus tower. Hurricanes also generate widespread thunderstorm activity. Thus, thunderstorms are related in some manner to all three types of cyclones mentioned here.

Thunderstorm Occurrence

Almost everyone has observed various small-scale phenomena that result from the vertical movements of relatively warm, unstable air. Perhaps you have seen a dust devil over an open field on a hot day, whirling its dusty load to great heights. Or maybe you have noticed a bird glide effortlessly skyward on an invisible thermal of hot air. These examples illustrate the dynamic thermal instability that occurs during the development of a thunderstorm.

A **thunderstorm** is a storm that generates lightning and thunder. Thunderstorms frequently produce gusty winds, heavy rain, and hail. A thunderstorm may be produced by a single cumulonimbus cloud and influence only a small area, or it may be associated with clusters of cumulonimbus clouds covering a large area.

Thunderstorms form when warm, humid air rises in an unstable environment. Various mechanisms can trigger the upward air movement needed to create thunderstorm-producing cumulonimbus clouds. One mechanism, the unequal heating of Earth's surface, significantly contributes to the formation *of air-mass thunderstorms*. These storms are associated with the scattered puffy cumulonimbus clouds that commonly form *within* maritime tropical air masses and produce scattered thunderstorms on summer days. Such storms are usually short-lived and seldom produce strong winds or hail.

Another type of thunderstorm not only benefits from uneven surface heating but is associated with the lifting of warm air, as occurs along a front or a mountain slope. Moreover, diverging winds aloft frequently contribute to the formation of these storms because they tend to draw air from lower levels upward beneath them. Some of the thunderstorms of this type may produce high winds, damaging hail, flash floods, and tornadoes. Such storms are described as *severe*.

At any given time, an estimated 2000 thunderstorms are in progress on Earth. As we would expect, the greatest number occur in the tropics, where warmth, plentiful moisture and instability are always present. About 45,000 thunderstorms take place each day, and more than 16 million occur annually around the world. The lightning from these storms strikes Earth 100 times each second (**Figure 14.17A**). Annually, the United States experiences about 100,000 thunderstorms and millions of lightning strikes. A glance at **Figure 14.17B** shows that thunderstorms

Did You Know?

Since the late 1980s, it has been possible to detect cloud-to-ground lightning flashes in real time across the country. Since 1989, the National Lightning Detection Network has recorded an average of about 25 million cloud-to-ground flashes per year in the contiguous 48 states. In addition, about half of all flashes have more than one ground-strike point so that more than 40 million points on the ground are struck in an average year. Besides cloud-to-ground flashes, there are roughly 5 to 10 times as many flashes within clouds.

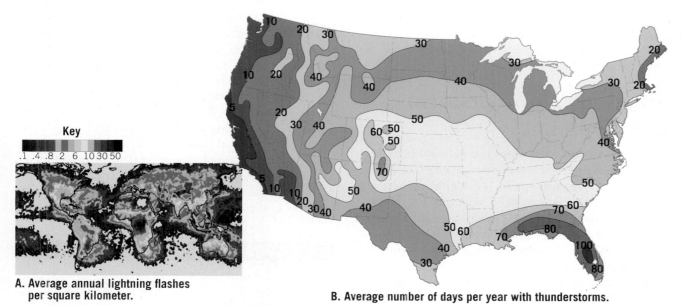

Key

.1 .4 .8 2 6 10 30 50

A. Average annual lightning flashes per square kilometer.

B. Average number of days per year with thunderstorms.

Figure 14.17 Occurrence of lightning and thunderstorms A. Data from space-based optical sensors show the worldwide distribution of lightning, with color variations indicating the average annual number of lightning flashes per square kilometer. The map includes data obtained from April 1995 to March 2000 from NASA's Optical Transient Detector and from December 1997 to November 2000 from NASA's Lightning Imaging Sensor. Both of these satellite-based sensors use high-speed cameras capable of detecting brief lightning flashes even under daytime conditions. (NASA) **B.** Average number of days each year with thunderstorms. The humid subtropical climate that dominates the southeastern United States receives much of its precipitation in the form of thunderstorms. Most of the Southeast averages 50 or more days each year with thunderstorms. (Environmental Data Service, NOAA)

Figure 14.18 Cumulus development A. Buoyant thermals often produce fair-weather cumulus clouds that soon evaporate into the surrounding air, making it more humid. As this process of cumulus development and evaporation continues, the air eventually becomes sufficiently humid so that newly forming clouds do not evaporate but continue to grow. (Photo by Henry Lansford/Photo Researchers, Inc.) **B.** This developing cumulonimbus cloud became a towering August thunderstorm over central Illinois. (Photo by E. J. Tarbuck)

are most frequent in Florida and the eastern Gulf coast region, where such activity is recorded between 70 and 100 days each year. The region on the eastern side of the Rockies in Colorado and New Mexico is next, with thunderstorms occurring on 60 to 70 days each year. Most of the rest of the nation experiences thunderstorms on 30 to 50 days annually. The western margin of the United States has little thunderstorm activity. The same is true for the northern tier of states and for Canada, where warm, moist, unstable mT air seldom penetrates.

Stages of Thunderstorm Development

All thunderstorms require warm, moist air, which, when lifted, releases sufficient latent heat to provide the buoyancy necessary to maintain its upward flight. This instability and associated buoyancy are triggered by a number of different processes, yet most thunderstorms have a similar life history.

Because instability and buoyancy are enhanced by high surface temperatures, thunderstorms are most common in the afternoon and early evening (**Figure 14.18A**). However, surface heating alone is not sufficient for the growth of towering cumulonimbus clouds. A solitary cell of rising hot air produced by surface heating could, at best, produce a small cumulus cloud, which would evaporate within 10 to 15 minutes.

The development of 12,000-meter (40,000-foot) (or, on rare occasions, 18,000-meter [60,000-foot]) cumulonimbus towers requires a continual supply of moist air (**Figure 14.18B**). Each new surge of warm air

rises higher than the last, adding to the height of the cloud (**Figure 14.19**). These updrafts occasionally reach speeds greater than 100 kilometers (60 miles) per hour based on the size of hailstones they are capable of carrying upward. Usually within an hour, the amount and size of precipitation that has accumulated is too much for the updrafts to support, and consequently downdrafts develop in one part of the cloud, releasing heavy precipitation. This represents the most active stage of the thunderstorm. Gusty winds, lightning, heavy precipitation, and sometimes hail are experienced.

Eventually the warm, moist air supplied by updrafts ceases as downdrafts dominate throughout the cloud. The cooling effect of falling precipitation, coupled with the influx of colder air aloft, marks the end of the thunderstorm activity. The life span of a typical cumulonimbus cell within a thunderstorm complex is only about an hour, but as the storm moves, fresh supplies of warm, water-laden air generate new cells to replace those that are dissipating.

14.4 CONCEPT CHECKS

1. Briefly compare and contrast midlatitude cyclones, hurricanes, and tornadoes. How are thunderstorms related to each?

2. What are the basic requirements for the formation of a thunderstorm?

3. Where are thunderstorms most common on Earth? In the United States?

4. Summarize the stages in the development of a thunderstorm.

During the cumulus stage, strong updrafts provide moisture that condenses and builds the cloud.

The mature stage is marked by heavy precipitation. Updrafts exist side by side with downdrafts and continue to enlarge the cloud.

When updrafts disappear, precipitation becomes light and then stops. Without a supply of moisture from updrafts, the cloud evaporates.

SmartFigure 14.19 Thunderstorm development Once a cloud passes beyond the freezing level, the Bergeron process begins producing precipitation. Eventually, the accumulation of precipitation in the cloud is too great for the updraft to support. The falling precipitation causes drag on the air and initiates a downdraft. Once downdrafts dominate, rainfall diminishes, and the cloud starts to dissipate.

14.5 Tornadoes

Summarize the atmospheric conditions and locations that are favorable to the formation of tornadoes. Discuss tornado destruction and tornado forecasting.

Tornadoes are local storms of short duration that rank high among nature's most destructive forces (**Figure 14.20**). Their sporadic occurrence and violent winds cause many deaths each year. The nearly total destruction in some stricken areas has led many to liken their passage to bombing raids during war.

During the very stormy spring of 2011, there were 753 confirmed tornadoes during April, setting a record for the number of tornadoes in a single month. The deadliest and most destructive outbreak occurred between April 25 and 28, when 326 confirmed tornadoes struck, mostly in the South. The loss of life and property were extraordinary. Estimated fatalities numbered between 350 and 400, and damages were in the billions of dollars. The tornado that struck Tuscaloosa, Alabama, was especially notable for the death and destruction it caused. A few weeks later, on May 22, another outbreak struck the Midwest. Joplin, Missouri, was in the direct path of a storm that took more than 150 lives (**Figure 14.21**).

Tornadoes, sometimes called *twisters* or *cyclones*, are violent windstorms that take the form of a rotating column of air, or *vortex*. Pressures within some tornadoes have been estimated to be as much as 10 percent lower than immediately outside the storm. Drawn by the much

Figure 14.20 A violent tornado near Manchester, South Dakota A tornado is a violently rotating column of air in contact with the ground. The air column is visible when it contains condensation or when it contains dust and debris. Often the appearance is a result of both. When the column of air is aloft and does not produce damage, the visible portion is properly called a *funnel cloud*. (Photo by Carsten Peter/ National Geographic Stock)

Figure 14.21 Tornado destruction at Joplin, Missouri A devastating multivortex tornado rated EF-5 struck this community in southwestern Missouri on May 22, 2011. There were more than 150 tornado deaths and nearly 1000 injured. (Photo by c51/ZUMA Press/Newscom)

lower pressure in the center of the vortex, air near the ground rushes into the tornado from all directions. As the air streams inward, it is spiraled upward around the core until it eventually merges with the airflow of the parent thunderstorm deep in the cumulonimbus tower. Because of the tremendous pressure gradient associated with a strong tornado, maximum winds can sometimes approach 480 kilometers (300 miles) per hour.

A tornado may consist of a single vortex, but within many stronger tornadoes are smaller whirls called *suction vortices* that rotate within the main vortex (**Figure 14.22**). Suction vortices have diameters of only about 10 meters (33 feet) and rotate very rapidly. This structure accounts for occasional observations of virtually total destruction of one building while another one, just 10 meters (33 feet) away, suffers little damage.

Figure 14.22 Multiple-vortex tornado Some tornadoes have multiple suction vortices. These small and very intense vortices are roughly 10 meters (30 feet) across and move in a counterclockwise path around the tornado center. Because of this multiple-vortex structure, one building might be heavily damaged and another one, just 10 meters away, might suffer little damage.

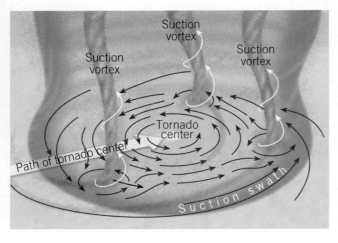

Tornado Occurrence and Development

Tornadoes form in association with severe thunderstorms that produce high winds, heavy (sometimes torrential) rainfall, and often damaging hail. Fortunately, fewer than 1 percent of all thunderstorms produce tornadoes. Nevertheless, a much higher number of thunderstorms must be monitored as potential tornado producers. A tornado is the product of the interaction between strong updrafts in a thunderstorm and the winds in the troposphere.

Tornadoes can form in any situation that produces severe weather, including cold fronts and tropical cyclones (hurricanes). The most intense tornadoes are usually those that form in association with huge thunderstorms called *supercells*. An important precondition linked to tornado formation in severe thunderstorms is the development of a *mesocyclone*—a vertical cylinder of rotating air, typically about 3 to 10 kilometers (2 to 6 miles) across, that develops in the updraft of a severe thunderstorm (**Figure 14.23**). The formation of this large vortex often precedes tornado formation by 30 minutes or so.

The formation of a mesocyclone does not necessarily mean that tornado formation will follow. Only about half of all mesocyclones produce tornadoes. Forecasters cannot determine in advance which mesocyclones will spawn tornadoes.

General Atmospheric Conditions Severe thunderstorms —and, hence, tornadoes—are most often spawned along the cold front of a midlatitude cyclone or in association with a supercell thunderstorm such as the one pictured in Figure 14.23D. Throughout spring, air masses associated with middle-latitude cyclones are most likely to have greatly contrasting conditions. Continental polar air from Canada may still be very cold and dry, whereas maritime tropical air from the Gulf of Mexico is warm, humid, and unstable. The greater the contrast when these air masses meet, the more intense the storm. The two contrasting air masses are most likely to meet in the central United States because there is no significant natural barrier separating the center of the country from the Arctic or the Gulf of Mexico. Consequently, this region generates more tornadoes than any other area of the country or, in fact, the world. The map in **Figure 14.24**, which depicts tornado incidence in the United States for a 27-year period, readily substantiates this fact.

Tornado Climatology An average of 1352 tornadoes were reported annually in the United States between 2003 and 2012. Still, the actual number that occurs from one year to the next varies greatly. During this 10-year span, for example, yearly totals ranged from a low of 1043 in 2012 to a high of 1820 in 2004.

Tornadoes occur during every month of the year. April through June is the period of greatest tornado frequency in the United States; the number is lowest during December and January (see Figure 14.24, graph inset).

Spinning along horizontal axis

Stronger winds

Weaker winds

A.

Thunderstorm forming

Updraft

B.

Clouds overshoot top of thunderstorm

Anvil

Mesocyclone (3 to 10 km diameter)

Tornado

Air inflow

C.

D.

SmartFigure 14.23 The formation of a mesocyclone often precedes tornado formation A. Winds are stronger aloft than at the surface (called speed wind shear), producing a rolling motion about a horizontal axis. **B.** Strong thunderstorm updrafts tilt the horizontally rotating air to a nearly vertical alignment. **C.** The mesocyclone, a vertical cylinder of rotating air, is established. **D.** If a tornado develops, it will descend from a slowly rotating wall cloud in the lower portion of the mesocyclone. (Photo by Corbis Premium RF/Alamy)

Of the 40,522 confirmed tornadoes reported over the contiguous 48 states during the 50-year period from 1950 through 1999, an average of almost 6 per day occurred during May. At the other extreme, a tornado was reported only about every other day in December and January.

Profile of a Tornado An average tornado has a diameter of between 150 and 600 meters (500 and 2000 feet), travels across the landscape at approximately 45 kilometers (30 miles) per hour, and cuts a path about 10 kilometers (6 miles) long.° Because tornadoes usually occur slightly ahead of a cold front, in the zone of southwest winds, most move toward the northeast.

Of the hundreds of tornadoes reported in the United States annually, more than half are comparatively weak and short-lived. Most of these small tornadoes have lifetimes of 3 minutes or less and paths that seldom exceed 1 kilometer (0.6 mile) in length and 100 meters (330 feet) in width. Typical wind speeds are on the order of 150 kilometers

° The 10-kilometer (6-mile) figure applies to documented tornadoes. Because many small tornadoes may go undocumented, the real average path of all tornadoes is unknown but is shorter than 10 kilometers.

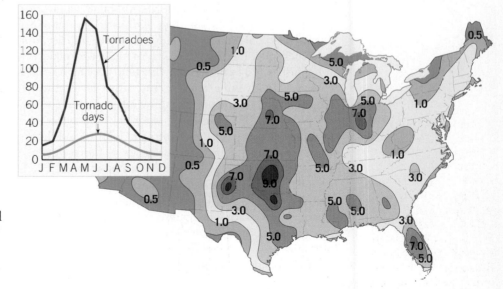

Figure 14.24 Tornado occurrence The map shows average annual tornado incidence per 26,000 square kilometers (10,000 square miles) for a 27-year period. The graph shows the average number of tornadoes and tornado days each month in the United States for the same period.

(90 miles) per hour or less. On the other end of the tornado spectrum are the infrequent and often long-lived violent tornadoes. Although large tornadoes constitute only a small percentage of the total reported, their effects are often devastating. Such tornadoes may exist for periods in excess of 3 hours and produce an essentially continuous damage path more than 150 kilometers (90 miles) long and perhaps a kilometer or more wide. Maximum winds range beyond 500 kilometers (310 miles) per hour.

Tornado Destruction and Loss of Life

The potential for tornado destruction depends largely on the strength of the winds generated by the storm. Because tornadoes generate the strongest winds in nature, they have accomplished many seemingly impossible tasks, such as driving a piece of straw through a thick wooden plank and uprooting huge trees. Although it may seem impossible for winds to cause some of the extensive damage attributed to tornadoes, tests in engineering facilities have repeatedly demonstrated that winds in excess of 320 kilometers (200 miles) per hour are capable of incredible feats (**Figure 14.25**).

Most tornado losses are associated with a few storms that strike urban areas or devastate entire small communities. The amount of destruction caused by such storms depends to a significant degree (but not completely) on the strength of the winds. A wide spectrum of tornado strengths, sizes, and lifetimes are observed. The commonly used guide to tornado intensity is the **Enhanced Fujita intensity scale**, or **EF-scale** for short (**Table 14.1**). Because tornado winds cannot be measured directly, a rating on the EF-scale is determined by assessing the worst damage produced by a storm. Although widely used, the EF-scale is not perfect. Estimating tornado intensity based on damage alone does not take into account the structural integrity of the objects hit by a tornado. A well-constructed building can withstand very high winds, whereas a poorly built structure can suffer devastating damage from the same or weaker winds.

Although the greatest part of tornado damage is caused by violent winds, most tornado injuries and deaths result from flying debris. The proportion of tornadoes that result in the loss of life is small. In most years, slightly fewer than 2 percent of all reported tornadoes in the United States are "killers." Although the percentage of tornadoes resulting in death is small, each tornado is potentially lethal. When tornado fatalities and storm intensities are compared, the results are quite interesting: The majority (63 percent) of tornadoes are weak (EF-0 and EF-1), and the number of storms decreases as tornado intensity increases. The distribution of tornado fatalities, however, is just the opposite. Although only 2 percent of tornadoes are classified as violent (EF-4 and EF-5), they account for nearly 70 percent of the deaths.

A.

B.

Figure 14.25 Tornado winds: the strongest in nature A. The force of the wind during a tornado near Wichita, Kansas, in April 1991 was enough to drive this piece of metal into a utility pole. (Photo by John Sokich/NOAA) **B.** The remains of a truck wrapped around a tree in Bridge Creek, Oklahoma, on May 4, 1999, following a major tornado outbreak. (AP Photo/L. M. Otero)

Tornado Forecasting

Because severe thunderstorms and tornadoes are small and relatively short-lived phenomena, they are among the most difficult weather features to forecast precisely. Nevertheless, the prediction, detection, and monitoring of such storms are among the most important services provided by professional meteorologists. Both the timely issuance and dissemination of watches and warnings are critical to the protection of life and property.

The Storm Prediction Center (SPC) located in Norman, Oklahoma, is part of the National Weather Service (NWS) and the National Centers for Environmental

TABLE 14.1   Enhanced Fujita Intensity Scale*

Scale	Wind Speed		Damage
	Km/Hr	Mi/Hr	
EF-0	105–137	65–85	*Light.* Some damage to siding and shingles.
EF-1	138–177	86–110	*Moderate.* Considerable roof damage. Winds can uproot trees and overturn single-wide mobile homes. Flagpoles bend.
EF-2	178–217	111–135	*Considerable.* Most single-wide homes destroyed. Permanent homes can shift off foundations. Flagpoles collapse. Softwood trees debarked.
EF-3	218–265	136–165	*Severe.* Hardwood trees debarked. All but small portions of houses destroyed.
EF-4	266–322	166–200	*Devastating.* Complete destruction of well-built residences, large sections of school buildings.
EF-5	>322	>200	*Incredible.* Significant structural deformation of mid- and high-rise buildings.

*The original Fujita scale was developed by T. Theodore Fujita in 1971 and put into use in 1973. The Enhanced Fujita intensity scale is a revision that was put into use in February 2007. Wind speeds are estimates (not measurements) based on damage, and represent 3-second gusts at the point of damage.

Prediction (NCEP). The mission of the SPC is to provide timely and accurate forecasts and watches for severe thunderstorms and tornadoes.

Severe thunderstorm outlooks are issued several times daily. *Day 1* outlooks identify the areas that are likely to be affected by severe thunderstorms during the next 6 to 30 hours, and *day 2* outlooks extend the forecast through the following day. Both outlooks describe the type, coverage, and intensity of the severe weather expected. Many local NWS field offices also issue severe weather outlooks that provide more local descriptions of the severe weather potential for the next 12 to 24 hours.

Tornado Watches and Warnings   **Tornado watches**
alert the public to the possibility of tornadoes over a specified area for a particular time interval. Watches serve to fine-tune forecast areas already identified in severe weather outlooks. A typical watch covers an area of about 65,000 square kilometers (25,000 square miles) for a 4- to 6-hour period. A tornado watch is an important part of the tornado alert system because it sets in motion the procedures necessary to deal adequately with detection, tracking, warning, and response. Watches are generally reserved for organized severe weather events where the tornado threat will affect at least 26,000 square kilometers (10,000 square miles) and/or persist for at least 3 hours. Watches typically are not issued when the threat is thought to be isolated and/or short-lived.

Whereas a tornado watch is designed to alert people to the possibility of tornadoes, a **tornado warning** is issued by local offices of the National Weather Service when a tornado has actually been sighted in an area or is indicated by weather radar. It warns of a high probability of imminent danger. Warnings are issued for much smaller areas than for watches, usually covering portions of a county or counties. In addition, they are in effect for much shorter periods, typically 30 to 60 minutes. Because a tornado warning may be based on an actual sighting, warnings are occasionally issued after a tornado has already developed. However, most warnings are issued prior to tornado formation, sometimes by several tens of minutes, based on Doppler radar data and/or spotter reports of funnel clouds.

If the direction and the approximate speed of the storm are known, an estimate of its most probable path can be made. Because tornadoes often move erratically, the warning area is fan-shaped downwind from the point where the tornado has been spotted. Improved forecasts and advances in technology have contributed to a significant decline in tornado deaths over the past 50 years.

Doppler Radar   Many of the difficulties that once limited the accuracy of tornado warnings have been reduced or eliminated by an advancement in radar technology called **Doppler radar**. Doppler radar not only performs the same tasks as conventional radar but also has the ability to detect motion directly (**Figure 14.26**). Doppler radar can detect the initial formation and subsequent development of a *mesocyclone*, the intense rotating wind system in the lower part of a thunderstorm that frequently precedes tornado development. Almost all

Reflectivity   Storm-relative velocity

Figure 14.26 Doppler radar This is a dual Doppler radar image of a violent tornado near Moore, Oklahoma, on May 3, 1999. The left image (reflectivity) shows precipitation in the supercell thunderstorm. The right image shows motion of the precipitation along the radar beam—that is, how fast rain or hail is moving toward or away from the radar. In this example, the radar was unusually close to the tornado—close enough to make out the signature of the tornado itself. (Most of the time only the weaker and larger mesocyclone is detected.) (After NOAA) mesocyclones produce damaging hail, severe winds, or tornadoes. Those that produce tornadoes (about 50 percent) can sometimes be distinguished by their stronger wind speeds and their sharper gradients of wind speeds.

It should also be pointed out that not all tornado-bearing storms have clear-cut radar signatures and that other storms can give false signatures. Detection, therefore, is sometimes a subjective process, and a given display could be interpreted in several ways. Consequently, trained observers continue to form an important part of the warning system.

The benefits of Doppler radar are many. As a research tool, it is not only providing data on the formation of tornadoes but also helping meteorologists gain new insights into thunderstorm development, the structure and dynamics of hurricanes, and air-turbulence hazards that plague aircraft. As a practical tool for tornado detection, Doppler radar has significantly improved our ability to track thunderstorms and issue warnings.

14.5 CONCEPT CHECKS

1. Why do tornadoes have such high wind speeds?
2. What general atmospheric conditions are most conducive to the formation of tornadoes?
3. During what months is tornado activity most pronounced in the United States?
4. Name the scale commonly used to rate tornado intensity. How is a rating on this scale determined?
5. Distinguish between a tornado watch and a tornado warning.

14.6 Hurricanes

Identify areas of hurricane formation on a world map and discuss the conditions that promote hurricane formation. List the three broad categories of hurricane destruction.

Most of us view the weather in the tropics with favor. Places such as the islands of the Caribbean are known for their lack of significant day-to-day variations. Warm breezes, steady temperatures, and rains that come as heavy but brief tropical showers are often the rule. It is ironic that these relatively tranquil regions sometimes produce the most violent storms on Earth.

Hurricanes are intense centers of low pressure that form over tropical oceans and are characterized by intense convective (thunderstorm) activity and strong cyclonic circulation (**Figure 14.27**). Sustained winds must equal or exceed 119 kilometers (74 miles) per hour. Unlike midlatitude cyclones, hurricanes lack contrasting air masses and fronts. Rather, the source of energy that produces and maintains hurricane-force winds is the huge quantity of latent heat liberated during the formation of the storm's cumulonimbus towers.

The vast majority of hurricane-related deaths and damage are caused by relatively infrequent, yet powerful, storms. Hurricane Sandy, shown in the chapter-opening photo, devastated coastal areas of New Jersey, New York, and Connecticut and caused billions of dollars in damages in late October 2012. The storm that pounded an unsuspecting Galveston, Texas, in 1900 was not just the deadliest U.S. hurricane ever but the deadliest natural disaster of *any kind* to affect the United States. The deadliest and most costly storm in recent memory occurred in August 2005, when Hurricane Katrina devastated the Gulf Coast of Louisiana, Mississippi, and Alabama and took an estimated 1800 lives. Although hundreds of thousands fled before the storm made landfall, thousands of others were caught by the storm. In addition to the human suffering and tragic loss of life that were left in the wake of Hurricane Katrina, the financial losses caused by the storm are practically incalculable.

The eye wall surrounds the eye and is the most intense part of the storm

Well-developed eye

100 km
N

Figure 14.27 Super Typhoon Jangmi In the western Pacific, hurricanes are called typhoons. This storm struck portions of Taiwan, China, and Japan in late September 2008. It was the strongest storm worldwide that year, with sustained winds that reached 270 kilometers (165 miles) per hour. The counterclockwise spiral of the clouds indicates that it is a Northern Hemisphere storm. (NASA)

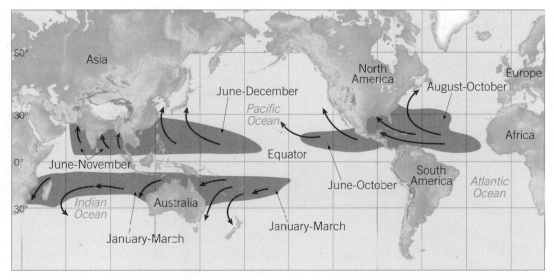

Figure 14.28 Regions where hurricanes form This world map shows the regions where most hurricanes form as well as their principal months of occurrence and the most common tracks they follow. Hurricanes do not develop within about 5° of the equator because the Coriolis effect is too weak. Because warm surface ocean temperatures are necessary for hurricane formation, they seldom form poleward of 20° latitude nor over the cool waters of the South Atlantic and the eastern South Pacific.

Profile of a Hurricane

Most hurricanes form between the latitudes of 5° and 20° over all the tropical oceans except the South Atlantic and the eastern South Pacific (**Figure 14.28**). The North Pacific has the greatest number of storms, averaging 20 each year. Fortunately for those living in the coastal regions of the southern and eastern United States, fewer than five hurricanes, on the average, develop annually in the warm sector of the North Atlantic.

These intense tropical storms are known in various parts of the world by different names. In the western Pacific, they are called *typhoons*, and in the Indian Ocean, including the Bay of Bengal and Arabian Sea, they are simply called *cyclones*. In the following discussion, these storms will be referred to as hurricanes. The term *hurricane* is derived from Huracan, a Carib god of evil.

Although many tropical disturbances develop each year, only a few reach hurricane status. By international agreement, a hurricane has wind speeds in excess of 119 kilometers (74 miles) per hour and a rotary circulation. Mature hurricanes average 600 kilometers (375 miles) across, although they can range in diameter from 100 kilometers (60 miles) up to about 1500 kilometers (930 miles). From the outer edge to the center, the barometric pressure has on occasion dropped 60 millibars, from 1010 millibars to 950 millibars. The lowest pressures ever recorded in the Western Hemisphere are associated with these storms.

A steep pressure gradient generates the rapid, inward-spiraling winds of a hurricane (**Figure 14.29**). As the air rushes toward the center of the storm, its velocity increases. This occurs for the same reason that skaters with their arms extended spin faster as they pull their arms in close to their bodies.

As the inward rush of warm, moist surface air approaches the core of the storm, it turns upward and ascends in a ring of cumulonimbus towers. This doughnut-shaped wall of intense convective activity surrounding the center of the storm is called the **eye wall**. It is here that the greatest wind speeds and heaviest rainfall occur. Surrounding the eye wall are curved bands of clouds that trail away in a spiral fashion. Near the top of the hurricane, the airflow is outward, carrying the rising air away from the storm center, thereby providing room for more inward flow at the surface.

At the very center of the storm is the **eye** of the hurricane. This well-known feature is a zone about 20 kilometers (12.5 miles) in diameter where precipitation ceases and winds subside. It offers a brief but deceptive break from the extreme weather in the enormous curving wall clouds that surround it. The air within the eye gradually descends and heats by compression, making it the warmest part of the storm. Although many people believe that the eye is characterized by clear blue skies, this is usually not the case because the subsidence in the eye is seldom strong enough to produce cloudless conditions. Although the sky appears much brighter in this region, scattered clouds at various levels are common.

Outflow of air at the top of the hurricane is important because it prevents the convergent flow at lower levels from "filling in" the storm.

Eye

Sinking air in the eye warms by compression.

Eye wall, the zone where winds and rain are most intense.

Tropical moisture spiraling inward creates rain bands that pinwheel around the storm center.

Cross Section of a Hurricane. Note that the vertical dimension is greatly exaggerated. (After NOAA)

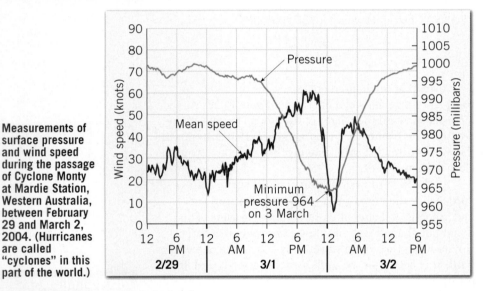

Measurements of surface pressure and wind speed during the passage of Cyclone Monty at Mardie Station, Western Australia, between February 29 and March 2, 2004. (Hurricanes are called "cyclones" in this part of the world.)

Pressure

Mean speed

Minimum pressure 964 on 3 March

Figure 14.29 Cross section of a hurricane

Hurricane Formation and Decay

A hurricane is a heat engine that is fueled by the latent heat liberated when huge quantities of water vapor condense. The amount of energy produced by a typical hurricane in just a single day is truly immense. The release of latent heat warms the air and provides buoyancy for its upward flight. The result is to reduce the pressure near the surface, which encourages a more rapid inward flow of air. To get this engine started, a large quantity of warm, moisture-laden air is required, and a continual supply is needed to keep it going.

Hurricane Formation Hurricanes develop most often in the late summer, when ocean waters have reached

temperatures of 27°C (80°F) or higher and thus are able to provide the necessary heat and moisture to the air (**Figure 14.30**). This ocean-water temperature requirement accounts for the fact that hurricanes do not form over the relatively cool waters of the South Atlantic and the eastern South Pacific. For the same reason, few hurricanes form poleward of 20° latitude. Although water temperatures are sufficiently high, hurricanes do not form within 5° of the equator because the Coriolis effect is too weak to initiate the necessary rotary motion.

Many tropical storms begin as disorganized arrays of clouds and thunderstorms that develop weak pressure gradients but exhibit little or no rotation. Such areas of low-level convergence and lifting are called *tropical disturbances.* Most of the time, these zones of convective activity die out. However, tropical disturbances occasionally grow larger and develop a strong cyclonic rotation.

What happens on occasions when conditions favor hurricane development? As latent heat is released from the clusters of thunderstorms that make up the tropical disturbance, areas within the disturbance get warmer. As a result, air density lowers and surface pressure drops, creating a region of weak low pressure and cyclonic circulation. As pressure drops at the storm center, the pressure gradient steepens. If you were watching an animated weather map of the storm, you would see the isobars get closer together. In response, surface wind speeds increase and bring additional supplies of moisture to nurture storm growth. The water vapor condenses, releasing latent heat, and the heated air rises. Adiabatic cooling of rising air triggers more condensation and the release

Sea Surface Temperature (°C)

−2   16.5   27.8   35

Figure 14.30 Sea-surface temperatures Among the necessary ingredients for a hurricane is warm ocean temperatures above 27°C (80°F). This color-coded satellite image from June 1, 2010, shows sea-surface temperatures at the beginning of hurricane season. (NASA)

of more latent heat, which causes a further increase in buoyancy. And so it goes.

Meanwhile, at the top of the storm, air is diverging. Without this outward flow up top, the inflow at lower levels would soon raise surface pressures (that is, fill in the low) and thwart storm development.

Other Tropical Storms Many tropical disturbances occur each year, but only a few develop into full-fledged hurricanes. By international agreement, lesser tropical cyclones are placed in different categories, based on wind strength. When a cyclone's strongest winds do not exceed 61 kilometers (38 miles) per hour, it is called a **tropical depression**. When winds are between 61 and 119 kilometers (38 and 74 miles) per hour, the cyclone is termed a **tropical storm**. It is during this phase that a name is given (Andrew, Katrina, Sandy, etc.). If the tropical storm becomes a hurricane, the name remains the same. Each year, between 80 and 100 tropical storms develop around the world. Of these, usually half or more eventually become hurricanes.

Hurricane Decay Hurricanes diminish in intensity whenever they (1) move over ocean waters that cannot supply warm, moist tropical air; (2) move onto land; or (3) reach a location where the large-scale flow aloft is unfavorable. When a hurricane moves onto land, it loses its punch rapidly. The most important reason for this rapid demise is the fact that the storm's source of warm, moist air is cut off. When an adequate supply of water vapor does not exist, condensation and the release of latent heat must diminish. In addition, friction from the increased roughness of the land surface rapidly slows surface wind speeds. This factor causes the winds to move more directly into the center of the low, thus helping to eliminate the large pressure differences.

Hurricane Destruction

A location only a few hundred kilometers from a hurricane—just 1 day's striking distance away—may experience clear skies and virtually no wind. Prior to the age of weather satellites, this situation made the task of warning people of impending storms very difficult.

The amount of damage caused by a hurricane depends on several factors, including the size and population density of the area affected and the shape of the ocean bottom near the shore. The most significant factor, of course, is the strength of the storm itself. By studying past storms, a scale has been established to rank the relative intensities of hurricanes. As **Table 14.2** indicates, a *category 5* storm is the worst possible, whereas a *category 1* hurricane is least severe.

During hurricane season, it is common to hear scientists and reporters use the numbers from the

TABLE 14.2	Saffir–Simpson Hurricane Scale				
Scale Number (category)	Central Pressure (millibars)	Winds (km/hr)	Storm Surge (meters)	Damage	
1	≥980	119–153	1.2–1.5	Minimal	
2	955–979	154–177	1.6–2.4	Moderate	
3	945–964	178–209	2.5–3.6	Extensive	
4	920–944	210–250	3.7–5.4	Extreme	
5	<920	>250	>5.4	Catastrophic	

Saffir–Simpson hurricane scale. When Hurricane Katrina made landfall, sustained winds were 225 kilometers (140 miles) per hour, making it a strong category 4 storm. Storms that fall into category 5 are rare. Damage caused by hurricanes can be divided into three categories: (1) storm surge, (2) wind damage, and (3) heavy rains and inland flooding.

Storm Surge The most devastating damage in the coastal zone is usually caused by storm surge (**Figure 14.31**). It not only accounts for a large share of coastal property losses but also is responsible for a high percentage of all hurricane-caused deaths. A **storm surge** is a dome

Figure 14.31 Storm surge destruction This is Crystal Beach, Texas, on September 16, 2008, 3 days after Hurricane Ike came ashore. At landfall the storm had sustained winds of 155 kilometers (105 miles) per hour. The extraordinary storm surge caused most of the damage shown here. (Photo by Earl Nottingham/Associated Press)

of water 65 to 80 kilometers (40 to 50 miles) wide that sweeps across the coast near the point where the eye makes landfall. If all wave activity were smoothed out, the storm surge would be the height of the water above normal tide level. In addition, tremendous wave activity is superimposed on the surge. The worst surges occur in places like the Gulf of Mexico, where the continental shelf is very shallow and gently sloping. In addition, local features such as bays and rivers can cause the surge height to double and increase in speed.

As a hurricane advances toward the coast in the Northern Hemisphere, storm surge is always most intense on the right side of the eye (viewed from the ocean), where winds are blowing *toward* the shore. In addition, on this side of the storm, the forward movement of the hurricane contributes to the storm surge. In **Figure 14.32**, assume that a hurricane with peak winds of 175 kilometers (109 miles) per hour is moving toward the shore at 50 kilometers (31 miles) per hour. In this case, the net wind speed on the right side of the advancing storm is 225 kilometers (140 miles) per hour. On the left side, the hurricane's winds are blowing opposite the direction of storm movement, so the net winds are *away* from the coast at 125 kilometers (78 miles) per hour.

Figure 14.32 An approaching hurricane Winds associated with a Northern Hemisphere hurricane that is advancing toward the coast. This hypothetical storm, with peak winds of 175 kilometers (109 miles) per hour, is moving toward the coast at 50 kilometers (31 miles) per hour. On the right side of the advancing storm (as viewed from the ocean), the 175-kilometer-per-hour winds are in the same direction as the movement of the storm (50 kilometers per hour). Therefore, the *net* wind speed on the right side of the storm is 225 kilometers (140 miles) per hour. On the left side, the hurricane's winds are blowing opposite the direction of storm movement, so the *net* winds of 125 kilometers (78 miles) per hour are away from the coast. Storm surge will be greatest along that part of the coast hit by the right side of the advancing hurricane.

Along the shore facing the left side of the oncoming hurricane, the water level may actually decrease as the storm makes landfall.

Wind Damage Destruction caused by wind is perhaps the most obvious of the classes of hurricane damage. Debris such as signs, roofing materials, and small items left outside become dangerous flying missiles in hurricanes. For some structures, the force of the wind is sufficient to cause total ruin. Mobile homes are particularly vulnerable. High-rise buildings are also susceptible to hurricane-force winds. Upper floors are most vulnerable because wind speeds usually increase with height. Recent research suggests that people should stay below the 10th floor but remain above any floors at risk for flooding. In regions with good building codes, wind damage is usually not as catastrophic as storm-surge damage. However, hurricane-force winds affect a much larger area than storm surge and can cause huge economic losses. For example, in 1992 it was largely the winds associated with Hurricane Andrew that produced more than $25 billion of damage in southern Florida and Louisiana.

Hurricanes sometimes produce tornadoes that contribute to the storm's destructive power. Studies have shown that more than half of the hurricanes that make landfall produce at least one tornado. In 2004 the number of tornadoes associated with tropical storms and hurricanes was extraordinary. Tropical Storm Bonnie and five landfalling hurricanes—Charley, Frances, Gaston, Ivan, and Jeanne—produced nearly 300 tornadoes that affected the southeastern and mid-Atlantic states.

Heavy Rains and Inland Flooding The torrential rains that accompany most hurricanes represent a third significant threat: flooding. Whereas the effects of storm surge and strong winds are concentrated in coastal areas, heavy rains may affect places hundreds of kilometers from the coast for up to several days after the storm has lost its hurricane-force winds.

In September 1999, Hurricane Floyd brought flooding rains, high winds, and rough seas to a large portion of the Atlantic seaboard. More than 2.5 million people evacuated their homes from Florida north to the Carolinas and beyond. It was the largest peacetime evacuation in U.S. history up to that time. Torrential rains falling on already saturated ground created devastating inland flooding. Altogether Floyd dumped more than 48 centimeters (19 inches) of rain on Wilmington, North Carolina, 33.98 centimeters (13.38 inches) in a single 24-hour span.

Tracking Hurricanes

Today we have the benefit of numerous observational tools for tracking tropical storms and hurricanes. Using input from satellites, aircraft reconnaissance, coastal radar, and remote data buoys in conjunction with sophisticated computer models, meteorologists monitor and forecast storm movements and intensity. The goal is to issue timely watches and warnings.

An important part of this process is the *track forecast*—the predicted path of the storm. The track forecast is probably the most basic information because accurate prediction of other storm characteristics (winds and rainfall) is of little value if there is significant uncertainty about where the storm is going. Accurate track forecasts are important because they can lead to timely evacuations from the surge zone, where the greatest number of deaths usually occur. Fortunately, track forecasts have been steadily improving. During the span 2001 to 2005, forecast errors were roughly half of what they were in 1990. During the very active 2004 and 2005 Atlantic hurricane seasons, 12- to 72-hour track forecast accuracy was at or near record levels. Consequently, the length of official track forecasts issued by the National Hurricane Center was extended from 3 days to 5 days (**Figure 14.33**). Current 5-day track forecasts are now as accurate as 3-day forecasts were 15 years ago.

Despite improvements in accuracy, forecast uncertainty still requires that hurricane warnings be issued for relatively large coastal areas. During the span 2000 to 2005, the average length of coastline under a hurricane warning in the United States was 510 kilometers (316 miles). This represents a significant improvement over the preceding decade, when the average was 730 kilometers (452 miles). Nevertheless, only about one-quarter of an average warning area experiences hurricane conditions.

14.6 CONCEPT CHECKS

1. Define *hurricane*. What other names are used for this storm?
2. In what latitude zone do hurricanes develop?
3. Distinguish between the eye and the eye wall of a hurricane. How do conditions differ in these zones?
4. What is the source of energy that drives a hurricane?
5. Why do hurricanes *not* form near the equator? Explain the lack of hurricanes in the South Atlantic and eastern South Pacific.
6. When do most hurricanes in the North Atlantic and Caribbean occur? Why are these months favored?
7. Why does the intensity of a hurricane diminish rapidly when it moves over land?
8. What are the three broad categories of hurricane damage?

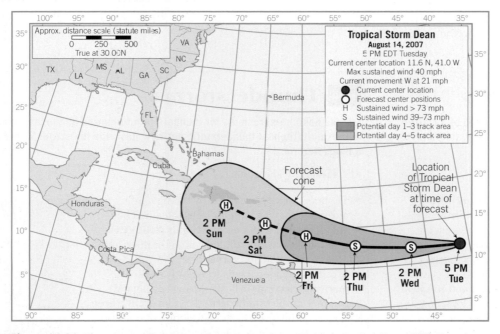

Figure 14.33 Five-day track forecast for Tropical Storm Dean issued at 5 P.M. EDT, Tuesday, August 14, 2007 When a hurricane track forecast is issued by the National Hurricane Center, it is termed a *forecast cone.* The cone represents the probable track of the center of the storm and is formed by enclosing the area swept out by a set of circles along the forecast track (at 12 hours, 24 hours, 36 hours, etc.). The size of each circle gets larger with time. Based on statistics from 2003 to 2007, the entire track of an Atlantic tropical cyclone can be expected to remain entirely within the cone roughly 60 to 70 percent of the time. (National Weather Service/National Hurricane Center)

CONCEPTS IN REVIEW: WEATHER PATTERNS AND SEVERE WEATHER

14.1 Air Masses

Discuss air masses, their classification, and associated weather.

KEY TERMS: air mass, air-mass weather, source region, polar (P) air mass, arctic (A) air mass, tropical (T) air mass, continental (c) air mass, maritime (m) air mass, lake-effect snow

- An *air mass* is a large body of air, usually 1600 kilometers (1000 miles) or more across, that is characterized by a sameness of temperature and moisture at any given altitude. When this air moves out of its region of origin, called the source region, it carries these temperatures and moisture conditions elsewhere, perhaps eventually affecting a large portion of a continent.
- Air masses are classified according to the nature of the surface in the source region and the latitude of the source region. Continental (c) designates an air mass of land origin, with the air likely to be dry; a maritime (m) air mass originates over water and, therefore, will be relatively humid. Polar (P) and arctic (A) air masses originate in high latitudes and are cold. Tropical (T) air masses form in low latitudes and are warm. According to this classification scheme, the four basic types of air masses are continental polar (cP), continental tropical (cT), maritime polar (mP), and maritime tropical (mT).
- Continental polar (cP) and maritime tropical (mT) air masses influence the weather of North America most, especially east of the Rocky Mountains. Maritime tropical air is the source of much, if not most, of the precipitation received in the eastern two-thirds of the United States.

14.2 Fronts

Compare and contrast typical weather associated with a warm front and a cold front. Describe an occluded front and a stationary front.

KEY TERMS: front, overrunning, warm front, cold front, stationary front, occluded front

- Fronts are boundary surfaces that separate air masses of different densities, one usually warmer and more humid than the other. As one air mass moves into another, the warmer, less dense air mass is forced aloft in a process called overrunning.
- Along a warm front, a warm air mass overrides a retreating mass of cooler air. As the warm air ascends, it cools adiabatically to produce clouds and, frequently, light to moderate precipitation over a large area.
- A cold front forms where cold air is actively advancing into a region occupied by warmer air. Cold fronts are about twice as steep as and move more rapidly than do warm fronts. Because of these two differences, precipitation along a cold front is generally more intense and of shorter duration than precipitation associated with a warm front.

? Identify each of these symbols used to designate fronts. On which side of each symbol are the warm air and the cool air?

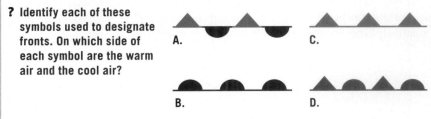

14.3 Midlatitude Cyclones

Summarize the weather associated with the passage of a mature midlatitude cyclone. Describe how airflow aloft is related to cyclones and anticyclones at the surface.

KEY TERM: midlatitude (middle-latitude) cyclone

- The primary weather producers in the middle latitudes are large centers of low pressure that generally travel from west to east, called middle-latitude cyclones. These bearers of stormy weather, which last from a few days to a week, have a counterclockwise circulation pattern in the Northern Hemisphere, with an inward flow of air toward their centers.
- Most middle-latitude cyclones have a cold front and frequently a warm front extending from the central area of low pressure. Convergence and forceful lifting along the fronts initiate cloud development and frequently cause precipitation. The particular weather experienced by an area depends on the path of the cyclone.
- Guided by west-to-east-moving jet streams, cyclones generally move eastward across the United States. Airflow aloft (divergence and convergence) plays an important role in maintaining cyclonic and anticyclonic circulation. In cyclones, divergence aloft supports the inward flow at the surface.

14.4 Thunderstorms

List the basic requirements for thunderstorm formation and locate places on a map that exhibit frequent thunderstorm activity. Describe the stages in the development of a thunderstorm.

KEY TERM: thunderstorm

- Thunderstorms are caused by the upward movement of warm, moist, unstable air. They are associated with cumulonimbus clouds that generate heavy rainfall, lightning, thunder, and occasionally hail and tornadoes.
- Air mass thunderstorms frequently occur in maritime tropical (mT) air during spring and summer in the middle latitudes. Generally, three stages are involved in the development of these storms: the cumulus stage, mature stage, and dissipating stage.

? Which stage in the development of a thunderstorm is shown in this sketch? Describe what is occurring. Is there a stage that follows this one? If so, describe what occurs during that stage.

14.5 Tornadoes

Summarize the atmospheric conditions and locations that are favorable to the formation of tornadoes. Discuss tornado destruction and tornado forecasting.

KEY TERMS: tornado, Enhanced Fujita intensity scale (EF-scale), tornado watch, tornado warning, Doppler radar

- Tornadoes are violent windstorms that take the form of a rotating column of air called a vortex that extends downward from a cumulonimbus cloud. Many strong tornadoes are multiple-vortex storms. Because of the tremendous pressure gradient associated with a strong tornado, maximum winds can approach 480 kilometers (300 miles) per hour.
- Tornadoes are most often spawned along the cold front of a midlatitude cyclone or in association with a supercell thunderstorm. Tornadoes also form in association with tropical cyclones (hurricanes). In the United States, April through June is the period of greatest tornado activity, but tornadoes can occur during any month of the year.
- Most tornado damage is caused by the tremendously strong winds. One commonly used guide to tornado intensity is the Enhanced Fujita intensity scale (EF-scale). A rating on the EF-scale is determined by assessing damages produced by the storm.
- Because severe thunderstorms and tornadoes are small and short-lived phenomena, they are among the most difficult weather features to forecast precisely. When weather conditions favor the formation of tornadoes, a tornado watch is issued. The National Weather Service issues a tornado warning when a tornado has been sighted in an area or is indicated by Doppler radar.

14.6 Hurricanes

Identify areas of hurricane formation on a world map and discuss the conditions that promote hurricane formation. List the three broad categories of hurricane destruction.

KEY TERMS: hurricane, eye wall, eye, tropical depression, tropical storm, Saffir–Simpson hurricane scale, storm surge

- Hurricanes, the greatest storms on Earth, are tropical cyclones with wind speeds in excess of 119 kilometers (74 miles) per hour. These complex tropical disturbances develop over tropical ocean waters and are fueled by the latent heat that is liberated when huge quantities of water vapor condense.
- Hurricanes form most often in late summer, when ocean-surface temperatures reach 27°C (80°F) or higher and thus are able to provide the necessary heat and moisture to the air. Hurricanes diminish in intensity when they move over cool ocean water that cannot supply adequate heat and moisture, move onto land, or reach a location where large-scale flow aloft is unfavorable.
- The Saffir–Simpson scale ranks the relative intensities of hurricanes. A 5 on the scale represents the strongest storm possible, and a 1 indicates the lowest severity. Damage caused by hurricanes is divided into three categories: (1) storm surge, (2) wind damage, and (3) heavy rains and inland flooding.

? This world map shows the tracks and intensities of thousands of hurricanes and other tropical cyclones. It was put together by the National Hurricane Center and the Joint Typhoon Warning Center. (a) What area has experienced the greatest number of category 4 and 5 storms? (b) Why do hurricanes not form in the very heart of the tropics, astride the equator? (c) Explain the absence of storms in the South Atlantic and the eastern South Pacific.

Saffir–Simpson Hurricane Intensity Scale

| Tropical depression | Tropical storm | 1 | 2 | 3 | 4 | 5 |

GIVE IT SOME THOUGHT

1. We know that during the winter, all polar (P) air masses are cold. Which is likely to be colder: a wintertime mP air mass or a wintertime cP air mass? Briefly explain. We expect tropical (T) air masses to be warm, but some are warmer than others. Which should be warmer: a summertime cT air mass or a summertime mT air mass? How did you figure this out?

2. Refer to Figure 14.4 to answer these questions.
   a. Thunder Bay and Marquette are both on the shore of Lake Superior, yet Marquette gets much more snow than Thunder Bay. Why is this the case?
   b. Notice the narrow, north–south zone of relatively heavy snow east of Pittsburgh and Charleston. This area is not affected by the Great Lakes. Speculate on the likely reason for the higher snowfall here. Does your answer explain the shape of this snowy area?

3. If you hear that a cyclone is approaching, should you immediately seek shelter? Why or why not?

4. Refer to the accompanying weather map to answer the following questions:
   a. What is a likely wind direction at each city?
   b. Identify the likely air mass that is influencing each city.
   c. Identify the cold front, warm front, and occluded front.
   d. What is the barometric tendency at city A and city C?
   e. Which one of the three cities is probably coldest? Which one is probably warmest?

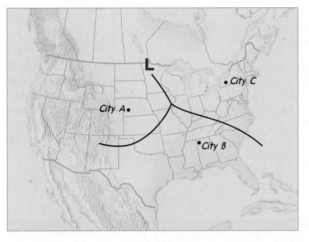

5. The accompanying diagrams show surface temperatures with isotherms labeled in degrees Fahrenheit for noon and 6 P.M. on January 29, 2008. On this day, a powerful front moved through Missouri and Illinois.
   a. What type of front passed through the Midwest?
   b. Describe how the temperature changed in St. Louis, Missouri, over the 6-hour period.
   c. Describe the likely shift in wind direction in St. Louis during this time span.

6. The accompanying table lists the number of tornadoes reported in the United States by decade. Propose a reason to explain why the totals for the 1990s and 2000s are so much higher than for the 1950s and 1960s.

Number of U.S. Tornadoes Reported, by Decade	
Decade	Number of Tornadoes Reported
1950–1959	4796
1960–1969	6613
1970–1979	8579
1980–1989	8196
1990–1999	12,138
2000–2009	12,914

7. The number of tornado deaths in the United States in the 2000s was less than 40 percent the number that occurred in the 1950s, even though there was a significant increase in population. Suggest a likely reason for the decline in the death toll.

8. A television meteorologist is able to inform viewers about the intensity of an approaching hurricane. However, the meteorologist can report the intensity of a tornado only after it has occurred. Why is this the case?

9. Refer to the graph in Figure 14.29. Explain why wind speeds are greatest when the slope of the pressure curve is steepest.

10. Assume that it is late September 2016, and that the eye of Hurricane Gaston, a category 5 storm, is projected to follow the path shown on the accompanying map of Texas. Answer the following questions:
    a. Name the stages of development that Gaston must have gone through to become a hurricane. At what stage did the storm receive its name?
    b. If the storm follows the projected path, will the city of Houston experience Gaston's fastest winds and greatest storm surge? Explain why or why not.
    c. What is the greatest threat to life and property if this storm approaches the Dallas-Fort Worth area?

MasteringGeology™

15

The Nature of the Solar System

FOCUS ON CONCEPTS

Each statement represents the primary learning objective for the corresponding major heading within the chapter. After you complete the chapter, you should be able to:

15.1 Explain the geocentric view of the solar system and describe how it differs from the heliocentric view.

15.2 List and describe the contributions to modern astronomy of Nicolaus Copernicus, Tycho Brahe, Johannes Kepler, Galileo Galilei, and Isaac Newton.

15.3 Describe the formation of the solar system according to the nebular theory. Compare and contrast the terrestrial and Jovian planets.

15.4 List and describe the major features of Earth's Moon and explain how maria basins were generated.

15.5 Outline the principal characteristics of Mercury, Venus, and Mars. Describe their similarities to and differences from Earth.

15.6 Compare and contrast the four Jovian planets.

15.7 List and describe the principal characteristics of the small bodies that inhabit the solar system.

°This chapter was revised with the assistance of Professors Mark Watry and Teresa Tarbuck.

View of the windblown Martian surface obtained by NASA rover *Curiosity*. The black rocks are volcanic and have a composition similar to the rock basalt found on the Hawaiian islands. (Photo courtesy of NASA)

The science of astronomy provides a rational way of knowing and understanding the origins of Earth, the solar system, and the universe. Earth was once thought to be unique, different in every way from everything else in the universe. However, through the science of astronomy, we have discovered that Earth and the Sun are similar to other objects in the universe and that the physical laws that apply on Earth seem to apply everywhere else in the universe.

How did our understanding of the universe change so drastically? In this chapter we examine the transformation from the ancient view of the universe, which focused on the positions and movements of celestial objects, to the modern perspective, which focuses on *understanding how* these objects came to be and *why* they move the way they do.

15.1 Ancient Astronomy

Explain the geocentric view of the solar system and describe how it differs from the heliocentric view.

Long before recorded history, people were aware of the close relationship between events on Earth and the positions of heavenly bodies. They realized that changes in the seasons and floods of great rivers such as the Nile in Egypt occurred when certain celestial bodies, including the Sun, Moon, planets, and stars, reached particular places in the heavens. Early agrarian cultures, whose survival depended on seasonal change, believed that if these heavenly objects could control the seasons, they could also strongly influence all Earthly events. These beliefs undoubtedly encouraged early civilizations to begin keeping records of the positions of celestial objects.

The ancient Chinese, Egyptians, and Babylonians are well known for their record keeping. These cultures recorded the locations of the Sun, the Moon, and the five visible planets, as these objects moved slowly against the background of "fixed" stars. Eventually, it was not enough to track the motions of celestial objects; predicting their future positions (to avoid getting married at an unfavorable time, for example) became important.

A study of Chinese archives shows that the Chinese recorded every appearance of the famous Halley's Comet for at least 10 centuries. However, because this comet appears only once every 76 years, they were unable to link these appearances to establish that what they saw was the same object multiple times. Like most other ancients, the Chinese considered comets to be mystical. Generally, comets were seen as bad omens and were blamed for a variety of disasters, from wars to plagues (**Figure 15.1**). In addition, the Chinese kept quite accurate records of "guest stars." Today we know that a "guest star" is a normal star, usually too faint to be visible, which increases its brightness as it explosively ejects gases from its surface, a phenomenon we call a *nova* (*novus* = new) or *supernova* (**Figure 15.2**).

Figure 15.1 The Bayeux Tapestry This tapestry, which hangs in Bayeux, France, shows the apprehension caused by Halley's Comet in A.D. 1066. This event preceded the defeat of King Harold by William the Conqueror. ("Sighting of a comet." Detail from Bayeux Tapestry. Musee de la Tapisserie, Bayeux. With special authorization of the City of Bayeux. Bridgeman-Giraudon/Art Resource, NY)

The Golden Age of Astronomy

The "Golden Age" of early astronomy (600 B.C.–A.D. 150) was centered in Greece. Although the early Greeks have been criticized for using purely philosophical arguments

Figure 15.2 The Chinese recorded the sudden appearance of a "guest star" in A.D. 1054 The scattered remains of that supernova is the Crab Nebula in the constellation Taurus. This image comes from the Hubble Space Telescope. (NASA, ESA, J. Hester and A. Loll/Arizona State University)

to explain natural phenomena, they employed observational data as well. The basics of geometry and trigonometry, which they developed, were used to measure the sizes of and distances to the largest-appearing bodies in the heavens—the Sun and the Moon.

The early Greeks held the incorrect **geocentric** (*geo* = Earth, *centric* = centered) view of the universe—which professed that Earth was a sphere that remained motionless at the center of the universe. Orbiting Earth were the seven wanderers (*planetai* in Greek): the Sun, the Moon, Mercury, Venus, Mars, Jupiter, and Saturn. Beyond the planets was a transparent, hollow **celestial sphere** to which the stars were attached and which keeps them in the same relative position to one another. The celestial sphere in turn traveled daily around Earth. Some early Greeks realized that the motion of the stars could be explained just as easily by a rotating

Earth, but they rejected that idea because Earth exhibits no sense of motion, and it seemed too large to be movable. In fact, proof that Earth rotates on its axis was not demonstrated until 1851.

The first Greek to profess a *Sun-centered*, or **heliocentric** (*helios* = Sun, *centric* = centered) universe was Aristarchus (312–230 B.C.). Aristarchus also used simple geometric relationships to calculate the relative distances from Earth to the Sun and the Moon. He later used these data to calculate their sizes. As a result of an observational error beyond his control, he came up with measurements that were much too small. However, he did discover that the Sun was many times more distant than the Moon and many times larger than Earth. The latter fact may have prompted him to suggest a Sun-centered universe.

Ptolemy's Model

Much of our knowledge of Greek astronomy comes from a 13-volume treatise, *Almagest* (the great work), that was compiled by Claudius Ptolemy in A.D. 141. In addition to presenting a summary of Greek astronomical knowledge, Ptolemy is credited with developing a model of the universe, known as the **Ptolemaic system** (**Figure 15.3**).

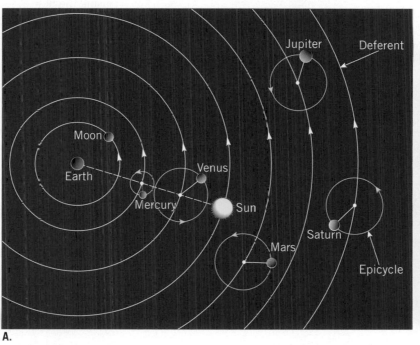

A.   **B.**

Figure 15.3 The universe according to Ptolemy, second century A.D. A. Ptolemy believed that the star-studded celestial sphere made a daily trip around a motionless Earth. In addition, he proposed that the Sun, Moon, and planets made trips of various lengths along individual orbits. **B.** A three-dimensional model of an Earth-centered system. Ptolemy likely utilized something similar to this to calculate the motions of the heavens. (Photo by Science Museum, London, UK/The Bridgeman Art Library)

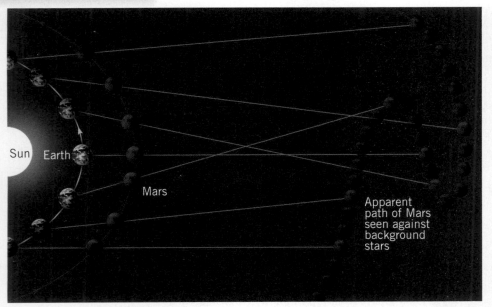

Figure 15.4 Retrograde (backward) motion of Mars, as seen against the background of distant stars When viewed from Earth, Mars moves eastward among the stars each day and then periodically appears to stop and reverse direction. This apparent westward drift is a result of the fact that Earth has a faster orbital speed than Mars and overtakes it. As this occurs, Mars appears to be moving backward—that is, it exhibits retrograde motion.

In the Greek tradition, the Ptolemaic model had the planets moving in perfect circular orbits around a motionless Earth. (The Greeks considered the circle to be a pure and perfect shape.) However, the motion of the planets, as seen against the background of stars, is not so simple. Each planet, if watched night after night, moves slightly eastward among the stars. Periodically, each planet appears to stop, reverse direction for a period of time, and then resume an eastward motion. The apparent westward drift is called **retrograde** (*retro* = to go back, *gradus* = walking) **motion**. This rather odd apparent

SmartFigure 15.5 Ptolemy's explanation of retrograde motion In Ptolemy's model, the planets move on small circles (epicycles) and orbit Earth on larger circles (deferents). Through trial and error, Ptolemy discovered the right combination of circles to produce the retrograde motion observed for each planet.

motion results from the combination of the motion of Earth and the planet's own motion around the Sun.

The retrograde motion of Mars is shown in **Figure 15.4**. Because Earth has a faster orbital speed than Mars, it overtakes its neighbor. While doing so, Mars *appears* to be moving backward, in retrograde motion. This is analogous to what a driver sees out the side window when passing a slower car. The slower planet, like the slower car, appears to be going backward, although its actual motion is in the same direction as the faster-moving body.

It is difficult to accurately represent retrograde motion using the incorrect Earth-centered model, but that is what Ptolemy was able to accomplish (**Figure 15.5**). Rather than use a single circle for each planet's orbit, he proposed that the planets orbited on small circles (*epicycles*), revolving along large circles (*deferents*). By trial and error, he found the right combination of circles to produce the amount of retrograde motion observed for each planet. (An interesting note is that almost any closed curve can be produced by the combination of two circular motions, a fact that can be verified by anyone who has used the Spirograph™ design-drawing toy.)

It is a tribute to Ptolemy's genius that he was able to account for the planets' motions as well as he did, considering that his model was incorrect. The precision with which he was able to predict planetary motion is attested to by the fact that his model went virtually unchallenged, in principle if not in detail, until the seventeenth century. When Ptolemy's predicted positions for the planets became out of step with the observed positions (which took 100 years or more), his model was simply recalibrated using the new observed positions as a starting point.

With the decline of the Roman Empire around the fourth century, much of the accumulated knowledge disappeared as libraries were destroyed. After the decline of Greek and Roman civilizations, the center of astronomical study moved east, to Baghdad where, fortunately, Ptolemy's work was translated into Arabic. It wasn't until sometime after the tenth century that the ancient Greeks' contributions to astronomy were reintroduced to Europe. The Ptolemaic model soon dominated Europe as the correct representation of the heavens, which created problems for anyone who found it lacking.

15.1 CONCEPT CHECKS

1. Why did the ancients believe that celestial objects had some influence over their lives?
2. What is the modern explanation of the "guest stars" that suddenly appeared in the night sky?
3. Explain the geocentric view of the universe.
4. In the Greek model of the universe, what were the seven wanderers, or planets? How were they different from stars?
5. Describe what produces the retrograde motion of Mars. What geometric arrangements did Ptolemy use to explain this motion?

15.2 The Birth of Modern Astronomy

List and describe the contributions to modern astronomy of Nicolaus Copernicus, Tycho Brahe, Johannes Kepler, Galileo Galilei, and Isaac Newton.

Ptolemy's Earth-centered universe was not discarded overnight. Modern astronomy's development was more than a scientific endeavor; it required a break from deeply entrenched philosophical and religious views that had been a basic part of Western society for thousands of years. Its development was brought about by the discovery of a new and much larger universe governed by discernible laws.

We examine the work of five noted scientists involved in this transition from an astronomy that merely describes what is observed to an astronomy that tries to explain what is observed and, more importantly, why the universe behaves the way it does. They include Nicolaus Copernicus, Tycho Brahe, Johannes Kepler, Galileo Galilei, and Sir Isaac Newton.

Nicolaus Copernicus

For almost 13 centuries after the time of Ptolemy, very few astronomical advances were made in Europe. Some were even lost, including the notion of a spherical Earth. The first great astronomer to emerge after the Middle Ages was Nicolaus Copernicus (1473–1543) from Poland (**Figure 15.6**). After discovering Aristarchus's writings, Copernicus became convinced that Earth is a planet, just like the other five then-known planets. The daily motions of the heavens, he reasoned, could be more simply explained by a rotating Earth.

Having concluded that Earth is a planet, Copernicus constructed a *heliocentric* model for the solar system, with the Sun at the center and the planets Mercury, Venus, Earth, Mars, Jupiter, and Saturn orbiting it. This was a major break from the ancient and prevailing idea that a motionless Earth lies at the center of all movement in the universe. However, Copernicus retained a link to the past and used circles to represent the orbits of the planets. Because of this, Copernicus was unable to accurately predict the future locations of the planets. Copernicus found it necessary to add smaller circles (epicycles) like those that Ptolemy used. The discovery that the planets actually have *elliptical* orbits occurred a century later and is credited to Johannes Kepler.

Like his predecessors, Copernicus also used philosophical justifications to support his point of view: "In the midst of all stands the Sun. For who could in this most beautiful temple place this lamp in another or better place than that from which it can at the same time illuminate the whole?"

Copernicus's monumental work *De Revolutionibus, Orbium Coelestium (On the Revolution of the Heavenly*

Figure 15.6 Polish astronomer Nicolaus Copernicus (1473–1543) Copernicus provided convincing evidence that Earth was just another planet that revolved around the Sun. (Detlev van Ravenswaay/Photo Researchers, Inc.)

Spheres), which set forth his controversial Sun-centered solar system, was published as he lay on his deathbed. Hence, he never suffered the criticisms that fell on many of his followers. Although Copernicus's model was a vast improvement over Ptolemy's, it did not attempt to explain how planetary motions occurred or why.

The greatest contribution of the Copernican system to modern science is its challenge of the primacy of Earth in the universe. At the time, many Europeans considered this heretical.

Tycho Brahe

Tycho Brahe (1546–1601) was born of Danish nobility 3 years after the death of Copernicus. Reportedly, Tycho became interested in astronomy while viewing a solar eclipse that astronomers had predicted. He persuaded King Frederick II to establish an observatory near Copenhagen, which Tycho headed. There he designed and built pointers (the telescope would not be invented for a few more decades), which he used for 20 years to systematically measure the locations of the heavenly bodies in

Figure 15.7 Tycho Brahe (1546–1601) in his observatory, in Uraniborg, on the Danish island of Hveen Tycho (central figure) and the background are painted on the wall of the observatory within the arc of the sighting instrument called a quadrant. In the far right, Tycho can be seen "sighting" a celestial object through the "hole" in the wall. Tycho's accurate measurements of Mars enabled Johannes Kepler to formulate his three laws of planetary motion. (Courtesy of Royal Geographical Society, London, UK/The Bridgeman Art Library)

an effort to disprove the Copernican theory (**Figure 15.7**). His observations, particularly of Mars, were far more precise than any made previously and are his legacy to astronomy.

Tycho did not believe in the Copernican model because he was unable to observe an apparent shift in the position of stars that should result if Earth traveled around the Sun. His argument went like this: If Earth orbits the Sun, the position of a nearby star, when observed from two locations in Earth's orbit 6 months apart, should shift with respect to the more distant stars. Tycho was correct, but his measurements did not have great enough precision to show any displacement. The apparent shift of the stars is called *stellar parallax*, and today it is used to measure distances to the nearest stars. (Stellar parallax is discussed in Appendix D, page 543.)

The principle of parallax is easy to visualize: Close one eye, and with your index finger vertical, use your eye to line up your finger with some distant object. Now, without moving your finger, view the object with your other eye and notice that the object's position appears to change. The farther away you hold your finger, the less the object's position seems to shift. Herein lay the flaw in Tycho's argument. He was right about parallax, but the distance to even the nearest stars is enormous compared to the width of Earth's orbit. Consequently, the shift that Tycho was looking for is too small to be detected without the aid of a telescope—an instrument that had not yet been invented.

With the death of his patron, the king of Denmark, Tycho was forced to leave his observatory. Known for his arrogance and extravagant nature, Tycho was unable to continue his work under Denmark's new ruler. As a result, Tycho moved to Prague in the present-day Czech Republic, where, in the last year of his life, he acquired an able assistant, Johannes Kepler. Kepler retained most of the observations made by Tycho and put them to exceptional use. Ironically, the data Tycho collected to refute the Copernican view of the solar system would later be used by Kepler to support it.

Johannes Kepler

If Copernicus ushered out the old astronomy, Johannes Kepler (1571–1630) ushered in the new (**Figure 15.8**). Armed with Tycho's data, a good mathematical mind, and, of greater importance, a strong belief in the accuracy of Tycho's work, Kepler derived three basic laws of planetary motion. The first two laws resulted from his inability to fit Tycho's observations of Mars to a circular orbit. Unwilling to concede that the discrepancies were a result of observational error, he searched for another solution. This endeavor led him to discover that the orbit of Mars is not a perfect circle but is slightly elliptical (**Figure 15.9**). About the same time, he realized that the orbital speed of Mars varies in a predictable way. As it approaches the Sun, it speeds up, and as it moves away, it slows down.

In 1609, after nearly a decade of work, Kepler proposed his first two laws of planetary motion:

1. The path of each planet around the Sun, while almost circular, is actually an ellipse, with the Sun at one focus (see Figure 15.9).
2. Each planet revolves so that an imaginary line connecting it to the Sun sweeps over equal areas in equal intervals of time (**Figure 15.10**). This *law of equal areas* geometrically expresses the variations in orbital speeds of the planets.

Figure 15.10 illustrates the second law. Note that in order for a planet to sweep equal areas in the same amount of time, it must travel more rapidly when it is nearer the Sun and more slowly when it is farther from the Sun.

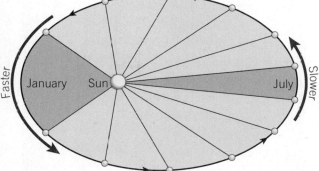

Figure 15.10 Kepler's law of equal areas A line connecting a planet (Earth) to the Sun sweeps out an area in such a manner that equal areas are swept out in equal times. This law accounts for the fact that Earth revolves slower when it is farther from the Sun (aphelion) and faster when it is closest (perihelion). The eccentricity of Earth's orbit is greatly exaggerated in this diagram.

Kepler was devout and believed that the Creator made an orderly universe and that this order would be reflected in the positions and motions of the planets. The uniformity he tried to find eluded him for nearly a decade. Then, in 1619, Kepler published his third law in *The Harmony of the Worlds*:

3. The orbital periods of the planets and their distances to the Sun are proportional.

In its simplest form, the orbital period is measured in Earth years, and the planet's distance to the Sun is expressed in terms of Earth's mean distance to the Sun. The latter "yardstick" is called the **astronomical unit (AU)** and is equal to about 150 million kilometers (93 million miles). Using these units, Kepler's third law states that the planet's orbital period squared is equal to its mean solar distance cubed. Consequently, the solar distances of the planets can be calculated when their periods of revolution are known. For example, Mars has an orbital period of 1.88 years, and 1.88 squared equals 3.54. The cube root of 3.54 is 1.52, and that is the average distance from Mars to the Sun, in astronomical units (**Table 15.1**).

Figure 15.8 German astronomer Johannes Kepler (1571–1630) A keen mathematician, Kepler helped establish the era of modern astronomy by deriving three laws of planetary motion. (Photo by Imagno/Getty Images)

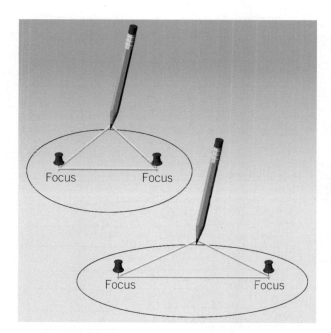

Figure 15.9 Drawing ellipses with various eccentricities Using two straight pins for foci and a loop of string, trace out a curve while keeping the string taut, and you will have drawn an ellipse. The farther apart the pins (the foci) are moved, the more flattened (more eccentric) is the resulting ellipse.

TABLE 15.1	Period of Revolution and Solar Distances of Planets		
Planet	Solar Distance (AU)*	Period (years)	Ellipticity 0 = Circle
Mercury	0.39	0.24	0.205
Venus	0.72	0.62	0.007
Earth	1.00	1.00	0.017
Mars	1.52	1.88	0.094
Jupiter	5.20	11.86	0.049
Saturn	9.54	29.46	0.057
Uranus	19.18	84.01	0.046
Neptune	30.06	164.80	0.011
*AU = astronomical unit			

Figure 15.11 Italian scientist Galileo Galilei (1564–1642) Galileo used a new invention, the telescope, to observe the Sun, Moon, and planets in more detail than had ever been done before. (Nimatallah/Art Resource N.Y.)

the Copernican theory of a Sun-centered solar system. Galileo's greatest contributions to science were his descriptions of the behavior of moving objects, which he derived from experimentation. The method of using experiments to determine natural laws had essentially been lost since the time of the early Greeks.

All astronomical discoveries before Galileo's time were made without the aid of a telescope. In 1609, Galileo heard that a Dutch lens maker had devised a system of lenses that magnified objects. Apparently without ever having seen a telescope, Galileo constructed his own, which magnified distant objects three times the size seen by the unaided eye. He immediately made others, the best having a magnification of about 30 (**Figure 15.12**).

With the telescope, Galileo was able to view the universe in a new way. He made many important discoveries that supported the Copernican view of the universe, including the following:

1. *The discovery of Jupiter's four largest satellites, or moons* (**Figure 15.13**). This finding dispelled the old

Kepler's laws assert that the planets revolve around the Sun and therefore support the Copernican theory. Kepler, however, did not determine the *forces* that act to produce the planetary motion he had so ably described. That task would remain for Galileo Galilei and Sir Isaac Newton.

Galileo Galilei

Galileo Galilei (1564–1642) was the greatest Italian scientist of the Renaissance (**Figure 15.11**). He was a contemporary of Kepler and, like Kepler, strongly supported

Figure 15.12 One of Galileo's telescopes Although Galileo did not invent the telescope, he built several—the largest of which had a magnification of 30. (Photo by Gianni Tortoli/ Photo Researchers, Inc.)

Figure 15.13 Sketch by Galileo of how he saw Jupiter and its four largest satellites through his telescope The positions of Jupiter's four largest Moons (drawn as stars) change nightly. You can observe these same changes with binoculars. (Yerkes Observatory Photograph/University of Chicago)

idea that Earth was the sole center of motion in the universe; for here, plainly visible, was another center of motion—Jupiter. It also countered the frequently used argument that the Moon would be left behind if Earth revolved around the Sun.

2. *The discovery that the planets are circular disks rather than just points of light, as was previously thought.* This indicated that the planets must be Earth-like as opposed to star-like.

3. *The discovery that Venus exhibits phases just as the Moon does and that Venus appears smallest when it is in full phase and thus is farthest from Earth* (**Figure 15.14B,C**). This observation demonstrates that Venus orbits its source of light—the Sun. In the Ptolemaic system, shown in **Figure 15.14A**, the orbit of Venus lies between Earth and the Sun, which means that only the crescent phases of Venus should ever be seen from Earth.

4. *The discovery that the Moon's surface is not a smooth glass sphere, as the ancients had proclaimed.* Rather, Galileo saw mountains, craters, and plains, indicating that the Moon is Earth-like. He thought the plains might be bodies of water, and this idea was strongly promoted by others, as we can tell from the names given to these features (Sea of Tranquility, Sea of Storms, etc.).

5. *The discovery that the Sun (the viewing of which may have caused the eye damage that later blinded him) had sunspots—dark regions caused by slightly lower temperatures.* He tracked the movement of these spots and estimated the rotational period of the Sun as just under a month. Hence, another heavenly body was found to have both "blemishes" and rotational motion.

Each of these observations eroded a bedrock principle held by the prevailing view on the nature of the universe.

In 1616, the Roman Catholic Church condemned the Copernican theory as contrary to Scripture because it did not put humans in their rightful place at the center of Creation, and Galileo was told to abandon this theory. Undeterred, Galileo began writing his most famous work, *Dialogue of the Great World Systems.* Despite poor health, he completed the project and in 1630 went to Rome, seeking permission from Pope Urban VIII to publish. Since the book was a dialogue that expounded

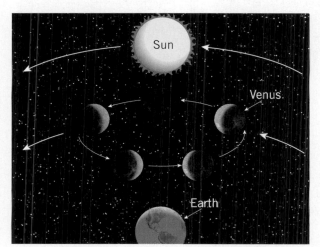

A. In the Ptolemaic (Earth-centered) system, the orbit of Venus lies between the Sun and Earth, as shown here. Thus, in an Earth-centered solar system, only the crescent phase of Venus would be visible from Earth.

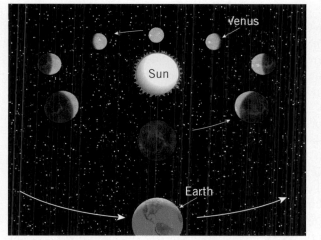

B. In the Copernican (Sun-centered) system, Venus orbits the Sun and hence all of the phases of Venus should be visible from Earth.

C. As Galileo observed, Venus goes through a series of Moonlike phases. Venus appears smallest during the full phase when it is farthest from Earth and largest in the crescent phase when it is closest to Earth. This led Galileo to conclude that the Sun was the center of the solar system.

SmartFigure 15.14 Using a telescope, Galileo discovered that Venus has phases just like the Moon. (Photo courtesy of Lowell Observatory)

both the Ptolemaic and Copernican systems, publication was allowed. However, Galileo's detractors were quick to realize that he was promoting the Copernican view at the expense of the Ptolemaic system. Sale of the book was quickly halted, and Galileo was called before the Inquisition. Tried and convicted of proclaiming doctrines contrary to religious teachings, he was sentenced to permanent house arrest, under which he remained for the last 10 years of his life.

Despite this restriction, and his grief following the death of his eldest daughter, Galileo continued to work. In 1637 he became totally blind, yet during the next few years he completed his finest scientific work, a book on the study of motion in which he stated that the natural tendency of an object in motion is to remain in motion. Later, as more scientific evidence in support of the Copernican system was discovered, the Church allowed Galileo's works to be published.

Sir Isaac Newton

Sir Isaac Newton (1642–1727) was born in the year of Galileo's death (**Figure 15.15**). His many accomplishments in mathematics and physics led a successor to say, "Newton was the greatest genius that ever existed."

Although Kepler and those who followed attempted to explain the forces involved in planetary motion, their explanations were less than satisfactory. Kepler believed that some force pushed the planets along in their orbits. Galileo, however, correctly reasoned that no force is required to keep an object in motion. Instead, Galileo proposed that the natural tendency for a moving object that is unaffected by an outside force is to continue moving at a uniform speed and in a straight line. Newton later formalized this concept, **inertia**, as his first law of motion.

The problem, then, was not to explain the force that keeps the planets moving but rather to determine the force that *keeps them from going in a straight line out into space*. It was to this end that Newton conceptualized the force of gravity. At the early age of 23, he envisioned a force that extends from Earth into space and holds the Moon in orbit around Earth. Although others had theorized the existence of such a force, he was the first to formulate and test the **law of universal gravitation**. It states:

> Every body in the universe attracts every other body with a force that is directly proportional to their masses and inversely proportional to the square of the distance between them.

Thus, gravitational force decreases with distance, so that two objects 3 kilometers apart have 3^2, or 9, times less gravitational attraction than if the same objects were 1 kilometer apart.

Figure 15.15 English scientist Sir Isaac Newton (1642–1727) Newton was the first to explain how gravity holds planets in orbit around the Sun. (The Granger Collection, NYC)

The law of gravitation also states that the greater the mass of an object, the greater its gravitational force. For example, the large mass of the Moon has a gravitational force strong enough to cause ocean tides on Earth, whereas the tiny mass of a communications satellite has very little effect on Earth.

With his laws of motion, Newton proved that the force of gravity—combined with the tendency of a planet to remain in straight-line motion—would result in a planet having an elliptical orbit, as established by Kepler. Earth, for example, moves forward in its orbit about 30 kilometers (18.5 miles) each second, and during the same second, the force of gravity pulls it toward the Sun about 0.5 centimeter (1/8 inch). Therefore, as Newton concluded, it is the combination of Earth's forward motion and its "falling" motion that defines its orbit (**Figure 15.16**). If gravity were somehow eliminated, Earth would move in a straight line out into space. Conversely, if Earth's forward motion suddenly stopped, gravity would pull it until it crashed into the Sun.

SmartFigure 15.16 Orbital motion of Earth and the other planets Diagram showing that the orbiting planet continues its forward motion because of inertia, while the Sun's gravity bends the planet's path into an elliptical orbit.

Diagram labels: Planet; A. Tendency of planet without the influence of gravity; B. Gravitational force; Net motion (A + B); Orbit; Sun

15.2 CONCEPT CHECKS

1. What major change did Copernicus make in the Ptolemaic system? Why was this change philosophically different?
2. What data did Tycho Brahe collect that was useful to Johannes Kepler in his quest to describe planetary motion?
3. Who discovered that planetary orbits are ellipses rather than circles?
4. Does Earth move faster in its orbit near perihelion (January) or near aphelion (July)?
5. Explain why Galileo's discovery of a rotating Sun supports the Copernican view of a Sun-centered universe.
6. Newton discovered that the orbits of the planets result from opposing forces. Briefly explain these forces.

15.3 Our Solar System: An Overview

Describe the formation of the solar system according to the nebular theory. Compare and contrast the terrestrial and Jovian planets.

The Sun is at the center of a revolving system, trillions of miles wide, consisting of eight planets, their satellites, and numerous smaller asteroids, comets, and meteoroids. An estimated 99.85 percent of the mass of our solar system is contained within the Sun. Collectively, the planets account for most of the remaining 0.15 percent. Starting from the Sun, the planets are Mercury, Venus, Earth, Mars, Jupiter, Saturn, Uranus, and Neptune (**Figure 15.17**). Pluto was recently reclassified as a member of a new class of solar system bodies called *dwarf planets.*

Tethered to the Sun by gravity, all the planets travel in the same direction on slightly elliptical orbits. Gravity causes objects nearest the Sun to travel fastest. Therefore, Mercury has the highest orbital velocity, 48 kilometers per second, and the shortest period of revolution around the Sun, 88 Earth-days. By contrast, the distant dwarf planet Pluto has an orbital speed of just 5 kilometers per second and requires 248 Earth-years to complete one revolution. Most large bodies orbit the Sun approximately in the same plane. The planets' inclination with respect to the Earth–Sun orbital plane, known as the *ecliptic,* is shown in **Table 15.2**.

Nebular Theory: Formation of the Solar System

The **nebular theory**, which explains the formation of the solar system, proposes that the Sun and planets formed from a rotating cloud of interstellar gases (mainly hydrogen and helium) and dust called the **solar nebula**. As the solar nebula contracted due to gravity, most of the material collected in the center to form the hot *proto-sun*. The remaining materials formed a thick, flattened, rotating disk, within which matter gradually cooled and condensed into grains and clumps of icy, rocky material. Repeated collisions resulted in most of the material clumping together into larger and larger chunks that eventually became asteroid-sized objects called **planetesimals**.

The composition of planetesimals was largely determined by their proximity to the protosun. As you might expect, temperatures were highest in the inner solar system and decreased toward the outer edge of the disk. Therefore, between the present orbits of Mercury and Mars, the planetesimals were composed of materials with high melting temperatures—metals and rocky substances. Then, through repeated collisions and accretion, these asteroid-sized rocky bodies combined to form the four **protoplanets** that eventually became Mercury, Venus, Earth, and Mars.

The planetesimals that formed beyond the orbit of Mars, where temperatures were low, contained high percentages of ices—water, carbon dioxide, ammonia, and methane—as well as small amounts of rocky and metallic debris. It was mainly from these planetesimals that the four outer planets eventually formed. The accumulation of ices accounts, in part, for the large sizes and low densities of the outer planets. The two most massive planets, Jupiter and Saturn, had surface gravities sufficient to

SmartFigure 15.17 Orbits of the planets A. Artistic view of the solar system in which planets are not drawn to scale. **B.** Positions of the planets shown to scale using astronomical units (AU), where 1 AU is equal to the distance from Earth to the Sun—150 million kilometers (93 million miles).

attract and retain large quantities of hydrogen and helium, the lightest elements.

It took roughly 1 billion years after the protoplanets formed for the planets to gravitationally accumulate most of the interplanetary debris. This was a period of intense bombardment as the planets cleared their orbits by collecting much of the leftover material. The "scars" of this period are still evident on the Moon's surface. Because of the gravitational effect of the planets, particularly Jupiter, small bodies were flung into planet-crossing orbits or into interstellar space. The small fraction of interplanetary matter that escaped this violent period became either

asteroids or comets. By comparison, the present-day solar system is a much quieter place, although many of these processes continue today at a reduced pace.

The Planets: Internal Structures and Atmospheres

The planets fall into two groups based on location, size, and density: the **terrestrial** (Earth-like) **planets** (Mercury, Venus, Earth, and Mars) and the **Jovian** (Jupiter-like) **planets** (Jupiter, Saturn, Uranus, and

TABLE 15.2   Planetary Data

Planet	Symbol	AU*	Mean Distance from Sun — Millions of Miles	Mean Distance from Sun — Millions of Kilometers	Period of Revolution	Inclination of Orbit	Orbital Velocity — mi/s	Orbital Velocity — km/s
Mercury	☿	0.39	36	58	88^d	7°00'	29.5	47.5
Venus	♀	0.72	67	108	225^d	3°24'	21.8	35.0
Earth	⊕	1.00	93	150	365.25^d	0°00'	18.5	29.8
Mars	♂	1.52	142	228	687^d	1°51'	14.9	24.1
Jupiter	♃	5.20	483	778	12^{yr}	1°18'	8.1	13.1
Saturn	♄	9.54	886	1427	30^{yr}	2°29'	6.0	9.6
Uranus	♅	19.18	1783	2870	84^{yr}	0°46'	4.2	5.8
Neptune	♆	30.06	2794	4497	165^{yr}	1°46'	3.3	5.3

Planet	Period of Rotation	Diameter — Miles	Diameter — Kilometers	Relative Mass (Earth = 1)	Average Density (g/cm³)	Polar Flattening (%)	Eccentricity†	Number of Known Satellites††
Mercury	59^d	3015	4878	0.06	5.4	0.0	0.206	0
Venus	243^d	7526	12,104	0.82	5.2	0.0	0.007	0
Earth	23^{h}56^{m}04^s	7920	12,756	1.00	5.5	0.3	0.017	1
Mars	24^{h}37^{m}23^s	4216	6794	0.11	3.9	0.5	0.093	2
Jupiter	9^{h}56^m	88,700	143,884	317.87	1.3	6.7	0.048	67
Saturn	10^{h}30^m	75,000	120,536	95.14	0.7	10.4	0.056	62
Uranus	17^{h}14^m	29,000	51,118	14.56	1.2	2.3	0.047	27
Neptune	16^{h}07^m	28,900	50,530	17.21	1.7	1.8	0.009	13

*AU = astronomical unit, Earth's mean distance from the Sun.

†Eccentricity is a measure of the amount an orbit deviates from a circular shape. The larger the number, the less circular the orbit.

††Includes all satellites discovered as of December 2012.

Neptune). Because of their relative locations, the four terrestrial planets are also known as *inner planets*, and the four Jovian planets are known as *outer planets*. A correlation exists between planetary locations and sizes: The inner planets are substantially smaller than the outer planets, also known as *gas giants*. For example, the diameter of Neptune (the smallest Jovian planet) is nearly four times larger than the diameter of Earth. Furthermore, Neptune's mass is 17 times greater than that of Earth or Venus.

Other properties that differ among the planets include densities, chemical compositions, orbital periods, and numbers of satellites. Variations in the chemical composition of planets are largely responsible for their density differences. Specifically, the average density of the terrestrial planets is about 5 times the density of water, whereas the average density of the Jovian planets is only 1.5 times that of water. In fact, Saturn has a density only 0.7 times that of water, which means that it would float in a sufficiently large tank of water. The outer planets are also characterized by long orbital periods and numerous satellites.

Internal Structures Shortly after Earth formed, segregation of material resulted in the formation of three major layers, defined by their chemical composition—the crust, mantle, and core. This type of chemical separation occurred in the other planets as well. However, because the terrestrial planets are compositionally different from the Jovian planets, the nature of these layers is different as well (**Figure 15.18**).

The terrestrial planets are dense, having relatively large cores of iron and nickel. The outer cores of Earth and Mercury are liquid, whereas the cores of Venus and Mars are thought to be only partially molten. This difference is attributable to Venus and Mars having lower internal temperatures than those of Earth and Mercury. Silicate minerals and other lighter compounds make up the mantles of the terrestrial planets. Finally, the silicate crusts of terrestrial planets are relatively thin compared to their mantles.

The two largest Jovian planets, Jupiter and Saturn, likely have small, solid cores consisting of iron compounds, like the cores of the terrestrial planets, and rocky material similar to Earth's mantle. Progressing outward, the layer above the core consists of liquid hydrogen that is under extremely high temperatures and pressures. There is substantial evidence that under these conditions, hydrogen behaves like a metal in that its electrons move freely about and are efficient conductors of both heat and electricity. Jupiter's intense magnetic field is thought to be

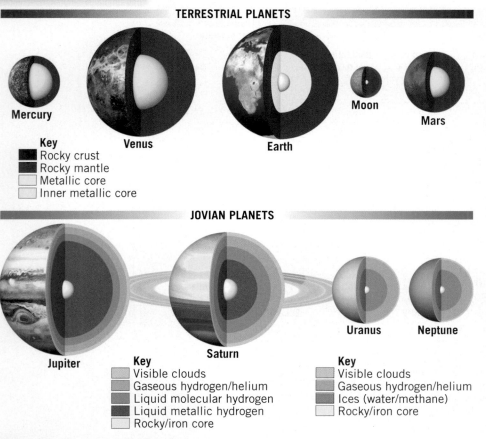

Figure 15.18 Comparing the internal structures of the planets

the result of electric currents flowing within a spinning layer of liquid metallic hydrogen. Saturn's magnetic field is much weaker than Jupiter's, due to its smaller shell of liquid metallic hydrogen. Above this metallic layer, both Jupiter and Saturn are thought to be composed of molecular liquid hydrogen that is intermixed with helium. The outermost layers are gases of hydrogen and helium, as well as ices of water, ammonia, and methane—which mainly account for the low densities of these giants.

Uranus and Neptune also have small iron-rich, rocky cores but their mantles are likely hot dense water and ammonia. Above their mantles, the amount of hydrogen and helium increases, but these gases exist in much lower concentrations than in Jupiter and Saturn.

The Atmospheres of the Planets The Jovian planets have very thick atmospheres composed mainly of hydrogen and helium, with lesser amounts of water, methane, ammonia, and other hydrocarbons. By contrast, the terrestrial planets, including Earth, have relatively meager atmospheres composed of carbon dioxide, nitrogen, and oxygen. Two factors—solar heating (temperature) and gravity—explain these significant differences (**Figure 15.19**). These variables determine what planetary gases, if any, were captured by planets during the formation of the solar system and which were ultimately retained.

During planetary formation, the inner regions of the developing solar system were too hot for ices and gases to condense. By contrast, the Jovian planets formed where temperatures were low and solar heating of planetesimals was minimal. This allowed water vapor, ammonia, and methane to condense into ices. Hence, the gas giants contain large amounts of these volatiles. As the planets grew, the largest Jovian planets, Jupiter and Saturn, also attracted large quantities of the lightest gases, hydrogen and helium.

How did Earth acquire water and other volatile gases? It seems that early in the history of the solar system, gravitational tugs by the developing protoplanets sent planetesimals into very eccentric orbits. As a result, Earth was bombarded with icy objects that originated beyond the orbit of Mars. This was a fortuitous event for organisms that currently inhabit our planet. Mercury, our Moon, and numerous other small bodies lack significant atmospheres even though they certainly would have been bombarded by icy bodies early in their development.

Airless bodies develop where solar heating is strong and/or gravities are weak. Simply stated, *small warm bodies* have a better chance of losing their atmospheres because gas molecules are more energetic and need less speed to escape their weak gravities. For example, warm bodies with small surface gravity, such as our Moon, are unable to hold even heavy gases such as carbon dioxide and nitrogen. Mercury is massive enough to hold trace amounts of hydrogen, helium, and oxygen gas.

The slightly larger terrestrial planets, Earth, Venus, and Mars, retain some heavy gases, including water vapor, nitrogen, and carbon dioxide. However, their atmospheres are miniscule compared to their total mass. Early in their development, the terrestrial planets probably had much thicker atmospheres. Over time, however, these primitive atmospheres gradually changed as light gases trickled away into space. For example, Earth's atmosphere continues to leak hydrogen and helium (the two lightest gases) into space. This phenomenon occurs near the top of Earth's atmosphere where air is so tenuous that nothing stops the fastest-moving particles from flying off into space. The speed required to escape a planet's gravity is called **escape velocity**. Because hydrogen is the lightest gas, it most easily reaches the speed needed to overcome Earth's gravity.

Planetary Impacts

Planetary impacts between solar system bodies have occurred throughout the history of the solar system. On bodies that have little or no atmosphere (like the Moon) and, therefore, no air resistance, even the smallest pieces of interplanetary debris (meteorites) can reach the surface. At high enough velocities, this debris can produce microscopic cavities on individual mineral grains. By contrast, large **impact craters** result from collisions with massive bodies, such as asteroids and comets.

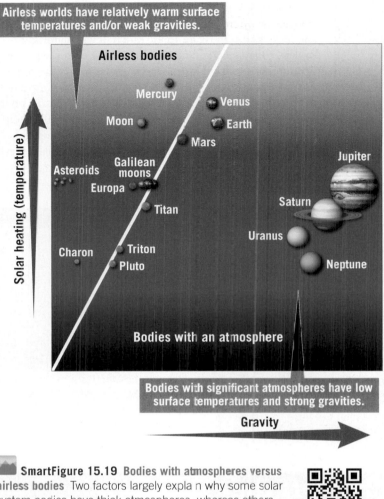

Airless worlds have relatively warm surface temperatures and/or weak gravities.

Airless bodies

Solar heating (temperature)

Mercury
Venus
Moon
Earth
Mars
Jupiter
Asteroids
Galilean moons
Europa
Saturn
Titan
Uranus
Charon
Triton
Pluto
Neptune

Bodies with an atmosphere

Bodies with significant atmospheres have low surface temperatures and strong gravities.

Gravity

 SmartFigure 15.19 Bodies with atmospheres versus airless bodies Two factors largely explain why some solar system bodies have thick atmospheres, whereas others are airless. Airless worlds have relatively warm surface temperatures and/or weak gravities. Bodies with significant atmospheres have low surface temperatures and strong gravities.

A. The energy of a rapidly moving body is transformed into heat and shock waves.

Compressional wave   Meteoroid

High-speed ejecta

Rebound

B. The rebound of over-compressed rock causes debris to be explosively ejected from the crater.

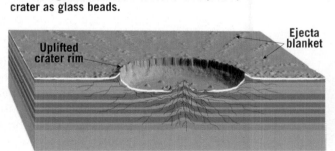

Melt   Central peak

Fractured rock

C. Heating melts some material that may be ejected from the crater as glass beads.

Uplifted crater rim   Ejecta blanket

D. Small secondary craters often form when the material "splashed" from the impact crater strikes the surrounding landscape.

Figure 15.20 Formation of an impact crater

Planetary impacts were considerably more common in the early history of the solar system than they are today. Following that early period of intense bombardment, the rate of cratering diminished dramatically and now remains essentially constant. Because weathering and erosion are almost nonexistent on the Moon and Mercury, evidence of their cratered past is clearly evident.

On larger bodies, thick atmospheres may cause the impacting objects to break up and/or decelerate. For example, Earth's atmosphere causes meteoroids with masses of less than 10 kilograms (22 pounds) to lose up to 90 percent of their speed as they penetrate the atmosphere. Therefore, impacts of low-mass bodies produce only small craters on Earth. Our atmosphere is much less effective in slowing large bodies; fortunately, they make very rare appearances.

The formation of a large impact crater is illustrated in **Figure 15.20**. The meteoroid's high-speed impact compresses the material it strikes, causing an almost instantaneous rebound, which ejects material from the surface. On Earth, impacts can occur at speeds that exceed 50 kilometers per second. Impacts at such high speeds produce shock waves that compress both the impactor and the material being impacted. Almost instantaneously, the over-compressed material rebounds and explosively ejects material out of the newly formed crater. This process is analogous to the detonation of an explosive device that has been buried underground.

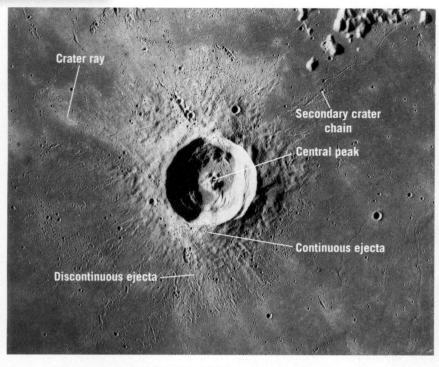

Figure 15.21 Lunar crater Euler This 20-kilometer-wide (12 mile-wide) crater is located in the southwestern part of Mare Imbrium. Clearly visible are the bright rays, central peak, secondary craters, and large accumulation of ejecta near the crater rim. (Courtesy of NASA)

Craters excavated by objects that are several kilometers across often exhibit a central peak, such as the one in the large crater in **Figure 15.21**. Much of the material expelled, called *ejecta*, lands in or near the crater, where it accumulates to form a rim. Large meteoroids may generate sufficient heat to melt and eject some of the impacted rock as glass beads. Specimens of glass beads produced in this manner, as well as melt breccia consisting of broken fragments welded by the heat of impact, have been collected on Earth, as well as the Moon, allowing planetary geologists to more fully understand such events.

15.3 CONCEPT CHECKS

1. Briefly outline the steps in the formation of our solar system, according to the nebular theory.
2. By what criteria are planets considered either terrestrial or Jovian?
3. What accounts for the large density differences between the terrestrial and Jovian planets?
4. Explain why the terrestrial planets have meager atmospheres compared to the Jovian planets.
5. Why are impact craters more common on the Moon than on Earth, even though the Moon is a much smaller target and has a weaker gravitational field?

15.4 Earth's Moon: A Chip Off the Old Block

List and describe the major features of Earth's Moon and explain how maria basins were generated.

The Earth–Moon system is unique because the Moon is the largest satellite relative to its planet. Mars is the only other terrestrial planet that has moons, but its tiny satellites are likely captured asteroids. Most of the 150 or so satellites of the Jovian planets are composed of low-density rock–ice mixtures, none of which resemble the Moon. As we will see later, our unique planet–satellite system is closely related to the mechanism that created it.

The diameter of the Moon is 3475 kilometers (2160 miles), about one-fourth of Earth's 12,756 kilometers (7926 miles). The Moon's surface temperature averages about 107°C (225°F) during daylight hours and –153°C (–243°F) at night. Because its period of rotation on its axis equals its period of revolution around Earth, the same lunar hemisphere always faces Earth. All of the landings of staffed *Apollo* missions were confined to the side of the Moon that faces Earth.

The Moon's density is 3.3 times that of water, comparable to that of *mantle* rocks on Earth but considerably less than Earth's average density (5.5 times that of water). The Moon's relatively small iron core is thought to account for much of this difference.

The Moon's low mass relative to Earth results in a lunar gravitational attraction that is one-sixth that of Earth. A person who weighs 150 pounds on Earth weighs only 25 pounds on the Moon, although the person's mass remains the same. This difference allows an astronaut to carry a heavy life-support system with relative ease. If not burdened with such a load, an astronaut could jump six times higher on the Moon than on Earth. The Moon's small mass (and low gravity) is the primary reason it was not able to retain an atmosphere.

How Did the Moon Form?

Until recently, the origin of the Moon—our nearest planetary neighbor—was a topic of considerable debate among scientists. Current models show that Earth is too small to have formed with a moon, particularly one so large. Furthermore, a captured moon would likely have an eccentric orbit similar to the captured moons that orbit the Jovian planets.

The current consensus is that the Moon formed as a result of a collision between a Mars-sized body and a youthful, semimolten Earth about 4.5 billion years ago. During this collision, some of the ejected debris was thrown into orbit around Earth and gradually coalesced to form the Moon. Computer simulations show that most of the ejected material would have come from the rocky mantle of the impactor, while its core was assimilated into the growing Earth. This *impact model* is consistent with the Moon having a proportionately smaller core than Earth's and, hence, a lower density.

The Lunar Surface When Galileo first pointed his telescope toward the Moon, he observed two different types of terrain: dark lowlands and brighter, highly cratered highlands (**Figure 15.22**). Because the dark regions appeared to be smooth, resembling seas on Earth, they were called **maria** (*mar* = sea, singular mare). The *Apollo 11* mission showed conclusively that the maria are exceedingly smooth plains composed of basaltic lavas. These vast plains are strongly concentrated on the side of the Moon facing Earth and cover about 16 percent of the lunar surface. The lack of large volcanic cones on these surfaces is evidence of high eruption rates of very fluid basaltic lavas similar to the Columbia Plateau flood basalts on Earth.

By contrast, the Moon's light-colored areas resemble Earth's continents, so the first observers dubbed them **terrae** (Latin for "lands") These areas are now generally referred to as the **lunar highlands** because they are elevated several kilometers above the maria. Rocks retrieved from the highlands are mainly breccias, pulverized by massive bombardment early in the Moon's history. The arrangement of terrae and maria account for the legendary "face" of the "man in the moon."

Some of the most obvious lunar features are impact craters. A meteoroid 3 meters (10 feet) in diameter can blast out a crater 50 times larger, or about 150 meters (500 feet) in diameter. The larger craters shown in Figure 15.22, such as Kepler and Copernicus (32 and 93 kilometers in diameter, respectively) were created from bombardment by bodies 1 kilometer or more in diameter. These two craters are thought to be relatively young because of the bright *rays* (light-colored ejected material) that radiate from them for hundreds of kilometers.

History of the Lunar Surface The evidence used to unravel the history of the lunar surface comes primarily from radiometric dating of rocks returned from *Apollo* missions and studies of crater densities—counting the number of craters per unit area. The greater the crater density, the older the feature is inferred to be. Such evidence suggests that, after the Moon coalesced, it passed through the following four phases: (1) formation of the original crust, (2) excavation of the large impact basins, (3) filling of maria basins, and (4) formation of rayed craters.

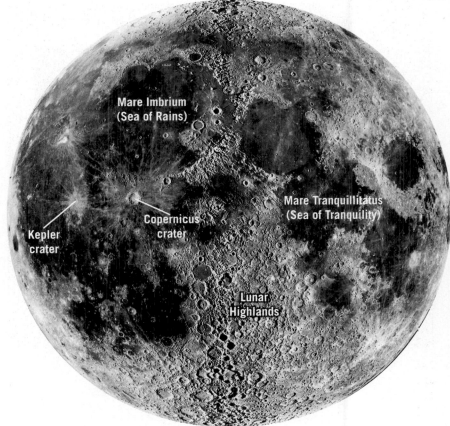

Figure 15.22 Telescopic view of the lunar surface The major features are the dark maria and the light, highly cratered highlands. (UCO/Lick Observatory Image)

During the late stages of its accretion, the Moon's outer shell was most likely completely melted—literally a magma ocean. Then, about 4.4 billion years ago, the magma ocean began to cool and underwent magmatic differentiation (see Chapter 3). Most of the dense minerals, olivine and pyroxene, sank, while less dense silicate minerals floated to form the Moon's crust. The highlands are made of these igneous rocks, which rose buoyantly like "scum" from the crystallizing magma. The most common highland rock type is *anorthosite*, which is composed mainly of calcium-rich plagioclase feldspar.

Once formed, the lunar crust was continually impacted as the Moon swept up debris from the solar nebula. During this time, several large impact basins were created. Then, about 3.8 billion years ago, the Moon, as well as the rest of the solar system, experienced a sudden drop in the rate of meteoritic bombardment.

The Moon's next major event was the filling of the large impact basins, which had been created at least 300 million years earlier (**Figure 15.23**). Radiometric dating of the maria basalts puts their age between 3.0 billion and 3.5 billion years, considerably younger than the lunar highlands.

Impact of an asteroid-size body produced a huge crater hundreds of kilometers in diameter and disturbed the lunar crust far beyond the crater.

Filling of the impact crater with fluid basalts, perhaps derived from partial melting deep within the lunar mantle.

Today these basins make up the lunar maria and a few similar large structures on Mercury.

SmartFigure 15.23 Formation and filling of large impact basins

The maria basalts are thought to have originated at depths between 200 and 400 kilometers. They were likely generated by a slow rise in temperature attributed to the decay of radioactive elements. Partial melting probably occurred in several isolated pockets, as indicated by the

diverse chemical makeup of the rocks retrieved during the *Apollo* missions. Recent evidence suggests that some mare-forming eruptions may have occurred as recently as 1 billion years ago.

Other lunar surface features related to this period of volcanism include small shield volcanoes (8–12 kilometers in diameter), evidence of pyroclastic eruptions, rilles (narrow winding valleys thought to be lava channels), and grabens (down-faulted valleys).

The last prominent features to form were rayed craters, as exemplified by the 90-kilometer-wide Copernicus crater shown in Figure 15.23. Material ejected from these craters blankets the maria surfaces and many older, rayless craters. The relatively young Copernicus crater is thought to be about 1 billion years old. Had it formed on Earth, weathering and erosion would have long since obliterated it.

Today's Lunar Surface The Moon's small mass and low gravity account for its lack of atmosphere and flowing water. Therefore, the processes of weathering and erosion that continually modify Earth's surface are absent on the Moon. In addition, tectonic forces are essentially inactive on the Moon, so moonquakes are weak and volcanic eruptions have ceased. Because the Moon is unprotected by an atmosphere, erosion is dominated by the impact of tiny particles from space (*micrometeorites*) that continually bombard its surface and gradually smooth the landscape. This activity has crushed and repeatedly mixed the upper portions of the lunar crust.

Both the maria and terrae are mantled with a layer of gray, unconsolidated debris derived from a few billion years of meteoric bombardment (**Figure 15.24**). This soil-like layer, properly called **lunar regolith** (*rhegos* = blanket, *lithos* = stone) is composed of igneous rocks, breccia, glass beads, and fine *lunar dust*. The lunar regolith is anywhere from 2 to 20 meters thick, depending on the age of the surface.

15.4 CONCEPT CHECKS

1. Briefly describe the origin of the Moon.
2. Compare and contrast the Moon's maria and highlands.
3. How are maria on the Moon similar to the Columbia Plateau in the Pacific Northwest?
4. How is crater density used in the relative dating of surface features on the Moon?
5. List the major stages in the development of the modern lunar surface.
6. Compare and contrast the processes of weathering and erosion on Earth with the same processes on the Moon.

Figure 15.24 Astronaut Harrison Schmitt sampling the lunar surface Notice the footprint (inset) in the lunar "soil," called regolith, which lacks organic material and is therefore not a true soil. (Courtesy of NASA)

15.5 Terrestrial Planets

Outline the principal characteristics of Mercury, Venus, and Mars. Describe their similarities to and differences from Earth.

The terrestrial planets in order from the Sun are Mercury, Venus, Earth, and Mars. Because most of the book focused on Earth, we will consider the three other Earth-like planets next.

Mercury: The Innermost Planet

Mercury, the innermost and smallest planet, revolves around the Sun quickly (88 days) but rotates slowly on its axis. Mercury's day–night cycle, which lasts 176 Earth-days, is very long compared to Earth's 24-hour cycle. One "night" on Mercury is roughly equivalent to 3 months on Earth and is followed by the same duration of daylight. Mercury has the greatest temperature extremes, from as low as −173°C (−280°F) at night to noontime temperatures exceeding 427°C (800°F), hot enough to melt tin

and lead. These extreme temperatures make life as we know it impossible on Mercury.

Mercury absorbs most of the sunlight that strikes it, reflecting only 6 percent into space, a characteristic of terrestrial bodies that have little or no atmosphere. The minuscule amount of gas that is present on Mercury may have originated from several sources, including ionized gas from the Sun, ices that vaporized during a recent comet impact, and/or outgassing of the planet's interior.

Although Mercury is small and scientists expected the planet's interior to have already cooled, *Mariner 10* measured Mercury's magnetic field in 2012. It found Mercury's magnetic field to be about 100 times less than Earth's, which suggests that Mercury has a large core that remains hot and fluid—a requirement for generating a magnetic field.

Mercury resembles Earth's Moon in that it has very low reflectivity, no sustained atmosphere, numerous

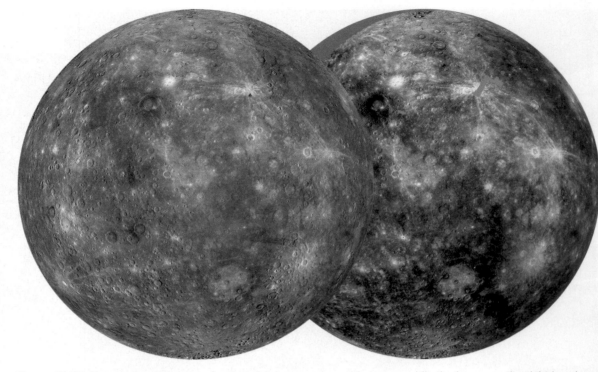

Figure 15.25 Two views of Mercury On the left is a monochromatic image, while the image on the right is color enhanced. These are high-resolution mosaics constructed from thousands of images obtained by the *Messenger* orbiter. (Courtesy of NASA)

volcanic features, and a heavily cratered terrain (**Figure 15.25**). The largest-known impact crater (1300 kilometers [800 mi] in diameter) on Mercury is Caloris Basin. Images and other data gathered by *Mariner 10* show evidence of volcanism in and around Caloris Basin and a few other smaller basins. Also like our Moon, Mercury has smooth plains that cover nearly 40 percent of the area imaged by *Mariner 10*. Most of these smooth areas are associated with large impact basins, including Caloris Basin, where lava partially filled the basins and the surrounding lowlands. Consequently, these smooth plains appear to be similar in origin to lunar maria. Recently, *Messenger* found evidence of volcanism by revealing thick volcanic deposits similar to those on Earth in the Columbia Basin. In addition, researchers were surprised by the recent detection of probable ice caps on Mercury.

Venus: The Veiled Planet

Venus, second only to the Moon in brilliance in the night sky, is named for the Roman Goddess of love and beauty. It orbits the Sun in a nearly perfect circle once every 225 Earth-days. However, Venus rotates in the opposite direction of the other planets at an agonizingly slow pace—1 Venus day is equivalent to about 244 Earth days. Venus has the densest atmosphere of the terrestrial planets, consisting mostly of carbon dioxide (97 percent)—the prototype for an extreme *greenhouse effect*. As a result, the surface temperature of Venus averages about 450°C (900°F) day and night.

Temperature variations at the surface are generally minimal because of the intense mixing within the planet's dense atmosphere. Investigations of the extreme and uniform surface temperature led scientists to more fully understand how the greenhouse effect operates on Earth.

The composition of the Venusian interior is probably similar to Earth's. However, Venus's weak magnetic field means its internal dynamics must be very different from Earth's. Mantle convection is thought to operate on Venus, but the processes of plate tectonics, which recycle rigid lithosphere, do not appear to have contributed to the present Venusian topography.

The surface of Venus is completely hidden from view by a thick cloud layer composed mainly of tiny sulfuric acid droplets. In the 1970s, despite extreme temperatures and pressures, four Russian spacecraft landed successfully and obtained surface images. As expected, however, all the probes were crushed by the planet's immense atmospheric pressure, approximately 90 times that on Earth, within an hour of landing. Using radar imaging, the unstaffed spacecraft *Magellan* mapped Venus's surface in stunning detail (**Figure 15.26**).

A few thousand impact craters have been identified on Venus—far fewer than on Mercury and Mars but more than on Earth. Researchers expected that Venus would show evidence of extensive cratering from the heavy bombardment period but found instead that a period of extensive volcanism was responsible for resurfacing Venus. The planet's thick atmosphere also

Figure 15.26 Global view of the surface of Venus This computer-generated image of Venus was constructed from years of investigations culminating with the *Magellan* mission. The twisting bright features that cross the globe are highly fractured ridges and canyons of the eastern Aphrodite highlands. (Courtesy of NASA/JPL)

Aphrodite highlands

Elevation of surface

Low ⟶ High

limits the number of impacts by breaking up large incoming meteoroids and incinerating most of the small debris.

About 80 percent of the Venusian surface consists of low-lying plains covered by lava flows, some of which traveled along lava channels that extend hundreds of kilometers. Venus's Baltis Vallis, the longest-known lava channel in the solar system, meanders 6800 kilometers (4255 miles) across the planet.

More than 1000 volcanoes larger than 20 kilometers (12 miles) wide have been identified on Venus. However, high surface pressures keep the gaseous components in lava from escaping. This retards the production of pyroclastic material and lava fountaining, phenomena that tend to steepen volcanic cones. In addition, because of Venus's high temperature, lava remains mobile longer and thus flows far from the vent. Both of these factors result in volcanoes that tend to be shorter and wider than those on Earth or Mars (**Figure 15.27**). Maat Mons, the largest volcano on Venus, is about 8.5 kilometers high and 400 kilometers wide. By comparison, Mauna Loa, the largest volcano on Earth, is about 9 kilometers high and only 120 kilometers wide.

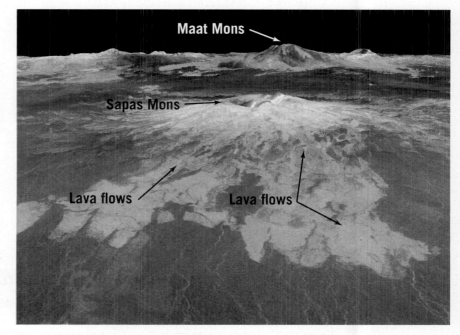

Maat Mons

Sapas Mons

Lava flows

Lava flows

Figure 15.27 Volcanoes on Venus Sapas Mons is a broad volcano, 400 kilometers (250 miles) wide. The bright areas in the foreground are lava flows. Another large volcano, Maat Mons, is in the background. (Courtesy of NASA/JPL)

Venus also has major highlands that consist of plateaus, ridges, and topographic rises that stand above the plains. The rises are thought to have formed where hot mantle plumes encountered the base of the planet's crust, causing uplift. Much like mantle plumes on Earth, abundant volcanism is associated with mantle upwelling on Venus. Recent data collected by the European Space Agency's *Venus Express* suggest that Venus's highlands contain silica-rich granitic rock. As such, these elevated landmasses resemble Earth's continents, albeit on a much smaller scale.

Mars: The Red Planet

Mars, the fourth planet from the Sun, is approximately one-half the diameter of Earth and revolves around the Sun in 687 Earth-days. Mean surface temperatures range from lows of −140°C (−220°F) at the poles in the winter to highs of 20°C (68°F) at the equator in the summer. Although seasonal temperature variations are similar to Earth's, daily temperature variations are greater due to the very thin atmosphere of Mars (only 1 percent as dense as Earth's). The tenuous Martian atmosphere consists primarily of carbon dioxide (95 percent), with small amounts of nitrogen, oxygen, and water vapor.

Topography Mars, like the Moon, is pitted with impact craters. The smaller craters are usually filled with wind-blown dust—confirming that Mars is a dry, desert world.

The reddish color of the Martian landscape is due to iron oxide (rust). Large impact craters provide information about the nature of the Martian surface. For example, if the surface is composed of dry dust and rocky debris, ejecta similar to that surrounding lunar craters is to be expected. But the ejecta surrounding some Martian craters has a different appearance—that of a muddy slurry that was splashed from the crater. Planetary geologists infer that a layer of permafrost (frozen, icy soil) lies below portions of the Martian surface and that impacts heated and melted the ice to produce the fluid-like appearance of these ejecta.

About two-thirds of the surface of Mars consists of heavily cratered highlands, concentrated mostly in its southern hemisphere (**Figure 15.28**). The period of extreme cratering occurred early in the planet's history and ended about 3.8 billion years ago, as it did in the rest of the solar system. Thus, the Martian highlands are similar in age to the lunar highlands.

Based on relatively low crater counts, the northern plains, which account for the remaining one-third of the planet, are younger than the highlands. If Mars once had abundant water, it would have flowed to the north, which is lower in elevation, forming an expansive ocean (shown in blue in Figure 15.28), indicating lower elevation. The relatively flat topography of the northern plains, possibly the smoothest surface in the solar system, is consistent with vast outpourings of fluid basaltic lavas. Visible on these plains are volcanic cones, some with summit pits (craters) and lava flows with wrinkled edges.

Two hemispheres of Mars

Figure 15.28 Two hemispheres of Mars Color represents height above (or below) the mean planetary radius: White is about 12 kilometers (7 miles) above average, and dark blue is 8 kilometers (5 miles) below average. (Courtesy of NASA/JPL)

Located along the Martian equator is an enormous elevated region, about the size of North America, called the *Tharsis bulge*. This feature, about 10 kilometers (6 mile) high, appears to have been uplifted and capped with a massive accumulation of volcanic rock that includes the solar system's largest volcanoes.

The tectonic forces that created the Tharsis region also produced fractures that radiate from its center, like spokes on a bicycle wheel. Along the eastern flanks of the bulge, a series of vast canyons called *Valles Marineris* (Mariner Valleys) developed. Valles Marineris is so vast it can be seen in the image of Mars in Figure 15.28. This canyon network was largely created by down-faulting, not by stream erosion, as is the case with Arizona's Grand Canyon. Thus, it consists of graben-like valleys similar to the East African Rift valleys. Once formed, Valles Marineris grew by water erosion and collapse of the rift walls. The main canyon is more than 5000 kilometers long, 7 kilometers deep, and 100 kilometers wide.

Other prominent features on of the Martian landscape are large impact basins. Hellas, the largest identifiable impact structure on the planet, is about 2300 kilometers (1400 miles) in diameter and has the planet's lowest elevation (see Figure 15.28). Debris ejected from this basin contributed to the elevation of the adjacent highlands. Other buried crater basins that are even larger than Hellas probably exist.

Volcanoes on Mars Volcanism has been prevalent on Mars during most of its history. The scarcity of impact craters on some volcanic surfaces suggests that the planet is still active. Mars has several of the largest-known volcanoes in the solar system, including the largest, Olympus Mons, which is about the size of Arizona and stands nearly three times higher than Mount Everest. This gigantic volcano was last active about 100 million years ago and resembles Earth's Hawaiian shield volcanoes (**Figure 15.29**).

How did the volcanoes on Mars grow so much larger than similar structures on Earth? The largest volcanoes on the terrestrial planets tend to form where plumes of hot rock rise from deep within their interiors. On Earth, moving plates keep the crust in constant motion. Consequently, mantle plumes tend to produce a chain of volcanic structures, like the Hawaiian Islands. By contrast, plate tectonics on Mars is absent, so successive eruptions accumulate in the same location. As a result, enormous volcanoes such as Olympus Mons form rather than a string of smaller ones.

Wind Erosion on Mars Currently, the dominant force shaping the Martian surface is wind erosion. Extensive

Caldera of
Olympus Mons

SmartFigure 15.29 Olympus Mons This massive inactive shield volcano on Mars covers an area about the size of the state of Arizona. (NASA/JPL)

dust storms, with winds up to 270 kilometers (170 miles) per hour, can persist for weeks. Dust devils have also been photographed. Most of the Martian landscape resembles Earth's rocky deserts, with abundant dunes and low areas partially filled with dust.

Water Ice on Mars Liquid water does not appear to exist anywhere on the Martian surface. However, poleward of about 30° latitude, ice can be found within 1 meter (3 feet) of the surface. In the polar regions, it forms small permanent ice caps. Current estimates place the maximum amount of water ice held by the Martian polar ice caps at about 1.5 times the amount covering Greenland.

Considerable evidence indicates that in the first 1 billion years of the planet's history, liquid water flowed on the surface, creating stream valleys and related features (**Figure 15.30**). One location where running water

This NASA image obtained by *Curiosity* rover shows rounded gravel fragments within a rock outcrop consistent with the sedimentary rock conglomerate. Weathered rock fragments can be seen below.

A typical sample of the sedimentary rock conglomerate that contains rounded gravel fragments deposited in a stream bed on Earth.

Figure 15.30 Similar rock outcrops on Mars and Earth This set of images compares a rock outcrop on Mars (left) with similar rocks on Earth. The rock outcrop on Earth formed in a stream bed, which suggests that the Martian rocks formed in a similar environment. Based on this finding, the chief scientist, John Grotzinger, concluded that there was once "a vigorous flow on the surface of Mars." (Courtesy of NASA/JPL)

the surface collapsed. The most likely source of water for these flood-created valleys was the melting of subsurface ice. If the meltwater was trapped beneath a thick layer of permafrost, pressure could mount until a catastrophic release occurred. As the water escaped, the overlying surface would collapse, creating the chaotic terrain.

Not all Martian valleys appear to have resulted from water released in this manner. Some exhibit branching, tree-like patterns that resemble dendritic drainage networks on Earth. In addition, the *Opportunity* rover investigated structures similar to features created by water on Earth. These included layered sedimentary rocks, playas (salt flats), and lake beds. Minerals that form only in the presence of water, such as hydrated sulfates, were also detected. Small spherical concretions of hematite, dubbed "blueberries," were found that probably precipitated from water to form lake sediments. Nevertheless, except in the polar regions, water does not appear to have significantly altered the topography of Mars for more than 1 billion years.

On August 6, 2012, the Mars rover *Curiosity* landed in Gale Crater, near what NASA calls Mount Sharp (officially known as Aeolis Mons). We can only imagine what *Curiosity* will uncover in its quest to study the habitability, climate, and geology of Mars.

15.5 CONCEPT CHECKS

1. What body in our solar system is most like Mercury?
2. Why are the surface temperatures so much higher on Venus than on Earth?
3. Venus was once referred to as "Earth's twin." How are these two planets similar? How do they differ from one another?
4. What surface features do Mars and Earth have in common?
5. Why are the largest volcanoes on Earth so much smaller than the largest ones on Mars?
6. What evidence suggests that Mars had an active hydrologic cycle in the past?

---

was involved in carving valleys can be seen in the Mars *Reconnaissance Orbiter* image in **Figure 15.31**. Notice the streamlike banks that contain numerous teardrop-shaped islands. These valleys appear to have been cut by catastrophic floods with discharge rates that were more than 1000 times greater than those of the Mississippi River. Most of these large flood channels emerge from areas of chaotic topography that appear to have formed when

---

15.6 Jovian Planets

Compare and contrast the four Jovian planets.

The four Jovian (Jupiter-like) planets in order from the Sun are Jupiter, Saturn, Uranus, and Neptune. Because of their location within the solar system and their size and composition they are also commonly called the *outer planets* and the *gas giants*.

Jupiter: Lord of the Heavens

The giant among planets, Jupiter has a mass 2.5 times greater than the combined mass of all other planets, satellites, and asteroids in the solar system. However, it

pales in comparison to the Sun, with only 1/800 of the Sun's mass.

Jupiter orbits the Sun once every 12 Earth-years, and it rotates more rapidly than any other planet, completing one rotation in slightly less than 10 hours. When viewed telescopically, the effect of this fast spin is noticeable. The bulge of the equatorial region and the slight flattening at the poles are evident (see the Polar Flattening column in Table 15.2).

Jupiter's appearance is mainly attributable to the colors of light reflected from its three main cloud layers (**Figure 15.32**). The warmest, and lowest, layer is composed mainly of water ice and appears blue-gray; it is generally not seen in visible-light images. The middle layer, where temperatures are lower, consists of brown to orange-brown clouds of ammonium hydrosulfide droplets. These colors are thought to be by-products of chemical reactions occurring in Jupiter's atmosphere. Near the top of its atmosphere lie white wispy clouds of ammonia ice.

Because of its immense gravity, Jupiter is shrinking a few centimeters each year. This contraction generates most of the heat that drives Jupiter's atmospheric circulation. Thus, unlike winds on Earth, which are driven by solar energy, the heat emanating from Jupiter's interior produces the huge convection currents observed in its atmosphere.

Jupiter's convective flow produces alternating dark-colored *belts* and light-colored *zones*, as shown in Figure 15.32. The light clouds (*zones*) are regions where warm material is ascending and cooling, whereas the dark belts represent cool material that is sinking and warming. This convective circulation, along with Jupiter's rapid rotation, generates the high-speed east–west flow observed between the belts and zones.

The largest storm on the planet is the Great Red Spot. This enormous anticyclonic storm that is twice the size of Earth has been known for 300 years. In addition to the Great Red Spot, there are various white and brown oval-shaped storms (see Figure 15.32). The white ovals are the cold cloud tops of huge storms many times larger than hurricanes on Earth. The brown storm clouds reside at lower levels in the atmosphere. Lightning in various white oval storms has been photographed by the *Cassini* spacecraft, but the strikes appear to be less frequent than on Earth.

Jupiter's Moons Jupiter's satellite system, consisting of 67 moons discovered thus far, resembles a miniature solar system. Galileo discovered the four largest satellites, referred to as Galilean satellites, in 1610 (**Figure 15.33**).

Figure 15.32 The structure of Jupiter's atmosphere The areas of light clouds (*zones*) are regions where gases are ascending and cooling. Sinking dominates the flow in the darker cloud layers (*belts*). This convective circulation, along with the rapid rotation of the planet, generates the high-speed winds observed between the belts and zones. (NASA/JPL/Space Science Institute)

A. Io is one of only three volcanically active bodies other than Earth known to exist in the solar system.

B. Europa, the smallest of the Galilean moons, has an icy surface that is crisscrossed by many linear features.

C. Ganymede, the largest Jovian satellite, exhibits cratered areas, smooth regions, and areas covered by numerous parallel grooves.

D. Callisto, the outermost of the Galilean satellites, is densely cratered, much like Earth's Moon

Figure 15.33 Jupiter's four largest moons These moons are often referred to as the Galilean moons because Galileo discovered them. (Courtesy of NASA/NGS Image Collection)

The two largest, Ganymede and Callisto, are roughly the size of Mercury, whereas the two smaller ones, Europa and Io, are about the size of Earth's Moon. The eight largest moons appear to have formed around Jupiter as the solar system condensed.

Jupiter also has many very small satellites (about 20 kilometers [12 miles] in diameter) that revolve in the opposite direction (*retrograde motion*) of the largest moons and have eccentric (elongated) orbits steeply inclined to the Jovian equator. These satellites appear to be asteroids or comets that passed near enough to be gravitationally captured by Jupiter or are remnants of the collisions of larger bodies.

The Galilean moons can be observed with binoculars or a small telescope and are interesting in their own right. Images from *Voyagers 1* and *2* revealed, to the surprise of most geoscientists, that each of the four Galilean satellites is a unique world (see Figure 15.33). The *Galileo* mission also unexpectedly revealed that the composition of each satellite is strikingly different, implying a different evolution for each. For example, Ganymede has a dynamic core that generates a strong magnetic field not observed in other satellites.

The innermost of the Galilean moons, Io, is perhaps the most volcanically active body in our solar system. In all, more than 80 active, sulfurous volcanic centers have been discovered. Umbrella-shaped plumes have been observed rising from Io's surface to heights approaching 200 kilometers (120 miles) (**Figure 15.34A**). The heat source for volcanic activity is tidal energy generated by a relentless "tug of war" between Jupiter and the other Galilean satellites—with Io as the rope. The gravitational field of Jupiter and the other nearby satellites pull and push on Io's tidal bulge as its slightly eccentric orbit takes it alternately closer to and farther from Jupiter. This gravitational flexing of Io is transformed into heat (similar to the back-and-forth bending of a piece of sheet metal) and results in Io's spectacular sulfurous volcanic eruptions. Moreover, lava, thought to be mainly composed of silicate minerals, regularly erupts on its surface (**Figure 15.34B**).

Jupiter's Rings   One of the surprising aspects of the *Voyager 1* mission was the discovery of Jupiter's ring system. More recently, the ring system was thoroughly investigated by the *Galileo* mission. By analyzing how these rings scatter light, researchers determined that the rings are composed of fine, dark particles that are similar in size to smoke particles. Furthermore, the faint nature of the rings indicates that these minute particles are widely dispersed. The main ring is composed of particles believed to be fragments blasted from the surfaces of Metis and Adrastea, two small moons of Jupiter. Impacts on Jupiter's moons Amalthea and Thebe are believed to be the source of the debris from which the outer gossamer ring formed.

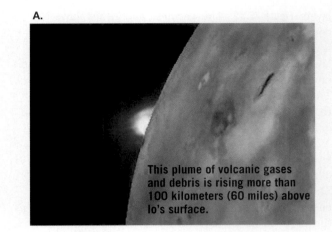

A.

This plume of volcanic gases and debris is rising more than 100 kilometers (60 miles) above Io's surface.

B.

The bright red area on the left side of the image (see arrow) is newly erupted lava.

Figure 15.34 A volcanic eruption on Jupiter's moon Io
(Courtesy of NASA)

Saturn: The Elegant Planet

Requiring more than 29 Earth-years to make one revolution, Saturn is almost twice as far from the Sun as Jupiter, yet their atmospheres, compositions, and internal structures are remarkably similar. The most striking feature of Saturn is its system of rings, first observed by Galileo in 1610 (**Figure 15.35**). Through his primitive telescope, the rings appeared as two small bodies adjacent to the planet. Their ring nature was determined 50 years later by Dutch astronomer Christian Huygens (1629–1695).

Saturn's atmosphere, like Jupiter's, is dynamic. Although the bands of clouds are fainter and wider near the equator, rotating "storms" similar to Jupiter's Great Red Spot occur in Saturn's atmosphere, as does intense lightning. Although the atmosphere is nearly 75 percent hydrogen and 25 percent helium, the clouds (or condensed gases) are composed of ammonia, ammonia hydrosulfide, and water, each segregated by temperature. As on Jupiter, the dynamics of Saturn's atmosphere are driven by the heat released by gravitational compression.

Figure 15.35 Saturn's dynamic ring system The two bright rings, called A ring (outer) and B ring (inner), are separated by the Cassini division. A second small gap (Encke gap) is also visible as a thin line in the outer portion of the A ring. (Courtesy of NASA)

Saturn's Moons The Saturnian satellite system consists of 62 known moons, of which 53 have been named. The moons vary significantly in size, shape, surface age, and origin. Twenty-three of the moons are "original" satellites that formed in tandem with their parent planet. At least 3 (Rhea, Dione, and Tethys) show evidence of tectonic activity, where internal forces have ripped apart their icy surfaces. Others, such as Hyperion, are so porous that impacts punch into their surfaces. Many of Saturn's smallest moons have irregular shapes and are only a few tens of kilometers in diameter.

Saturn's largest moon, Titan, is larger than Mercury and is the second-largest satellite in the solar system. Titan and Neptune's Triton are the only satellites in the solar system known to have substantial atmospheres. Titan was visited and photographed by the *Huygens* probe in 2005. The atmospheric pressure at Titan's surface is about 1.5 times that at Earth's surface, and the atmospheric composition is about 98 percent nitrogen and 2 percent methane, with trace organic compounds. Titan has Earth-like geologic landforms and geologic processes, such as dune formation and stream-like erosion caused by methane "rain." In addition, the northern latitudes appear to have lakes of liquid methane.

Enceladus is another unique satellite of Saturn—one of the few where active eruptions have been observed (**Figure 15.36**). The outgassing, comprised mostly of water, is thought to be the source that replenishes the material in Saturn's E ring. The volcanic-like activity occurs in areas called "tiger stripes" that consists of four large fractures with ridges on either side.

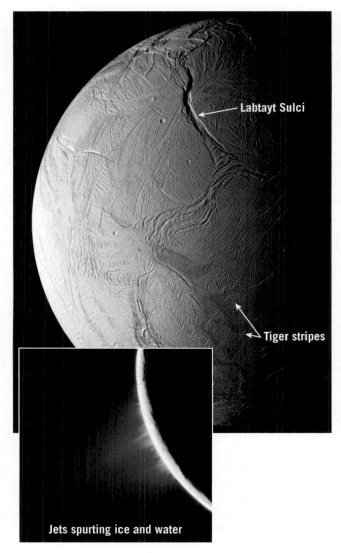

Figure 15.36 Enceladus, Saturn's tectonically active, icy satellite The northern hemisphere contains a 1-kilometer-deep chasm, and linear features, called tiger stripes, are visible in the lower right. Inset image shows jets spurting ice particles, water vapor, and organic compounds from the area of the tiger stripes. (Courtesy of NASA/JPL)

Labtayt Sulci

Tiger stripes

Jets spurting ice and water

A. Pan is a small moon about 30 kilometers in diameter that orbits in the Encke gap, located in the A ring. It is responsible for keeping the Encke gap open by sweeping up any stray material that may enter.

B. Prometheus, a potato-shaped moon, acts as a ring shepherd. Its gravity helps confine the moonlets in Saturn's thin F ring.

Figure 15.37 Two of Saturn's ring moons (Courtesy of NASA/JPL)

Saturn's Ring System In the early 1980s, the nuclear-powered *Voyagers 1* and *2* explored Saturn within 100,000 miles of its surface. More information was collected about Saturn in that short time than had been acquired since Galileo first viewed this "elegant planet" in the early 1600s. More recently, observations from ground-based telescopes, the Hubble Space Telescope, and the *Cassini-Huygens* spacecraft have added to our knowledge of Saturn's ring system. In 1995 and 1996, when the positions of Earth and Saturn allowed the rings to be viewed edge-on, Saturn's faintest rings and satellites became visible. (The rings were visible edge-on again in 2009.)

Saturn's ring system is more like a large rotating disk of varying density and brightness than a series of independent ringlets. Each ring is composed of individual particles—mainly water ice, with lesser amounts of rocky debris—that circle the planet while regularly impacting one another. There are only a few gaps; most of the areas that look like empty space either contain fine dust particles or coated ice particles that are inefficient reflectors of light.

Most of Saturn's rings fall into one of two categories, based on density. Saturn's main (bright) rings, designated A and B, are tightly packed and contain particles that range in size from a few centimeters (pebble-size) to tens of meters (house-size), with most of the particles being roughly the size of a large snowball (see Figure 15.35). In the dense rings, particles collide frequently as they orbit the planet. Although Saturn's main rings (A and B) are 40,000 kilometers wide, they are very thin, only 10 to 30 meters from top to bottom.

At the other extreme are Saturn's faint rings. Saturn's outermost ring (E ring), not visible in Figure 15.35, is

composed of widely dispersed, tiny particles. Recall that volcanic-like activity on Saturn's satellite Enceladus is thought to be the source of material for the E ring.

Studies have shown that the gravitational tugs of nearby moons tend to shepherd the ring particles by gravitationally altering their orbits (**Figure 15.37**). For example, the F ring, which is very narrow, appears to be the work of satellites located on either side that confine the ring by pulling back particles that try to escape. On the other hand, the Cassini division, a clearly visible gap in Figure 15.35, arises from the gravitational pull of Mimas, one of Saturn's moons.

Some of the ring particles are believed to be debris ejected from the moons embedded in them. It is also possible that material is continually recycled between the rings and the ring moons. The ring moons gradually sweep up particles, which are subsequently ejected by collisions with large chunks of ring material, or perhaps by energetic collisions with other moons. It seems, then, that planetary rings are not the timeless features that we once thought; rather, they are continually recycled.

The origin of planetary ring systems is still being debated. Perhaps the rings formed simultaneously and from the same material as the planets and moons—condensing from a flattened cloud of dust and gases that encircled the parent planet. Or perhaps the rings formed later, when a moon or large asteroid was gravitationally pulled apart after straying too close to a planet. Yet another hypothesis suggests that a foreign body collided catastrophically with one of the planet's moons, and the fragments jostled one another and formed a flat, thin ring. Researchers expect more light to be shed on the origin of planetary rings as the *Cassini-Huygens* spacecraft continues its tour of Saturn.

Uranus and Neptune: Twins

Although Earth and Venus have many similar traits, Uranus and Neptune are perhaps more deserving of being called "twins." They are nearly equal in diameter (both about four times the size of Earth), and they are both bluish in appearance, as a result of methane in their atmospheres. Their days are nearly the same length, and their cores are made of rocky silicates and iron—similar to the other gas giants. Their mantles, made mainly of water, ammonia, and methane, are thought to be very different from those of Jupiter and Saturn. One of the most pronounced differences between Uranus and Neptune is the time they take to complete one revolution around the Sun—84 and 165 Earth-years, respectively.

Uranus: The Sideways Planet

Unique to Uranus is the orientation of its axis of rotation. Whereas the other planets resemble spinning toy tops as they circle the Sun, Uranus is like a top that has been knocked on its side but remains spinning (**Figure 15.38**). This unusual characteristic of Uranus is likely due to one or more impacts essentially knocking the planet sideways from its original orientation early in its evolution.

Uranus shows evidence of huge storm systems equivalent in size to the United States. Recent photographs from the Hubble Space Telescope also reveal banded clouds composed mainly of ammonia and methane ice—similar to the cloud systems of the other gas giants.

Uranus's Moons

Spectacular views from *Voyager 2* showed that Uranus's five largest moons have varied terrains. Some have long, deep canyons and linear scars, whereas others possess large, smooth areas on otherwise crater-riddled surfaces. Studies conducted at California's Jet Propulsion Laboratory suggest that Miranda, the innermost of the five largest moons, was recently geologically active—most likely driven by gravitational heating, as occurs on Io.

Uranus's Rings

A surprise discovery in 1977 showed that Uranus has a ring system. The discovery was made as Uranus passed in front of a distant star and blocked its view, a process called *occultation* (occult = hidden.) Observers saw the star "wink" briefly five times (meaning five rings) before the primary occultation and again five times afterward (see Figure 15.38). More recent ground- and space-based observations indicate that Uranus has at least 10 sharp-edged, distinct rings orbiting its equatorial region. Interspersed among these distinct structures are broad sheets of dust.

Neptune: The Windy Planet

Because of Neptune's great distance from Earth, astronomers knew very little about this planet until 1989. Twelve years and nearly 3 billion miles of *Voyager 2* travel provided investigators an amazing opportunity to view the outermost planet in the solar system.

Neptune has a dynamic atmosphere, much like that of the other Jovian planets (**Figure 15.39**). Record wind

Figure 15.38 Uranus, surrounded by its major rings and a few of its known moons Also visible in this image are cloud patterns and several oval storm systems. This false-color image was generated from data obtained by Hubble's Near Infrared Camera. (Image by Hubble Space Telescope, courtesy of NASA)

Great dark spot

Cirrus-like clouds

Dark spot with bright core

Figure 15.39 Neptune's dynamic atmosphere (Courtesy of NASA/JPL)

speeds exceeding 2400 kilometers (1500 miles) per hour encircle the planet, making Neptune one of the windiest places in the solar system. In addition, Neptune exhibits large dark spots that are thought to be rotating storms similar to Jupiter's Great Red Spot. However, Neptune's storms appear to have comparatively short life spans—usually only a few years. Another feature that Neptune has in common with the other Jovian planets is layers of white, cirrus-like clouds (probably frozen methane) about 50 kilometers (30 miles) above the main cloud deck.

Neptune's Moons   Neptune has 13 known satellites, the largest of which is the moon Triton; the remaining 12 are small, irregularly shaped bodies. Triton is the only large moon in the solar system that exhibits retrograde motion, indicating that it most likely formed independently and was later gravitationally captured by Neptune.

Triton and a few other icy moons erupt "fluid" ices—an amazing manifestation of volcanism. **Cryovolcanism** (from the Greek *kryos*, meaning "frost") describes the eruption of magmas derived from the partial melting of ice instead of silicate rocks. Triton's icy magma is a mixture of water ice, methane, and probably ammonia. When partially melted, this mixture behaves as molten rock does on Earth. In fact, upon reaching the surface, these magmas can generate quiet outpourings of ice lavas or, occasionally, produce explosive eruptions. An explosive eruptive column can generate the ice equivalent of volcanic ash. In 1989, *Voyager 2* detected active plumes on Triton that rose 8 kilometers (5 miles) above the surface and were blown downwind for more than 100 kilometers (60 miles). In other environments, ice lavas develop that can flow great distances from their source—similar to the fluid basaltic flows on Hawaii.

Neptune's Rings   Neptune has five named rings; two of them are broad, and three are narrow, perhaps no more than 100 kilometers (60 miles) wide. The outermost ring appears to be partially confined by the satellite Galatea. Neptune's rings are most similar to Jupiter's in that they appear faint, which suggests that they are composed mostly of dust-size particles. Neptune's rings also display red colors, indicating that the dust is composed of organic compounds.

15.6 CONCEPT CHECKS

1. What is the nature of Jupiter's Great Red Spot?
2. Why are the Galilean satellites of Jupiter so named?
3. What is distinctive about Jupiter's satellite Io?
4. Why are many of Jupiter's small satellites thought to have been captured?
5. How are Jupiter and Saturn similar to one another?
6. What two roles do ring moons play in the nature of planetary ring systems?
7. How are Saturn's satellite Titan and Neptune's satellite Triton similar to one another?
8. Name three bodies in the solar system other than Earth that exhibit active volcanism.

15.7 Small Solar System Bodies

List and describe the principal characteristics of the small bodies that inhabit the solar system.

There are countless chunks of debris in the vast spaces separating the eight planets and in the outer reaches of the solar system. Some of these objects formed in the young solar nebula and are thought to have changed little since their formation. In 2006, the International Astronomical Union organized solar system objects not classified as planets or moons into two broad categories: **small solar system bodies**, that include *asteroids*, *comets*, and *meteoroids*, and **dwarf planets**. The newest grouping, dwarf planets, includes Ceres, the largest known object in the asteroid belt, and Pluto, a former planet.

Asteroids and meteoroids are composed of rocky and/or metallic material with compositions somewhat like the terrestrial planets. They are distinguished according to size: Asteroids are larger than 100 meters (60 miles) in diameter, whereas meteoroids have diameters less than 100 meters. Comets, on the other hand, are composed mainly of ices, with lesser amounts of dust and small rocky particles, and formed in the outer reaches of the solar system.

Asteroids: Leftover Planetesimals

Asteroids are small bodies (planetesimals) remaining from the formation of the solar system, which means they are about 4.6 billion years old. Most asteroids orbit the Sun between Mars and Jupiter in the region known as the **asteroid belt** (**Figure 15.40**). Only five asteroids are more than 400 kilometers (250 miles) in diameter, but the solar system hosts an estimated 1 to 2 million asteroids larger than 1 kilometer and many millions that are smaller. Some travel along eccentric orbits that take them very near the Sun, and others regularly pass close to Earth and the Moon (Earth-crossing asteroids). Many

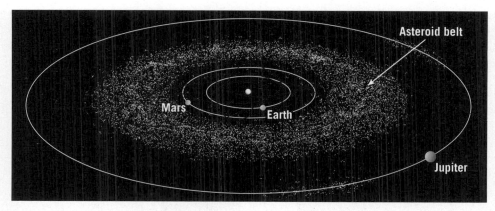

Figure 15.40 The asteroid belt The orbits of most asteroids lie between Mars and Jupiter. Shown in red are the orbits of a few known near-Earth asteroids.

of the recent large-impact craters on the Moon and Earth probably resulted from collisions with asteroids. There are an estimated 1000 to 2000 Earth-crossing asteroids that are more than 0.6 kilometer in diameter. Inevitably, Earth–asteroid collisions will occur again.

Asteroids are rocky and/or metallic remnants of the formation of the solar system. They are not spherical and have varied compositions and histories. Some have low densities, suggesting that they are porous bodies, like "piles of rubble," loosely bound together (**Figure 15.41**). Others belong to groups that came from larger parent bodies that were shattered in past collisions with other asteroids.

In February 2001, an American spacecraft became the first visitor to an asteroid. Although it was not designed for landing, *NEAR Shoemaker* landed successfully on Eros and collected information that has planetary

geologists both intrigued and perplexed. Images obtained as the spacecraft drifted toward the surface of Eros revealed a barren, rocky surface composed of particles ranging in size from fine dust to boulders up to 10 meters (30 feet) across. Researchers unexpectedly discovered that fine debris tends to concentrate in the low areas, where it forms flat deposits resembling ponds. Surrounding the low areas, the landscape is marked by an abundance of large boulders.

One of several hypotheses to explain the boulder-strewn topography is seismic shaking, which would cause the boulders to move upward as the finer materials sink. This is analogous to what happens when a jar of sand and various-sized pebbles is shaken: The larger pebbles rise to the top, while the smaller sand grains settle to the bottom (sometimes referred to as the Brazil nut effect).

Indirect evidence from meteorites suggests that some asteroids might have been heated by a large impact event. A few large asteroids may have completely melted, causing them to differentiate into a dense iron core and a rocky mantle. In November 2005, the Japanese probe *Hayabusa* landed on a small near-Earth asteroid named 25143 Itokawa; it returned to Earth in June 2010. Analyzed samples suggest that Itokawa was probably made up of interior fragments of a large asteroid that broke apart. *Hayabusa 2* is scheduled to launch in 2014, to eventually expose subsurface samples by blasting a crater in asteroid 1999 JU3.

Comets: Dirty Snowballs

Comets, like asteroids, are leftover material from the formation of the solar system. They are loose collections of rocky material, dust, water ice, and frozen gases (ammonia, methane, and carbon dioxide), thus the nickname "dirty snowballs." Recent space missions to comets have shown their surfaces to be dry and dusty, which indicates that their ices are hidden beneath a rocky layer.

Most comets reside in the outer reaches of the solar system and take hundreds of thousands of years

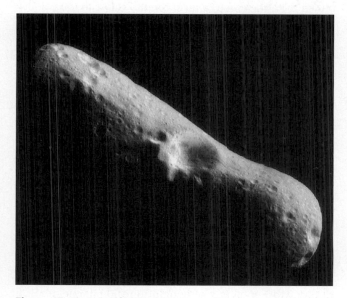

Figure 15.41 Asteroid 25143 Itokawa The barren rocky surface of asteroid Itokawa appears to be a pile of rubble held together by the asteroid's weak gravitational field. (Courtesy of JAXA)

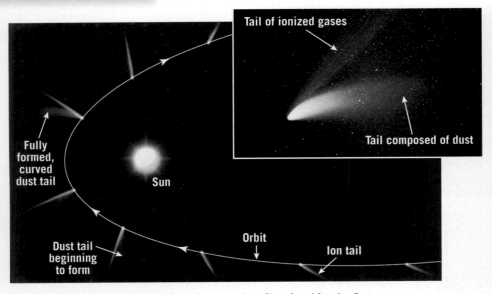

Figure 15.42 Changing orientation of a comet's tail as it orbits the Sun

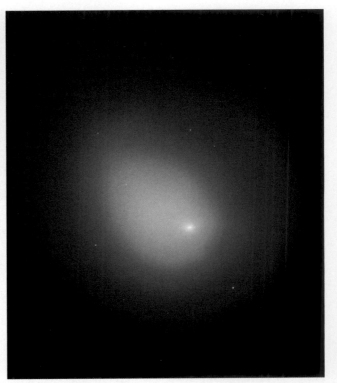

Figure 15.43 Coma and nucleus of Comet Holmes Comet Holmes, which orbits the Sun every 6 years, was uncharacteristically active during its most recent entry into the inner solar system. The nucleus is the small bright spot within the reddish-orange coma. (Image by the Spitzer Space Telescope, courtesy of NASA)

to complete a single orbit around the Sun. However, a smaller number of *short-period comets* (those having orbital periods of less than 200 years), such as the famous Halley's Comet, make regular encounters with the inner solar system (**Figure 15.42**). The shortest-period comet (Encke's Comet) orbits around the Sun once every 3 years.

Structure and Composition of Comets All of the phenomena associated with comets come from a small central body called the **nucleus**. These structures are typically 1 to 10 kilometers in diameter, but nuclei 40 kilometers (25 miles) across have been observed. When comets reach the inner solar system, solar radiation (sunlight) begins to vaporize their ices. The escaping gases carry dust from the comet's surface, producing a highly reflective halo called a **coma** (**Figure 15.43**). Within the coma, the much smaller glowing nucleus can sometimes be detected.

As comets approach the Sun, most develop tails that can extend for millions of kilometers. The tail of a comet points away from the Sun in a slightly curved manner (see Figure 15.42), which led early astronomers to believe that the Sun has a repulsive force that pushes away particles of the coma to form the tail. Scientists have identified two solar forces known to contribute to tail formation. One is *radiation pressure* caused by radiant energy (light) emitted by the Sun, and the second is the *solar wind*, a stream of charged particles ejected from the Sun. Sometimes a single tail composed of both dust and ionized gases is produced, but two tails are often observed. The heavier dust particles produce a slightly curved tail, whereas the extremely light ionized gases are "pushed" directly away from the Sun, forming a straight tail. The inset image in Figure 15.42 clearly shows the orientation of both tails.

As a comet's orbit carries it away from the Sun, the gases forming the coma re-condense, the tail disappears, and the comet returns to cold storage. Material that was blown from the coma to form the tail is lost to the comet forever. When all the gases are expelled, the inactive comet, which closely resembles an asteroid, continues its orbit without a coma or tail. Scientists believe that most comets are active for a few hundred close orbits of the Sun, but this varies based on the size of the nucleus and average solar distance.

The first samples from a comet's coma (Comet Wild 2) were returned to Earth in January 2006 by NASA's *Stardust* spacecraft (**Figure 15.44**). Images from *Stardust* show that the comet's surface was riddled with flat-bottomed depressions and appeared dry, although at least 10 gas jets were active. Laboratory studies revealed that the coma contained a wide range of organic compounds and substantial amounts of silicate crystals.

The Realm of Comets: The Kuiper Belt and Oort Cloud Most comets are thought to reside in one of two regions: the *Kuiper belt* or the *Oort cloud*. Named in honor of astronomer Gerald Kuiper, who predicted its existence, the **Kuiper belt** hosts comets that orbit

Figure 15.44 Comet Wild 2, as seen by NASA's *Stardust* spacecraft The inset is an artist's depiction of jets of gas and dust erupting from Comet Wild 2. (Courtesy of NASA)

Meteoroids: Visitors to Earth

Nearly everyone has seen **meteors**, commonly (but inaccurately) called "shooting stars." These streaks of light can be observed in as little as the blink of an eye or can last as "long" as a few seconds. They occur when a small solid particle, a **meteoroid**, enters Earth's atmosphere from interplanetary space. Heat, created by friction between the meteoroid and the air, produces the light we see trailing across the sky. Most meteoroids originate from one of three sources: (1) interplanetary debris missed by the gravitational sweep of the planets during formation of the solar system, (2) material that is continually being ejected from the asteroid belt, or (3) the rocky and/or metallic remains of comets that once passed through Earth's orbit. A few meteoroids are probably fragments of the Moon, Mars, or possibly Mercury, ejected by a violent asteroid impact. Before *Apollo* astronauts brought Moon rocks back to Earth, meteorites were the only extraterrestrial materials that could be studied in the laboratory.

Meteoroids less than about 1 meter in diameter generally vaporize before reaching Earth's surface. Some, called *micrometeorites*, are so tiny and their rate of fall so slow that they drift to Earth continually as space dust. Researchers estimate that thousands of meteoroids enter Earth's atmosphere every day. After sunset on a clear, dark night, many are bright enough to be seen with the naked eye from Earth.

Meteor Showers Occasionally, meteor sightings increase dramatically to 60 or more per hour. These displays, called **meteor showers**, result when Earth encounters a swarm of meteoroids traveling in the same direction and at nearly the same speed as Earth. The close association of these swarms to the orbits of some short-term comets strongly suggests that they represent material lost by these comets (**Table 15.3**). Some swarms, not associated with the orbits of known comets, are probably the scattered remains of the nucleus of a long-defunct

in the outer solar system, beyond Neptune (see Figure 15.17). This doughnut-shaped structure, which is similar to, but much larger than, the asteroid belt, is thought to contain about 1 billion objects over 1 kilometer in size. However, most comets are too small and too distant to be observed from Earth, even using the Hubble Space Telescope.

Like the asteroids in the inner solar system, most Kuiper belt comets move in slightly elliptical orbits that lie roughly in the same plane as the planets. However, a group of comets have highly eccentric orbits that carry them more than 100 AU away from the Sun and then back toward the orbit of Neptune. Most comets that enter the inner solar system likely belong to this dynamic group of comets. A chance collision between two comets or the gravitational influence of one of the Jovian planets occasionally alters their orbits sufficiently to send them into our view.

Named for Dutch astronomer Jan Oort, the **Oort cloud** is thought to consist of comets that are distributed in all directions from the Sun, forming a spherical shell around the solar system. Most Oort cloud comets orbit the Sun at distances between about 5000 and 100,000 times the Earth–Sun distance. The gravitational effect of a distant passing star may send an occasional Oort cloud comet into a highly eccentric orbit that carries it toward the Sun. However, only a tiny fraction of Oort cloud comets have orbits that bring them into the inner solar system.

TABLE 15.3	Major Meteor Showers	
Shower	**Approximate Dates**	**Associated Comet**
Quadrantids	January 4–6	Unknown
Lyrids	April 20–23	Comet 1861 I
Eta Aquarids	May 3–5	Halley's Comet
Delta Aquarids	July 30	Unknown
Perseids	August 12	Comet 1862 III
Draconids	October 7–10	Comet Giacobini-Zinner
Orionids	October 20	Halley's Comet
Taurids	November 3–13	Comet Encke
Andromedids	November 14	Comet Biela
Leonids	November 18	Comet 1866 I
Geminids	December 4–16	Unknown

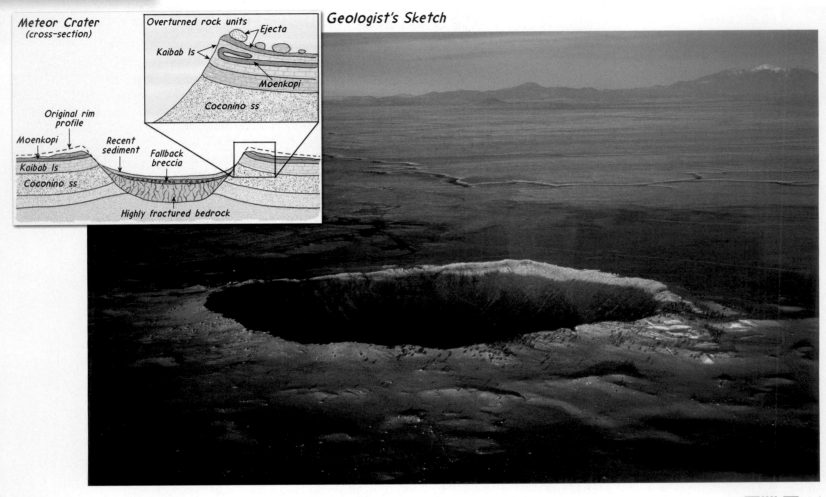

Geologist's Sketch

Meteor Crater
(cross-section)

Overturned rock units — Ejecta
Kaibab ls
Moenkopi
Coconino ss

Original rim
profile
Moenkopi
Recent
sediment
Fallback
breccia
Kaibab ls
Coconino ss
Highly fractured bedrock

SmartFigure 15.45 Meteor Crater, near Winslow, Arizona This cavity is about 1.2 kilometers (0.75 mile) across and 170 meters (560 feet) deep. The solar system is cluttered with asteroids and comets that can strike Earth with explosive force. (Photo by Michael Collier)

comet. The notable *Perseid meteor shower* that occurs each year around August 12 is likely material ejected from the comet Swift-Tuttle on previous approaches to the Sun.

Most meteoroids large enough to survive passage through the atmosphere to impact Earth probably originate among the asteroids, where chance collisions or gravitational interactions with Jupiter modify their orbits and send them toward Earth. Earth's gravity does the rest.

A few very large meteoroids have blasted craters on Earth's surface that strongly resemble those on our Moon. At least 40 terrestrial craters exhibit features that could be produced only by an explosive impact of a large asteroid, or perhaps even a comet nucleus. More than 250 others may be of impact origin. Notable among them is Arizona's Meteor Crater, a huge cavity more than 1 kilometer (0.7 mile) wide and 170 meters (560 feet) deep, with an upturned rim that rises above the

surrounding countryside (**Figure 15.45**). More than 30 tons of iron fragments have been found in the immediate area, but attempts to locate the main body have been unsuccessful. Based on the amount of erosion observed on the crater rim, the impact likely occurred within the past 50,000 years.

Types of Meteorites The remains of meteoroids, when found on Earth, are referred to as **meteorites**. Classified by their composition, meteorites are either (1) *irons*, mostly aggregates of iron with 5–20 percent nickel (**Figure 15.46**); (2) *stony* (also called *chondrites*), silicate minerals with inclusions of other minerals; or (3) *stony–irons*, mixtures of the two. Although stony meteorites are the most common, irons are found in large numbers because metallic meteorites withstand impacts better, weather more slowly, and are easily distinguished from terrestrial rocks. Iron meteorites are probably fragments of once-molten cores of large asteroids or small planets.

The Holsinger Meteorite is the largest discovered fragment of the 150-foot (45-meter) meteor that created Meteor Crater.

Figure 15.46 Iron meteorite found near Meteor Crater, Arizona (Courtesy of M2 Photography/Alamy)

One type of stony meteorite, called a *carbonaceous chondrite*, contains organic compounds and occasionally simple amino acids, which are some of the basic building blocks of life. This discovery confirms similar findings in observational astronomy, which indicate that numerous organic compounds exist in interstellar space.

Data from meteorites have been used to ascertain the internal structure of Earth and the age of the solar system. If meteorites represent the composition of the terrestrial planets, as some planetary geologists suggest, our planet must contain a much larger percentage of iron than is indicated by surface rocks. This is one reason that geologists think Earth's core is mostly iron and nickel. In addition, radiometric dating of meteorites indicates that the age of our solar system is about 4.6 billion years. This "old age" has been confirmed by data obtained from lunar samples.

Dwarf Planets

Since its discovery in 1930, Pluto has been a mystery to astronomers who were searching for another planet in order to explain irregularities in Neptune's orbit. At the time of its discovery, Pluto was thought to be the size of Earth—too small to significantly alter Neptune's orbit. Later, estimates of Pluto's diameter, adjusted because of improved satellite images, indicated that it was less than half Earth's diameter. Then, in 1978, astronomers realized that Pluto appeared much larger than it really is because of the brightness of its newly discovered satellite, Charon. Most recently, calculations based on images obtained by the Hubble Space Telescope show that Pluto's diameter is 2300 kilometers (1430 miles), about one-fifth the diameter of Earth and less than half that of Mercury (long considered the solar system's "runt"). In fact, seven moons in the solar system, including Earth's, are larger than Pluto.

Even more attention was given to Pluto's status as a planet when astronomers discovered another large icy body in orbit beyond Neptune. Soon, more than 1000 of these *Kuiper belt objects* were discovered to be forming a band of objects—a second "asteroid belt" but located at the outskirts of the solar system. The Kuiper belt objects are rich in ices and have physical properties similar to those of comets. Many other planetary objects, some perhaps larger than Pluto, are thought to exist in this belt of icy worlds beyond Neptune's orbit. Researchers soon

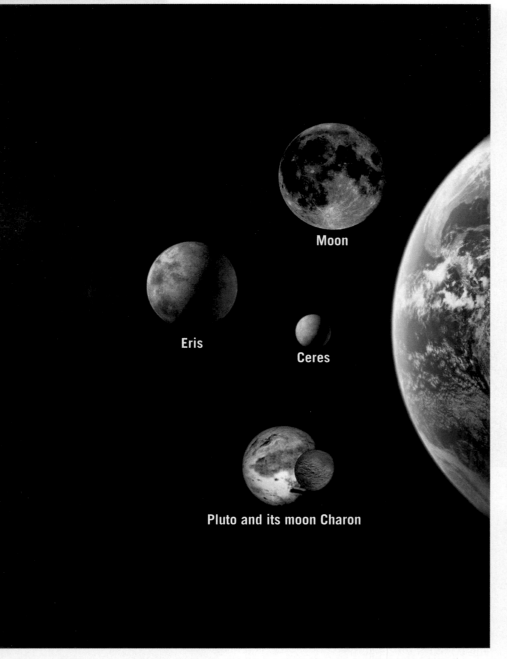

Moon

Eris

Ceres

Pluto and its moon Charon

Figure 15.47 Artist's drawing of the relative sizes of the best-known dwarf planets compared to Earth and its moon Eris, the largest known dwarf planet, has an eccentric orbit that takes it as far as 100 AU away from the Sun. Both Eris and Pluto are composed mainly of ices of water, methane, and ammonia. Ceres is the only identified dwarf planet in the asteroid belt and is located between the orbits of Mars and Jupiter. (Courtesy of NASA)

recognized that Pluto was unique among the planets—completely different from the four rocky, innermost planets, as well as the four gaseous giants.

In 2006, the International Astronomical Union, the group responsible for naming and classifying celestial objects, voted to designate a new class of solar system objects called *dwarf planets*. These are celestial bodies that orbit the Sun and are essentially spherical due to their own gravity but are not large enough to sweep their orbits clear of other debris. By this definition, Pluto is recognized as a dwarf planet and the prototype of this new category of planetary objects. Other dwarf planets include Eris, a Kuiper belt object, and Ceres, the largest-known asteroid (**Figure 15.47**).

Pluto's reclassification was not the first such "demotion." In the mid-1800s, astronomy textbooks listed as many as 11 planets in our solar system, including the asteroids Vesta, Juno, Ceres, and Pallas. Astronomers continued to discover dozens of other "planets," a clear signal that these small bodies represent a class of objects separate from the planets.

The new classification will give a home to the hundreds of additional dwarf planets astronomers assume exist in the solar system. *New Horizons*, the first spacecraft designed to explore the outer solar system, was launched in January 2006. As of September 2012, *New Horizons* was half way between the orbits of Uranus and Neptune. Scheduled to fly by Pluto in July 2015 and later explore the Kuiper belt, *New Horizons* carries tremendous potential for aiding researchers in further understanding the solar system.

15.7 CONCEPT CHECKS

1. Where are most asteroids found?
2. Compare and contrast asteroids and comets.
3. What do you think would happen if Earth passed through the tail of a comet?
4. Where are most comets thought to reside? What eventually becomes of comets that orbit close to the Sun?
5. Differentiate between the following solar system bodies: meteoroid, meteor, and meteorite.
6. What are the three main sources of meteoroids?
7. Why was Pluto demoted from the ranks of the officially recognized planets?

CONCEPTS IN REVIEW: THE NATURE OF THE SOLAR SYSTEM

15.1 Ancient Astronomy

Explain the geocentric view of the solar system and describe how it differs from the heliocentric view.

KEY TERMS: geocentric, celestial sphere, heliocentric, Ptolemaic system, retrograde motion

- Early agrarian cultures, whose survival depended on seasonal change, believed that if the heavenly objects could control the seasons, they must also strongly influence all other Earth events.
- The early Greeks held a geocentric ("Earth-centered") view of the universe, believing that Earth is a sphere that stays motionless at the center of the universe. They believed that the Moon, the Sun, and the known planets—Mercury, Venus, Mars, Jupiter, and Saturn—orbited Earth.
- The early Greeks believed that the stars traveled daily around Earth on a transparent, hollow celestial sphere. In A.D. 141, Claudius Ptolemy documented this geocentric view, called the Ptolemaic system, which became the dominant view of the solar system for over 15 centuries.

15.2 The Birth of Modern Astronomy

List and describe the contributions to modern astronomy of Nicolaus Copernicus, Tycho Brahe, Johannes Kepler, Galileo Galilei, and Isaac Newton.

KEY TERMS: astronomical unit (AU), inertia, law of universal gravitation

- Modern astronomy evolved during the 1500s and 1600s. The scientists involved in the transition from an astronomy that merely describes what is observed to an astronomy that tries to explain why the universe behaves the way it does included Nicolaus Copernicus, Tycho Brahe, Johannes Kepler, Galileo Galilei, and Isaac Newton.
- Nicolaus Copernicus (1473–1543) reconstructed the solar system with the Sun at the center and the planets orbiting around it, but he erroneously continued to use circles to represent the orbits of planets. His Sun-centered view was rejected by the establishment of his day.
- Tycho Brahe's (1546–1601) observations of the planets were far more precise than any made previously and are his legacy to astronomy.
- Johannes Kepler (1571–1630) used Tycho Brahe's observations to usher in a new astronomy with the formulation of his three laws of planetary motion.
- After constructing his own telescope, Galileo Galilei (1564–1642) made many important discoveries that supported the Copernican view of a Sun-centered solar system. This included charting the movement of Jupiter's four largest moons, proving that Earth was not the center of all planetary motion.
- Sir Isaac Newton (1642–1727) demonstrated that the orbit of a planet is a result of a planet's inertia (the tendency of a planet to move in a straight line) and the Sun's gravitational attraction, which bends the planet's path into an elliptical orbit.

15.3 Our Solar System: An Overview

Describe the formation of the solar system according to the nebular theory. Compare and contrast the terrestrial and Jovian planets.

KEY TERMS: nebular theory, solar nebula, planetesimal, protoplanet, terrestrial planet, Jovian planet, escape velocity, impact crater

- Our Sun is the most massive body in a solar system that includes planets, dwarf planets, moons, and other small bodies. The planets all orbit in the same direction and at speeds proportional to their distance from the Sun, with inner planets moving faster, and outer planets moving more slowly.
- The solar system's formation is described by the nebular theory, which proposes that the system began as a solar nebula before condensing due to gravity. While most of the matter ended up in the Sun, some material formed a thick disc around the young Sun and later clumped together into larger and larger bodies. Planetesimals collided to form protoplanets, and protoplanets grew into planets.
- The four terrestrial planets are enriched in rocky materials, whereas the Jovian planets have a higher proportion of ice and gas. The terrestrial planets are relatively dense, with thin atmospheres, while the Jovian planets are less dense and have thick atmospheres.
- Smaller planets have less gravity to retain gases in their atmosphere. It's easier for lightweight gases such as hydrogen and helium to escape, which makes the atmospheres of the terrestrial planets enriched in heavier gases such as water vapor, carbon dioxide, and nitrogen.

15.4 Earth's Moon: A Chip Off the Old Block

List and describe the major features of Earth's Moon and explain how maria basins were generated.

KEY TERMS: maria, lunar highlands (terrae), lunar regolith

- Earth's Moon is the largest moon relative to its planet, and its composition is unique in the solar system, approximately the same as the composition of Earth's mantle (density = 3.3 g/cm^3). The Moon likely formed due to a collision between a Mars-sized protoplanet and the early Earth. The bulk of the protoplanet's iron core was incorporated into Earth, while its rocky mantle material spun off to make the Moon.
- Two types of topography dominate the lunar surface: (1) light-colored lunar highlands (or terrae) dominated by old breccias pulverized by massive bombardments and (2) darker lowlands called maria, which are blanketed by younger flood basalts. Both the terrae and maria are partially covered by a layer of fine material called lunar regolith, produced by micrometeorite bombardment.

? **Briefly describe the formation of our Moon and how its formation accounts for its low density compared to that of Earth.**

15.5 Terrestrial Planets

Outline the principal characteristics of Mercury, Venus, and Mars. Describe their similarities to and their differences from Earth.

- Mercury is the planet closest to the Sun. Because it has a very thin atmosphere and its rate of rotation is extremely slow, the temperature on the surface varies from less than −173°C (−280°F) at night to 427°C (800°F) during daylight hours.
- Venus, the second planet from the Sun, has a relatively dense atmosphere that is dominated by carbon dioxide. The resulting extreme greenhouse effect produces surface temperatures around 450°C (900°F). The topography of Venus has been resurfaced by active volcanism.
- Mars is the fourth planet from the Sun. It has about 1 percent as much atmosphere as Earth, and is relatively cold (−140°C to 20°C [−220°F to 68°F]). Mars appears to be the closest planetary analogue to Earth, showing surface evidence of rifting, volcanism, and surface modification by flowing water. Volcanoes on Mars, such as Olympus Mons, are much larger than volcanoes on Earth because of the lack of plate motion on Mars: The lava accumulates in a single cone rather than forming a long chain of cones, as exemplified by the Hawaiian Islands.

? As you can see from the graph below, Mercury's temperature varies a lot from "day" to "night," but Venus's temperature is relatively constant "around the clock." Suggest a reason for this difference.

Idealized graph comparing the daily temperature variations on Venus and Mercury.

15.6 Jovian Planets

Compare and contrast the four Jovian planets.

KEY TERM: cryovolcanism

- Jupiter is the fifth planet from the Sun. It is very large: several times greater than the masses of everything else in the solar system combined, except for the Sun. Convective flow among its three layers of clouds produces its characteristic banded appearance. Persistent giant rotating storms exist between these bands. Many moons orbit Jupiter, including Io, which shows active volcanism, and Europa, with its icy shell.
- Saturn is the sixth planet from the Sun. Like Jupiter, it is big, gaseous, and endowed with dozens of moons. Some of these moons show evidence of tectonics, while Titan has its own atmosphere. Saturn's well-developed rings are made of numerous particles of water ice and rocky debris.
- Uranus is the seventh planet from the Sun. Like its "twin" Neptune, it has a blue atmosphere dominated by methane, and its diameter is about four times greater than Earth's. Uranus rotates sideways relative to the plane of the solar system. It has a relatively thin ring system and at least five moons.
- Neptune, the eighth planet from the Sun, has an active atmosphere, with fierce winds and giant storms. It has one large moon, Triton, which shows evidence of cryovolcanism, as well as a dozen smaller moons and a ring system.

? Prepare and label a sketch comparing the typical characteristics of terrestrial planets and those of Jovian planets.

15.7 Small Solar System Bodies

List and describe the principal characteristics of the small bodies that inhabit the solar system.

KEY TERMS: small solar system body, dwarf planet, asteroid, asteroid belt, comet, nucleus, coma, Kuiper belt, Oort cloud, meteor, meteoroid, meteor shower, meteorite

- Small solar system bodies include rocky asteroids and icy comets. Both are remnants from the formation of the solar system or fragments from later impacts.
- Most asteroids are concentrated in a wide belt between the orbits of Mars and Jupiter. Some are rocky, some are metallic, and some are basically "piles of rubble" loosely held together by their own weak gravity.
- Comets are dominated by ices, "dirtied" by rocky material and dust. They originate in either the Kuiper belt, located beyond Neptune, or the Oort cloud (a spherical shell around the outermost solar system). When the orbit of a comet brings it through the inner solar system, solar radiation (sunlight) causes its ices to begin to vaporize, generating the coma (gaseous envelope around the comet's nucleus) and its characteristic "tail."
- A meteoroid is debris that enters Earth's atmosphere, flaring briefly as a meteor before either burning up or striking Earth's surface to become a meteorite. Asteroids and material lost from comets as they travel through the inner solar system are the most common source of meteoroids.
- Dwarf planets include Ceres (located in the asteroid belt), the dwarf planet Pluto, and Eris, a Kuiper belt object. They are spherical bodies that orbit the Sun but are not massive enough to have cleared their orbits of debris.

? Shown here are four small solar system bodies. Identify each and explain the differences among them.

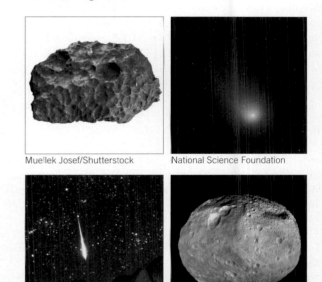

Muellek Josef/Shutterstock   National Science Foundation

Jerry Schad/Science Source/   NASA/JPL
Photo Researchers, Inc.

GIVE IT SOME THOUGHT

1. Use Kepler's third law to answer the following questions:
   a. Determine the period (the time it takes to orbit the Sun) of an imaginary planet that orbits the Sun at a distance of 10 AU.
   b. Determine the distance between the Sun and a planet with a period of 5 years.
   c. Imagine two bodies, one twice as large as the other, orbiting the Sun at the same distance. Which of the bodies, if either, would complete its rotation around the Sun sooner than the other? Explain.

2. Galileo used his telescope to observe the planets and moons in our solar system. These observations allowed him to determine the positions and relative motions of the Sun, Earth, and other objects in the solar system. Refer to Figure 15.14A, which shows an Earth-centered solar system, and Figure 15.14B, which shows a Sun-centered solar system, to complete the following:
   a. Using the Earth-centered model of the solar system, describe the phases of Venus that an observer on Earth would see.
   b. Using the Sun-centered model of the solar system, describe the phases of Venus that an observer on Earth would see.
   c. Explain how Galileo used his observations of the phases of Venus shown in Figure 15.14C to determine the correct model for our solar system.

3. Refer to the accompanying diagram, which shows three pairs of asteroids (A, B, and C). (Assume that all of these asteroids are composed of the same material.) Use Newton's universal law of gravitation to answer the following:
   a. Which asteroid pair exerts the strongest mutual gravitation?
   b. Which asteroid pair exerts the weakest mutual gravitation?

4. Assume that a solar system has been discovered in a nearby region of the Milky Way Galaxy. The accompanying table shows data that have been gathered about three of the planets orbiting the central star of this newly discovered solar system. Using Table 15.2 as a guide, classify each planet as either Jovian, terrestrial, or neither. Explain your reasoning.

	Planet 1	Planet 2	Planet 3
Relative Mass (Earth = 1)	1.2	15	0.1
Diameter (km)	15,000	52,000	5000
Mean Distance from Star (AU)	1.4	17	35
Density (g/cm3)	4.8	1.22	5.3
Orbital Eccentricity	0.01	0.05	0.23

5. In order to conceptualize the size and scale of Earth and Moon as they relate to the rest of the solar system, complete the following:
   a. Approximately how many Moons (diameter 3475 kilometers [2160 miles]) would fit side-by-side across the diameter of Earth (diameter 12,756 kilometers [7926 miles])?
   b. Given that the Moon's orbital radius is 384,798 kilometers, approximately how many Earths would fit side by side between Earth and the Moon?
   c. Approximately how many Earths would fit side by side across the Sun, whose diameter is about 1,390,000 kilometers.
   d. Approximately how many Suns would fit side by side between Earth and the Sun, a distance of about 150,000,000 kilometers?

6. Compare and contrast the lunar highlands' oldest rocks (about 4.4 billion years old) to the oldest rocks on Earth (about 4.0 billion years old). What processes formed, modified, and erased the earliest rocks on both bodies?

7. The accompanying graph shows the temperature at various distances from the Sun during the formation of our solar system. Use it to answer the following:
   a. Which planet(s) formed at locations where the temperature in the solar system was hotter than the boiling point of water?
   b. Which planet(s) formed at locations where the temperature in the solar system was cooler than the freezing point of water?

Condition	Temperature (Farenheit)	Temperature (Kelvins)
Water freezes	32	199
Room temp.	72	296
Human body	98.6	310
Water boils	212	373

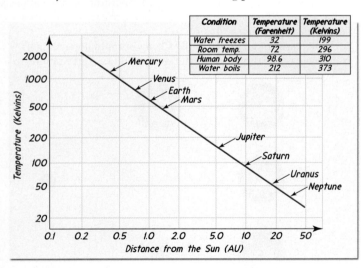

8. The accompanying sketch shows four primary craters (A, B, C, and D). The impact that produced crater A produced two secondary craters (labeled "a") and three rays. Crater D has one secondary crater (labeled "d"). Rank the four primary craters from oldest to youngest and explain your ranking.

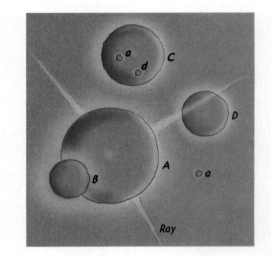

9. The accompanying diagram shows two of Uranus's moons, Ophelia and Cordelia, which act as shepherd moons for the Epsilon ring. Explain what would happen to the Epsilon ring if a large asteroid struck Ophelia, knocking it out of the Uranian system.

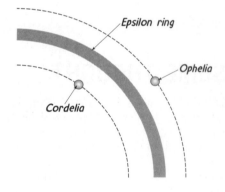

10. It has been estimated that Halley's Comet, which has an orbital period of 76 years, has a mass of 100 billion tons. Furthermore, it is estimated to lose about 100 million tons of material when its orbit brings it close to the Sun. Calculate the maximum remaining life span of Halley's Comet.

11. The accompanying diagram shows a comet traveling toward the Sun at the first position, where it has both an ion tail and a dust tail. Refer to this diagram to complete the following:
    a. For each of the three numbered sites, indicate whether the comet will have no tail, one tail, or two tails. If one tail or two tails are present, in what direction will they point?
    b. Would your answers to the preceding question change if the Sun's energy output were to increase significantly? If so, how would they change?
    c. If the solar wind suddenly ceased, how would it affect this comet and its tails?

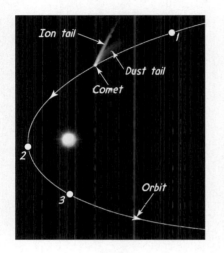

12. Assume that three irregularly shaped planet-like objects, each smaller than our Moon, have just been discovered orbiting the Sun at a distance of 35 AU. One of your friends argues that the objects should be classified as planets because they are large and orbit the Sun. Another friend argues that the objects should be classified as dwarf planets, such as Pluto. State whether you agree or disagree with either or both of your friends. Explain your reasoning.

MasteringGeology™

16

Beyond Our Solar System*

*Revised with the assistance of Professors Teresa Tarbuck and Mark Watry.

FOCUS ON CONCEPTS

Each statement represents the primary learning objective for the corresponding major heading within the chapter. After you complete the chapter, you should be able to:

16.1 Define *cosmology* and describe Edwin Hubble's most significant discovery about the universe.

16.2 Explain why interstellar matter is often referred to as a stellar nursery. Compare and contrast bright and dark nebulae.

16.3 Define *main-sequence star*. Explain the criteria used to classify stars as giants.

16.4 List and describe the stages in the evolution of a typical Sun-like star.

16.5 Compare and contrast the final state of Sun-like stars to the remnants of the most massive stars.

16.6 List the three major types of galaxies. Explain the formation of large elliptical galaxies.

16.7 Describe the big bang theory. Explain what it tells us about the universe.

The Orion Nebula, a huge mass of glowing interstellar gas and dust. (© Anglo-Australian Observatory/Royal Observatory, photo by David Malin)

Astronomers and cosmologists study the nature of our vast universe, attempting to answer questions such as these: Is our Sun a typical star? Do other stars have solar systems with planets similar to Earth? Are galaxies distributed randomly, or are they organized into groups? How do stars form? What happens when stars expend their fuel? If the early universe consisted mostly of hydrogen and helium, how did other elements come into existence? How large is the universe? Did it have a beginning? Will it have an end? This chapter explores these as well as other questions.

16.1 The Universe

Define *cosmology* and describe Edwin Hubble's most significant discovery about the universe.

The universe is more than a collection of dust clouds, stars, stellar remnants, and galaxies (**Figure 16.1**). It is an entity with its own properties.

Cosmology is the study of the universe, including its properties, structure, and evolution. Over the years, cosmologists have developed a comprehensive theory that describes the structure and evolution of the universe. Some of the questions cosmologists seek to answer with this theory include these: How did the universe evolve to its present state? How long has it existed, and how will it end? Modern cosmology addresses these important issues and helps us understand the universe we inhabit.

How Large Is It?

For most of human existence, our universe was thought to be Earth centered, containing only the Sun, Moon, 5 wandering stars, and about 6000 fixed stars. Even after the Copernican view of a Sun-centered universe became widely accepted, the entire universe was believed to consist of a single galaxy, the Milky Way, composed of innumerable stars, along with many faint "fuzzy patches," thought to be clouds of dust and gases.

In the mid-1700s, German philosopher Immanuel Kant proposed that many of the telescopically visible

Figure 16.1 The Trifid Nebula, in the constellation Sagittarius This colorful nebula is a cloud consisting mostly of hydrogen and helium gases. These gases are excited by light emitted by hot, young stars within and produce a reddish glow. (Courtesy of National Optical Astronomy Observatories)

fuzzy patches of light scattered among the stars were actually distant galaxies similar to the Milky Way. Kant described them as "island universes." Each galaxy, he believed, contained billions of stars and was a universe in itself. In Kant's time, however, the weight of opinion favored the hypothesis that the faint patches of light occurred within our galaxy. Admitting otherwise would have implied a vastly larger universe, thereby diminishing the status of Earth and, likewise, humankind.

In 1919 Edwin Hubble arrived at the observatory at Mount Wilson, California, to conduct research using a 2.5-meter (100-inch) telescope, then the world's largest and most advanced astronomical instrument. Armed with this modern tool, Hubble set out to solve the mystery of the "fuzzy patches." At that time, the debate was still raging as to whether the fuzzy patches were "island universes," as Kant had proposed more than 150 years earlier, or clouds of dust and gases (nebulae). To accomplish this task, Hubble studied a group of pulsating stars known as *Cepheid variables*—extremely bright variable stars that increase and decrease in brightness in a repetitive cycle. This group of stars is significant because their "true" brightness, called **absolute magnitude**, can be determined by knowing the rate at which they pulsate (see Appendix D). When the absolute magnitude of a star is compared to its observed brightness, a reliable approximation of distance can be established. (This is similar to how we judge the distance of an oncoming vehicle when driving at night.) Thus, Cepheid variables are important because they can be used to measure large astronomical distances.

Using the telescope at Mount Wilson, Hubble found several Cepheid variables embedded in one of the fuzzy patches. However, because these intrinsically bright stars appeared faint, Hubble concluded that they must lie outside the Milky Way. Indeed, one of the objects Hubble observed lies more than 2 million light-years away and is now known as the *Andromeda Galaxy* (**Figure 16.2**).

Based on his observations, Hubble determined that the universe extended far beyond the limits of our

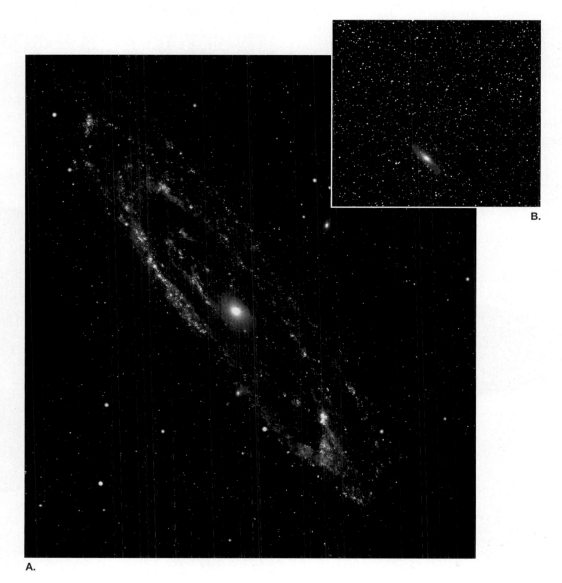

Figure 16.2 Andromeda: A nearby galaxy that is larger than our Milky Way A. Photo of Andromeda Galaxy taken by an optical telescope aboard GALEX Orbiter. (NASA) **B.** Andromeda Galaxy as it appears under low magnification. When viewed with the naked eye, Andromeda appears as a fuzzy patch surrounded by stars. (European Southern Observatory/ESO)

B.

A.

imagination. Today, we know that there are hundreds of billions of galaxies, each containing hundreds of billions of stars. For example, researchers estimate that a million galaxies exist in the portion of the sky bounded by the cup of the Big Dipper. There are literally more stars in the heavens than grains of sand in all the beaches on Earth.

A Brief History of the Universe

Large telescopes can actually "look back in time," which accounts for much of the knowledge astronomers have acquired regarding the history of the universe. Light from celestial objects that are great distances from Earth require millions or even billions of years to reach Earth. The distance light travels in 1 year is called a **light-year** (slightly less than 10 trillion kilometers [6 trillion miles]). Therefore, the farther out telescopes can "see," the farther back in time astronomers are able to study. Even the closest large galaxy, the Andromeda Galaxy, is a staggering 2.5 million light-years away. Light that left Andromeda Galaxy 2.5 million years ago is just now reaching Earth, allowing scientists to observe Andromeda as it was 2.5 million years ago. Light from the farthest-known objects, about 13 billion light-years away, came from stars that have long since burned out.

The time line for the history of the universe, shown in **Figure 16.3**, highlights some of the major events in the evolution of matter and energy. The model that most accurately describes the birth and current state of the universe is the **big bang theory**. According to this theory, all of the energy and matter of the universe existed in an incomprehensibly hot and dense state. About 13.7 billion years ago, our universe began as a cataclysmic explosion, which continued to expand, cool, and evolve to its current state.

In the earliest moments of this expansion, only energy and quarks (subatomic particles that are the building blocks of protons and neutrons) existed. Not until 380,000 years after the initial expansion did the universe cool sufficiently for electrons and protons to combine to form hydrogen and helium atoms—the lightest elements in the universe. For the first time, light traveled through space. Eventually, temperatures decreased sufficiently to allow clumps of matter to collect. This material formed massive clouds of dust and gases (nebulae), which quickly evolved into the first stars and galaxies. Our Sun and planetary system, having formed about 5 billion years ago (nearly 9 billion years after the big bang), is a latecomer to the universe.

16.1 CONCEPT CHECKS

1. What is cosmology?
2. Explain how Edwin Hubble used Cepheid variables to change our view of the structure of the universe.
3. When do cosmologists think the universe began?
4. What two elements were the first to form?

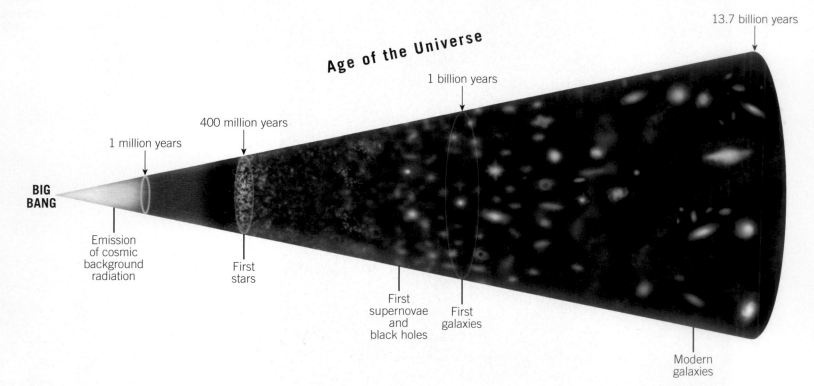

Figure 16.3 Time line for the evolution of the universe According to the big bang theory, the universe began 13.7 billion years and has been expanding ever since.

16.2 Interstellar Matter: Nursery of the Stars

Explain why interstellar matter is often referred to as a stellar nursery. Compare and contrast bright and dark nebulae.

As the universe expanded, gravity caused matter to accumulate into large "clumps" and "strands" of **interstellar matter** known as **nebulae** (meaning "clouds"; the singular of *nebulae* is *nebula*). In addition, considerable amounts of interstellar matter once resided in the interiors of stars and were subsequently returned to space. Some stars ejected matter as part of their normal life cycle, some exploded when they died, and some formed black holes that ejected stream of matter through structures called jets.

Interstellar matter resides between stars within the galaxies and consists of roughly 90 percent hydrogen and 9 percent helium. The remainder, *interstellar dust*, is composed of atoms, molecules, and larger dust grains of the heavier elements. These huge concentrations of interstellar dust and gases are extremely *diffuse*, similar to fog, with no distinct edges or boundaries. However, because nebulae are enormous, their total mass is many times that of the Sun. If a nebula is dense enough, it will contract due to gravity, leading to processes that form stars and planets.

When nebulae are in close proximity to very hot (blue) stars, they glow and are called **bright nebulae**. By contrast, when clouds of interstellar material are too far from bright stars to be illuminated, they are referred to as **dark nebulae**.

Bright Nebulae

There are three main types of bright nebulae: *emission*, *reflection*, and *planetary nebulae*. Emission and reflection nebulae consist of the material from which new stars are born—stellar nurseries. By contrast, planetary nebulae form when stars die and expel material into space. Thus, all stars form from nebulae and many ultimately return to the material of nebulae when their life cycle ends.

Emission Nebulae Glowing clouds of hydrogen gas, called **emission nebulae**, are produced in active star-forming regions of galaxies. Energetic ultraviolet light emitted from hot, young stars ionizes the hydrogen atoms in the nebulae. Because these gases exist under extremely low pressure, they radiate, or emit, their energy as less energetic *visible light*. The conversion of ultraviolet light to visible light is known as *fluorescence*—the same phenomenon that causes neon lights to glow. Hydrogen emits much of its energy in the red portion of the spectrum, which accounts for the red glow from emission nebulae (**Figure 16.4**).

Figure 16.4 Emission nebulae Lagoon Nebulae is a large emission nebula composed mainly of hydrogen. Its red color is attributed to ionized gases, which are excited by the energetic light emitted from young, hot stars embedded in the nebula. (National Optical Astronomy Observatories)

Figure 16.5 Reflection nebulae This blue reflection nebula, located in the Pleiades star cluster, is the result of the scattering of starlight by relatively large molecules and interstellar dust. The Pleiades star cluster, barely visible to the naked eye in the constellation Taurus, is spectacular when viewed through binoculars or a small telescope. (Palomar Observatories/California Institute of Technology)

Figure 16.6 Planetary nebulae The Helix Nebula, the nearest planetary nebula to our solar system. A planetary nebula is the ejected outer envelope of a Sun-like star that formed during the star's collapse from a red giant to a white dwarf. (© Anglo-Australian Observatory, photo by David Malin)

When elements other than hydrogen are ionized, the glowing nebula may exhibit a broader range of colors. An emission nebula that is easily seen with binoculars is located in the constellation Orion, in the sword of the hunter.

Reflection Nebulae As the name implies, **reflection nebulae** merely reflect the light of nearby stars (**Figure 16.5**). Reflection nebulae are likely composed of significant amounts of comparatively large debris, including grains of carbon compounds. This view is supported by the fact that atomic gases with low densities could not reflect light sufficiently to produce the glow observed. Reflection nebulae are usually blue because blue light (shorter wavelength) is scattered more efficiently than red light (longer wavelength)—a process that also produces the blue color of the sky. The blue wisps in Pleiades, shown in Figure 16.5, are reflection nebulae.

Planetary Nebulae **Planetary nebulae** are generally not as diffuse as other nebulae, and they originate from the remnants of dying Sun-like stars (**Figure 16.6**). Planetary nebulae consist of glowing clouds of dust and hot gases that have been expelled near the end of a star's life. When first viewed through primative optical telescopes, planetary nebulae resembled giant planets such as Jupiter, hence their name. A good example of a planetary nebula is the Helix Nebula in the constellation Aquarius (see Figure 16.6).

Dark Nebulae

Recall that when clouds of interstellar material are too distant from bright stars to be illuminated, they are referred to as **dark nebulae**. Exemplified by the Horsehead Nebula in Orion, dark nebulae appear as opaque objects silhouetted against bright backgrounds (**Figure 16.7**). In addition, dark nebulae can also easily be seen as starless regions—"holes in the heavens" when viewing the Milky Way (see Figure 16.14). Although dark nebulae often appear dense, they are made of the same matter as bright nebulae and consist of thinly scattered matter.

Figure 16.7 Dark nebulae The Horsehead Nebula is a dark nebula in the constellation Orion. (Courtesy of the European Southern Observatory)

16.2 CONCEPT CHECKS

1. Why in the phrase "nursery of stars" an appropriate way of describing interstellar matter (nebulae)?
2. Compare and contrast bright and dark nebulae.
3. Why are reflection nebulae generally blue?
4. How are planetary nebulae different from other types of bright nebulae?

16.3 Classifying Stars: Hertzsprung–Russell Diagrams (H-R Diagrams)

Define *main-sequence star*. Explain the criteria used to classify stars as giants.

Early in the twentieth century, Einar Hertzsprung and Henry Russell independently studied the relationship between the true brightness (absolute magnitude) of stars and their respective temperatures. Their work resulted in the development of a graph, called a **Hertzsprung–Russell diagram (H-R diagram)**. By studying H-R diagrams, we can learn a great deal of information about the relationships among the sizes, colors, and temperatures of stars (see Appendix D). For example, we learned that the hottest stars are blue in color and the coolest are red.

To produce an H-R diagram, astronomers survey a portion of the sky and plot each star according to its absolute magnitude (stellar brightness) and temperature (**Figure 16.8**). Notice that the stars in Figure 16.8 are not uniformly distributed. Rather, about 90 percent of all stars fall along a band that runs from the upper-left corner to the lower-right corner of the H-R diagram. These "ordinary" stars are called **main-sequence stars**. As you can see in Figure 16.8, the hottest main-sequence stars are intrinsically the brightest, and, conversely, the coolest are the dimmest.

The absolute magnitude of main-sequence stars is also related to their mass. The hottest (blue) stars are about 50 times more massive than the Sun, whereas the coolest (red) stars are only 1/10 as massive. Therefore, on the H-R diagram, the main-sequence stars appear in decreasing order, from hotter, more massive blue stars to cooler, less massive red stars.

Note the location of the Sun in Figure 16.8. The Sun is a yellow main-sequence star with an absolute magnitude, or "true" brightness, of about 5 (see Appendix D). Because the vast majority of main-sequence stars have magnitudes between –10 and 20, the Sun's midpoint position in this range results in its classification as an "average star." (Note that stellar magnitudes are measured so that the lower the number the brighter the star, and vice versa.)

Just as all humans do not fall into the normal size range, some stars differ significantly from main-sequence stars. Above and to the right of the main sequence stars lies a group of very luminous stars called *giants*, or, on the basis of their color, **red giants** (see Figure 16.8). The size of these giants can be estimated by comparing them

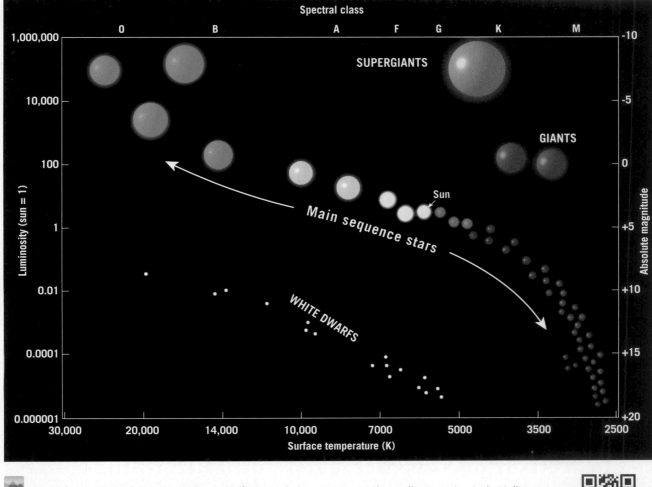

Spectral class

Main sequence stars

SUPERGIANTS

GIANTS

Sun

WHITE DWARFS

SmartFigure 16.8 Hertzsprung–Russell diagram Astronomers use these diagrams to study stellar evolution by plotted stars according to their temperatures and luminosities (absolute magnitudes).

with stars of known size that have the same surface temperature. Scientists have discovered that objects having equal surface temperatures radiate the same amount of energy per unit area. Any difference in the brightness of two stars having the same surface temperature can be attributed to their relative sizes. Therefore, if one red star is 100 times more luminous than another red star, it must have a surface area that is 100 times larger. Stars with large radiating surfaces appear in the upper-right position of an H-R diagram and are appropriately called *giants*.

Some stars are so immense that they are called **supergiants**. Betelgeuse, a bright red supergiant in the constellation Orion, has a radius about 800 times that of the Sun. If this star were at the center of our solar system, it would extend beyond the orbit of Mars, and Earth would find itself buried inside this supergiant!

In the lower portion of the H-R diagram, opposite conditions are observed. These stars are much fainter than main-sequence stars of the same temperature and likewise are much smaller. Some probably

approximate Earth in size. These stars are called *white dwarfs*.

H-R diagrams have proved to be an important tool for interpreting stellar evolution: The stages in which stars, similar to living things, are born, age, and die. Considering that almost 90 percent of stars lie on the main sequence, we can be relatively certain that they spend most of their active years as main-sequence stars. Only a few percent are giants, and perhaps 10 percent are white dwarfs.

16.3 CONCEPT CHECKS

1. On an H-R diagram, where do stars spend most of their life span?
2. How does the Sun compare in size and brightness to other main-sequence stars?
3. Describe how the H-R diagram is used to determine which stars are "giants."

16.4 Stellar Evolution

List and describe the stages in the evolution of a typical Sun-like star.

The idea of describing how a star is born, ages, and dies may seem a bit presumptuous, for most stars have life spans that exceed billions of years. However, by studying stars of different ages, at different points in their life cycles, astronomers have been able to assemble a model for stellar evolution.

The method used to create this model is analogous to how an alien, upon reaching Earth, might determine the developmental stages of human life. By observing large numbers of humans, this stranger would witness the onset of life, the progression of life in children and adults, and the death of the elderly. From this information, the alien could put the stages of human development into their natural sequence. Based on the relative abundance of humans in each stage of development, it would even be possible to conclude that humans spend more of their lives as adults than as toddlers. Similarly, astronomers have pieced together the life story of stars.

The first stars probably formed about 300 million years after the big bang. The most massive nebulae were the birthplace of the first stars because their immense gravity caused them to collapse quickly. As a result, the stars that formed early in the history of the universe, often referred to as *first-generation stars*, were very massive. First-generation stars consisted mostly of hydrogen, with lesser amounts of helium, the primary elements formed during the big bang. Massive stars have relatively short lifetimes followed by violent, explosive deaths. These explosions create the heavier elements and expel them into space. Some of this ejected matter is incorporated into subsequent generations of stars such as our Sun.

Every stage of a star's life is ruled by gravity. The mutual gravitational attraction of particles in a thin, gaseous nebula causes the cloud to collapse on itself. As the cloud is squeezed to unimaginable pressures, its temperature increases, eventually igniting its nuclear furnace, and a star is born. A star is a ball of very hot gases, caught between the opposing forces of gravity trying to contract it and thermal nuclear energy trying to expand it. Eventually, all of a star's nuclear fuel will be exhausted, and gravity will prevail, collapsing the stellar remnant into a small, dense body.

Stellar Birth

The birthplaces of stars are interstellar clouds, rich in dust and gases (see Figure 16.4). In the Milky Way, these gaseous clouds are about 92 percent hydrogen, 7 percent helium, and less than 1 percent heavier elements.

If these thin gaseous clouds become sufficiently concentrated, they begin to gravitationally contract (**Figure 16.9**). A mechanism that may trigger star

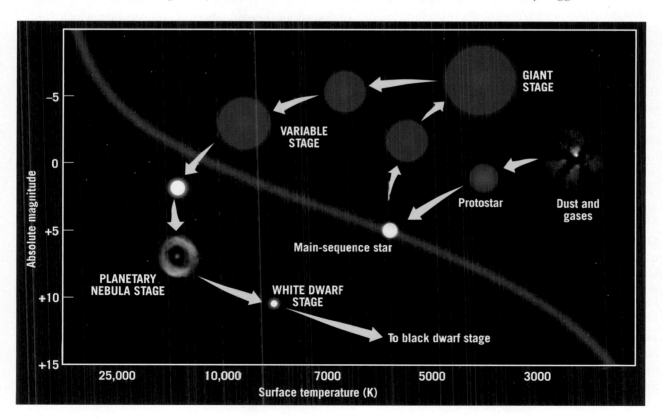

Figure 16.9 H-R diagram showing stellar evolution for a star about as massive as the Sun

formation is a shock wave from a catastrophic explosion (supernova) of a nearby star. Slow dissipation of thermal energy is also thought to cause nebulae to collapse. Regardless of how the process is initiated, once it begins, mutual gravitational attraction of the particles causes the cloud to contract, pulling every particle toward the center.

As the cloud collapses, gravitational energy is converted into energy of motion, or thermal energy, causing the contracting gases to gradually increase in temperature. When the temperature of these gaseous bodies increases sufficiently, they begin to radiate energy in the form of long-wavelength red light. Because these large red objects are not hot enough to engage in nuclear fusion, they are not yet stars. The name **protostar** is applied to these bodies (see Figure 16.9).

Protostar Stage

During the protostar stage, gravitational contraction continues, slowly at first and then much more rapidly. This collapse causes the core of the developing star to heat much faster than its outer envelope. (Stellar temperatures are expressed in Kelvins (K); see Appendix A.) When the core reaches a temperature of 10 million K, the pressure within is so intense that groups of four hydrogen nuclei (through a several-step process) fuse into single helium nuclei. Astronomers refer to this nuclear reaction, in which hydrogen nuclei are fused into helium, as **hydrogen fusion**.

The immense amount of heat released by hydrogen fusion causes the gases inside stars to move with increased vigor, raising the internal gas pressure. At some point, the increased atomic motion produces an outward force (gas pressure) that balances the inward-directed force of gravity. Upon reaching this balance, the stars become *stable main-sequence stars* (see Figure 16.9). In other words, a main-sequence star is one in which the force of gravity, in an effort to squeeze the star into the smallest possible ball, is precisely balanced by gas pressure created by hydrogen fusion in the star's interior.

Main-Sequence Stage

During the main-sequence stage, stars experience minimal changes in size or energy output. Hydrogen is continually being converted into helium, and the energy released maintains the gas pressure sufficiently high to prevent gravitational collapse.

How long can stars maintain this balance? Hot, massive blue stars radiate energy at such an enormous rate that they substantially deplete their hydrogen fuel in only a few million years, approaching the end of their main-sequence stage rapidly. By contrast, the smallest (red) main-sequence stars may take hundreds of billions of years to burn their hydrogen; they live practically forever. A yellow star, such as the Sun, typically remains a main-sequence star for about 10 billion years. Because the solar system is about 5 billion years old, the Sun is expected to remain a stable main-sequence star for another 5 billion years.

An average star spends 90 percent of its life as a hydrogen-burning main-sequence star. Once the hydrogen fuel in a star's core is depleted, it evolves rapidly and dies. However, with the exception of the least-massive stars, death is delayed when another type of nuclear reaction is triggered and the star becomes a *red giant* (see Figure 16.9).

Red Giant Stage

Evolution to the red giant stage begins when the usable hydrogen in a star's interior is consumed, leaving a helium-rich core. Although hydrogen fusion is still progressing in the star's outer shell, no fusion is taking place in the core. Without a source of energy, the core no longer has the gas pressure necessary to support itself against the inward force of gravity. As a result, the core begins to contract.

The collapse of a star's interior causes its temperature to rise rapidly as gravitational energy is converted into thermal energy. Some of this energy is radiated outward, initiating a more vigorous level of hydrogen fusion in the shell surrounding the star's core. The additional heat from the accelerated rate of hydrogen fusion expands the star's outer gaseous shell enormously. Sun-like stars become bloated **red giants**, while the most massive stars become *supergiants*, which are thousands of times larger than their main-sequence size.

As a star expands, its surface cools, which explains the star's color: Relatively cool objects radiate more energy as long-wavelength radiation (nearer the red end of the visible spectrum). Eventually the star's gravitational force stops this outward expansion, and the two opposing forces—gravity and gas pressure—once again achieve balance. The star enters a stable state but is much larger in size. Some red giants overshoot the equilibrium point and instead rebound like an overextended spring. These stars, which alternately expand and contract, and never reach equilibrium, are known as **variable stars**.

While the outer envelope of a red giant expands, the core continues to collapse, and the internal temperature eventually reaches 100 million K. This astonishing temperature triggers another nuclear reaction, in which helium is converted to carbon. At this point, a red giant consumes both hydrogen and helium to produce energy. In stars more massive than the Sun, other thermonuclear reactions occur that generate the elements on the periodic table up to and including number 26, iron.

Burnout and Death

What happens to stars after the red giant phase? We know that stars, regardless of their size, must eventually exhaust their usable nuclear fuel and collapse in response to their immense gravity. Because the gravitational field of a star is dependent on its mass, low-mass stars and high-mass stars have different fates.

Death of Low-Mass Stars Stars less than one-half the mass of the Sun (0.5 solar mass) consume their fuel at relatively low rates (**Figure 16.10A**). Consequently, there are many small, *cool red stars* that may remain stable for as long as 100 billion years. Because the interiors of low-mass stars never attain sufficiently high temperatures and pressures to fuse helium, their only energy source is hydrogen fusion. Thus, a low-mass star never becomes a bloated red giant. Rather, it remains a stable main-sequence star until it consumes

its usable hydrogen fuel and collapses into a hot, dense *white dwarf*.

Death of Medium-Mass (Sun-Like) Stars Stars with masses ranging between one-half and eight times that of the Sun have a similar evolutionary history (**Figure 16.10B**). During their red giant phase, Sun-like stars fuse hydrogen and helium fuel at accelerated rates. Once this fuel is exhausted, these stars (like low-mass stars) collapse into Earth-size bodies of great density—white dwarfs. Without a source of nuclear energy, white dwarfs become cooler and dimmer as they continually radiate thermal energy into space.

During their collapse from red giants to white dwarfs, medium-mass stars cast off their bloated outer atmosphere, creating an expanding spherical cloud of gas. The remaining hot, central white dwarf heats the gas cloud, causing it to glow. Recall that these spectacular spherical clouds are called *planetary nebulae* (see Figure 16.6).

SmartFigure 16.10
Evolutionary stages of stars having various masses

Figure 16.11 Crab Nebula in the constellation Taurus This spectacular nebula is thought to be the remains of the supernova of A.D. 1054. (Courtesy of NASA)

telescope, although Tycho Brahe and Johannes Kepler each recorded one, about 30 years apart, late in the sixteenth century. Chinese astronomers recorded an even brighter supernova in A.D. 1054. Today, the remnant of that great outburst is the Crab Nebula, shown in **Figure 16.11**.

A supernova event is triggered when a massive star has consumed most of its nuclear fuel. Without a heat engine to generate the gas pressure required to balance its immense gravitational field, it collapses. This implosion is enormous, resulting in a shock wave that rebounds out from the star's interior. This energetic shock wave blasts the star's outer shell into space, generating the supernova event.

Theoretical work predicts that during a supernova, the star's interior condenses into an incredibly hot object, possibly no larger than 20 kilometers (12 miles) in diameter. These incomprehensibly dense bodies have been named *neutron stars*. Some supernova events are thought to produce even smaller and more intriguing objects called *black holes*. We consider the nature of neutron stars and black holes in the following section.

Death of Massive Stars In contrast to Sun-like stars, which expire nonviolently, stars that are more than eight times the mass of the Sun have relatively short life spans and terminate in brilliant explosions called **supernovas** (**Figure 16.10C**). During supernova events, these stars become millions of times brighter than they were in prenova stages. If a star located near Earth produced such an outburst, its brilliance would surpass that of the Sun. Fortunately for us, supernovae are relatively rare events; none have been observed in our galaxy since the advent of the

16.4 CONCEPT CHECKS

1. What element is the fuel for main-sequence stars?
2. Describe how main-sequence stars become giants.
3. Why are less massive stars thought to age more slowly than more massive stars, despite the fact they have much less "fuel"?
4. List the steps thought to be involved in the evolution of Sun-like stars.

16.5 Stellar Remnants

Compare and contrast the final state of Sun-like stars to the remnants of the most massive stars.

Eventually, all stars consume their nuclear fuel and collapse into one of three celestial objects—*white dwarfs*, *neutron stars*, or *black holes*. How a star's life ends, and what final form it takes, depends largely on the star's mass during its main-sequence stage (**Table 16.1**).

White Dwarfs

After low- and medium-mass stars consume their remaining fuel, gravity causes them to collapse into **white dwarfs**. The density of these Earth-sized objects, having masses roughly equal to the Sun, may be 1 million times greater than water. A spoonful of such matter would weigh several tons on Earth. Densities of this magnitude are possible only when electrons are displaced inward from their regular orbits around an atom's nucleus. Material in this state is called **degenerate matter**.

The atoms in degenerate matter have been squeezed together so tightly that the electrons are pushed very

TABLE 16.1 Summary of Evolution for Stars of Various Masses

Initial Mass of Main-Sequence Star (Sun = 1)*	Main-Sequence Stage	Giant Phase	Evolution After Giant Phase	Terminal State (Final Mass)
0.001	None (Planet)	No	Not applicable	Planet (0.001)
0.1	Red	No	Not applicable	White dwarf (0.1)
1–3	Yellow	Yes	Planetary Nebula	White dwarf (<1.4)
8	White	Yes	Supernova	Neutron star (1.4–3)
25	Blue	Yes (Supergiant)	Supernova	Black hole (>3)

*These mass numbers are estimates.

close to the nucleus. However, the electrical repulsion that occurs between the negatively charged electrons supports the star against complete gravitational collapse.

As main-sequence stars contract into white dwarfs, their surfaces become extremely hot, sometimes exceeding 25,000 K. Without sources of energy, main-sequence stars slowly cool and eventually become small, cold, burned-out embers called *black dwarfs*. However, our galaxy is not yet old enough for any black dwarfs to have formed.

Neutron Stars

A study of white dwarfs produced a surprising conclusion: The *smallest white dwarfs* are the *most massive*, and the *largest* are the *least massive*. Researchers have discovered that more massive stars, because of their greater gravitational fields, are squeezed into smaller, more densely packed objects than less massive stars. Thus, the smallest white dwarfs were produced from the collapse of larger, more massive main-sequence stars than are the largest white dwarfs.

This conclusion led to the prediction that stellar remnants even *smaller* and *more massive* than white dwarfs must exist. Named **neutron stars**, these objects are the remnants of explosive supernova events. In white dwarfs, the electrons are pushed close to the nucleus, whereas in neutron stars, the electrons are forced to combine with protons in the nucleus to produce neutrons (hence the name *neutron star*). A pea-size sample of this matter would weigh 100 million tons. This is approximately the density of an atomic nucleus; thus, a neutron star can be thought of as a large atomic nucleus, composed entirely of neutrons.

During a supernova implosion, the outer envelope of the star is ejected, while the core collapses into a very hot star that is only about 20 to 30 kilometers (12 to 18 miles) in diameter. Although neutron stars have high surface temperatures, their small size greatly limits their luminosity, making them nearly impossible to locate visually.

Theoretical models predict that neutron stars have a very strong magnetic field and a high rate of rotation. As stars collapse, they rotate faster, for the same reason ice skaters rotate faster as they pull their arms in as they spin. Radio waves generated by the rotating magnetic fields of neutron stars are concentrated into two narrow zones that align with the star's magnetic poles. Consequently, these stars resemble rapidly rotating beacons emitting strong radio waves. If Earth happened to be in the path of these beacons, the star would appear to blink on and off, or pulsate, as the waves swept past.

In the early 1970s, a source that radiates short pulses of radio energy named a **pulsar** (*pulsating radio source*) was discovered in the Crab Nebula (**Figure 16.12**). Visual inspection of this radio source indicated that it was coming from a small star centered in the nebula. The pulsar found in the Crab Nebula is most likely the remains of the supernova of A.D. 1054 (see Figure 16.11).

Figure 16.12 Crab Pulsar: A young neutron star centered in the Crab Nebula This is the first pulsar to have been associated with a supernova. The energy emitted from this star illuminates the Crab Nebula. (Courtesy of NASA)

Black Holes

Although neutron stars are extremely dense, they are not the densest objects in the universe. Stellar evolutionary theory predicts that neutron stars cannot exceed three times the mass of the Sun. Above this mass, not even tightly packed neutrons can withstand the star's gravitational pull. Following supernova explosions, if the core of a remaining star exceeds three solar masses, gravity prevails, and the stellar remnant collapses into an object more dense than a neutron star. (The pre-supernova mass of such a star was likely 25 times that of our Sun.) The incredible objects, or celestial phenomena, created by such a collapse are called **black holes**.

Einstein's theory of general relativity predicts that even though black holes are extremely hot, their surface gravity is so immense that even light cannot escape. Consequently, they literally disappear from sight. Anything that moves too close to a black hole can be swept in and devoured by its immense gravitational field.

How do astronomers find objects whose gravitational field prevents the escape of all matter and energy? Theory predicts that as matter is pulled into a black hole, it should become extremely hot and emit a flood of x-rays before it is engulfed. Because *isolated* black holes do not have a source of matter to engulf, astronomers decided to look at binary-star systems for evidence of matter emitting x-rays while being rapidly swept into a region of apparent nothingness.

The first black hole to be identified, Cygnus X-1, orbits a massive supergiant companion once every 5.6 days. The gases that are pulled from the companion form an *accretion disk* that spirals around a "void" while emitting a steady stream of x-rays (**Figure 16.13**). Recent observations have determined that pairs of jets extend outward from these accretion disks and are thought to return some of this material back to space.

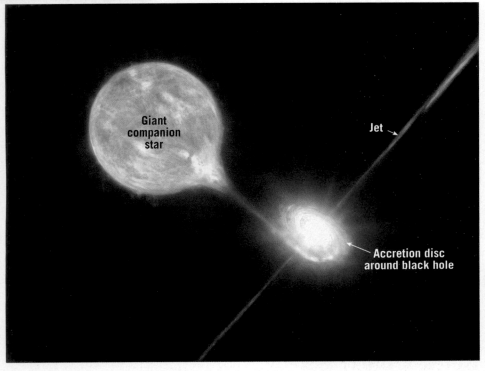

Figure 16.13 Artist's view of a black hole and a giant companion star Note the accretion disk surrounding the black hole. (Courtesy of European Southern Observatory/L. Calcada/M. Kornmesser)

Cygnus X-1, which is about 8 or 9 times as massive as our Sun, probably formed from a star of approximately 40 solar masses. Since the discovery of Cygnus X-1, many other x-ray sources have been discovered that are assumed to be black holes.

Astronomers have established that black holes are common objects in the universe and vary considerably in size. Small black holes have masses approximately 10 times that of our Sun but are only about 32 kilometers (20 miles) across, less than the distance of a marathon course. Intermediate black holes have masses 1000 times our Sun, and the largest black holes (*supermassive black holes*), found in the centers of galaxies, are estimated to be millions of solar masses. Because the earliest stars were thought to be massive, their deaths could have provided the seeds that eventually formed the supermassive black holes at the centers of galaxies.

16.5 CONCEPT CHECKS

1. Describe degenerate matter.
2. What is the final state of a medium-mass (Sun-like) star?
3. How do the "lives" of the most massive stars end? What are the two possible products of this event?
4. Explain how it is possible for the *smallest* white dwarfs to be the *most* massive.
5. Black holes are thought to be abundant, yet they are hard to find. Explain why.

16.6 Galaxies and Galactic Clusters

List the three major types of galaxies. Explain the formation of large elliptical galaxies.

On a clear and moonless night away from city lights, you can see a truly marvelous sight—a band of light stretching from horizon to horizon. With his telescope, Galileo discovered that this band of light was composed of countless individual stars. Today, we realize that the Sun is actually a part of this vast system of stars, the Milky Way Galaxy (**Figure 16.14**).

Galaxies (*galaxias* = milky) are collections of interstellar matter, stars, and stellar remnants that are gravitationally bound (see Figure 16.14). Recent observational data indicate that supermassive black holes may exist at the centers of most galaxies. In addition, spherical halos of very tenuous gas and numerous star clusters (*globular clusters*) appear to surround many of the largest galaxies.

The first galaxies were small and composed mainly of massive stars and abundant interstellar matter. These galaxies grew quickly by accreting nearby interstellar matter and by colliding and merging with other galaxies. In fact, our galaxy is currently absorbing at least two tiny satellite galaxies.

Figure 16.14 View of our Milky Way Galaxy at sunset The dark patches in the "milky" band of light are caused by the presence of dark nebulae. (Courtesy of European Southern Observatory)

Types of Galaxies

Among the hundreds of billions of galaxies, three basic types have been identified: *spiral*, *elliptical*, and *irregular.* Within each of these categories are many variations, the causes of which are still a mystery.

Spiral Galaxies Our Milky Way Galaxy is an example of a large **spiral galaxy** (**Figure 16.15**). Spiral galaxies are flat, disk-shaped objects that range from 20,000 to about 125,000 light-years in diameter. Typically, spiral galaxies have a greater concentration of stars near their centers, but there are numerous variations. As shown in **Figure 16.16**, spiral galaxies have arms (usually two) extending from the central nucleus. Spiral galaxies rotate rapidly in the center, while the outermost stars rotate more slowly, which gives these galaxies the appearance of a fireworks pinwheel. Generally, the central bulge contains older stars that give it a yellowish color, while younger hot stars are located in the arms. The young, hot stars in the arms are found in large groups that appear as bright patches of blue and violet light shown in Figure 16.16.

Many spiral galaxies have a band of stars extending outward from the central bulge that merges with the spiral arms. These are known as **barred spiral galaxies** (**Figure 16.17**). Recent investigations have found evidence that our galaxy probably has a bar structure. What produces these bar-shaped structures is a matter of ongoing research.

Figure 16.15 Dramatic image of the spiral galaxy Messier 83 Although smaller, Messier 83 is thought to be very similar to the Milky Way Galaxy. (European Space Observatory)

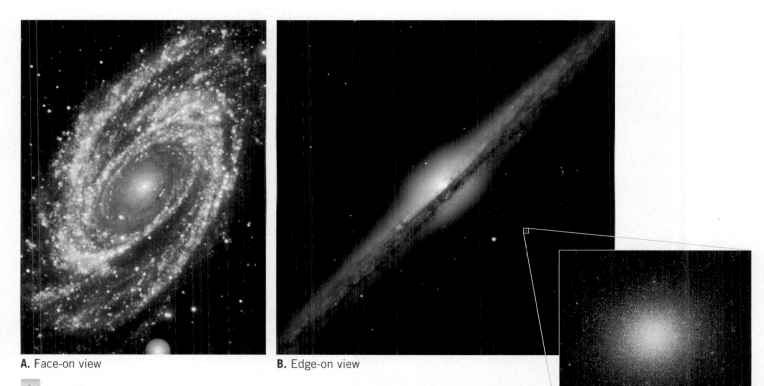

A. Face-on view

B. Edge-on view

C. Globular cluster

SmartFigure 16.16 Spiral galaxies A. Spiral galaxies typically have a greater concentration of older stars near their center, which gives the central bulge its yellowish color. By contrast, the arms of spiral galaxies contain numerous hot, young stars that give these structures a bluish or violet tint. **B.** Edge-on view showing the central bulge. **C.** Surrounding most large galaxies are spherical halos of very tenuous gases and groups of stars called globular clusters. This large globular cluster contains an estimated 10 million stars. (Image A courtesy of NASA images B and C courtesy of European Southern Observatory)

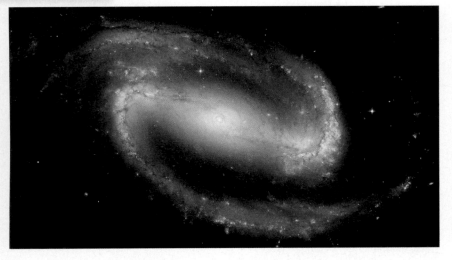

Figure 16.17 Barred spiral galaxy (Courtesy of NASA)

Elliptical Galaxies As the name implies, **elliptical galaxies** have an ellipsoidal shape that can be nearly spherical, and they lack arms (**Figure 16.18**). Some of the largest and the smallest galaxies are elliptical. The smallest of these are known as **dwarf galaxies**. The two small companions of Andromeda shown in Figure 16.2 are dwarf galaxies.

The very largest known galaxies (1 million light-years in diameter) are also elliptical. For comparison, the Milky Way, a large spiral galaxy, is about one-half the

diameter of a large elliptical galaxy. Most large elliptical galaxies are believed to result from the merger of two or more smaller galaxies.

Large elliptical galaxies tend to be composed of older, low-mass stars (red) and have minimal amounts of interstellar matter. Thus, unlike the arms of spiral galaxies, they have low rates of star formation. As a result, elliptical galaxies appear yellow to red in color, as compared to the bluish tint emanating from the young, hot stars in the arms of spiral galaxies.

Irregular Galaxies Approximately 25 percent of known galaxies show no symmetry and are classified as **irregular galaxies**. Some were once spiral or elliptical galaxies that were subsequently distorted by the gravity of a larger neighbor. Two well-known irregular galaxies, the *Large* and *Small Magellanic Clouds*, are named for explorer Ferdinand Magellan, who observed them when he circumnavigated the globe in 1520. They are among our nearest galactic neighbors.

Recent images of the Large Magellanic Cloud revealed a central bar-shaped structure. Thus, the Large Magellanic Cloud was once a barred spiral galaxy that was subsequently distorted by gravity exerted by another galaxy as it passed by.

Galactic Clusters

Once astronomers discovered that stars occur in groups (galaxies), they set out to determine whether galaxies also occur in groups or whether they are just randomly distributed. They found that galaxies are grouped into gravitationally bound clusters (**Figure 16.19**). Some large **galactic clusters** contain thousands of galaxies. Our own galactic cluster, called the **Local Group**, consists of more than 40 galaxies and may contain many undiscovered dwarf galaxies. Of the Local Group galaxies, three are large spiral galaxies, including the Milky Way and Andromeda Galaxies.

Galactic clusters also reside in huge groups called *superclusters*. There are possibly 10 million superclusters; our Local Group is found in the Virgo Supercluster. From visual observations, it appears that superclusters are the largest entities in the universe.

Galactic Collisions

Within galactic clusters, interactions between galaxies, often driven by one galaxy's gravity disturbing another, are common. For example, a large galaxy may engulf a dwarf satellite galaxy. In this case, the larger galaxy will retain its form, while the smaller galaxy will be torn apart and assimilated into the larger galaxy. Recall that two dwarf satellite galaxies are currently merging with the Milky Way.

Galactic interactions may also involve two galaxies of similar size that pass through one another without merging. It is unlikely that the individual stars within

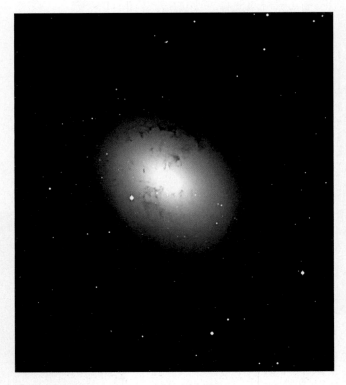

Figure 16.18 Large elliptical galaxy belonging to the Fornax Cluster Dark clouds of interstellar matter are visible within the central nucleus of this galaxy. Some of the star-like objects in this image are large groups of stars (globular clusters) that belong to the galaxy. (Courtesy of European Southern Observatory)

these galaxies will collide because they are so widely dispersed. However, the interstellar matter will likely interact, triggering an intense period of star formation.

In an extreme case, two large galaxies may collide and merge into a single large galaxy (**Figure 16.20**). Many of the largest elliptical galaxies are thought to have been produced by the merger of two large spiral galaxies. Some studies have predicted that in 2 to 4 billion years, there is a 50 percent probability that the Milky Way and Andromeda Galaxies will collide and merge.

Figure 16.19 The Fornax Galaxy Cluster, one of the nearest groupings of galaxies to our Local Group Although many of the galaxies shown are elliptical, an elegant barred spiral galaxy is visible in the lower right. (Courtesy of European Southern Observatory/ J. Emerson/VISTA)

16.6 CONCEPT CHECKS

1. Compare the three main types of galaxies.
2. What type of galaxy is our Milky Way?
3. Describe a possible scenario for the formation of a large elliptical galaxy.

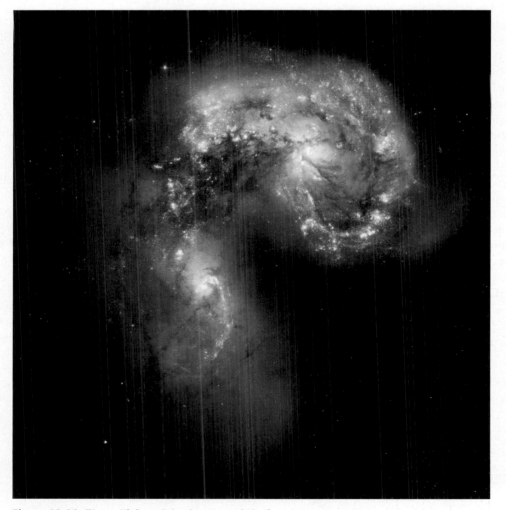

Figure 16.20 The collision of the Antennae Galaxies When two galaxies collide, the stars generally do not. However, the clouds of dust and gas that are common to both collide. During these galactic encounters, there is a rapid birth of millions of stars, shown as the bright regions in this image. (Courtesy of NASA, ESA, and the Hubble Heritage Team (STScI/AURA)-ESA/Hubble Collaboration)

16.7 The Big Bang Theory

Describe the big bang theory. Explain what it tells us about the universe.

The *big bang theory* describes the birth, evolution, and fate of the universe. According to the big bang theory, the universe was originally in an extremely hot, supermassive state that expanded rapidly in all directions. Based on astronomers' best calculations, this expansion began about 13.7 million years ago. What scientific evidence exists to support this theory?

Evidence for an Expanding Universe

In 1912 Vesto Slipher, while working at the Lowell Observatory in Flagstaff, Arizona, was the first to discover that galaxies exhibit motion. The motions he detected were twofold: Galaxies rotate, and galaxies move relative to each other. Slipher's efforts focused on the shifts in the spectra of the light emanating from galaxies (see the information on the Doppler effect in Appendix D). When a source of light is moving away from an observer, the spectral lines shift toward the red end of the spectrum (longer wavelengths). Conversely, when celestial objects approach the observer, the spectral lines shift to the blue end of the spectrum (shorter wavelengths).

In 1929, a study of galaxies conducted by Edwin Hubble expanded the groundwork established by Slipher. Hubble noticed that most galaxies have spectral shifts toward the red end of the spectrum—which occurs when an object emitting light is receding from an observer (**Figure 16.21**). Therefore, all galaxies (except those in the Local Group) appear to be moving away from the Milky Way. These patterns were later named **cosmological red shifts** because the movement they revealed was the result of the expansion of the universe.

Recall that Hubble had also found a way to measure galactic distances. By comparing the distance to a galaxy with Vesto Slipher's measurements of its red shift, Hubble made an unexpected discovery: He discovered

Standard spectral lines (unshifted)

Red shift moves spectral lines to longer wavelengths

Figure 16.21 Red shift Illustration of the shift in spectral lines toward the red end of the spectrum, which occurs when an object emitting light is receding from an observer.

that the red shifts of galaxies increase with distance and that the most distant galaxies are receding from the Milky Way at the fastest rate. This concept, now called **Hubble's law**, states that *galaxies recede at speeds proportional to their distances from the observer*.

This discovery surprised Hubble because conventional wisdom was that the universe was unchanging and would likely remain unchanged. What cosmological theory could explain Hubble's findings? Researchers concluded that an expanding universe accounts for the observed red shifts.

To help visualize why Hubble's law implies an expanding universe, imagine a batch of raisin bread dough that has been set out to rise for a few hours (**Figure 16.22**). In this analogy, the raisins represent galaxies and the dough represents space. As the dough doubles in size, so does the distance between all of the raisins. The distance between raisins that were originally 2 centimeters apart will become 4 centimeters, while the distance between raisins originally 6 centimeters apart will increase to

SmartFigure 16.22 Raisin bread analogy for an expanding universe As the dough rises, raisins originally farthest apart travel greater distances than those located closer together. Thus, in an expanding universe (as with the raisins), more space is created between two objects that are farther apart than between two objects that are closer together.

A. Raisin bread dough before it rises.

B. Raisin bread dough a few hours later.

12 centimeters. The raisins that were originally farthest apart traveled greater distances than those located closer together. Therefore, in an expanding universe, as in our analogy, more space is created between two objects located farther apart than between two objects that are closer together.

Another feature of the expanding universe can be demonstrated using the same bread analogy. Regardless of which raisin you look at, it will move away from all the other raisins. Likewise, at any point in the universe, all other galaxies (except those in the same cluster) are receding from an observer at that location. Hubble's law implies a centerless universe that is expanding uniformly. The Hubble Space Telescope is named in honor of Edwin Hubble's invaluable contributions to the scientific understanding of the universe.

Predictions of the Big Bang Theory

Recall from "The Introduction " that in order for a hypothesis to become an accepted component of scientific knowledge (a theory), it must incorporate predictions that can be tested. One prediction of the big bang model is that if the universe was initially unimaginaby hot, then researchers should be able to detect the remnant of that heat. The electromagnetic radiation (light) emitted by a white-hot universe would have extremely high energy and short wavelengths. However, according to the big bang theory, the continued expansion of the universe would have stretched the waves so that by now they should be detectable as long-wavelength radio waves called microwave radiation. Scientists began to search for this "missing" radiation, which they named *cosmic microwave background radiation*. As predicted, in 1965, this microwave radiation was detected and found to fill the entire visible universe.

Detailed observations of the cosmic microwave background radiation since its original discovery have confirmed many theoretical details of the big bang theory, including the order and timing of important events in the early history of the universe.

What Is the Fate of the Universe?

Cosmologists have developed different scenarios for the ultimate fate of the universe. In one scenario, the stars will slowly burn out, being replaced by invisible degenerate matter and black holes that travel outward through an endless, dark, cold universe. Another possibility is that the outward flight of the galaxies will slow and eventually stop. Gravitational contraction would follow, causing all matter to eventually collide and coalesce into the high-energy, high-density state from which the universe began. This fiery death of the universe, the big bang operating in reverse, has been called the "*big crunch*."

Whether the universe will expand forever or eventually collapse upon itself is contingent on its density. If the average density of the universe were more than an amount known as its *critical density* (about one atom for every cubic meter), gravitational attraction would be sufficient to stop the outward expansion and cause the universe to collapse. On the other hand, if the density of the universe is less than the critical value, it will continue to expand forever. Current estimates place the known matter in the universe below the critical density, which predicts an ever-expanding, or *open*, universe. Additional support for an open universe is based on research which indicates that the universe is expanding faster now than it did in the past. Hence, the view currently favored by most cosmologists is an endless, expanding universe.

It should be noted, however, that the methods used to determine the density of the universe are subject to substantial uncertainties. It is possible that previously undetected matter called **dark matter** exists in great quantities in the universe. Although the concept of *dark matter* may sound foreboding, it simply allows for the possibility that matter exists that does not interact with electromagnetic radiation. Recall that most knowledge about the universe comes to us via light. If there is a form of matter that does not interact with light, we will not be able to "see" it—hence the term *dark matter*. However, that is not the end of the story. It is possible that dark matter will show itself through gravitational interactions with types of matter with which we are familiar. Many astronomers are presently looking for these interactions. If there is enough dark matter, the universe could, in fact, collapse in the "big crunch." Consider the following as astronomers search for dark matter: "Absence of evidence is not evidence of absence."

16.7 CONCEPT CHECKS

1. In your own words, explain how astronomers determined that the universe is expanding.
2. What did the big bang theory predict that was finally confirmed years after it was formulated?
3. Which view of the fate of the universe is currently favored: a "big crunch" or an endless, ever-expanding universe?
4. What property does the universe possess that will determine its final state?

CONCEPTS IN REVIEW: BEYOND OUR SOLAR SYSTEM

16.1 The Universe

Define *cosmology* and describe Edwin Hubble's most significant discovery about the universe.

KEY TERMS: cosmology, absolute magnitude, light-year, big bang theory

- Cosmology is the study of the universe, including its properties, structure, and evolution.
- The universe consists of hundreds of billions of galaxies, each containing billions of stars.
- The model that most accurately describes the birth and current state of the universe is the big bang theory. According to this model, the universe began about 13.7 billion years ago, in a cataclysmic explosion, and then it continued to expand, cool, and evolve to its current state.

16.2 Interstellar Matter: Nursery of the Stars

Explain why interstellar matter is often referred to as a stellar nursery. Compare and contrast bright and dark nebulae.

KEY TERMS: interstellar matter, nebula, bright nebula, dark nebula, emission nebula, reflection nebula, planetary nebula, dark nebula

- New stars are born out of enormous accumulations of dust and gases, called nebulae, which are scattered between existing stars.
- Emission nebulae derive their visible light from nearby hot stars or stars that are embedded in them. Reflection nebulae contain comparatively large debris, including grains of carbon compounds that reflect the light of nearby stars. Planetary nebulae consist of glowing clouds of dust and hot gases that have been expelled near the end of a star's life. Nebulae that are too distant from bright stars to be illuminated are referred to as dark nebulae.

? **Based on color, what type of nebula is likely shown in the accompanying image?**

National Optical Astronomy Observatories

16.3 Classifying Stars: Hertzsprung–Russell Diagrams (H-R Diagrams)

Define *main-sequence star*. Explain the criteria used to classify stars as giants.

KEY TERMS: Hertzsprung–Russell diagram (H-R diagram), main-sequence star, red giant, supergiant

- Hertzsprung–Russell diagrams are constructed by plotting the absolute magnitudes and temperatures of stars on graphs. Considerable information about stars and stellar evolution has been discovered through the use of H-R diagrams.
- Stars are positioned within H-R diagrams as follows: (1) Main-sequence stars, 90 percent of all stars, are in the band that runs from the upper-left corner (massive, hot blue stars) to the lower-right corner (low-mass, red stars); (2) red giants and supergiants, very luminous stars with large radii, are located in the upper-right position; and (3) white dwarfs, small, dense stars, are located in the lower portion.

? **Identify the type of stars located in the positions labeled A, B, and C on the accompanying H-R diagram.**

16.4 Stellar Evolution

List and describe the stages in the evolution of a typical Sun-like star.

KEY TERMS: protostar, hydrogen fusion, red giant, variable star, supernova

- Stars are born when their nuclear furnaces are ignited by the unimaginable pressures and temperatures generated during the collapse of nebulae.
- Red star-like objects not yet hot enough for nuclear fusion are called protostars. When the core of a protostar reaches a temperature of about 10 million K, a process called hydrogen fusion begins, marking the birth of the star. Hydrogen fusion involves the conversion of four hydrogen nuclei into a single helium nuclei and the release of thermal nuclear energy.
- Two opposing forces act on a star: gravity, which tries to contract it into the smallest possible ball, and gas pressure (created by thermal nuclear energy), which tries to blow it apart. When the forces are balanced, the star becomes a stable main-sequence star.
- Medium- and high-mass stars experience another type of nuclear fusion that causes their outer envelopes to expand enormously (hundreds to thousands of times larger), making them red giants or supergiants. When a star exhausts all its usable nuclear fuel, gravity takes over, and the stellar remnant collapses into a small, dense body.

16.5 Stellar Remnants

Compare and contrast the final state of Sun-like stars to the remnants of the most massive stars.

KEY TERMS: white dwarf, degenerate matter, neutron star, pulsar, black hole

- The final fate of a star is determined by its mass.
- Stars with less than one-half the mass of the Sun collapse into hot, dense white dwarfs.
- Medium-mass stars, like the Sun, become red giants, collapse, and end up as white dwarfs, often surrounded by expanding spherical clouds of glowing gas called planetary nebulae.
- Massive stars terminate in a brilliant explosion called a supernova. Supernova events can produce small, extremely dense neutron stars, composed entirely of neutrons, or smaller, even denser black holes—objects that have such immense gravity that light cannot escape their surface.

16.6 Galaxies and Galactic Clusters

List the three major types of galaxies. Explain the formation of large elliptical galaxies.

KEY TERMS: spiral galaxy, barred spiral galaxy, elliptical galaxy, dwarf galaxy, irregular galaxy, galactic cluster, Local Group

- The various types of galaxies include (1) irregular galaxies, which lack symmetry and account for about 25 percent of the known galaxies; (2) spiral galaxies, which are disk shaped and have a greater concentration of stars near their centers and arms extending from their central nucleus; and (3) elliptical galaxies, which have an ellipsoidal shape that may be nearly spherical.
- Galaxies are grouped in galactic clusters, some containing thousands of galaxies. Our own, called the Local Group, contains at least 40 galaxies.

? What type of galaxy is shown in the accompanying image?

(NASA)

16.7 The Big Bang Theory

Describe the big bang theory. Explain what it tells us about the universe.

KEY TERMS: cosmological red shift, Hubble's law, dark matter

- Evidence for an expanding universe came from the study of red shifts in the spectra of galaxies. Edwin Hubble concluded that the observed red shifts, called cosmological red shifts, result from the expansion of space. This evidence strongly supports the big bang model of an expanded universe.
- One question that remains is whether the universe will expand forever or gravitationally contract in the "big crunch." Most cosmologists favor an endless, ever-expanding universe.

GIVE IT SOME THOUGHT

1. Assume that NASA is sending a space probe to each of the following locations:
   a. Polaris (the North Star)
   b. A comet near the outer edge of our solar system
   c. Jupiter
   d. The far edge of the Milky Way Galaxy
   e. The near side of the Andromeda Galaxy
   f. The Sun
   List the locations in order, *from nearest to farthest.*

2. Use the information provided below about three main-sequence stars (A, B, and C) to complete the following and explain your reasoning:

   • Star A has a main-sequence life span of 5 billion years.
   • Star B has the same luminosity (absolute magnitude) as the Sun.
   • Star C has a surface temperature of 5000 K.

   a. Rank the mass of these stars from *greatest to least.*
   b. Rank the energy output of these stars from *greatest to least.*
   c. Rank the main-sequence life span of these stars from *longest to shortest.*

3. The masses of three clouds of gas and dust (nebulae) are provided below. Imagine that each cloud will collapse to form a single star. Use this information to complete the following and explain your reasoning.

   • Cloud A is 60 times the mass of the Sun.
   • Cloud B is 7 times the mass of the Sun.
   • Cloud C is 2 times the mass of the Sun.

   a. Which cloud or clouds, if any, will evolve into a red main-sequence star?
   b. Which of the stars that will form from these clouds, if any, will reach the giant stage?
   c. Which of the stars that will form from these clouds, if any, will go through the supernova stage?

4. Refer to the accompanying images (A, B, C, and D) to complete the following:

   A.   B.

   C.   D.

   a. Which of these nebulae, if any, is an emission nebula?
   b. Which of these nebulae, if any, formed near the end of a star's lifetime?
   c. Which of these nebulae, if any, is a reflection nebula?

5. The accompanying photo shows the Trifid Nebula, which can be easily observed with a small telescope. What unique properties does this nebula exhibit?

   ESO image

6. How a star evolves is closely related to its mass as a main-sequence star. Complete the accompanying diagram by labeling the evolutionary stages for the three groups of main-sequence stars shown.

   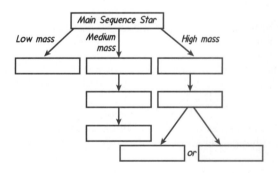

7. Refer to the accompanying photos of an elliptical galaxy and a spiral galaxy to complete the following:
   a. Which image (A or B) is an elliptical galaxy?
   b. Which of these galaxies appears to contain more young, hot massive stars? How did you determine your answer?
   c. When stars are born from a cloud of dust and gases, both large and small stars form at about the same time. Which group of stars, the large or the small, will die out first? Over time, how will this affect the color of the light we observe coming from this group of stars? Based on your response, which of these galaxies appears to be older? Explain.

A.   B.

8. Consider these three characteristics of the universe:
   - It does not have a center.
   - It does not have edges.
   - Its galaxies are all moving away from each other.

   a. Which of the three characteristics of the universe is/are depicted in the raisin bread dough analogy (see Figure 16.22)?
   b. Which of the three characteristics of the universe is/are not accurately depicted in Figure 16.22?

MasteringGeology™

www.masteringgeology.com

Looking for additional review and test prep materials? Visit the Study Area in MasteringGeology to enhance your understanding of this chapter's content by accessing a variety of resources, including **Self-Study Quizzes,** **Geoscience Animations, GEODe Tutorials, RSS feeds, flashcards, web links, self-study quizzes,** and an optional **Pearson eText.**

Appendix A Metric and English Units Compared

Units

1 kilometer (km)	= 1000 meters (m)
1 meter (m)	= 100 centimeters
1 centimeter (cm)	= 0.39 inch (in.)
1 mile (mi)	= 5280 feet (ft)
1 foot (ft)	= 12 inches (in.)
1 inch (in.)	= 2.54 centimeters (cm)
1 square mile (mi^2)	= 640 acres (a)
1 kilogram (kg)	= 1000 grams (g)
1 pound (lb)	= 16 ounces (oz)
1 fathom	= 6 feet (ft)

Conversions

Length

When you want to convert:	multiply by:	to find:
inches	2.54	centimeters
centimeters	0.39	inches
feet	0.30	meters
meters	3.28	feet
yards	0.91	meters
meters	1.09	yards
miles	1.61	kilometers
kilometers	0.62	miles

Area

When you want to convert:	multiply by:	to find:
square inches	6.45	square centimeters
square centimeters	0.15	square inches
square feet	0.09	square meters
square meters	10.76	square feet
square miles	2.59	square kilometers
square kilometers	0.39	square miles

Volume

When you want to convert:	multiply by:	to find:
cubic inches	16.38	cubic centimeters
cubic centimeters	0.06	cubic inches
cubic feet	0.028	cubic meters
cubic meters	35.3	cubic feet
cubic miles	4.17	cubic kilometers
cubic kilometers	0.24	cubic miles
liters	1.06	quarts
liters	0.26	gallons
gallons	3.78	liters

Masses and Weights

When you want to convert:	multiply by:	to find:
ounces	28.35	grams
grams	0.035	ounces
pounds	0.45	kilograms
kilograms	2.205	pounds

Temperature

When you want to convert degrees Fahrenheit (°F) to degrees Celsius (°C), subtract 32 degrees and divide by 1.8.

When you want to convert degrees Celsius (°C) to degrees Fahrenheit (°F), multiply by 1.8 and add 32 degrees.

When you want to convert degrees Celsius (°C) to kelvins (K), delete the degree symbol and add 273. When you want to convert kelvins (K) to degrees Celsius (°C), add the degree symbol and subtract 273.

Figure A.1 Temperature scales.

Appendix B — Earth's Grid System

A glance at any globe reveals a series of north–south and east–west lines that together make up Earth's grid system, a universally used scheme for locating points on Earth's surface. The north–south lines of the grid are called **meridians** and extend from pole to pole (**Figure B.1**). All are halves of great circles. A **great circle** is the largest possible circle that can be drawn on a globe; if a globe were sliced along one of these circles, it would be divided into two equal parts called **hemispheres**. By viewing a globe or Figure B.1, you can see that meridians are spaced farthest apart at the equator and converge toward the poles. The east–west lines (circles) of the grid are called **parallels**. As their name implies, these circles are parallel to one another (see Figure B.1). All meridians are parts of great circles, but all parallels are not. In fact, only one parallel, the equator, is a great circle.

Latitude and Longitude

Latitude may be defined as distance, measured in degrees, *north* and *south* of the equator. Parallels are used to show latitude. Because all points that lie along the same parallel are an identical distance from the equator, they all have the same latitude designation. The latitude of the equator is 0°, whereas the North and South Poles lie 90° north and 90° south, respectively.

Longitude is defined as distance, measured in degrees, *east* and *west* of the zero meridian, or *prime meridian*. Because all meridians are identical, the choice of a zero line is obviously arbitrary. However, the meridian that passes through the Royal Observatory at Greenwich, England, is universally accepted as the reference meridian. Thus, the longitude for any place on the globe is measured east or west from this line. Longitude can vary from 0° along the prime meridian to 180°, halfway around the globe.

It is important to remember that when a location is specified, directions must be given—that is, north or south latitude and east or west longitude (**Figure B.2**). If this is not done, more than one point on the globe is being designated. The only exceptions, of course, are places that lie along the equator, the prime meridian, or the 180° meridian. It should also be noted that while it is not incorrect to use fractions, 1° of latitude or longitude is usually divided into minutes and seconds. A minute (') is 1/60 of 1°, and a second (') is 1/60 of a minute. When locating a place on a map, the degree of exactness depends on the scale of the map. When using a small-scale world map or globe, it may be difficult to estimate latitude and longitude to the nearest whole degree or two. On the other hand, when a large-scale map of an area is used, it is often possible to estimate latitude and longitude to the nearest minute or second.

Distance Measurement

The length of 1° of longitude depends on where the measurement is taken. Along the equator, which is a great circle, 1° of east–west distance is equal to approximately 111 kilometers (69 miles). This figure is found by dividing Earth's circumference—40,075 kilometers (24,900 miles)—by 360. However, with an increase in latitude, the parallels become smaller, and the length of 1° of longitude diminishes (see **Table B.1**). Thus, at about latitude 60° north and south, 1° of longitude has a value equal to about half of what it is at the equator.

As all meridians are halves of great circles, 1° of latitude is equal to about 111 kilometers (69 miles), just as 1° of longitude along the equator is. However, Earth is not a perfect sphere but is slightly flattened at the poles and bulges slightly at the equator. Because of this, there are small differences in the length of 1° of latitude.

Determining the shortest distance between two points on a globe can be done easily and fairly accurately by using the globe-and-string method. It should be noted here that the arc of a great circle is the shortest distance between two points on a sphere. To determine the great circle distance (as well as observe the great circle route) between two places, stretch a piece of string between the locations in question. Then measure the length of the string along the equator (since it is a great circle with degrees marked on it) to determine the number of degrees between the two points. To calculate the distance in kilometers or miles, simply multiply the number of degrees by 111 or 69, respectively.

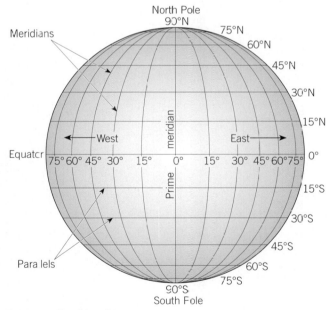

Figure B.1 Earth's grid system

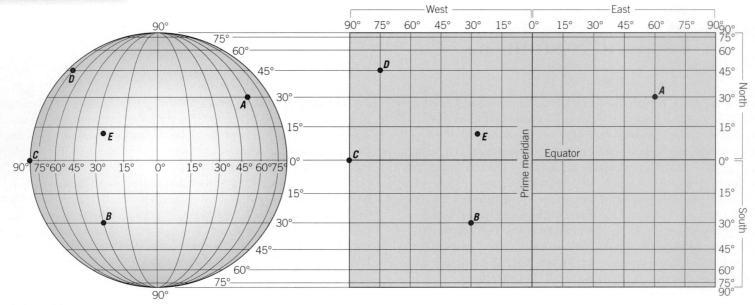

Figure B.2 Locating places using the grid system For both diagrams: Point *A* is latitude 30° north, longitude 60° east; Point *B* is latitude 30° south, longitude 30° west; point *C* is latitude 0°, longitude 90° west; Point *D* is latitude 45° north, longitude 75° west; Point *E* is approximately latitude 10° north, longitude 25° west.

TABLE B.1   Longitude as Distance

	Length of 1° Longitude			Length of 1° Longitude			Length of 1° Longitude	
°Latitude	km	miles	°Latitude	km	miles	°Latitude	km	miles
0	111.367	69.172	30	96.528	59.955	60	55.825	34.674
1	111.349	69.161	31	95.545	59.345	61	54.131	33.622
2	111.298	69.129	32	94.533	58.716	62	52.422	62.560
3	111.214	69.077	33	93.493	58.070	63	50.696	31.488
4	111.096	69.004	34	92.425	57.407	64	48.954	30.406
5	110.945	68.910	35	91.327	56.725	65	47.196	29.314
6	110.760	68.795	36	90.203	56.027	66	45.426	28.215
7	110.543	68.660	37	89.051	55.311	67	43.639	27.105
8	110.290	68.503	38	87.871	54.578	68	41.841	25.988
9	110.003	68.325	39	86.665	53.829	69	40.028	24.862
10	109.686	68.128	40	85.431	53.063	70	38.204	23.729
11	109.333	67.909	41	84.171	52.280	71	36.368	22.589
12	108.949	67.670	42	82.886	51.482	72	34.520	21.441
13	108.530	67.410	43	81.575	50.668	73	32.662	20.287
14	108.079	67.130	44	80.241	49.839	74	30.793	19.126
15	107.596	66.830	45	78.880	48.994	75	28.914	17.959
16	107.079	66.509	46	77.497	48.135	76	27.029	16.788
17	106.530	66.168	47	76.089	47.260	77	25.134	15.611
18	105.949	65.807	48	74.659	46.372	78	23.229	14.428
19	105.337	65.427	49	73.203	45.468	79	21.320	13.242
20	104.692	65.026	50	71.727	44.551	80	19.402	12.051
21	104.014	64.605	51	70.228	43.620	81	17.480	10.857
22	103.306	64.165	52	68.708	42.676	82	15.551	9.659
23	102.565	63.705	53	67.168	41.719	83	13.617	8.458
24	101.795	63.227	54	65.604	40.748	84	11.681	7.255
25	100.994	62.729	55	64.022	39.765	85	9.739	6.049
26	100.160	62.211	56	62.420	38.770	86	7.796	4.842
27	99.297	61.675	57	60.798	37.763	87	5.849	3.633
28	98.405	61.121	58	59.159	36.745	88	3.899	2.422
29	97.481	60.547	59	57.501	35.715	89	1.950	1.211
30	96.528	59.955	60	55.825	34.674	90	0.000	0.000

Appendix C — Relative Humidity and Dew-Point Tables

TABLE C.1 Relative Humidity (percent)

Dry bulb (°C)	Depression of Wet-Bulb Temperature (Dry-Bulb Temperature Minus Wet-Bulb Temperature = Depression of the Wet Bulb)																					
	1	2	3	4	5	6	7	8	9	10	11	12	13	14	15	16	17	18	19	20	21	22
−20	28																					
−18	40																					
−16	48	0																				
−14	55	11																				
−12	61	23																				
−10	66	33	0																			
−8	71	41	13																			
−6	73	48	20	0																		
−4	77	54	32	11																		
−2	79	58	37	20	1																	
0	81	63	45	28	11																	
2	83	67	51	36	20	6																
4	85	70	56	42	27	14																
6	86	72	59	46	35	22	10	0														
8	87	74	62	51	39	28	17	6														
10	88	76	65	54	43	33	24	13	4													
12	88	78	67	57	48	38	28	19	10	2												
14	89	79	69	60	50	41	33	25	16	8	1											
16	90	80	77	62	54	45	37	29	21	74	7	1										
18	91	81	72	64	56	48	40	33	26	19	12	6	0									
20	91	82	74	66	58	51	44	36	30	23	17	11	5									
22	92	83	75	68	60	53	46	40	33	27	21	15	10	4	0							
24	92	84	76	69	62	55	49	42	36	30	25	20	14	9	4	0						
26	92	85	77	70	64	57	51	45	39	34	28	23	18	13	9	5						
28	93	86	78	71	65	59	53	45	42	36	31	26	21	17	12	8	4					
30	93	86	79	72	66	61	55	49	44	39	34	29	25	20	16	12	8	4				
32	93	86	80	73	68	62	56	51	46	41	36	32	27	22	19	14	11	8	4			
34	93	86	81	74	69	63	58	52	48	43	38	34	30	26	22	18	14	11	8	5		
36	94	87	81	75	69	64	59	54	50	44	40	36	32	28	24	21	17	13	10	7	4	
38	94	87	82	76	70	66	60	55	51	46	42	38	34	30	26	23	20	16	13	10	7	5
40	94	89	82	76	71	67	61	57	52	48	44	40	36	33	29	25	22	19	16	13	10	7

(Relative Humidity Values)

*To determine the relative humidity, or dew point, find the air (dry-bulb) temperature on the vertical axis (far left) and the depression of the wet bulb on the horizontal axis (top). Where the two meet, the relative humidity (TABLE C.1) or dew point (TABLE C.2) is found. For example, when the dry-bulb temperature is 20°C and a wet-bulb temperature is 14°C, the depression of the wet bulb is 6°C (20°–14°C). From TABLE C.1, the relative humidity is 51 percent, and from TABLE C.2 the dew point is 10°C.

TABLE C.2   Dew-Point Temperature (°C)*

Dry bulb (°C)	(Dry-Bulb Temperature Minus Wet-Bulb Temperature = Depression of the Wet Bulb)																					
	1	2	3	4	5	6	7	8	9	10	11	12	13	14	15	16	17	18	19	20	21	22
−20	−33																					
−18	−28																					
−16	−24																					
−14	−21	−36																				
−12	−18	−28																				
−10	−14	−22																				
−8	−12	−18	−29																			
−6	−10	−14	−22																			
−4	−7	−12	−17	−29																		
−2	−5	−8	−13	−20																		
0	−3	−6	−9	−15	−24																	
2	−1	−3	−6	−11	−17																	
4	1	−1	−4	−7	−11	−19																
6	4	1	−1	−4	−7	−13	−21															
8	6	3	1	−2	−5	−9	−14															
10	8	6	4	1	−2	−5	−9	−14	−18													
12	10	8	6	4	1	−2	−5	−9	−16													
14	12	11	9	6	4	1	−2	−5	−10	−17												
16	14	13	11	9	7	4	1	−1	−6	−10	−17											
18	16	15	13	11	9	7	4	2	−2	−5	−10	−19										
20	19	17	15	14	12	10	7	4	2	−2	−5	−10	−19									
22	21	19	17	16	74	12	10	8	5	3	−1	−5	−10	−19								
24	23	21	20	18	16	14	12	10	8	6	2	−1	−5	−10	−18							
26	25	23	22	20	18	17	15	13	11	9	6	3	0	−4	−9	−18						
28	27	25	24	22	27	19	17	16	14	11	9	7	4	1	−3	−9	−16					
30	29	27	26	24	23	21	19	18	16	14	12	70	8	5	1	−2	−8	−15				
32	31	29	28	27	25	24	22	21	19	17	15	13	11	8	5	2	−2	−7	−14			
34	33	31	30	29	27	26	24	23	21	20	18	16	14	12	9	6	3	−1	−5	−12	−29	
36	35	33	32	31	29	28	27	25	24	22	20	19	17	15	13	10	7	4	0	−4	−10	
38	37	35	34	33	32	30	29	28	26	25	23	21	19	17	15	13	11	8	5	1	−3	−9
40	39	37	36	35	34	32	31	30	28	37	25	24	22	20	18	16	14	12	9	6	2	−2

Dew-Point Values

Appendix D Stellar Properties

Measuring Distances to the Closest Stars

Measuring the distance to stars is difficult. Nevertheless, astronomers have developed some direct as well as indirect methods to measure stellar distances. One simple measurement, called *stellar parallax*, is effective in determining the distances to only the closest stars.

Stellar parallax is the slight back-and-forth shift of the apparent position of a nearby star due to the orbital motion of Earth around the Sun. The principle of parallax is easy to visualize. Close one eye, and with your index finger in a vertical position, use your open eye to align your finger with some distant object. Without moving your finger, view the object with your other eye and notice that its position appears to have changed. Now repeat the exercise, this time holding your finger farther away and notice that the farther away you hold your finger, the less its position seems to shift. This method of measuring stellar distances is elementary and was practiced by ancient Greek astronomers.

Modern cosmologists determine parallax by photographing a nearby star against the background of distant stars. Then, when Earth has moved halfway around its orbit 6 months later, the same star is photographed again. When these two photographs are compared, the position of the nearby star appears to have shifted with respect to the background stars. **Figure D.1** illustrates this shift, and the parallax angle that is determined from it. The nearest stars have the largest parallax angles, whereas the parallax angles of distant stars are much too small to measure.

In practice, conducting parallax measurements is complex because of the miniscule angles being measured. The process is further complicated because both the Sun and the star being measured are moving relative to each other. The first accurate stellar parallax was not determined until 1838. Even today, parallax angles for only a few thousand of the nearest stars are known with certainty; nearly all others have such small parallax shifts that accurate measurements are not possible.

Fortunately, other methods have been developed to estimate distances to more distant stars. In addition, the Hubble Space Telescope, which is not hindered by Earth's light-distorting atmosphere, has obtained accurate parallax distances for many more stars.

Stellar Brightness

The oldest means of classifying stars is based on their *brightness*, also called *luminosity* or *magnitude*. Three factors control the brightness of a star as seen from Earth: *how large* it is, *how hot* it is, and its *distance* from Earth. The stars in the night sky come in a grand assortment of sizes, temperatures, and distances, so their apparent brightness varies widely.

Apparent Magnitude

Stars have been classified according to their apparent brightness since at least the second century B.C., when Hipparchus placed about 850 stars into six categories, based on his ability to see differences in brightness. Because he could only reliably see six different brightness levels, he created six categories. These categories were later called *magnitudes*, with first magnitude being the brightest and sixth magnitude the dimmest. Because some stars may appear dimmer than others only because they are farther away, a star's brightness, *as it appears when viewed from Earth*, is called its *apparent magnitude*. With the invention of the telescope, many stars fainter than the sixth magnitude were discovered.

In the mid-1800s, a method was developed to standardize the magnitude scale. An absolute comparison was made between the light coming from stars of the first magnitude and from stars of the sixth magnitude. It was determined that a first-magnitude star was about 100 times brighter than a sixth-magnitude star. On the scale that was devised, any

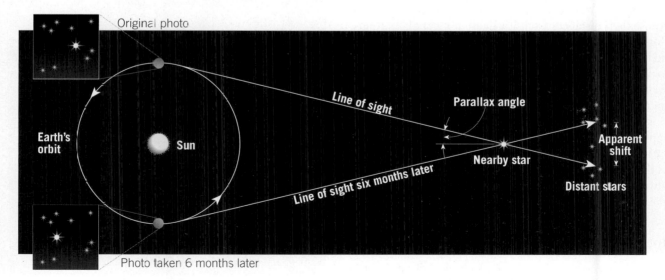

Figure D.1 Geometry of stellar parallax The parallax angle shown here is enormously exaggerated to illustrate the principle. Because distances to even the nearest stars are thousands of times greater than the Earth–Sun distance, the triangles that astronomers work with are extremely long and narrow, making the angles that are measured very small.

TABLE D.1   Ratios of Star Brightness

Difference in Magnitude	Brightness Ratio
0.5	1.6:1
1	2.5:1
2	6.3:1
3	16:1
4	40:1
5	100:1
10	10,000:1
20	100,000,000:1

Calculations: $2.512 \times 2.512 \times 2.512 \times 2.512 \times 2.512$, or 2.512 raised to the fifth power, equals 100.

TABLE D.2   Distance, Apparent Magnitude, and Absolute Magnitude of Some Stars

Name	Distance (light-years)	Apparent Magnitude	Absolute Magnitude*
Sun	NA	−26.7	5.0
Alpha Centauri	4.27	0.0	4.4
Sirius	8.70	−1.4	1.5
Arcturus	36	−0.1	−0.3
Betelgeuse	520	0.8	−5.5
Deneb	1600	1.3	−6.9

* The more negative, the brighter; the more positive, the dimmer.

two stars that differ by five magnitudes have a ratio in brightness of 100 to 1. Hence, a third-magnitude star is 100 times brighter than an eighth-magnitude star. It follows, then, that the brightness ratio of two stars differing by only one magnitude is about 2.5.° A star of the first magnitude is about 2.5 times brighter than a star of the second magnitude. **Table D.1** shows how differences in magnitude correspond to brightness ratios.

Because some celestial bodies are brighter than first-magnitude stars, zero and negative magnitudes were introduced. On this scale, the Sun has an apparent magnitude of −26.7. At its brightest, Venus has a magnitude of −4.3. At the other end of the scale, the Hubble Space Telescope can view stars with an apparent magnitude of 30, more than 1 billion times dimmer than stars that are visible to the unaided eye.

Absolute Magnitude

Apparent magnitudes were good approximations of the true brightness of stars when astronomers thought that the universe was very small—containing no more than a few thousand stars, all at very similar distances from Earth. However, we now know that the universe is unimaginably large and contains innumerable stars at wildly varying distances. Because astronomers are interested in the "true" brightness of stars, they devised a measure called *absolute magnitude*.

Stars of the same apparent magnitude usually do not have the same brightness because their distances from Earth are not equal. Astronomers correct for distance by determining what brightness (magnitude) the stars would have if they were at a standard distance—about 32.6 light-years. For example, if the Sun, which has an apparent magnitude of −26.7, were located 32.6 light-years from Earth, it would have an absolute magnitude of about +5. Thus, stars with absolute magnitudes greater than 5 (smaller numerical value) are intrinsically brighter than the Sun but appear much dimmer because of their distance from Earth. **Table D.2** lists the absolute and apparent magnitudes of some stars as well as their distances from Earth. Most stars have an absolute magnitude between −5 (very bright) and 15 (very dim). The Sun is near the midpoint of this range.

Stellar Color and Temperature

The next time you are outside on a clear night, look carefully at the stars and note their colors (**Figure D.2**). Because human eyes do not respond well to color in low-intensity light (when it is very dark, we see in only

black and white), we tend to look at the brightest stars. Some stars that are quite colorful can be found in the constellation Orion. Of the two brightest stars in Orion, Rigel (β Orionis) appears blue, and Betelgeuse (α Orionis) is definitely red.

Very hot stars with surface temperatures above 30,000 K emit most of their energy in the form of short-wavelength light and therefore appear blue. On the other hand, cooler red stars, with surface temperatures generally less than 3,000 K, emit most of their energy as longer-wavelength red light. Stars, such as the Sun, with surface temperatures between 5,000 and 6,000 K, appear yellow. Because color is primarily a manifestation of a star's surface temperature, this characteristic provides astronomers with useful information. As you will see, combining temperature data with stellar magnitude tells us a great deal about the size and mass of stars.

Binary Stars and Stellar Mass

One of the night sky's best-known constellations, the Big Dipper, appears to consist of seven stars. But those with good eyesight can see that the second star in the handle is actually two stars. In the early nineteenth century, careful examination of numerous star pairs by William Herschel showed that many stars found in pairs actually orbit one another. In such cases, the two stars are in fact united by their mutual gravitation. These

Figure D.2 Time-lapse photograph of stars in the constellation Orion These star trails show some of the various star colors. It is important to note that the human eye sees color somewhat differently than photographic film. (Courtesy of National Optical Astronomy Observatories)

° The more negative, the brighter; the more positive, the dimmer.

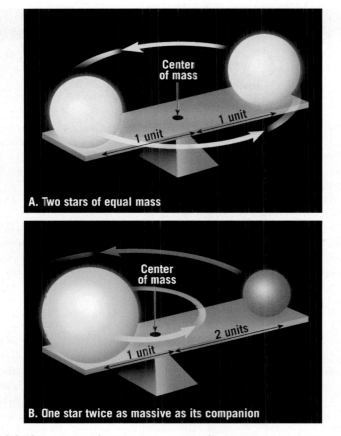

Figure D.3 **Binary stars orbit each other around their common center of mass**
A. For stars of equal mass, the center of mass lies exactly halfway between them.
B. If one star is twice as massive as its companion, it is twice as close to the two stars' common center of mass. Therefore, more massive stars have proportionately smaller orbits than their less massive companions.

pairs, in which the members are far enough apart to be telescopically identified as two stars, are called *visual binaries* (*binaries* = double). The idea of one star orbiting another may seem unusual, but many stars in the universe exist in pairs or multiples.

Binary stars can be used to determine the star property that is most difficult to calculate—its *mass*. The mass of a body can be established if it is gravitationally attached to a partner. Binary stars orbit each other

around a common point called the *center of mass* (**Figure D.3**). For stars of equal mass, the center of mass lies exactly halfway between them. When one star is more massive than its partner, their common center will be located closer to the more massive one. Thus, if the sizes of their orbits can be observed, their individual masses can be determined. You can experience this relationship on a seesaw by trying to balance a person who has a much greater (or smaller) mass.

For illustration, when one star has an orbit half the size (radius) of its companion, it is twice as massive as its companion. If their combined masses are equal to three solar masses, then the larger will be twice as massive as the Sun, and the smaller will have a mass equal to that of the Sun. Most stars have a mass that falls in a range between 1/10 and 50 times the mass of the Sun.

The Doppler Effect

The positions of the bright and dark lines in the spectra shift when the source of energy moves relative to the observer. This effect is observed for all types of waves. Think about the change in pitch you hear in a car horn or an ambulance siren as it passes by. When it is approaching, the sound seems to have a higher-than-normal pitch, and when it is moving away, the pitch sounds lower than normal. This effect, first explained by Christian Doppler in 1842, is called the *Doppler effect*. The reason for the difference in pitch is that it takes time for the wave to be emitted. If the source is moving away, the beginning of the wave is emitted nearer to you than the end, which stretches the wave—that is, gives it a longer wavelength (**Figure D.4**). The opposite is true for an approaching source.

In the case of light, when a source is moving away, its light appears redder than it actually is because its waves are lengthened. Objects approaching have their light waves shifted toward the blue (shorter-wavelength) end of the spectrum. Thus, if a source of red light approached you at a very high speed (near the speed of light), it would actually appear blue. The same effect is produced if you are moving and the light remains stationary.

The Doppler effect is important because it reveals whether Earth is approaching or receding from a star or another celestial body. In addition, the amount of shift allows us to calculate the rate at which the relative movement is occurring. Large Doppler shifts indicate high velocities; small Doppler shifts indicate low velocities.

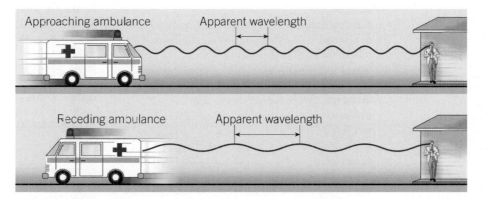

Figure D.4 **The Doppler effect** This image illustrates the apparent lengthening and shortening of wavelengths caused by the relative motion between a source and an observer.

Glossary

Aa flow A type of lava flow that has a jagged, blocky surface.

Ablation A general term for the loss of ice and snow from a glacier.

Abrasion The grinding and scraping of a rock surface by the friction and impact of rock particles carried by water, wind, or ice.

Absolute instability The condition of air that has an environmental lapse rate that is greater than the dry adiabatic rate (1°C per 100 meters).

Absolute magnitude The apparent brightness of a star if it were viewed from a distance of 10 parsecs (32.6 light-years). Used to compare the true brightness of stars.

Absolute stability The condition of air that has an environmental lapse rate that is less than the wet adiabatic rate.

Abyssal plain A very level area of the deep-ocean floor, usually lying at the foot of the continental rise.

Accretionary wedge A large wedge-shaped mass of sediment that accumulates in subduction zones. Here sediment is scraped from the subducting oceanic plate and accreted to the overriding crustal block.

Active continental margin A margin that is usually narrow and consists of highly deformed sediments. Such margins occur where oceanic lithosphere is being subducted beneath the margin of a continent.

Adiabatic temperature change Cooling or warming of air that occurs because air is allowed to expand or is compressed, not because heat is added or subtracted.

Advection fog A fog formed when warm, moist air is blown over a cool surface.

Aerosols Tiny solid and liquid particles suspended in the atmosphere.

Air The mixture of gases and particles that make up Earth's atmosphere. Nitrogen and oxygen are most abundant.

Air mass A large body of air that is characterized by a sameness of temperature and humidity.

Air-mass weather The conditions experienced in an area as an air mass passes over it. Because air masses are large and fairly homogeneous, air-mass weather will be fairly constant and may last for several days.

Air pressure Force exerted by the weight of the air above.

Albedo The reflectivity of a substance, usually expressed as a percentage of the incident radiation reflected.

Alluvial fan A fan-shaped deposit of sediment formed when a stream's slope is abruptly reduced.

Alluvium Unconsolidated sediment deposited by a stream.

Alpine glacier A glacier confined to a mountain valley, which in most instances was previously a stream valley. Also known as a *valley glacier*.

Andesitic composition See *Intermediate composition*.

Aneroid barometer An instrument for measuring air pressure that consists of evacuated metal chambers that are very sensitive to variations in air pressure.

Angular unconformity An unconformity in which the strata below dip at an angle different from that of the beds above.

Annual mean temperature An average of the 12 monthly means.

Annual temperature range The difference between the highest and lowest monthly means.

Anticline A fold in sedimentary strata that resembles an arch.

Anticyclone A high-pressure center characterized by a clockwise flow of air in the Northern Hemisphere.

Apparent magnitude The brightness of a star when viewed from Earth.

Aquicludes Impermeable beds that hinder or prevent groundwater movement.

Aquifer Rock or soil through which groundwater moves easily.

Arête A narrow knifelike ridge separating two adjacent glaciated valleys.

Artesian well A well in which the water rises above the level where it was initially encountered.

Asteroid One of thousands of small planetlike bodies, ranging in size from a few hundred kilometers to less than 1 kilometer, whose orbits lie mainly between those of Mars and Jupiter.

Asthenosphere A subdivision of the mantle situated below the lithosphere. This zone of weak material exists below a depth of about 100 kilometers and in some regions extends as deep as 700 kilometers. The rock within this zone is easily deformed.

Astronomical unit (AU) The average distance from Earth to the Sun; 1.5×10^8 kilometers (93×10^6 miles).

Astronomy The scientific study of the universe; it includes the observation and interpretation of celestial bodies and phenomena.

Atmosphere The gaseous portion of a planet; the planet's envelope of air. One of the traditional subdivisions of Earth's physical environment.

Atoll A continuous or broken ring of coral reef surrounding a central lagoon.

Atom The smallest particle that exists as an element.

Atomic number The number of protons in the nucleus of an atom.

Autumnal equinox See *Equinox*.

Backswamp A poorly drained area on a floodplain that results when natural levees are present.

Bajada An apron of sediment along a mountain front created by the coalescence of alluvial fans.

Barograph A recording barometer.

Barometer An instrument that measures atmospheric pressure.

Barometric tendency See *Pressure tendency*.

Barred spiral A galaxy that has straight arms extending from its nucleus.

Barrier island A low, elongated ridge of sand that parallels the coast.

Basaltic composition A compositional group of igneous rocks in which the rock contains substantial dark silicate minerals and calcium-rich plagioclase feldspar. Also known as *mafic composition*.

Base level The level below which a stream cannot erode.

Basin A circular downfolded structure.

Batholith A large mass of igneous rock that formed when magma was emplaced at depth, crystallized, and subsequently exposed by erosion.

Bathymetry The measurement of ocean depths and the charting of the topography of the ocean floor.

Baymouth bar A sandbar that completely crosses a bay, sealing it off from the open ocean.

Beach drift The transport of sediment in a zigzag pattern along a beach caused by the uprush of water from obliquely breaking waves.

Beach nourishment The addition of large quantities of sand to a beach system to offset losses caused by wave erosion.

Bed load Sediment that is carried by a stream along the bottom of its channel.

Bergeron process A theory that relates the formation of precipitation to supercooled clouds, freezing nuclei, and the different saturation levels of ice and liquid water.

Big bang theory A theory which proposes that the universe originated as a single mass, which subsequently exploded.

Biochemical sedimentary rock A sedimentary rock composed of material that was extracted from water by organisms to create hard parts such as shells.

Biogenous sediment Seafloor sediment consisting of material of marine-organic origin.

Biosphere The totality of life on Earth; the parts of the lithosphere, hydrosphere, and atmosphere in which living organisms can be found.

Black hole A massive star that has collapsed to such a small volume that its gravity prevents the escape of all radiation.

Blowout (deflation hollow) A depression excavated by the wind in easily eroded deposits.

Bowen's reaction series A concept proposed by N. L. Bowen that illustrates the relationships between magma and the minerals crystallizing from it during the formation of igneous rocks.

Breakwater A structure that protects a near-shore area from breaking waves.

Bright nebula A cloud of glowing gas excited by ultraviolet radiation from hot stars.

Brittle deformation The loss of strength by a material, usually in the form of sudden fracturing.

Caldera A large depression typically caused by collapse or ejection of the summit area of a volcano.

Capacity The total amount of sediment a stream is able to transport.

Catastrophism The concept that Earth was shaped by catastrophic events of a short-term nature.

Cavern A naturally formed underground chamber or series of chambers most commonly produced by solution activity in limestone.

Celestial sphere An imaginary hollow sphere on which the ancients believed the stars were hung and were carried around Earth.

Cenozoic era A time span on the geologic calendar beginning about 66 million years ago, following the Mesozoic era.

Chemical bond A strong attractive force that exists between atoms in a substance. It involves the transfer or sharing of electrons that allows each atom to attain a full valence shell.

Chemical sedimentary rock Sedimentary rock consisting of material that was precipitated from water by either inorganic or organic means.

Chinook A wind that blows down the leeward side of a mountain and that warms by compression.

Cinder cone A rather small volcano built primarily of pyroclastics ejected from a single vent. Also known as a *scoria cone*.

Circle of illumination The great circle that separates daylight from darkness.

Cirque An amphitheater-shaped basin at the head of a glaciated valley produced by frost wedging and plucking.

Cirrus One of three basic cloud forms; also one of the three high cloud types. These thin, delicate ice-crystal clouds often appear as veil-like patches or thin, wispy fibers.

Cleavage The tendency of a mineral to break along planes of weak bonding.

Climate A description of aggregate weather conditions; the sum of all statistical weather information that helps describe a place or region.

Cloud A form of condensation best described as a dense concentration of suspended water droplets or tiny ice crystals.

Cloud of vertical development A cloud that has its base in the low height range but extends upward into the middle or high altitudes.

Coarse-grained texture An igneous rock texture in which the crystals are roughly equal in size and large enough that individual minerals can be identified with the unaided eye.

Cold front A front along which a cold air mass thrusts beneath a warmer air mass.

Collision–coalescence process A theory of raindrop formation in warm clouds (above 0°C [32°F]) in which large cloud droplets (giants) collide and join together with smaller droplets to form raindrops. Opposite electrical charges may bind the cloud droplets together

Color An obvious mineral characteristic that is often unreliable as a diagnostic property.

Columnar joints A pattern of cracks that form during cooling of molten rock to generate columns that are generally six sided.

Comet A small body that generally revolves around the Sun in an elongated orbit.

Competence A measure of the largest particle a stream can transport; a factor dependent on velocity.

Composite volcano A volcano composed of both lava flows and pyroclastic material. Also known as a *stratovolcano*.

Compound A substance formed by the chemical combination of two or more elements in definite proportions and usually having properties different from those of its constituent elements.

Compressional mountains Mountains in which great horizontal forces have shortened and thickened the crust. Most major mountain belts are of this type.

Concordant A term used to describe intrusive igneous masses that form parallel to the bedding of the surrounding rock.

Condensation The change of state from a gas to a liquid.

Condensation nuclei Tiny bits of particulate matter that serve as surfaces on which water vapor condenses.

Conditional instability The condition of moist air with an environmental lapse rate between the dry and wet adiabatic rates.

Conduction The transfer of heat through matter by molecular activity. Energy is transferred through collisions from one molecule to another.

Conduit A pipelike opening through which magma moves toward Earth's surface. It terminates at a surface opening called a vent.

Cone of depression A cone-shaped depression in the water table immediately surrounding a well.

Confined aquifer An aquifer that has impermeable layers (aquitards) both above and below.

Conformable Having layers of rock deposited without interruption.

Contact metamorphism Changes in rock caused by the heat from a nearby magma body. Also known as *thermal metamorphism*.

Continental (c) air mass An air mass that forms over land; it is normally relatively dry.

Continental drift theory A theory which originally proposed that the continents are rafted about. It has essentially been replaced by the plate tectonics theory.

Continental margin The portion of the seafloor adjacent to the continents. It may include the continental shelf, continental slope, and continental rise.

Continental rift A linear zone along which continental lithosphere stretches and pulls apart. Its creation may mark the beginning of a new ocean basin.

Continental rise The gently sloping surface at the base of the continental slope.

Continental shelf The gently sloping submerged portion of the continental margin that extends from the shoreline to the continental slope.

Continental slope The steep gradient that leads to the deep-ocean floor and marks the seaward edge of the continental shelf.

Continental volcanic arc Mountains formed in part by igneous activity associated with the subduction of oceanic lithosphere beneath a continent. Examples include the Andes and the Cascades.

Convection The transfer of heat by the movement of a mass or substance. It can take place only in fluids.

Convergence The condition that exists when the distribution of winds within a given area results in a net horizontal inflow of air into the area. Since convergence at lower levels is associated with an upward movement of air, areas of convergent winds are regions favorable to cloud formation and precipitation.

Convergent plate boundary A boundary in which two plates move together, causing one of the slabs of lithosphere to be consumed into the mantle as it descends beneath an overriding plate.

Coral reef A structure formed in a warm, shallow, sunlit ocean environment that consists primarily of the calcite-rich remains of corals as well as the limy secretions of algae and the hard parts of many other small organisms.

Core The innermost layer of Earth, based on composition. It is thought to be largely an iron–nickel alloy, with minor amounts of oxygen, silicon, and sulfur.

Coriolis effect (force) The deflective force of Earth's rotation on all free-moving objects, including the atmosphere and oceans. Deflection is to the right in the Northern Hemisphere and to the left in the Southern Hemisphere.

Correlation The establishment of the equivalence of rocks of similar age in different areas.

Cosmological red shift Changes in the spectra of galaxies which indicate that they are moving away from the Milky Way as a result of the expansion of space.

Crater The depression at the summit of a volcano, or a depression that is produced by a meteorite impact.

Creep The slow downhill movement of soil and regolith.

Crevasse A deep crack in the brittle surface of a glacier.

Cross-bedding Relatively thin layers that are inclined at an angle to the main bedding. Formed by currents of wind or water.

Cross-cutting A principle of relative dating which states that a rock or fault is younger than any rock or fault through which it cuts.

Crust The very thin outermost layer of Earth.

Crystal settling The downward movement of minerals during the crystallization of magma. The earlier-formed minerals are denser than the liquid portion and settle to the bottom of the magma chamber.

Crystal shape See *Habit*.

Crystallization The formation and growth of a crystalline solid from a liquid or gas.

Cumulus One of three basic cloud forms; also the name given to one of the clouds of vertical development. Cumulus are billowy individual cloud masses that often have flat bases.

Cup anemometer An instrument used to determine wind speed.

Curie point The temperature above which a material loses its magnetization.

Cut bank The area of active erosion on the outside of a meander.

Cutoff A short channel segment created when a river erodes through the narrow neck of land between meanders

Cyclone A low-pressure center characterized by a counterclockwise flow of air in the Northern Hemisphere.

Daily mean temperature A statistic that is determined by averaging the hourly readings or, more commonly, by averaging the maximum and minimum temperatures for a day.

Daily temperature range The difference between the maximum and minimum for a day.

Dark matter Undetected matter that is thought to exist in great quantities in the universe.

Dark nebula A cloud of interstellar dust that obscures the light of more distant stars and appears as an opaque curtain.

Dark silicate mineral A silicate mineral that contains ions of iron and/or magnesium in its structure. It is dark in color and has a higher specific gravity than a light silicate mineral.

Decompression melting Melting that occurs as rock ascends due to a drop in confining pressure.

Deep-ocean basin The portion of the seafloor that lies between the continental margin and the oceanic ridge. This region comprises almost 30 percent of Earth's surface.

Deep-ocean trench A narrow, elongated depression on the floor of the ocean.

Deep-sea fan A cone-shaped deposit at the base of the continental slope. The sediment is transported to the fan by turbidity currents that follow submarine canyons.

Deflation The lifting and removal of loose material by wind.

Deformation General term for the processes of folding, faulting, shearing, compression, or extension of rocks.

Degenerate matter Incomprehensibly dense material formed when stars collapse and form a white dwarf.

Delta An accumulation of sediment formed where a stream enters a lake or an ocean.

Dendritic pattern A stream system that resembles the pattern of a branching tree.

Density The weight per unit volume of a particular material.

Deposition The process by which water vapor is changed directly to a solid, without passing through the liquid state.

Desalination The removal of salts and other chemicals from seawater.

Desert One of the two types of dry climate; the driest of the dry climates.

Desert pavement A layer of coarse pebbles and gravel created when wind removes the finer material.

Detrital sedimentary rock Rock formed from the accumulation of material that originated and was transported in the form of solid particles derived from both mechanical and chemical weathering.

Dew point The temperature to which air has to be cooled in order to reach saturation. Also known as *dew-point temperature*.

Dike A tabular-shaped intrusive igneous feature that cuts through the surrounding rock.

Dip-slip fault A fault in which the movement is parallel to the dip of the fault.

Discharge The quantity of water in a stream that passes a given point in a period of time.

Disconformity A type of unconformity in which the beds above and below are parallel.

Discordant A term used to describe plutons that cut across existing rock structures, such as bedding planes.

Dissolved load The portion of a stream's load that is carried in solution.

Distributary A section of a stream that leaves the main flow.

Diurnal tidal pattern A tidal pattern exhibiting one high tide and one low tide during a tidal day; a daily tide.

Divergence The condition that exists when the distribution of winds within a given area results in a net horizontal outflow of air from the region. In divergence at lower levels, the resulting deficit is compensated for by a downward movement of air from aloft; hence, areas of divergent winds are unfavorable to cloud formation and precipitation.

Divergent plate boundary A region where the rigid plates are moving apart, typified by the mid-ocean ridges. Also known as a *spreading center*.

Divide An imaginary line that separates the drainage of two streams; often found along a ridge.

Dome A roughly circular upfolded structure similar to an anticline.

Doppler effect The apparent change in wavelength of radiation caused by the relative motions of the source and the observer.

Doppler radar A new generation of radar that can handle the tasks performed by conventional radar and also detect motion directly. It greatly improves tornado and severe storm warnings.

Drainage basin The land area that contributes water to a stream.

Drawdown The difference in height between the bottom of a cone of depression and the original height of the water table.

Drift See *Glacial drift*.

Drumlin A streamlined asymmetrical hill composed of glacial till. The steep side of the hill faces the direction from which the ice advanced.

Dry adiabatic rate The rate of adiabatic cooling or warming in unsaturated air. The rate of temperature change is 1°C per 100 meters.

Dry climate A climate in which yearly precipitation is less than the potential loss of water by evaporation.

Ductile deformation A type of solid-state flow that produces a change in the size and shape of a rock body without fracturing. Occurs at depths where temperatures and confining pressures are high.

Dune A hill or ridge of wind-deposited sand.

Dwarf galaxy A very small galaxy, usually elliptical and lacking spiral arms.

Dwarf planets Celestial bodies that orbit stars, massive enough to be spherical but have not cleared their neighboring regions of planetesimals.

Earthquake Vibration of Earth produced by the rapid release of energy.

Echo sounder An instrument used to determine the depth of water by measuring the time interval between emission of a sound signal and the return of its echo from the bottom.

Economic mineral A mineral used extensively in the manufacture of products.

Elastic deformation Rock deformation in which the rock will return to nearly its original size and shape when the stress is removed.

Elastic rebound The sudden release of stored strain in rocks that results in movement along a fault.

Electron A negatively charged subatomic particle that has a negligible mass and is found outside an atom's nucleus.

Element A substance that cannot be decomposed into simpler substances by ordinary chemical or physical means.

Elements of weather and climate Quantities or properties of the atmosphere that are measured regularly and that are used to express the nature of weather and climate.

Elliptical galaxy A galaxy that is round or elliptical in outline. It contains little gas and dust, no disk or spiral arms, and few hot, bright stars.

Emergent coast A coast where land that was formerly below sea level has been exposed either because of crustal uplift or a drop in sea level or both.

Emission nebula A gaseous nebula that derives its visible light from the fluorescence of ultraviolet light from a star in or near the nebula.

End moraine A ridge of till that marks a former position on the front of a glacier.

Enhanced Fujita intensity scale (EF-scale) A scale originally developed by Theodore Fujita for classifying the severity of a tornado, based on the correlation of wind speed and the degree of destruction.

Entrenched meander A meander cut into bedrock when uplifting rejuvenated a meandering stream.

Environmental lapse rate The rate of temperature decrease with increasing height in the troposphere.

Eon The largest time unit on the geologic time scale, next in order of magnitude above era.

Ephemeral stream A stream that is usually dry because it carries water only in response to specific episodes of rainfall. Most desert streams are of this type.

Epicenter The location on Earth's surface that lies directly above the forces of an earthquake.

Epoch A unit of the geologic calendar that is a subdivision of a period.

Equatorial low A belt of low pressure that lies near the equator and between the subtropical highs.

Equatorial system A method of locating stellar objects much like the coordinate system used on Earth's surface.

Equinox (spring or autumnal) The time when the vertical rays of the Sun are striking the equator. At equinox, the length of daylight and darkness is equal at all latitudes.

Era A major division on the geologic calendar; eras are divided into shorter units called periods.

Erosion The incorporation and transportation of material by a mobile agent, such as water, wind, or ice.

Eruption column A buoyant plume of hot, ash-laden gases that can extend thousands of meters into the atmosphere.

Escape velocity The initial velocity an object needs to escape from the surface of a celestial body.

Esker A sinuous ridge composed largely of sand and gravel deposited by a stream flowing in a tunnel within or beneath a glacier near its terminus.

Estuary A funnel-shaped inlet of the sea that formed when a rise in sea level or subsidence of land caused the mouth of a river to be flooded.

Evaporation The process of converting a liquid to a gas.

Evaporite A sedimentary rock formed of material deposited from solution by evaporation of the water.

Evapotranspiration The combined effect of evaporation and transpiration.

Exfoliation dome A large, dome-shaped structure, usually composed of granite, formed by sheeting.

External process A process such as weathering, mass wasting, or erosion that is powered by the Sun and transforms solid rock into sediment.

Extrusive Igneous activity that occurs outside the crust. Also known as *volcanic*.

Eye A zone of scattered clouds and calm averaging about 20 kilometers in diameter at the center of a hurricane.

Eye wall The doughnut-shaped area of intense cumulonimbus development and very strong winds that surrounds the eye of a hurricane.

Fault A break in a rock mass along which movement has occurred.

Fault-block mountain A mountain formed by the displacement of rock along a fault.

Fault creep Gradual displacement along a fault. Such activity occurs relatively smoothly and with little noticeable seismic activity.

Fault scarp A cliff created by movement along a fault. It represents the exposed surface of the fault prior to modification by weathering and erosion.

Felsic composition See *Granitic composition*.

Fetch The distance that the wind has traveled across the open water.

Fine-grained texture A texture of igneous rocks in which the crystals are too small for individual minerals to be distinguished with the unaided eye.

Fiord A steep-sided inlet of the sea that formed when a glacial trough was partially submerged.

Fissure A crack in rock along which there is a distinct separation.

Fissure eruption An eruption in which lava is extruded from narrow fractures or cracks in the crust.

Flood The overflow of a stream channel that occurs when discharge exceeds the channel's capacity. A flood is the most common and destructive geologic hazard.

Flood basalts Flows of basaltic lava that issue from numerous cracks or fissures and that commonly cover extensive areas to thicknesses of hundreds of meters.

Floodplain A flat, low-lying portion of a stream valley that is subject to periodic inundation.

Focus (earthquake) The zone within Earth where rock displacement produces an earthquake. Also known as a *hypocenter*.

Fog A cloud with its base at or very near Earth's surface.

Fold A bent rock layer or series of layers that were originally horizontal and subsequently deformed.

Foliated texture A texture of metamorphic rocks that gives the rock a layered appearance.

Foliation A term for a linear arrangement of textural features that metamorphic rocks often exhibit.

Footwall block The rock surface below a fault.

Foreshocks Small earthquakes that often precede a major earthquake.

Fossil magnetism See *Paleomagnetism*.

Fossil succession The definite and determinable order in which fossil organisms succeed, which allows any time period to be recognized by its fossil content.

Fossils The remains or traces of organisms preserved from the geologic past.

Fracture Any break or rupture in rock along which no appreciable movement has taken place.

Fracture zone A linear zone of irregular topography on the deep-ocean floor that follows transform faults and their inactive extensions.

Freezing The change of state from a liquid to a solid.

Freezing nuclei Solid particles that serve as cores for the formation of ice crystals.

Front The boundary between two adjoining air masses that have contrasting characteristics.

Frontal fog Fog formed when rain evaporates as it falls through a layer of cool air.

Frontal wedging Lifting of air that results when cool air acts as a barrier over which warmer, lighter air will rise.

Frost wedging The mechanical breakup of rock caused by the expansion of freezing water in cracks and crevices.

Fumarole A vent in a volcanic area from which fumes or gases escape.

Galactic cluster A system of galaxies that contains from several to thousands of member galaxies.

Geocentric The concept of an Earth-centered universe.

Geologic time The span of time since the formation of Earth.

Geologic time scale The division of Earth history into blocks of time, such as eons, eras, periods, and epochs. The time scale was originally created using relative dating principles.

Geology The science that examines Earth, its form and composition, and the changes it has undergone and is undergoing.

Geostrophic wind A wind, usually above a height of 600 meters (2000 feet), that blows parallel to the isobars.

Geothermal gradient The gradual increase in temperature with depth in the crust. The average is 30°C per kilometer in the upper crust.

Geyser A fountain of hot water ejected periodically.

Giant (star) A luminous star of large radius.

Glacial budget The balance, or lack of balance, between ice formation at the upper end of a glacier and ice loss in the zone of wastage.

Glacial drift An all-embracing term for sediments of glacial origin, no matter how, where, or in what shape they were deposited. Also known as *drift*.

Glacial erratic An ice-transported boulder that was not derived from bedrock near its present site.

Glacial striations Scratches and grooves on bedrock caused by glacial abrasion.

Glacial trough A mountain valley that has been widened, deepened, and straightened by a glacier.

Glacier A thick mass of ice that originates on land from the compaction and recrystallization of snow and that shows evidence of past or present flow.

Glassy texture A term used to describe the texture of certain igneous rocks, such as obsidian, that contain no crystals.

Glaze A coating of ice on objects that is formed when supercooled rain freezes on contact.

Graben A valley formed by the downward displacement of a fault-bounded block.

Gradient The slope of a stream; generally measured in feet per mile.

Granitic composition A compositional group of igneous rocks in which the rock is made up almost entirely of light-colored silicates. Also known as *felsic composition*.

Greenhouse effect The transmission of short-wave solar radiation by the atmosphere coupled with the selective absorption of longer-wavelength terrestrial radiation, especially by water vapor and carbon dioxide.

Groin A short wall built at a right angle to the shore to trap moving sand.

Ground moraine An undulating layer of till deposited as the ice front retreats.

Groundwater Water in the zone of saturation.

Guyot A submerged flat-topped seamount. Also known as a *tablemount*.

Gyre The large, circular surface-current pattern found in each ocean.

Habit The common or characteristic shape of a crystal or aggregate of crystals. Also known as *crystal shape*.

Hail Nearly spherical ice pellets having concentric layers and formed by the successive freezing of layers of water.

Half-life The time required for one-half of the atoms of a radioactive substance to decay.

Hanging valley A tributary valley that enters a glacial trough at a considerable height above its floor.

Hanging wall block The rock surface immediately above a fault.

Hard stabilization An artificial structure built to protect a coast or to prevent the movement of sand along a beach. Examples include groins, jetties, breakwaters, and seawalls.

Hardness The resistance a mineral offers to scratching.

Heliocentric The view that the Sun is at the center of the solar system.

Hertzsprung-Russell (H-R) diagram A plot of stars according to their absolute magnitudes and spectral types.

High clouds Clouds that normally have their base above 6000 meters; the base may be lower in winter and at high-latitude locations.

High-pressure center See *Anticyclone*.

Horn A pyramid-like peak formed by glacial action in three or more cirques surrounding a mountain summit.

Horst An elongated, uplifted block of crust bounded by faults.

Hot spot A concentration of heat in the mantle that is capable of producing magma, which in turn extrudes onto Earth's surface. The intraplate volcanism that produced the Hawaiian islands is one example.

Hot spring A spring in which the water is 6° to 9°C (10° to 15°F) warmer than the mean annual air temperature of its locality.

Hubble's law A law that relates the distance to a galaxy and its velocity.

Humidity A general term referring to water vapor in the air but not to liquid droplets of fog, cloud, or rain.

Hurricane A tropical cyclonic storm that has winds in excess of 119 kilometers (74 miles) per hour.

Hydrogen burning The conversion of hydrogen through fusion to form helium.

Hydrogen fusion A nuclear reaction in which hydrogen nuclei are fused into helium nuclei.

Hydrogenous sediment Seafloor sediment consisting of minerals that crystallize from seawater. An important example is manganese nodules.

Hydrologic cycle The unending circulation of Earth's water supply. The cycle is powered by energy from the Sun and is characterized by continuous exchanges of water among the oceans, the atmosphere, the geosphere, and the biosphere.

Hydrosphere The water portion of our planet; one of the traditional subdivisions of Earth's physical environment.

Hygrometer An instrument designed to measure relative humidity.

Hygroscopic nuclei Condensation nuclei that have a high affinity for water, such as salt particles.

Hypocenter See *Focus (earthquake)*.

Hypothesis A tentative explanation that is tested to determine whether it is valid.

Ice cap A mass of glacial ice covering a high upland or plateau and spreading out radially.

Ice sheet A very large, thick mass of glacial ice flowing outward in all directions from one or more accumulation centers.

Iceberg A mostly submerged mass of floating ice created by the calving of a glacier.

Igneous rock A rock formed by the crystallization of molten magma.

Immature soil A soil lacking horizons.

Inclination of the axis The tilt of Earth's axis from the perpendicular to the plane of Earth's orbit.

Inclusion A piece of one rock unit contained within another. Inclusions are useful in relative dating.

Inclusions (principle of) A principle that uses pieces of rock contained within another to determine a relative date. According to the principle of inclusions, the rock mass that provided the inclusion is older than the rock mass containing the inclusion.

Index fossil A fossil that is associated with a particular span of geologic time.

Infrared Radiation with a wavelength from 0.7 to 200 micrometers.

Inner core The solid innermost layer of Earth, about 1300 kilometers (800 miles) in radius.

Interface A common boundary where different parts of a system interact.

Interior drainage A discontinuous pattern of intermittent streams that do not flow to the ocean.

Intermediate composition A compositional group of igneous rocks, in which the rock contains at least 25 percent dark silicate minerals. The other dominant mineral is plagioclase feldspar. Also known as *andesitic composition*.

Internal process A process such as volcanic activity, earthquake, and mountain building that derives its energy from Earth's interior.

Interstellar matter Dust and gases found between stars.

Intertropical convergence zone (ITCZ) The zone of general convergence between the Northern and Southern Hemisphere trade winds.

Intraplate volcanism Igneous activity that occurs within a tectonic plate away from plate boundaries.

Intrusion See *Pluton*.

Intrusive Igneous rock that formed below Earth's surface. Also known as *plutonic*.

Iron meteorite One of the three main categories of meteorites. This group is composed largely of iron, with varying amounts of nickel (5–20 percent). Most meteorite finds are irons.

Irregular galaxy A galaxy that lacks symmetry.

Island arc See *Volcanic island arc*.

Isobar A line drawn on a map that connects points of equal atmospheric pressure, usually corrected to sea level.

Isotherms Lines that connect points of equal temperature.

Isotopes Varieties of the same element that have different mass numbers; their nuclei contain the same number of protons but different numbers of neutrons.

Jet stream Swift, high-altitude winds (120–240 kilometers per hour).

Jovian planet The Jupiter-like planets Jupiter, Saturn, Uranus, and Neptune. These planets have relatively low densities.

Kame A steep-sided hill composed of sand and gravel that originates when sediment collects in openings in stagnant glacial ice.

Karst topography A topography that consists of numerous depressions called sinkholes.

Kettle A depression created by a block or blocks of ice becoming lodged in a glacial deposit and subsequently melting.

Kuiper belt A region outside the orbit of Neptune where most short-period comets are thought to originate.

Laccolith A massive igneous body intruded between preexisting strata.

Lahar Mudflows on the slopes of volcanoes that result when unstable layers of ash and debris become saturated and flow downslope, usually following stream channels.

Land breeze A local wind that blows from land toward the water during the night in coastal areas.

Latent heat The energy absorbed or released during a change in state.

Lateral moraine A ridge of till along the sides of an alpine glacier composed primarily of debris that fell to the glacier from the valley walls.

Lava Magma that reaches Earth's surface.

Light silicate mineral A silicate mineral that lacks iron and/or magnesium. It is generally lighter in color and has a lower specific gravity than a dark silicate mineral.

Lightning A sudden flash of light generated by the flow of electrons between oppositely charged parts of a cumulonimbus cloud or between the cloud and the ground.

Light-year The distance light travels in a year; about 6 trillion miles.

Liquefaction A phenomenon, sometimes associated with earthquakes, in which soils and other unconsolidated materials containing abundant water are turned into a fluidlike mass that is not capable of supporting buildings.

Lithification The process, generally cementation and/or compaction, of converting sediments to solid rock.

Lithosphere The rigid outer layer of Earth, including the crust and upper mantle.

Lithospheric plate A coherent unit of Earth's rigid outer layer that includes the crust and uppermost mantle. Also known as a *tectonic plate*.

Local Group The cluster of 20 or so galaxies to which our galaxy belongs.

Local wind A small-scale wind produced by a locally generated pressure gradient.

Localized convective lifting Unequal surface heating that causes localized pockets of air (thermals) to rise because of their buoyancy.

Loess Deposits of windblown silt that are lacking visible layers, generally buff-colored, and capable of maintaining a nearly vertical cliff.

Longitudinal profile A cross section of a stream channel along its descending course from the head to the mouth.

Longshore current A near-shore current that flows parallel to the shore.

Low clouds Clouds that form below a height of about 2000 meters.

Low-pressure center See *Cyclone*.

Lower mantle See *Mesosphere (geology)*.

Lunar regolith A thin, gray layer on the surface of the Moon, consisting of loosely compacted, fragmented material believed to have been formed by repeated meteoric impacts.

Luster The appearance or quality of light reflected from the surface of a mineral.

Mafic composition See *Basaltic composition*.

Magma A body of molten rock found at depth, including any dissolved gases and crystals.

Magnetic reversal A change in the polarity of Earth's magnetic field that occurs over time intervals of roughly 200,000 years.

Magnetic time scale The history of magnetic reversals through geologic time.

Magnetometer A sensitive instrument used to measure the intensity of Earth's magnetic field.

Magnitude (earthquake) The total amount of energy released during an earthquake.

Magnitude (stellar) A number given to a celestial object to express its relative brightness.

Main-sequence stars A sequence of stars on the Hertzsprung-Russell diagram, containing the majority of stars, that runs diagonally from the upper left to the lower right.

Mantle The 2900-kilometer- (1800-mile-) thick layer of Earth located below the crust.

Mantle plume Structures that originate at great depth and, upon reaching the crust, spread laterally, creating a localized volcanic zone called a hot spot. A mantle plume is a source of some intraplate basaltic magma.

Maria The Latin name for the smooth areas of the Moon formerly thought to be seas.

Marine terrace A wave-cut platform that has been exposed above sea level.

Maritime (m) air mass An air mass that originates over the ocean. Maritime air masses are relatively humid.

Mass number The number of neutrons and protons in the nucleus of an atom.

Mass wasting The downslope movement of rock, regolith, and soil under the direct influence of gravity.

Massive An igneous pluton that is not tabular in shape.

Mean solar day The average time between two passages of the Sun across the local celestial meridian.

Meander A looplike bend in the course of a stream.

Medial moraine A ridge of till formed when lateral moraines from two coalescing alpine glaciers join.

Megathrust fault The plate boundary separating a subducting slab of oceanic lithosphere and the overlying plate.

Melting The change of state from a solid to a liquid.

Mercury barometer A mercury-filled glass tube in which the height of the mercury column is a measure of air pressure.

Mesosphere (atmosphere) The layer of the atmosphere immediately above the stratosphere and characterized by decreasing temperatures with height.

Mesosphere (geology) The part of the mantle that extends from the core–mantle boundary to a depth of 660 kilometers. Also known as the *lower mantle*.

Mesozoic era A time span on the geologic calendar between the Paleozoic and Cenozoic eras—from about 245 to 66.4 million years ago.

Metallic bond A chemical bond that is present in all metals that may be characterized as an extreme type of electron sharing in which the electrons move freely from atom to atom.

Metamorphic rock Rock formed by the alteration of preexisting rock deep within Earth (but still in the solid state) by heat, pressure, and/or chemically active fluids.

Metamorphism The changes in mineral composition and texture of a rock subjected to high temperatures and pressures within Earth.

Meteor The luminous phenomenon observed when a meteoroid enters Earth's atmosphere and burns up; popularly called a "shooting star."

Meteor shower Many meteors appearing in the sky that occurs when Earth intercepts a swarm of meteoritic particles.

Meteorite Any portion of a meteoroid that survives its traverse through Earth's atmosphere and strikes Earth's surface.

Meteoroid Small, solid particles that have orbits in the solar system.

Meteorology The scientific study of the atmosphere and atmospheric phenomena; the study of weather and climate.

Microcontinent A relatively small fragment of continental crust that may lie above sea level, such as the island of Madagascar, or that may be submerged, as exemplified by the Campbell Plateau located near New Zealand.

Middle clouds Clouds that occupy the height range from 2000 to 6000 meters.

Middle-latitude (midlatitude) cyclone A large low-pressure center with a diameter often exceeding 1000 kilometers (600 miles) that moves from west to east and may last from a few days to more than a week and usually has a cold front and a warm front extending from the central area of low pressure.

Mid-ocean ridge A continuous elevated zone on the floor of all the major ocean basins and varying in width from 500 to 5000 kilometers (300 to 3000 miles). The rifts at the crests of these ridges represent divergent plate boundaries. Also known as an *oceanic ridge*.

Mineral A naturally occurring, inorganic crystalline material that has a unique chemical composition.

Mineral resources All discovered and undiscovered deposits of a useful mineral that can be extracted now or at some time in the future.

Mixed tidal pattern A tidal pattern exhibiting two high tides and two low tides per tidal day, with a large inequality in high water heights, low water heights, or both. Coastal locations that experience such a tidal pattern may also show alternating periods of diurnal and semidiurnal tidal patterns. Also called *mixed semidiurnal*.

Model A term often used synonymously with hypothesis but that is less precise because it is sometimes used to describe a theory as well.

Modified Mercalli Intensity scale A 12-point scale developed to evaluate earthquake intensity based on the amount of damage to various structures.

Mohs hardness scale A series of 10 minerals used as a standard in determining hardness.

Moment magnitude A more precise measure of earthquake magnitude than the Richter scale that is derived from the amount of displacement that occurs along a fault zone.

Monsoon A seasonal reversal of wind direction associated with large continents, especially Asia. In winter, the wind blows from land to sea; in summer, from sea to land.

Monthly mean temperature The mean temperature for a month that is calculated by averaging the daily means for the month.

Mountain breeze The nightly downslope winds commonly encountered in mountain valleys.

Natural hazard A natural phenomenon that represents a potential danger to people, such as an earthquake, a volcanic eruption, or a hurricane.

Natural levees The elevated landforms that parallel some streams and act to confine their waters, except during flood stage.

Neap tide The lowest tidal range, occurring near the times of the first and third quarters of the Moon.

Nebula A cloud of interstellar gas and/or dust.

Nebular theory A model for the origin of the solar system that supposes a rotating nebula of dust and gases that contracted to form the Sun and planets.

Neutron A subatomic particle found in the nucleus of an atom. A neutron is electronically neutral and has a mass approximately that of a proton.

Neutron star A star of extremely high density that is composed entirely of neutrons.

Nonconformity An unconformity in which older metamorphic or intrusive igneous rocks are overlain by younger sedimentary strata.

Nonfoliated texture Metamorphic rocks that do not exhibit foliation.

Nonrenewable resource A resource that forms or accumulates over such long time spans that it must be considered as fixed in total quantity.

Normal fault A fault in which the rock above the fault plane has moved down relative to the rock below.

Normal polarity A magnetic field that is the same as that which exists at present.

Nucleus (atomic) The small, heavy core of an atom that contains all of its positive charge and most of its mass.

Nucleus (comet) The small central body of a comet, typically 1 to 10 kilometers in diameter.

Nuée ardente Incandescent volcanic debris buoyed up by hot gases that moves downslope in an avalanche fashion.

Numerical date A date that specifies the actual number of years that have passed since an event occurred.

Oceanic ridge See *Mid-ocean ridge*.

Oceanography The scientific study of the oceans and oceanic phenomena.

Octet rule A rule which states that atoms combine in order that each may have the electron arrangement of a noble gas—that is, so the outer energy level contains eight neutrons.

Oort cloud A spherical shell composed of comets that orbit the Sun at distances generally greater than 10,000 times the Earth–Sun distance.

Ore Usually a useful metallic mineral that can be mined at a profit. The term is also applied to certain nonmetallic minerals such as fluorite and sulfur.

Organic matter Carbon-rich substances derived from living things that are in various stages of decomposition.

Organic sedimentary rock Sedimentary rock composed of organic carbon from the remains of plants that died and accumulated on the floor of a swamp. Coal is the primary example.

Original horizontality Layers of sediments generally deposited in a horizontal or nearly horizontal position.

Orogenesis The processes that collectively result in the formation of mountains.

Orographic lifting Mountains acting as barriers to the flow of air and forcing the air to ascend. The air cools adiabatically, and clouds and precipitation may result.

Outer core A layer beneath the mantle that is about 2200 kilometers (1364 miles) thick and has the properties of a liquid.

Outgassing The escape of gases that were once dissolved in magma

Outlet glacier A tongue of ice normally flowing rapidly outward from an ice cap or ice sheet, usually through mountainous terrain to the sea.

Outwash plain A relatively flat, gently sloping plain consisting of materials deposited by meltwater streams in front of the margin of an ice sheet.

Overrunning Warm air gliding up a retreating cold air mass.

Oxbow lake A curved lake produced when a stream cuts off a meander.

Ozone A molecule of oxygen that contains three oxygen atoms.

P wave See *Primary wave*.

Pahoehoe flow A lava flow with a smooth-to-ropey surface.

Paleoclimatology The study of ancient climates; the study of climate and climate change prior to the period of instrumental records using proxy data.

Paleomagnetism The natural remnant magnetism in rock bodies. The permanent magnetization acquired by rock that can be used to determine the location of the magnetic poles and the latitude of the rock at the time it became magnetized. Also known as *fossil magnetism*.

Paleozoic era A time span on the geologic calendar between the Precambrian and Mesozoic eras—from about 570 million to 245 million years ago.

Pangaea A proposed supercontinent that 200 million years ago began to break apart and form the present landmasses.

Parasitic cone A volcanic cone that forms on the flank of a larger volcano.

Parcel An imaginary volume of air enclosed in a thin elastic cover. Typically it is considered to be a few hundred cubic meters in volume and is assumed to act independently of the surrounding air.

Partial melting The process by which most igneous rocks melt. Since individual minerals have different melting points, most igneous rocks melt over a temperature range of a few hundred degrees. If the liquid is squeezed out after some melting has occurred, a melt with a higher silica content results.

Passive continental margin A margin that consists of a continental shelf, continental slope, and continental rise. These margins are not associated with plate boundaries and therefore experience little volcanism and few earthquakes.

Perched water table A localized zone of saturation above the main water table that is created by an impermeable layer (aquiclude).

Period A basic unit of the geologic calendar that is a subdivision of an era. Periods may be divided into smaller units called epochs.

Permeability A measure of a material's ability to transmit water.

Perturbation The gravitational disturbance of the orbit of one celestial body by another.

Phanerozoic eon The part of geologic time represented by rocks containing abundant fossil evidence. The eon extending from the end of the Proterozoic eon (570 million years ago) to the present.

Physical environment The part of the environment that encompasses water, air, soil, and rock, as well as conditions such as temperature, humidity, and sunlight.

Piedmont glacier A glacier that forms when one or more alpine glaciers emerge from the confining walls of mountain valleys and spread out to create a broad sheet in the lowlands at the base of the mountains.

Pipe A vertical conduit through which magmatic materials have passed.

Planetary nebula A shell of incandescent gas expanding from a star.

Plate One of numerous rigid sections of the lithosphere that move as a unit over the material of the asthenosphere.

Plate tectonics A theory which proposes that Earth's outer shell consists of individual plates that interact in various ways and thereby produce earthquakes, volcanoes, mountains, and the crust itself.

Playa lake A temporary lake in a playa.

Pleistocene epoch An epoch of the Quaternary period beginning about 1.6 million years ago and ending about 10,000 years ago. Best known as a time of extensive continental glaciation.

Plucking (quarrying) The process by which pieces of bedrock are lifted out of place by a glacier.

Pluton A structure that results from the emplacement and crystallization of magma beneath the surface of Earth. Also known as an *intrusion*.

Plutonic See *Intrusive*.

Pluvial lake A lake formed during a period of increased rainfall. During the Pleistocene epoch, this occurred in some nonglaciated regions during periods of ice advance elsewhere.

Point bar A crescent-shaped accumulation of sand and gravel deposited on the inside of a meander.

Polar (P) air mass A cold air mass that forms in a high-latitude source region.

Polar easterlies In the global pattern of prevailing winds, winds that blow from the polar high toward the subpolar low. Unlike the trade winds, however, these winds should not be thought of as persistent winds.

Polar front A stormy frontal zone that separates air masses of polar origin from air masses of tropical origin.

Polar high An anticyclone that is assumed to occupy the inner polar regions and is believed to be thermally induced, at least in part.

Polar wandering A theory based on paleomagnetic studies in the 1950s, in which researchers proposed that either the magnetic poles migrated greatly through time or the continents had gradually shifted their positions.

Porosity The volume of open spaces in rock or soil.

Porphyritic texture An igneous texture consisting of large crystals embedded in a matrix of much smaller crystals.

Positive-feedback mechanism As used in climatic change, any effect that acts to reinforce an initial change.

Pothole A depression formed in a stream channel by the abrasive action of the water's sediment load.

Precambrian All geologic time prior to the Paleozoic era.

Precession A slow motion of Earth's axis that traces out a cone over a period of 26,000 years.

Precipitation fog Fog formed when rain evaporates as it falls through a layer of cool air.

Pressure gradient The amount of pressure change occurring over a given distance.

Pressure tendency The nature of the change in atmospheric pressure over a period of several hours. It can be a useful aid in short-range weather prediction. Also known as *barometric tendency*.

Prevailing wind A wind that consistently blows from one direction more than from any other.

Primary pollutants Pollutants emitted directly from identifiable sources.

Primary (P) wave A type of seismic wave that involves alternating compression and expansion of the material through which it passes.

Proglacial lake A lake created when a glacier acts as a dam, blocking the flow of a river or trapping glacial meltwater. The term refers to the position of such lakes just beyond the outer limits of a glacier.

Proton A positively charged subatomic particle found in the nucleus of an atom.

Proton–proton chain A chain of thermonuclear reactions by which nuclei of hydrogen are built up into nuclei of helium.

Protostar A collapsing cloud of gas and dust that is destined to become a star.

Psychrometer A device consisting of two thermometers (wet-bulb and dry-bulb) that is rapidly whirled and, with the use of tables, yields the relative humidity and dew point.

Ptolemaic model An Earth-centered system of the universe.

Pulsar A variable radio source of small size that emits radio pulses in very regular periods.

Pumice A light-colored, glassy vesicular rock that commonly has a granitic composition.

Pycnocline A layer of water in which there is a rapid change of density with depth.

Pyroclastic flow A highly heated mixture, largely of ash and pumice fragments, traveling down the flanks of a volcano or along the surface of the ground.

Pyroclastic material The volcanic rock ejected during an eruption, including ash, bombs, and blocks.

Pyroclastic texture An igneous rock texture resulting from the consolidation of individual rock fragments that are ejected during a violent eruption.

Quaternary period The most recent period on the geologic time scale. It began about 2.6 million years ago and extends to the present.

Radial pattern A system of streams running in all directions away from a central elevated structure, such as a volcano.

Radiation (electromagnetic radiation) The transfer of energy (heat) through space by electromagnetic waves.

Radiation fog Fog resulting from radiation heat loss by Earth.

Radiation pressure The force exerted by electromagnetic radiation from an object such as the Sun.

Radioactive decay (Radioactivity) The spontaneous decay of certain unstable atomic nuclei.

Radiometric dating The procedure of calculating the absolute ages of rocks and minerals that contain radioactive isotopes.

Radiosonde A lightweight package of weather instruments fitted with a radio transmitter and carried aloft by a balloon.

Rain Drops of water that fall from a cloud and have a diameter of at least 0.5 millimeter.

Rainshadow desert A dry area on the lee side of a mountain range. Many middle-latitude deserts are of this type.

Rectangular pattern A drainage pattern characterized by numerous right-angle bends that develops on jointed or fractured bedrock.

Red giant A large, cool star of high luminosity; a star occupying the upper-right portion of the Hertzsprung-Russell diagram.

Reflecting telescope A telescope that concentrates light from distant objects by using a concave mirror.

Reflection nebula A relatively dense dust cloud in interstellar space that is illuminated by starlight.

Refracting telescope A telescope that employs a lens to bend and concentrate the light from distant objects.

Refraction The process by which the portion of a wave in shallow water slows, causing the wave to bend and tend to align itself with the underwater contours. Also known as *wave refraction*.

Regional metamorphism Metamorphism associated with large-scale mountain-building processes.

Regolith The layer of rock and mineral fragments that nearly everywhere covers Earth's land surface.

Rejuvenation A change, often caused by regional uplift, that causes the force of erosion to intensify.

Relative dating Rocks placed in their proper sequence or order. Relative dating determines only the chronologic order of events.

Relative humidity The ratio of the air's water-vapor content to its water-vapor capacity.

Renewable resource A resource that is virtually inexhaustible or that can be replenished over relatively short time spans.

Reserve Identified deposits from which minerals can be extracted profitably.

Retrograde motion The apparent westward motion of the planets with respect to the stars.

Reverse fault A fault in which the material above the fault plane moves up in relation to the material below.

Reverse polarity A magnetic field opposite that which exists at present.

Revolution The motion of one body about another, as Earth about the Sun.

Richter scale A scale of earthquake magnitude based on the motion of a seismograph.

Ridge push A mechanism that may contribute to plate motion. It involves the oceanic lithosphere sliding down the oceanic ridge under the pull of gravity.

Rift valley A region of Earth's crust along which divergence is taking place.

Rime A thin coating of ice on objects that is produced when supercooled fog or cloud droplets freeze on contact.

Ring of Fire The zone of active volcanoes surrounding the Pacific Ocean.

Rock A consolidated mixture of minerals.

Rock cycle A model that illustrates the origin of the three basic rock types and the interrelatedness of Earth materials and processes.

Rock flour Ground-up rock produced by the grinding effect of a glacier.

Rock-forming minerals The few dozen most abundant minerals that make up most common rocks. Minerals rich in oxygen and silicon (silicate minerals) dominate this group.

Rotation The spinning of a body, such as Earth, about its axis.

S wave See *Secondary wave*.

Saffir–Simpson scale A scale from 1 to 5 that is used to rank relative intensities of hurricanes.

Salinity The proportion of dissolved salts to pure water, usually expressed in parts per thousand (‰).

Santa Ana The local name given to a chinook wind in southern California.

Saturation The maximum possible quantity of water vapor that the air can hold at any given temperature and pressure.

Scoria Vesicular ejecta that is the product of basaltic magma.

Scoria cone See *Cinder cone*.

Sea arch An arch formed by wave erosion when caves on opposite sides of a headland unite.

Sea breeze A local wind blowing from the sea during the afternoon in coastal areas.

Sea ice Frozen seawater that is associated with polar regions. The area covered by sea ice expands in winter and shrinks in summer.

Sea stack An isolated mass of rock standing just offshore, produced by wave erosion of a headland.

Seafloor spreading The process of producing new seafloor between two diverging plates.

Seamount An isolated volcanic peak that rises at least 1000 meters (3000 feet) above the deep-ocean floor.

Seawall A barrier constructed to prevent waves from reaching the area behind the wall. Its purpose is to defend property from the force of breaking waves.

Secondary (S) wave A seismic wave that involves oscillation perpendicular to the direction of propagation.

Sediment Unconsolidated particles created by the weathering and erosion of rock, by chemical precipitation from solution in water, or from the secretions of organisms and transported by water, wind, or glaciers.

Sedimentary rock Rock formed from the weathered products of preexisting rocks that have been transported, deposited, and lithified.

Seismic sea wave A rapidly moving ocean wave generated by earthquake activity that is capable of inflicting heavy damage in coastal regions.

Seismic waves A form of elastic energy released during an earthquake that causes vibrations in the materials that transmit them.

Seismogram A record made by a seismograph.

Seismograph An instrument that records earthquake waves. Also known as a *seismometer*.

Seismology The study of earthquakes and seismic waves.

Seismometer See *Seismograph*.

Semidiurnal tidal pattern A tidal pattern exhibiting two high tides and two low tides per tidal day, with small inequalities between successive highs and successive lows; a semidaily tide.

Shield volcano A broad, gently sloping volcano built from fluid basaltic lavas.

Silicate mineral Any one of numerous minerals that have the silicon–oxygen tetrahedron as their basic structure.

Silicon–oxygen tetrahedron A structure composed of four oxygen atoms surrounding a silicon atom that constitutes the basic building block of silicate minerals.

Sill A tabular igneous body that was intruded parallel to the layering of preexisting rock.

Sink hole (sink) A depression produced in a region where soluble rock has been removed by groundwater.

Slab pull A mechanism that contributes to plate motion in which cool, dense oceanic crust sinks into the mantle and "pulls" the trailing lithosphere along.

Slab suction One of the driving forces of plate motion, which arises from the drag of the subducting plate on the adjacent mantle. It is an induced mantle circulation that pulls both the subducting and overriding plates toward the trench.

Sleet Frozen or semifrozen rain that forms when raindrops freeze as they pass through a layer of cold air.

Slip face The steep, leeward slope of a sand dune; it maintains an angle of about 34 degrees.

Small solar system bodies Objects in the solar system that are not planets or dwarf planets. Examples include comets and asteroids.

Snow A solid form of precipitation produced by sublimation of water vapor.

Solstice (summer or winter) The time when the vertical rays of the Sun are striking either the Tropic of Cancer or the Tropic of Capricorn. Solstice represents the longest or shortest day (length of daylight) of the year.

Sonar An instrument that uses acoustic signals (sound energy) to measure water depths. Sonar is an acronym for *so*und *na*vigation and *r*anging.

Sorting The process by which solid particles of various sizes are separated by moving water or wind. Also the degree of similarity in particle size in sediment or sedimentary rock.

Source region The area where an air mass acquires its characteristic properties of temperature and moisture.

Specific gravity The ratio of a substance's weight to the weight of an equal volume of water.

Specific humidity The weight of water vapor compared with the total weight of the air, including the water vapor.

Spiral galaxy A flattened, rotating galaxy with pinwheel-like arms of interstellar material and young stars winding out from its nucleus.

Spit An elongated ridge of sand that projects from the land into the mouth of an adjacent bay.

Spreading center See *Divergent plate boundary*.

Spring A flow of groundwater that emerges naturally at the ground surface.

Spring equinox See *Equinox*.

Spring tide The highest tidal range. Occurs near the times of the new and full moons.

Stable air Air that resists vertical displacement. If it is lifted, adiabatic cooling will cause its temperature to be lower than the surrounding environment; if it is allowed to do so, it will sink to its original position.

Stalactite An icicle-like structure that hangs from the ceiling of a cavern.

Stalagmite A columnlike rock formation that grows upward from the floor of a cavern.

Steam fog Fog that has the appearance of steam; produced by evaporation from a warm-water surface into the cool air above.

Stellar parallax A measure of stellar distance.

Steppe One of the two types of dry climate. A marginal and more humid variant of the desert that separates it from bordering humid climates.

Stock A pluton similar to but smaller than a batholith.

Stony meteorite One of the three main categories of meteorites. Such meteorites are composed largely of silicate minerals, with inclusions of other minerals.

Stony-iron meteorite One of the three main categories of meteorites. This group is a mixture of iron and silicate minerals.

Storm surge The abnormal rise of the sea along a shore as a result of strong winds.

Strata Parallel layers of sedimentary rock.

Stratified drift Sediments deposited by glacial meltwater.

Stratosphere The layer of the atmosphere immediately above the troposphere, characterized by increasing temperatures with height due to the concentration of ozone.

Stratovolcano See *Composite volcano*.

Stratus One of three basic cloud forms; also the name given to one type of low cloud. They are sheets or layers that cover much or all of the sky.

Streak The color of a mineral in powdered form.

Strike-slip fault A fault along which the movement is horizontal.

Subduction erosion A process at subduction zones in which sediment and rock are scraped off the bottom of the overriding plate and transported into the mantle.

Subduction zone A long, narrow zone where one lithospheric plate descends beneath another.

Sublimation The conversion of a solid directly to a gas, without passing through the liquid state.

Submarine canyon A canyon carved into the outer continental shelf, slope, and rise by turbidity currents.

Submergent coast A coast with a form that is largely a result of the partial drowning of a former land surface either because of a rise of sea level or subsidence of the crust or both.

Subpolar low Low pressure located at about the latitudes of the Arctic and Antarctic Circles. In the Northern Hemisphere, the low takes the form of individual oceanic cells; in the Southern Hemisphere, there is a deep and continuous trough of low pressure.

Subtropical high Not a continuous belt of high pressure but rather several semipermanent, anticyclonic centers characterized by subsidence and divergence located roughly between latitudes 25° and 35°.

Summer solstice See *Solstice*.

Supergiant A very large star with high luminosity.

Supernova An exploding star that increases in brightness many thousands of times.

Superposition (principle of) A principle of relative dating which uses the fact that in any underformed sequence of sedimentary rocks, each bed is older than the layers above and younger than the layers below.

Surf A collective term for breakers; also the wave activity in the area between the shoreline and the outer limit of breakers.

Surface waves Seismic waves that travel along the outer layer of Earth.

Suspended load The fine sediment carried within the body of flowing water.

Syncline A linear downfold in sedimentary strata; the opposite of anticline.

Tablemount See *Guyot*.

Tabular A term used to describe a feature such as an igneous pluton having two dimensions that are much longer than the third; a pluton that is thin in one dimens

Tectonic plate See *Lithospheric plate.*

Temperature control A factor that causes temperature to vary from place to place or from time to time.

Temperature gradient The amount of temperature change per unit of distance.

Terrane A crustal block bounded by faults, whose geologic history is distinct from the histories of adjoining crustal blocks.

Terrestrial planet Any of the Earthlike planets, including Mercury, Venus, Mars, and Earth.

Terrigenous sediment Seafloor sediment derived from weathering and erosion on land.

Texture The size, shape, and distribution of the particles that collectively constitute a rock.

Theory A well-tested and widely accepted view that explains certain observable facts.

Thermal metamorphism See *Contact metamorphism.*

Thermocline A layer of water in which there is a rapid change in temperature in the vertical dimension.

Thermohaline circulation Movements of ocean water caused by density differences brought about by variations in temperature and salinity.

Thermosphere The region of the atmosphere immediately above the mesosphere, which is characterized by increasing temperatures due to absorption of very shortwave solar energy by oxygen.

Thrust fault A low-angle reverse fault.

Thunderstorm A storm produced by a cumulonimbus cloud and always accompanied by lightning and thunder. It is of relatively short duration and usually accompanied by strong wind gusts, heavy rain, and sometimes hail.

Tidal current The alternating horizontal movement of water associated with the rise and fall of the tide.

Tidal delta A deltalike feature created when a rapidly moving tidal current emerges from a narrow inlet and slows, depositing its load of sediment.

Tidal flat A marshy or muddy area that is covered and uncovered by the rise and fall of the tide.

Tide Periodic change in the elevation of the ocean surface.

Till Unsorted sediment deposited by a glacier.

Tombolo A ridge of sand that connects an island to the mainland or to another island.

Tornado A small, very intense cyclonic storm with exceedingly high winds, most often produced along cold fronts in conjunction with severe thunderstorms.

Tornado warning A warning issued when a tornado has actually been sighted in an area or is indicated by radar.

Tornado watch A forecast issued for areas of about 65,000 square kilometers (25,000 square miles), indicating that conditions are such that tornadoes may develop; it is intended to alert people to the possibility of tornadoes.

Trade winds Two belts of winds that blow almost constantly from easterly directions and are located on the equatorward sides of the subtropical highs.

Transform boundary A boundary in which two plates slide past one another without creating or destroying lithosphere.

Transform fault A major strike-slip fault that cuts through the lithosphere and accommodates motion between two plates.

Transpiration The release of water vapor to the atmosphere by plants.

Trellis pattern A system of streams in which nearly parallel tributaries occupy valleys cut in folded strata.

Trench An elongated depression in the seafloor produced by bending of oceanic crust during subduction.

Trigger (mass wasting) An event, such as an earthquake or heavy rainfall, that initiates a mass wasting process.

Tropic of Cancer The parallel of latitude, $23\frac{1}{2}°$ north latitude, marking the northern limit of the Sun's vertical rays.

Tropic of Capricorn The parallel of latitude, $23\frac{1}{2}°$ south latitude, marking the southern limit of the Sun's vertical rays.

Tropical (T) air mass A warm-to-hot air mass that forms in the subtropics.

Tropical depression By international agreement, a tropical cyclone with maximum winds that do not exceed 61 kilometers (38 miles) per hour.

Tropical storm By international agreement, a tropical cyclone with maximum winds between 61 and 119 kilometers (38 and 74 miles) per hour.

Troposphere The lowermost layer of the atmosphere. It is generally characterized by a decrease in temperature with height.

Tsunami A rapidly moving ocean wave generated by earthquake activity that is capable of inflicting heavy damage in coastal regions.

Turbidity current A downslope movement of dense, sediment-laden water created when sand and mud on the continental shelf and slope are dislodged and thrown into suspension.

Ultramafic composition A compositional group of igneous rocks in which the rock contains mostly olivine and pyroxene.

Ultraviolet radiation Radiation with a wavelength from 0.2 to 0.4 micrometer

Unconformity A surface that represents a break in the rock record, caused by erosion or nondeposition.

Uniformitarianism The concept that the processes that have shaped Earth in the geologic past are essentially the same as those operating today.

Unsaturated zone The area above the water table where openings in soil, sediment, and rock are not saturated with water but filled mainly with air.

Unstable air Air that does not resist vertical displacement. If it is lifted, its temperature will not cool as rapidly as the surrounding environment, and so it will continue to rise on its own.

Upslope fog Fog created when air moves up a slope and cools adiabatically.

Upwelling The rising of cold water from deeper layers to replace warmer surface water that has been moved away.

Valley breeze The daily upslope winds commonly encountered in a mountain valley.

Valley glacier See *Alpine glacier.*

Valley train A relatively narrow body of stratified drift deposited on a valley floor by meltwater streams that issue from a valley glacier.

Vapor pressure The part of the total atmospheric pressure that is attributable to water-vapor content.

Variable stars Red giants that overshoot equilibrium and then alternately expand and contract.

Vent A conduit that connects a magma chamber to a volcanic crater.

Vesicular texture A term applied to fine-grained igneous rocks that contain many small cavities called vesicles, which are openings on the outer portion of a lava flow that were created by escaping gases.

Viscosity A measure of a fluid's resistance to flow.

Visible light Radiation with a wavelength from 0.4 to 0.7 micrometer.

Volatiles Gaseous components of magma dissolved in the melt. Volatiles readily vaporize (form a gas) at surface pressures.

Volcanic See *Extrusive.*

Volcanic bomb A streamlined pyroclastic fragment ejected from a volcano while molten.

Volcanic cone A cone-shaped structure built by successive eruptions of lava and/or pyroclastic materials.

Volcanic island arc A chain of volcanic islands generally located a few hundred kilometers from a trench where there is active subduction of one oceanic plate beneath another. Also known simply as *island arc.*

Volcanic neck An isolated, steep-sided, erosional remnant consisting of lava that once occupied the vent of a volcano.

Volcano A mountain formed of lava and/or pyroclastics.

Warm front A front along which a warm air mass overrides a retreating mass of cooler air.

Water table The upper level of the saturated zone of groundwater.

Wave height The vertical distance between the trough and crest of a wave.

Wave period The time interval between the passage of successive crests at a stationary point.

Wave refraction See *Refraction.*

Wave-cut cliff A seawater-facing cliff along a steep shoreline formed by wave erosion at its base and mass wasting.

Wave-cut platform A bench or shelf in the bedrock at sea level, cut by wave erosion.

Wavelength The horizontal distance separating successive crests or troughs.

Weather The state of the atmosphere at any given time.

Weathering The disintegration and decomposition of rock at or near the surface of Earth.

Well An opening bored into the zone of saturation.

Westerlies The dominant west-to-east motion of the atmosphere that characterizes the regions on the poleward side of the subtropical highs.

Wet adiabatic rate The rate of adiabatic temperature change in saturated air. The rate of temperature change is variable, but it is always less than the dry adiabatic rate.

White dwarf A star that has exhausted most or all of its nuclear fuel and has collapsed to a very small size; believed to be near its final stage of evolution.

Wind Air flowing horizontally with respect to Earth's surface.

Wind vane An instrument used to determine wind direction.

Winter solstice See *Solstice.*

Xenolith An inclusion of unmelted country rock in an igneous pluton.

Yazoo tributary A tributary that flows parallel to the main stream because a natural levee is present.

Zone of accumulation The part of a glacier characterized by snow accumulation and ice formation. Its outer limit is the snowline.

Zone of saturation The zone where all open spaces in sediment and rock are completely filled with water.

Zone of wastage The part of a glacier beyond the zone of accumulation, where all the snow from the previous winter melts, as does some of the glacial ice.

Index

Note: Boldfaced entries are key terms. References to figures are marked *f*, to tables marked *t*.